ISBN 978-0-260-05137-0
PIBN 10924606

JOURNAL OF AGRICULTURAL RESEARCH

VOLUME XI

OCTOBER 1—DECEMBER 31, 1917

PUBLISHED BY AUTHORITY OF THE SECRETARY OF AGRICULTURE,
WITH THE COOPERATION OF THE ASSOCIATION OF AMERICAN
AGRICULTURAL COLLEGES AND EXPERIMENT STATIONS

EDITORIAL COMMITTEE OF THE
UNITED STATES DEPARTMENT OF AGRICULTURE AND
THE ASSOCIATION OF AMERICAN AGRICULTURAL
' COLLEGES AND EXPERIMENT STATIONS

CONTENTS

iii

2094

ERRATA

Page 1, line 11, "(*Juniperus occidentalis* Hook)" should read "(*Thuya plicata* D. Don)."
Page 76, line 18 from bottom, "Jadin and Astruc (4)" should read "Jadin and Astruc (5) "
Page 79, Table I, " Phosphoric" should read " Phosphoric acid."
Page 144, Table V, column 5, "Coefficient of variation" should read "Coefficient of variability "
Page 147, "Kline" should read "Klein."
Page 193, lines 8, 21, "*Agaricus oreades*" should read "*Agaricus oreades*."
Page 195, Table I, column 2, "*Agaricus caryophylleus*" should read "*Marasmius caryophylleus*."
Page 301, last line, legend of figure 17, "September 29" should read "September 9."
Page 461, line 10 from bottom, "Meek" should read "Mech."
Page 513, legend of figure 2, "during 1916" should be omitted.
Page 556, Table IV, heading, "years" should read "ears."
Page 580, lines 10 and 11, "detaining" should read "destaining."
Page 588, line 30, "(Pl. 44. C)" should read "(Pl. 44. C, a) "
Page 699, line 17, "E. R. Brooks" should read "E. W. Brooks."

ILLUSTRATIONS

PLATES

TEXT FIGURES

JOURNAL OF AGRICULTURAL RES

VOL. XI　　　WASHINGTON, D. C., OCTOBER 1, 1917　　　No. 1

NATURAL REPRODUCTION FROM SEED STORED IN THE FOREST FLOOR

By J. V. HOFMANN,

Forest Examiner in Charge of Wind River Experiment Station, Forest Service, United States Department of Agriculture

PURPOSE OF THE STUDY

On many burns and cut-over areas in the Douglas fir and western white-pine region of northwestern Idaho, Washington, and Oregon there are found dense and irregular stands of young growth, the origin of which can not be traced in any way to the seed trees left after cutting or burning. The effort to find the true source of seed of these stands began with a study [1] to determine the efficiency of seed trees in restocking the ground and the distances to which seed is disseminated.

Studies on the Kaniksu National Forest in northern Idaho of cut-over areas, some with single seed trees and others with groups of seed trees, brought out the fact that western white pine (*Pinus monticola* Dougl.), Douglas fir (*Pseudotsuga taxifolia* [Lam.] Butler), western red cedar (*Juniperus occidentalis* Hook), and western hemlock (*Tsuga heterophylla* [Rof.] Sargent) were producing satisfactory stands of young growth for a distance of not more than from 2 to 5 chains from the seed trees. Instead of being blown abundantly over large areas, the seed was cast in sufficient stand-producing quantities within a radius of only a few chains. This is illustrated in figure 1, which is based on an area studied two years after cutting.

The cut-over area shown in figure 1 contains 160 acres. It was clearcut, blocks of seed trees (A, B, C, and D) containing from 2 to 4 acres each being left to reseed the area. The slash was left broadcast and burned the same year of the cutting. A very severe slash fire burned over all of the cutting except area E in the northwestern part and area F in the southwestern part. The whole cutting was studied by running

[1] The writer began the study of seed viability and factors influencing germination at the College of Forestry, University of Minnesota. The results of these studies are to be published as a bulletin of the University of Minnesota. He wishes gratefully to acknowledge the assistance given by Mr. D. R. Brewster, of the Forest Service, in this work; also the valuable aid of Mr C. J Kraebel, Forest Assistant, Wind River Experiment Station, in analyzing and interpreting the field data.

transects 2½ chains apart.   The striking point brought out by the study
was the very limited distance to which seed was being carried by the

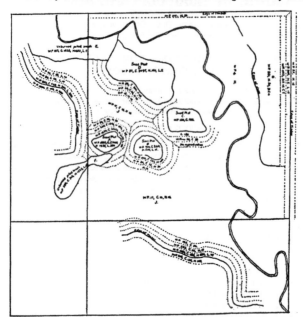

FIG. 1.—Migration chart: Graphic representation of the number of seedlings per acre in each chain distance
from seed trees. T. 57 N., R. 5 W., Boise M., sec. 26. Kaniksu National Forest. Scale, 16 inches=1
mile. W. P.=white pine, C =cedar; H =hemlock; L=larch; D. F.=Douglas fir.

wind. The distance of seeding and the effect of the slash fire are shown
in Table I.

TABLE I —*Average number of seedlings per acre at various distances from seed trees on
burned and unburned slash*

Species.	Burned slash.				Unburned slash.
	First chain.	Second chain.	Third chain.	All areas beyond 3 chains.	Distances of 3 or more chains.
Western white pine ...	615	187	40	17	274
Western red cedar..	5,600	266	24	16	924
Western hemlock..	1,438	195	20	16	854

It is apparent from this table that the seed trees are effective for a distance of only 2 chains on this area, and that reproduction occurs in unburned slash without relation to the distance from seed trees. The limited migration from seed trees threw doubt upon the accepted theory that wind-blown seed is responsible for the dense stands of reproduction occurring in burns at distances of a mile or more from seed-bearing trees. If the seed did not come from the seed trees, then it must have been produced before the burn or the cutting took place, and must have been stored in the duff of the forest floor or in cones. Attention therefore was directed to the leaf litter, and an effort was made to ascertain whether the forest floor of a virgin forest contains any germinable seed of western white pine and Douglas fir. It was found that the duff contains a large number of germinable seed, which might remain dormant there for a number of years and which evidently germinates and results in a dense stand of young growth as soon as the forest is cut down or burned over and light and heat are admitted to the ground.

These facts led to a comprehensive study of large burns, particularly in the Douglas-fir region of Washington, and of cut-over lands in the Puget Sound region. The burns studied were divided into two classes: Areas burned only once were classed as "single burns" and areas burned twice or more were classed as "repeated burns."

The results of the 5-year study from 1912 to 1916 are based on surveys of burns comprising about 750,000 acres, of which 68 acres have been actually examined by the transect method, and about 7,780 acres of cut-over land, of which 22.25 acres have been examined by the transect-and-plot method.

RESTOCKING OF BURNS BY SEED STORED IN THE FOREST FLOOR

THE COLUMBIA BURN, A SINGLE BURN ON WHICH THE DUFF WAS ONLY PARTLY DESTROYED BY THE FIRE

The Columbia burn (locally known as the Yacolt burn), extends northward from the Columbia River on the Columbia National Forest in southern Washington. The Columbia fire burned over an area of 604,000 acres in the foothills on the west side of the Cascade Mountains at elevations varying from 500 to 4,000 feet (Pl. 1). At the lower altitudes the forest traversed by the fire was the well-known Douglas-fir type, which includes the associates, western hemlock, western red cedar, western white pine, and lowland or "grand" fir (*Abies grandis* Lindl.). Above 1,100 feet, silver fir (*Abies amabilis* Forbes) makes its appearance, and then noble fir (*Abies nobilis* Lindl.), until at about 3,000 or 3,500 feet the forest developes the true fir type, composed almost entirely of noble and silver fir with a slight admixture of western white pine and Douglas fir. Pacific yew (*Taxus brevifolia* Nutt.) is distributed almost throughout the forest, avoiding only the subalpine summits of the higher ridges. Dwarf juniper (*Juniperus communis* L.), on the other hand, is restricted to the subalpine summits.

The fire occurred from September 8 to 12, 1902, following an exceptionally dry season, and, driven by a dry southeast wind, traveled from southeast to northwest. So far as can be determined from local information, it traveled at perhaps a maximum rate of 8 miles an hour during the time it was doing the most damage. No portion of the area has been burned over by a second fire.

The main feature that is now of interest on the burn is the good stand of young growth almost uniformly covering the area and of the same species as that which made up the burned forest The presence of this reproduction is very obvious to anyone passing through the area, but the reason for its appearance after so severe a fire is a question that has always been open to conjecture. The problem, then, was to determine the history of the reproduction and, so far as possible, to account for its distribution.

An arbitrary section, chosen to include Lookout Mountain, was studied intensively by a gridiron system of east and west belt transects (Pl. 2, B). These were run 2½ chains apart over the entire section. Then, with this section as a hub, an arbitrary township surrounding it was studied extensively. For this study eight transects were run radially from the centers of the four sides and from the four corners of the section to the corresponding points in the township. Wherever a solid body of green timber was encountered, the transect was discontinued at that point. These belt transects served effectively to disclose the distribution of reproduction over the entire township. The plan of the survey is shown in figure 2. The lines radiating from the center section represent the transects which were run in making the township study.

Each species found on the Columbia burn is discussed separately.

SEED TREES LEFT ON THE BURN ONLY A MINOR FACTOR IN REPRODUCTION AFTER THE FIRE

DOUGLAS FIR.—The Douglas-fir records bring out the fact that there is scarcely any young growth on the south and east slopes of Lookout Mountain, and such as does occur is confined to the draws below the barren slopes. On the north and west slopes, however, it is uniformly scattered. This distribution of the young growth is due to the local topography. The fire approached the mountain from the southeast and naturally swept up these slopes with unusual intensity; also the south and east slopes were hot, dry sites after the fire, and were consequently unfavorable to the establishment of seedlings. On the other hand, on the north and west slopes the opposite of these conditions obtained—that is, the fire was less intense, and the site was inherently more propitious for seedling growth.

OF NATURAL REPRODUCTION

OF

DOUGLAS FIR

MADE IN 1913 BY

D RIVER EXPERIMENT STATION

ON THE

URN OF 1902 . COLUMBIA NAT'L FOREST

WASHINGTON

e.
Wernstedt & Kraebel

Contour Interval 100'
Mapped by Cal R.Paul

Legend

ld
dicated — 6,7,8,9,10 Years old
Each 100 per Acre indicated

11 Years old
Each 100 per Acre indicated

d
re — 6,7,8,9,10 Years old
Over 500 per Acre

11 Years old
Over 500 per Acre

ir — P White Pine — △ Over 100 yrs old
ɔ Fir — H Western Hemlock — Q Under 100 yrs old
Fir — C Western Red Cedar

The most significant facts brought out by the study of the young growth on this section have to do with the distribution of the age classes and their relative proportions, as shown in Table II. This table shows that the 11-year-old class includes 25 per cent of all Douglas-fir seedlings found; the 6- to 10-year-old class, 65 per cent; and the 0- to 5-year-old class, 10 per cent.

TABLE II.—*Classification of young growth on section of Columbia burn studied* a

Species.	Distance from seed trees.	Percentage of total area examined included in each distance.	Percentage of total number of seedlings found within each distance.			Grand total percentage of all seedlings found		
			Over 10 years.	6-10 years.	0-5 years.	Over 10 years.	6-10 years.	0-5 years.
	Chains.							
Douglas fir..............	Over 10....	79. 0	61. 1	38. 2	0. 7	25. 3	64. 6	10. 1
	6-10.......	9. 7	48. 1	33. 3	18. 6			
	0-5........	11. 3	67. 6	14. 6	17. 8			
Western white pine....	Over 10....	100. 0	15. 8	84. 2	......	15. 8	84. 2	......
	6-10.......	......	......	......	......			
	0-5........	......	......	......	......			
Noble fir...	Over 10....	64. 2	96. 0	3. 3	. 7	93. 9	3. 5	2. 6
	6-10.......	20. 4	90. 0	5. 4	4. 6			
	0-5........	15. 4	76. 5	9. 4	14. 1			
Silver fir..............	Over 10....	56. 6	87. 9	10. 2	1. 9	80. 1	13. 2	6. 7
	6-10.......	24. 4	71. 5	18. 7	9. 8			
	0-5........	19. 0	55. 1	22. 9	22. 0			
Western hemlock........	Over 10....	92. 0	36. 8	57. 8	5. 4	21. 8	65. 6	12. 6
	6-10.......	8. 0	......	76. 9	23. 1			
	0-5........	......	......	......	......			

a Total area examined=78.6 acres.

The proportion of age classes tells the history of the reproduction, showing that only a comparatively small percentage of the total Douglas-fir reproduction on this area started the first year after the fire, by far the greater part of the seed germinated during a period of from 2 to 6 years, and very little germination occurred in the period from 7 to 11 years. The decrease of germination of seed throughout the section after the first few years subsequent to the burn and the very small percentage of the young age classes found at more than five chains from seed trees indicated that the remaining seed trees had not been a factor in the restocking of the area.

These facts are supported by the records of the township study (Table III). Although the Douglas fir which germinated the first year after the fire was more than half the total Douglas-fir germination, the small percentage of germination after the sixth year following the fire was evidence that the remaining seed trees were not casting seed over the burn, because germination conditions were still favorable when examined 11 years after the fire.

TABLE III.—*Classification of young growth on township of Columbia burn studied* [a]

Species.	Distance from seed trees.	Percentage of total area examined included in each distance.	Percentage of total number of seedlings found within each distance.			Grand total percentage of all seedlings found.		
			Over 10 years.	6–10 years.	0–5 years.	Over 10 years.	6–10 years.	0–5 years.
	Chains.							
Douglas fir	Over 10....	89.9	80.6	7.0	12.4	} 55.0	33.2	11.8
	6–10.......	4.6	68.5	14.4	17.1			
	0–5........	5.5	54.9	13.4	31.7			
Western white pine....	Over 10....	98.0	25.3	74.7	.....	} 24.0	76.0	.....
	6–10.......	1.0	20.0	80.0	.....			
	0–5........	1.0	20.0	80.0	.....			
Noble fir	Over 10....	89.7	94.1	3.0	2.9	} 91.4	3.7	4.9
	6–10.......	5.5	90.9	2.2	6.9			
	0–5........	4.8	67.2	8.8	24.0			
Silver fir	Over 10....	85.2	93.5	5.1	1.4	} 92.0	6.8	1.2
	6–10.......	5.5	75.0	25.0	.....			
	0–5........	9.3	80.0	20.0	.....			
Western hemlock	Over 10....	78.4	26.5	53.0	20.5	} 24.4	51.7	23.9
	6–10.......	9.3	25.0	45.8	29.2			
	0–5........	12.3	19.3	51.6	29.1			

[a] Total area examined = 32.25 acres.

The section and township records demonstrate that the occurrence of the age class 0 to 5 years is limited to the vicinity of seed trees or to localities which are topographically within the influence of these trees. Trees on a hillside above a canyon disseminated their seed over a wider range of territory than trees on level land, since the seed could be blown both down into the canyon over the nearer slope and also across to the opposite slope.

There are no Douglas-fir seed trees on the section, except a single broken-topped one, but there are Douglas-fir seed trees within two or three chains of the northeast boundary of the section on the northwest slope of Little Lookout Mountain. These are from 100 to 300 feet above the areas containing the 0- to 5-year-old class of reproduction; hence, it is entirely possible that they are responsible for the occurrence of this age class in the northern part of the section. The limited distribution of the 0- to 5-year-old class, with reference to seed trees and topography, is consistent throughout the section and is particularly conspicuous at several points in the township (fig. 2).

The transect from the west-central point of the section passes within two chains of green timber. The influence of this timber is shown in figure 2 by the occurrence of the 0- to 5-year-old class for a distance of a few chains in the immediate vicinity of the seed trees in Texas Gulch. The remainder of the transect has a scattered stand of reproduction of the older classes, a very dense stand occurring in Poison Gulch almost 1 mile from the nearest seed trees. This same condition is illustrated

on the southwest transect, where the young growth of the older classes
is very heavy along the north slope of Bear Creek Canyon at a distance
of more than a mile from the nearest Douglas fir timber. The occurrence
in this locality of a small group of living trees (noble fir) demonstrates
the abatement of the fire as a feature coincident with the subsequent
occurrence of dense reproduction. The importance of the relation
between the severity of the fire and the amount of subsequent reproduction is developed later. As the transect approaches the timber at the
top of the ridge, older seedlings are again very scattered, and it is only
very close to the edge of the timber that the younger seedlings begin to
appear at all. This peculiar distribution of the reproduction can be
observed on all of the transects, and shows very definitely that the
green timber remaining after the fire has had very little influence on the
general occurrence of the Douglas fir reproduction over the burn.

NOBLE FIR.—The noble-fir records for the section bring out the same
points as the Douglas fir records, but in a more striking way because
the site was more suited to noble fir than to Douglas fir. It is noticeable
that the reproduction of this species is distributed in good stands with
remarkable uniformity over most of the section, in spite of the fact
that there are two well-defined groups of seed trees within the section.
One departure from this uniformity is the scarcity of seedlings on the
high south and east slopes of Lookout Mountain, which is undoubtedly
due to the dryness of the site and the direction of the fire. On the west
and south slope of the mountain extremely dense stands occur, amounting in several places to 20,000 or 30,000 seedlings per acre.

There is another departure from the uniformity of the noble-fir reproduction which is inconspicuous but very significant. This is the restricted
distribution of the age class 0 to 5 years, which is limited either to the
proximity of the groups of seed trees or to favorable situations downhill
from these trees. Even in such places the young class forms only a
small percentage of the total reproduction, and decreases rapidly with
the increase in distance from the seed trees. These facts can not be
construed otherwise than to indicate that the seed-tree groups are not
casting their seed for any great distance, in most cases only from 2 to
5 chains.

On the other hand, the older classes, especially the 11-year class,
occur at distances of from 10 to 50 chains from the groups of seed trees.
Over the whole section almost 94 per cent of all noble-fir reproduction
started the first year following the fire, less than 4 per cent in the next
five years following the fire, and only 2.5 per cent later than six years,
most of this latter being within 5 chains of seed trees (Table II).

The township study shows that noble fir occurs on all of the areas
on which it grew before the fire. Some parts of the burn are below
the noble-fir zone, and for that reason do not show any noble-fir reproduction. Wherever the species does occur in the township, it was

found that the proportion of age classes is practically the same as in the central section, over 90 per cent being in the 11-year-old class. The distribution of the age classes with reference to seed trees was found to exhibit the same peculiarities as were found in the section— namely, the seedlings under 5 years old are found only in the vicinity of seed trees, seedlings from 6 to 10 years old are more widely distributed over the burn, seedlings 11 years old are found everywhere in the burn where noble fir formerly grew, often forming dense stands at great distances from seed trees that survived (Table III).

These facts show clearly that all the young growth could not have come up from the seed scattered by the seed trees which have survived the fire. The restricted distribution of the youngest seedlings, which so definitely points to their having originated from the seed trees, makes it impossible to attribute all of the older classes to the same source. Had both classes come from the same seed source they would of necessity have exhibited the same distribution.

SILVER FIR.—The silver-fir records resemble in every respect those of the noble fir. In the same groups with the noble fir there are some silver-fir seed trees, the influence of which is about the same as described for the noble fir. The silver fir is found on all sites in the township where it occurred in the forest before the fire, although the silver-fir zone is still more limited than that of the noble fir to higher elevations and cooler slopes.

WESTERN HEMLOCK.—The area studied as a whole is above the western hemlock zone, but young growth occurs in several localities, although usually in sparse stands. The important point about the hemlock reproduction is that burned hemlock trees were found in every locality where stands of hemlock reproduction occurred, and the relative proportion of hemlock reproduction to other species is approximately the same as the proportion of hemlock in the original forest. For example, the transect running from the northwest corner of the section goes through a dense stand of hemlock reproduction at about half a mile from the nearest green timber. This is a favorable hemlock site, and the burned forest consisted of approximately 10 per cent of hemlock. This same circumstance is again evident on the southeast transect at approximately half a mile from the nearest seed trees. On this transect it is also important to note that the heavy stand of reproduction does not continue to the green timber; on the contrary, there is practically no reproduction for 20 chains from the green timber. These facts argue strongly against the theory that the seed source of such isolated reproduction is the green timber which escaped the fire.

WESTERN WHITE PINE.—Western white pine does not naturally form pure stands in the Cascade Mountains of Washington; but individual trees or very small groups of this species do regularly occur in the forests throughout the region. The section and township data for white pine show a fairly uniform but scant distribution of reproduction of this species over the entire burn, repeating with remarkable fidelity the distribution of the species in the forest before the fire.

A striking fact is brought out by a study of the proportions of the age classes. There was a total absence in the burn of seedlings under 5 years old (Tables II and III). The older classes, however, although sparse in numbers, were uniformly distributed over the township without any relation to the location of possible seed trees. These facts show clearly enough that germination and establishment of white pine in the open burn ceased altogether after the year 1908. This cessation of the appearance of white-pine reproduction was not abrupt but gradual, and is shown graphically in figure 3 by a simple curve based upon the averages of the age-class percentages for white pine in both section and township.

The stop in the appearance of white-pine seedlings occurred in spite of the fact that the possible white-pine seed trees in the township continued to produce seed during the four years from 1909 to 1912, and were carrying a crop of cones at the time they were examined in August, 1913. Moreover, the conditions for germi-

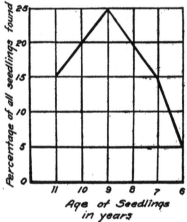

FIG. 3.—Curve showing the germination of white-pine seed in the Columbia burn 11 years after the fire. Note that 15 per cent of the seeds germinated the first year after the fire as shown by the 11-year-old class; the maximum germination of 25 per cent of the seed occurred the third year after the fire, shown by the 9-year-old class, and germination ceased entirely six years after the fire.

nation and establishment on white-pine sites over the entire area were still favorable in 1913, and had no doubt been equally favorable during the previous four years. If, then, the existing white-pine reproduction in the burn is attributed to seed transported from surrounding seed trees, why, in the light of the foregoing facts, was there no continued migration and establishment during 1909 to 1913?

In this connection it should be remembered that there were very few living white-pine trees within the township, and that these potential seed trees were always situated within stands of mature timber so that the chance for seed distribution by wind was extremely small. Moreover, with this species, as with all others, the reproduction is uniformly scattered over the township without any relation to the nearness or distance of the possible seed trees. Especially noticeable in this connection are the west, southwest, and south township transects, where white-pine reproduction over 5 years old occurs at a distance of 120 chains from the nearest seed trees. The exceptional quantity of white-pine reproduction near the end of the southeast township transect occurs on a very favorable white-pine site, where the presence of white-pine snags proved that this species had formed an unusual percentage of the original forests. In this particular instance the mature forest contained about 5 per cent of white pine, while the reproduction now constitutes 10 per cent of the stand.

It is obvious that some source of seed other than the seed trees which escaped the fire is responsible for the occurrence of white-pine reproduction in the burn.

PACIFIC YEW AND DWARF JUNIPER.—In the case of Pacific yew and dwarf juniper the theory of wind migration of seed to great distances is still more open to question. The important fact emphasized by the study of these heavy-seeded species is that reproduction is found only in localities where the trees occurred in the original forest. Before the fire Pacific yew was scattered throughout the area, and after the fire the reproduction was found always directly under or beside fire-killed yew snags. Dwarf juniper on the summit of Lookout Mountain grows close to the gnarled mats of dead juniper shrubs and could be found nowhere else in the entire township.

This peculiar and consistent limitation of occurrence argues against the possibility of regarding such reproduction as the result of seed deposited by birds. If birds were the agents of migration, the reproduction would have been found elsewhere than in the immediate vicinity of yew snags or dead juniper shrubs. Careful examination of the fire-killed specimens of both yew and juniper determined beyond doubt that they did not live through and produce seed after the fire; moreover, there are no green trees of seeding age of either species left within the burned areas. Neither of these sources of seed, therefore, can be held responsible for the reproduction of yew and juniper in the Columbia burn.

THE DUFF, THE PRINCIPAL FACTOR IN REPRODUCTION AFTER THE FIRE

The foregoing facts first cast a doubt upon the long-accepted theory of the restocking of large forest burns by the process of wind migration and finally proved it untenable. As the study progressed and this fact grew steadily more convincing, there arose naturally the question, "What was the source of seed for all this reproduction?" The answer to this question also developed naturally through the accumulation of evidence throughout the burn. It was found that the reproduction most often occurred, not in a solid unbroken cover, but in various-sized patches with very irregular and ramifying boundaries. Where the reproduction was lacking, the ground was covered with grasses, herbaceous plants, and shrubs, evidencing an uninterrupted growth since the burn was formed. The occurrence of these two types of cover made an interlaced pattern resembling mosaic over the entire burn, although each type often expanded solidly over a slope or basin many acres in extent. Everywhere the feature that was most striking was the sharp line of demarcation between the reproduction and the grass areas. For all its tortuous windings the boundary was always distinct. Obviously such a condition could not have resulted from any process of overhead seeding, but must rather have been produced by some action on the surface of the ground itself. The idea of ground fire suggested itself. One who has seen ground fire burning in forest duff will remember that it burns very irregularly, here leaving an island and there forming a deep bay between two points of unburned ground. When at length the smoldering fire is stopped, the result is just such a mosaic of burned and unburned territory as has been described for the reproduction and grass territory. A representative spot in the Columbia burn is shown on Plate 2, A, and the sketch in figure 4 is a graphic portrayal of the same spot.

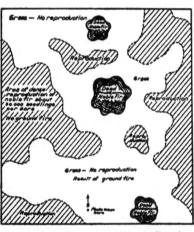

Fig. 4.—Ground plan of the spot shown in Plate 2, A. The conformance of this irregular boundary line with the erratic margin of ordinary ground fire is clearly evident. Constant repetition of this feature over the entire Columbia burn could lead to but one conclusion, that where ground fire burned there resulted no reproduction, excepting in proximity to living seed trees.

The likelihood that any part of the forest floor will burn depends on a number of varying factors, such as the quantity and kind of humus soil and its moisture content. Favorable conditions for a ground fire are shown in Plate 2, C, and the results of ground fire in Plate 2, D. Most severe ground fires occur on dry sites, provided those sites have a sufficient quantity of duff to carry fire at all. Accordingly the reproduction occurs most densely in the moist sites and is open or lacking on exposed dry sites, although this may be partly due to the fact that these dry sites are very unfavorable to the establishment of seedlings even though germination may take place. Furthermore, irregularity in the areas of young growth occurred on all sorts of sites. This could lead to but one conclusion: wherever ground fire occurred no reproduction appeared, except close to seed trees where seed could be cast upon the burned ground after the fire.

From this it was but a step to the complete explanation: wherever the duff and litter were burned out of the forest floor, there developed an area barren of reproduction; wherever the duff and litter were not burned out of the forest floor, there developed an area of more or less dense reproduction. Therefore, the duff must be the controlling element: the duff must be the storage medium of the seed, and that seed must have been produced and stored in the forest floor before the fire and have retained its viability through the fire.

Before this conclusion is accepted, however, another possible source of seed must be considered. Is it not possible that cones carried through the fire on the crowns of trees severely burned or killed furnished the seed from which the young growth originated? After the fire these cones may possibly have opened and dispersed their seed, becoming in that way an overhead source for the restocking of the burn. In fact, a very small percentage of germination of white-pine, noble-fir, and cedar seed has been secured from seed which passed through a crown fire. But even though this source does contribute some seed, it does not explain the great mass of reproduction, which, by its mosaic occurrence, demonstrates conclusively the impossibility of its having come from overhead seed distribution subsequent to the fire. The principal factor in reproduction after fire must be the seed stored in the duff.

UPPER CISPUS BURN, A SINGLE BURN ON WHICH THE DUFF WAS COM-
PLETELY DESTROYED BY THE FIRE

In an area on the Upper Cispus River, Rainier National Forest, burned in 1910, there was no reproduction except in the vicinity of seed trees; and the seedlings were more numerous close to the edge of blocks of seed trees than at a distance, which is invariably the condition where migration occurs from wind-blown seed. This is a typical example of an area made barren by a single fire which completely destroys the duff and thereby precludes the possibility of reproduction from seed stored in the forest floor. The complete destruction of the duff may have been due not so much to an unusually intense fire as to the extreme lightness of the dry volcanic soil of the region, which is easily "burned out" by ground fire. The conditions found on this area corroborate the conclusions reached in the study of the Columbia burn. Where the duff is completely burned, reproduction is found only in inverse numerical proportion to the distance from seed trees.

AREAS, PART OF WHICH ARE SINGLE BURNS AND PART REPEATED BURNS

The studies of the Columbia burn in 1913 established the fact that most of the reproduction which follows a single burn is due to seed which was stored in the forest floor before the fire and which retained its viability through the fire. Very often, however, burned areas are again burned, and the second fire creates many complications of the factors which influence reproduction subsequent to a single burn. In order to find out the reason for the presence or absence of reproduction following repeated burns, areas on the Rainer National Forest burned over in 1864, 1874, 1892, 1902, and 1910 were studied.

COWLITZ BURN —The Cowlitz area was burned in 1902, and the northeastern part was reburned in 1910. The original forest consisted of Douglas fir, noble fir, western red cedar, western hemlock, and western white pine.

The most striking feature of this burn is the fairly uniform distribution of reproduction of all age classes and species on the portion of the area burned only once, as shown by Table IV.

TABLE IV.—*Classification of young growth on Cowlitz area after the burn of 1902a*

Species.	Distance from seed trees.	Percentage of total area examined included in each distance.	Percentage of total number of seedlings found within each distance.			Grand total percentage of all seedlings found.		
			Over 10 years.	6–10 years.	0–5 years.	Over 10 years.	6–10 years.	0–5 years.
	Chains.							
Douglas fir	Over 10....	37. 5	16. 5	45. 5	38. 0			
	6–10......	52. 9	19. 6	45. 3	35. 1	35. 1	56. 8	8. 1
	0–5........	9. 6	24. 2	36. 3	39. 5			
Western white pine....	Over 10....	100	4. 8	67. 0	28. 2			
	6–10......	....	....	....	....	4. 8	67. 0	28. 2
	0–5........	....	....	....	....			
Noble fir	Over 10....	36. 2	33. 3	40. 9	25. 8			
	6–10......	57. 2	40. 0	36. 1	23. 9	37. 7	37. 4	24. 9
	0–5........	6. 6	20. 0	40. 0	40. 0			
Western hemlock	Over 10....	36. 2	18. 7	34. 7	46. 6			
	6–10......	58. 1	19. 2	20. 0	60. 8	17. 6	26. 3	56. 1
	0–5........	5. 7	10. 5	30. 7	58. 8			
Western red cedar	Over 10....	50. 5	23. 4	53. 2	23. 4			
	6–10......	42. 9	22. 2	49. 1	28. 7	20. 8	47. 4	31. 8
	0–5........	6. 6	7. 4	27. 8	64. 8			

a Total area examined = 5 15 acres.

The uniformity of distribution of each age class and the large percentages of the 0-to-5 and 6-to-10-year-age classes seemed at first to indicate restocking by the remaining seed trees after the fire. White pine snags were found on the burned area, and green white-pine trees were found in canyons and at elevations varying from 300 to 400 feet below the area studied and well inside the mature stand of timber; yet the white-pine reproduction is found up to the highest points of the burn, which are from 800 to 1,000 feet above where the green white-pine trees were found (Pl. 4, A). It is very evident, therefore, that the white-pine reproduction could not be due to seed produced by the seed trees remaining after the fire, but must have sprung up from the seed in the ground that survived the fire. Seed trees of the other species were found on the edge of the remaining green timber at higher elevations than the burn, and therefore could bring about the restocking of the ground by these species. Since, however, the influence of seed trees on the 1910 burn, which reburned a portion of this 1902 burn, was found to be very local, the part which the seed trees played after the first burn in restocking the ground becomes questionable (Pl. 4, B). The effect of the seed trees of the different species on the second burn is shown in Table V. As can be seen from this table, the influence of seed trees extended only for from 2 to 4 chains.

TABLE V.—*Distribution of seedlings on Cowlitz area after the burn of 1910*

Species.	Number of seedlings per acre.			
	Average of 1 and 2 chains from seed trees.	Average of 3 and 4 chains from seed trees.	Average of 5 and 6 chains from seed trees.	Average of 7 and 8 chains from seed trees.
Douglas fir.............................	280	240	40	40
Noble fir................................	80	0	0	20
Western hemlock........................	1,640	120	80	0
Western red cedar.......................	2,120	200	40	0

None of the white-pine seed trees left on the 1902 burn which might have influenced the area of the 1910 burn were killed by the 1910 fire, though it burned closer to them than the 1902 fire. Yet no white-pine seedlings followed the 1910 burn, while white pine reproduction of all age classes came in on the 1902 burn. This would indicate that the seed trees had little influence on the restocking of the ground after the 1902 fire.

MOUNT ADAMS BURN.—Northward and westward from Mount Adams there lies an extensive stretch of burned country, embracing altogether no less than the equivalent of three full townships, and probably much more. The greater part of this area was apparently only once burned and is now a standing forest of weather-bleached snags. The rest of the area represents a much older burn. It bears evidence of being irregularly reburned and includes large barren stretches of semiarid appearance with only decayed stumps and the crumbling remains of logs to suggest the existence of a former forest.

On the area burned only once in 1892 almost all of the timber was killed. Occasional patches in such places as near a marsh or spring were left unburned. The fire-killed trees still stand, most of them retaining their branches and some of them even fragments of their bark. These are indications of a killing but not a consuming fire. The area has very much the appearance of the Columbia burn, with one striking and consistent difference. Where the Columbia burn has a fairly dense and generally satisfactory cover of reproduction, the Mount Adams burn has very sparse and inadequate reproduction. The average per acre of all species, based on 80 contiguous acres, was 436 seedlings per acre on the Columbia burn and 56 seedlings per acre on the Mount Adams burn. This lack of seedlings in the Mount Adams burn is apparently due to the light volcanic soil, which causes a drier forest floor and consequently a more complete consumption of the litter and duff by an average fire.

The composition of the reproduction actually found in the Mount Adams burn is as follows:

			Per cent.
True firs	66	Noble	35
		Silver	31
		Whitebark (*Pinus albicaulis* Engelm.)	9
Pines	23.6	Western white	8
		Lodgepole (*Pinus murrayana*)	5
		Western yellow (*Pinus ponderosa* Laws)	1.6
Other species	10.4	Western hemlock	9
		Engleman spruce (*Picea engelmanni* Engelm.)	1
		Douglas fir	0.4

There is every reason to believe that the composition of the forest before the fire was closely in accord with these figures, although it was not possible to check this absolutely. The distribution of this reproduction over the burn was fairly even, with the natural exception that in the vicinity of green timber where the fire had obviously abated there was the usual fringe of dense reproduction.

That the distribution of reproduction was found to be independent of the location of possible seed trees indicates that here, as on the Columbia burn, the source of much of that reproduction must have been seed stored in the forest floor previous to the fire. This fundamental point is especially emphasized in the case of whitebark pine, which was found throughout the burn.

Whitebark pine produces a heavy winged seed, but the wing is retained on the cone scale when the seed is liberated, hence the distance of wind migration is very limited. However, seedlings were found at distances of more than a mile from the nearest seed trees.

A peculiar feature of the occurrence of these whitebark pine seedlings gave an early clue to a convincing explanation of their presence so far from seed trees. Throughout the burn they were found most often occurring in little clumps containing from 3 to 15 seedlings, and occasionally as many as 18 were found in a clump. The seedlings in such

clumps were always closely bunched, growing usually within a space of
only 2 or 3 inches in diameter.   They were most often found springing
from between the spreading roots of snags close to the trunks.   In age
the seedlings varied from 1 to 16 years, but always the individuals of a
clump were of the same age or within a year of the same age.   From the
location of many of the clumps at the bases of snags and from the fact
that the seed is known to be a favorite food of rodents, it appears that
the seeds were buried by rodents.[1]   Since the seedlings occur often at
distances of over a mile from seed trees of this species and since rodents
would not be likely to carry seeds to all parts of the burn for such distances, it is believed that the seeds were buried either before or immediately after the fire.

The young seedlings from 1 to 5 years old must therefore be remarkable
examples of delayed germination.   Having been buried approximately
in 1892 and having germinated in 1913 these seeds of whitebark pine
exhibit the retention of viability through a storage period of 21 years.
Several features of the cone and seed are calculated to aid in this delayed
germination.   The hard, thick-scaled cone is admirably adapted as a
storehouse and, when sealed by the heat of a forest fire, no doubt serves
effectively to preserve the seed.   Before the seed can germinate, the cone
must first open or disintegrate.   When buried deeply (as by rodents) in
a sealed cone, the seed will naturally be most favorably placed for pro-
tracted storage.   Finally, the seed itself is large, and is well protected
by a thick, hard seed coat.   It is very well equipped both to resist the
agents of disintegration from without and to retain life within.   Of all
the species dealt with, it should prove, and has proved, to be among
those which are able to retain their germinability for the longest period.

The reburned area was strikingly different from the area burned only
once.   It was apparent that the fire here had been much more severe,
because the area was noticeably bare of snags.   Such snags, stumps, and
logs as were present were also in a much more advanced state of decay.
The reproduction in this region was markedly confined to the areas only
once burned, or at most to those which were visited by one severe fire
followed after a long interval by one or more surface fires.   The reproduction on such areas was very open, but was occasionally dense enough
to make a satisfactory cover.   Named in the order of their numerical
importance, the species represented were lodgepole pine, hemlock (western
and mountain), western white pine, whitebark pine, noble fir, and others.
These covered a range of ages from 1 to 38 years, with possibly some older
trees which were seeding vigorously (Pl. 5, B, C).   It was impossible to
fix the age of the original fire closer than between 39 and 40 years, which
would place it approximately in the year 1875

---

[1] Caching of whitebark-pine seeds by blue jays has been noted by C. G. Bates at the Fremont Experiment Station in Colorado.

The reason for such sparse reproduction and for its location in patchy areas must be sought in the fact that in 1892 a second severe fire swept through the burn of 1875 in an irregular width, as most fires do.

The reproduction on the area swept over by fire for the second time averages less than one seedling per square rod, and even this figure is exaggerated by the occurrence of whitebark pine seedlings in clumps of from 6 to 14 in a single spot. There are many areas several acres in extent which are absolutely bare of coniferous growth.

Although these areas are now dry and inhospitable to tree growth, the occasional tall, standing snags which have escaped destruction attest by their size that a forest of real value once occupied the ground.

Similar conditions were found on the East Canyon Creek area where a 1910 burn followed a 1902 fire. On this area the source of scattered seedlings 1 to 3 years old was demonstrated to be seed produced by a few young trees which had escaped the previous fire and had now just reached seeding age. One such tree is shown in Plate 5, A.

The reproduction found on scattered areas which have escaped the second fire suggest that after the first fire the entire burned-over area, wherever the duff was not completely burned out, must have had a similar reproduction. Since the second fire occurred 17 years later or before the young stand could produce a large amount of seed and before the ground provided favorable conditions for seed storage, there was no stored seed from which a new stand could come up, and therefore the area covered by the second fire resulted in a barren. This point is more clearly brought out in the Tower Rock burn.

TOWER ROCK BURN.—The Cispus fire of 1902 burned in two distinct age classes of forest, one a mature Douglas-fir forest previously untouched by fire and the other a second-growth stand of pole size of almost pure Douglas fir about 40 years old. The results which followed in the two cases were strikingly different.

In the mature Douglas-fir forest the conditions after the burn repeated practically the conditions found on the Columbia burn. There was here the same occurrence and distribution of dense and sparse reproduction independent of the position of seed trees; there was the same proportion of age classes in the reproduction, evidencing the greatest germination during one or two years after the fire; there was the same alternation of patches of the original brush cover and reproduction, outlining the limits of irregular ground fire. There was, moreover, abundant proof of the inadequacy of survived seed trees to restock a burned forest area of large extent. There were left on this burn numerous groups of living seed trees of Douglas fir and minor species ideally placed for a study of their seeding influence. In spite of the most favorable conditions of site for the germination and establishment of seedlings, it was invariably found that 1-, 2-, and 3-year-old seedlings were limited to a radius of 3 and

4 chains from such seed trees. On the other hand, seedlings 10 and 11 years old were found everywhere, even at the maximum distance from seed trees attainable in the area, 12 chains (Pl. 6, A).

The destruction of a portion of the pole-sized stand of Douglas fir (Pl 6, B) by the fire of 1902 produced peculiar conditions. When examined in September, 1914, the ground was occupied by a dense cover of deciduous shrubs and fern which grew up through a network of dead-down logs of small diameter. Reproduction occurred on this area in spite of the heavy brush cover, but it was comparatively scarce. The reproduction, in its struggle against the surrounding brush, developed the tall, lank "shade form" common to any thicket or forest-grown plant (Pl. 7, A). The older seedlings occurred independently of the position of seed trees, but showed a tendency to be limited to moist sites, such as draws and hollows, where the germination and establishment conditions were most favorable immediately after the fire. The younger seedlings (2 to 5 years old), however, showed an increase in density in the neighborhood of seed trees. Several seedlings of this class were found dead as a result of excessive drouth and heat in open spots, but they were rarely found either living or dead at distances over 6 chains from seed trees.

The occurrence of two distinct age classes of seedlings and their position in the burn with reference to seed trees indicate here again the action of two sources of seed—namely, survived seed trees and the unburned duff of the forest floor. The reasons for the failure of seed trees to restock the burn have been discussed before. The scarcity of seedlings from duff-stored seed, however, presented a problem peculiar to this burn. The dense cover of shrubby and herbaceous growth on the area might serve to account for a scarcity or absence of seedlings from "wind-blown" seed, but it will not account for the scarcity of stored-seed seedlings. Reproduction from stored seed as a rule starts at the same time as the brush on the burn and does not come in after the brush has taken possession of the ground. Hence, it has merely to hold its own in height growth with the brush; and this it has frequently proved itself able to do.

The reason for the scarcity of reproduction from duff-stored seed must be sought in the condition of the stand. The fire burned, not in an old virgin forest, but in a very young forest, which was itself successor to a burned mature forest. The earlier fire (1860) killed most of the veteran trees which were the chief source of seed stored in the forest floor. From the stored seed of the old forest there resulted a thrifty second growth of almost pure Douglas fir. How complete this stand was could not be determined; it is likely that there were openings resulting from ground fire in parts of the original burn. In 1902, at an age when this young forest was seeding but long before it had been able to re-establish most favorable conditions of the forest floor for the duff storage of its seed, this forest, too, was destroyed by fire. In the hot fire of 1902 some of the duff was undoubtedly burned, and consequently some of the stored seed. The result of this combination of circumstances was the occurrence, as was found, of scattered, inadequate reproduction amid a rank growth of deciduous brush and weeds. The reproduction was more plentiful in moist or depressed spots. These had an accumulation of duff which in a large measure escaped the ground fire and provided most favorable conditions for the germination and establishment of seedlings after the fire.

The findings on the burns studied on the Rainier National Forest corroborated those established on the Columbia burn. They brought out also additional facts regarding the effect of the conditions of the original stand upon the reproduction that follows after the fire.

When reproduction which follows a burn is destroyed by fire before it reaches seeding age, a denuded area results if the area lies outside the influence of seed trees.

Where a stand of young growth is destroyed by fire shortly after it has reached seeding age, the result is, as a rule, a thin stand of reproduction. If the fire is severe enough to destroy all seed stored in the forest floor or in cones, then it results also in a barren. The reason for the light reproduction is found in the small accumulation of duff since the previous fire and the greater liability to drouth and consumption by fire, and in the small accumulation of seed in the duff owing to the limited seed production by the young trees.

The sparseness or entire lack of reproduction after a second fire in a young stand of timber leads to the conclusion that the source of seed from which the reproduction originated after the first fire was already on the area and did not come from outside; it points also to the great danger of fires in young stands of timber before they have begun to bear plentifully and before normal leaf litter has accumulated on the ground as a storage reservoir for the seed.

GERMINATION OF SEED STORED IN DUFF

The accumulation of seed in the forest floor is no longer a theory, but has been found to be an actual condition by an analysis of the duff, which revealed from ½ to 2 full seeds of Douglas fir, western white pine, and associated species per square foot. Of course, the age of these seeds could not be determined, and the depth to which they were buried can not be taken as an index as to how long they had been there. Very probably the greatest factor in storing seed is rodent activity, by which seeds are buried at various depths and forgotten.

The accumulation of seed over a period of years can prove advantageous only if the stored seed retains its viability. With regard to this, each species possesses a dormancy habit of its own, and these habits have been growing more familiar to the forester through nursery observation and experiment. Douglas fir sown in seed spots has been found to yield higher germination during the third season than during the first and second seasons combined. In the Wind River Nursery seed bed, germination of Douglas fir has continued through three seasons. Western white pine has often produced better germination during the second season after sowing, even under the best germinating conditions.[1] Juniper is difficult to germinate without special treatment. To some extent the rate of germination in nursery seed beds is controlled by the depth of soil cover.

With conditions such as these obtaining in regular nursery practice, it is not at all surprising that germination should be delayed under the forest cover. The cool shaded layers of leaf mold and general duff of the forest floor, which in the virgin Cascade forests seldom feel the warmth of the sun, constitute an ideal natural storage medium. Under conditions so unfavorable to germination and so favorable to its retardation, it can easily be imagined that the germination of forest tree seeds can be delayed to the limits of their various powers of dormancy.

On the Columbia burn, owing to the probability of seed accumulation for several years and to the possibility of there having been a general seed year just previous to the fire, it was not possible to determine absolutely how long the seed lay dormant. However, from the year of the fire until the date of germination (deduced from the age of seedlings found) the various species showed unquestionably the following periods of delayed germination :[2]

	Years		Years.
Douglas fir	6	Western white pine	6
Noble fir	3	Pacific yew	10
Silver fir	5	Dwarf juniper	10
Western hemlock .	6		

Delayed germination is believed to explain the proportions of age classes of the reproduction found in the open burns. Occasionally, however, single seedlings from 1 to 4 or 5 years old were found growing at great distances from seed trees in the midst of uniformly older reproduction. The presence of these "erratics" was at first a disturbing element, until it was decided that their presence could be accounted for by some one of the factors generally conceded to be responsible for the occurrence of scattered individuals of a species far beyond the normal range of that species, such as the occurrence of occasional western

---

[1] In this connection attention may well be called to the delayed germination experienced at the Feather River Forest Experiment Station and the Pilgrim Creek Nursery in California with the seeds of redwood (*Sequoia sempervirens* [Lamb] Endl.), white fir, incense cedar (*Libocedrus decurrens* Torr), and sugar pine (*Pinus lambertiana* Dougl.) when sown in the nursery. Often the second season has produced more germination than the first, when the seeds were spring sown.

According to Mr. S. B. Show, seed spotting in northern California with sugar pine, Jeffrey pine (*Pinus jeffreyi*), and western yellowpine has produced very little or no germination the first season, whereas the second season a number of seedlings were found in the seed spots, and sugar pine has produced a good stand from germination the third season after sowing.

[2] The ability of seed to retain its viability is shown further by the results of Beal (1–5); Ewart (6); Harrington (7); Nobbe and Haenlein (8); Rees (9), and others, although their work was not with forest tree seed.

Reference is made by number to "Literature cited," p 26.

yellow pines in the true Douglas-fir type of the west slope. These agencies need only be stated to be understood.

Whirlwinds.—Sudden powerful gusts of wind with unusual lifting power carry a few seeds of almost any species to great distances.

Birds.—Birds deposit undigested seeds at long distances from the parent trees. This is especially true of migratory birds, which move when seeds are ripe and most likely to be eaten.

Animals.—Animals transport seeds in their fur or caked in the mud of their feet.

The influence of these agencies is recognized by such authorities as Strassburger, Clements, and by ecologists generally. The common occurrence of juniper (*Juniperus virginiana* L.) and black elder (*Sambucus canadensis*) along the fence rows of the Eastern States is a typical example of seeds distributed by birds. Moreover, an added possibility, in the light of the conclusions of this study, is that occasional individual seeds may possess very unusual powers of dormancy and so cause the appearance of 1- and 2-year-old seedlings as late as 10 or 15 years after their deposit in the forest floor.

A storage test in the forest floor, under the forest canopy, of seeds of Douglas fir, silver fir, noble fir, western red cedar, western white pine, and western hemlock is in progress at the Wind River Forest Experiment Station. These tests have passed through one season with no germination under the forest canopy, which indicates favorable storage conditions. At the beginning of the second season germination tests showed that the seeds of all the species used were still viable. This is conclusive evidence that reproduction, which follows a burn of all these species, is not dependent on the seed crop which is produced during the season in which the fire occurs.

Seeds recovered from the forest floor have not been germinated by the writer. However, Mr. D. R. Brewster,[1] in charge of the Priest River Experiment Station, in northern Idaho, has recovered western-white-pine seed from the forest floor in about the same amount and has secured a germination of from 2 to 20 per cent of the total number of seeds found. The striking thing about the germination is that even with ideal greenhouse conditions the larger percentage occurred during the second year. Seeds gathered as mentioned above showed the probability of from 20,000 to 80,000 seeds per acre and a germination of from 400 to 16,000 per acre; more than this number of seedlings has been actually found on burns. These facts readily explain the occurrence of dense stands of reproduction regardless of seed trees.

With this stored seed in the forest floor, let there come a forest fire which destroys completely the forest cover over a vast area and leaves no seed trees, but in its rapid sweep fails to cover all of that area with ground fire. The main storage bulwark, the chief refrigerant, the tree canopy itself with its secondary undergrowth has been removed; and the stark, dead forest of snags offers just sufficient protection from the sun's heat and excessive evaporation to make of the forest floor an ideal germinating bed. The dormant seeds, bedded in the unburned duff, are stirred to life and respond to the stimulus. Germination proceeds for each species in a manner peculiar to itself; the true firs crowd over 90 per cent of their total germination into the first season after the removal of the forest; white pine reaches its highest figure during the third season; hemlock shows a similar tendency; and Douglas fir germinates probably less than half its seed during the first season.

A few years after the fire the greening of the hills in the Columbia burn, with a mantle of reproduction, arrested the attention and aroused the wonder of those who had taken for granted that this was to be a "denuded" burn, because of its large extent and the thoroughness of the burning. The occupation of the burn by reproduction was not a gradual creeping out from the surrounding bodies of green timber, but a sudden taking possession of the entire area. In ordinary wind migration 11 years (the age of this burn) would have sufficed for only the beginning of a process of reforestation that would take generations to complete. Yet within 5 years the duff-stored seed had clothed fully 80 per cent of the Columbia burn with a satisfactory tree cover.

CUT-OVER AREAS. THE EFFECT UPON RESTOCKING OF BURNING SLASH AND OF LEAVING SLASH UNBURNED

.   The conclusions reached with regard to the source of seed from which reproduction starts on burns have been corroborated by the study of cut-over lands about Puget Sound during 1915 and 1916. In the Douglas-fir region covered by this study the slash is left broadcast and almost invariably burned. Thus, a study of cut-over areas is only another step in the study of burns, and covers a fire which is more severe on the surface because of the débris left by logging operations.

On one area cut over in 1914, burned in the spring of 1915, and left covered with blackened logs and ashes, cones of cedar and hemlock were found almost everywhere buried in the unburned duff. The cones were papery and almost decayed, but they were still intact in shape and contained seeds with firm endosperm. Throughout this burn there were small islands which had escaped burning and which contained large numbers of cedar and Douglas-fir seedlings germinated in the fall of 1915. These islands of reproduction demonstrate the presence of seed over the area previous to the 1915 fire, and show that if the slash had not been burned the area would have been covered with a good stand of reproduction of the same species as the original forest and in about the same proportion.

Reproduction forming a satisfactory stand—that is, 300 or more seedlings per acre—was not found more than 4 chains from green timber, and usually only 2 chains in localities where the stand depended entirely upon seed produced after the burn.   On one area which had burned over twice no reproduction followed the second burn.   The patches which escaped the second fire have good reproduction of Douglas fir, cedar, and hemlock of age classes which indicate they germinated shortly after the cutting, while the area reburned supports only occasional seedlings. These seedlings very likely are due to seeds carried in by winds or animals after the second burn.

At Esperance, Wash., south of Martha Lake, a 20-year-old stand of good reproduction was found following a single slash fire (Pl. 7, C).   The débris was only partially consumed, and evidences of a rather light burn were apparent.   At Bitter Lake, south of Esperance, a good stand of reproduction 9 years old occurred in a slash area covered by a single fire   The remaining stumps and logs are only fire scarred and burned lightly—evidences of a light slash fire.   This stand of reproduction occurred at distances of 8 chains and more from green seed trees.   Another area in this locality burned by a severe slash fire had a dense stand of reproduction at 8 or 10 chains from seed trees

Excellent stands of reproduction 20 years old, Douglas fir principally, with scattered western red cedar and western hemlock, occurred along the Edmonds Road about ½ mile east of the Pacific Highway.   These are examples of reproduction in unburned slash following cuttings. The stands are great distances from green seed trees, and the even-aged class establishes the fact that they all followed within a few seasons after cutting.   The fact that no evidences of a slash fire and no seed trees were found at the time of the study indicates that no seed trees were in the area after cutting.   Had there been any seed trees left after cutting, they would still be there, since only a small percentage of trees left after cutting are wind-thrown in this region.

The results secured by the study of cut-over areas were not so definite as those obtained on burns in virgin stands.   The exact history of the areas could not be obtained, without which the interpretation of results was difficult and in many instances impossible.   The history of burns in virgin forests can usually be definitely determined; the complex conditions generally prevailing in cut-over areas as a result of man's operations, are as a rule difficult to analyze and often impossible of interpretation.   Final results in this field can be obtained only by means of the permanent plot, the complete history of which is known.

CONCLUSIONS

The study of burns and cut-over areas in the Douglas-fir region of the Pacific Northwest has brought out the following facts:

(1) The distance to which seed trees are capable of restocking the ground is limited to from 150 to 300 feet.   They can not, therefore, account for the restocking of the large burned areas.

(2) The irregular, dense stands of young growth are due to seed stored in the forest floor or in cones. This seed retains its viability through the fire and is responsible for the dense reproduction that springs up after the first fire.

(3) The even-aged stands of reproduction immediately following a fire, regardless of location of remaining seed trees, the irregular alternation of dense stands of reproduction with grass areas, and the failure of reproduction on areas burned over by a second fire before the stand reaches seeding age or by consuming all of the duff and precluding any possibility of seed remaining after the fire, all point to the seed stored in the duff as the principal source of seed responsible for the restocking.

(4) The ability of the seed to retain its viability when stored in the duff or when retained in cones during fires has been further demonstrated by recovering and germinating seed from duff under forest conditions and by recovering and germinating seed from cones which passed through a crown fire.

PRACTICAL VALUE OF THE RESULTS

The fundamental facts of seed storage and viability brought out by this study have thrown new light on the whole question of natural reproduction.

The even-aged second growth stands of Douglas fir which are the general rule throughout the Pacific Northwest are due to seed stored in the ground. It has been clearly shown that these even-aged stands could not be produced by the seed scattered by the remaining seed trees. The distance of effective migration during each generation is usually only from 150 to 300 feet, varying with the adaptation of the seed for wind distribution. Migration due to other agencies than wind is responsible for individual seedlings far from the parent tree. Such individual seedlings, however, are of no practical significance in restocking large areas.

The establishment of a forest by means of wind-disseminated seed is a slow process; and with only this means of succession practically all of the large burns would be denuded areas or would support but a few scattered trees, until several generations of these trees, grown within these areas, could again restock them. This would inevitably produce an uneven-aged and irregular forest; yet the stands which follow most of the burns are even-aged. When a forest is destroyed by fire or cutting and is replaced simultaneously over large areas, the succession depends upon the seed produced at the time of or before the destruction of the forest and the ability of the seed to retain its viability through the period of destruction, whether by fire or cutting. This form of succession is the replacement of a forest almost immediately by the same species which comprised the original stand and usually in the same proportions.

Since the seed must be produced by the stand before it is destroyed, the age at which different species begin to produce seed is of the utmost importance. It varies greatly, and this variation alone is often the controlling factor in determining the composition of the second growth. For example, when western white pine, Douglas fir, and knobcone pine (*Pinus attenuata* Lemmon.) appear in a mixed stand which is destroyed by fire, all of these species may again appear in the next stand; but if this second growth is destroyed by fire when it is from 10 to 12 years old the next stand will consist principally, if not wholly, of knobcone pine. The knobcone pine begins producing seed when it is 6 years old and is producing good crops of seed at 10 years; while the white pine and Douglas fir bear only occasional cones at ages under 12 years. Therefore the knobcone pine is the only species which has any seed present to produce a forest stand following the second fire. Instances of such types are the knobcone pine types on the Siskiyou National Forest.

For this same reason, fires which destroy young growth before it is of seeding age invariably produce a denuded area by precluding any possibility of immediate reproduction; hence, the great need for fire protection in young stands of timber. Areas of immature stands when destroyed by a second fire can be restocked only by the slow process of wind disseminated seed. Hence, the viability and delayed germination of forest seeds serves to explain many conditions in the virgin forest and its reestablishment after one or more fires. The nature and severity of the fire and the age of the forest are shown to be the principal factors in determining whether reproduction will or will not follow a burn in a virgin forest.

The practical value of stored seed in the forest floor for restocking cut-over areas depends on the condition in which the forest floor is left after cutting. As has been shown, the duff must not be burned or all of the seed is destroyed. In order to avoid the destruction of the duff, it is necessary to pile the slash and burn the piles when the fire will not spread over the entire surface of the ground. This method has proved satisfactory in the white-pine region of Idaho, where piling of slash is feasible. In the Douglas-fir region of the coast the piling of slash would not be practicable because of the large amount of débris and the consequent cost. The large percentage of the surface which would of necessity be burned over even by pile burning would reduce the value of the operation for conserving seed stored in the duff, which is usually all destroyed by broadcast burning of slash.

In regions where broadcast burning of slash is the only practical method of disposing of the débris left after logging, the seed trees after all may have to be depended upon to reseed the area, in which case it will be necessary to know just how much seeding can be expected from such seed trees.

Areas with about two seed trees left for every 3 acres have been studied, but the results are questionable because the age classes and distribution of seedlings which came in after the slash fire indicated that the seedling originated from stored seed. In order to determine the effect of seed trees, the complete history of the area will have to be obtained by the permanent-plot method.

LITERATURE CITED

(1) BEAL, W. J.
    1884. THE VITALITY OF SEEDS BURIED IN THE SOIL. *In* 23d Ann Rpt. State
        Bd Agr. Mich., 1883/84, p. 332-334. (Mich. Agr Exp. Sta. Bul 5.)
(2) ——
    1889. VITALITY AND GROWTH OF SEEDS BURIED IN SOIL. *In* Proc 10th Ann.
        Meeting Soc Prom. Agr. Sci. 1889, p. 15-16, illus.
(3) ——
    1894 THE VITALITY OF SEEDS BURIED IN THE SOIL. *In* Agr. Sci., v 8, no. 6/9,
        p 283-284
(4) ——
    1899. THE VITALITY OF SEEDS TWENTY YEARS IN THE SOIL. *In* Columbus Hort.
        Soc. Quart. Jour. Proc , v. 14 (Ann. Rpt. 1899), no. 3, p 143-144.
(5) ——
    1911. THE VITALITY OF SEEDS BURIED IN THE SOIL. *In* Proc. 31st Ann. Meeting
        Soc. Prom. Agr. Soc. 1910, p. 21-23.
(6) EWART, E. J.
    1908. ON THE LONGEVITY OF SEEDS. *In* Proc. Roy. Soc. Victoria n s v. 21,
        pt. 1, p 2-210, pl. 1
(7) HARRINGTON, G. T.
    1916. AGRICULTURAL VALUE OF IMPERMEABLE SEEDS. *In* Jour. Agr. Research,
        v. 6, no. 20, p. 761-796, 6 fig , pl 106.
(8) NOBBE, Friedrich, and HAENLEIN, Hermann.
    1877. UEBER DIE RESISTENZ VON SAMEN GEGEN DIE ÄUSSEREN FACTOREN DER
        KEIMUNG. *In* Landw. Vers. Stat., Bd. 20, p. 71-96, 13 fig.
(9) REES, Bertha.
    1911. LONGEVITY OF SEEDS AND STRUCTURE AND NATURE OF SEED COAT. *In*
        Proc. Roy. Soc. Victoria, n s. v. 23, pt. 2, p. 393-414, pl. 79-81.

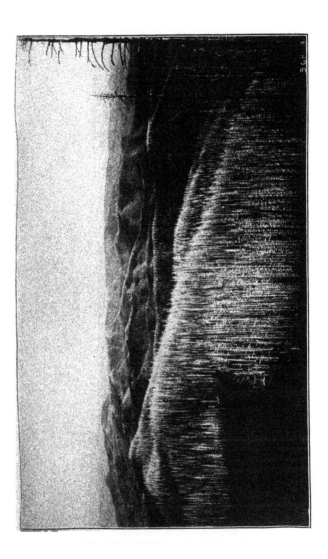

A.—"Interrupted" reproduction of noble fir on the west slope of Lookout Mountain. Note the clear-cut line between dense reproduction and grass areas. The 11-year-old seedlings here shown average about 30,000 per acre, yet these dense patches are contiguous with areas absolutely barren of seedlings. The grass areas mark the extent of ground fire; the seedlings grow where ground fire did not work. Photographed in Columbia National Forest, 1913.

B.—The picture shows the method of measuring the height of the seedlings by means of a graduated stick which was used also in determining the width of the transect, 8¼ feet. A representative stand of noble- and silver-fir reproduction on the northeast slope of Lookout Mountain. Photographed in Columbia National Forest, 1913.

C.—A fire line cut through 18 inches of litter and duff, Foss River fire, September, 1914, Snoqualmie National Forest.

D.—This picture was taken just inside the fire line, about 15 feet from the upper picture. At the time of the fire, September 2, the forest floor was very dry and the duff was completely consumed in many places Many trees fell after the fire because of the exposure and burning of the roots. Reproduction does not occur on such areas except near seed trees.

PLATE 4

A.—Reproduction in a 1902 burn of the Cowlitz area, about 10 chains from the nearest seed trees. White-pine seed trees are nearly ½ mile distant and are 800 feet below this area. Species shown are white pine (left), Douglas fir (center), noble fir (right). The stand on this slope is uniformly open, but will be sufficient ultimately to develop a satisfactory forest.

B.—Results of a 1910 fire following a 1902 burn. A practically denuded area. The reproduction which followed the fire had not reached seeding age when it was destroyed by the second fire. Note the dead young stand killed in 1910. Photographed in Rainier National Forest, July, 1914, on Cowlitz area.

7765°—17——3

PLATE 5

A.—Douglas fir, 16 years old, which has been producing seed since its fourteenth year. Photographed on Rainier National Forest on East Canyon Creek, 1914.

B.—A clump of reproduction which escaped a 1910 fire (Upper Cispus, Rainier National Forest). The white pine, bearing seven cones, was 15 years old. Photographed in 1914.

C.—A clump of true firs which escaped all fires, surrounded by a small area of reproduction which resulted from 1875 fire, but was injured, and some of it destroyed by subsequent fires, notably that of 1892. Photographed in Rainier National Forest, 1914.

D.—A barren waste left where the fire of 1892 followed the fire of 1875. A striking illustration of the limit of seeding distance of green timber into an open burn. The camera station is well beyond the farthest wind-cast seed. Seedlings from such seed were not encountered until within 5 chains of the timber shown. Photographed in Rainier National Forest, 1914.

PLATE 6

A.—Reproduction resulting from a single fire in a typical mature Douglas fir forest. This is an example of what is usually found on good sites where ground fire has not been at work during the general conflagration. A fine stand of Douglas fir and associate reproduction in the "old forest" portion of the Tower Rock burn of 1902. Photographed in Rainier National Forest, 1914.

B.—Homesteaders clearing in an alder "bottom" surrounded by second growth Douglas fir. This even-aged uniform stand of Douglas fir followed a virgin forest fire of about 1860. The charred stubs in the foreground and the tall, bare snags rising above the second growth are remnants of the former forest The burned area of this young forest in the 1902 fire at Tower Rock was not succeeded by dense reproducion as was the burned area of the virgin stand. Photographed in Rainier National Forest,

PLATE 7

A —Douglas-fir reproduction in the brush thickets of the Cispus burn, Rainier National Forest. In their struggle to keep abreast of the surrounding brush, the coniferous seedlings make tremendous height growth and very little diameter growth. The result is the slender "shade-form" reproduction here shown. These trees are holding their own and will soon overreach the crowding brush; in the end they will dominate and completely subdue it. The deciduous growth shown includes Oregon maple, vine maple, dogwood, alder, brake fern, and fireweed. Photographed in 1914.

B.—A 1902 fire followed by a 1910 fire, photographed in 1915, Oregon National Forest, near Summit R. S. Note the sharp boundary line of the second fire The dense reproduction succeeded the fire of 1902 and proved an effective barrier to the fire of 1910. This stoppage of fire by dense reproduction is a matter of frequent occurrence.

The picture shows also a striking comparison of duff-stored versus wind-blown seed as a source of reproduction. The dense, even-aged reproduction at the left resulted from duff-stored seed following a first burn in virgin forest. The scattered few, uneven-aged, young seedlings to the right resulted from seed blown out since 1910 from a group of seed trees from beneath which this picture was taken The scrubby reproduction to the left was seeded from the same source, since 1902, in an opening left barren of "stored-seed reproduction" on account of ground fire. It can be readily appreciated how slow will be the reforestation of the reburned area, which is now altogether dependent upon the few seed trees which have survived both fires. When the 13-year-old stand reaches seeding age it will, of course, contribute its seed to the same purpose, within the limited range of its influence. Such a process could not conceivably have produced the beautiful uniformity in density and age which is exhibited by the 13-year-old stand here pictured.

C.—Reproduction 20 years old, at Esperance, Wash. The stand shown followed a light slash fire. In the immediate foreground a second fire occurred The area covered by the second fire has no reproduction. Puget Sound cut-over studies. Photographed in 1915.

PLATE 7

A NEW PARASITIC NEMA FOUND INFESTING COTTON AND POTATOES

By N. A. Cobb,

Technologist in Charge, Office of Agricultural Technology, Bureau of Plant Industry, United States Department of Agriculture

INTRODUCTION

The new species of Tylenchus described below first came to the attention of the writer in 1911, when it was found infesting rootstocks of violets (*Viola* spp.) in New York. Practically all the morphological data given herein were obtained at that time; later examinations have, for the most part, been merely confirmatory. Since its discovery this nema has again and again come to notice, first on one plant, then on another, until it is now known to infest violets, cotton roots (*Gossypium* spp.), potatoes (*Solanum tuberosum*), and the roots of camphor (*Cinnamomum camphora*) over an area in the United States extending from Florida to New York and as far west as Michigan.

It injures all the plants upon which it has been found, and, though the extent of the possible damage is not yet fully ascertained, it has been sufficient to cause growers to make inquiry concerning the origin of the injuries they were suffering. The occurrence of this parasite on potatoes is of unusual interest, for the reason that any pest found upon potatoes—at any rate one whose lesions are similar to those of *Tylenchus penetrans*—is likely to become widely disseminated unless special preventive measures are adopted. Experience emphasizes the very grave danger of spreading serious pests in seed potatoes. If noninfested land be seeded with infested potatoes, the result is, in effect, the planting out of the pest under conditions very favorable to its increase.

Unfortunately our knowledge of nemas, especially those attacking the roots of plants, is relatively very meager, and in the possession of but few. First or last, the existence of root diseases caused by these organisms comes to the attention of numerous observers, who, not having the necessary literature at their command and not knowing to whom to send specimens, often let the matter drop without adequate investigation. As it is very desirable to know more about the frequency of this pest, its host plants, the extent of the damage done by it, on what plants and at what seasons the damage is greatest, and in what kind of soil it is most active, this preliminary note aims to enlist the services of such observers.

Journal of Agricultural Research,
Washington, D. C.

iy

Vol. XI, No. 1
Oct. 1, 1917
Key No. G—121

The illustrations have been very carefully prepared, and with their aid almost any careful microscopist should be able to identify the species. For this purpose it is advisable to use living specimens under sufficient pressure to prevent active motion. This method will give somewhat exaggerated width measurements, for which due allowance should be made. To see all the features shown, requires the use of an immersion lens under favorable conditions.

FIG. 1.—*Tylenchus penetrans:* Adult female. It is believed that the glands, *gl sal*, empty into the pharynx via the esophagus, as in *Tylenchus similis*, but the full details of the structures have not been worked out in *T. penetrans*. *ac*, accessory piece to spicula; *al*, wing of cuticle; *an*, anus; *blb oe*, median esophageal bulb; *blb on*, bulbs of the spear; *brs*, bursa; *cav som*, body cavity; *cl ren*, cell of the renette; *cl ut*, one of the cells of uterine wall; *cst brs*, ribs of bursa; *cut*, cuticle; *dct ren*, duct of the renette; *ex p*, excretory pore; *gl sal*, "salivary" gland; *int*, intestine; *int grn*, granule of the intestine; *lb*, lip region; *lum oe*, lumen of the esophagus; *lum on*, lumen of the spear; *msc blb*, radial muscles of bulb; *msc on*, protruding muscles of the spear; *msc som*, somatic muscle; *ncl nrv*, nucleus of nerve cell; *ncl ren*, nucleus of renette cell; *nrv r*, nerve-ring; *oe*, esophagus; *on*, spear; *on dir*, guiding apparatus for spear; *or*, mouth; *ov*, ovary; *rcl*, rectum; *sp*, spiculum; *sub cut*, subcuticle; *trm*, terminus; *ut*, uterus; *ut post rud*, posterior rudimentary uterus; *vag*, vagina; *vlv*, vulva. This specimen is mature, but has not yet begun to deposit eggs. The empty uterus indicates that the specimen is probably a virgin female.

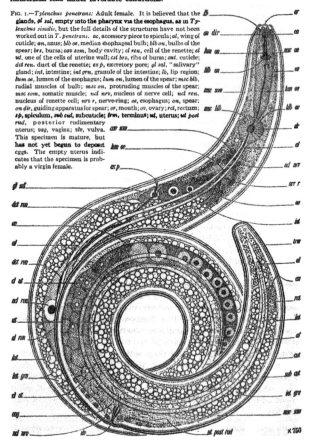

×750

APPEARANCES CAUSED BY THE INFESTATION

1. ON POTATOES

On potatoes the lesions are as follows: The tuber is spotted here and there, or over its whole surface, by pimples—slightly elevated, somewhat conical areas, 2 to 5 mm. across, having a roundish, broadly elliptical or quadrangular contour. Where one or more pimples run together, the contour may be considerably elongated. Near the center of each diseased area there arises a minute opening through which the young nemas escape to found new colonies. After the potatoes have been dug and have shrunken somewhat, the diseased areas may become surrounded by shallow channels, owing to the collapse of the immediately adjacent tissues, and this shrinkage may be sufficient to cause the whole of a diseased area to present somewhat the appearance of a depression (fig. 2).

× 3

FIG. 2.—Potato pustules caused by the nema *Tylenchus penetrans* At the left a double-headed pustule. These are from a potato that had been stored, and show the peripheral channel mentioned in the text.

2. ON VIOLETS

The rootstocks of violets infested with *Tylenchus penetrans* display brownish, somewhat collapsed areas. As in the potato, the tissues of these collapsed areas are inhabited by both sexes of the nema, and the infestation may be so intense that areas only a few millimeters in length harbor scores of specimens. The violets are very seriously injured, and as a result of careful examination, it has become quite evident that the nemas are the main cause, and in all probability the sole cause, of the injury.

3. ON COTTON ROOTS

Cotton plants infested by *Tylenchus penetrans* have been found in Georgia and North Carolina, the nemas being discovered in considerable numbers in small, diseased, colored areas on the roots. At the time the discoveries were made, the plants were not suspected of being infested; they were being examined in a general way in an effort to find out something about the nemas associated with the roots of cotton plants. No definite evidence exists that *T. penetrans* is a serious pest of the cotton plant.

4. ON THE ROOTS OF CAMPHOR TREES

Thus far no definite evidence exists that *Tylenchus penetrans* seriously injures the camphor tree. It was found in small diseased areas on the feeding roots, but not in such numbers as to seem to be the cause of trouble.

NUMBER OF HOST PLANTS

As a rule a given parasite infests but few hosts. In the genus Tylenchus and a few related genera, there are some marked exceptions to this rule, one of the most astounding of which, *Heterodera radicicola*, infests nearly 500 different species of plants. Not long ago in this Journal [1] attention was called to another species, *Tylenchus similis*, which infests such widely different plants as the banana and sugar cane. *Tylenchus penetrans* is another addition to the already rather formidable list of Tylenchi that infest a number of widely different plants; and a glance at the list of hosts just recorded warns us that we need not be surprised to find it infesting almost any of our crop plants.

IDENTIFICATION

As there are a number of Tylenchi closely resembling *Tylenchus penetrans*, no nema should be assigned to this species unless its anatomical details clearly correspond with all those given herein. The features that should be examined with special care in making comparisons are: Lip region; spear and median bulb of the esophagus; location of the excretory pore; the almost complete absence of wings;[2] the form of the tail, especially that of the male; and the size and general proportions of the body. These latter, however, may vary to some extent in accordance with the age of the specimen, and in the case of the females, according as they are or are not in a gravid condition. Some nemas, while keeping the relative proportions of many of the different parts, vary considerably in size, so that some specimens may be twice as long as others.

LOSSES

No definite information exists as to the full extent of the injury caused by this parasite upon cotton and camphor. In the case of violets the losses consist of failures, and poor growth resulting in inferior bloom. Such losses may reach an amount that constitutes a very serious handicap.

Present information indicates that the losses in the case of potatoes may sometimes be serious. The potatoes that reach a salable size often have an inferior appearance that detracts roughly 10 to 20 per cent from their market value, or even more. The infested tubers are inferior in size

[1] COBB, N. A. TYLENCHUS SIMILIS, THE CAUSE OF A ROOT DISEASE OF SUGAR CANE AND BANANA. *In* Jour. Agr. Research, v 4. no 6, p. 561-568, 2 fig. 1915.

[2] These are too strongly emphasized in figure 1.

as well as in quality, so that the yield is materially lessened. When the parasite infests the roots, as well as the tubers, it may cause an actual falling off in the number of tubers that are formed.

MUSCULAR POWERS: PENETRATION

When removed under the surface of water from potato pustules, the nemas sometimes lie so loosely in the tissues that they fall out three or four at a time with every movement of the dissecting needle. The structure and general appearance of this parasite discountenance the idea that its body movements are very active, at any rate after entering the host tissues. Specimens removed from potatoes exhibit only slight and weak movements.

The oral spear and the median bulb of *Tylenchus penetrans* are so well developed as to indicate clearly that the parasite is one readily capable of forcing its way into the tissues of the host plant. This fact will be appreciated by comparing these features with the corresponding features of other destructive species.

CONTROL

One important preventive measure is at once suggested by the presence of this parasite in the tubers of the potato—namely, great care in the selection of seed potatoes.

So far as potatoes are concerned, the problem of combating *Tylenchus penetrans* is very similar to that of combating the ravages due to *Heterodera radicicola;* the essence of it may be summed up in the old saying that "an ounce of prevention is better than a pound of cure." In this particular case it might almost be said that "an ounce of prevention is better than any amount of cure;" or rather, attempt at cure, for there is no assurance that land once infested with these nemas can ever be cleared of them. In the case of the average parasite, infesting but a single host, rotation of crops offers a solution in many cases at once effective and comparatively inexpensive. When, however, the parasite is capable of sustaining itself on a variety of hosts, especially if common weeds are among these hosts, the rotation of crops may not prove a thoroughly efficient remedy. It may be that no paying crop that is immune can be found. Even if such a crop exists there is the ever present menace that the parasite may maintain itself in the roots of weeds, difficult and expensive to exterminate. These facts strongly emphasize to potato growers the advisability of being very careful to plant nothing but perfectly healthy tubers.

An examination of potatoes treated with mercuric chlorid for scab showed that this treatment has a certain value in killing *Tylenchus penetrans* in the tubers. The scabby potatoes were also infested with *T. penetrans.* The vitality of the nema was much reduced and many of them appeared to have been killed by the mercuric chlorid.

SPECIFIC CHARACTERS OF THE PARASITE

Tylenchus penetrans, n. sp.—* $\frac{3.9}{2.1}...\frac{11.}{4.9}\infty\frac{30.}{76.5}\frac{92.701}{4.1}\cdot\frac{91.}{2.1}\cdot^9$ ■   The thin, color-
less, transparent, naked cuticle is traversed by about 400 plain transverse striæ,
resolvable with moderate powers. There are six amalgamated lips

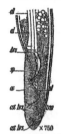

on which the cuticle is thickened in such a way as to have the
appearance of a frontal cap. This cap is apparently built from about
three ordinary striations. There are no labial papillæ. The spear
is about twice as long as the lip region is wide, and is composed
of two distinct regions. The posterior half is a cylindrical shaft
about 1½ μ in diameter. At the very base this portion expands
into a 3-lobed bulb having a diameter twice as great as that of
the shaft itself. The anterior half of the spear tapers. Nothing
is known concerning the lateral organs. There are no eye spots.
The esophagus may be traced backward from the median bulb a
distance approximately equal to the corresponding body diameter, and ends more or less indefinitely. The anterior ovary is
elongated and outstretched and ends at a point as far behind

Fig. 3.—*Tylenchus*
penetrans. Profile
of the tail end and
bursa of the male.
The two posterior
pairs of bursal ribs
are very faint and
easily escape ob-
servation. Com-
pare with figure 4.

the median bulb as the latter is behind the anterior extremity.
The eggs occur one at a time in the uterus, and segmentation begins
before they are deposited. The eggs are thin shelled, elongated,
fully twice as long as the body is wide, and about one-third as
wide as long. Eggs found deposited by the nemas in the tissues
of the host-plant measured 78 by 25 μ, and contained fully formed
embryos with well-developed spears three-fourths as long as those
of the adults (fig. 1).

* $\frac{3.9}{1.}...\frac{11.}{4.9}\infty\frac{16.}{7.6}\cdot\frac{-1.}{4.5}\cdot\frac{94.9}{2.1}r^9$ ■   The bursa surrounds the tail completely,
springing from near the lateral lines at a point opposite the proximal ends of the two
equal arcuate spicula. The general contour of the tail end of the male when viewed
dorsoventrally shows no enlargement, the distance across the

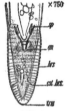

bursa being nowhere greater than the diameter of the body
immediately in front of the anus (fig. 3 and 4).

Habitat.—Parasitic in the roots of violets, Rhinebeck, New
York; roots of camphor trees, Orlando, Florida; roots of Upland
cotton, Millington, North Carolina, and Statesboro, Georgia,
in potatoes, vicinity of Kalamazoo, Michigan.

Fig. 4.—*Tylenchus pene-*
trans: Ventral view of
the tail end of the male.
Compare with figure 3.
The two pairs of ribs
near the terminus were
nearly invisible in this
specimen.

SUMMARY

(1) A parasitic nema, *Tylenchus penetrans*, n. sp.,
has been discovered infesting the tubers of the potato,
the feeding roots of camphor, the rootstocks of violets,
and the roots of Upland cotton.

(2) External indications of the presence of the
nema are the existence on the roots or tubers of
small, abnormal-looking areas, a few millimeters
across, sometimes in the form of pimples, but more
often in the form of slightly sunken, discolored areas.
Each of these diseased areas when fully developed contains up to about
50 specimens of *T. penetrans* in various stages of growth.

(3) The geographical distribution of the pest is suggested by its presence in Florida, Georgia, North Carolina, New York, and Michigan.

(4) The occurrence of this disease under such different climatic conditions and in such a diversity of hosts makes it certain that the nema causing it is another species which, like some other destructive members of its genus, can adapt itself to widely varying conditions.

(5) As yet too little is known about this parasite to accurately estimate the damage done by it.

(6) The occurrence of the parasite in the tubers of the potato is a peculiarly significant fact and again points to the necessity of being particularly careful to plant only perfectly healthy potatoes. The mercuric-chlorid treatment of potatoes, as for scab, decreases the vitality of the nemas.

APPENDIX

Figure 5 is a diagram illustrating the decimal formula used for nematodes; 6, 7, 8, 10, 6 are the transverse measurements, while 7, 14, 28, 50, 88 are the corresponding longitudinal measurements. A formula assembling these measurements appears just below the diagram. The unit of measurement is the hundredth part of the length of the body, whatever that may be. The measurements become, therefore, percentages of the length. The absolute length of the nema is given in millimeters as a final term, in this case 1 mm.

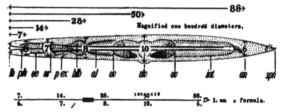

Fig. 5.—Diagram showing measurements of a nema and the corresponding formula.

The measurements are taken with the animal viewed in profile; the first are taken at the base of the pharynx, the second at the nerve ring, the third at the cardiac constriction or end of the neck, the fourth at the vulva in females and at the middle (M) in males, the fifth at the anus. The presence, situation, form, and extent of various organs is indicated in the formula by self-explanatory marks; in this case, the excretory pore, wings, spinneret, and sexual organs are so indicated. The formula thus becomes a sort of conventionalized sketch of the organism.

In this article the writer has adopted the improvement suggested by Micoletsky, and have split the measurement indicating the extent of the female sexual organs so as to show the relative proportions of the anterior and posterior branches. Thus, 50-f-5 indicates that the entire apparatus, consisting of two outstretched parts, occupies 55 per cent of the length of the nema, but that the posterior branch is a mere rudiment only five one-hundredths as long as the body.

ENDROT OF CRANBERRIES[1]

By C. L. SHEAR,

Plant Pathologist, Fruit Disease Investigations, Bureau of Plant Industry, United States Department of Agriculture

INTRODUCTION

In the writer's early studies of diseases of the cranberry (*Oxycoccus macrocarpus*) occasional cultures of an undetermined organism of a rather striking and characteristic color appeared. These cultures, however, failed to produce spores, and their identity was not determined at the time. Since beginning cooperative experiments with the Massachusetts Agricultural Experiment Station on Cape Cod, this organism has been found to be the most frequent fungus causing a rot in Late Howe berries from the experimental plots and checks on the Massachusetts State Cranberry Experiment Station bog at East Wareham. The fungus has also been frequently found in diseased berries, especially the Late Howe, from Maine, New Jersey, Michigan, Wisconsin, Washington, and Oregon. Although this rot has been found thus far most abundantly in the Late Howe cranberry, further investigation may show that it is equally serious with some other varieties. It has been isolated from the following varieties: Bennett Jumbo, Cape Cod Beauty, Early Black, Early Ohio, Mathews, McFarlin, Perry Red, Prolific, Searles Jumbo, Selected Howe, Vose Pride, and several unknown varieties. It no doubt occurs in others that have not yet been examined. Sufficient observations have not been made at present to determine the relative susceptibility of these varieties. It is evident from the amount of this rot found in various samples of fruit from all the cranberry-growing sections the past two or three years that it is the cause of much loss. Whether the recent increase of this rot indicates that the disease is becoming more serious or has only appeared more frequently the past year or two on account of unusually favorable conditions for its development and distribution is not known.

[1] Grateful acknowledgment is due Miss A M Beckwith and Mr Bert A Rudolph, of the Bureau of Plant Industry, for assistance in making cultures and sections of the fungus, and to Dr H J Franklin, of the Massachusetts State Cranberry Experiment Station, who carried out the spraying experiments in Massachusetts.

Journal of Agricultural Research,   Vol. XI, No 2
Washington, D. C   Oct. 8, 1917
In   Key No. G—122

(35)

CHARACTERISTICS OF THE DISEASE

In most cases in which the endrot organism has been isolated from berries, the fruit first showed signs of rot at the blossom end. Plate A, figure 1, shows the characteristic appearance of Late Howe berries in which the rot started at the blossom end. This was so generally the case in the early collections that the name "blossom end rot" was suggested by Dr. Franklin. Later a quantity of rotten fruit was received from Mr. O. G. Malde, of the Wisconsin Cranberry Station at Grand Rapids, Wis. These berries were affected with what he called "stem end rot," because it usually first appeared at the stem end of the berry. Isolations from this lot of fruit gave almost entirely the endrot fungus. Typical cases of the endrot beginning at the stem end are shown in Plate A, figures 2, 3, 4, and 5. Further studies have shown that rot caused by this organism usually begins at one end or the other of the fruit and rarely, if ever, at the side. The common name "endrot" has therefore been adopted to distinguish this disease in a general way from scald, anthracnose, and other rots which usually begin on some other part of the berry. Unfortunately most of the rots of the cranberry are difficult or impossible to distinguish by the superficial appearance of the diseased fruit.

Endrot first appears as a softening of the tissues accompanied by a slightly yellowish or brownish yellow watery discoloration of the skin. The diseased part is lighter colored than the sound portion of the berry, the various colors and shades usually occurring as shown in Plate A. This discoloration spreads as the rot develops, until the whole fruit is involved and becomes soft and elastic to the touch, but remains turgid.

CAUSAL ORGANISM

The tissues of the fruit affected with endrot are found to contain numerous hyalin, branched fungus filaments which when transferred to favorable culture media produce the characteristic growth and fructifications of one of the Sphaeropsidaceae which appears to be undescribed. In the present state of knowledge of this group of fungi it is difficult to assign the organism satisfactorily to a particular genus. For the present, or until its full life history is positively determined, it may be referred to the genus Fusicoccum as *F. putrefaciens*, n. sp., and characterized as follows:

Fusicoccum putrefaciens, n. sp.

Pycnidia subglobose to pyriform, rather thick-walled, more or less roughened, tawny to tawny brownish, embedded at first, becoming erumpent or subsuperficial when mature, sessile, or subsessile, simple or irregularly chambered forms 300 to 400 μ in diameter, larger chambered forms 400 to 450 μ; pycnospores elliptic to fusiform, hyalin or very faintly yellowish in mass, continuous or pseudoseptate, 8 to 18 by 2 to 3 μ, mostly 10 to 12 by 2.5 μ; sporophores simple or branched, cylindrical or somewhat tapering above, 20 to 36 by 2 μ

Type specimen.—No. 2918 on rotten prolific cranberries grown near Walton Junction, Michigan, 1916.

Distribution.—Maine to New Jersey and west to Oregon and Washington.

The pycnidia in culture are ovoid to globose or depressed globose with simple or irregular labyrinthiform chambers, erumpent or subsuperficial, 500 to 800 μ in diameter. The inner wall of the pycnidium is sordid yellowish and the pycnospores somewhat more variable in their extreme measurements than in specimens found on old cranberries, under natural conditions, but averaging about the same size. Figure 1 shows sections of pycnidia and spores from a pure culture of the fungus. When mature, the apical

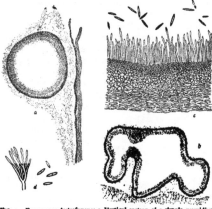

Fig. 1.—*Fusicoccum putrefaciens*: *a*, Vertical section of a simple pycnidium grown in pure culture on sterile sweet clover stems (*Melilotus alba*); *b*, vertical section of a chambered pycnidium from a corn-meal-agar culture, X 43.5; *c*, part of a section of *a* showing sporophores and spores, *d*, separate sporophores and spores from pure culture, X 450.

portion of the pycnidium ruptures more or less irregularly, permitting the escape of the pycnospores which are expelled in a mucilaginous whitish mass. Later more or less of the top breaks away as shown in figure 2, *a*.

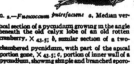

Fig. 2.—*Fusicoccum putrefaciens* *a*, Median vertical section of a pycnidium growing in the angle beneath the old calyx lobe of an old rotten cranberry, X 43.5; *b*, similar section of a two-chambered pycnidium, with part of the apical portion gone, X 43.5; *c*, portion of inner wall of a pycnidium, showing simple and branched sporophores with immature spores; *d*, separate pseudoseptate spores, X 450.

The pycnidia appear to be produced but rarely under natural conditions in the field, as they have been found only twice, although thorough search of diseased bogs and all parts of the cranberry plant has frequently been made. In a careful examination of a large number of dried rotten berries from a pile of discarded fruit about a year and a half old at the Massachusetts Cranberry Station, a considerable number of mummied berries were found, having a dirty yellowish color suggestive of that produced by the endrot fungus. On one of these berries four or five pycnidia bearing spores were found hidden beneath the old shriveled calyx lobes. Figure 2 shows sections of two of these pycnidia, *a* the simple form and *b* the chambered form. Sporophores with immature spores from the same specimen are shown in *c* and mature spores in *d*.

In April, 1917, a box of spoiled cranberries was received from Sioux City, Iowa, where they had been stored in the chaff during the winter. This fruit was of the Prolific variety and was grown near Walton Junction, Mich. A considerable part of the berries was affected with a fungus rot having the external appearance of endrot. Two hundred berries of this kind were selected and cultures made from each by transferring a portion of the pulp from the interior of the diseased berries. Of the 200 cultures made, 157 produced the typical growth of the endrot fungus. A considerable number of the entirely rotten and crushed berries from this lot bore pycnidia on and about the blossom end. These fructifications were larger and thicker-walled than those from the old dry fruit illustrated in figure 2. Figure 3 shows sections of pycnidia from the Michigan berries. In these specimens the pycnidia are at first entirely buried in the tissue of the fruit, as shown in figure 3, *a*. As they develop, they break through the epidermis and finally become erumpent or subsuperficial, as shown in figure 3, *b*. When a portion of the sporogenous tissue of the pycnidium is crushed out so that the individual sporophores are separated, it is found that they are frequently branched as shown in figure 3, *c*. The spores often appear septate. No real septum, however, has been demonstrated, and the lack of uniformity in the division of the protoplasm seems to indicate that they are only pseudoseptate.

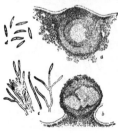

FIG. 3 —*Fusicoccum putrefaciens a,* Vertical section of a young pycnidium embedded in the tissue of a recently rotted cranberry, *b*, vertical section of a nearly mature pycnidium showing two chambers, × 45; *c*, separate branched and simple sporophores with immature spores; *d*, separate spores, showing variations in form and size, × 420.

LIFE HISTORY OF THE FUNGUS

Only pycnospores of the form described and illustrated have thus far been produced in cultures. On old rotten, dried, and mummied berries from 1 to 1½ years old, which had been apparently destroyed by the endrot fungus and had been lying on the ground exposed to the weather during the preceding winter, a discomycetous fungus has been found which is suspected of being the perfect form of this plant. The circumstantial evidence is as follows: The apothecia are found on and about the blossom end of the mummied berries which have the peculiar dirty-yellow color characteristic of the endrot fungus. They are also associated with old or obsolete pycnidia of *Fusicoccum putrefaciens*. In section they show the same character and color of the mycelium at the base and the same relation to the tissues of the host. Unfortunately as yet no cultures from ascospores have been obtained, as all the specimens collected have been too old or poorly developed. This fungus agrees very closely with *Cenangium urceolatum* Ellis (1, p. 9) [1] found on twigs of *Clethra alnifolia* in New Jersey and distributed in Fungi Columbiani No. 742 and North American Fungi Exsiccati No. 990. The author states (1, p. 9) that *Sphaeronema clethrincola* Ellis is the stylosporous form of this species. This statement is apparently based on the close association of the two forms on the same specimens, which may easily be misleading. The writer's suggestion made above in regard to the relation of the endrot fungus to this Discomycete may, however, have but little more value than that of Ellis. It is recorded here for the purpose of calling attention to this fungus, and perhaps thereby securing fresh material with viable ascospores from which pure culture studies may be made, and its life history positively determined.

CULTURAL CHARACTERS

The endrot organism grows well on corn-meal agar, potato agar, steamed corn meal, and stems of *Melilotus alba*. It fruits more frequently on stems than on the other media mentioned.

The superficial growth of the fungus is at first pure white, soon becoming pale green-yellow [2] to pale yellow-green, passing through La France pink to Mars orange, as indicated in Plate A, figure 8, which shows the appearance of a culture 39 days old on cranberry agar-gelatin.

On steam-sterilized stems of *Melilotus alba* the growth passes through the same shades during its early development, but finally becomes nopal-red or even garnet-brown in spots, as indicated in Plate A, figures 6 and 7, which show the appearance of cultures 46 days old

On corn-meal agar in petri dishes colonies 12 days old observed by transmitted light are white at the margin, passing through pale lemon-yellow and lemon-chrome to orange-citrine at the center and when older on slant agar tubes of the same medium produce colors very similar to those shown in Plate A, figure 8.

CONTROL OF THE DISEASE

As already stated, endrot does not usually show much evidence of its presence in the fruit before picking. The time and manner of infection have not yet been positively determined. From the fact that spraying during the growing season greatly reduces the rot, which usually develops after picking, it seems probable that infection occurs before the fruit is mature and that the fungus is in the berries in a dormant or slowly incubating condition when they are picked.

---

[1] Reference is made by number to "Literature cited," p 41

[2] The nomenclature of colors in the description of the cultures is according to Ridgway (RIDGWAY, ROBERT. COLOR STANDARDS AND COLOR NOMENCLATURE. 43 p., 53 col. pl. Washington, D. C., 1912.)

Under ordinary commercial conditions of handling after picking, sometimes 25 to 50 per cent of the Late Howe variety, which is one of the most seriously affected, will show endrot when screened and sorted for shipment two or three months after harvesting. In spraying experiments with Bordeaux mixture, carried on for several years in cooperation with the Massachusetts Agricultural Experiment Station at East Wareham, it has been found that in all cases there was a decided improvement in the keeping quality of the sprayed fruit and in some cases two or three times as much endrot developed in the unsprayed fruit as in the sprayed fruit. Details of the results of these experiments are given by the published reports of Franklin (3, 4, 5) and of the writer (6, 7).

The following formula and dates of applications followed in the general treatment of other cranberry rots have been found effective in reducing loss from this disease: Bordeaux mixture (4_{-3-50}) plus 2 pounds of commercial resin-fishoil soap. The first application should be made just before the blossoms open, the second just as soon as most of the fruit is set, the third about 10 days later, and the fourth when the fruit is about three-fourths grown. Even though spraying is practiced, in order to reduce the loss from endrot to the minimum, it is necessary to take all practicable precautions to pick and handle the fruit with care and place it in a cool, well-ventilated place as soon as possible. A low temperature is very essential in preventing the development of this rot in storage, as it has been found to develop to some extent at $_0°$ C. $_{(32°}$ F.$)$.

Whether or not it will be profitable to practice spraying for this disease will depend largely upon the amount of loss from decayed fruit usually sustained in any individual case. The greatly improved keeping and shipping quality of the sprayed fruit should, of course, be given due consideration in this connection.

It may also be mentioned that under the conditions obtaining on the sprayed plots at the Massachusetts State bog, some injury to the vines is apparent and, associated with it except for the second year (2, 5), was a reduction in the amount of fruit produced. The effect on the vines is not a burning of the new foliage as in ordinary injury by Bordeaux mixture, but a reddening of the old foliage. The cause of this is not at present understood. No injury of this sort has been observed in New Jersey, where spraying with Bordeaux mixture has been practiced on the same bogs for 8 or 10 years in succession. Further investigation and experiments in Massachusetts are necessary to determine if possible whether this case is due to some peculiar local condition or combination of conditions of this bog, and if so, what they are. No injury to vines has been reported or observed on other Massachusetts bogs which have been sprayed with this mixture for several years in succession.

SUMMARY

Endrot is a disease of the cranberry caused by a fungus which does not appear to have been described heretofore.

It has been found to occur in all of the cranberry-growing sections of the United States and in the past few years has occasioned considerable loss especially to the Late Howe variety.

The rot usually starts at either the blossom or stem end of the berry, finally producing complete softening of the fruit.

The causal organism is a sphaeropsidaceous fungus which is here described and named *"Fusicoccum putrefaciens*, n. sp."

Only the pycnidial form has been produced in culture. It is suspected on circumstantial evidence of being genetically related to a species of Cenangium resembling *C. urceolatum* Ellis.

Cultures produce a characteristic series of colors in the mycelium, and pycnidia and pycnospores are produced especially on stems of *Melilotus alba*.

Spraying experiments in Massachusetts have shown that this rot can be largely prevented by the use of Bordeaux mixture.

Some injury to the cranberry vines has been associated with the application of Bordeaux mixture on the experimental plots at the Cranberry Experiment Station in Massachusetts, but not elsewhere. The cause is being investigated.

LITERATURE CITED

(1) COOKE, M. C., and ELLIS, J. B.
　　1877. NEW JERSEY FUNGI. *In* Grevillea, v. 6, no. 37, p. 1–18, pl. 95–96. (Continued article.)

(2) FRANKLIN, H. J.
　　1913. STATE BOG REPORT. *In* Rpt. 26th Ann. Meeting Cape Cod Cranberry Growers' Assoc., 1913, p. 17–47.

(3) ———
　　1915. REPORT OF CRANBERRY SUBSTATION FOR 1914. Mass. Agr. Exp. Sta. Bul. 160, p. 91–117.

(4) ———
　　1916. REPORT OF CRANBERRY SUBSTATION FOR 1915. Mass. Agr. Exp. Sta. Bul. 168, 48 p.

(5) ——— and MORSE, F. W.
　　1914. REPORTS ON EXPERIMENTAL WORK IN CONNECTION WITH CRANBERRIES. Mass. Agr. Exp. Sta. Bul. 150, p. 37–68.

(6) SHEAR, C. L.
　　1912. REPORT OF CO-OPERATIVE CRANBERRY SPRAYING EXPERIMENTS IN MASSACHUSETTS FOR 1911. *In* Rpt. 25th Ann. Meeting Cape Cod Cranberry Growers' Assoc., 1912, p. 9–13.

(7) ———
　　1913. CRANBERRY SPRAYING EXPERIMENTS IN MASSACHUSETTS IN 1912. *In* Rpt. 26th Ann. Meeting Cape Cod Cranberry Growers' Assoc., 1913, p. 9–14.

PLATE A

1.—Five Late Howe cranberries showing typical appearance of endrot starting at the blossom end

2, 3.—Cranberries from Wisconsin, showing the appearance of the endrot starting at the stem end.

4.—Entirely rotten berry affected with endrot.

5.—Appearance of endrot on the lighter-colored fruit, beginning at the stem end. Cultures of *Fusicoccum putrefaciens* were isolated from the diseased tissues of all these berries.

6, 7.—Cultures of the endrot fungus 46 days old on stems of *Melilotus alba*, showing the various color changes in the fungus from the youngest to the oldest growth.

8.—Culture of the endrot fungus 39 days old on cranberry agar-gelatin, showing the various colors produced.

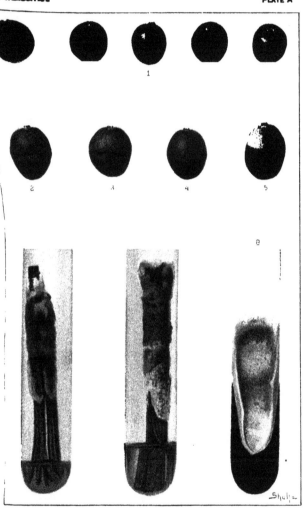

SOME FACTORS AFFECTING NITRATE-NITROGEN ACCUMULATION IN SOIL

By P. L. Gainey, *Soil Bacteriologist*, and L. F. Metzler, *Research Student, Research Laboratory in Soil Biology, Kansas Agricultural Experiment Station*

INTRODUCTION

The literature of nitrification in soils impresses one with the varied technic employed in such studies. It is an exception to find identical methods in use in any two laboratories. In many instances the differences are so slight that they would seem to have little effect upon nitrification, but in other instances the differences are too great, it would seem, not to influence bacterial activity. The variations in volume of soil, depth of column, ratio of volume to surface exposed, shape, size, and type of container, degree of compactness, and methods of preventing contamination and evaporation suggest that the variations which might be produced in bacterial activity could in part explain the wide variations in results often reported from different laboratories.

PRELIMINARY EXPERIMENT

Some preliminary experiments designed to test the validity of such a conception were conducted. No attempt was made to duplicate identically the various methods that have been used or suggested, but the principal variable factors were varied sufficiently to produce different results, provided differences actually arose from such variations. In Table I are reported the variations to which a soil was submitted, together with the results such variations produced upon nitrate accumulation.

In these and most of the succeeding experiments an Oswego silt loam of the following mechanical analysis was used: Fine gravel, none; coarse sand, 0.6 per cent; medium sand, 0.3 per cent; fine sand, 1.5 per cent; very fine sand, 6.9 per cent; silt, 71.8 per cent; clay, 17.8 per cent.

Ammonium sulphate was added at the rate of 60 mgm. of nitrogen per 100 gm. of soil; calcium carbonate, 1 gm. per 100 gm. of soil; water, 27 gm. per 100 gm. of soil. Incubation was for four weeks at room temperature. Nitrate nitrogen, as in all succeeding experiments, was determined by the phenoldisulphonic-acid method as modified by Lipman and Sharp (9). The results in this experiment are reported in milligrams of nitrogen recovered as nitrates per 100 gm. of soil. Ammonia was tested qualitatively by Nessler's reagent. Where reported as "abundant," a heavy red precipitate was formed. Where reported as "some,"

Journal of Agricultural Research,
Washington, D. C
hb

Vol. XI, No. 2
Oct. 8, 1917
Key No. Kans.—10

a strong yellow color without precipitate was formed. Where reported as "slight" or "trace" only a very light yellow color developed upon the addition of the reagent.

TABLE I —*Variations in the nitrates and ammonia produced in Orwego silt loam under different conditions of incubation*

Sample No.	Container.	Quantity of soil.	Nitrogen as NO₃.[a]	Ammonia (NH₃).
		Gm.	*Mgm.*	
1	100-c. c. Erlenmeyer flask, cotton-plugged	50	52. 8	Slight.
2	....do..............................	100	52. 8	Trace.
3	350-c. c. Erlenmeyer flask, cotton-plugged	50	48. 4	Slight.
4	. .do..............................	100	52. 8	Trace.
5	....do..	200	54. 2	Do.
6	500-c. c. Erlenmeyer flask, cotton-plugged..	50	50. 4	Do.
7	....do..............................	100	54. 0	Do.
8	....do..............................	200	54. 0	Do.
9	...:.do..............................	400	56. 9	Do.
10	1,000-c. c. Erlenmeyer flask, cotton-plugged.	50	53. 5	Slight.
11	....do..............................	100	54. 2	Trace.
12	....do..............................	200	54. 2	Do.
13	....do..............................	400	54. 8	Do.
14	....do..............................	800	55. 0	Do.
15	100-c. c. bottle, cotton-plugged..............	50	54. 0	Slight.
16	....do..............................	100	54. 2	Trace.
17	250-c. c. bottle, cotton-plugged..............	50	54. 0	Slight.
18	....do..............................	100	55. 0	Trace.
19	....do..............................	200	52. 8	Do.
20	500-c. c. bottle, cotton-plugged............	50	44. 9	Abundant.
21	....do..............................	100	54. 2	Slight.
22A	....do..............................	200	53. 0	Trace.
22B	....do..............................	200	53. 5	Do.
23A	....do..............................	400	....	Do.
23B	....do..............................	400	54. 8	Do.
23C	....do..............................	400	52. 8	Do.
23D	....do..............................	400	52. 8	Do.
24X	500-c. c. bottle, cotton-plugged............	[a] 100	19. 0	Abundant.
24Y	....do..............................	[a] 100	52. 8	Slight.
25XA	....do..............................	[a] 400	9. 9	Abundant.
25XB	....do..............................	[a] 400	1. 8	Do.
25XC	....do..............................	[a] 400	Trace.	Do.
25XD	....do..............................	[a] 400	Trace.	Do.
25YA	....do..............................	[a] 400	44. 0	Do.
25YB	....do..............................	[a] 400	35. 2	Do.
25YC	....do..............................	[a] 400	35. 2	Do.
25YD	....do..............................	[a] 400	35. 2	Do.
26	500-c. c. beaker, cotton-plugged............	[a] 100	50. 0	Slight.
27	500-c. c. beaker, petri-dish cover...........	100	52. 8	Do.
28	Tumbler, cotton-plugged............	100	52. 8	Do.
29	Tumbler, petri-dish cover..................	100	52. 8	Trace.
30	1-pint fruit jar, cotton-plugged............	100	52. 8	Do.
31	1-pint fruit jar, petri-dish cover...........	100	56. 9	Do.
32	....do..............................	200	55. 0	Do.
33	....do..............................	500	54. 2	Do.
34A	500-c. c. bottle, corked......................	300	18. 2	Abundant.
34B	....do..............................	300	18. 2	Do.
34C	....do..............................	300	18. 2	Do.
35	500-c. c. bottle, petri-dish cover...........	100	55. 0	Trace.

[a] The figures reported are averages of closely agreeing duplicates. Where duplicates varied widely, they are both reported as X and Y. A, B, C, etc., have reference to the depth from which samples were taken for analysis. When no letter is given, the whole sample was analyzed or thoroughly mixed and a 100-gm. sample taken. A, is in all cases the surface 100 gm.; B, second 100 gm.; C, third 100 gm., etc.

TABLE I.—*Variations in the nitrates and ammonia produced in Oswego silt loam under different conditions of incubation*—Continued

Sample No.	Container.	Quantity of soil.	Nitrogen as NO₃.	Ammonia (NH₃)
		Gm.	*Mgm.*	
36	Large petri dish, open....................	100	54. 2	Slight.
37	...do...............................	ᵃ 100	54. 2	Trace.
38A	20-inch cylinder, cotton-plugged	ᵃ 900	52. 8	Do.
38B.....	.....do............	ᵃ 900	58. 0	Do.
38C.....	.....do............	ᵃ 900	55. 0	Do.
38D.....	.....do............	ᵃ 900	57. 2	Do.
38E.....	.....do............	ᵃ 900	58. 0	Do.
38F.....	.....do............	ᵃ 900	59. 8	Do.
38G.....	.....do............	ᵃ 900	58. 0	Do.
38H.....	.....do............	ᵃ 900	55. 0	Do.
39A.....	.....do............	ᵃ 1, 400	52. 8	Some.
39B.....	.....do............	ᵃ 1, 400	49. 0	Do.
39C.....	.....do............	ᵃ 1, 400	37. 4	Do.
39D.....	.....do............	ᵃ 1, 400	26. 4	Abundant.
39E.....	.....do............	ᵃ 1, 400	7. 0	Do.
39F.....	.....do............	ᵃ 1, 400	.9	Do.
39G.....	.....do............	ᵃ 1, 400	Trace.	Do.
39H ᵇ...	.....do............	ᵇ 1, 400	00. 0	Do.
40......	500-c. c. bottle, sealed..................	100	55. 0	Trace.
41......	Large Fernbach flask	100	55. 0	Do.
42......	.....do...............................	1, 000	58. 3	Do.

ᵃ Packed. ᵇ Samples below this depth were not analyzed

The data recorded in Table I support the following observations:

(*a*) The quantity of soil varied from 50 to 1,000 gm., with but little, if any, effect upon nitrate accumulation. Apparently there were slightly larger quantities of nitrate nitrogen in the large volumes of soil and also slight quantities of ammonia nitrogen in the smaller volumes. However, the differences in both instances are probably within the limit of error. There is no evidence that increasing the volume of soil decreases nitrate formation.

(*b*) The depth of the column of soil varied from ¼ inch (No. 42) to 20 inches (No. 39) without producing any appreciable effect upon nitrate accumulation, provided the soil was left loose. There is also no evidence that nitrate formation 20 inches below the surface was any less vigorous than at the surface, provided the soil was left unpacked.

(*c*) Packing the soil in a thin layer (No. 38) was without effect. If, however, the depth of the column was appreciably increased, packing (reducing volume from 14 to 9) decreased very markedly the nitrate accumulation (No. 40), the accumulation becoming negative only a few inches below the surface. In view of the data to follow, it should be borne in mind that, with a water content of 27 c. c. per 100 gm. of soil, reducing the volume of this soil from 14 to 9 increased the percentage saturation from one-half to practically complete saturation.

(*d*) Decreasing the ratio of surface exposed per 100 gm. of soil from 314 square centimeters (No. 42) to 2 square centimeters (No. 39) was without effect upon the nitrate accumulation.

(e) The shape and size of the container, as well as methods of preventing evaporation and contamination, were without effect except when the container was stoppered tightly, and the volume of inclosed air was relatively small in proportion to soil volume.

The principal vital factor which the above variations would seem to affect is the access of oxygen. The fact that the factors influencing the access of oxygen could be varied as widely as in the experiments just recorded without in turn effecting the aerobic processes, suggested the advisability of further study along similar lines. The subsequent experiments herein reported were therefore designed to ascertain the effect produced upon the accumulation of nitrate nitrogen in soils by varying some of the more important factors controlling aeration.

According to Buckingham (1), aeration is a direct function of porosity or free pore space. This being true, the factors influencing porosity are the factors influencing aeration. In any given soil porosity is a direct function of compactness and moisture content, being in an inverse ratio to either one considered alone. In addition to these two variables, it is also essential to take into consideration the length of column or depth to which air or oxygen must diffuse. Accordingly these were the factors varied in the succeeding experiments. Typical experiments of those conducted will now be reported.

EXPERIMENTAL DATA

The data reported in Table II were secured under the following conditions: All samples contained 50 mgm. of nitrogen as ammonium sulphate and 0.5 gm. of calcium carbonate per 100 gm. of soil. Samples were incubated in a moist chamber at room temperaure for the periods indicated in the table. Such loss in moisture as occurred was replaced from time to time. Samples 1 and 2 were contained in large, open petri dishes. Samples 3 and 4 were incubated in wide-mouthed 500-c. c. bottles plugged with cotton. Samples 5 and 6 were in glass cylinders 24 inches high and 2½ inches in diameter. The moisture content of all samples was 20 c. c. per 100 gm. of soil. The samples marked "loose" were placed in the containers and the latter gently tapped on the bottom to cause the soil to settle into the larger openings. Those marked "compact" were so tamped that the same weight of soil occupied only two-thirds the volume that the "loose" soil occupied or in other words there were 1½ times by weight as much soil in same containers.

From the data contained in Table II the following facts seem evident: (1) The "compact" samples in the major portion of possible comparisons show a higher nitrate gain than do the "loose" samples. The greater gains in "loose" samples are so small as to be negligible. (2) The deeper soil columns show higher nitrate gains than do the shallow. (Samples 5 and 6 are exceptions, but for some reason these failed to show active nitrification. Other data recorded in this paper show that this low nitrification could not be due to the depth of column. Also, if access of oxygen were influencing nitrification these samples should have shown greater nitrate accumulation in upper layers.)   (3) The lower-lying strata of all samples show higher nitrate gains than do the upper layers.

TABLE II —*Effect of variation in depth of column and compactness of soil upon nitrate accumulation*

[Results are expressed as gain or loss in milligrams of NO₃ per 100 gm. of soil]

Sample No. a	Ratio, diameter to depth	Depth of column.	Sample from depth of—	Incubation.					
				10 days.		20 days.		30 days.	
				Loose.	Compact.	Loose.	Compact.	Loose.	Compact.
		Inches.	*Inches*						
1........	14:1	0. 4	0–4	0. 05	0. 60	....	4. 50	12. 60	......
2........	7. 5:1	1. 25	0–1. 25	−2. 00	2. 90	6. 20	10. 20	12. 70	26. 00
3A........	3. 5:1	2. 5	0–1. 25	− . 22	8. 60	7. 30	28. 80	26. 80	36. 00
3B........	3. 5:1	2. 5	1. 25–2. 5	. 59	5. 70	7. 10	28. 20	27. 40	35. 50
4A........	1. 75:1	5. 0	0–1	4. 60	4. 50	31. 60	32. 50	35. 60	41. 40
4B........	1. 75:1	5. 0	2–3	6. 62	6. 40	31. 50	41. 50	44. 90	51. 80
4C........	1. 75:1	5. 0	4–5	7. 08	4. 40	32. 60	44. 60	47. 00	55. 90
5A........	1:2. 5	18. 0	0–2	....	...	....	....	− . 20	. 50
5B........	1:2. 5	18. 0	8–10	....	...	....	....	3. 80	3. 40
5C........	....	18. 0	16–18	....	. . . .	....	....	2. 40	10. 80
6A........	1:9	24. 0	0–2	. 70	−4. 30	....	....	− . 40	4. 70
6B........	1:9	24. 0	7–9	− . 20	− . 50	....	....	3. 80	3. 50
6C.	1:9	24. 0	10–12	1. 80	1. 80	....	....	....	......
6D ..	1:9	24. 0	14–16	− . 80	. 20	....	....	10. 80	23. 40
6E........	1:9	24. 0	18–20	− . 60	. 60	....	....	10. 70	26. 70
6F . . .	1:9	24. 0	22–24	1. 20	2. 30	....	....	20. 20	23. 40

a A, B, C, etc. are different layers of same column.

The data reported in Table III were secured from samples containing 50 mgm. of nitrogen as ammonium sulphate and 0.5 gm. of calcium carbonate per 100 gm. of soil. They were incubated at room temperature for three weeks. Water lost by evaporation was replaced from time to time. Soil was incubated in 500-c. c. wide-mouthed bottles. The water content varied as indicated in Table III. The samples marked "loose" were as in previous experiments. Those marked "medium" were tamped to two-thirds the "loose" volume. Those marked "compact" were packed to four-sevenths the original volume. It was difficult to reduce further the volume of this soil. In some of the following experiments it was possible by continuous pounding to reduce some samples to one-half the "loose" volume, but this was exceedingly difficult either with high or low moisture contents. With low-water contents such tightly packed samples would burst the heavy glass cylinders upon slight change in temperature. The above phenomenon, together with the fact that an unprotected column of such packed soil 6 inches long would support considerable stress applied at right angle to the length without breaking, gives some idea of how tightly the soil in this and succeeding experiments was packed.

TABLE III.—*Effect of variations in depths of column, moisture content, and compactness of a soil upon nitrate accumulation*

[Results are expressed as gain or loss in milligrams of NO₃ per 100 gm. of soil]

Sample No a	Depth of column.	Sample from depth of—	Water per 100 gm. of soil.	Loose.	Medium.	Compact.
	Inches.	*Inches.*	*C. c.*			
1A...................	5	0-1	15	0. 34	0. 84	13. 40
1B...................	5	4-5	15	2. 36	6. 50	10. 70
2A...................	5	0-1	20	7. 80	21. 40	....
2B...................	5	4-5	20	6 30	35. 70	31 60
3A...................	5	0-1	30	76. 50	25. 80	—10. 70
3B...................	5	4-5	30	123. 80	— 7. 60	b —14. 30
4A...................	5	0-1	38	185. 70	— 8. 20	b —14. 30
4B...................	5	4-5	38	114. 80	b —14. 30	b —14. 30

a A and B are different layers of same column.   b Complete loss of NO₃ initially present.

Here again the evidence shows that with a water content of 15 c. c. per 100 gm. of soil the nitrate accumulation increased with an increasing compactness and also with an increasing depth of column. With a moisture content of 20 c. c. per 100 gm. of soil the medium and compact samples both showed a greater nitrate gain than the loose. There was, however, slight difference between the two. Again, there is no evidence of lack of aeration in the lower-lying strata, the medium compact sample showing a much higher gain at the bottom than at the surface.

At moisture contents of 30 and 38 c. c. per 100 gm. of soil, compacting inhibited nitrate accumulation in all except the surface of medium compact samples containing 30 c. c. of water. However, at both these moisture contents the compact samples were saturated. The medium samples were saturated at 38 c. c. and approached very closely to saturation at 30 c. c. per 100 gm. of soil; hence, nitrification could not be expected to take place in them. At 30 c. c. of moisture per 100 gm. of soil the loose samples showed marked increases in the lower layers over the surface layers, while at 38 c. c., which is slightly above the optimum moisture content for this soil in loose condition, the surface soil showed the largest net gain.

In the experiments reported in Table IV combinations of the three variable factors are presented. The calcium-carbonate content was the same as in previous experiments, while the nitrogen added was only 20 mgm. per 100 gm. of soil. The samples were contained in 500-c. c. bottles and tall glass cylinders, and were incubated in a moisture chamber at room temperature for three weeks. In the samples marked "medium" the volume was reduced to three-fourths that of the "loose," while the samples marked "compact" were reduced to one-half that of the "loose." The ammonia was tested qualitatively in order to show whether or not nitrification was complete.

TABLE IV.—*Effect of variations in depth of column, moisture content, and compactness upon nitrate accumulation*

[Results are expressed as gain or loss in milligrams of NO_3 per 100 gm. of soil]

Sample No.[a]	Depth of column.	Sample from depth of—	12 c.c. of water per 100 gm. of soil.						24 c.c. of water per 100 gm. of soil.						36 c.c. of water per 100 gm. of soil.[c]			
			Loose.		Medium.		Compact.		Loose.		Medium.		Compact.		Loose.		Medium.	
			NO₃	NH₄[b]	NO₃	NH₄[b]	NO₃	NH₄[b]	NO₃	NH₄[b]	NO₃	NH₄[b]	NO₃	NH₄[b]	NO₃	NH₄[b]	NO₃	NH₄[b]
	Inches.	*Inches.*																
1A	2	0– 2	13.0	Ab	22.0	Ab	16.4	Ab	72.0	T	88.0	T	–1.0	Ab	68.0	T	14.6	Ab
2A	4	0– 2	19.8	Ab	22.0	Ab	11.3	Ab	88.0	T	88.0	Ab	–1.0	Ab	68.0	T	2.0	Ab
		2– 4	48.0	Ab	28.0	Ab	16.4	Ab	78.0	T	44.0	Ab	–1.3	Ab	73.0	T	–0.6	Ab
3A	6	0– 2	48.0	Ab	28.0	Ab	16.4	Ab	88.0	S	88.0	Ab	–1.0	Ab	73.0	T	–0.6	X
3B	6	2– 4			28.0	Ab	19.8	Ab			–1.0	Ab	X	X	73.0	T	–0.6	X
3C	6	4– 6	48.0	Ab	31.4	Ab	14.6	Ab	88.0	T	98.0	Ab	22.0	Ab	23.0		6.0	Ab
3D	6	0– 2	31.4	Ab	30.4	Ab	14.3	Ab	70.0	S	76.0	Ab	10.8	Ab	84.0	T	–1.0	X
4A	16	0– 2			22.0	Ab	15.2	Ab	70.0		88.0	T	–1.0	Ab				X
4B	16	2– 4			28.0	Ab	15.2	Ab			96.0	T						X
4C	16	4– 6			28.0	Ab	18.4	Ab			76.0	T	X	X				X
4D	16	4– 5			38.0	Ab					84.0	T						X
4E	16	4– 6			31.4	Ab	22.0	Ab	78.0	T	84.0	T	X	X	78.0	T		X
4F	16	6– 7	48.0	Ab	31.4	Ab					78.0	T						X
4G	16	7– 9			34.8	Ab					84.0	T						X
4H	16	9–10			38.0	Ab	22.0	Ab	88.0	T	78.0	T	X	X				X
4I	16	10–11			31.4	Ab					84.0	T						X
4J	16	11–13			34.8	Ab					84.0	T						X
4K	16	13–14			30.4	Ab	22.0	Ab			84.0	T	X	X				X
4L	16	14–15			48.0	Ab					84.0	T						X
4M	16	13–14			48.0	Ab					84.0	T						X
4N	16	14–15			38.0	Ab					84.0	T						X
4O	16	15–16	31.4	Ab	41.6	Ab	22.0	Ab	80.0	T	73.0	T	X	X	73.0	T		X

[a] A, B, C, etc. are successive layers of the same column.
[b] Ab= NH_4 abundant; S—some NH_4, T—NH_4 trace; X— sample not analyzed, because sample above showed complete loss of NO
[c] Compact samples not set up.

p.

	10 c. c. of water per 100 gm. of soil.a			15 c. c. of water per 100 gm. of soil.				20 c. c. of water per 100 gm. of soil.				25 c. c. of wa[ter] per 100 gm. of soil.	
	Medium.			Loose.		Compact.				Compact.		Loose.	
	cNH3	NO2	cNH3	NO2	cNH3	NO2	cNH3			NO2	cNH3	NO2	cNH3
	6 Ab	1.4	Ab	13	4	5.8	Ab	13.0 Ab	62.2	4	43.0 Ab	98.0	Ab 1
						8.0	Ab				48.0 Ab		
		0.6	Ab			0.0	Ab	19.2 Ab	88.0		48.0 Ab		
						0.4	Ab				40.9 Ab		
	0 Ab	.06	Ab	28	0	0.4	Ab	23.8 Ab	76.0	0	34.0 Ab	118.0	T 1
						3.6	Ab		79.6		30.1 Ab		
		0.6	Ab			6.0	Ab	21.4 Ab			30.1 Ab		
						6.0	Ab			6	24.4 Ab		
	0 Ab	0.6	Ab	33	0	8.0	Ab	22.5 Ab	76.0		24.4 Ab	136.0	T 1

b A, B, C, etc.=different layers of the same column.

c Ab=abundant; S=some; T [...]

With a moisture content of 12 c. c. per 100 gm. of soil the medium packed gave a greater nitrate accumulation than the loose. The compact, however, showed a somewhat smaller accumulation. In all three instances increasing the depth of column increased the nitrate accumulation, indicating no lack of aeration in the lower layers. Even at 16 inches below the surface in the compact sample the nitrate content was much greater than at the surface.

With a moisture content of 24 c. c. per 100 gm. of soil, the nitrification of the added nitrogen was complete, both in loose and medium-packed soils, except samples 2 and 3 of the medium. These samples probably became puddled on the surface in the course of preparation. The compact samples at this moisture content were saturated; hence, they showed no nitrate accumulation except at the surface of No. 4.

At a moisture content of 36 c. c. per 100 gm. of soil, the nitrification was complete in loose samples at all depths. The medium-packed samples were saturated at this moisture content, and only slight nitrification took place, even in surface soil. It was impossible to set up the compact sample with this high moisture content.

The very compact samples 3 and 4, containing 24 c. c. of water per 100 gm. of soil, and the medium compact samples 2, 3, and 4, containing 36 c. c. water per 100 gm. of soil, all showed heavy gas formation, which resulted in the breaking and pushing upward of the columns at different places, the former two near the bottom, the latter three near the top. It is evident from this that there was little or no gaseous exchange even at the high gas pressure necessary to move the tightly packed columns of soil. Water stood on top of the medium-packed samples 2, 3, and 4.

In Table V are reported other experiments in which the moisture content and degree of compactness were varied. These samples were contained in glass cylinders and incubated in moist chamber at room temperature for 16 days. The period was reduced in order that an analysis might be made before nitrification was complete. Nitrification was so active, however, that even in this short period it was complete in many instances. To all samples 0.5 gm. of calcium carbonate and approximately 25 mgm. of nitrogen as ammonium sulphate were added per 100 gm. of soil.

The quantity of nitrate nitrogen found in those samples containing 10 c. c. of water per 100 gm. of soil was too small to be of any significance. With a moisture content of 15 c. c. per 100 gm. of soil, compacting slightly decreased the nitrate accumulation; however, in all three degrees of compactness the nitrate content increased with increasing depth of column. With 20 c. c. of water per 100 gm. of soil, the nitrification was complete in the lower layers of both loose and medium packed, while there was an abundance of ammonia in the surface soil of both. In the compact sample there was a maximum nitrification at 2 to 4 inches deep, below which the accumulation decreased with the increasing depth.

When the moisture content was increased to 25 c. c. per 100 gm. of soil, the total nitrate accumulation in the loose and medium samples was greatly increased, indicating the possibility of less loss of ammonia by volatilization with the higher moisture content. Here again the nitrification was incomplete in the surface soil of both the loose and the medium samples. The nitrate content also increased as the depth increased, except in the lower strata of the medium packed, where the nitrification apparently was not so active. In the compact samples the nitrate accumulation was inhibited except in surface soil. This soil, however, was in a saturated condition.

The evidence thus far submitted seems sufficient to show that increasing the depth of a column of soil, up to at least 2 feet, increases the formation or rather the accumulation of nitrate nitrogen from ammonium sulphate, provided the moisture content is not so high that aeration is prevented. In the type of soil under study aeration will be sufficient for optimum nitrification, at least up to two-thirds the total amount of water the soil will retain with any degree of compactness tried. Packing, of course, decreases the water-holding power almost in the same proportion as the volume of soil is decreased; hence, the quantity of water necessary to limit aeration is correspondingly less. In as compact condition as it was possible to obtain by the methods used and with a moisture content at least up to 15 c. c. per 100 gm. of soil, nitrification was more active a foot below than nearer the surface. With 20 c. c. of moisture per 100 gm. of soil the nitrate accumulation a foot below the surface on very compact soil was far in excess of any nitrification ever observed under field conditions. Compacting this soil type to two-thirds its loose volume increased the nitrate accumulation with any moisture content up to 25 c. c. per 100 gm. of soil.

From the data available one can estimate roughly the free pore space necessary to supply oxygen in sufficient quantities for maximum nitrate accumulation at the depths to which these experiments were carried out. They have not been conducted to greater depths because under our field conditions practically all the nitrate formation takes place in the first foot of soil. One hundred gm. of this soil in loose condition occupies approximately 116 c. c. of space. The specific gravity of this soil is 2.4. Therefore 100 gm. of soil actually occupy approximately 41 c. c , thus leaving 75 c. c. of space to be occupied by water and air. From Tables III and IV it appears that 36 c. c. of water (which is two-thirds that the soil will retain) is approximately the highest moisture content permitting sufficient aeration for optimum nitrification in loose condition. Thirty-eight c. c. did not permit sufficient aeration, while 36 c. c. permitted complete nitrification of added nitrogen. Accordingly 39 c. c., or 33.6 per cent, of pore space were required for optimum aeration.

On applying similar calculations to the medium-packed samples, 100 gm. of soil occupied approximately 80 c. c. of space. The actual volume of soil is, of course, the same as in the loose condition—that is, 41 c. c. This leaves 39 c. c. to be occupied by air and moisture. From Tables IV and V it appears that 25 c. c. of water gave approximately optimum moisture conditions for nitrification. This leaves 14 c. c., or 17.5 per cent, of free pore space necessary for efficient aeration. Similarly 100 gm. of compact soil occupied approximately 60 c. c. of space. Subtracting the 41 c. c. of soil volume leaves 19 c. c. for air and water. On referring to Table V it is observed that 20 c. c. of water per 100 gm. of soil did not permit sufficient aeration in the lower strata, while 15 c. c. did. Therefore, with an air content of 4 c. c. per 100 gm. of soil, or 6.33 per cent of free pore space, oxygen was supplied in sufficient quantities for maximum nitrification at a depth of 12 inches. Stoklasa (13) says that if the free pore space falls below 2 per cent anaerobic processes set in. Even though these figures are meager and only approximations, it is evident that increasing the moisture content of a soil not only decreases aeration by actually decreasing pore space, but also increases the demands for oxygen, and, hence, the free pore space necessary to supply sufficient oxygen for maximum aerobic processes. This is probably due to a number of factors. According to Rahn (10), the soil-moisture film should be 10 to 20 μ thick for optimum aerobic conditions. Therefore, increasing the moisture content until this thickness is reached should increase aerobic activity, provided saturation does not exclude aeration. Rahn's views are in accord with daily observations that increasing moisture up to a certain point increases bacterial activity. In the soil under study saturation will result before this optimum film thickness is reached. One would expect, therefore, that the more moisture added until saturation inhibits aeration, the greater the aerobic activity will be, and consequently the greater the demands for oxygen. The writers observed that increasing the moisture content of a similar soil in loose condition from 12 to 20 c. c. per 100 gm. of soil, or from 20 to 33 per cent of saturation, increased nitrate accumulation 4.8 times. Increasing the same from 20 to 30 c. c., or from 33 to 50 per cent of saturation, increased NO_3 accumulation 2.6 times that which occurred at 20 c. c. A further increase of from 30 to 42 c. c., or from 50 to 70 per cent, increased the nitrate accumulation 1.4 times. There is no reason for not believing that other aerobic processes are increased in similar proportions.

Then, too, increasing moisture content decreases aeration at a greater rate than it does porosity; especially is this true with relative high moisture contents. This is due to the fact that, as the water content increases the thickness of the films, the continuous air tubes upon which rapid exchange of gases is dependent are closed at the narrow points much faster than they are actually filled with water. When all such passages are closed and all available oxygen consumed, aeration will be inhibited regardless of the free pore space remaining.

The writers can offer no satisfactory reason as to why compacting increases the nitrate accumulation, but suggest the explanation that bringing the soil particles into closer contact increases the thickness of the moisture films by pressing the moisture out of those areas actually in contact, and thus reducing the area to be covered by film. Also by transforming the films surrounding individual soil grains into films surrounding clusters the area to be covered by the film is decreased. The same factors operating in increasing nitrification in compact soil probably operate in increasing nitrification in lower layers. Nitrates accumulate faster in the lower layers of a column, even if evaporation from the surface be stopped.

The question will very naturally arise as to whether the oxygen utilized in the lower depths of such samples as are here reported actually diffused there, or was not already contained in the freshly aerated soil. From the data given above it is very easy to calculate the maximum amount of nitrogen that could be oxidized to NO_3 condition, provided all the inclosed oxygen was utilized in this process. If the soil is saturated, only that oxygen actually dissolved in the water or adsorbed by the soil itself would be available. As is well known, this is insufficient for a measurable accumulation of nitrate nitrogen; and it can therefore be eliminated from consideration. In the very compact soil, with a moisture content of 15 c. c. per 100 gm. of soil, the air content could not exceed approximately 5 c. c., the oxygen content of which could not form more than approximately 1.85 mgm. of NO_3. In Tables IV and V it will be observed that under such conditions the NO_3 formation was more than 10 times this amount. With 20 c. c., in which there must have been a negligible quantity of air, the NO_3 accumulation was in some instances more than 20 times that calculated for 15 c. c. of moisture. Similar calculations can be made for the other degrees of compactness, and in all instances it will be found that the NO_3 formed was many times in excess of that possible from the inclosed oxygen.

The above calculations have been verified experimentally, in that controls have been run with each experiment to show that the nitrate gains could not be due to the oxygen contained in the soil when the experiment was set up. The results reported in Table VI are typical.

Here is given, in parallel columns, the quantity of NO_3 accumulation in open and stoppered bottles when both were filled with soil. With low moisture content and loose conditions nitrification, to a limited extent, is possible in absence of aeration. However, with a very compact soil even containing low quantities of water or with a high moisture content even in loose condition, the quantity of inclosed oxygen is insufficient to prevent decreases in NO_3. In no instance does the quantity formed under inclosed conditions equal that when aeration was possible.

TABLE VI.—*Nitrate accumulation in soil in a sealed container compared with soil with its surface exposed to the atmosphere*

[Results expressed as gain or loss in milligrams of NO_3 per 100 gm. of soil]

Water per 100 gm. of soil.	Condition.	Gain or loss.	
		Open bottle	Stoppered bottle.
C. c.			
10	Loose	0. 60	0. 25
	Medium	1. 40	0. 25
	Compact..		.
15	Loose	13. 40	0. 82
	Medium ...	13. 00	7. 00
	Compact .	5. 80	— 2. 00
20	Loose .	54. 40	10. 00
	Medium. .	62. 20	28. 00
	Compact	43. 00	— 2. 00
25	Loose......	98. 00	2. 50
	Medium....	122. 00	— 2. 00
	Compact .	43 00	....
28	Loose.....	....	—14. 40
	Medium. .	. ..	—14. 40
	Compact...	. ...	. ....

It is evident, therefore, that the nitrate accumulation herein reported was possible only through the atmospheric oxygen diffusing into those depths wherein such accumulations have occurred. The results above reported were somewhat surprising, and it was deemed advisable to attempt a somewhat similar study of undisturbed soil columns, it being possible that the manipulations through which the soil was carried might facilitate aeration even after such vigorous repacking. Accordingly columns were taken of the same soil as used above to a depth of 12 inches and incubated in glass cylinders at room temperature, as indicated in Table VII. This soil had been thoroughly cultivated the previous season to a depth of 7 inches.

TABLE VII.—*Nitrate accumulation in unbroken soil columns as compared with broken columns*

[Results expressed as gain or loss in milligrams of NO_3 per 100 gm. of soil]

Sample No.	Condition.	Water.		Sample from depth of—	Gain or loss.
		Initial	Final		
1.....	Natural position, unbroken..........	28	26	*Inches.* Surface	25. 65
				8	5. 78
				12	— 1. 38
2.....	Inverted, unbroken...................	28	26	Surface	3. 57
				8	— 0. 11
				12	— 2. 64
3.....	Natural position, sieved.............	28	24	Surface	11. 09
				8	4. 56
				12	2. 63
4.....	Inverted, sieved......................	28	22	Surface	2. 30
				8	8. 65
				12	12. 33

It will be observed that nitrification was much more vigorous in unbroken soil in natural position down to a depth of 8 inches than in the
sieved soil. Below this depth there was a decrease, indicating insufficient
aeration. This did not occur in sieved soil. It should be remembered,
though, that this soil at such a depth has never been cultivated and also
that the percentage of organic matter is low and the percentage of clay
high. These factors, coupled with the fact that the soil contained 28 c. c.
of water per 100 gm. of soil, produced practically saturated conditions
in the lower layers. When the column was inverted, air could not
penetrate the lower layers in sufficient quantity to prevent denitrification
a few inches below the surface. It will also be observed that the sieved
soil column when inverted gave a higher nitrate accumulation than it
did in its normal position. All these facts are in accord with the experiments already given.

In addition to the experiments just mentioned, others have been
carried out upon a soil that has not been cultivated for seven years.
Columns of soil 5 inches in diameter and 7 inches long were taken by
driving a sharp edged cylinder into the soil. These were easily transferred, without even cracking, to anaerobic jars through which a current
of moist air could be drawn. At the same time 100 gm. samples of the
same soil were incubated under as favorable conditions as possible.
The quantities of nitrate nitrogen accumulating under such conditions
are, of course, small, because of the absence of added nitrogen. Nevertheless the gains were greater in the unbroken column in every instance.

In addition, in order to see whether oxygen could diffuse through such
soil columns in excess of that required by the normal activities of the
organisms in the column, the following experiments were planned and
carried out Below an unbroken column 5 inches long were placed
2 inches of sieved soil containing ammonium sulphate and calcium
carbonate. The column was then pressed down upon the loose soil
and the space surrounding it filled with melted paraffin.[1] This ran into
all cracks, etc., made in transferring the column from field to jar, thus
making conditions such that all oxygen reaching the loose soil below had
to pass through the column. Another 7-inch-long column was well
moistened over the bottom and lower sides with a solution of ammonium
sulphate and then dusted with calcium carbonate The column was
then placed in a jar and melted paraffin poured around to fill all space
at bottom and sides. In this instance there was no method of controlling
the concentration of ammonium sulphate in the outer layers of soil,
and it is possible that it was sufficient to retard nitrification. In
both instances the gain in nitrate was far in excess of that in untreated

[1] Since these experiments were carried out, the senior writer has called attention to the danger of the
use of paraffin in such experiments. (GAINEY, P. L. EFFECT OF PARAFFIN ON THE ACCUMULATION OF
AMMONIA AND NITRATES IN THE SOIL. In Jour. Agr. Research, v. 10, no. 7, p. 355–364. Literature cited, p.
363–364. 1917.) In these experiments, however, the paraffin would probably tend to eliminate rather
than to accentuate the differences to which attention is called

soil as well as in excess of any accumulations observed under field conditions.

The moisture content per 100 gm. of soil, incubation period, etc., are given below, along with results reported as milligrams of NO_3 per 100 gm. of soil gained or lost. A sample containing the same quantities of ammonium sulphate and calcium carbonate was also incubated in tightly stoppered bottles, to simulate conditions of loose soil below the unbroken column, provided no oxygen diffused through.

TABLE VIII.—*Nitrate accumulation in unbroken and sieved columns of soil*

[Results expressed as gain or loss in milligrams of NO_3 per 100 gm. of soil]

Sample No.	Depth of column.	Condition.	Moisture content.	Gain or loss.
	Inches.		*C. c.*	
1........	7	Unbroken. Incubated 7 weeks	25	9. 60
2........	7	Sieved, incubated 7 weeks.................	25	6. 50
3........		Unbroken. Incubated 4 weeks............	24	. 32
4........	7	Sieved, incubated 4 weeks	24	. 07
5........	5	Unbroken. Loose soil below incubated 4 weeks.................................	24	39. 80
6........	7	Moistened with ammonium sulphate, incubated 4 weeks.	24	14. 50
7........	7	Sieved, moistened with ammonium sulphate, incubated 4 weeks.	24	30. 00
8........	7	Sieved, moistened with ammonium sulphate, sealed, incubated 4 weeks.	24	-. 90
9........	7	Unbroken. Incubated 12 weeks...........	24	2. 50
10.......	7	Sieved, incubated 12 weeks................	24	1. 60

It will be observed that the oxygen contained in the loose soil was insufficient to prevent denitrification, also that oxygen passed to the loose soil below the 5-inch column in sufficient quantity to enable a more rapid accumulation of NO_3 than occurred in loose soil freely exposed to the air. The accumulation in the 7-inch column moistened with ammonium sulphate is also far in excess of that where no nitrogen was added.

DISCUSSION OF RESULTS

The results presented above are certainly sufficient to open the question of whether or not the beneficial effect upon bacterial activity, observed to follow cultivation, can be attributed to increased aeration. The question is not altogether new, for Leather (8) has pointed out that the beneficial effect of cultivation upon crop production must be attributed to some factor other than aeration.

The available data relative to the effect of different degrees of aeration upon nitrate formation is confined almost wholly to the effect that cultivation produces. We do not believe, however, that one is justified in attributing the major effect of cultivation upon nitrification to better aeration. There is no question but that cultivation may beneficially influence nitrate formation other than through increasing aeration. Be this as it may, it has recently been shown by Call and Sewell (2), of this Station, that just as vigorous nitrate accumulation can take place in noncultivated as in cultivated soil. The following results are taken from their paper. The figures represent pounds of nitrates gained or lost between the given dates. The noncultivated plot was kept bare of weeds by scraping the surface.

Year.	Period.	Not cultivated.	Cultivated 3 inches deep.	Cultivated 6 inches deep.
		Pounds.	*Pounds.*	*Pounds.*
1914.................	Apr. 5 to Nov. 2..............	620	84	369
1915.................	Apr. 15 to Sept. 8...........	419	180	207
1916.................	May 29 to Sept. 23..........	283	281	485
Average........................		441	182	354

King and Whitson (5) conducted somewhat similar experiments under greenhouse conditions. In one experiment, carried out in the greenhouse, soil in cylinders in which various crops had been grown, was found after 91 days to show an average gain in nitric nitrogen per acre-foot (4,000,000 pounds) in all cultivated soils of 105.62 pounds. Soil in the noncultivated cylinders showed a gain of 162.42 pounds. In another series, which ran for 258 days, the average gain in soil in noncultivated cylinders was 325.48 pounds, and in the soil of all cultivated cylinders, 303.68 pounds per acre-foot. In summarizing they say (p. 9):

The increase of nitric nitrogen has been greater at all depths, as a rule, where the soils have not been cultivated than where they have been cultivated. As a general rule, there has been the highest increase of nitric nitrogen in the surface foot, and the increase in the third foot has generally exceeded that in the second foot.

King and Whitson (6) also carried out the only experiments found reported wherein an effort was made under field conditions to increase aeration without cultivation. Holes were bored to a depth of 4 feet and left open but protected in order to prevent drying out and also to prevent surface drainage. These were so arranged that other samples could be taken for nitrate analysis at definite distances from the original holes. Duplicate sets were arranged and analyzed both after 4 and 10 weeks' standing. Below are given the average nitrate contents per foot to a depth of 4 feet at the 10-week analysis.

Holes on number of sides......	4	4	1	2
Distance from holes, inches..	6	8.5	12	24
Nitrates....................	26.24	23.94	27.00	27.01

If the figures from any depth, even third and fourth foot layers, are studied, there is little to indicate that increasing aeration, as such a procedure must have done, increased the nitrate formation. The moisture content of this soil varied from about 12 to 25 c. c. per 100 gm. of soil, averaging approximately 20 c. c.

The writers have also endeavored to increase aeration by forcing air daily into horizontal perforated tubes placed underneath both cultivated and uncultivated plots of ground. Such experiments have been conducted for two seasons without the differences in nitrate accumulation exceeding the experimental error.

Such data as are available, therefore, upon the effect of increasing aeration under conditions more nearly approaching normal field conditions are in accord with the laboratory results.

There is another angle from which the question of availability of oxygen in soils of different types, different depths, etc., under natural conditions can be approached. One can readily measure, with a fair degree of accuracy, the oxygen actually potentially available for bacterial metabolism. Such data, together with data as to the quantity of oxygen essential for normal aerobic activity, should be very valuable in connection with the problem under consideration.

Fortunately Schlœsing (12), in his original classic researches upon nitrification, secured very definite data, which, so far as the writers are aware, has never been refuted, on the effect of varying the oxygen content of the atmosphere upon nitrate accumulation. Working with a fertile soil, rich in humus—containing 0.263 per cent of nitrogen and of the following mineral composition: Clay (*argile*), 14.6 per cent; chalk (*calcaire fin*), 19.5; silicious sand (*sable siliceux*), 48; calcareous sand (*sable calcaire*), 17.7—and with a moisture content of 15.9 parts per 100, he secured the results given below:

Oxygen, per cent..	1.5	6	11	16	21
Nitric acid formed, mgm.... .	45.7	95.7	132.5	246.6	162.6

Incubation was for four weeks at temperature of $21°$ to $29°$ C.

Duplicating this experiment, except that one sample received no oxygen, the moisture content was 24 parts per 100 of soil (as much as it would absorb), and incubation eight instead of four months, he secured the following results:

Oxygen, per cent........	0	6	11	16	21
Nitric acid formed, mgm .	−64	199	222	203	225

In this instance, with the capacity of the soil to hold air reduced very low and an oxygen content in the air of only 6 per cent, practically as much nitric acid was formed as with normal atmospheric oxygen content, while with 11 per cent of oxygen the quantity was equal to that of normal air.

It is also interesting to note that the carbon dioxid evolved, a factor, as the evidence shows, more or less dependent upon oxygen content, was practically as great at 6 per cent of oxygen as with 21 per cent.

Below are given the relative averages of carbon dioxid evolved in the two experiments just cited:

Oxygen, per cent................... 0   1.6   6   11   16   21
Carbon dioxid evolved, mgm. per
   kgm. of soil in 24 hours   9.03   10.4   16.25   16.05   15.85   17.5

It is evident, therefore, that for either the process of carbon dioxid formation (partially aerobic) or nitric acid formation (obligate aerobic) the oxygen content of the soil atmosphere can be reduced to one-half normal atmospheric oxygen content or below without any appreciable effect upon the processes. Also, the total air content of the soil can be reduced very low without any detrimental effect upon the bacterial activities in question, a fact entirely in accord with the data submitted in this paper. With this information available accurate analysis of soil atmosphere, both as to quantity present and composition, should give us valuable information as to the probability of sufficient quantities of oxygen being available under that particular condition. Fortunately there are a number of recent carefully executed investigations upon soil gases, particularly those of Lau (7), Russell and Appleyard (11), Djarenko (3), Jodidi and Wells (4), and Leather (8)

Lau (7), working with three very different types of soil (sand, loam, and peat) at Rostock, Germany, found, upon analyzing the soil atmosphere monthly for a period of one year, that the carbon-dioxid content varied from 0.11 to 0.143 per cent at a depth of 6 inches; 0.157 to 0.65 per cent at 12 inches; and 0.29 to 2.09 per cent at 24 inches. The oxygen varied from 20.35 to 20.77 per cent at a depth of 6 inches; 19.99 to 20.63 per cent at 12 inches; and 18.42 to 20.19 per cent at 24 inches. Manuring a sandy soil and cropping to potatoes increased the carbon-dioxid from 0.11 to 0.57 per cent and decreased the oxygen content from 20 79 to 20.22 per cent

Russell and Appleyard (11) analyzed the soil gases under the following conditions at the Rothamsted Station, Broadbalk wilderness (uncultivated); Boradbalk wheat field, both manured and unmanured; and Hoos field, both under wheat and fallow. Analyses were made on an average of three times a month for periods in most cases extending over approximately two years. Below are given the extreme and average results of their analysis when air was drawn from a depth of 6 inches:

Constituent.	Minimum.	Maximum.	Average.
	Per cent.	*Per cent.*	*Per cent.*
Carbon dioxid......................	0.02 to 0.28	0.38 to 2.27	0.30 to 0.52
Oxygen........................	17.61 to 20.44	20.81 to 21.30	20.19 to 20.46
Nitrogen........................	78.49 to 79.01	79.38 to 80.48	79.25 to 79.30

They also found very slight differences in the composition of air drawn from depths of 6 inches and 18 inches

Djarenko (3), investigating the effect of different methods of fallowing practiced in Russia upon the composition of the soil atmosphere, secured the results given below.   He analyzed the soil gases 14 times between April 20 and September 1.

	Black fallow			April fallow.			June fallow.			Peasant fallow.		
	Min.	Max.	Ave.	Min.	Max.	Ave.	Min.	Max.	Ave.	Min.	Max.	Ave.
Oxygen..........per cent..	17.1	20.7	18.9	19.2	21.5	19.9	18.2	20.0	19.4	10.6	19.1	15.5
Air per liter of soil....c. c..	290	410	332	240	420	348	200	330	264	150	280	208
Oxygen per liter of soil.c.c..	50	81.5	64.5	48	84	67.8	36.5	65.9	51.3	16	48	32.7

Black fallow is worked in autumn and again the following April and June.   April fallow is worked in April and June.   June fallow is first worked in June.   Cattle are allowed to roam upon peasant fallow in the spring, thus giving it a much-trodden effect.   It is first cultivated in June.   All are seeded on August 1.

Jodidi and Wells (4), analyzing soil gases from 22 plots at the Iowa Station daily from April to August, inclusive, found the lowest, highest, and average monthly analyses to be 0.04, 0.82, and 0.25 per cent for carbon dioxid and 19.36, 21.20, and 20.5 per cent, respectively, for oxygen.   The 5-month average of individual plots varied from 0.18 to 0.44 per cent for carbon dioxid and 20.18 to 20.62 per cent for oxygen. Samples were drawn from 7 inches below the surface.

Leather (8), in his work upon soil gases at Pusa, India, used very different methods of analysis, which render it difficult to compare his results with those given above.   From a definite volume of soil all gases that could be extracted with a Topler pump were removed for analysis. Thus, not only the air existing in free pore space but also much that was held in solution was removed.   Owing to the much greater solubility of carbon dioxid than oxygen and nitrogen in water, the percentage present in dissolved air is much higher and as a result the percentage in air extracted from the soil by this method is much higher. Sufficient data are not given for making the necessary corrections in order to place the figures upon the same basis as those already given. By assuming, though, that the five instances in which such corrections are made by the author are representative, the carbon dioxid figures reported below are approximately five times too high, while the oxygen figures should be increased by approximately 15 per cent.   According to Leather (8, p. 106)—

When the total carbon dioxide is much less than 10 per cent and the soil is not particularly dry, it is nearly all in solution.

With these facts in mind are quoted the following figures from Leather's paper. These data were secured from experiments with an unmanured fallow soil and show the variations in analysis for a number of dates.

Depth.	Per cent oxygen.	Per cent CO₂.	Per cent gas in volume.
1st 12 inches..........................	12. 7 to 18. 6	2. 9 to 12	18. 6 to 41. 2
2d 12 inches	9. 2 to 18. 5	3. 9 to 8. 5	14. 8 to 39. 0
3d 12 inches	9. 7 to 17. 7	1. 9 to 6. 9	23. 8 to 42. 2
4th 12 inches..........................	7. 6 to 17. 8	1. 4 to 7. 0	21. 2 to 45. 3
5th 12 inches..........................	11. 9 to 14. 9	2. 7 to 8. 5	3. 8 to 28. 2
6th 12 inches..........................	12. 7 (one analysis).	6.1 (one analysis).	22. 6 to 25. 3
7th 12 inches..........................	15. 5 (one analysis).	8.2 (one analysis).	....

The low oxygen figures followed the heavy monsoon rains, while the high figures followed prolonged droughts.

As a general summary of the above-reported soil atmospheric analysis, coming from as wide a variety of localities, conditions, etc., as they do, one may say:

(a) The percentage of air by volume in soil rarely ever falls below 20 even several feet below the surface where soil has never been artifically disturbed. Excessive rains or excessive trampling of cattle when the soil is moist may reduce the percentage.

(b) The percentage of oxygen contained in the free soil atmosphere is normally only slightly below that in the atmospheric air. This is true to such depths as one should expect appreciable nitrification to take place. The exceptional conditions just cited, together with heavy applications of green or stable manure, as well as the presence of certain growing plants, may cause a marked reduction in the oxygen content. Such exceptional reductions rarely ever cause a reduction of more than one-half the atmospheric content, even several feet below the surface.

(c) The carbon-dioxid content of the soil air is usually materially higher than that of the atmospheric air. Any factor tending to increase the bacterial activity or to decrease the exchange of gases tends to increase the carbon-dioxid content at the expense of oxygen. The proportion of carbon dioxid present in free soil air rarely ever exceeds 2 per cent, and usually is only a few tenths of 1 per cent. Owing to the greater solubility of carbon dioxid, the dissolved soil atmosphere is usually high in this constituent. According to Leather, the major portion of this is probably in loose chemical combination as an acid carbonate.

Since the composition of the soil atmosphere is directly controlled by the rate of exchange of gases between it and the atmospheric air, it is evident from slight variations in composition reported above that the exchange of gases between the two must take place very rapidly. This is true under practically all conditions and to such depths for which data are available.

So far as the writers are aware, there are no data available invalidating the assumption that the results reported by Schlœsing (12) are applicable to soils in general. If this be true, it would certainly seem with the experimental data now available, that conditions are rarely ever met with where there is not sufficient oxygen potentially available to insure maximum nitrification. Furthermore, Stoklasa (13) mentions 2 per cent of porosity as the point below which anaerobic processes become active, 3.8 per cent is the lowest found recorded for field conditions, and seldom does it fall below 15 to 20 per cent The writers have shown that in a tightly compact soil with a moisture content of 15 c c. per 100 gm. of soil 6.33 per cent of porosity were sufficient for maximum nitrate accumulation at considerable depths below the surface. Therefore these experimental data and theoretical considerations from an entirely different point of view all tend to the same conclusions as do the original experimental data herein reported.

SUMMARY

In the experimental data submitted in this paper it has been shown · that:

(1) As the moisture content of a soil decreases, increasing the compactness from a very loose condition will increase the accumulation of nitrate nitrogen.

(2) With any degree of compactness tested the optimum moisture content will be reached when the soil contains approximately two-thirds the total amount of moisture it will retain.

(3) Aeration will be sufficient to the depth of 1 foot with any degree · of compactness, provided the moisture content does not exceed the above relation.

(4) Increasing the depth of column up to 2 feet does not, as far as tested, alter the above relations. In fact, the accumulation of nitrate nitrogen increases with increasing depth down to 2 feet, so long as the moisture does not exceed approximately two-thirds saturation.

(5) Nitrate nitrogen accumulates more rapidly in unbroken soil columns than in pulverized soil. Aeration in a column of soil uncultivated for seven years is far in excess of that required to maintain aerobic conditions.

It has also been pointed out that such experimental data as is available, regarding oxygen relations in normal field soils, indicate that obligate aerobic conditions almost universally exists within the first foot of surface. Therefore such beneficial effect as cultivating may have upon biological activity can not be attributed to increased aeration.

LITERATURE CITED

(1) BUCKINGHAM, Edgar
　　1904. CONTRIBUTIONS TO OUR KNOWLEDGE OF THE AERATION OF SOILS. U. S.
　　　　Dept Agr. Bur. Soils Bul 25, 52 p.
(2) CALL, L. E , and SEWELL, M. C.
　　1917. THE SOIL MULCH. *In* Jour. Amer Soc. Agron , v. 9, no. 2, p. 49–61.
(3) DJARENKO, A. G.
　　1915. THE AERATION OF CULTIVATED SOILS. (Abstract.) *In* Intern. Inst.
　　　　Agr. [Rome] Bul Agr Intel. and Plant Diseases, year 6, no. 10, p.
　　　　1299–1302. (Original article in Izv Moskov Selskoch. Inst. (Ann.
　　　　Inst Agron Moscou), year 21, v 1, p 1–41. Original not seen)
(4) JODIDI, S. L , and WELLS, A A.
　　1911. INFLUENCE OF VARIOUS FACTORS ON DECOMPOSITION OF SOIL ORGANIC
　　　　MATTER. *In* Iowa Agr Exp. Sta Res. Bul. 3, p 135–154, 4 fig.
(5) KING, F H , and WHITSON, A R
　　1901. DEVELOPMENT AND DISTRIBUTION OF NITRATES AND OTHER SOLUBLE
　　　　SALTS IN CULTIVATED SOILS. Wis Agr Exp Sta. Bul. 85, 48 p.,
　　　　11 fig
(6) ——— ———.
　　1902. DEVELOPMENT AND DISTRIBUTION OF NITRATES IN CULTIVATED SOILS.
　　　　Second paper　Wis Agr. Exp. Sta Bul. 93, 39 p., 6 fig.
(7) LAU, Erich
　　1906 BEITRÀGE ZUR KENNTNIS DER ZUSAMMENSETZUNG DER IM ACKERBODEN
　　　　BEFINDLICHEN LUFT. 34 p , illus., 3 fold. tab. Rostock.　Inaugural
　　　　Dissertation.
(8) LEATHER, J W.
　　1915. SOIL GASES. *In* Mem Dept Agr India, Chem Ser , v. 4, no. 3, p.
　　　　85–134, 4 fig.
(9) LIPMAN, C B , and SHARP, L T.
　　1912. STUDIES ON THE PHENOLDISULPHONIC ACID METHOD FOR DETERMINING
　　　　NITRATES IN SOILS　*In* Univ Cal. Pub. Agr , v 1, no 2, p. 21–37
(10) RAHN, Otto.
　　1912 DIE BAKTERIENTÁTIGKEIT IM BODEN ALS FUNKTION VON KORNGRÖSSE
　　　　UND WASSERGEHALT　*In* Centbl Bakt [etc], Abt. 2, Bd 35, No.
　　　　17/19, p. 429–465
(11) RUSSELL, E. J , and APPLEYARD, Alfred.
　　1915. THE ATMOSPHERE OF THE SOIL; ITS COMPOSITION AND THE CAUSES OF
　　　　VARIATION　*In* Jour Agr Sci , v. 7, pt. 1, p 1–48, 17 fig
(12) SCHLŒSING, Théophile
　　1873. ÉTUDE DE LA NITRIFICATION DANS LES SOLS. *In* Compt Rend Acad·
　　　　Sci [Paris], t 77, no 3, p. 203–207, no. 5, p. 353–356
(13) STOKLASA, Julius
　　1912. METHODEN ZUR BIOCHEMISCHEN UNTERSUCHUNG DES BODENS *In*
　　　　Abderhalden, Emil. Handbuch der Biochemischen Arbeitsmetho-
　　　　den, Bd. 5, T. 2, p. 843–910, fig. 215–221.　Berlin, Wien.

EFFECT OF SULPHUR ON DIFFERENT CROPS AND SOILS [1]

By O. M. SHEDD,[2]

Chemist, Kentucky Agricultural Experiment Station

INTRODUCTION

There has recently been some discussion as to the importance and supply for plant growth of sulphur in its various compounds in soils, and whether or not it may be a limiting element in crop production While it is one of the essential elements, the amounts found by the old method of ashing plants were so low in most cases that it was generally assumed there was an abundant supply of its compounds in soils for all crop requirements. More recently, however, it has been demonstrated by improved methods of analysis that most plants contain much more sulphur than was formerly thought to be the case, owing to the fact that in many instances by the old method the bulk of the sulphur was lost on ashing the plant, and therefore was overlooked The question then arose as to whether there is an ample supply of sulphur compounds in soils for crop needs and especially for the best growth of those which are now known to have a high sulphur content.

Hart and Peterson (4),[3] of Wisconsin, later the writer (5), and afterwards Brown and Kellogg (2), of Iowa, Ames and Boltz (1), of Ohio, and others have found that the sulphur content of many soils is low and that there has been a decided loss of sulphur in some soils which have been cultivated for a long term of years when compared with the corresponding virgin soils.

Some investigators contend that, although some soils are low in sulphur, lower in many cases than in phosphorus, this is compensated for by the amount brought down in the rainfall; and as a result it will never be a limiting element in crop production. There are others, however,

[1] Approved for publication in the Journal of Agricultural Research by A M. Peter, Acting Director, Kentucky Agricultural Experiment Station, July 25, 1917.

[2] The writer desires to express his thanks to Dr. Peter, for his helpful suggestions in the preparation of the manuscript, and to the county agents who assisted in the collection of the soils

[3] Reference is made by number to "Literature cited," p. 103.

who maintain that the sulphur brought down in the rainfall will not equal the loss of this element in the drainage, and their point seems to be established by the limited work that has been done along this line. They contend that the application of sulphur compounds may be beneficial for the maximum production of crops high in sulphur when they are grown on soils that are low in this element.

To establish this point, numerous experiments have been carried on, at first abroad and later in this country; and many of these have been mentioned in former publications (1–6). Since then other work has been published or started along this line, but it is not thought necessary to describe these experiments here further than to state that in some cases decidedly beneficial results have been obtained by applying sulphur or its compounds to some soils and crops; in other instances no benefits were obtained; and in still other experiments an injury was noted. These variations are to be expected when different soils and crops are employed in experiments.

In demonstrating the effect of a given element on plant growth it is sometimes extremely difficult to arrange the experiment so as to prove the desired point, for the reason that the element is generally applied in combination with other elements; consequently the results are not always easy of interpretation.

EXPERIMENTAL WORK

In an effort to avoid the above difficulty the writer used flowers of sulphur mixed with soil to which the necessary fertilizing ingredients were added, together with calcium carbonate. For this purpose eight surface soils, taken to a depth of 6⅔ inches, each representing a distinct type in Kentucky, were selected; and on these were grown five different crops. These soils had been in cultivation for a number of years and, with one or two exceptions, had had little manure or fertilizer added. Their locations and descriptions are as follows:

No. 892.—Eastern coal-field area, Lawrence County, from the farm of Ernest Shannon, 3 miles west of Louisa and 20 rods north of the Louisa-Bussyville road. This land had been cultivated over 60 years; no fertilizer used; produced about 20 bushels of corn per acre. The land appeared to be somewhat worn.

No. 893.—St. Louis-Chester area, Warren County, from the Alderson farm, 1 mile west of Bowling Green. This land had been cultivated for 60 years; no fertilizer used; produced about 20 bushels of corn and 15 bushels of wheat per acre.

No. 894.—Cincinnatian area, Mason County, from the farm of John M. Chambers, about 2½ miles northwest of Washington. This land had been continuously cropped for nearly 100 years; very little fertilizer added, was worn out and produced very poor crops.

No. 895.—Western coal-field area, Muhlenburg County, from the Station experiment field, 2½ miles southwest of Greenville. This land had been cultivated for 40 years, and some fertilizer and manure used. It produced 30 bushels of corn, 12 bushels of wheat, and 1 ton of hay per acre.

No. 896.—Keokuk-Waverly area, Barren County, from the Underwood farm near Roseville. The land had been long in cultivation, was worn out; had little fertilizer used and produced 20 bushels corn, 8 bushels wheat, and 12 bushels oats per acre.

No. 897.—Quaternary area, McCracken County, from the Station experiment field, about 3 miles southwest of Paducah, at Lone Oak. A part of this land had been cultivated for 15 years, the remainder for 30 years Some stable manure had been added. It produced about 15 bushels of corn, 8 bushels of wheat, and 1 ton of hay per acre.

No. 898.—Devonian area, Madison County, from the farm of Mark Settle, northwest of Big Hill. This land had been cultivated for 50 years, was much worn, no fertilizers used. It produced 15 bushels of corn, 10 bushels of wheat, and ½ ton of hay per acre.

No. 899.—Silurian area, Jefferson County, from a farm adjoining the Fair Grounds woods at Fern Creek. The land had been cultivated for many years; was much worn; no commercial fertilizers and little stable manure used. It produced light crops.

Before the experiments were started the soils were partially analyzed and found to have the following composition, in pounds per acre, on assuming that the surface 6⅔ inches weighs 2,000,000 pounds (Table I):

TABLE I.—*Composition of experimental Kentucky soils*

Soil number.	Total phosphorus [a]	Total sulphur [b]	Sulphate sulphur.[c]	Calcium carbonate requirement.[d]
892	360	180	160	393
893	540	400	140	129
894	1,540	520	180	3,176
895	520	420	200	107
896	400	700	240	143
897	520	400	200	36
898	400	240	200	1,335
899	600	380	200	107

[a] Magnesium-nitrate method (Wiley, H W., ed. Official and provisional methods of analysis, Association of Official Agricultural Chemists. As compiled by the Committee on Revision of Methods. U. S. Dept. Agr. Bur. Chem. Bul. 107 (rev.), p. 2. 1908)
[b] Sodium-peroxid method
[c] Hydrochloric-acid (sp (g). digestion (Wiley, H. W. Idem, p. 14).
[d] Hopkins method (Wiley, H. W) Idem, p. 20).

These soils were air-dried, put through a coarse sieve, and 15 pounds (in triplicate) were placed in 2-gallon glazed earthenware jars, supplied with drainage and thoroughly mixed with the following materials, added at the rate of pounds per acre of soil on the above assumption of soil weight:

> 500 pounds of tricalcium phosphate, C. P., precipitated.
> 200 pounds of potassimum nitrate, C. P.
> 8,000 pounds of calcium carbonate, C. P.
> 100 pounds of flowers of sulphur; or
> 200 pounds of flowers of sulphur.

Triplicate jars like the above were also prepared of each soil, except that the sulphur was omitted; and these were used as controls.

SERIES I: SOYBEANS.—Fifteen uninoculated soybean seeds (*Soja max*) were planted in each jar in the greenhouse on November 2, 1914, and watered in all cases with the same amount of distilled water. When the plants were about 3 inches in height, they were thinned to 6 plants of average size in each jar. After the pods had matured, which took place at different times, according to the soil, the plants were cut close to the ground, air-dried, and weighed. They were cut on the following dates: No. 894 and 895 on December 31; No. 892 and 893 on January 9; and the remainder on January 19, 1915.

SERIES II: CLOVER.—After the plants in Series I had been cut, the soil in each jar was sifted, and 25 inoculated red-clover seeds (*Trifolium pratense*) planted. After germinating, these were thinned to 15 average-sized plants in each jar. The plants were allowed to grow until they had attained a height of about 10 inches. Four cuttings were made at a distance of about 2 inches from the crown. When cut in this manner and not allowed to blossom, the plants grew faster, and this procedure was followed until July 14, when the weather became too warm for greenhouse experiments.

SERIES III: OATS.—The soil in each jar of Series II was pulverized and stirred to a depth of 6 inches, and on December 23, 1915, 36 Burt oat seeds (*Avena sativa*) were planted in each. Two weeks later they were thinned to 25 average-sized plants in each jar, and potassium nitrate was then added in solution at the rate of 100 pounds per acre and repeated on February 15. On May 12, 1916, after the seed had matured, the plants were cut close to the ground, air-dried, and weighed.

SERIES IV: ALFALFA.—Similar jars of the same soils used in Series I to III were prepared, and, in addition, jars representing 200 pounds of sulphur per acre were included in triplicate. Thirty-five inoculated alfalfa seeds (*Medicago sativa*) were planted on October 14, 1914, in each of the jars containing soil from Warren, Mason, Muhlenburg, and McCracken Counties, and a like number on November 6 in the remainder. On November 17 the plants were thinned to 15 average-sized plants, and cuttings were made at intervals—whenever the plants attained a height of about 10 inches—just as was done in Series II. Five cuttings were made from the Lawrence, Barren, Madison, and Jefferson County soils, and six from the remainder. Final cuttings on all were made on July 9, 1915, when the experiments were stopped.

SERIES V: WHEAT.—The soils in Series IV were stirred and pulverized to a depth of 6 inches and on December 23, 1915, were planted with 36 Jersey Fultz wheat seeds (*Triticum aestivum*) in each jar. The plants were thinned to 25 average-sized plants, and potassium nitrate was added in the same amounts and at the same time as in Series III. On June 13, 1916, after the seeds had matured, the plants were cut close to the ground, air-dried, and weighed.

The weight in grams of the total air-dry materials in Series I–V, together with the percentage gains or losses are given in Table II.

TABLE II.—*Weight (in grams) and percentage gains or losses of the total air-dried material of Series I to V*

LAWRENCE COUNTY SOIL, EASTERN COAL-FIELD AREA

Quantity of sulphur per acre.	Soybeans.		Clover.	Oats.		Alfalfa.	Wheat.	
	Hay.	Grain.	Hay.	Straw.	Grain.	Hay.	Straw.	Grain.
Control	6.9	0.5	24.1	28.7	11.3	23.1	26.6	1.5
Do.	4.7	1.3	15.4	30.8	12.3	22.6	23.4	1.6
Do.	4.9	1.1	17.3	25.3	12.7	21.0	24.2	1.9
Total	16.5	2.9	56.8	84.8	36.3	66.7	74.2	5.0
100 pounds	6.0	.9	25.4	29.6	13.4	22.5	29.9	3.2
Do.	6.4	1.1	18.6	28.8	13.2	17.3	25.4	1.6
Do.	5.2	1.0	23.6	30.3	12.7	21.5	25.1	1.9
Total	17.6	3.0	67.6	88.7	39.3	61.3	80.4	6.7
200 pounds	.....	.....	.....	.....	.....	21.0	28.6	2.4
Do.	.....	.....	.....	.....	.....	23.6	24.3	1.7
Do.	.....	.....	.....	.....	.....	22.7	32.0	2.0
Total	.....	.....	.....	.....	.....	67.3	84.9	6.1

WARREN COUNTY SOIL, ST LOUIS-CHESTER AREA

Quantity of sulphur per acre.	Soybeans Hay.	Soybeans Grain.	Clover Hay.	Oats Straw.	Oats Grain.	Alfalfa Hay.	Wheat Straw.	Wheat Grain.
Control	8.7	.8	33.0	20.9	9.1	34.1	23.6	4.4
Do	7.6	.8	27.8	26.1	8.9	40.2	24.7	5.3
Do.	7.2	1.6	26.0	22.5	9.6	38.9	21.9	3.1
Total	23.5	3.2	86.8	69.5	27.6	113.2	70.2	12.8
100 pounds	5.2	1.0	27.2	25.5	9.5	39.8	22.2	2.9
Do.	6.5	1.2	24.1	28.9	9.1	41.1	28.0	4.0
Do.	7.3	1.1	28.1	22.6	8.4	39.1	25.0	5.0
Total	19.0	3.3	79.4	77.0	27.0	120.0	75.2	11.9
200 pounds	.....	.....	.....	.....	.....	40.3	32.8	6.2
Do.	.....	.....	.....	.....	.....	41.2	26.5	5.5
Do.	.....	.....	.....	.....	.....	41.4	26.4	4.6
Total	.....	.....	.....	.....	.....	122.9	85.7	16.3

MASON COUNTY SOIL, CINCINNATIAN AREA

Quantity of sulphur per acre.	Soybeans Hay.	Soybeans Grain.	Clover Hay.	Oats Straw.	Oats Grain.	Alfalfa Hay.	Wheat Straw.	Wheat Grain.
Control	9.5	4.6	27.7	17.5	6.5	36.1	24.0	5.0
Do.	8.5	4.4	19.0	17.3	7.7	39.0	26.5	3.5
Do.	8.2	4.1	19.5	23.3	8.7	37.9	27.0	4.0
Total	26.2	13.1	66.2	58.1	22.9	113.0	77.5	12.5
100 pounds	9.1	4.8	15.4	18.6	8.4	43.0	26.0	4.0
Do.	8.2	4.1	13.9	19.8	8.2	47.1	29.0	5.0
Do.	9.2	4.9	17.2	23.8	7.2	40.2	25.3	4.7
Total	26.5	13.8	46.5	62.2	23.8	130.3	80.3	13.7
200 pounds	.....	.....	.....	.....	.....	36.2	31.4	5.6

TABLE II.—*Weight (in grams) and percentage gains or losses of the total air-dried material of Series I to V*—Continued

MUHLENBURG COUNTY SOIL, WESTERN COAL-FIELD AREA

Quantity of sulphur per acre.	Soybeans.		Clover.	Oats.		Alfalfa.	Wheat.	
	Hay.	Grain.	Hay.	Straw.	Grain.	Hay.	Straw.	Grain.
Control.	11. 8	7. 7	27. 2	23. 7	10. 3	38. 3	30. 0	7. 0
Do.	11. 6	6. 3	29. 0	21. 7	9. 3	39. 2	30. 2	6. 8
Do.	12. 2	6. 8	26. 0	25. 9	9. 2	40. 1	27. 0	5. 0
Total.	35. 6	20. 8	82. 2	71. 3	28. 8	117. 6	87. 2	18. 8
100 pounds.	11. 0	6. 8	24. 6	25. 3	10. 7	42. 5	31. 3	5. 7
Do.	9. 4	3. 8	22. 5	24. 3	9. 7	41. 1	25 6	6. 4
Do.	12. 5	5. 6	20. 5	24. 7	8. 3	39. 7	29. 7	7. 3
Total.	32. 9	16. 2	67. 6	74. 3	28. 7	123. 3	86. 6	19. 4
200 pounds.						29. 9	26. 0	5. 0
Do.						44. 1	27. 3	2. 7
Do.						39. 2	26. 9	5. 1
Total.						113. 2	80. 2	12. 8

BARREN COUNTY SOIL, KEOKUK-WAVERLY AREA

Control.	8. 0	1. 0	28. 7	31. 2	12. 9	27. 8	31. 6	5. 4
Do.	5. 4	2. 4	28. 1	27. 4	14. 6	26. 4	31. 5	6. 5
Do.	7. 1	2. 3	17. 9	29. 5	13. 5	28. 0	33. 2	5. 8
Total.	20. 5	5. 7	74. 7	88. 1	41. 0	82. 2	96. 3	17. 7
100 pounds.	7. 4	2. 4	28. 9	28. 6	12. 4	24. 0	30. 8	6. 2
Do.	5. 0	1. 8	20. 5	25. 8	12. 3	30. 9	33. 7	6. 3
Do.	6. 4	3. 3	22. 6	31. 3	12. 7	28. 4	33. 9	6. 1
Total.	18. 8	7. 5	72. 0	85. 7	37. 4	83. 3	98. 4	18. 6
200 pounds.						25. 1	32. 5	6. 5
Do.						25. 6	28. 7	3. 3
Do.						27. 4	34. 2	6. 8
Total.						78. 1	95. 4	16. 6

MCCRACKEN COUNTY SOIL, QUATERNARY AREA

Control.	9. 9	2. 5	28. 4	21. 9	12. 1	45. 8	29. 5	6. 5
Do.	9. 1	2. 7	32. 3	32. 9	14. 1	46. 5	32. 1	6. 9
Do.	8. 6	3. 0	14. 8	29. 1	13. 9	49. 3	32. 7	6. 3
Total.	27. 6	8. 2	75. 5	83. 9	40. 1	141. 6	94. 3	19. 7
100 pounds.	9. 5	2. 4	18. 8	32. 1	15. 9	48. 4	30. 6	5. 4
Do.	8. 6	2. 4	22. 7	32. 9	14. 1	50. 8	29. 1	5. 9
Do.	10. 2	2. 2	23. 6	28. 3	12. 7	51. 4	28. 9	7. 1
Total.	28. 3	7. 0	65. 1	93. 3	42. 7	150. 6	88. 6	18. 4
200 pounds.						46. 9	30. 7	8. 3
Do.						49. 8	32. 9	7. 1
Do.						53. 2	31. 6	6. 4
Total.						149. 9	95. 2	21. 8

TABLE II.— *Weight (in grams) and percentage gains or losses of the total air-dried material of Series I to V*—Continued

MADISON COUNTY SOIL, DEVONIAN AREA

Quantity of sulphur per acre.	Soybeans.		Clover.	Oats.		Alfalfa.		Wheat.
	Hay.	Grain.	Hay.	Straw.	Grain.	Hay.	Straw.	Grain.
Control.......................	7.0	2.3	27.2	23.0	13.0	19.9	33.4	8.6
Do........................	6.8	2.2	27.1	26.8	12.2	30.2	23.0	3.1
Do........................	6.7	2.4	24.8	22.3	12.7	16.9	33.3	5.7
Total.................	20.5	6.9	79.1	72.1	37.9	67.0	89.7	17.4
100 pounds....................	9.2	2.6	24.1	23.5	12.5	18.8	26.8	3.3
Do........................	7.6	1.7	30.0	25.4	13.6	30.7	25.1	4.9
Do........................	8.6	3.1	24.0	23.1	15.0	28.7	26.4	5.6
Total.................	25.4	7.4	78.1	72.0	41.1	78.2	78.3	13.8
200 pounds....................	......	......	......	......	......	27.4	30.7	7.3
Do........................	......	......	......	......	......	28.3	19.5	2.5
Do........................	......	......	......	......	......	30.5	32.3	6.7
Total.................	......	......	......	......	......	86.2	82.5	16.5

JEFFERSON COUNTY SOIL, SILURIAN AREA

Quantity of sulphur per acre.	Soybeans.		Clover.	Oats.		Alfalfa.		Wheat.
	Hay.	Grain.	Hay.	Straw.	Grain.	Hay.	Straw.	Grain.
Control.......................	10.4	2.6	13.6	24.5	11.5	35.7	33.4	8.6
Do........................	8.8	2.8	21.3	26.7	13.3	34.9	33.8	8.2
Do........................	9.4	3.1	13.6	25.9	13.1	38.2	34.5	7.5
Total.................	28.6	8.5	48.5	77.1	37.9	108.8	101.7	24.3
100 pounds....................	10.9	3.2	16.1	24.9	13.1	37.1	34.3	8.7
Do........................	11.8	3.4	19.0	22.7	13.3	41.0	32.6	8.4
Do........................	10.5	3.5	15.7	24.0	13.0	38.9	33.4	7.6
Total.................	33.2	10.1	50.8	71.6	39.4	117.0	100.3	24.7
200 pounds....................	......	......	......	......	......	34.8	32.0	7.0
Do........................	......	......	......	......	......	33.8	31.4	8.6
Do........................	......	......	......	......	......	35.1	31.7	8.3
Total.................	......	......	......	......	......	103.7	95.1	23.9

TABLE II.—*Weight (in grams) and percentage gains or losses of the total air-dried material of Series I to V*—Continued.

PERCENTAGE GAIN OR LOSS

County.	Quantity of sulphur added per acre.	Soybeans.		Clover hay.	Oats.		Alfalfa hay.	Wheat.	
		Hay.	Grain.		Straw.	Grain.		Straw.	Grain.
	Pounds.								
Lawrence....	100	+ 6.9	+ 3.4	+19.0	+ 4.6	+8.3	— 8.1	+ 8.4	+34.0
Do......	200	......	......	......	......	......	+ .9	+14.4	+22.0
Warren........	100	—19.1	+ 3.1	— 8.5	+10.8	—2.2	+ 6.0	+ 7.1	— 7.0
Do......	200	......	......	......	......	......	+ 8.6	+22.1	+27.3
Mason........	100	+ 1.1	+ 5.3	—29.8	+ 7.1	+3.9	+15.3	+ 3.6	+ 9.6
Do......	200	......	......	......	......	......	— 4.0	+21.7	+33.3
Muhlenburg	100	— 7.6	—22.1	—17.3	+ 4.2	— .03	+ 4.8	— .7	+ 3.2
Do......	200	......	......	......	......	......	— 3.7	— 8.0	—31.9
Barren........	100	— 8.3	+31.6	— 3.6	— 2.7	—8.8	+ 1.3	+ 2.2	+ 5.1
Do......	200	......	......	......	......	......	— 5.0	— .9	— 6.2
McCracken...	100	+ 2.5	—14.6	—13.7	+11.2	+6.5	+ 6.4	— 6.0	— 6.6
Do......	200	......	......	......	......	......	+ 5.9	+ .9	+10.7
Madison......	100	+23.9	+ 7.2	— 1.3	— .1	+8.4	+16.7	—12.7	—20.7
Do......	200	......	......	......	......	......	+28.7	— 8.0	— 5.2
Jefferson. .	100	+16.1	+18.8	+ 4.7	— 7.1	+4.0	+ 7.5	— 1.4	+ 1.6
Do......	200	......	......	......	......	......	— 4.7	— 6.5	— 1.6

EFFECT OF SULPHUR ON THE TOTAL AND SULPHATE-SULPHUR CONTENT OF SOYBEANS, CLOVER, AND ALFALFA

The air-dried plants in Series I, II, and IV, were finely ground for this work, a composite sample being made of the triplicates in each case, and the soybean seeds were ground with the corresponding sample of hay.

The total sulphur determinations were made by the sodium-peroxid method [1] and the sulphate sulphur was determined by the following procedure:

Ten gm. of material were digested in 400 c. c. of water on the water bath for several hours, with frequent stirring. It stood overnight, and was made to 500 c. c. volume, first deducting the volume occupied by the sample. This was then filtered, and a 5-gm. aliquot used, to which was added 1 c. c. of hydrochloric acid (1:1) to precipitate the protein and organic matter. The whole was then heated on the water bath several hours longer, allowed to stand overnight, filtered, and washed. The filtrate was made slightly acid with hydrochloric acid, heated, and barium sulphate precipitated by adding, hot, 10 per cent barium-chlorid solution, and letting it stand overnight.

The results obtained on the above samples are given in Table III.

[1] A slight modification was used, which is now the official method of the Association of Official Agricultural Chemists (Wiley, H. W., ed. *Op. cit.*, p. 23.)

TABLE III.—*Percentage of total and sulphate sulphur in air-dry soybeans, clover, and alfalfa*

SOYBEANS

Quantity of sulphur to the acre.	Lawrence County.	Warren County.	Mason County.	Muhlenburg County.	Barren County.	McCracken County.	Madison County.	Jefferson County.
Control:								
Total sulphur	0.207	0.301	0.252	0.260	0.209	0.223	0.198	0.202
Sulphate sulphur	.025	.038	.079	.068	.039	.049	.042	.056
Residual sulphur	.182	.263	.173	.192	.170	.174	.156	.146
100 pounds:								
Total sulphur	.302	.376	.298	.278	.229	.247	.237	.254
Sulphate sulphur	.037	.044	.096	.086	.035	.054	.056	.083
Residual sulphur	.265	.332	.202	.192	.194	.193	.181	.171

CLOVER

	Lawrence County.	Warren County.	Mason County.	Muhlenburg County.	Barren County.	McCracken County.	Madison County.	Jefferson County.
Control:								
Total sulphur	0.238	0.276	0.275	0.304	0.219	0.213	0.279	0.231
Sulphate sulphur	.097	.140	.125	.143	.084	.081	.148	.071
Residual sulphur	.141	.136	.150	.161	.135	.132	.131	.160
100 pounds:								
Total sulphur	.267	.281	.321	.379	.248	.291	.340	.351
Sulphate sulphur	.130	.152	.178	.218	.100	.107	.198	.195
Residual sulphur	.137	.129	.143	.161	.148	.184	.142	.156

ALFALFA

	Lawrence County.	Warren County.	Mason County.	Muhlenburg County.	Barren County.	McCracken County.	Madison County.	Jefferson County.
Control:								
Total sulphur	0.410	0.457	0.352	0.409	0.334	0.306	0.387	0.284
Sulphate sulphur	.168	.223	.142	.209	.152	.123	.204	.093
Residual sulphur	.242	.234	.210	.200	.182	.183	.183	.191
100 pounds:								
Total sulphur	.511	.480	.425	.451	.397	.398	.467	.435
Sulphate sulphur	.265	.256	.205	.265	.206	.166	.267	.248
Residual sulphur	.246	.224	.220	.186	.191	.232	.200	.187
200 pounds:								
Total sulphur	.517	.503	.528	.462	.422	.409	.474	.412
Sulphate sulphur	.286	.301	.349	.295	.213	.212	.298	.218
Residual sulphur	.231	.202	.179	.167	.209	.197	.176	.194

FORMATION OF SULPHATE IN SEEDS ON GERMINATION

As the results in Table III indicate that the excess of sulphur in those plants which have been grown in soil to which this element has been added exists in the form of sulphate, it was thought that it might be of interest to ascertain whether sulphate is formed in seeds from their sulphur compounds when they are allowed to germinate.

As a control the sulphate was determined in the finely ground ungerminated seed by the same method used for the work in Table III, and about the same weight of the seeds were then allowed to germinate in covered dishes between cheesecloth kept moistened with distilled water, after which the sample was ground in a mortar and the sulphate determined. As a precaution all precipitates of barium sulphate were fused with sodium carbonate and reprecipitated and blanks made on the reagents.

The results of these experiments are given in Table IV.

TABLE IV.—*Percentage of sulphur existing as sulphate in seeds before and after germination*

Variety.	Before germination.	After germination.	Increase during germination.	Period of germination.
				Days.
Corn	None.	None.	None.	6
Beans	None.	0.0003	0.0003	6
Cowpeas	None.	.0079	.0079	6
Alfalfa	None.	.0144	.0144	6
Millet	None.	.0223	.0223	6
Oats	None.	.0312	.0312	6
Soybeans	0.0007	.0034	.0027	6
Wheat	.0007	.0219	.0212	6
Hemp	.0014	.0107	.0093	6
Timothy	.0034	.0316	.0282	6
Rye	.0048	.0168	.0120	6
Tobacco	.0072	.0261	.0189	13
Peas	.0117	.0212	.0095	6
Onion	.0220	.0566	.0346	6
Bluegrass	.0258	.0309	.0051	13
Clover	.0483	.0447	a .0006	6

a Loss.

GENERAL DISCUSSION

While a few of the duplicates in Table II vary widely, yet on the whole they agree fairly well, considering work of this character. If an allowance of 10 per cent, compared with the controls, which is a safe amount, is made for unavoidable factors, then we find from an examination of Table II that applications of sulphur have affected the crops grown on the soils from the different counties as follows:

LAWRENCE.—Beneficial, clover from the smaller application and wheat grain and straw from the larger application. Injurious, none.

WARREN.—Beneficial, oat straw from the smaller application, and wheat, both in grain and straw, from the larger application. Injurious, soybean hay.

MASON.—Beneficial, alfalfa and wheat, both in grain and straw, from the larger application. Injurious, clover.

MUHLENBURG.—Beneficial, none. Injurious, soybean grain, clover, and wheat grain from larger application.

BARREN.—Beneficial, soybean grain. Injurious, none.

McCRACKEN.—Beneficial, oat straw and wheat grain from larger application. Injurious, soybean grain and clover.

MADISON.—Beneficial, soybean hay and alfalfa. Injurious, wheat, both in grain and straw, from smaller application.

JEFFERSON.—Beneficial, soybeans, both in hay and grain. Injurious, none.

From the foregoing we find that the sulphur has affected the crops differently, depending on the soil. Some undoubtedly were benefited, others were injured, while in many cases no effect was apparent. On the whole, there is a preponderance of gains from the sulphur, although generally small.

Some observers have found that sulphur had a more marked effect on certain crops when applied to soils fairly well supplied with organic matter. What the effect would have been if such had been the case here or if the other fertilizing ingredients had been omitted is not easy to forecast, for, as stated before, it is difficult to have all soil conditions ideal in order to prove a certain point. Soil fertility involves so many factors that its study is very complicated. The question of mineral plant food has occupied considerable attention, and rightly so, but oftentimes another important side has been overlooked—namely, the organic matter, involving, as it does, all bacterial activities of the soil. If a bacterial study was carried on in mineral-nutrition work, probably different deductions would be drawn then where each is considered alone. In this connection it might be of interest to state that Fred and Hart (3) have been found that soluble phosphates have a more marked effect on promoting the bacterial activity of a soil than sulphates; and for this reason, while sulphates are important and as low in amount in most soils as phosphates, they will not in all probability have the same crop-producing power as the phosphates.

From an examination of Tables III, it will be found that applications of sulphur increased the total and sulphate-sulphur content of the plant; and the larger the application, the greater the increase. Furthermore, it will also be observed that in the clover and alfalfa, the sulphur marked "residual" more nearly approaches a constant figure, regardless of whether sulphur was applied or not. This does not hold true with respect to the soybeans in most cases, however, and this indicates that the excess sulphur in the clover and alfalfa plants exists as sulphate, while part of the excess in the soybeans is in a form other than sulphate.

As sulphur is combined with the protein of plants, it was thought that in the soybeans protein determinations might show that where an increased sulphur content is shown, owing to the sulphur applied, a correspondingly larger protein content might be found; but such is not always the case, as will be seen in Table V. For these determinations the same materials used for the sulphur work in Table III were employed.

TABLE V.—*Protein in air-dry soybeans, tops and seed*

	Protein.			
County.	Controls.		Sulphur, 100 pounds per acre.	
	Per cent.	Weight (in grams).	Per cent.	Weight (in grams).
Lawrence	24. 3	4. 7	24. 5	5. 0
Warren	25. 5	6. 8	24. 9	5. 6
Mason	23. 6	9. 3	21. 0	8. 5
Muhlenburg	23. 2	13. 1	21. 5	10. 6
Barren	26. 5	6. 9	26. 9	7. 1
McCracken	25. 1	9. 0	26. 5	9. 4
Madison	29. 8	8. 2	28. 5	9. 3
Jefferson	25. 7	9. 5	23. 6	10. 2

The results in Table IV are interesting in showing that some seeds contain no sulphate soluble in water; others contain small amounts; while some are fairly well supplied. Furthermore, these results show that in most cases more or less sulphate is formed from the reserve sulphur compounds in the seed on germinating, but there are exceptions—namely, corn and clover—and the latter is of particular interest, since it possessed the highest sulphate content and seemed to show a slight loss on germination.

SUMMARY

(1) Soybeans, clover, oats, alfalfa, and wheat were grown in the greenhouse on eight soils, each taken from a different county and representing a distinct type in Kentucky. They were more or less impoverished by cultivation. To these soils applications of flowers of sulphur at the rate of 100 and 200 pounds per acre, together with the calcium carbonate and other fertilizing ingredients, were added.

(2) The results show that the sulphur increased the production of some crops, had no effect on others, and on some was injurious, depending on the crop and the soil on which it was grown. There was a preponderance of gains, however, from the sulphur application, but these were generally small.

(3) Analyses of some of the crops show that the sulphur increased the total and sulphate-sulphur content of the plant, and the greater the application, the greater the increase.

(4) Where sulphur was applied to clover and alfalfa, the excess sulphur in those plants was in the form of sulphate, while in soybeans part of the excess was in another form.

(5) In the soybeans which showed an increased sulphur content, no corresponding increased protein content was always found. In five instances out of eight, however, soybeans grown in soil where sulphur was added show an increase in the total weight of protein.

(6) It was found that, of the 16 varieties of field and garden seeds examined, some contain sulphates, while others do not, but that, on germinating, all except two form a greater or less amount of sulphate. The highest sulphate content obtained in the ungerminated seed was 0.048 per cent, in clover, and the increase due to germination varied from none, in corn, to 0.035 per cent, in the onion. There was a slight loss in only one sample, clover.

LITERATURE CITED

(1) AMES, J. W., and BOLTZ, G. E.
1916. SULPHUR IN RELATION TO SOILS AND CROPS. Ohio Agr. Exp. Sta. Bul. 292, p. 221-256. References, p. 255-256.

(2) BROWN, P. E., and KELLOGG, E. H.
1914. SULFOFICATION IN SOILS. Iowa Agr. Exp. Sta. Research Bul 18, p. 49-111.

(3) FRED, E. B., and HART, E. B.
1915. THE COMPARATIVE EFFECT OF PHOSPHATES AND SULPHATES ON SOIL BACTERIA. Wis. Agr. Exp. Sta. Research Bul. 35, p. 35-66, 6 fig.

(4) HART, E. B., and PETERSON, W. H.
1911. SULPHUR REQUIREMENTS OF FARM CROPS IN RELATION TO THE SOIL AND AIR SUPPLY. Wis. Agr. Exp. Sta. Research Bul. 14, 21 p.

(5) SHEDD, O. M.
1913. THE SULFUR CONTENT OF SOME TYPICAL KENTUCKY SOILS. Ky. Agr. Exp. Sta. Bul. 174, p. 269-306. References, p. 306.

(6) ——
1914. THE RELATION OF SULFUR TO SOIL FERTILITY. Ky. Agr. Exp. Sta. Bul. 188, p. 595-630.

A STATISTICAL STUDY OF SOME INDIRECT EFFECTS OF CERTAIN SELECTIONS IN BREEDING INDIAN CORN

By H. L. Rietz, *Professor of Mathematical Statistics and Statistician*, and L. H. Smith, *Chief in Plant Breeding, Illinois Agricultural Experiment Station*

INTRODUCTION

At the Illinois Agricultural Experiment Station a considerable number of experiments are being conducted in the breeding of Indian corn (*Zea mays*) by the selection of parents with respect to particular characters. One of the best known of these experiments is concerned with selections with regard to chemical composition. This corn now consists of four strains, known as the "high-protein," the "low-protein," the "high-oil," and the "low-oil" strains. Other selection experiments are concerned with such characters as the height of ears on stalks, the number of ears per stalk, and the inclination of the ear on the stalk.

The primary purpose of these experiments is to determine what progress, if any, can be made in regard to particular characters by selecting for parents from year to year with respect to such characters. It is the purpose of the present paper to report the results of an investigation into what may be called the indirect effects of some of the above-mentioned selections, by a statistical investigation of the changes that have taken place in certain physical characters of the ears of corn. The physical characters treated are length, circumference, weight of ears, and the number of rows of kernels on the ears. A part of this work on physical characters of ears has been in progress on each of 11 generations of corn. Each year we have either measured for each strain all the measureable ears produced, or a random sample of sufficient size to insure fairly small probable errors. To give a notion of the extent of the data used and of the statistical analysis, it may be stated that this analysis involves the preparation of 476 distinct frequency distributions. It would hardly be feasible to exhibit each of these frequency distributions, but we shall present in Table I, as a representative illustration, the 16 distributions of the four chemical composition strains for the crop of 1914.

Journal of Agricultural Research,
Washington, D. C.
kz

Vol. XI, No 4
Oct. 22, 1917
Key No. Ill.—6

TABLE I.—*Frequency distributions with respect to physical characters of the ears of four strains of corn bred for chemical composition—crop of 1914*

HIGH-PROTEIN STRAIN

Length of ears		Weight of ears.		Circumference of ears.		Rows of kernels	
Inches.	Frequency.	Ounces.	Frequency.	Inches.	Frequency.	Number.	Frequency.
3.0	1	2	4	4.00	1	8	1
3.5	6	3	19	4.25	9	10	25
4.0	8	4	43	4.50	26	12	236
4.5	13	5	66	4.75	71	14	241
5.0	28	6	135	5.00	113	16	74
5.5	36	7	144	5.25	158	18	13
6.0	42	8	96	5.50	142		
6.5	67	9	28	5.75	73		
7.0	95	10	6	6.00	27		
7.5	118			6.25	5		
8.0	100			6.50	8		
8.5	89			6.75	5		
9.0	43			7.00	0		
9.5	17			7.25	1		
10.0	3						

LOW-PROTEIN STRAIN

Length of ears		Weight of ears.		Circumference of ears.		Rows of kernels	
3.0		2	4	4.00		8	
3.5	3	3	10	4.25		10	9
4.0	4	4	17	4.50	1	12	117
4.5	6	5	21	4.75	6	14	179
5.0	11	6	38	5.00	27	16	124
5.5	17	7	59	5.25	36	18	41
6.0	12	8	78	5.50	58	20	7
6.5	34	9	97	5.75	99	22	1
7.0	34	10	72	6.00	113		
7.5	66	11	49	6.25	98		
8.0	62	12	15	6.50	54		
8.5	93	13	4	6.75	15		
9.0	83	14	1	7.00	9		
9.5	54			7.25	2		
10.0	34						
10.5	1						
11.0	2						

HIGH-OIL STRAIN

Length of ears		Weight of ears.		Circumference of ears.		Rows of kernels	
2.5	2	2	16	4.00	2	8	
3.0	4	3	44	4.25	15	10	2
3.5	7	4	88	4.50	31	12	78
4.0	17	5	145	4.75	57	14	158
4.5	23	6	114	5.00	103	16	171
5.0	48	7	61	5.25	96	18	49
5.5	71	8	23	5.50	117	20	3
6.0	59	9	6	5.75	84	22	3
6.5	116			6.00	34		
7.0	85			6.25	11		
7.5	77			6.50	4		
8.0	34						
8.5	27						
9.0	10						
9.5	2						
10.0	1						

TABLE I —*Frequency distributions with respect to physical characters of the ears of four strains of corn bred for chemical composition—crop of 1914*—Continued

LOW-OIL STRAIN

Length of ears		Weight of ears.		Circumference of ears.		Rows of kernels	
Inches.	Frequency.	Ounces.	Frequency.	Inches.	Frequency.	Number.	Frequency.
2.5	....	2	....	4.00	1	8	2
3.0	3	3	4	4.25	0	10	59
3.5	2	4	19	4.50	3	12	287
4.0	10	5	40	4.75	7	14	185
4.5	22	6	83	5.00	19	16	30
5.0	25	7	126	5.25	20	18	4
5.5	29	8	97	5.50	46	......	......
6.0	62	9	100	5.75	77	......	......
6.5	106	10	55	6.00	110	......	......
7.0	100	11	23	6.25	117	......	......
7.5	116	12	9	6.50	130	......	......
8.0	95	13	4	6.75	57	......	......
8.5	60	......	......	7.00	36	......	......
9.0	18	......	......	7.25	13		
9.5	7	......	......	7.50	3		
10.0	2	......	......	7.75	1		

The character of the frequency distributions may be described in a general way as not far from a symmetrical form. Occasionally a distribution has presented an appearance that seemed pretty decidedly skew, and we have followed such cases into succeeding generations without good evidence that such decided skew form persisted in later generations. On this account it seems likely that the decided skew condition was due mainly to soil differences on a given plot. There is, however, some tendency for the distributions as to length and weight of ears to be slightly skew, with the mode greater than the mean.

We have tested a number of the distributions, and found some of the normal type, and some differing significantly from the normal type. Taken as a whole, it seems, however, that these distributions are sufficiently near the normal type, to be well described for our purposes by the means, standard deviations, and coefficients of variability.

THE FOUR STRAINS WITH RESPECT TO CHEMICAL COMPOSITION

These four strains were produced from the same original stock (163 ears) and each seed selection was made from the highest (or lowest) ears within the strain with respect to the character in question. For example, the high-protein strain is the offspring of those ears showing the highest protein content in the original stock, and was developed by the selection, for successive generations of planting, of the highest protein ears, always within the high-protein strain. In a similar manner, each of the other three strains was produced. Attention [1] has been called to some indi-

---

[1] DAVENPORT, Eugene. PRINCIPLES OF BREEDING, p. 447. Boston, 1907.

rect effects of selection as shown in the crop of 1905, after selection for chemical composition had been made since 1897. It was there shown that the four strains exhibit some significant differences with respect to physical characters of ears. As we have measured each ear of a fair-sized random sample of ears from each strain for crops grown from 1905–1915, we are now in a position to determine whether the differences existing in 1905 have persisted pretty regularly under changed conditions of soil and season, or if some further indirect influence of the selection for chemical composition is to be noted in considering the four physical characters on which we have collected data. Table II gives the means, empirical modes, standard deviations, and coefficients of variability of the length of ears, the circumference of ears, the weight of ears, and the number of rows of kernels on ears, for each of the four strains bred for chemical composition, for the 11 crops from 1905 to 1915, inclusive.

TABLE II.—_Type and variability of the four chemical-composition strains of Indian corn for 11 crops, 1905 to 1915_

CROP OF 1905

Character and strain.	Mean.	Empirical mode.	Standard deviation.	Coefficient of variability.
Length of ear, in inches:				
High-protein...........	7.21±0.04	7.5	1.27±0.03	17.6±0.4
Low-protein...........	7.80±.04	8.5	1.54±.03	19.7±.4
High-oil..............	6.87±.04	7.0	1.39±.03	20.2±.4
Low-oil...............	7.48±.04	8.0	1.30±.03	17.4±.4
Circumference of ear, in inches:				
High-protein...........	5.71±.01	5.7	.44±.01	7.6±.2
Low-protein...........	6.51±.02	6.3	.61±.01	9.4±.2
High-oil..............	6.05±.01	6.3	.53±.01	8.8±.2
Low-oil...............	6.65±.02	6.9	.59±.01	8.9±.2
Weight of ear, in ounces:				
High-protein...........	7.53±.04	8.0	2.50±.03	33.2±.4
Low-protein...........	9.66±.10	10.0	3.30±.07	34.2±.7
High-oil..............	7.79±.07	9.0	2.43±.05	31.2±.6
Low-oil...............	9.84±.08	11.0	2.87±.06	29.2±.7
Rows of kernels on ears:				
High-protein...........	13.72±.03	14.0	1.85±.02	13.5±.2
Low-protein...........	14.71±.06	14.0	1.94±.04	13.7±.3
High-oil..............	15.65±.06	16.0	2.08±.04	13.3±.3
Low-oil...............	12.80±.05	12.0	1.77±.04	13.8±.3

TABLE II.—*Type and variability of the four chemical-composition strains of Indian corn for 11 crops, 1905 to 1915*—Continued

CROP OF 1910

Character and strain.	Mean.	Empirical mode.	Standard deviation.	Coefficient of variability.
Length of ear, in inches:				
High-protein............	7.771±0.039	8.0	1.581±0.027	20.34±0.37
Low-protein	8.476± .051	8.5	1.678± .036	19.80± .44
High-oil	7.315± .030	7.5	1.019± .021	13.93± .30
Low-oil......	8.094± .034	9.0	1.233± .024	15.23± .31
Circumference of ear, in inches:				
High-protein............	6.034± .012	6.0	.469± .009	7.77± .15
Low-protein............	6.730± .016	6.5	.546± .012	8.12± .17
High-oil................	6.104± .014	6.0	.442± .011	7.24± .15
Low-oil................	6.859± .014	7.0	.510± .010	7.43± .15
Weight of ear, in ounces:				
High-protein............	8.415± .071	9.0	2.835± .051	33.69± .67
Low-protein............	10.036± .090	11.0	3.460± .090	34.48±1.00
High-oil................	7.600± .061	7.0	1.960± .043	25.82± .61
Low-oil................	10.986± .090	13.0	3.060± .064	27.85± .62
Rows of kernels on ears:				
High-protein............	13.345± .039	14.0	1.546± .028	11.58± .21
Low-protein..........	14.114± .059	14.0	1.995± .042	14.13± .30
High-oil................	14.750± .060	14.0	1.936± .042	13.13± .29
Low-oil................	13.280± .047	12.0	1.672± .033	14.82± .30

CROP OF 1911

Character and strain.	Mean.	Empirical mode.	Standard deviation.	Coefficient of variability.
Length of ear, in inches:				
High-protein............	6.302±0.038	6.5	1.402±0.027	22.25±0.45
Low-protein............	7.888± .045	8.5	1.411± .032	17.89± .42
High-oil................	6.744± .034	6.5	1.216± .024	18.03± .37
Low-oil................	6.486± .039	7.0	1.376± .027	21.21± .43
Circumference of ear, in inches:				
High-protein............	5.356± .014	5.5	.510± .010	9.52± .19
Low-protein............	6.026± .014	6.0	.502± .010	8.96± .19
High-oil................	5.769± .014	6.0	.504± .010	8.74± .18
Low-oil................	6.094± .014	6.5	.515± .010	8.45± .18
Weight of ear, in ounces:				
High-protein............	4.980± .056	5.0	1.854± .040	37.24± .91
Low-protein............	8.402± .071	10.0	2.336± .051	27.80± .65
High-oil ,...............	6.209± .050	6.0	1.672± .036	26.93± .63
Low-oil................	6.323± .070	6.0	2.038± .050	32.23± .85
Rows of kernels on ears:				
High-protein............	12.920± .046	12.0	1.541± .033	11.93± .26
Low-protein............	13.700± .052	14.0	1.744± .037	10.08± .22
High-oil................	15.350± .060	14.0	1.980± .042	12.89± .28
Low-oil................	12.721± .060	12.0	1.693± .043	13.31± .33

TABLE II.—*Type and variability of the four chemical-composition strains of Indian corn for 11 crops, 1905 to 1915*—Continued

CROP OF 1908

Character and strain.	Mean	Empirical mode.	Standard deviation.	Coefficient of variability.
Length of ear, in inches:				
High-protein...........	7 625±0. 035	7. 5	1. 195±0. 025	15. 67±0. 34
Low-protein...........	8. 496± .040	9. 0	1. 363± .028	16. 04± .34
High-oil...............	7. 334± .027	7. 5	.983± .019	13. 40± .27
Low-oil...............	8. 194± .041	8. 0	1. 407± .029	17. 17± .36
Circumference of ear, in inches:				
High-protein...........	5. 452± .014	5. 5	.460± .010	8. 44± .18
Low-protein...........	6. 338± .015	6 2	.490± .011	7. 76± .17
High-oil...............	5. 774± .012	6. 0	.438± .009	7. 59± .15
Low-oil...............	6 336± .016	6. 5	.548± .012	8. 65± .18
Weight of ear, in ounces:				
High-protein...........	6. 301± .058	6. 0	1. 804± .041	28. 63± .65
Low-protein...........	9. 783± .077	10. 0	2. 288± .054	23. 39± .59
High-oil...............	7 566± .046	8. 0	1. 616± .033	21. 36± .47
Low-oil...............	9. 944± .089	11. 0	2. 935± .063	29. 52± .68
Rows of kernels on ears:				
High-protein.	13. 276± .052	14. 0	1. 631± .037	12. 27± .28
Low-protein...........	13 984± .050	14. 0	1. 606± .035	11. 51± .26
High-oil...............	15. 010± .056	14. 0	1. 970± .040	13. 12± .27
Low-oil...............	12. 326± .050	14. 0	1. 591± .035	12. 90± .30

CROP OF 1909

Length of ear, in inches				
High-protein...........	7. 680±0. 035	7. 5	1. 127±0. 025	14. 67±0. 35
Low-protein...........	7. 672± .034	8. 0	1. 221± .024	15. 91± .32
High-oil...............	7 050± .031	7. 0	1. 140± .022	16. 17± .32
Low-oil...............	7. 748± .028	8. 0	1. 123± .020	14. 49± .27
Circumference of ear, in inches:				
High-protein...........	5. 693± .013	5. 75	.424± .009	7. 45± .17
Low-protein...........	5 967± .015	5. 75	.560± .011	9. 38± .19
High-oil...............	5 914± .013	6. 00	.489± .009	8. 27± .16
Low-oil...............	6 660± .013	6. 50	.517± .009	7. 76± .15
Weight of ear, in ounces:				
High-protein...........	6. 982± .054	7. 0	1. 694± .039	24. 26± .60
Low-protein...........	7 005± .068	7. 0	2. 407± .048	34. 36±1. 08
High-oil...............	6 837± .052	7. 0	1. 880± .037	27. 50± .82
Low-oil...............	9 244± .059	9. 0	2. 275± .042	24. 61± 48
Rows of kernels on ears:				
High-protein.... ...	13. 020± .050	12. 0	1. 554± .035	11. 93± .28
Low-protein...........	13. 789± .048	14. 0	1. 571± .034	11. 39± .27
High-oil.......... ..	14. 770± .051	14. 0	1. 758± .036	11. 90± .25
Low-oil...............	12. 806± .042	12. 0	1. 593± .030	12. 44± .24

Table II.—*Type and variability of the four chemical-composition strains of Indian corn for 11 crops, 1905 to 1915*—Continued

CROP OF 1910

Character and strain.	Mean.	Empirical mode.	Standard deviation.	Coefficient of variability.
Length of ear, in inches:				
High-protein	7.771±0.039	8.0	1 581±0.027	20.34±0.37
Low-protein	8.476±.051	8.5	1.678±.036	19.80±.44
High-oil	7.315±.030	7.5	1.019±.021	13.93±.30
Low-oil	8.094±.034	9.0	1 233±.024	15.23±.31
Circumference of ear, in inches:				
High-protein	6.034±.012	6.0	.469±.009	7.77±.15
Low-protein	6.730±.016	6.5	.546±.012	8.12±.17
High-oil	6.104±.014	6.0	.442±.011	7.24±.15
Low-oil	6.859±.014	7.0	.510±.010	7.43±.15
Weight of ear, in ounces:				
High-protein	8.415±.071	9.0	2.835±.051	33.69±.67
Low-protein	10.036±.090	11.0	3.460±.090	34.48±1.00
High-oil	7.600±.061	7.0	1.960±.043	25.82±.61
Low-oil	10.986±.090	13.0	3.060±.064	27.85±.62
Rows of kernels on ears:				
High-protein	13.345±.039	14.0	1.546±.028	11.58±.21
Low-protein	14.114±.059	14.0	1.995±.042	14.13±.30
High-oil	14.750±.060	14.0	1.936±.042	13.13±.29
Low-oil	13.280±.047	12.0	1.672±.033	14.82±.30

CROP OF 1911

Character and strain.	Mean.	Empirical mode.	Standard deviation.	Coefficient of variability.
Length of ear, in inches:				
High-protein	6.302±0.038	6.5	1 402±0.027	22.25±0.45
Low-protein	7.888±.045	8.5	1.411±.032	17.89±.42
High-oil	6.744±.034	6.5	1 216±.024	18.03±.37
Low-oil	6.486±.039	7.0	1 376±.027	21.21±.43
Circumference of ear, in inches:				
High-protein	5.356±.014	5.5	.510±.010	9.52±.19
Low-protein	6.026±.014	6.0	.502±.010	8.96±.19
High-oil	5.769±.014	6.0	.504±.010	8.74±.18
Low-oil	6.094±.014	6.5	.515±.010	8.45±.18
Weight of ear, in ounces:				
High-protein	4.980±.056	5.0	1.854±.040	37.24±.91
Low-protein	8.402±.071	10.0	2.336±.051	27.80±.65
High-oil	6.209±.050	6.0	1.672±.036	26.93±.63
Low-oil	6.323±.070	6.0	2.038±.050	32.23±.85
Rows of kernels on ears:				
High-protein	12.920±.046	12.0	1.541±.033	11.93±.26
Low-protein	13.700±.052	14.0	1.744±.037	10.08±.22
High-oil	15.350±.060	14.0	1.980±.042	12.89±.28
Low-oil	12.721±.060	12.0	1.693±.043	13.31±.33

TABLE II —*Type and variability of the four chemical-composition strains of Indian corn for 11 crops, 1905 to 1915*—Continued

CROP OF 1912

Character and strain.	Mean.	Empirical mode.	Standard deviation.	Coefficient of variability.
Length of ear, in inches:				
High-protein...........	7. 992±0. 028	8. 5	1. 356±0. 028	16. 97±0. 36
Low-protein...........	8. 550± 038	9. 0	1. 490± . 027	17. 43± . 34
High-oil..............	7. 210± . 031	8. 0	1. 232± . 022	17. 09± . 31
Low-oil..............	7 752± . 032	11. 3	1. 180±. 023	15. 22± . 32
Circumference of ear, in inches:				
High-protein...........	5. 850± . 014	6. 0	. 475± . 010	8. 12± . 17
Low-protein...........	6. 443± . 014	6. 50	. 549± . 010	8. 52± . 16
High-oil..............	6. 066± . 011	6. 25	. 459± . 008	7. 57± . 14
Low-oil..............	6. 619± . 032	6. 75	. 575± . 011	8. 66± . 17
Weight of ear, in ounces:				
High-protein...........	8. 012±. 059	10. 0	1. 985± . 042	24. 77± . 57
Low-protein...........	10. 962± . 071	12. 0	2. 640± . 051	24. 08± . 45
High-oil..............	7. 852± . 049	9. 0	1. 941± . 035	24. 72± . 45
Low-oil....... . .	10. 369± . 073	10. 0	2. 561± . 052	24. 70± . 54
Rows of kernels on ears:				
High-protein...........	12. 866± . 052	12. 0	1. 745± . 037	13. 56± . 29
Low-protein...........	13. 794± . 059	14. 0	1. 801± . 042	13. 05± . 26
High-oil..............	14. 700± . 045	14. 0	1. 788± . 032	12. 16± . 22
Low-oil..............	12. 448± . 053	12. 0	1. 817± . 038	14. 60± . 3

CROP OF 1913

Character and strain.	Mean.	Empirical mode.	Standard deviation.	Coefficient of variability.
Length of ear, in inches:				
High-protein...........	7. 438±0. 035	8. 0	1. 214±0. 025	16. 32±0. 3
Low-protein...........	8. 051± . 041	8. 5	1. 397± . 029	17. 35± . 3
High-oil..............	6. 582± . 033	6. 5	1. 178± . 023	17. 90± . 3
Low-oil..............	7. 026± . 033	7. 5	1. 263± . 023	17. 98± . 3
Circumference of ear, in inches:				
High-protein...........	5. 593± . 019	5. 5	. 589± . 013	10. 53± . 2
Low-protein...........	5. 956± . 017	6. 0	. 578± . 012	9. 70± . 2
High-oil..............	5. 395± . 014	5. 5	. 493± . 010	9. 14± . 1
Low-oil..............	6. 380± . 016	6. 5	. 604± . 011	9. 47± . 1
Weight of ear, in ounces:				
High-protein...........	6 754± . 051	6. 0	1. 708± . 036	25. 29± . 5
Low-protein...........	8. 221± . 060	9 0	1. 890± . 042	22. 99± . 5
High-oil..............	5 604± . 043	6. 0	1. 437± . 030	25. 64± . 5
Low-oil..............	7. 492± . 057	8. 0	2. 000± . 040	26. 70± . 5
Rows of kernels on ears:				
High-protein...........	13. 008± . 052	12. 0	1. 782± . 037	13. 70± . 5
Low-protein........ .	14. 260± . 060	14. 0	1. 914± . 042	13. 42± . 4
High-oil..............	14. 512± . 057	14. 0	1. 862± . 040	12. 83± . 5
Low-oil..............	13. 010± . 053	12. 0	1. 842± . 038	14. 16± . 5

TABLE II.—*Type and variability of the four chemical-composition strains of Indian corn for 11 crops, 1905 to 1915*—Continued.

CROP OF 1914

Character and strain.	Mean.	Empirical mode.	Standard deviation.	Coefficient of variability.
Length of ear, in inches:				
High-protein.............	7. 245±0. 033	7. 5	1. 279±0. 024	17. 65±0. 33
Low-protein.............	8. 037± . 041	8. 5	1. 368± . 029	17. 02± . 37
High-oil................	6. 423± . 035	6. 5	1. 248± . 025	19. 43± . 40
Low-oil.................	6. 985± . 032	7. 5	1. 200± . 022	17. 18± . 33
Circumference of ear, in inches:				
High-protein.............	5. 287± . 012	5. 3	. 441± . 008	8. 34± . 17
Low-protein.............	5. 918± . 014	6. 0	. 474± . 010	8. 01± . 17
High-oil................	5. 282± . 013	5. 5	. 466± . 009	8. 82± . 18
Low-oil.................	6. 162± . 015	6. 5	. 544± . 010	8. 83± . 17
Weight of ear, in ounces:				
High-protein.............	6. 396± . 044	6. 0	1. 499± . 031	23. 44± . 51
Low-protein.............	8. 284± . 068	9. 0	2. 170± . 048	26. 19± . 62
High-oil................	5. 207± . 043	5. 0	1. 417± . 030	27. 21± . 63
Low-oil.................	7. 711± . 054	7. 0	1. 863± . 038	24. 16± . 52
Rows of kernels on ears:				
High-protein.............	13. 360± . 046	14. 0	1. 668± . 033	12. 48± . 25
Low-protein.............	14. 402± . 063	14. 0	2. 048± . 045	14. 22± . 32
High-oil................	14. 489± . 061	16. 0	1. 936± . 043	13. 00± . 29
Low-oil.................	12. 684± . 044	12. 0	1. 552± . 031	12. 24± . 25

CROP OF 1915

Character and strain.	Mean.	Empirical mode.	Standard deviation.	Coefficient of variability.
Length of ear, in inches:				
High-protein.............	7. 176±0. 029	7. 5	1. 173±0. 021	16. 35±0. 29
Low-protein.............	7. 859± . 034	8. 0	1. 231± . 024	15. 66± . 31
High-oil................	7. 358± . 028	8. 0	1. 071± . 020	14. 56± . 28
Low-oil.................	7. 103± . 035	8. 0	1. 148± . 025	16. 16± . 35
Circumference of ear, in inches:				
High-protein.............	5. 674± . 010	5. 75	. 412± . 007	7. 26± . 13
Low-protein.............	6. 125± . 016	6. 50	. 583± . 011	9. 52± . 19
High-oil................	5. 781± . 012	5. 75	. 468± . 009	8. 10± . 16
Low-oil.................	6. 455± . 017	6. 50	. 592± . 012	9. 17± . 19
Weight of ear, in ounces:				
High-protein.............	7. 076± . 044	7. 0	1. 669± . 031	23. 59± . 47
Low-protein.............	8. 685± . 071	10. 0	2. 395± . 050	27. 58± . 62
High-oil................	7. 361± . 049	7. 0	1. 801± . 035	24. 47± . 50
Low-oil.................	8. 073± . 072	10. 0	2. 294± . 051	28. 42± . 68
Rows of kernels on ears:				
High-protein.............	13. 986± . 041	14. 0	1. 588± . 029	11. 35± . 21
Low-protein.............	15. 038± . 053	16. 0	1. 896± . 037	12. 61± . 25
High-oil................	15. 360± . 047	16. 0	2. 262± . 047	14. 73± . 31
Low-oil.................	14. 034± . 053	14. 0	1. 712± . 037	12. 20± . 27

MEAN LENGTH OF EARS

In the first year (1905) we find that the means in length for any two of the strains are significantly different. The smallest difference that exists is 0.27±0.056. We thus note that the smallest difference is about five times the probable error. The average difference in the means is about nine times the probable error of any difference between two means.

Having thus found in the 1905 crop significant differences when the differences are judged by probable errors, we very naturally inquire how persistently these differences are maintained and to what extent they are accentuated under continued selection for chemical composition. In these considerations we must not lose sight of the differences in soil conditions, such as we have on Experiment Station plots, and

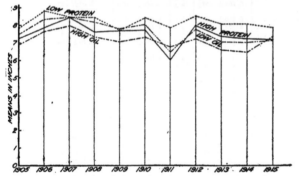

FIG. 1.—Graphs showing the mean length of the ears in the four Illinois strains of Indian corn for 11 crops, 1905-1915.

differences in seasonal conditions—that is to say, it should be kept in mind in considering these results—that the various strains have occupied different plots of ground in different years and that there has been no special attempt to maintain uniform or comparable soil conditions among these breeding plots.

In figure 1 is shown graphically the mean values of length of ears in the four strains. It may be noted from these graphs that the order of magnitude of 1905 was maintained with slight exception till 1911. The year 1911 was an abnormal one characterized by lack of moisture at the time when the corn was much in need of moisture. The high-protein and the low-oil strains changed positions in 1912, and these two strains have maintained the new order through 1915. In 1915 the high-oil surpassed the high-protein and the low-oil strain, but it is not unlikely that this is only a seasonal fluctuation. While there have been marked seasonal fluctuations in the mean length of ears for each of these strains, there is no marked regular change in the length of the ear for any strain. From the graphs in figure 1 there is, however, at least some evidence that, on the whole, the mean length of ears has been smaller during the second half of the period 1905–1915, than during the first half of the period. But it seems in explanation of this, that in the second half of the period there were at least three years in which the length of ears showed decidedly the effects of abnormal seasons.

There has been one progressive change worth noting, in that the mean lengths of ears for the high-protein strain for the past four years maintained a position above the low-oil strain. The differences between the mean lengths of ears in the latter half of the period 1905–1915 are slightly larger, but hardly significantly larger than those that existed in the earlier part of the period.

Since the probability that a given order of four elements in a trial under pure chance is 1 to 24, we note that there is a decided tendency to maintain order in mean length of ears, although there are a few exceptions.

MEAN CIRCUMFERENCE OF EARS

In the first year (1905) we find significant differences in the means except possibly for the case of the low-protein and the low-oil strains. For this pair we find a difference of 0.14 ± 0.03. While, under random sampling, we should probably ascribe at least some slight significance to such a difference, it is doubtful whether the difference is large enough to persist regularly under differences of seasons and soil. But, based on 11 years' experience, we find that the low-protein strain has always shown a smaller mean circumference than the low-oil strain, except in the one year (1908) when the two were substantially equal.

We note from figure 2 the regularity with which the order of 1905 has been maintained. The order for circumference of ears has been at least as well maintained as that for the length of ears, and perhaps a little more persistently. Figure 2 shows that the mean circumference of the high-oil strain was lower than that of the high-protein strain in the year 1913 only.

As to progressive changes, it is true, on the whole, that each of these strains shows a smaller mean circumference in the second half of the period 1905–1915 than in the first half, but it is not unlikely that this is due to seasonal fluctuations. What is more important for our purposes is to note that the mean differences between mean values are no more pronounced during the second half than during the first half of the period. That is to say, the differences existing in 1905 have well been maintained, considering the changes of soils and seasons, but they have not been decidedly increaeds since 1905 by the selections for chemical composition.

MEAN WEIGHT OF EARS

In 1905 there appeared to be no significant difference in weight of ears between the high-protein and the high-oil strains, and no significant difference between the low-protein and the low-oil. The high-protein and the high-oil were significantly different in weight of ears from the low-protein and the low-oil. These two classes seem to persist well under differences of season and soil, as shown by the graphs of figure 3.

In regard to progressive changes in weight, there is no doubt that, on the whole, the mean weight of the ears for each strain have been less

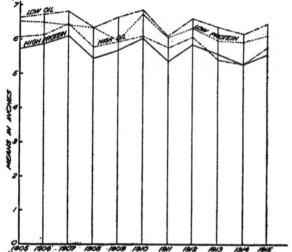

Fig. 2.—Graphs showing the mean circumference of the ears in the four Illinois strains of Indian corn for 11 crops, 1905–1915.

during the second half of the period 1905–1915 than during the first half of the period; that this is, however, merely in response to abnormal seasons is shown by the fact that the ears are particularly light for the crops of 1911, 1913, and 1914. It is important to note that the selection for chemical composition has probably neither increased nor decreased significantly the differences in the mean weight of ears during the period 1905–1915.

MEAN NUMBER OF ROWS OF KERNELS ON EARS

In the year 1905 there was a significant difference in the number of rows of kernels on ears. Figure 4 shows that the same order of magnitude of number of rows of kernels has been invariably maintained throughout 11 years, except the years 1913 and 1915, when the low-oil changed

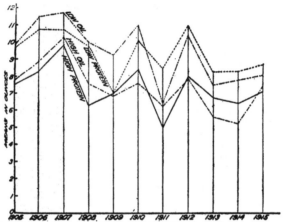

Fig. 3.—Graphs showing the mean weight of the ears in the four Illinois strains of Indian corn for 11 crops, 1905-1915.

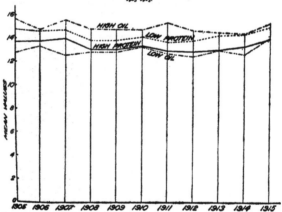

Fig. 4.—Graphs showing the mean number of rows of kernels in the four Illinois strains of Indian corn for 11 crops, 1905-1915.

places with the high-protein strain. The differences, however, are insignificant in these years. As the character (rows of kernels) is determined by inherent tendencies and is less susceptible to external or environmental influences than the other characters considered, we should expect smaller seasonal fluctuations than in the case of the other characters.

As to progressive changes, we may note that while the low-oil strain seemed to tend toward a 12-row type for some years, it appears that in the 1910, 1913, and 1915 crops 14 rows of kernels are the modal value. That is to say, there is no longer good evidence that the low-oil strain is approaching a condition where the ears have 12 rows of kernels as a modal value. The differences existing in 1905 have been well maintained, except in the case of the high-protein and low-oil strains and we have indicated above that these two strains did not differ significantly in several recent crops. It is fair to say in a general way that the differences in the mean number of rows of kernels existing in 1905 have not been significantly accentuated by the selection for chemical composition from 1905 to 1915, although the mean of the differences is slightly greater in the latter half of the period.

GENERAL STATEMENT ABOUT DISTINCT TYPES

Taken as a whole, it is found that distinct types, as shown by the mean length of ears, the circumference of ears, the weight of ears, and the number of rows of kernels, are so well established that in most cases we may assign an order of magnitude that persists well in such changes of environment as have been experienced in the 11 years of planting from 1905 to 1915. While certain progressive changes have been noted above, the selections for chemical composition from 1905 to 1915 have not increased the difference in the mean values to such an extent that we are able to assert that the strains differ more with respect to these characters during the second half of the period than during the first half, although they certainly do not as a whole differ less during the second half than during the first half. In fact, as stated above, the differences of the mean length of ears and the number of rows of kernels are larger, by an insignificant amount, during the second half than during the first half of the period.

STANDARD DEVIATIONS

In general, the differences in standard deviations for the different strains compared to the corresponding probable errors are much smaller than the differences in the mean values for the different strains compared to their probable errors.

Many of the differences in standard deviations do not persist well from year to year under differences of soils and seasons. Perhaps we may properly infer from figure 5, that, on the whole, of the four strains the low-protein strain has the largest standard deviation in length of ears, and the high-oil strain has the smallest standard deviation; but there are

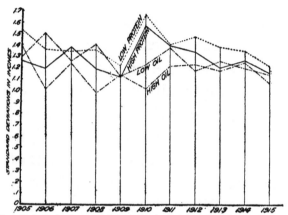

FIG. 5.—Graphs showing the standard deviation of the length of the ears in the four Illinois strains of Indian corn for 11 crops, 1905–1915.

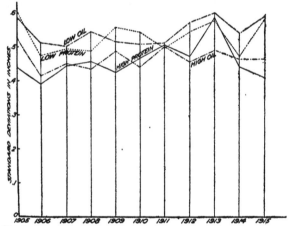

FIG. 6.—Graphs showing the standard deviation of the circumference of the ears in the four Illinois strains of Indian corn for 11 crops, 1905–1915.

some exceptions to this statement.   At least we may s
protein strain has for each of the 11 crops a larger stan
length of ears than the high-oil strain.

As shown graphically on figure 6, we find that on th
oil and the low-protein strains are more variable in
ears than the high-protein and the high-oil strains.

From figure 7 it is pretty clear that the low-protei
strains are decidedly more variable in weight of ears t
two strains.   When judged by probable errors, thes
decidedly significant, and of such magnitude that we
persist from year to year if soil and seasonal conditions
different.

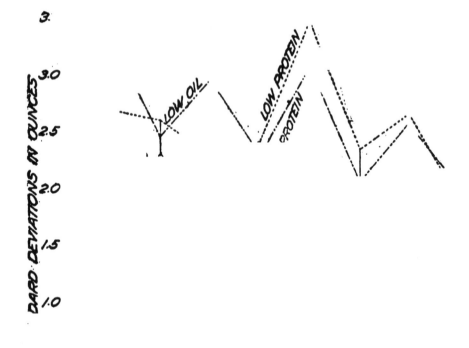

figures 1 to 4 and as the standard deviations have the smaller probable errors, we very naturally inquire for a plausible explanation of these results. Shall we conclude that the mean value is more stable and characterizes the strain better than the standard deviation does? To be sure, we often give a mean value to characterize a type; but in accord with what is stated above about these four strains of corn, the ratio of their differences of means to the probable errors of such differences are in general much greater than the ratio of the differences of their standard deviations to corresponding probable errors. In more definite form it is found that the ratio of the average difference of means in a crop to the average of the corresponding probable errors varies from 12 in the case of length of ears to 23 in the case of circumference of ears. On the other hand, the ratio of average differences in standard deviations in a crop to the average of the corresponding probable errors varies from 3.5 in the

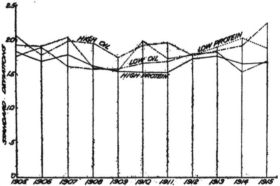

Fig. 1.—Graphs showing the standard deviation of the rows of kernels on the ears in the four Illinois strains of Indian corn for 11 crops, 1905–1915

case of rows of kernels to 7 in the case of weight of ears. Incidentally, in the latter case, we find that the standard deviation is strikingly higher for the low-protein and the low-oil than for the high-protein and the high-oil strains, and that this difference persists without exception for the 11 crops. We should recognize, therefore, that the standard deviations of the four strains for a given year differ much less when compared with their probable errors than do the corresponding means. Hence, we should expect much less tendency to maintain an order of values of standard deviations than of means. If, on the other hand, the mean values were close together, and the standard deviations were significantly different, we could characterize strains better by standard deviations than by means.

Stated in another form, the differences of standard deviations of the four strains when taken as a whole are not large enough to separate the strains nearly so well as the differences of means separate them, although the standard deviations separate the strains well with regard to some of the four characters. For example, we distinguish the high-protein and the high-oil strains very effectively from the low-protein and the low-oil strains by differences of standard deviations in weight of ears. The fact that differences in standard deviations do not enable us to distinguish the strains so well as the differences of means does not mean that the standard deviations are less stable from season to season for a given strain than are the means. In fact, taken as a whole, the standard deviations are considerably less variable from season to season than the corresponding means. This is in accord with the fact that the probable errors of standard deviations are only about 0.7 those of corresponding means   We are justified in putting the above statement a little stronger by saying that the ratio of fluctuations of standard deviations actually experienced to the probable errors of standard deviations are less for these four strains than the corresponding ratios of fluctuations of means to the probable errors in the means.

COEFFICIENTS OF VARIABILITY

The coefficients of variability do not in any marked way maintain a definite order of values from year to year. We prepared graphic representations of the coefficients, but hesitate to take the space for their publication, since the conclusions drawn consist largely in the statement of the existence of insignificant differences that may be obtained by inspection of the tables. For length of ears these graphs gave a fair illustration of as many crossings as we should expect under pure chance. On the other hand, in circumference of ears the low-protein and low-oil strains have in general somewhat higher coefficients of variability than the high-oil strain.

When we compare differences of coefficients of variability with their probable errors, the differences in a fair number of cases are significant when judged by the usual requirements in regard to random sampling. But these differences are generally insufficient to withstand changes of season and soil. Further, we find that the ratio of the average of the differences of coefficients of variability in a season to the average of the probable errors in such differences varies from 2.7 in the case of rows of kernels to 3.8 in the case of weight of ears. The corresponding ratios of 12 and 23 given above for means show a striking contrast. This accounts for the maintenance of a pretty regular order for mean values, while no considerable degree of regularity is maintained among coefficients of variability. It seems that to maintain any considerable regularity from one season to another the differences must be larger, compared to probable errors, than those which exist in the case of these coefficients of variability.

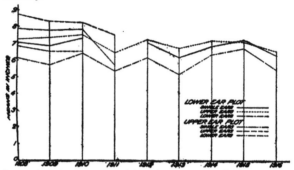

Fɪɢ. 9.—Graphs showing the mean length of the ears in the two-ear strains of Indian corn, 1908-1916.

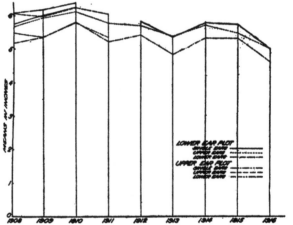

Fɪɢ. 10.—Graphs showing the mean circumference of the ears in the two-ear strains of Indian corn, 1908-1916.

TWO-EAR STRAINS

Table III gives the means, standard deviations, coefficients of variability, and corresponding probable errors, for length of ears, circumference of ears, weight of ears, and number of rows of kernels on ears for the two-ear strains.

7768°—17——3

COMPARISON OF UPPER AND LOWER EARS ON THE UPPER- AND LOWER-
EAR PLOTS

The "upper-ear plots" were started by selecting seed from upper ears
for the purpose primarily of determining the ratio of the number of

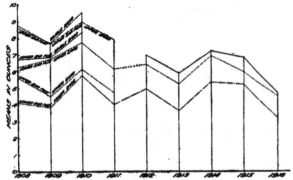

FIG. 11.—Graphs showing the mean weight of the ears in the two-ear strains of Indian corn, 1908–1916.

stalks with single ears to the number that have two ears in the offspring
of such parents. In succeeding years, the upper ears continued to be

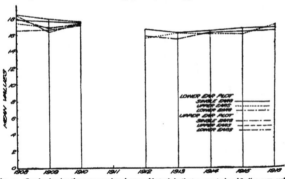

FIG. 12.—Graphs showing the mean number of rows of kernels in the two-ear strains of Indian corn, 1908–1916.

selected for planting. This perhaps defines sufficiently what is called
the "upper-ear plots." Similarly on the "lower-ear plots" the lower
ears were selected for planting. In the present paper we are not concerned with the primary purposes of the experiment, but with the effects

of the selections on the four physical characters of ears mentioned above.

Our data on the upper-ear plots cover four crops, 1908 to 1911. Our data on the lower-ear plots cover nine crops, 1908 to 1916, with the one exception of the single ears for 1911, owing to the fact that the records are not available for that case.

DIFFERENCES IN MEANS

[Table III and fig. 9 to 12]

It is true in every case that the upper ears have a significantly larger mean value in length, weight, and circumference than have the corresponding lower ears of the same stalks. This fact stands out prominently in the graphic representations of figures 9 to 12.

In mean length of ears these differences range from 5 to 30 per cent of the mean length of the shorter group of ears.

In mean weight of ears the differences range from 14 to 70 per cent of the mean length of the lighter group of ears.

In mean circumference of ears the differences range from a little less than 5 to 13 per cent of the mean circumference of the smaller group of ears.

TABLE III.—*Type and variability of the two-ear strains of Indian corn*

UPPER-EAR PLOT, CROP OF 1908

Character and strain.	Mean.	Empirical mode.	Standard deviation.	Coefficient of variability.
Length of ear, in inches:				
Single ears.............	7.870±0.054	9.0	1.750±0.038	22.24±0.51
Upper ears.............	8.739± .052	8.5	.956± .037	10.94± .42
Lower ears.............	6.840± .072	7.5	1.344± .052	19.65± .77
Circumference of ear, in inches:				
Single ears.............	6.067± .020	6.0	.630± .014	10.38± .24
Upper ears.............	6.060± .025	5.75	.454± .018	7.49± .29
Lower ears.............	5.479± .030	5.50	.533± .021	9.73± .40
Weight of ear, in ounces:				
Single ears.............	8.64 ± .11	10.0	2.986± .071	34.57± .97
Upper ears.............	8.786± .098	9.0	1.658± .070	18.88± .81
Lower ears....	5.69 ± .13	6.0	2.034± .091	35.75±1.68
Rows of kernels on ear:				
Single ears.............	18.35 ± .086	18.0	2.516± .060	13.71± .34
Upper ears.............	18.03 ± .14	18.0	2.55 ± .10	14.05± .56
Lower ears.............	18.25 ± .20	18.0	2.74 ± .14	15.04± .78

TABLE III.—*Type and variability of the two-ear strains of Indian corn*—Continued

UPPER-EAR PLOT, CROP OF 1909

Character and strain.	Mean.	Empirical mode.	Standard deviation.	Coefficient of variability.
Length of ear, in inches:				
Single ears.............	7.775±0.043	8.0	1.085±0.030	13.96±0.40
Upper ears	8.325± .062	8.5	.897± .043	10.77± .53
Lower ears.............	6.512± .097	6.5	1.349± .065	20.72±1.10
Circumference of ears, in inches:				
Single ears......... .	6.183± .020	6.0	.512± .014	8.42± .24
Upper ears..	5.976± .028	6.0	.408± .020	6.83± .34
Lower ears.............	5.373± .036	5.5	.485± .025	9.03± .48
Weight of ear, in ounces:				
Single ears.............	7.719± .087	8.0	2.166± .061	28.06± .86
Upper ears.............	7.69 ± .12	8.0	1.622± .084	21.09±1.10
Lower ears.............	4.51 ± .14	4.0	1.810± .098	40.11±2.50
Rows of kernels on ears:				
Single ears.............	17.91 ± .10	18.0	2.482± .071	13.86± .41
Upper ears.............	16.76 ± .16	16.0	2.30 ± .11	13.72± .72
Lower ears.............	16.32 ± .18	18.0	2.41 ± .13	12.93± .80

UPPER-EAR PLOT, CROP OF 1910

Character and strain.	Mean.	Empirical mode.	Standard deviation.	Coefficient of variability.
Length of ear, in inches:				
Single ears.............	7.820±0.057	9.0	1.592±0.040	20.36±0.53
Upper ears.............	8.268± .056	8.0	1.021± .040	12.35± .48
Lower ears.............	6.522± .079	7.0	1.364± .056	20.91± .89
Circumference of ear, in inches:				
Single ears.............	6.376± .019	6.5	.526± .014	8.25± .22
Upper ears.............	6.260± .025	6.0	.459± .018	7.33± .18
Lower ears.............	5.784± .032	5.5	.529± .023	9.15± .40
Weight of ear, in ounces:				
Single ears.............	9.57 ± .11	11.0	2.957± .079	30.91± .90
Upper ears.............	8.99 ± .12	9.0	2.205± .087	24.5 ±1.0
Lower ears.............	6.13 ± .14	6.0	2.128± .099	34.7 ±1.8
Rows of kernels on ear:				
Single ears.	17.542± .091	18.0	2.366± .064	13.49± .37
Upper ears.............	17.26 ± .14	16.0	2.54 ± .10	14.73± .60
Lower ears.............	17.31 ± .15	16.0	2.27 ± .11	13.12± .64

UPPER-EAR PLOT, CROP OF 1911

Character and strain.	Mean.	Empirical mode.	Standard deviation.	Coefficient of variability.
Length of ear, in inches:				
Single ears.............				
Upper ears.............	7.537±0.010	7.5	1.319±0.070	17.50±0.94
Lower ears..	5.715± .012	6.0	1.298± .083	22.71± .13
Circumference of ear, in inches:				
Single ears.............				
Upper ears...	6.066± .041	6.0	.535± .029	8.82± .48
Lower ears	5.339± .049	5.0	.603± .035	11.29± .67
Weight of ear, in ounces:				
Single ears.............				
Upper ears.............	7.942± .015	9.0	1.941± .011	24.44± .15
Lower ears.............	4.716± .022	3.0	2.031± .016	43.10± .40

TABLE III.—*Type and variability of the two-ear strains of Indian corn*—Continued

LOWER-EAR PLOT, CROP OF 1908

Character and strain.	Mean.	Empirical mode.	Standard deviation.	Coefficient of variability.
Length of ear, in inches:				
Single ears............	7. 056±0. 043	8. 0	1. 295±0. 030	18. 35±0. 44
Upper ears............	7. 238± . 097	7. 5	1. 227± . 068	16. 95± . 96
Lower ears............	6. 14 ± . 10	6. 0	1. 188± . 071	19. 3 ±1. 2
Circumference of ear, in inches:				
Single ears............	5. 741± . 020	6. 0	. 574± . 014	10. 00± . 25
Upper ears............	5. 671± . 051	5. 75	. 575± . 036	10. 14± . 64
Lower ears............	5. 192± . 046	5. 25	. 514± . 032	10. 09± . 64
Weight of ear, in ounces:				
Single ears............	6. 740± . 078	7. 0	2. 176± . 055	32. 28± . 90
Upper ears............	6. 26 ± . 20	5. 0	2. 24 ± . 14	35. 7 ±2. 5
Lower ears............	4. 11 ± . 17	3. 0	1. 79 ± . 12	43. 6 ±3. 4
Rows of kernels on ear:				
Single ears............	17. 79 ± . 10	18. 0	2. 55 ± . 071	14. 32± . 41
Upper ears............	17. 35 ± . 20	16. 0	2. 16 ± . 14	12. 45± . 82
Lower ears............	16. 49 ± . 23	16. 0	2. 20 ± . 16	13. 3 ±1. 0

LOWER-EAR PLOT, CROP OF 1909

Character and strain.	Mean.	Empirical mode.	Standard deviation.	Coefficient of variability.
Length of ear, in inches:				
Single ears............	6. 834±0. 039	8. 0	1. 143±0. 028	16. 73±0. 42
Upper ears............	7. 350± . 061	7. 5	. 901± . 043	12. 26± . 59
Lower ears..	5. 701± . 078	5. 5	1. 176± . 055	20. 6 ±1. 0
Circumference of ear, in inches:				
Single ears............	6. 014± . 019	6. 0	. 552± . 013	9. 18± . 23
Upper ears............	5. 932± . 032	6. 0	. 464± . 022	7. 82± . 37
Lower ears............	5. 356± . 041	5. 5	. 546± . 029	10. 19± . 55
Weight of ear, in ounces:				
Single ears............	6. 87 ± . 11	7. 0	2. 228± . 078	32. 4 ±1. 3
Upper ears............	6. 763± . 14	7. 0	1. 994± . 099	29. 5 ±1. 5
Lower ears............	3. 88 ± . 15	2. 0	1. 89 ± . 11	48. 7 ±3. 2
Rows of kernels on ear:				
Single ears............	17. 63 ± . 094	16. 0	2. 496± . 066	14. 16± . 38
Upper ears............	16. 93 ± . 16	16. 0	2. 28 ± . 11	13. 48± . 67
Lower ears............	16. 60 ± . 28	16. 0	2. 75 ± . 20	16. 5 ±1. 2

LOWER-EAR PLOT, CROP OF 1910

Character and strain.	Mean.	Empirical mode.	Standard deviation.	Coefficient of variability.
Length of ear, in inches:				
Single ears............	7. 314±0. 051	8. 0	1. 416±0. 036	19. 36±0. 26
Upper ears............	7. 517± . 039	8. 0	1. 088± . 020	14. 48± . 38
Lower ears............	6. 405± . 045	7. 0	1. 195± . 032	18. 66± . 51
Circumference of ear, in inches:				
Single ears............	6. 272± . 022	6. 0	. 587± . 016	9. 36± . 25
Upper ears............	6. 122± . 018	6. 0	. 476± . 012	7. 78± . 20
Lower ears............	5. 816± . 019	6. 0	. 480± . 013	8. 25± . 23
Weight of ear, in ounces:				
Single ears............	8. 359± . 12	5. 0	2. 853± . 085	34. 1 ±1. 1
Upper ears............	7. 770± . 087	7. 0	2. 124± . 061	27. 32± . 85
Lower ears...... .	5. 793± . 081	5. 0	1. 903± . 057	32. 8 ±1. 1
Rows of kernels on ear:				
Single ears............	17. 44 ± . 11	16. 0	2. 644± . 076	15. 15± . 45
Upper ears............	17. 24 ± . 10	16. 0	2. 674± . 071	15. 51± . 42
Lower ears............	17. 07 ± . 11	16. 0	2. 584± . 076	15. 14± . 45

TABLE III —*Type and variability of the two-ear strains of Indian corn*—Continued

LOWER-EAR PLOT, CROP OF 1911

Character and strain.	Mean.	Empirical mode.	Standard deviation.	Coefficient of variability.
Length of ear, in inches:				
Single ears.............	....	.....	....	....
Upper ears.............	6. 462±0. 087	5. 5	1. 357±0. 062	21. 00±1. 00
Lower ears.............	5. 358± .091	5. 5	1. 358± .065	25. 35±1. 30
Circumference of ear, in inches:				
Single ears... ...	....	.....	....	....
Upper ears.............	5. 790± .078	6. 0	. 532± .056	9. 19± .50
Lower ears.............	5. 216± .091	5. 25	. 561± .065	10. 75± .65
Weight of ear, in ounces:				
Single ears.............	....	.....	....	....
Upper ears.............	6. 18 ± .170	6. 0	1. 764± . 120	28. 5 ±1. 4
Lower ears.............	4. 04 ± . 130	3. 0	1. 677± .089	41. 5 ±2. 6

CROP OF 1912

Character and strain.	Mean.	Empirical mode.	Standard deviation.	Coefficient of variability.
Length of ear, in inches:				
Single ears.............	7. 056±0. 048	7. 5	1. 233±0. 034	17. 47±0. 49
Upper ears.............	7. 250± .066	7. 0	1. 900± .047	26. 21± .67
Lower ears.............	6. 176± .075	6. 5	2. 100± .053	34. 00± .84
Circumference of ear, in inches:				
Single ears.............	5. 853± .020	6. 0	. 523± .014	8. 94± .24
Upper ears.............	5. 700± .015	5. 75	. 417± .010	7. 32± .19
Lower ears	5. 427± .014	5. 5	. 392± .010	7. 22± .19
Weight of ear, in ounces:				
Single ears.............	7. 004± .088	6. 0	2. 053± .062	29. 31± .96
Upper ears.............	6. 429± .052	6. 0	1. 406± .037	21. 87± .65
Lower ears.............	4. 977± .052	5. 0	1. 245± .037	25. 02± .79
Rows of kernels on ear:				
Single ears.............	16. 560± .088	16. 0	2. 212± .062	13. 61± .39
Upper ears.............	15. 560± .069	16. 0	1. 990± .049	12. 79± .32
Lower ears.............	15. 822± .023	16. 0	1. 870± .016	11. 82± .10

CROP OF 1913

Character and strain.	Mean.	Empirical mode.	Standard deviation.	Coefficient of variability.
Length of ear, in inches:				
Single ears.............	6. 202±0. 061	6. 0	1. 394±0. 043	22. 48±0. 75
Upper ears.............	6. 741± .064	6. 5	1. 238± .046	18. 37± .72
Lower ears.............	5. 192± .069	5. 0	1. 257± .049	24. 2 ±1. 0
Circumference of ear, in inches:				
Single ears.............	5. 359± .025	5. 0	. 581± .018	10. 84± .35
Upper ears	5. 374± .028	5. 5	. 528± .020	9. 82± .39
Lower ears.............	4. 844± .025	4. 75	. 454± .018	9. 37± .39
Weight of ear, in ounces:				
Single ears.............	5. 908± .093	6. 0	1. 641± .066	27. 8 ±1. 3
Upper ears.............	5. 280± .083	5. 0	1. 411± .059	26. 7 ±1. 2
Lower ears.............	3. 655± .093	3. 0	1. 173± .060	32. 1 ±1. 8
Rows of kernels on ears:				
Single ears.............	16. 19 ± . 13	16. 0	2. 486± .089	15. 35± .58
Upper ears.............	16. 10 ± . 13	16. 0	2. 252± .090	13. 99± .58
Lower ears.............	15. 42 ± . 15	14. 0	2. 07 ± . 10	13. 42± .69

TABLE III.—*Type and variability of the two-ear strains of Indian corn*—Continued

CROP OF 1914

Character and strain.	Mean.	Empir-ical mode.	Standard devia-tion.	Coefficient of varia-bility.
Length of ears, in inches:				
Single ears..............	6.855±0.061	8.0	1.473±0.043	21.49±0.71
Upper ears.............	7.190± .031	7.5	1.052± .022	14.63± .32
Lower ears.............	6.340± .031	6.5	1.015± .022	16.01± .36
Circumference of ear, in inches:				
Single ears.............	5.789± .024	5.75	.574± .017	9.92± .30
Upper ears.............	5.696± .014	5.75	.456± .010	8.01± .17
Lower ears.............	5.427± .014	5.50	.440± .010	8.11± .18
Weight of ear, in ounces:				
Single ears.............	7.23 ± .10	7.0	2.246± .070	31.1 ±1.0
Upper ears.............	7.060± .048	7.0	1.553± .034	22.00± .51
Lower ears.............	5.376± .049	5.0	1.439± .035	26.77± .69
Rows of kernels on ear:				
Single ears......... . .	16.434± .097	16.0	2.245± .069	13.66± .42
Upper ears.............	16.004± .065	16.0	2.128± .046	13.30± .29
Lower ears.............	16.140± .066	16.0	2.060± .047	12.76± .30

CROP OF 1915

Character and strain.	Mean.	Empir-ical mode.	Standard devia-tion.	Coefficient of varia-bility.
Length of ears, in inches:				
Single ears..............	7.220±0.075	7.5	1.288±0.053	17.84±0.79
Upper ears.............	7.107± .024	7.5	.900± .017	12.66± .24
Lower ears.............	6.718± .025	6.5	.939± .018	13.98± .27
Circumference of ears, in inches:				
Single ears.............	5.718± .028	5.75	.470± .020	8.22± .35
Upper ears.............	5.485± .011	5.5	.404± .008	7.37± .14
Lower ears.............	5.295± .011	5.5	.414± .008	7.82± .15
Weight of ear, in ounces:				
Single ears.............	6.860± .115	8.0	1.878± .081	27.4 ±1.3
Upper ears.............	5.981± .033	6.0	1.228± .024	20.53± .41
Lower ears.............	5.243± .036	5.0	1.285± .025	24.51± .50
Rows of kernels on ear:				
Single ears.............	16.69 ± .13	16.0	2.250± .095	13.48± .60
Upper ears.............	15.990± .055	16.0	2.086± .039	13.05± .25
Lower ears.............	16.282± .054	16.0	2.026± .038	12.44± .24

CROP OF 1916

Character and strain.	Mean.	Empir-ical mode.	Standard devia-tion.	Coefficient of varia-bility.
Length of ears, in inches:				
Single ears..............	6.327±0.042	6.5	0.934±0.030	14.77±0.49
Upper ears.............	6.545± .047	6.5	1.057± .033	16.15± .52
Lower ears.............	5.441± .050	5.5	1.111± .035	20.42± .67
Circumference of ear, in inches:				
Single ears.............	4.995± .020	5.0	.444± .014	8.89± .29
Upper ears.............	5.011± .016	4.75	.369± .012	7.35± .23
Lower ears.............	4.615± .018	4.75	.413± .013	8.95± .25
Weight of ear, in ounces:				
Single ears.............	4.762± .087	4.0	1.776± .062	37.29±1.46
Upper ears.............	4.549± .054	4.0	1.143± .038	25.13± .87
Lower ears.............	3.270± .053	4.0	1.035± .037	31.64±1.25
Rows of kernels on ear:				
Single ears.............	16.249± .090	16.0	1.849± .063	11.37± .42
Upper ears.............	16.327± .095	16.0	2.070± .067	12.68± .42
Lower ears.............	16.14 ± .10	16.0	2.123± .074	13.15± .47

With respect to the number of rows of kernels on ears, no significant difference is shown to exist between the upper and the lower ears. While the upper-ear plots were continued for only four years, it seems that it may be of value to compare for these four years the upper ears of the upper-ear plots with the upper ears of the lower-ear plots and the lower ears of the upper-ear plots with the lower ears of the lower-ear plots. From such a comparison (see fig. 9 to 12) we may assert that in mean length of ears, circumference of ears, and weight of ears, the upper ears of the upper-ear plot are larger in each case than the corresponding upper ears of the lower-ear plot. The lower ears of the upper-ear plots exceed in mean length and weight the lower ears of the lower-ear plots, but in circumference no significant difference exists.

DIFFERENCES IN VARIABILITY

[Table III and fig. 13 to 16]

The standard deviations of length of ears have in general been only slightly different for the upper and lower ears of the lower-ear plots. In circumference of ears for the lower-ear plots it seems that the differences in standard deviation are insufficient to persist from season to season. In the weight of ears the standard deviations of the upper ears of the lower-ear plots have generally been significantly greater than those of the lower ears of the lower-ear plots. The year 1915 gives an exception to this rule. The standard deviations of the upper ears on the upper-ear plots for length of ears, weight of ears, and circumference of ears are in general smaller than the corresponding standard deviations of the lower ears on these plots.

The coefficients of variability for length and weight of the lower ears of the lower-ear plots are decidedly larger than the coefficients of variability for these characters in the upper ears of the lower-ear plots. Furthermore the coefficients of variability of length, circumference, and weight of the upper ears are larger for the lower-ear plots than for the corresponding upper-ear plots.

COMPARISON OF SINGLE EARS ON UPPER- AND LOWER-EAR PLOTS WITH THE UPPER AND LOWER EARS

[Table III and fig 9 to 16]

DIFFERENCES OF MEANS

[Fig. 9 to 12]

The means of single ears for weight, length, and circumference are larger in each case than the corresponding means of the lower ears.

In length of ears the mean values of the upper ears are larger than those of the corresponding single ears in 10 out of 11 cases. To judge by probable errors, no more than half of these differences are significant.

But on the whole there seems to be some tendency for the upper ears to exceed the single ears in length.

In the circumference of ears the means of the single ears exceed the means of the upper ears in 8 out of 11 cases, but only 4 of these differences are significant when judged by probable errors.

In weight of ears, the means for the single ears exceed those for the upper ears in every case but one. Thus, it appears in a general way that single ears tend to be relatively heavier and larger in circumference, but shorter than upper ears.

In rows of kernels on ears the mean number for the single ears is larger in each case but one than that for the upper and lower ears, but three of these differences are so small as hardly to be regarded as significant when compared with probable errors.

DIFFERENCES IN VARIABILITY

[Table III and fig. 13 to 16]

With respect to the length of ears nothing very definite is to be inferred. In cir- cumference of ears, with a single exception in 22 cases, the standard deviation of the single ears exceeds that of upper and lower ears,and the difference is insignificant in this exceptional case.

In weight of ears the single ears have a greater standard deviation than the upper or lower ears in 21 out of 22 cases. In fact, so great is this difference that in some cases it looks as if the two frequency distributions of upper and lower ears might be combined to form the frequency distribution of single ears.

In length and weight of ears the coefficients of variability of the lower ears exceed those of the corresponding upper ears. In circumference of ears, the coefficient of variability of the lower ears either exceeds in each case that of the corresponding upper ears or differs from it by an insignificant amount.

ERECT- AND DECLINING-EAR STRAINS

Table IV gives the means, standard deviations, coefficients of variability, and the corresponding probable errors for the characters in question of the erect- and declining-ear strains.

Although we exhibit graphically the results on the erect- and declining-ear strains and those on the high- and low-ear strains on the same figures, we do so to save space, and not to make comparisons of the erect- and declining-ear strains with the high- and low-ear strains. Comparisons are made only between the erect- and declining-ear strains and between the high- and low-ear strains.

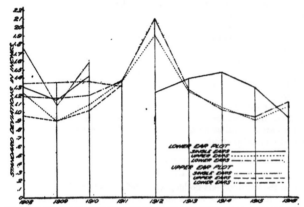

Fig. 13.—Graphs showing the standard deviation of the length of the ears in the two-ear strains of Indian corn, 1908-1916.

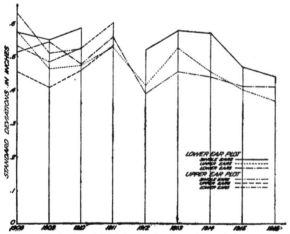

Fig. 14.—Graphs showing the standard deviation of the circumference of the ears in the two-ear strains of Indian corn, 1908-1916.

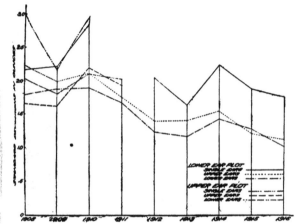

Fig. 15.—Graphs showing the standard deviation of the weight of the ears in the two-ear strains of Indian corn, 1908–1916.

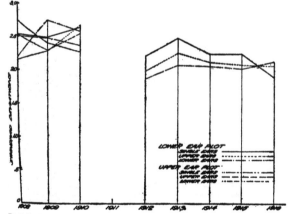

Fig. 16.—Graphs showing the standard deviation of the number of rows of kernels in the two-ear strains of Indian corn.

DIFFERENCES IN MEANS

[Fig 17 to 20]

In 6 of the 10 years of this experiment we find that the mean lengths of the erect ears exceed those of the declining ears. In 7 of the 10 years,

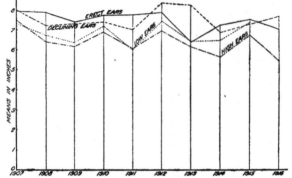

Fig. 17.—Graphs showing the mean length of the ears of the erect- and declining-ear strains of Indian corn.

the means of weights of erect ears exceed those of declining ears. In mean number of rows of kernels on ears, there is no significant difference

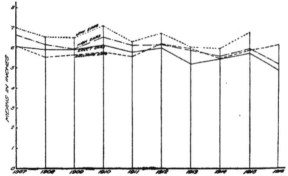

Fig. 18.—Graphs showing the mean circumference of the ears of the erect- and declining-ear strains of Indian corn.

between erect and declining ears. Taken as a whole it seems that no differences in means are large enough to persist well from season to season. The fact that the erect ears are on the whole, at least as heavy as the declining ears is of special interest because of a natural suggestion that the ears decline on account of their weight.[1]

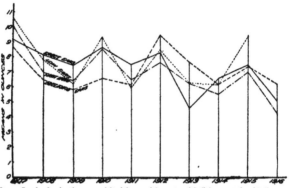

Fig. 19.—Graphs showing the mean weight of the ears of the erect- and declining-ear strains of Indian corn.

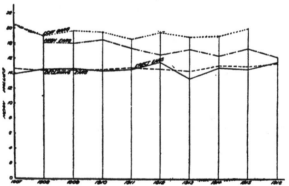

Fig. 20.—Graphs showing the mean number of rows of kernels on the ears of the erect- and declining-ear strains of Indian corn.

DIFFERENCES IN VARIABILITY

[Fig. 17 to 20]

In regard to the standard deviations it seems that no inference of value is to be drawn in regard to differences between erect and declining-ear strains. It seems that in about one-half of the cases the erect ears are

---

[1] Smith, L. H. The effect of selection upon certain physical characters in the corn plant. Ill. Agr. Exp. Sta. Bull. 132, p. 47-62, 5 p. 1909.

more variable than the declining, while the reverse is true for the remainder of the cases. The differences are in general of no significance. The same may be said in regard to the coefficients of variability, of which we prepared graphic representations but hesitate to take the space to exhibit them, since the conclusions drawn can be readily obtained from the values in Table IV.

TABLE IV.—*Type and variability of the erect- and declining-ear strains of Indian corn*

CROP OF 1907

Character and strain.	Mean.	Empirical mode.	Standard deviation.	Coefficient of variability.
Length of ear, in inches:				
Erect ears.............	7. 925±0. 037	9. 0	1. 380±0. 026	17. 41±0. 35
Declining ears..... .	7. 968± . 053	9. 0	1. 575± . 037	19. 77± . 50
Circumference of ears, in inches:				
Erect ears.............	6. 053± . 014	6. 0	. 519± . 010	8. 58± . 17
Declining ears.........	6. 086± . 020	6. 5	. 568± . 014	9. 33± . 23
Weight of ears, in ounces:				
Erect ears.............	9. 191± . 075	10. 0	2. 688± . 053	29. 25± . 64
Declining ears.........	8. 660± . 110	9. 0	2. 953± . 071	33. 98± . 99
Rows of kernels on ears:				
Erect ears...	13. 918± . 054	14. 0	1. 900± . 038	13. 65± . 28
Declining ears.........	14. 688± . 072	14. 0	1. 983± . 052	13. 50± . 35

CROP OF 1908

Character and strain.	Mean.	Empirical mode.	Standard deviation.	Coefficient of variability.
Length of ear, in inches:				
Erect ears.............	7. 850±0. 067	8. 0	1. 243±0. 024	15. 84±0. 31
Declining ears. .	7. 183± . 031	7. 0	1 109± 022	15. 44± . 31
Circumference of ear, in inches:				
Erect ears.............	5. 900± . 014	6. 0	. 500± . 010	. 848± . 17
Declining ears	5. 511± . 013	5 5	. 447± . 009	8. 11± . 16
Weight of ear, in ounces:				
Erect ears.............	8 128± . 067	8. 0	2. 132± . 047	26. 23± . 62
Declining ears.........	6. 485± . 053	6 0	1. 803± . 037	27. 80± . 62
Rows of kernels on ears:				
Erect ears.	14 560± . 051	14. 0	1. 754± . 036	12. 05± . 25
Declining ears..... .	14. 370± 056	14 0	1. 880± . 039	· 13. 08± . 28

CROP OF 1909

Character and strain.	Mean.	Empirical mode.	Standard deviation.	Coefficient of variability.
Length of ear, in inches:				
Erect ears....	7 366±0 033	8. 0	1 246±0 023	16. 92±0. 32
Declining ears.........	7 278± . 034	7 0	1 298± 024	17. 83± . 36
Circumference of ear, in inches:				
Erect ears.........	5 911± 013	6 0	. 512± . 009	866± 17
Declining ears.........	5 647± . 016	5 5	. 570± . 012	10. 09± . 20
Weight of ear, in ounces:				
Erect ears.............	7. 442± . 058	8. 0	2. 186± . 041	29. 37± . 60
Declining ears.........	5. 836± . 056	6 0	1. 916± . 040	32. 83± . 75
Rows of kernels on ears:				
Erect ears.. ..	14. 623± . 053	14. 0	1. 927± . 038	13. 07± . 26
Declining ears.....	14 429± . 057	14 0	1. 911± . 040	13. 24± . 28

TABLE IV—*Type and variability of the erect- and declining-ear strains of Indian corn*—Continued

CROP OF 1910

Character and strain.	Mean.	Empirical mode.	Standard deviation.	Coefficient of variability.
Length of ear, in inches:				
Erect ears............	7. 691±0. 031	8. 0	1. 065±0. 022	13. 85±0. 29
Declining ears.... .	7. 350± . 035	8. 0	1. 143± . 024	15. 55± . 34
Circumference of ear, in inches:				
Erect ears............	6. 135± . 013	6. 0	. 455± . 009	7. 42± . 16
Declining ears.........	5 796± . 015	6. 0	. 484± . 011	8. 35± . 18
Weight of ear, in ounces:				
Erect ears............	8. 643± . 068	9. 0	2. 267± . 048	26. 23± . 59
Declining ears.........	6. 598± . 061	7. 0	1. 875± . 043	28. 42± . 71
Rows of kernels on ears:				
Erect ears............	14. 314± . 050	14. 0	1. 676± . 035	11. 71± . 25
Declining ears.........	14. 520± . 056	14. 0	1. 765± . 040	12. 16± . 28

CROP OF 1911

Character and strain.	Mean.	Empirical mode.	Standard deviation.	Coefficient of variability.
Length of ear, in inches:				
Erect ears............	7. 708±0. 036	8. 5	1. 418±0. 026	18. 39±0. 34
Declining ears.........	6 985± . 036	7. 5	1. 312± . 026	18. 78± . 38
Circumference of ear, in inches:				
Erect ears............	5. 785± . 011	5. 75	. 457± . 008	7. 90± . 15
Declining ears.........	5. 575± . 015	5. 50	. 510± . 011	9. 15± . 19
Weight of ear, in ounces:				
Erect ears............	7. 480± . 053	8. 0	1. 925± . 038	25. 74± . 54
Declining ears.........	6. 154± . 069	6. 0	1. 700± . 049	27. 62± . 81
Rows of kernels on ears:				
Erect ears .	14. 538± . 049	14. 0	1. 877± . 035	12. 91± . 26
Declining ears... .	14. 758± . 077	14. 0	1. 900± . 055	12. 87± . 36

CROP OF 1912

Character and strain.	Mean.	Empirical mode.	Standard deviation.	Coefficient of variability.
Length of ear, in inches:				
Erect ears............	7. 821±0. 037	8. 5	1. 379±0. 021	17. 63±0. 29
Declining ears.........	8. 294± . 033	9. 0	1. 314±0. 024	15. 84± . 29
Circumference of ear, in inches:				
Erect ears............	5. 998± . 010	6. 0	. 467±0. 007	7. 79± . 12
Declining ears.........	6. 220± . 012	6. 0	. 479±0. 009	7. 70± . 15
Weight of ear, in ounces:				
Erect ears............	8. 255± . 051	9. 0	2. 135±0. 036	25. 86± . 47
Declining ears	9. 436± . 059	9. 0	2. 218±0. 042	23. 51± . 47
Rows of kernels on ears:				
Erect ears... . .	15. 506± . 044	14. 0	1. 844±0. 031	12. 15± . 20
Declining ears... ...	14. 546± . 048	14. 0	1. 845±0. 034	12. 86± . 24

TABLE IV.—*Type and variability of the erect- and declining-ear strains of Indian corn*—Continued

CROP OF 1913

Character and strain.	Mean.	Empirical mode.	Standard deviation.	Coefficient of variability.
Length of ear, in inches:				
Erect ears..............	6. 338±0. 044	6. 5	1. 311±0. 031	20. 68±0. 51
Declining ears.........	8. 182± . 034	8. 5	1. 277± . 024	15. 61±0. 30
Circumference of ear, in inches:				
Erect ears..............	5. 205± . 016	5. 0	0. 480± . 011	9. 22±0. 22
Declining ears.........	5. 969± . 012	6. 0	0. 437± . 008	7. 32±0. 14
Weight of ear, in ounces:				
Erect ears..............	4. 565± . 065	4. 0	1. 543± . 046	33. 80±1. 12
Declining ears.........	7. 661± . 050	8. 0	1. 772± . 035	23. 13±0. 49
Rows of kernels on ears:				
Erect ears.. . .	13. 380± . 065	12. 0	1. 598± . 046	11. 94±0. 35
Declining ears.........	14. 310± . 169	14. 0	1. 811± . 119	12. 65±0. 91

CROP OF 1914

Character and strain.	Mean.	Empirical mode.	Standard deviation.	Coefficient of variability.
Length of ear, in inches:				
Erect ears....... ...	7. 197±0. 038	7. 5	1. 198±0. 027	16. 65±0. 39
Declining ears.........	6. 849± . 030	7. 5	1. 112± . 022	16. 24±0. 32
Circumference of ear, in inches:				
Erect ears..............	5. 464± . 017	5. 25	0. 524± . 012	9. 590±0. 23
Declining ears.........	5. 482± . 013	5. 50	0. 454± . 009	8. 281±0. 17
Weight of ear, in ounces:				
Erect ears..............	6. 567± . 057	7. 0	1. 665± . 041	25. 35±0. 66
Declining ears.........	6. 145± . 049	6. 0	1. 555± . 035	25. 31±0. 61
Rows of kernels on ears:				
Erect ears..............	14. 796± . 033	14. 0	1. 970± . 083	13. 314±0. 16
Declining ears.........	15. 100± . 061	14. 0	1. 936± . 043	12. 821±0. 30

CROP OF 1915

Character and strain.	Mean.	Empirical mode.	Standard deviation.	Coefficient of variability.
Length of ear, in inches:				
Erect ears..............	7. 455±0. 030	8. 0	1. 220±0. 021	16. 36±0. 29
Declining ears.........	7. 212± . 025	7. 5	1. 020± . 018	14. 14±0. 25
Circumference of ear, in inches:				
Erect ears..............	5. 728± . 012	5. 75	0. 496± . 009	8. 66±0. 15
Declining ears.........	5. 856± . 012	6. 00	0. 481± . 008	8. 22±0. 15
Weight of ear, in ounces:				
Erect ears..............	7. 449± . 049	8. 0	1. 889± . 035	25. 36±0. 50
Declining ears.........	7. 344± . 044	7. 0	1. 640± . 031	22. 33±0. 45
Rows of kernels on ears:				
Erect ears..............	14. 616± . 044	14. 0	1 768± . 031	12. 10±0. 22
Declining ears.........	14. 930± . 070	14. 0	2. 116± . 049	14. 17±0. 34

TABLE IV.—*Type and variability of the erect- and declining-ear strains of Indian corn*—Continued

CROP OF 1916

Character and strain.	Mean.	Empirical mode.	Standard deviation.	Coefficient of variability.
Length of ear, in inches:				
Erect ears............	6. 985±0. 041	7. 0	1. 290±0. 029	18. 47±0. 43
Declining ears.........	7. 639± .034	7. 5	1. 141± .024	14. 94±0. 32
Circumference of ear, in inches:				
Erect ears............	4. 892± .014	5. 0	.409± .010	8. 37±0. 20
Declining ears.........	5. 599± .013	5. 5	.438± .009	7. 82±0. 17
Weight of ear, in ounces:				
Erect ears............	5. 019± .061	5. 0	1. 459± .043	29. 07±1. 25
Declining ears.........	6. 178± .048	7. 0	1. 417± .034	22. 78±0. 57
Rows of kernels on ears:				
Erect ears........	15. 573± .072	14. 0	1. 679± .051	11. 52±0. 35
Declining ears.........	15. 420± .059	16. 0	1. 767± .059	11. 46±0. 27

HIGH- AND LOW-EAR STRAINS

Table V gives the means, standard deviations, coefficients of variability, and corresponding probable errors for the characters in question of the high- and low-ear strains.

DIFFERENCES IN MEANS

[Fig. 17 to 20]

The mean lengths of ears of the low-ear strain exceed in general slightly but significantly the mean lengths of ears of the high-ear strain, but there are two exceptions in nine years. The mean circumference of ears from the low-ear strain exceeds decidedly in each year the mean circumference of ears of the high-ear strain. The mean weight of ears of the low-ear strain exceeds the mean weight of the high-ear strain in seven of the nine years. Some of these differences would be regarded as unquestionably significant in the light of their probable errors, but the excess of the mean weights of low ears is not so well fixed but that in two seasons of the nine the mean weights of high ears exceed slightly those of the low ears. In mean number of rows of kernels on ears the low ears exceed the high ears in every case. This difference is decidedly significant.

DIFFERENCES IN VARIABILITY

[Fig. 21 to 24]

In length of ears it is doubtful if there is any significant difference in variability between the high- and low-ear strains. In circumference of ears the low-ear strain shows in general the greater variability. In weight of ears it is doubtful whether there is any significant difference

7768°—17——4

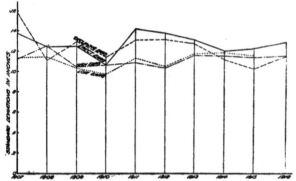

Fig. 21.—Graphs showing the standard deviation of the length of the ears of the high- and low-ear strains of Indian corn.

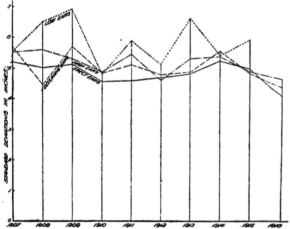

Fig. 22.—Graphs showing the standard deviation of the circumference of the ears of high- and low-ear strains of Indian corn.

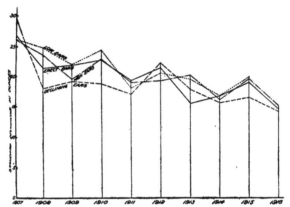

Fig. 25.—Graphs showing the standard deviation of the weight of the ears of high- and low-ear strains of Indian corn.

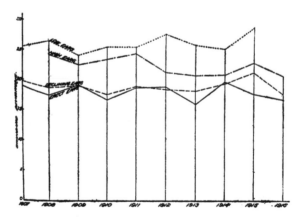

.... 20. 577± .083   2. 559± .059   12. 45± ·

in inches:					
....	6. 380±0. 039	6. 5	1. 251±0. 027	19. 61±0.
....	6. 700± .035	6. 5	1. 148± .024	17. 13± ·
of ear, in				
....	6. 148± .018	6. 25	.560± .013	9. 11± ·
....	6. 540± .020	6. 5	.654± .014	10. 00± ·
in ounces:				
....	6. 980± .080	6. 0	2. 358± .057	33. 78± ·
....	7. 795± .080	9. 0	2. 477± .056	31. 78± ·
ls on ears:				
....	18. 09 ± .091	18. 0	2. 424± .064	13. 39± ·
....	18. 96 ± .088	18. 0	2. 650± .062	13. 98± ·

in inches:				
....	6. 128±0. 030		1. 042±0. 022	76±0.
....	6. 322± .027		1. 032± .022	32± ·
of ear, in				
....	5. 942± .015		.529± .011	90± ·
....	6. 522± .021	6. 25	.693± .015	62± ·
in ounces:				
....	6. 432± .057		1. 952± .040	30. 35± ·
....	6. 234± .070		2. 204± .050	·
ls on ears:				
....	17. 986± .076	18. 0	2. 258± .054	12. 55± ·3
....	19. 744± .097		2. 410± .069	12. 21± ·3

TABLE V.—*Type and variability of high- and low-ear strains of Indian corn*—Continued

CROP OF 1910

Character and strain.	Mean.	Empirical mode.	Standard deviation.	Coefficient of variability.
Length of ear, in inches:				
High ears...............	6. 844±0. 027	7. 0	1. 064±0. 019	15. 55±0. 29
Low ears...............	7. 152± .048	8. 0	. 978± .039	13. 67± .24
Circumference of ear, in inches:				
High ears...............	6. 543± .013	6. 5	. 487± .009	7. 44± .14
Low ears...............	7. 118± .012	7. 0	. 481± .008	6. 76± .12
Weight of ear, in ounces:				
High ears...............	8. 516± .060	9. 0	2. 281± .042	26. 77± .53
Low ears...............	9. 353± .061	10. 0	2. 435± .043	26. 03± .49
Rows of kernels on ears:				
High ears...............	18. 422± .065	18. 0	2. 342± .046	12. 71± .25
Low ears...............	19. 500± .067	20. 0	2. 546± .048	13. 06± .25

CROP OF 1911

Character and strain.	Mean.	Empirical mode.	Standard deviation.	Coefficient of variability.
Length of ear, in inches:				
High ears...............	6. 044±0. 029	6. 0	1. 088±0. 021	17. 99±0. 37
Low ears...............	5. 991± .024	6. 0	1. 133± .017	18. 91± .30
Circumference of ear, in inches:				
High ears...............	6. 130± .015	6. 25	. 545± .011	8. 89± .18
Low ears...............	6. 317± .013	6. 5	. 588± .009	9. 31± .15
Weight of ear, in ounces:				
High ears...............	6. 488± .057	6. 0	1. 892± .041	29. 16± .68
Low ears...............	5. 913± .045	6. 0	1. 800± .032	30. 44± .58
Rows of kernels on ears:				
High ears...............	17. 308± .087	16. 0	2. 437± .062	14. 08± .36
Low ears...............	18. 510± .074	18. 0	2. 546± .053	13. 75± .29

CROP OF 1912

Character and strain.	Mean.	Empirical mode.	Standard deviation.	Coefficient of variability.
Length of ear, in inches:				
High ears...............	6. 898±0. 031	7. 5	1. 035±0. 022	15. 00±0. 33
Low ears...............	7 396± .029	7. 5	1. 052± .021	14. 12± .29
Circumference of ear, in inches:				
High ears...............	6. 202± .014	6. 5	. 459± .010	7. 40± .16
Low ears...............	6. 745± .014	6. 75	. 510± .010	7. 56± .16
Weight of ear, in ounces:				
High ears...............	7. 609± .062	7. 0	1. 919± .044	25. 22± .61
Low ears...............	8. 428± .060	9. 0	2. 055± .043	24. 38± .55
Rows of kernels on ears:				
High ears...............	16. 500± .071	16. 0	2. 130± .051	12. 91± .32
Low ears...............	19. 459± .083	18 0	2. 766± .059	14. 21± .31

TABLE V.—*Type and variability of high- and low-ear strains of Indian corn*—Continued

CROP OF 1913

Character and strain.	Mean.	Empirical mode.	Standard deviation.	Coefficient of variation.
Length of ear, in inches:				
High ears.............	6. 107±0. 037	6. 5	1. 157±0. 026	18. 95±0. 44
Low ears..............	6. 362± .032	6. 5	1. 172± .023	18. 42± .37
Circumference of ear, in inches:				
High ears.............	5. 895± .017	6. 0	.529± .012	8. 97± .21
Low ears..............	6. 093± .017	6. 0	.660± .012	10. 83± .21
Weight of ear, in ounces:				
High ears.............	6. 224± .068	6. 0	2. 008± .048	32. 26± .84
Low ears..............	6. 250± .057	6. 0	1. 942± .040	31. 07± .69
Rows of kernels on ears:				
High ears.............	17. 190± .076	16. 0	2. 080± .054	12. 10 ± .32
Low ears..............	18. 854± .089	18. 0	2. 576± .063	13. 663± .34

CROP OF 1914

Character and strain.	Mean.	Empirical mode.	Standard deviation.	Coefficient of variation.
Length of ear, in inches:				
High ears.............	5. 581±0. 035	6. 5	1. 149±0. 025	20. 588±0. 46
Low ears..............	6. 410± .030	7 0	1. 183± .022	18. 456± .35
Circumference of ear, in inches:				
High ears.............	5. 594± .017	5. 75	.536± .012	9. 582± .22
Low ears..............	5. 992± .014	6. 0	.540± .009	9. 012± .17
Weight of ear, in ounces:				
High ears.............	5. 491± .053	5. 0	1. 605± .038	29. 230± .35
Low ears..............	6. 105± .047	6. 5	1. 669± .033	27. 338± .35
Rows of kernels on ears:				
High ears.............	16. 350± .075	16. 0	2. 098± .053	12. 832± .33
Low ears..............	18. 992± .075	18. 0	2. 510± .053	13. 216± .28

CROP OF 1915

Character and strain.	Mean.	Empirical mode.	Standard deviation.	Coefficient of variation.
Length of ear, in inches:				
High ears.............	6. 693±0. 034	6. 5	1. 137±0. 024	16. 990±0. 37
Low ears..............	7. 325± .032	8. 0	1. 151± .023	15. 710± .33
Circumference of ear, in inches:				
High ears.............	5. 963± .015	6. 0	.486± .011	8. 150± .18
Low ears..............	6. 787± .017	6. 75	.590± .012	8. 690± .17
Weight of ear, in ounces:				
High ears.........	6. 991± .063	8. 0	1. 946± .045	27. 840± .69
Low ears........... ..	9. 341± .065	10. 0	2 178± .046	23. 320± .53
Rows of kernels on ears:				
High ears.........	17. 328± .072	16. 0	2. 280± .051	13. 150± .30
Low ears...........	19. 992± .084	20. 0	2. 874± .059	14. 380± .30

CROP OF 1916[a]

Character and strain.	Mean.	Empirical mode.	Standard deviation.	Coefficient of variation.
Length of ear, in inches:				
High ears.............	5. 399±0. 039	5. 5	1. 147±0. 028	21. 24±0. 54
Circumference of ear, in inches:				
High ears.............	5. 178± .014	5. 25	.465± .010	8. 99± .20
Weight of ear, in ounces:				
High ears.............	4. 221± .058	4. 0	1. 511± .041	35. 79±1. 09
Rows of kernels on ears:				
High ears.............	16. 631± .090	16. 0	2. 075± .063	12. 48± .39

SUMMARY

THE FOUR STRAINS WITH RESPECT TO CHEMICAL COMPOSITION

(1) It is found that four distinct types of corn as regards length, circumference, weight of ears, and number of rows of kernels on ears are so well established that we may assign orders of values to the means of these characters that persist with but few exceptions in such changes of environment as have been experienced in 11 years of planting, from 1905 to 1915.

(2) While a few slight but significant progressive changes have been noted, the selections for chemical composition from 1905 to 1915 have not changed decidedly the differences in mean values of these characters. In fact, we are unable to assert with any high degree of probability that the strains differ more or less with respect to these characters during the second half of the period 1905 to 1915 than during the first half.

(3) The standard deviations of the strains do not differ nearly so much compared to their probable errors as do the means, and it is not in general nearly so easy to discriminate among strains by the differences of standard deviations as by the use of means. There is one marked exception to this, in that we easily distinguish high-protein and high-oil from low-protein and low-oil strains by the differences in the standard deviations in weights of ears.

(4) No progressive change of consequence has taken place in standard deviations.

(5) The coefficients of variability, in comparison to their probable errors, differ still less in a given season than the standard deviations, and there is no very general tendency for the coefficients of variability to maintain a definite order of values. That is to say, the differences of coefficients of variability of the four strains seem to be fairly well described, with certain exceptions noted in the paper, as random fluctuations.

TWO-EAR STRAINS

(6) The upper ears have a significantly larger mean value in length, weight, and circumference than have the lower ears on the same stalks.

(7) The means with respect to weight, length, and circumference of single ears are in each case larger than the corresponding means for the lower ears of the same plot. The means with respect to weight and circumference are also in general larger than these means for upper ears of the same plot. However, strange as it may appear, the mean lengths of single ears are on the whole less than those of upper ears.

(8) A striking fact in the comparison of the single ears with the upper and lower ears is the greater standard deviation in the weight of single ears.

ERECT- AND DECLINING-EAR STRAINS

(9) Taken as a whole, there are no significant differences in these two strains with respect to the characters considered. In view of the suggestion that ears are declining because of their greater weight, it is a fact of special interest that the declining ears are not on the whole heavier than erect ears.

HIGH- AND LOW-EAR STRAINS

(10) The ears of the low-ear strain are on the whole significantly larger in mean length, circumference, and weight than those of the high-ear strain, but there are a few exceptions. In each of the eight years considered the mean number of rows of kernels on ears is larger for the low-ear strain than for the high-ear strain.

(11) The standard deviation of number of rows of kernels in each year is distinctly greater for the low ears than for the high ears, and the standard deviation of circumstance of ears is in general larger for the low-ear strain than for the high-ear strain.

SOME NOTES ON THE DIRECT DETERMINATION
OF THE HYGROSCOPIC COEFFICIENT [1]

By Frederick J. Alway, *Chief of Division of Soils, Agricultural Experiment Station, University of Minnesota*, Millard A. Kline, *formerly Assistant in Chemistry, Nebraska Agricultural Experiment Station*, and Guy R. McDole, *Assistant in Soils, Agricultural Experiment Station, University of Minnesota*

INTRODUCTION

The hygroscopic coefficient expresses the percentage of moisture contained in a soil which, in a dry condition, has been brought into a saturated atmosphere, kept at a constant temperature, and allowed to remain until approximate equilibrium with this atmosphere has been attained (12, p. x; 14, p. 76; 15, p. 196).[2] It has a twofold significance, both serving as a single-valued expression of the relative fineness of texture (12, p. xi), and, in soil-moisture studies, permitting the approximate estimation of the maximum amount of water available for growth and for the maintenance of life in the case of ordinary crop plants—the difference between the total amount of water and the hygroscopic coefficient (3, p. 121).

The error is sometimes made (20, p. 209) of confusing Mitscherlich's "*Hygroskopizität*" with the hygroscopic coefficient as above defined. The former is determined by allowing the exposed soils to come into equilibrium with an atmosphere in contact with a 10 per cent sulphuric-acid solution (24, p. 56) instead of with water, Mitscherlich holding that the determination by Hilgard's method gives results much too high on account of the condensation of moisture on the exposed samples (24, p 156). The values obtained by his method are much lower than the hygroscopic coefficients, although either will serve to indicate the relative hygroscopicity and fineness of texture of the samples, as may also the much simpler determination of the hygroscopic moisture contained in air-dried soils exposed freely side by side to the air of an ordinary room for some time (4, p. 351), or that of the moisture equivalent (7, p. 140).

The term "hygroscopic coefficient" was introduced by Hilgard in 1874 (13, p. 9), but as early as 1859 he had developed the method for its determination which he continued to use for more than 50 years, referring to the value, previous to the date mentioned, as the hygroscopic moist-

---

[1] The work reported in this paper was carried out in 1910 to 1913 at the Nebraska Agricultural Experiment Station, where the authors were, respectively, Chemist, Assistant in Chemistry, and Research Assistant in Chemistry.

[2] Reference is made by number to "Literature cited," p. 165-166.

Journal of Agricultural Research,
Washington, D C.
ki

Vol XI, No 4
Oct. 22, 1917
Key No Minn —20

(147)

ure absorbed at $x°$ C., x being the temperature at which the absorption boxes had been exposed. The conception of making such a determination did not originate with Hilgard, but with Schubler as early as 1830. The method (27, p. 80) of the latter, however, did not insure the absorption of the maximum amount of moisture, too thick a layer of soil being exposed. Hilgard's method is a refinement of that of Schübler, providing for a higher humidity of the atmosphere, a shallower layer of soil, thus permitting a more rapid saturation, and lastly the avoidance of fluctuations in temperature, which cause a precipitation of dew upon the soil.

The hygroscopic coefficient of soils has been but rarely determined during the past 40 years, although during the previous four or five decades it had received considerable recognition. In the early days of soil investigation undue importance was attached to the relative absorbent powers of soils with respect to atmospheric moisture, it being supposed that it served to some extent as an index of fertility. Thus, Davy (9, p. 184) states:

I have compared the absorbent power of many soils with respect to atmospheric moisture, and I have always found it greatest in the most fertile soils; so that it affords one method of judging of the productiveness of land.

He also considered that in the case of soils with a high hygroscopic coefficient plants were supplied with moisture in periods of drouth, the hygroscopic power of the soil enabling it to render the water of the atmosphere available to the plants (9, p. 183). Schübler accepted the second of Davy's views but not the first, mentioning (27, p. 82) that he had found that a

pure, infertile clay absorbs in 12 hours   . more than the very fertile garden soil.—Translation.

Liebig (17. p. 48) also accepted Davy's view as to the value of the moisture absorbed from the air, and it continued to be indorsed by the foremost investigators until 1875, when the work of Sachs, Wilhelm, Reisler, Heinrich, and Mayer finally proved the falsity of it. The reaction from the old view carried to the opposite extreme, and it became almost universally accepted that a knowledge of the hygroscopic power of different soils is practically valueless. Apparently the question of the usefulness of a knowledge of the relative hygroscopicity of a soil was confused with the distinct question of the usefulness to plants of the hygroscopic moisture. Mayer appears largely responsible for this confusion. In 1871 (21, p. 130) he fully indorsed the older view:

The absorptive power of the soil for water . . . is under all circumstances a *useful soil property*, as it becomes active only when there is an actual scarcity of water in the soil and so acts as a regulator . . . The water thus condensed in porous solid bodies conducts itself exactly like other capillary held water and can, for example, when a field soil has in this way condensed water, be taken up by a plant's roots just like that which has entered the small interstices of the soil from the rain or by watering.— Translation.

Four years later, having completed his proof of the falsity of this view, he strongly discouraged determining the hygroscopicity of soils on the ground that it would cause confusion if very hygroscopic soils were indicated as retentive of an especially high percentage of water in a non-available form, when this was not an actual disadvantage, for the reason that the most hygroscopic soils were the very ones that were able to hold the largest amounts of total water (22, p. 248):

By a remarkable coincidence the most strongly condensing soils usually are those with the highest water capacity and it would probably only cause confusion if we were to mark these soils with a blemish in regard to the supply of water on account of a not very productive correction —Translation.

However, he still clearly recognized that the amount of water remaining in a soil on the wilting of a plant was closely related to the maximum amount of moisture which the soil was able to absorb from a nearly saturated atmosphere.

During the last quarter of the nineteenth century Hilgard and Loughridge appear to have been the only investigators who continued to attach any importance to the determination of the hygroscopic coefficient in connection with soil-moisture studies. The general attitude of soil investigators during that period is well illustrated by the numerous publications of one of the foremost, the late Dr. F. H. King, in which the determination, the importance, and the use of the hygroscopic coefficient are ignored.

At the beginning of the present century Mitscherlich began to call attention to the importance of the relative hygroscopicity of soils (23) to which he attached extreme importance. In Europe the importance of a knowledge of the hygroscopicity of soils is still practically ignored except by Mitscherlich and by Hall (11, p. 84–88).

Naturally, under such circumstances, the particular method developed by Hilgard received little critical consideration. It appears that no one except himself had investigated the method with the object of determining the experimental errors involved, previous to our beginning the use of it on an extensive scale for the interpretation of field-moisture data. We had originally no intention of studying the method itself, and our observations have resulted from difficulties encountered in its use, part of them due to attempts to expedite the determination without lessening its accuracy.

Lipman and Sharp (18), while associated with Hilgard, studied the effect of a rise or fall of the temperature and also of the thickness of the exposed layer of soil upon the amount of hygroscopic moisture absorbed.

Briggs and Shantz (8, p. 64) studying the relationship between the moisture equivalent and the hygroscopic coefficient found that the determination of the latter "unless carried out with special precautions, is not very exact."

HILGARD'S METHOD

Details of this were published earliest by Loughridge (19), who had been a student under Hilgard at the University of Mississippi. Later (14, p. 76) it was described by the latter as follows:

The fine earth is exposed to an atmosphere saturated with moisture for about twelve hours at the ordinary temperature (60° F.) of the cellar in which the box should be kept. For this it is sifted in a layer of about 1 mm. thickness upon glazed paper, on a wooden table in a small water-tight covered box (12 by 9 by 8 inches) in which there is about an inch of water; the interior sides and cover of the box should be lined with blotting paper, kept saturated with water, to insure the saturation of the air. . . .

After eight to twelve hours the earth is transferred as quickly as possible, in the cellar, to a weighed drying-tube and weighed. The tube is then placed in a paraffin bath, the temperature gradually raised to 200° C. and kept there 20 to 30 minutes (rapidity of raising temperature depending upon the amount of moisture in the soil), a current of dry air passing continually through the tube. It is then weighed again, and the loss in weight gives the hygroscopic moisture in saturated air.

Some time later, to avoid the decomposition of the organic matter of surface soils, Hilgard modified the method to the extent of using an air bath, raising the temperature to only 110° C, keeping the sample in for an hour, weighing, drying again, and continuing the process until a practically constant weight was obtained.[1]

Under "fine earth" was included the material finer than 0.5 mm. resulting from the sample being reduced by a rubber pestle, instead of by one of steel or porcelain, to prevent any crushing of soil particles. He sometimes digested clayey soils with distilled water until fully disintegrated, after which the muddy water was evaporated with the soil slush and the whole thoroughly mixed. As the determination of the hygroscopic coefficient was, with Hilgard, almost always incidental to a chemical or physical analysis of the soil, this rather tedious preliminary preparation of the sample was no serious matter in view of the great amount of time and labor involved in the detailed analyses to follow.

Hilgard studied the effect of differences in temperature upon the amount of moisture absorbed by a soil and found that it increased with the temperature, provided that the saturation of the air was maintained. In this view he was not confirmed by the work of Knop (16), Schlœsing (26), Ammon (6), and Von Dobeneck (10), who found a lower absorption at higher temperatures. He attributed their findings to their failure to provide for the saturation of the air inclosed within the absorption vessel.

More recently Patten and Gallagher (25, p. 31-35) have reported some data confirming the view of Knop, etc., but, as Lipman and Sharp have pointed out (18, p. 716), their description of their methods makes it evident

---

[1] HILGARD, E. W. METHODS OF PHYSICAL AND CHEMICAL SOIL ANALYSIS. Cal. Agr. Exp. Sta. Circ. 6, p. 17. 1903.

that the conditions of their experiments were not such as to insure that the exposed soils became saturated with hygroscopic moisture. The work of Patten and Gallagher (25), confirming that of the earlier investigators mentioned, has unfortunately had the effect of leading to the generally accepted view that, in his conclusions, Hilgard was in error (20, p. 367).

WORK OF LIPMAN AND SHARP

As a reply to the criticisms of Patten and Gallagher, Lipman and Sharp (18) report a study of the effect of the thickness of the layer of exposed soil and of the temperature of exposure, using boxes 12 by 18 by 19 inches, similar to those employed by Hilgard in his later work (15, p. 198), but longer and wider than those mentioned in the above quotation (14, p. 76). They also used glazed paper placed on wooden tables with the tops only 1 inch above the surface of the water. They compared three depths (1.5, 4, and 8.5 mm.) of an adobe soil exposed for various intervals, ranging from 1.5 to 455 hours, and at temperatures from $12°$ to 34.75 °C. With two other soils the periods of exposure were quite similar, but the range in temperature was less, while the thickness of the soil layer did not exceed 2 mm.

Lipman and Sharp found in general that a greater amount of moisture was absorbed at the higher temperatures, and emphasize the necessity of using very thin layers of soil, having found about 1 mm. the best. They conclude that 8 to 10 hours' exposure of air-dried soils is sufficient when the depth of the layer is only 1 mm. and the boxes are of the dimensions they used. They call attention to the difficulty of securing the saturation of the air at any considerable distance from the water surface.

These authors mention that duplicate samples agreed remarkably well. However, if they were run side by side, this agreement, as pointed out below, does not serve as evidence of the accuracy of the determination.

EXPERIMENTAL WORK

Two of us had previously (1, p. 590; 2; 5, p. 277–281) employed the method on a very limited scale, using an improvised place for the absorption boxes. We found the loose sheets of glazed paper so inconvenient that we introduced a modification that greatly facilitated the work, substituting light pasteboard trays 7 inches long, 5 inches wide, and 0.75 inch high, lined with glazed paper. On starting the determinations at the University of Nebraska we again used this modification, having had a large supply of the pasteboard trays lined with the glazed paper manufactured by a local box company. As these one by one became somewhat limp from exposure to the damp atmosphere maintained in the absorption boxes, they were discarded and new ones substituted.

We used wooden boxes of the dimensions given by Hilgard, lining both the boxes and the covers with blotting paper as specified. To prevent the warping of the sides and cover, it was found desirable first thoroughly to paraffin these.   To prevent leakage of water a galvanized-iron tray, 3 to 4 inches high and made to fit snugly, was placed in the bottom of the wooden box, the blotting paper extending from the top of the sides of the wooden box to the bottom of the metal tray.   By means of metal supports a wooden table was held in position 1 to 2 inches above the surface of the water and from 0.5 to 1 inch from each paper-lined side of the box. In the later determinations a table of galvanized-wire screen was substituted, as it occupied less space.

To provide a constant temperature room, we inclosed with alternate layers of building paper and boards a portion of the cellar of the Experiment Station building that was windowless and not in contact with any outside wall.   An electric light on the outside of a double window furnished the necessary illumination without heating the room.   The operator opened and closed the door as quickly as possible on entering the room daily to change the soil in the trays.   A maximum and minimum thermometer was hung in the room, and the daily readings were recorded. Later a recording hygrometer was added.   No sudden changes in temperature or humidity were observed.

We employed 24 absorption boxes, which were placed on low shelves against the heavy inside brick wall.   From the first we had exposed samples of two control soils in different absorption boxes.   At first the object of this was to find the variation in the coefficient of the same soil from day to day, but later to indicate whether any abnormal conditions had prevailed in the room, the boxes, or the drying oven.   Of the two control soils used on the same day the one was a subsoil with a coefficient of about 5 to 6, while the other was a finer-textured soil or subsoil with a coefficient between 10 and 22.   The two control soils were exposed, each in duplicate, from day to day along with a large number of the samples under investigation, the latter also being in duplicate.   The time of exposure varied from 15 to 16 hours, the boxes being filled at 4 or 5 o'clock in the afternoon and opened shortly after 8 o'clock on the following morning.   The results both with the duplicates exposed on the same day and with the controls from day to day were so concordant that the method was accepted as reliable, and some 2,000 determinations were thus made. We did not suspect the unreliability of the results until, through force of circumstances, one set had to be left in the boxes for 48 hours.   On drying these samples it was found that the two control soils had absorbed much more moisture than on any previous occasion.   We then investigated the cause of this and finally discarded all the data and made new determinations.   It was while trying to devise such modifications of the method described by Hilgard as would permit the large number of determinations in connection with our field moisture studies being made as rapidly as possible without lessening the accuracy that we secured the data reported below on the influence of the material of the trays, the effect of the temperature, previous drying, etc.

RAPID LOSS OF HYGROSCOPIC MOISTURE

When soils carrying the maximum amount of hygroscopic moisture—that is, an amount equal to the hygroscopic coefficient—are exposed to an ordinarily dry air, they lose water at first as rapidly as though placed in a desiccator containing sulphuric acid. This makes it important that the transfer of the exposed soil from the trays to the weighing bottles should be made as rapidly as possible.

Fifty-gm. portions of two soils, A and B, which had previously been exposed in hygroscopic boxes until the moisture content was approximately equal to the hygroscopic coefficient were placed in each of 10 weighing bottles. The depth of soil was about 5 cm. Five bottles of each were placed in a desiccator containing concentrated sulphuric acid, while the other five were exposed on a shelf in the laboratory. All were weighed at intervals until those of the latter set ceased to lose weight, four weeks being necessary. Those in the desiccators were exposed nearly five months longer, and at the end of this time they were almost oven-dry. The changes in the case of the individual members of each of the four sets were so similar that only the averages need be reported (Table I). In a similar desiccator containing sulphuric acid and in weighing bottles of the same kind we exposed distilled water. The loss of water between February 12 and March 19 amounted to about 45 gm., or 90 per cent of the weight of soil A contained in similar weighing bottles.

TABLE I.—*Percentage of water in 50-gm. portions of soils exposed to the air of the laboratory compared with that in a desiccator with sulphuric acid*

Date.	Soil A.		Soil B.	
	Exposed to air.	In desiccator.	Exposed to air.	In desiccator.
Feb. 12	5. 26	5. 26	11. 00	11. 00
18	2. 27	2. 45	6. 18	6. 70
26	1. 60	1. 30	4. 07	4. 04
Mar. 5	1. 79	. 96	4. 09	2. 95
19	1. 82	. 61	4. 17	1. 91
Apr. 1		. 44		1. 49
8		. 40		1. 35
May 21		. 25		. 88
Aug. 10		. 12		. 43

At the end of two weeks both soils appeared to have reached equilibrium with the air of the laboratory.

In the case of the absorption boxes and trays, an operator, after a little practice, will be able to open a box, transfer the samples from the two trays to weighing bottles, and stopper these within the space of 30 seconds.

.... o. 5
.... 4. 8
.... 18. 7

SUITABILITY OF TRAYS OF VARIOUS MATERIALS

four control soils, A, C, F, and G, we compared trays of alu
zinc, tin plate, glass, vulcanized rubber, granite ware, an
board described above.   Part of the last had been saturat
before being lined with glazed paper.   In some, paraffined
ormed the sides, but the bottom was cut out, thus making
the glazed paper lining the bottom of the tray.   The meta
n   es long, 5 inches wide, and 0.75 inch deep.   The trays o
ware, and vulcanized rubber were photographers' trays,
the same dimensions as the preceding.   To distribute th
dry soil in the trays, an aluminum salt shaker was foun
ient.

der to determine the approximate maximum amount of m
s likely to condense on the layers of soil independent of its ˙
, we exposed coarse quartz sand of 0.5 to 1 mm. diam
te in all the different kinds of trays, both for 16- and for 3
.   The amount of absorbed moisture was independent of bc
tray and the time of exposure, varying from 0.02 to o.
ith an average of 0.06, an amount quite negligible in conr
il-moisture studies.

TABLE III.—*Comparison of trays of different materials in the determination of the hygroscopic coefficient*

	Hygroscopic coefficients found after—				
Material of tray.	24 hours' exposure.				48 hours' exposure.
	Soil C.	Soil A.	Soil F.	Soil G.	Soil G.
Pasteboard lined with glazed paper..............	0. 6	4. 5	......	15. 5	16. 6
Paraffined pasteboard lined with glazed paper	. 7	5. 1	......	18. 2	17. 4
Paraffined pasteboard sides, but no bottom, lined with glazed paper............................	......	5. 4	......	20. 3	......
Glass..	. 6	5. 6	......	19. 4	19. 2
Vulcanized rubber................................	. 6	5. 7	17. 7	......	......
Graniteware.......................................	. 6	5. 4	17. 9	......	......
Tin plate..	. 6	5. 6	......	......	......
Zinc...	. 5	5. 7	17. 9	20. 0	19. 7
Copper...	. 7	5. 5	17. 7	19. 2	18. 9
Aluminum..	. 7	5. 6	17. 7	19. 5	19. 9

The data reported in Table III show that any of the materials mentioned, except the pasteboard, may safely be used. Even the trays with paraffined pasteboard sides, but with only a glazed paper bottom, appear allowable. The low amount of moisture absorbed by the soils when placed on glazed paper in the pasteboard trays with pasteboard bottoms is probably due to the competition of the pasteboard, which, itself, is hygroscopic and absorbs moisture from a saturated atmosphere. The somewhat low results obtained with the glazed paper used in paraffined pasteboard trays may be due to the competition of the lower unglazed surface of the glazed paper, which, when placed directly upon the table in the absorption box is able to absorb moisture from the air reaching it from below.

That the saturation with hygroscopic moisture is practically complete at the end of 24 hours with all the trays except those of pasteboard is to be seen from the failure of the soil G to absorb additional moisture during the second 24 hours. All the samples of this soil were placed in the hygroscopic boxes at the same time, but part were removed at the end of the first 24 hours. During the first day the temperature in the hygroscopic room fell from $17°$ to $16°$ C. and during the next day from $16°$ to $14°$ C.

The trays of tin plate readily rusted, and those of zinc soon corroded to such an extent that the soil adhered. On the whole, those of aluminum or copper were found the most satisfactory, and in our later work we used trays made of the former metal, with the dimensions given above.

In handling a large number of absorption box exposure necessitates a working day in the laboratory on account of the additional time required in the m samples in the trays and in the afternoon to transfe bottles. Twelve hours, the maximum exposure use is inconvenient, as usually it will necessitate an extra tory. We tried various intervals. The soils expo trays, unlike those on the glazed paper used by Hil continued to gain in weight during the first 18 ho both the 8- and 12- hour exposures too brief. As, un from 22 to 24 hours was the most convenient inter this with longer exposures. Except in the case of· soils like I, after the first 20 hours there was gene increase in weight (Table IV). As under our wo was not possible to maintain the temperature co periods, this increase with such very hygroscopic so caused by dew formation resulting from a depressi ture after the atmosphere had become saturated. exposures there was a slight decrease in weight durin of the period.

TABLE IV.—*Influence of the length of the time of exposure upon absorbed*

Experiment No.	Time of exposure in hygro-scopic boxes.	Tempera-ture.
		°C.
I................	hours................	18.

In the case of one of the control soils, I, the values shown in Table V were obtained by 24 hours' exposure at different temperatures. Another control soil, A, was used in both summer and winter. Higher values (Table VI) were found during the warmer weather. When any distinct change was caused by an elevation of the temperature, it appeared in all cases to be an increase rather than a decrease in the amount of moisture absorbed. This is in accordance with the conclusions of Hilgard and of Lipman and Sharp.

TABLE V —*Influence of the temperature upon the amount of moisture absorbed by Soil I*

Date.	Temperature.		Hygroscopic coefficient.				
	Maxi-mum.	Mini-mum.	Deter-mina-tion 1.	Deter-mina-tion 2.	Deter-mina-tion 3.	Deter-mina-tion 4.	Aver-age.
	°C.	°C.					
Jan. 17–18	21.5	21	24.0	22.3	24.7	25.3	24.1
Feb. 27–28	16.0	15	20.7	20.4	20.2	......	20.4
Feb. 29–Mar. 1	14.5	12	19.7	19.6	......	......	19.6

TABLE VI —*Concordance of values found on successive days for the hygroscopic coefficients of control soils*

Date.	Temperature of room.		Soil A.			Soil B.		
	Maxi-mum.	Mini-mum.	Determination.			Determination.		
			1	2	Aver-age.	1	2	Aver-age.
	°C.	°C.						
Jan. 22	16	15	5.6	6.0	5.8	10.2	10.0	10.1
23	16	15	5.9	5.9	5.9	10.5	10.0	10.2
24	17	17	5.7	5.7	5.7	10.2	10.0	10.1
25	18	18	5.5	5.5	5.5	10.4	10.0	10.2
26	19	18	5.5	5.6	5.5	10.0	10.0	10.0
27	21	16	5.6	5.6	5.6	9.9	9.9	9.9
28	21	16	5.6	5.8	5.7	10.2	......	10.2
29	20	16	5.5	5.5	5.5	10.0	10.0	10.0
Feb. 2	20	16	5.5	5.6	5.5	10.1	10.0	10.0
4	17	15	5.9	6.1	6.0	10.2	10.4	10.3
Average	......	......	......	......	5.7	......	......	10.1

			Soil A.			Soil I.		
June 12	22	22	5.9	5.4	5.6	20.0	20.1	20.0
13	22	21	5.7	5.9	5.8	20.0	20.0	20.0
14	23	23	6.3	6.1	6.2	21.6	21.6	21.6
15	23	22	6.0	6.4	6.2	22.9	22.3	22.6
16	23	22	6.2	6.3	6.2	22.0	22.5	22.2
17	23	22	6.0	6.3	6.1	23.0	23.0	23.0
19	22	22	6.0	5.8	5.9	22.0	22.1	22.0
20	23	22	6.2	6.1	6.1	20.4	20.9	20.5
21	23	22	6.4	....	6.4	22.0	22.6	22.3
23	23	23	6.4	....	6.4	20.0	21.4	20.7
Average	......	......	......	......	6.1	......	......	21.5

EFFECT OF GREAT FLUCTUATIONS IN THE TEMPERATURE OF THE ROOM

To obtain some idea as to how great a fluctuation in temperature is permissible we exposed soil F for 17 hours simultaneously in a laboratory and in a greenhouse as well as in the basement room used regularly for such work. In the plant house where the range of temperature was greatest the amount of moisture absorbed (Table VII) was the highest. In the laboratory where the temperature was much higher during the day than during the night the amount of moisture absorbed was lower than in the constant temperature room.

TABLE VII —*Influence of variations in the temperature of the room of exposure upon the amount of moisture absorbed*

	Hygroscopic coefficient.		
Determination.	In constant temperature room (11° to 11° C.).	In laboratory (9° to 16° C.).	In plant house (13° to 29° C.).
1	17.4	16.5	19.7
2	17.0	16.5	20.1

EFFECT OF OVEN DRYING

In field moisture studies it is usually much more convenient to determine the total moisture content of the sample and then to use a portion of this dried material for the determination of the hygroscopic coefficient than to air dry a portion of the original sample. If the previous drying should appreciably alter the value of the coefficient, it would become necessary to air dry a separate portion of the sample.

With four subsoils (Table VIII) we determined the coefficient, using the air-dried form of each. as well as portions dried at 100°, 110°, 150°, and 175° C.

TABLE VIII.—*Effect of drying soils at different temperatures previous to their exposure in the absorption boxes*

	Hygroscopic coefficient.				
Soil.	Air-dried.	Dried at 100° C.	Dried at 110° C.	Dried at 150° C.	Dried at 175° C.
A	0.7	....	....	0.7	0.7
C	5.6	5.3	5.2	5.2	5.4
J	12.0	11.6	11.7	12.1	....
I	22.3	....	....	21.3	20.8

The above would indicate that in the case of a mineral subsoil a temperature as high as $175°$ C. may be permissible. If the subsoil should be one containing any considerable amount of organic matter, such a temperature should not be employed, but one of $105°$ to $110°$ C. we found to have no distinct influence upon the values found for silt-loam surface soils containing 2 per cent of organic carbon. We made no experiments with peat soils or with mineral soils very rich in organic matter.

EFFECT OF GRINDING

As many fine-textured samples when first removed from the drying oven are so bricklike that it is difficult to reduce them with a rubber pestle so that they will pass a 1-mm. sieve it is desirable to know whether the hygroscopicity is appreciably increased by freely grinding them in a steel mortar until they will pass through such a sieve. The samples we used were four control soils, three of which, unless they had first been moistened and then oven-dried, needed no grinding to pass a 1-mm. sieve, while the fourth did not need it even then. A portion of each was ground in a steel mortar until it passed through a silk bolting cloth with openings of 0.25 mm. (Table IX.)

TABLE IX.—*Influence of grinding upon the amount of moisture absorbed*

| | Hygroscopic coefficient. | |
Soil.	Unground (2-1 mm.).	Finely ground (below 0.25 mm.).
C.	0.8	1.4
A.	5.8	5.7
B.	10.9	10.5
I.	22.0	21.7

Only in the case of the very coarse-textured soil, C, which in practice would not need grinding, did we find any increase in the hygroscopicity. If a ball mill were employed, the reduction in size of particles might easily be so great as to affect seriously the value found.

It would appear that the dried samples, if very hard, may safely be reduced in a steel mortar until they pass a 1-mm. sieve.

INFLUENCE OF SIZE OF ABSORPTION BOXES AND THE NUMBER OF TABLES

With two soils, H and S, larger absorption boxes were also tried, these being lined with blotting paper as with the smaller ones. We used one 23.25 inches long, 16 inches wide, and 6.25 inches high, and another 12 inches square and 30 inches high, these being the inside dimensions. In the first an exposure of about 72 hours was necessary to bring the absorbed moisture to as high a point as was found in the smaller boxes at the end of 24 hours, while in the second even this was not a sufficiently long exposure.

In the small absorption boxes we also tried using two tables, one placed . 1 to 2 inches above the surface of the water and the other 2 inches still higher. In this manner four trays, two on each table, could be placed in a single absorption box. In all cases the results were unsatisfactory, the soil in the trays on the lower table absorbing less moisture than those on the upper.

CONCORDANCE OF DETERMINATIONS IN PRACTICE

An illustration of the variation in the values found for a particular soil from day to day is afforded by a series of 30 samples of loess soils used in cylinder experiments (3, p. 78–115). On four consecutive days a single determination was made on each (Table X), using as controls two soils, H and K, with coefficients of 6.1 and 14.2, respectively. Three trays with the former and two with the latter were exposed each day. It will be observed that the values obtained for the different samples rose and fell together day by day and that the control soils indicate both the direction and in general the magnitude of this fluctuation. Both controls showed the highest values on the second day and the next highest on the first, while on both the third and the fourth they were slightly lower than on either of the previous days. Of the 30 other soils 28 showed the highest values on the second day, 25 the next highest on the first, while 22 were third highest on the third day, and 14 on the fourth day. As the values obtained for the controls on the last two days were nearest those commonly found, we used the averages of the values found on these two days for the 30 other soils in estimating the amount of free water (3, p. 97).

TABLE X —*Concordance of determinations of the hygroscopic coefficient on successive days*

Source (Nebraska).	Depth.	Hygroscopic coefficient.				
		1st day.	2d day.	3d day.	4th day.	Average.
	Feet.					
Wauneta	1	10.0	10.3	9.4	9.0	9.6
Do.	2	10.4	10.7	9.5	9.5	10.0
Do.	3	10.3	10.9	10.0	9.7	10.2
Do.	4	10.3	10.7	10.2	10.0	10.3
Do.	5	9.7	10.2	9.5	9.4	9.7
Do.	6	8.7	9.0	8.8	8.5	8.7
McCook	1	9.7	10.5	9.5	9.3	9.7
Do.	2	11.0	11.9	11.1	10.8	11.2
Do.	3	10.1	10.9	9.8	9.5	10.1
Do.	4	9.4	9.6	9.0	9.0	9.2
Do.	5	9.4	9.6	8.8	8.7	9.1
Do.	6	9.4	9.3	8.6	8.2	8.8
Holdrege	1	10.7	11.5	10.0	10.3	10.6
Do.	2	12.2	13.2	11.7	11.7	12.2
Do.	3	12.2	13.8	12.0	11.9	12.5
Do.	4	11.1	12.0	10.6	10.6	11.0
Do.	5	10.0	11.1	9.8	9.6	10.1
Do.	6	10.0	10.8	9.5	9.6	9.9
Hastings	1	11.1	12.0	11.0	10.4	11.1
Do.	2	13.7	15.0	13.2	13.2	13.7
Do.	3	12.5	13.0	12.0	11.5	12.2
Do.	4	11.6	12.9	11.6	11.5	11.9
Do.	5	10.7	11.9	10.7	11.0	11.1
Do.	6	10.9	11.7	10.3	10.7	10.9
Lincoln	1	12.3	12.7	12.3	12.0	12.3
Do.	2	15.5	16.6	15.3	15.3	15.7
Do.	3	15.3	16.0	14.6	14.6	15.1
Do.	4	15.2	15.1	13.2	13.3	14.2
Do.	5	14.5	14.7	13.1	13.2	13.8
Do.	6	15.0	15.2	13.6	13.5	14.3
Control H, determination 1	......	6.7	7.1	6.1	6.3	......
Control H, determination 2	......	6.2	6.7	6.2	6.0	......
Control H, determination 3	......	6.3	6.7	6.1	6.0	......
Control K, determination 1	......	14.5	15.0	14.2	14.3	......
Control K, determination 2	......	14.8	15.0	14.0	14.2	......
Maximum temperature ....°C.	......	23	24	22	23	......
Minimum temperature ....°C.	......	22	23	22	22	......

TABLE XI.—*Concordance of single determination of the hygroscopic coefficient made on different days with the average of 10 previous determinations*[a]

Source (Nebraska).	Depth.	Hygroscopic coefficient.				Departure of individual determinations from average of 10 previous determinations.		
		Average of 10 previous determinations.	Individual determination.			Det. 1.	Det. 2.	Det. 3.
			1st day.	2d day.	3d day.			
	Foot.							
Wauneta	1	9.1	10.1	8.6	9.2	1.0	−0.5	0.1
Do	2	9.6	10.2	9.1	9.3	.6	−.5	−.3
Do	3	9.7	10.5	9.4	9.2	.8	−.3	−.5
Do	4	9.9	10.2	9.1	9.2	.3	−.8	−.7
Do	5	9.0	10.0	8.5	8.2	1.0	−.5	−.8
Do	6	8.3	8.3	8.1	7.8	0.0	−.2	−.5
McCook	1	10.0	10.4	9.8	9.5	0.4	−.2	−.5
Do	2	10.9	9.6	11.0	10.0	−.3	−.1	−.9
Do	3	10.7	11.1	9.9	9.9	.4	−.8	−.8
Do	4	9.7	10.3	9.9	9.0	.6	.2	−.7
Do	5	9.1	9.2	8.7	8.6	.1	−.4	−.5
Do	6	9.1	9.3	8.5	8.7	.2	−.6	−.4
Holdrege	1	10.1	10.9	11.0	*8.0*	.8	.9	−2.1
Do	2	11.2	11.1	12.0	11.9	−.1	.8	0.7
Do	3	11.3	11.0	12.0	*14.0*	−.3	.7	2.7
Do	4	10.2	11.1	10.5	10.3	.9	.3	.1
Do	5	9.6	10.3	9.6	9.6	.7	.0	.0
Do	6	9.4	10.3	9.4	8.3	.9	.0	−1.1
Hastings	1	9.6	10.2	10.3	9.6	.6	.7	.0
Do	2	11.6	12.6	12.8	12.9	1.0	1.2	1.3
Do	3	12.4	13.3	13.2	11.5	0.9	.8	−0.9
Do	4	11.1	12.3	11.8	11.2	1.3	.7	.1
Do	5	10.7	11.5	11.1	10.8	0.8	.4	.1
Do	6	10.7	11.6	10.7	10.5	.9	.0	−.2
Lincoln	1	12.0	12.3	12.5	12.2	.3	.5	.2
Do	2	14.4	14.8	14.4	14.0	.4	.0	−.4
Do	3	13.6	*10.2*	14.5	13.7	2.0	.9	.1
Do	4	13.0	13.6	13.5	13.8	.6	.5	.8
Do	5	12.8	13.5	12.9	12.7	.7	.1	.1
Do	6	12.7	12.7	12.5	12.0	.0	−.2	−.7
Weeping Water	1	12.1	12.5	12.3	12.1	.4	.2	.0
Do	2	13.7	14.5	*10.8*	13.4	.8	−2.9	−.3
Do	3	13.9	14.3	14.3	13.4	.4	.4	−.5
Do	4	13.0	13.6	13.5	12.5	.6	.5	−.5
Do	5	12.6	13.2	13.1	12.1	.6	.5	−.5
Do	6	12.5	13.0	12.6	11.9	.5	.1	−.6
Control A, Determination 1			5.8	5.5	5.5			
Control A, Determination 2			5.9	5.7	5.5			
Control A, Determination 3			5.8	5.7	5.4			
Control A, average		5.6	5.8	5.7	5.5	.2	.1	.0
Maximum temperature...°C			19	19	19			
Minimum temperature...°C			17	18	16			

[a] The data in italics would in practice be rejected, but serve to illustrate the importance of duplicate determinations.

Another illustration is afforded by the data in Table XI on distinct sets of samples taken from the same localities as those appearing in the preceding table.

In the case of 30 sets, each consisting of 6 samples, hygroscopic-coefficient determinations had been made in duplicate on each. The sets had been taken from five fields at each of six different places, and to secure a single set of composite samples representing each area we mixed equal weights of soil from the corresponding depths of the five fields, making allowance for the unequal amounts of hygroscopic moisture at the time. Accordingly the average of the values for the various depths in each area represents the mean of 10 determinations. These data are given in the third column of Table XI. Later, samples of each of the 30 composites were exposed singly on three consecutive days. Only the one control soil, A, was used; on the first day the average value for this was found slightly higher and on the third slightly lower than on the second day. Of the 30 other soils 27 showed the highest value on the first day and 22 the lowest on the third day.

From the above the advantage of using samples of control soils will be evident. The mere concordance of determinations on different days and even by different operators is in itself no evidence that such data express the correct hygroscopic coefficients. Especially striking illustrations of this were afforded by the data obtained using pasteboard trays described above—data that we later wholly discarded. The development of an indirect method for the determination of the hygroscopic coefficient from the content of hygroscopic moisture was suggested by the parallel rise and fall of the two control soils from day to day (4, p. 348–359).

RELIABILITY OF THE METHOD DESCRIBED BY HILGARD

To determine this, we followed his method exactly, except that we used a table of paraffined wood, which he does not specify. To give the method the most severe test we each day employed fresh and, accordingly, air-dry sheets of glazed paper. Four soils, two of them of high hygroscopicity, were employed, they being exposed in duplicate for 12 hours and also for 24 hours (Table XII). During the whole of the time the temperature of the room was between $14°$ and $15°$ C.

TABLE XII —*Results showing the reliability of Hilgard's method when carried out as he has described it*

Soil.	Hygroscopic coefficient.		
	On glazed paper.		In aluminum trays (24 hours' exposure).
	12 hours' exposure.	24 hours' exposure.	
Control soil L	15.4	15.7	16.0
Control soil I	19.5	20.0	21.5
Control soil A	5.2	5.3	5.6
Control soil C	.6	.6	.6

The data in Table XII show that in the case of soils of medium hygroscopicity a 12-hour exposure gives results so close to those obtained by one of longer duration or by the use of metal trays that the differences may be considered within the range of experimental error. However, in the case of soils of very high hygroscopicity, such as I, it is not certain that the results are not slightly too low; in the case of these it is especially important to have the layer of soil very shallow. With L and I we used only from 2 to 5 gm., 8 to 16 gm. giving a decidedly lower result, while with soil A 10 gm. was not too much and with C as much as 80 gm. could be used without lessening the value found.

It is highly important that the method of Hilgard, carried out exactly as he described it gives reliable results, as this makes available as trustworthy material for investigational purposes the great number of data on the hygroscopic coefficient which he has reported along with physical and chemical analyses of arid and humid soils.

MODIFICATIONS OF METHOD WHEN SAMPLE CONTAINS GRAVEL OR PEBBLES

On many soil types the sample in which moisture has been determined contains gravel or even small pebbles. These can not as such be included in the material exposed in the absorption boxes, as the quantity so used is so small that the coarse particles would constitute either a much smaller or much larger proportion of these than they did in the moisture sample. Accordingly they must either be separated and a correction introduced or be reduced to a state sufficiently fine to permit of a uniform sample being secured while still coarse enough to have practically no absorptive capacity. As a coarse sand has a hygroscopic coefficient of only about 0.5, the pebbles and gravel if crushed just sufficiently to pass a 2-mm. sieve will cause no difficulty on the trays and introduce no appreciable errors.

In practice we have found it most convenient to crush the sample in a steel mortar, sifting out the fine material at frequent intervals with a 2-mm. sieve.

CONCLUSIONS

The amount of hygroscopic moisture absorbed increases with the rise of temperature.

Drying of mineral soils at temperature of $100°$ to $110°$ C. does not appreciably decrease their hygroscopicity.

Intractable samples may be reduced in a steel mortar to pass a 1-mm. sieve without appreciably affecting their hygroscopicity.

Twelve hours' exposure in the absorption boxes is sufficient only when the soil layer is very shallow. In practice a longer interval is found more convenient, 20 to 24 hours proving very satisfactory. An exposure of more than 24 hours gives higher values in the case of only very fine textured soils.

A soil containing the amount of moisture corresponding to its hygroscopic coefficient loses water very rapidly when exposed to an ordinarily dry atmosphere, but in determining the hygroscopic coefficient the time necessary to transfer the soils from the absorption boxes to weighing bottles is so brief that the loss during the transfer is too small to appreciably affect the accuracy of the results.

Hilgard's method for the determination of the hygroscopic coefficient, carried out exactly as he describes it, gives reliable results. However, the loose sheets of glazed paper thus involved are very inconvenient when many determinations are to be made and may advantageously be replaced by shallow trays, either of aluminum or of copper. Trays of glass, graniteware, and vulcanized rubber give satisfactory results, but are less convenient, while those of tin plate or zinc, although satisfactory at first, soon corrode. Pasteboard trays lined with glazed paper give results much too low, unless the period of exposure be greatly prolonged, and even those of paraffined pasteboard lined with glazed paper give somewhat low results. Any considerable increase in the size of the absorption boxes over that recommended by Hilgard or the use of a larger number of exposed samples within the boxes of the same size cause too low results, unless the time of exposure be greatly increased.

LITERATURE CITED

(1) ALWAY, F. J.

1906. STUDIES ON THE SOILS OF THE NORTHERN PORTION OF THE GREAT PLAINS REGION: THE SECOND STEPPE. *In* Amer. Chem. Jour., v. 36, no. 6, p. 580-594.

(2) ——

1908. STUDIES OF SOIL MOISTURE IN THE "GREAT PLAINS" REGION. *In* Jour. Agri. Sci., v. 2, pt 4, p 333-342

(3) ——

1913. STUDIES ON THE RELATION OF THE NON-AVAILABLE WATER OF THE SOIL TO THE HYGROSCOPIC COEFFICIENT. Nebr. Agr. Exp. Sta. Research Bul. 3, 122 p., 37 fig.

(4) —— and CLARK, V. L.

1916. USE OF TWO INDIRECT METHODS FOR THE DETERMINATION OF THE HYGROSCOPIC COEFFICIENTS OF SOILS. *In* Jour. Agr. Research, v. 7, no. 8, p. 345-359, 1 fig. Literature cited, p. 359.

(5) —— and McDOLE, G. R.

1907. STUDIES ON THE SOILS FROM THE NORTHERN PORTION OF THE GREAT PLAINS REGION: THE DISTRIBUTION OF CARBONATES ON THE SECOND STEPPE. *In* Amer. Chem. Jour., v. 37, no. 3, p. 275-283.

(6) AMMON, Georg.

1879. UNTERSUCHUNGEN ÜBER DAS CONDENSATIONSVERMÖGEN DER BODEN-CONSTITUENTEN FÜR GASE. *In* Forsch. Geb. Agr. Phys., Bd. 2, p. 1-46.

(7) BRIGGS, L. J., and McLANE, J. W.

1911. MOISTURE EQUIVALENT DETERMINATIONS AND THEIR APPLICATION. *In* Proc. Amer. Soc. Agron., v. 2, 1910, p. 138-147, pl. 6.

      1906. SOILS . . . 593 p., illus., diagr.   New York, London
(16) KNOP, W.

      1868. LEHRBUCH DER AGRIKULTUR-CHEMIE.   v. 2.   Leip
(17) LIEBIG, Justus von.

      1859. LETTERS ON MODERN AGRICULTURE.   ed. by John
            London.
(18) LIPMAN, C. B., and SHARP, L. T.

      1911. A CONTRIBUTION TO THE SUBJECT OF THE HYGROSCOP
            In Jour. Phys. Chem., v. 15, no. 8, p. 709–722.
(19) LOUGHBRIDGE, R. H.

      1874. ON THE INFLUENCE OF STRENGTH OF ACID AND TIME
            EXTRACTION OF SOILS.   *In* Amer. Jour. Sci., s. 3,
(20) LYON, T. L., FIPPIN, E. O., and BUCKMAN, H. O.

      1915. SOILS, THEIR PROPERTIES AND MANAGEMENT.   764
(21) MAYER, Adolph.

      1871. LEHRBUCH DER AGRIKULTURCHEMIE.   v. 2.
(22) ———

      1877. STUDIEN ÜBER DIE WASSERVERDICHTUNG IN DEI
            stract.)   *In* Centbl. Agr. Chem., Bd. 11, p. 243–2
            in Fühling's Landw. Ztg., Jahrg. 24, p. 87–97.
(23) MITSCHERLICH, Alfred.

      1901. UNTERSUCHUNGEN ÜBER DIE PHYSIKALISCHEN BC
            In Landw. Jahrb., Bd. 30, Heft 3, p. 361–445,
(24) ———

      1905. BODENKUNDE FÜR LAND– UND FORSTWIRTE.   364
(25) PATTEN, H. E., and GALLAGHER, F. E.

      1908. ABSORPTION OF VAPORS AND GASES BY SOILS.
            Soils Bul. 51, 50 p., 19 fig.
(26) SCHLŒSING, Théophile.

      1884. INFLUENCE DE LA TEMPÉRATURE SUR L'HYGROSCC
            VÉGÉTALE.   *In* Compt. Rend. Acad. Sci. [Paris],
(27) SCHÜBLER, G.

      1830. GRUNDSÄTZE DER AGRIKULTUR-CHEMIE.   T. 2, 272 p

INHERITANCE STUDIES IN PISUM. IV. INTERRELATION OF THE GENETIC FACTORS OF PISUM

By ORLAND E. WHITE,

Curator of Plant Breeding, Brooklyn Botanic Garden; Collaborator, Office of Forage Crop Investigations and Office of Horticultural and Pomological Investigations, Bureau of Plant Industry, United States Department of Agriculture

INTRODUCTION

The present paper describes and discusses the number of demonstrated factors in Pisum, their modifying effects upon each other's expression, the modification of their expression by different environments, and their relation to one another in inheritance, whether independent or linked.

MATERIAL AND METHODS

As pointed out by Mendel, Correns, Lock, and many others, the pea plant is in many respects especially favorable material for studying such genetic problems as these because of the ease with which it grows, its short life cycle, its practically complete self-pollination (1 to 10,000 in this locality) in outdoor cultures, its comparatively small number of chromosomes (seven pairs), the sharp cleavage in its multitudinous character differences, and the slight trouble involved in making successful crosses. On the other hand, back-crossing for various purposes, such as Morgan and his coworkers do, is practically out of the question with peas, because of the very small amount (2 to 4) of seed obtained from each cross for the amount of work required.

The material upon which the original and confirmatory data in this study are based consists of a collection of over 200 different varieties and species of peas gathered from all over the world through the help of the Offices of Foreign Seed and Plant Introduction and Forage-Crop Investigations; Sutton & Sons, Reading, England; P. Vilmorin and Vilmorin et Cie., Paris, France; W. Bateson and C. Pellew; Haage & Schmidt, Erfurt, Germany; several botanic gardens in Europe and Asia; and from many other sources. The material particularly used consists of the varieties listed in Table I, for which the factors directly concerned in this study are also given. All the strains have been inbred for at least two generations, and they breed true to the factors designated for them. All the pedigree cultures to be compared with one another were grown as nearly as practicable under the same environmental conditions—for example, the parents and the F_1 and F_2 generations are often grown side by side; crosses are practically all made in the greenhouse in winter; and the seed for growing the F_3 progeny are also grown under these conditions, as a greater quantity and better quality of seed can be secured than in field cultures. Under such conditions the bagging of emasculated flowers is absolutely unnecessary. Under extremely favorable conditions three generations a year can be grown, especially by using dwarf short-cycled

Journal of Agricultural Research,
Washington, D. C.
hd

Vol. XI, No. 4
Oct. 22, 1917
Key No. G—123

(167)

types, such as P21 The forepart of the winter and late fall, however, is poor pea weather, because of the short daylight hours, so that generally at the Brooklyn Botanic Garden only two crops a year are matured.

TABLE I.—*Designation, source, and factorial composition of varieties of Pisum studied in determining the relation of factors to each other in inheritance*

No.	Variety.	Factors studied.[a]	Source.
1	Mummy..............	Ab(fa)G(Gp)I(Le)OR (Tl)	H. Eckford, Wem, England.
11	Quite Content..........	aB(Fa)(Le)	Vaughan Seed Co., New York.
12	Dwarf Grey Sugar......	AB(le)T..............	Do.
13	Tall Grey Sugar........	AB(Fa)GI(Le)pR......	Do.
14	Black-eyed Marrowfat...	G(Gp)I..............	Do.
21	Nott's Excelsior........	aB(Fa)Gi(le)(Tl)........	Peter Henderson & Co., New York.
22	First of All	aBGI(Le)OR............	Do.
23	White-eyed Marrowfat...	aB(Fa)I..............	Do.
26	Telephone.............	aB(Fa)Gi............	Do.
27	Laxtonian.............	aB(Fa)Gi(le)...........	Do.
28	Aldermann.............	aB(Fa)Gi............	Haage & Schmidt, Erfurt, Germany.
29	Späte Gold.............	aB(Fa)GI	Do.
30	Goldkönig.............	aB(Fa)g(Gp)i(Le)or(Tl).	Do.
32	Wachs Schwert.........	aB(bt)(Fa)G(gp)I(Le) OpR.	Do.
35	"Market Split"..........	aB(Fa)G(Gp)iOR......	New York City Markets.
36	Selection of MS (P35)...	AB(Fa)................	
38	Acacia D..............	aB(Fa)Gi(le)rO(tl)....	W. Bateson and C. Pellew.
39	Acacia T..............	aB(Bt)(Fa)Gi(Le)PR(tl)	Do.
40	*P. Jomardi?*...........	AB(Fa)GI(Le)OR(Tl)	Cambridge (England) Botanic Garden.
41	*P. elatius*..............	ABGI(Le)OPR(Tl) . ..	Do.
43	*P. quadratum*...........	AB(Fa)I.	W. Bateson and C. Pellew.
47	*P. sat. umbellatum* .	aB(fa).	United States Department of Agriculture.
57	Quite Content...........	GiOr.................	Stumpp and Walters, New York.
58	Early White Dwarf Sugar.	aB(bt)(le).. ..	Do.
59	Velocity...............	aBG(Le)R.	Vaughan Seed Co., New York.
62	Everbearing.. .	aBi(le). . .	Peter Henderson & Co , New York.
65	Yorkshire Hero.........	aBi(le)	Thorburn & Co , New York.
67	Advancer...............	aBGi(le)Pr.	Peter Henderson & Co , New York.
72	Acacia V..............	aBGi(le)r(tl)..	Vilmorin & Cie, Paris. France.
91	"Chenille"............	aB(bl)s...	P. Vilmorin, Paris, France.
94	Little Marvel ...	a(le)u....	Sutton & Sons, Reading, England.
132	Abyssinian	ABI(Le)RU	Bureau of Plant Industry.
138	Benton................	ABGIOR	United States Department of Agriculture.
198	Scotch Beauty . .	aBGiOR.	Do.

Peas are easily crossed and the flowers but little affected by mutilation. The set of seed in good weather is, at least, 60 to 75 per cent. Of 200 or more varieties grown in the greenhouse in winter in 4-inch pots and in the open bench, none have failed to blossom and set seed.[1]

THE FACTORS OF PISUM

Mendel (1866)[2], Tschermak (1911), Correns (1902), Bateson (1909), Darbishire (1911), Lock (1905), Pellew (1913), Keeble (1910), Vilmorin (1911, 1913), Hoshino (1915), White (1916), and others have demonstrated, on the presence and absence hypothesis, the existence of over 35 factors in the genus Pisum, these obviously forming only a small part of the full factorial Pisum complex. If the experimental data secured by the above-mentioned investigators are interpreted on the "paired allelomorph" hypothesis, the "absence" of each factor is to be regarded as a recessive factor, and the number of demonstrated factors is thus increased to over 70. In Table II the factors are listed according to the "presence and absence" hypothesis. The presence and absence of these 35 factors gives rise, under certain specific environmental conditions, to over 70 differential Pisum characters.

The detailed experimental data by which these factors were demonstrated and the references to the literature, together with much other summarized data on the genetics of peas, are given in No. II of this series of studies on Pisum[3]; hence need not be repeated here. Suffice it to say that in many cases the existence of these factors has been proved by a large amount of experimental data, including large numbers of F_2 and F_3 generation individuals, results from the back-crossing of F_1 and homozygous F_2 and F_3 individuals with the ancestral "parent" types and with each other, reciprocal crosses, etc. With the exception of factors 11, 15, 17, 26, 27, and 33, the data are especially complete and satisfactory. In several cases studies on a single factorial difference have involved observations on over 20,000 individuals. In a large number of the cases the studies involve a thousand or more individuals. Most of these factors (except factors 13 and 24) have been isolated by other workers, but in many cases the writer has repeated their experiments with similar results.

In collecting these factors different symbols from those originally assigned them by their discoverers have often had to be given, in order to make them intelligible (Table II). Since the letters of the alphabet

[1] The writer is much indebted to his assistant, Miss M. Mann, and to Miss S Streeter, a graduate student of the Brooklyn Botanic Garden, for assistance in collecting these data   Mr. Montague Free, of the Garden staff has also helped collect and classify the data.

[2] Bibliographic citations in parentheses refer to "Literature cited," p. 178–180.

[3] WHITE, O. E.   INHERITANCE STUDIES IN PISUM.   II. THE PRESENT STATE OF KNOWLEDGE OF HEREDITY AND VARIATION IN PEAS.   Brooklyn Botanic Garden Contrib. 19.   (Compilation and correlation of all the genetic data on Pisum up to 1916, with critical discussion and full bibliography.   Read by title before the American Philosophical Society, May 4, 1917.   Not yet published.   Manuscript copy in Brooklyn Botanic Garden Library.)

are too few, double-letter expressions of a descriptive nature have been used, one capital and one lower-case letter, the two being joined by a parenthesis, as "(Gc)." In some cases, where two factors are responsible for the same result, the same letter, subtended by different numerals has been used by Tschermak (1911), and these are retained. The same factor often expresses itself in several distinct regions of the plant, such as the inflorescence, the leaf axils, etc. When more is known concerning the genetic behavior of some of these factors, two or more of them may be found to be absolutely coupled such as (1) 1 and (5), in which case matters are simplified by representing them by one factor symbol.

TABLE II.—*List of Pisum factors, alphabetically arranged, and their corresponding character expressions*

No.	Factor.	Expression.
1	A......	Salmon-pink or rose flower color. With CD gives reddish leaf axils.
2	B......	Purpling factor plus A gives purple flowers. With CD plus A gives purplish leaf axils and stem bases.
3	(Bl)	Glaucous foliage, stems and pods (with W); "bloom."
4	(Bt)....	Pods with blunt apex.
5	C[A]....	With D gives leaf axil and stem color.
6	D......	With C gives leaf axil and stem color.
7	E[A]....	With F and B gives purple dotting on seed coats, in the absence of B gives reddish dots.
8	(Ef)....	Modifies the expression of (Lf) toward earlier flowering.
9	F......	With E and B gives purple dotting on seed coats; in the absence of B gives reddish dots.
10	(Fa)....	Axillary flowers, round stems, regular phyllotaxy
11	(Fn)....	1 to 2 flowers per peduncle.
12	(Gc)[A].	Yellowish green to grayish brown seed-coat color (weak chromogen factor), brown hilum.
13	G......	Green cotyledon pigment.
14	(Gp)...	Green pod color.
15	H......	Brightener or inhibitor of expression of Gc.
16	I......	Factor which causes green cotyledon color to fade.
17	J......	With (Gc) gives dark-brown seed-coat color.
18	L_1[A]...	With L_2 gives indent peas.
19	L_2.....	With L_1(A) gives indent or dimpled peas.
20	(Le)....	Long internodes; with T gives tall plants.
21	(Lf) ...	Primarily responsible for late flowering.
22	M......	Brown or maple mottling on seed coat; or "ghost mottling" in absence of A.
23	N......	Violet eye on seeds.
24	O......	Green foliage, stems, and pods.
25	P......	Inflated, parchmented, nonedible pods with V.
26	P_1.....	With P_2 gives purple pods.
27	P_2.....	With P_1 gives purple pods.
28	(Pl) ...	Black-eyed seed-coat pattern.
29	R......	Round, smooth seeds with simple, oval starch grains, low water content and with excellent powers of germination under unfavorable weather conditions
30	S......	Pods with seeds separated or free
31	T......	Tall, robust plants, large number of internodes.
32	(Tl)...	Leaves with tendrils.
33	U......	Dark self-colored purple seed coat.
34	V......	With P gives parchmented, smooth pods.
35	W......	With (Bl) gives glaucous foliage, pods.

[1] For key to numbers 1 to 35 in parentheses, see Table II.

Factors A, C, E, (Gc), and L_1, so far as the evidence forthcoming up to now is concerned, appear absolutely coupled, and it is much simpler to regard them all as one factor with many separate expressions. A could be substituted in each case where C, E, (Gc), and L_1 are used.

MODIFICATION OF THE EXPRESSION OF PISUM FACTORS BY DIFFERENT ENVIRONMENTS AND BY EACH OTHER

Most of these data have been compiled from other papers, but this brief review also represents for the most part, first-hand confirmatory knowledge by the writer. References to literature and credit are given in the earlier paper previously referred to.[1]

The factor expressions given in Table II represent only the usual or common expression, and changes in environment may quite radically modify these expressions. Weather conditions, especially prolonged damp or rainy weather, often wash out or suppress the development of the purple flower color, so that it resembles the pink, and pinks are modified by the same causes to white. Under greenhouse conditions, where proper control is exercised, this never happens. Factors 3 and 35, either independently of each other or when combined, produce the glaucous waxy covering of peas. Factor 3 is arbitrarily regarded as the common one. Plants from which either or both of these are absent are called "emeralds," and lack this covering. Such plants are very subject to death from disease and other external causes; hence, are generally deficient in number in the F_2 ratios from crosses of glaucous with emerald varieties. The usual expression of factor (Bt) (4) depends partly on whether the pod is well filled with peas. Factors C and D (5, 6) are dependent on plenty of sunlight; otherwise the color is either completely absent or faint. The anthocyanin pigment of the seedcoat pattern, brought about by the presence of factors E and F (7, 9) is soluble in water and may be washed out in rainy weather. Factors (Ef) and (Lf) (8, 21) are modified by a large number of environmental conditions, so that studies on these should always be carried on with parents, F_1, F_2s, etc., growing side by side from plantings made at the same time. In crosses between certain normal and fasciated varieties of peas, dominance may be partially reversed by cloudy weather and conditions very favorable for rank growth. Unfasciated stem is usually dominant over fasciated stem, but in such crosses as noted above, the F_1 plants have been slightly fasciated under the conditions mentioned, while under other conditions, F_1 plants of the same cross have remained normal.

A character similar to that produced by the absence of factor (Fa) (10) may be brought about in varieties of peas in which (Fa) is present by very damp weather, lack of sufficient sunlight, and crowding and twisting of the stems (White, 1916b). Various environmental causes

---

1 WHITE, O. E   Op. cit.

bring about fasciation in plants, but in the case of such varieties of peas as the Mummy (P1), the cause is innate or hereditary.

The expression of (Gc) (12) changes to a deeper brown, resembling that of (Gc) (12) and J (17) combined, if the ripe seeds remain exposed to the weather or even ripen under conditions of much alternation of sun and rain. Under greenhouse conditions water rarely touches the ripe seeds. The expression of factor G (13) fades on prolonged exposure to sunlight and wet weather, and seeds of wrinkled-seeded varieties fade more quickly than those of green round-seeded forms. Lack of sufficient sunlight, immaturity due to prolonged growth of the vine, and other environmental factors effect the action of factor I (16) on green cotyledon pigment, especially in such varieties as Spàte Gold (P29). Pitting or "spurious indent" resembling the expression of factors L_1 (18) and L_2 (19) is found as an environmental effect in peas from which these factors are absent. This is commonly the case in field-grown seeds in this locality of such white-flowered, round-seeded varieties as Tall Acacia (P39), First of All (P22), and Black-Eyed Marrowfat (P14). Factor (Le) (20) is generally responsible for the difference between dwarf and tall peas, and this difference may be noted as soon as the plants are 2 to 3 weeks old. Factor T (31), aside from the fluctuations caused by differences in environment, is responsible for the number of internodes. Pea varieties may be classified into three general types as regards height—dwarfs, half dwarfs, and talls. Dwarfs have 8 to 15 short internodes. Half dwarfs are at least of two general types—plants with 10 to 18 long internodes or 20 to 30 short ones. Tall peas are made up of a large number of long internodes (35 to 60). Other hereditary elements not yet isolated, together with environmental conditions such as disease, very poor soil, and dry, hot weather, modify the expression of these factors, especially T (31). The absence of factor (Fa) (10), bringing about fasciation, modifies height by shortening the internodes, though possibly the number of internodes is increased. Factor M (22) expresses itself very faintly ("ghost mottling") in the absence of factor A (1) for flower color. Prolonged exposure to sun and rain after the seeds are mature darkens the brown pigment.

The expression of factor O (24) may be modified or almost entirely suppressed by certain diseases or the lack of sufficient salts, such as potassium. Sickly yellow foliage results, somewhat resembling varieties from which O is absent, though whiter and duller. Crosses involving factors G, O, and R (13, 24, 29) apparently give complex results. Either the presence of the shape factor R (29) causes no zygotes to be formed in which R is present and G and O are absent, or the presence of R in some way brings about the production of green pigment, even though G (13) is absent. More data are being accumulated, and it is hoped these particular problems will be solved shortly through their aid. The expression of factor S (30), in the absence of which pea seeds are stuck together when mature, is regarded by the Hagedoorns (1914) as being modified by the factors for flower color and glaucous foliage and stems. In working over Vilmorin's data (1913), the present writer concluded that these data were more simply explained by regarding factors (Bl) (3) and S (30) as partially coupled or linked and by assuming certain of the F_2 combinations to be adversely affected by weather conditions. The variety with stuck-together seeds ("chenille") has emerald foliage, but emerald varieties are known with free seeds.

RELATION OF PISUM FACTORS TO EACH OTHER IN INHERITANCE, WHETHER LINKED (COUPLED) OR INDEPENDENT

The studies of Bateson (1909), Pellew (1913), Surface (1916), Tanaka (1913), Emerson (1911), Vilmorin (1913), Gregory (1911), Doncaster (1913), Pearl (1912), Altenburg (1916), and especially of Morgan and his coworkers and students (1915), Sturtevant (1915), Bridges (1914), and Muller (1916) make it increasingly evident that the various hereditary characteristics of organisms and their determiners or factors are inherited in more or less closely linked groups, instead of independently, as was supposed in the early Mendelian studies, these groups being held together, perhaps, by chromosomes. As shown by Cannon (1903), Strasburger (1911), and others, at least some varieties of the common pea have seven pairs of chromosomes; and a wide cytological investigation, now in progress, will determine how true this is for all species of Pisum. On the assumption that seven pairs of chromosomes are the number usually present in all varieties, the number of independently inherited groups in Pisum, on the chromosome hypothesis as developed by Morgan and his students (1915), should be seven. Should accurately checked data show that more than seven of these groups occur, for instance, eight or nine, either the varieties used possess more than seven pairs of chromosomes or the chromosome theory as now held, so far as peas are concerned, would need revision or would be disproved. The chromosomes of peas, as compared with those of *Drosophila* spp., however, are long; and from this fact one might assume the linear distances between many of the factors to be greater. This being true, greater opportunities for crossing over would occur, and when the number of cross-overs approached 50 per cent, greater difficulty would be found in securing sufficiently accurate data to distinguish between ratios involving linkage and those showing only independent inheritance.

Table III shows the totals of F_2 ratios from crosses involving the presence and absence of eight factors [A, B, (Fa), I, (Le), G, R, (Tl)], two of which [R and (Tl)] belong to one group. Each factor has been tested out in crosses involving all its possible combinations with the other seven, the results indicating seven independently inherited groups in Pisum. Hence, in Table III all the factors of one group have a common number, as, for example, R and (Tl) meaning that both R and (Tl) belong to group 7. The F_2 generation totals in most cases are fairly large and dependable, but in some cases still more data are needed, in view of the fact that the linked groups may be less closely bound together in Pisum than in Drosophila, because of difference in chromosome dimensions. In several cases, involving small totals, the approximation between the classes actually obtained and those theoretically expected is not very close. As soon as more data are obtainable probable errors for these ratios will be calculated. In Table III, the ratios actually obtained are given first and the calculated expectation directly beneath. Each factorial combination is separated by a line and the combinations themselves (somatic) are expressed by descriptive character symbols, explanations of which are given in a note below the table. Factors (Tl) and R are used interchangeably, as they are closely linked. The totals represent the F_2 populations from crosses of different varieties, though each of the two varieties involved in a single cross differed in the designated factors. The data from the individual crosses with the numbered designations of the varieties involved, are given in the appendix. Each group of crosses involving one type of factorial difference is designated so as to correspond with the designation in Table III, as *exempli gratia* (1), (2), etc. All data except those involving group 7 are from the Brooklyn Botanic Garden cultures. The data for group 7 are the combined results from the studies of Vilmorin (1911), Bateson (1909), Bateson et al. (1905), Pellew (1913), as well as from the writer's own studies.

The employment of data from crosses involving many different varieties to demonstrate the independent inheritance of specific factors, such as those given in Table III, as compared with data from crosses involving only two varieties with the desired factorial differences, is open to some criticism on the hypothesis of multiple factors. Nilsson-Ehle, (1908, 1909), East (1910), Morgan et al. (1915), and many others have secured F_2 ratios involving only one character difference, which, however, gives a di-, tri-, or poly-hybrid ratio.

Apparently the same characters in plants and animals of the same species in other experiments have given an ordinary monohybrid 3 : 1 or 1 : 2 : 1 F_2 ratio. These results (dark-brown glume color in oats, yellow endosperm color in maize, pink eye color in Drosophila, capsule-shape in shepherd's-purse (*Bursa bursa-pastoris*), etc., are interpreted by their discoverers as showing the presence of two (in case of a dihybrid ratio) or more factors, each of which by itself gives rise to practically the same effect, expression or character, as when both are concerned. Further, as the ratio indicates, they are inherited independently of each other, and according to Morgan, Sturtevant, and their colleagues (1915), they are located in different chromosomes. By the method in use in the present study, different factors in different varieties giving the same or very nearly the same "somatic" expression could not be distinguished; hence, one can not be certain that he is always experimenting with the same factor.

TABLE III.—*Frequency distribution actually obtained and theoretically expected of the various factorial combinations of F_2 populations of peas derived from crosses involving eight pairs of factorial differences, two of which are partially linked*

No.	Factors.	F_2 frequency ratio.[a]
1	$A^1, (Fa)^5$ Total 32.	25 RpN 7 RpF 6 WN : 1 WF. 22 RpN 7 RpF 7 WN 2 4 WF.
2	$A^1, B^3, (Fa)^5$ Total 1,936.	823 RpN 264 RpF 260 PN 68 PF 401 WN 120 WF. 817 RpN 272 RpF : 272 PN : 90 PF : 363 WN : 121 WF.
3	$B^3, (Fa)^5$ Total 550.	323 RpN 94 RpF 100 PN 33 PF 309 RpN 103 RpF 103 PN 34 PF
4	$A^1, B^3, I^4.$ Total 777.	332 RpY 103 RpG 112 PY 26 PG 150 WY 54 WG. 327 RpY 109 RpG 109 PY 36 PG . 145 WY : 48 WG.
5	$(Fa)^5, I^4$. Total 608.	411 YN 129 YF 115 GN 43 GF 392 YN 131 YF 131 GN 44 GF
6	$A^1, B^3, (Le)^6.....$ Total 416.	179 RpT 64 RpD 50 PT 19 PD 73 WT 31 WD. 175 5 RpT 58 5 RpD 58.5 PT 19 5 PD 78 WT : 26 WD.
7	$A^1, (Le)^6.$ Total 1,182.	750 RpT 270 RpD 276 WT 84 WD. 776 7 RpT 259 RpD 259 WT 86 3 WD.
8	$(Fa)^5, (Le)^6$. Total 416.	233 NT 82 ND 69 FT 32 FD 234 NT 78 ND : 78 FT 26 FD.
9	$I4, (Le)^6...$. Total 1,232.	669 YT 255 YD 225 GT : 83 GD. 693 YT 231 YD 231 GT 77 GD
10	A^1, B^3, I^4, G^6 Total 167.	80 RpY 13 RpG 28 PY 5 PG 30 WY 11 WG. 76 0 RpY . 27 3 RpG 25 3 PY . 5.8 PG 33.8 WY : 7.8 WG.
11	$A^1, I^4, G^6.$ Total 265.	168 CY 32 CG : 52 WY 13 WG 160 CY 36 9 CG 53 3 WY 12 3 WG.
12	$(Fa)^5, G^6.$ Total 167.	102 YN : 46 YF 25 GN 4 GF 101.4 YN : 34 YF . 23.4 GN : 7.8 GF.
13	$I4, G^6$. Total 1,410.	1,142 Y 268 G. 1,145.3 Y : 264.3 G.
14	$(Le)^6, G^6.$. Total 39.	22 GT 6 GD 7 YT 1 YD. 20.25 GT . 6 75 GD . 6.75 YT · 2.25 YD.
15	$A^1, R^7.$ Total 579.	317 RRp 106 RW 113 WrRp 43 WrW. 324 9 RRp 108.3 RW 108 3 WrRp 36 WrW.
16	$A^1 (Tl)^7$ Total 404.	233 TlRp 77 TlW 68 ARp 26 AW 226 8 TlRp 75 6 TlW 75 6 ARp 25 2 AW.
17	$B^3, (Tl)^7$. Total 72.	53 TlB 10 AB 14 Tlb 2 Ab 45 TlB 15 AB 15 Tlb 5 Ab
18	$(Fa)^5, (Tl)^7$ Total 72.	(No coupling; figures not available (Pellew). 50+1? TNL : 5+1? NA 16 TlF 5+1? AF. 45 TlN 15 NA 15 TlF 5 AF
19	$I4, R^7$ Total 9,217.	5,241 RV 1,729 RG 1,757 WrV 510 WrG. 5,195 RV : 1,731 RG 1,731 WrV 577 WrG.
20	$I4, (Tl)^7$ Total 481.	275 TlY 94 TlG 82 AY 32 AG 270 2 TlY 90 2 TlG 90 2 AY 30 2 AG
21	$(Le)^6, (Tl)^7$ Total 96.	(No coupling; figures not available (Pellew). 51 TlT 25 TlD 13 AT 7 AD 54 TlT 18 TlD 18 AT 6 AD
22	$(Le)^6, R^7$. Total 143.	73 RT 34 RD 28+1? WrT 7 WrD. 80 1 RT 26 7 RD 26 7 WrT 8 9 WrD.
23	$G^6, (Tl)^7$. Total 66.	31 GTl 13 GA 19 YTl 5 YA 37 8 GTl 12 6 GA 12 6 YTl 4 2 YA

a Rp=purple flower color; P=pink flower color; W=white flower color; C=colored flowers. N=normal stems; F=fasciated stems. V=yellow cotyledon color; G=green cotyledon color. T=long internodes (tall plants); D=short internodes (dwarf plants). R=round cotyledons; Wr=wrinkled cotyledons. Tl=leaves with tendrils; A=leaves without tendrils (acacia). B and b=with and without factor B.

Two such multiple factors as A and A_1, each capable of producing pink flower color, each located in a separate chromosome, and each independently inherited from a third such factor A_2, having the same expression, could bring about considerable confusion in such an attempt as the present to determine the factorial groups of Pisum. With data from crosses of two varieties having seven or more factorial differences, no such difficulties would be encountered, and perhaps even with the method used above the chance of encountering more than one such factor, is comparatively rare. As soon as two true breeding varieties differing in seven or more suitable factors can be secured from crossing experiments the independence in inheritance of the factors listed in Table III will again be tested.

From among 16,384 or more F_2 progeny of such a cross, if each factor proved again to be independently inherited, there should be over 128 (2^7) visibly distinct forms, representing 2,187 distinct factorial combinations. Provided the cross involved 8 factor differences, each inherited independently of the other, 256 distinct forms would be theoretically expected and if a cross were made involving 25 such factor differences, each independently inherited, the F_2 generation, if it were practicable to grow a large enough population (4^{25}), should consist of at least (2^{25}) 33,554,432 visibly distinct forms. The classification of such a population, even when obtained, would be impracticable, if not impossible.

It is brought into the present discussion to emphasize the enormous amount of variation possible through crossing, for even on the chromosome-crossing-over-linkage hypothesis of Morgan and his coworkers (1915), plants and animals are known with more than enough factor differences and chromosomes to have 25 independently inherited groups. For example, *Nicotiana tabacum*, the commercial tobacco, as observed by the writer (1913), has 24 chromosome pairs and the recent lists of Tischler (1916) and Ishikawa (1916) cite plants with a much higher number. As pointed out by East (1915), if the facts and theories regarding inheritance in *Drosphila* spp. are found to be true for organisms in general, it is very important that plant breeders should know the number of chromosome pairs in the plants with which they are experimenting. When one contrasts the number of forms that can be derived from such crosses as mentioned above with the number of recognized plant species (approximately 225,000) and considers that to the plant breeder and horticulturist many of these forms are as distinct as taxonomic species, one is not surprised at all that a certain well-known taxonomist and student of phylogeny, Lotsy (19), should advocate that all evolutionary change of a hereditary nature is due to crossing.

Other factors in addition to those listed in Table III have been tested out with each other, but only part of the possible combinations with those given in Table III have been made, so these data are reserved for a later paper.

LINKAGE IN PISUM

In Table IV the available data on linkage of factors in Pisum are given. As in Table III, only totals are listed, the detailed results from which the totals were compiled being given in the appendix, the tables being numbered to correspond with the linked factor groups in Table IV. Linked group 24 in Table IV is based on the writer's own interpretation of Vilmorin's (1911) data from F_2 hybrid populations and F_3 progeny from heterozygotes similar to the F_1. The deficiency of EA individuals is, as previously stated, probably due to insufficient ability to resist disease. At least, that is the writer's own experience with emerald varieties. The calculation of crossover to non-crossover gametes is based on Castle's table (1916).

Group 25 is based on the studies of Lock (1905), Tschermak (1911), and Hoshino (1915), supplemented by the writer's own studies, though none of his own data are incorporated. The figures and interpretation are taken from Hoshino's paper. As previously stated, factors A, C, E, (Gc), and L_1, are absolutely coupled and most simply interpreted as expressions of A, instead of as five distinct factors, as held by Tschermak and others. Group 26 is based on the data and the interpretation of Vilmorin and Bateson (1911), and Pellew (1913), though the discovery was first made by Vilmorin (1911). The results of the writer from numerous crosses, among which was one, the F_2 seed of which was kindly sent me by Prof. Bateson, completely confirm the previous studies. Group 27 is based entirely on the writer's own data, but they are too few as yet to determine the ratio of non-crossover to crossover gametes. Groups 25, 26, and 27 are inherited independently of each other, but, as shown in Table III, they belong to three of the seven independently inherited groups already demonstrated. The relation of group 24 to these, so far as the writer is aware, is undetermined.

TABLE IV.—*Linked groups of Pisum* [a]

No.	Factors.	Linkage.
24	(Bl), S ... Total 340....	265 GlFs : 6 GlAs : 12 EFs : 57 EAs. 226 GlFs : 17 GlAs : 17 EFs : 64 EAs (8:1). 189.9 GlFs : 63.3 GlAs : 63.3 EFs : 21.1 EAs (no coupling).
25	A (Lf)........ Total 2,636....	513 WE : 163 WL : 141 RpE : 1813 RpL. 505 WE : 154 WL : 154 RpE : 1823 RpL (7:1). 164.3 WE : 492.9 WL : 492.9 RpE : 1478.7 RpL (no coupling).
25	A, C, E (Gc), and L_1	Absolutely coupled, so may be regarded as expressions of a single factor A.
26	R (Tl)........ Total 2,065....	1,466 RT : 20 RA : 15 WT : 564 WA. 1,471 RT : 15 RA : 15 WT : 561 WA (63:1). 1,161 RT : 387 RA : 387 WT : 120 WA (no coupling).
27	G, O......... Total 196....	134 GcGf : 6 GcYf (?) : 16 YcGf (?) : 40 YcYf. 109.7 GcGf : 36.3 GcYf (?) : 36.3 YcGf (?) : 12.2 YcYf (no coupling).

[a] Actually obtained F_2 classes and frequencies given first, followed (1) by approximated amount of crossing-over and (2) by theoretical expectancy on a non-linked or no-coupling basis.
Gl = glaucous, E = emerald or non-glaucous Fs = free seeds, As = adhering or "chenille" seeds

SUMMARY

(1) Thirty-five genetic factors of the genus Pisum together with their common expressions, are listed and discussed in the present paper. The "presence and absence" of these 35 factors are mainly responsible for 70 or more differential Pisum characters.

(2) The modifying effects of the expression of one factor upon that of another, and the effects of external environmental conditions upon the expression of these factors, so far as known, are presented and discussed.

(3) Data, involving many thousand F_2 generation progeny, indicate the factors A, B, (Fa), I, (Le), G, and R to be independently inherited— that is, not linked, unless the linkage is very loose.

(4) Data for four linked groups are presented, three of which involve the factors mentioned under (3), and one of which the relations to the above seven are still undetermined.

LITERATURE CITED

ALTENBURG, Edgar.
    1916. LINKAGE IN PRIMULA SINENSIS. *In* Genetics, v. 1, no. 4, p. 354-366. Bibliography, p. 366.
BATESON, William.
    1909. MENDEL'S PRINCIPLES OF HEREDITY. 396 p., illus. Bibliography, p. 369-385.
—— SAUNDERS, Edith R., and PUNNETT, R. C.
    1905. EXPERIMENTAL STUDIES IN THE PHYSIOLOGY OF HEREDITY. *In* Roy. Soc. [London] Rpts. Evol. Com., Rpt. 2, 154 p.
BRIDGES, C. B.
    1914. THE CHROMOSOME HYPOTHESIS OF LINKAGE APPLIED TO CASES IN SWEET PEAS AND PRIMULA. *In* Amer. Nat., v. 48, no. 573, p. 524-534.
CANNON, W. A.
    1903. STUDIES IN PLANT HYBRIDS: THE SPERMATOGENESIS OF HYBRID PEAS. *In* Bul. Torrey Bot. Club, v. 30, no. 10, p. 519-543, pl. 17-19. Bibliography, p. 534-541.
CASTLE, W. E.
    1916. TABLES OF LINKAGE INTENSITIES. *In* Amer. Nat., v. 50, no. 597, p. 575-576, 2 tab.
CORRENS, C F. J. E.
    1900. G MENDEL'S REGEL ÜBER DAS VERHALTEN DER NACHKOMMENSCHAFT DER RASSENBASTARDE. *In* Ber. Deut. Bot. Gesell., Bd. 18, Heft 4, p. 158-168.
————
    1900. GREGOR MENDEL'S "VERSUCHE ÜBER PFLANZEN-HYBRIDEN" UND DIE BESTÄTIGUNG IHRER ERGEBNISSE DURCH DIE NEUESTEN UNTERSUCHUNGEN. *In* Bot. Ztg., Jahrg. 58, Abt. 2, No. 15, p. 229-235.
————
    1902. UEBER DEN MODUS UND DEN ZEITPUNKT DER SPALTUNG DER ANLAGEN BEI DEN BASTARDEN VON ERBSEN-TYPUS. *In* Bot. Ztg., Jahrg. 60, Abt. 2, No. 5/6, p. 65-82.
DARBISHIRE, A. D.
    1911. BREEDING AND THE MENDELIAN DISCOVERY. 282 p., 34. pl. (partly col., partly fold.). London, New York.
DONCASTER, L.
    1913. ON SEX-LIMITED INHERITANCE IN CATS, AND ITS BEARING ON THE SEX-LIMITED TRANSMISSION OF CERTAIN HUMAN ABNORMALITIES. *In* Jour. Genetics,

EAST, E. M.
1910. A MENDELIAN INTERPRETATION OF VARIATION THAT IS APPARENTLY CONTINUOUS. *In* Amer. Nat , v. 44, no. 518, p. 65-82.

1915. THE CHROMOSOME VIEW OF HEREDITY AND ITS MEANING TO PLANT BREEDERS. *In* Amer. Nat., v. 49, no. 584, p. 451-494, 5 fig. Literature cited, p. 488-494.

EMERSON, R. A.
1911. GENETIC CORRELATION AND SPURIOUS ALLELOMORPHISM IN MAIZE. *In* Nebr. Agr. Exp. Sta., 24th Ann. Rpt. [1910], p. 59-90, 9 fig.

GREGORY, R. P.
1911. ON GAMETIC COUPLING AND REPULSION IN PRIMULA SINENSIS. *In* Proc. Roy. Soc. [London], s. B, v. 84, no. 568, p. 12-15.

HAGEDOORN, A. L., and HAGEDOORN, (Mrs.) A. C.
1914. STUDIES ON VARIATION AND SELECTION. *In* Ztschr. Indukt Abstamm. u. Vererb., Bd. 11, Heft 3, p. 145-183, 4 fig.

HOSHINO, Yuzo.
1915 ON THE INHERITANCE OF THE FLOWERING TIME IN PEAS AND RICE *In* Jour Col. Agr. Tohoku Imp. Univ., v. 6, pt. 9, p 229-288, fold. tab pl. 12-15 (fold.). Literature cited, p. 285-286.

ISHIKAWA, Mitsuharu.
1916. A LIST OF THE NUMBER OF CHROMOSOMES. *In* Bot. Mag. [Tokyo], v. 30, no. 360, p. 404-448, 32 fig

KEEBLE, Frederick, and PELLEW, Caroline.
1910 THE MODE OF INHERITANCE OF STATURE AND OF TIME OF FLOWERING IN PEAS (PISUM SATIVUM). *In* Jour. Genetics, v. 1, no. 1, p. 47-56.

LOCK, R. H.
1905. STUDIES IN PLANT BREEDING IN THE TROPICS. II. EXPERIMENTS WITH PEAS. *In* Ann. Roy. Bot. Gard. Peradeniya, v. 2, pt. 3, p. 357-414.

LOTSY, J. P.
1916. EVOLUTION BY MEANS OF HYBRIDIZATION. 166 p , illus. The Hague.

MENDEL, Gregor.
1866. VERSUCHE ÜBER PFLANZEN-HYBRIDEN. *In* Verhandl. Naturf. Ver. Brünn, Bd. 4, 1865, Abhandl., p. 3-47.

MORGAN, T. H., STURTEVANT, A. H., MULLER, H. J., and BRIDGES, C. B.
1915 THE MECHANISM OF MENDELIAN HEREDITY. 262 p., illus. New York.

MULLER, H. J.
1916 THE MECHANISM OF CROSSING OVER. I-IV. *In* Amer. Nat., v. 50, no. 592, p. 193-221, 5 fig., no. 593, p. 284-305, 4 fig.; no. 594, p. 350-366; no. 595, p 421-434, 4 fig.

NILSSON-EHLE, Herman.
1908 EINIGE ERGEBNISSE VON KREUZUNGEN BEI HAFER UND WEIZEN. *In* Bot Notiser, 1908, Häftet 6, p. 257-294.

1909 KREUZUNGSUNTERSUCHUNGEN AN HAFER UND WEIZEN. Lunds Univ. Årsskr., N. F., Afd. 2, Bd. 5, No. 2, 122 p.

PEARL, Raymond.
1912 THE MODE OF INHERITANCE OF FECUNDITY IN THE DOMESTIC FOWL. *In* Jour. Exp. Zool , v. 13, no. 2, p. 153-268, 3 fig. Literature cited, p. 266-268.

PELLEW, Caroline.
1913 NOTE ON GAMETIC REDUPLICATION IN PISUM. *In* Jour. Genetics, v. 3, no. 2, p 105-106.

SHULL, G H.
1914. DUPLICATE GENES FOR CAPSULE-FORM IN BURSA BURSA-PASTORIS *In* Ztschr. Indukt. Abstamm. u. Vererb., Bd. 12, Heft 2, p. 97-149, 7 fig. Literature cited, p. 148-149.

Genetics, v. 1, no. 3, p. 252-286, pl. 2-3.   Literature

TANAKA, Yoshimaro.

1913. GAMETIC COUPLING AND REPULSION IN THE SILKWORM,
Jour. Col. Agr. Tohoku Imp. Univ., v. 5, pt. 5, p. 1
Literature cited, p. 146-147.

TISCHLER, G.

1916. CHROMOSOMENZAHL, -FORM UND -INDIVIDUALITÄT IM PFI
Prog. Rei Bot., Bd. 5, Heft 2, p. 164-284.   Citierte Li

TSCHERMAK, Erich von.

1911. UEBER DIE VERERBUNG DER BLÜTEZEIT BEI ERBSEN.   I
Ver. Brünn, Bd. 49, 1910, p. 169-191, 9 fig.

VILMORIN, Philippe de.

1913. ÉTUDE SUR LE CARACTÈRE "ADHÉRENCE DES GRAINS
LE POIS "CHENILLE."   In IV. Conf. Intern. Genetiqu
Rap., 1911, p. 368-372, 2 fig.

—— and BATESON, William.

1911. A CASE OF GAMETIC COUPLING IN PISUM.   In Proc. Roy
B, v. 84, no. 568, p. 9-11, 1 fig.

WHITE, O. E.

1913. THE BEARING OF TERATOLOGICAL DEVELOPMENT IN NICOT
OF HEREDITY.   In Amer. Nat., v. 47, no. 556, p. 206-2

——

1916a. INHERITANCE STUDIES IN PISUM.   I. INHERITANCE OF
In Amer. Nat., v. 50, no. 597, p. 530-547.   Literature

——

1916b. STUDIES OF TERATOLOGICAL PHENOMENA IN THEIR RE
TION AND THE PROBLEMS OF HEREDITY.   II. THE NA
TRIBUTION, AND INHERITANCE OF FASCIATION, WITH SI
TO ITS OCCURRENCE IN NICOTIANA.   In Ztschr. In
Vererb., Bd. 15, fig. 1-28, tab. A-F, 1-26.

APPENDIX

[See pages 173-177]

TABLE 1.—*Inheritance of factors A and (Fa)*

TABLE 2 —*Inheritance of factors A, B, and (Fa)*

Cross.	F₂ classes and frequencies.						Total.
	RpN.	RpF.	PN.	PF.	WN	WF.	
(P1−4×P23−4)−1.	45+3?	15	12	2	24	7	108
(P1−4×P32−5)−1.	4	2	1	0	3	1	11
(P1−4×P32−5)−3.	0	1	1	0	1	0	3
(P1−4×P32−5)−4.	13+4?	6	3	0	11	3 8	40
(P1−4×P39−3)−1.	34	9	11	5	12	8	79
(P1−6×P29−1)−1.	3	1	4	1	4	1	14
(P11−1×P1−1)−1.	33+1?	3	7	5	10	1	60
(P11−1×P1−1)−2,							
−3, −4..........	34	15	13	2	17	5	86
(P11−1×P1−2)−1.	21	12	6	3	18	4	64
(P11−1×P1−1)−1,							
etc..............	172	54	44	18	70	25	383
(P13−1×P1−2)−1,							
−2..............	80	17+1?	22	5	41+1?	9	176
(P16−1×P1−1)−1,							
−2,−3,−4......	38	12	16	5	21	5	97
(P17−1×P1−1)−1.	15	2	7	0	7	2	33
(P18−1×P1−1)−1,							
−2.............							
(P18−2×P1−2)−1,	66	26	30	6	36	13	177
−2.............							
(P29−1×P1−4)−2,							
−3, −4.........	47	19	10	2	25	11	114
(P30−1×P1−1)−1,							
etc..............	69	24	27	6	31	10	167
(P31−1×P1−4)−1,							
−2.............	117	38	39	7	56	13	270
(P35−3×P1−4)−1,							
....	24	7	7	1	13	2	54
Grand total...	815+8?	263+1?	260	68	400+1?	120	1,936
Theoretically expected ...	816.7	272.2	272.2	90.7	363	121	1,936
Ratio.........	27	9	9	3	12	4	......

TABLE 3.—*Inheritance of factors B and (Fa)*

	F₂ classes and frequencies.				Total.
	RpN.	RpF.	PN.	PF.	
h−1×Red-flowered Normal	52	12+2?	14+1?	3	84
Do.............................	100+2?	29+1?	38+1?	11+2?	134
(P13−1×P1−1)..................	79+5?	26+1?	27	11	149
(P13−3×P1−3)−1, −2.............	81+3?	20+2?	19	6	131
(P30 1×P1−6)−1.................	1	1	0	0	2
Grand total...................	313+10?	88+6?	98+2?	31+2?	550
Theoretically expected.........	308.7	102.9	102.9	34.3	550
....	9	3	3	1	......

F$_2$ classes and frequencies.

	RpY.	RpG.	PY.	PG.	WY.	WG.	Total.
. . . .	28	15	11				79
. . . .	176	50	53				383
−4 . . .	36	7	13				84
. . . .	73	19					177
. . . .	19	12					54
. . . .	332		112	26	150	54	777
. . . .	326. 7	108. 9	108. 9	36. 3	145. 2	48. 4	777
. . . .	27	9	9	3	12	4	. . . .

F$_2$ classes and frequencies.

				GN.		Total.
. . . .						383
4						84
. . . .						177
. . . .						54
. . . .	411	129	115		43	698
. . . .	392. 4	130. 8	130. 8		43. 6	698
. . . .	9	3	3		1	. . . .

TABLE 7 —*Inheritance of factors A and (Le)*

Cross.	F₂ classes and frequencies.				Total.
	RpT.	RpD.	WT.	WD.	
(P₂₁−1×P₁−1)−1, etc.	167	59	65	30	321
(P₂₇−1×P₁−1)−1	12	5	8	1	26
(P₄₀−1×P₅₈−2)−1, −2	25	8	12	2	47
(P₄₁−1×P₂₁−1)−1, −2	122	36	40	8	206
P₄₁−3×P₅₈−3)−1	23	6	5	1	35
(P₄₈−3×P₄₁−5)−1, −2	82	21	34	4	141
(P₆₂−1×P₄₁−3)−1, −2	67	16	20	4	107
(P₆₅−1×P₄₁−3)−1	37	22	10	4	73
P₆₇−6×P₁₃−3)−1, −2, −3	39	15	26	4	84
P₇₂−1×P₄₁−5)−1, −2	176	82	56	26	340
Grand total	750	270	276	84	1,380
Theoretically expected	776.7	258.9	258.9	86.3	1,380
Ratio	9	3	3	1	......

TABLE 8 —*Inheritance of (Fa) and (Le)*

Cross.	F₂ classes and frequencies.				Total.
	NT.	ND.	FT.	FD.	
(P₂₁−1×P₁−1)−1, etc.	210	76	65	32	383
(P₂₇−1×P₁−1)−1	23	6	4	0	33
Grand total	233	82	69	32	416
Theoretically expected	234	78	78	26	416
Ratio	9	3	3	1	......

TABLE 9.—*Inheritance of factors I and (Le)*

Cross.	F₂ classes and frequencies				Total.
	YT.	YD.	GT.	GD.	
(P₂₁−1×P₁−1)−1, etc.	221	79	54	29	383
(P₄₁−1×P₂₁−1)−1, −2	121	34	42	11	208
(P₆₂−1×P₄₁−3)−1, −2	71	20	29	5	125
(P₆₅−1×P₄₁−3)−1	38	20	17	9	84
(P₆₇−6×P₁₃−3−3)−1, −2, −3	51	15	15	6	87
(P₇₂−1×P₄₁−5)−1, −2	167	87	68	23	345
Grand total	669	255	225	83	1,232
Theoretically expected	693	231	231	77	1,232
Ratio	9	3	3	1	......

TABLE 10 —*Inheritance of factors A, B, I, and G*

Cross.	F₂ classes and frequencies.						Total.
	RpY.	RpG.	PY.	PG.	WY.	WG.	
(P30−1✕P1−1)−1, etc	80	13	28	5	30	11	167
Theoretically expected	76. 0	17. 5	25. 3	5. 8	33. 8	7. 8	167
Ratio.......................	117	27	39	9	52	12	......

TABLE 11 —*Inheritance of factors A, I, and G*

Cross.	F₂ classes and frequencies				Total
	CY	CG	WY	WG	
(P30−1✕P1−1)−1, etc............................	108	18	30	11	167
(P40−2✕P30−4)−1............................	60	14	22	2	98
Grand total............................	168	32	52	13	265
Theoretically expected . .	160	36. 9	53. 3	12. 3	265
Ratio....................................	39	9	13	3	......

TABLE 12.—*Inheritance of factors (Fa) and G*

Cross.	F₂ classes and frequencies.				Total.
	YN	YF	GN	GF	
(P30−1✕P1−1)−1, etc	102	36	25	4	167
Theoretically expected....................	101. 4	33. 8	23. 4	7. 8	167
Ratio....................................	39	13	9	3	......

TABLE 13.—*Inheritance of factors G and I*

Cross.	F₂ classes and frequencies.		Total.
	Y	G	
Dom Y (GGII)×rec. Y (ggii)[a]			
Total data transferred from table 3b (1916)	457	87+22?	566
(P29-2-1×P30—A—10)—1.	78	17	95
(P29-2-1×P30—A—10)—2.	155	40	195
(P29-2-1×P30—A—10)—3.	75	19	94
(P29-2-2×P30—A—11)—1.	108	25	133
(P138-1×P30—4—1)—1.	28	10	38
(P138-2×P30—4—1)—1.	140	30	170
(P138-2×P30—4—1)—2.	72	16	88
(P71-2-2—1×P30—4—3—1)—1.	4	0	4
(P70-3-1×P30—4—3—1)—1.	11	1	12
(P70-3-1×P30—4—3—1)—2.	9	1	10
(P100-3-1×P30—4—3—1)—2.	5	0	5
Grand total.	1, 142	246+22?	1, 410
Theoretically expected	1, 145. 3	264. 3	1, 410
Ratio.	13	3	. . . .
Dom Y×G cot. (GGii)[a]			
For crosses, *etc.*, see Durbishire (1913, p. 71) and White (1916, pp. 537–538).	43, 764	14, 490	58, 254
Theoretically expected	43, 690. 5	14, 563	58. 254
Ratio.	3	1	. . . .
Rec. Y×G cot.			
(P21-15-1×P30—A—2)—1.	5	22	27
(P21-15-1×P30—A—2)—2.	8	23	31
(P70-5-4×P38—20—1P)—1.	15	49	64
(P70-5-4×P38—20—1P)—2.	11	41	52
(P70-5-4×P38—20noP)—1.	6	26	32
(P70-5-1×P38—20—1)—1.	12	31	43
(P71-2×P30—5—7)—1.	3	13	16
(P71-2×P30—5—7)—2.	3	12	15
Grand total.	63	217	280
Theoretically expected.	70	210	280
Ratio.	1	3	. . . .

[a] Factorial composition.

TABLE 14.—*Inheritance of factors (Le) and G*

Cross.	F₁ classes and frequencies.				Total.
	GT	GD	YT	YD	
(P21—15—1×P30—A—2)—1,—2................	22	6	7	1	36
Theoretically expected	20. 2	6. 7	6. 7	2. 2	36
Ratio....................................	9	3	3	1	......

TABLE 15.—*Inheritance of factors A and R*

Cross.	F₂ classes and frequencies.				Total.
	RRp	RW	WrRp	WrW	
(P40—2×P30—4)—1	55	20	19	4	98
(P41—1×P30—6)—1............	32	10	7	2	51
(P67—6×P13—3—3)—1,—2,—3.	45	21	12	9	87
(P72—1×P41—5)—1,—2....... .	185	55	75	28	343
Grand total ...	317	106	113	43	579
Theoretically expected .	324. 9	108. 3	108. 3	36. 1	579
Ratio.....................	9	3	3	1	....

TABLE 16 —*Inheritance of factors A and (Tl)*

Cross.	F₁ classes and frequencies.				Total.
	TlRp	TlW	ARp	AW	
(P72—1×P41—5)—1,—2........	197	60	61	22+1?	....
(P1—4×P39—3)—1.............	35+1?	17	5+2?	3	....
Grand total	232+1?	77	66+2?	25+1?	404
Theoretically expected .	226. 8	75. 6	75. 6	25. 2	404
Ratio.....................	9	3	3	1	....

TABLE 17 —*Inheritance of factors B and (Tl)*

Cross.	F₂ classes and frequencies.				Total.
	TlB.	AB.	Tlb.	Ab.	
(P1—4×P39—3)—1	52+1?	8+2?	14	2	79
Theoretically expected .	45	15	15	5	79
Ratio.....................	9	3	3	1	........I..

TABLE 18 —*Inheritance of factors (Fa) and (Tl)*

Cross.	F₂ classes and frequencies.				Total.
	TlN.	AN.	TlF.	AF.	
(P1−4×P39−3)−1...............	50+1?	5+1?	16	5+1?	79
Theoretically expected....	45	15	15	5	79
Ratio......................	9	3	3	1	....

TABLE 19.—*Inheritance of factors I and R*

Cross.	F₂ classes and frequencies.				Total.
	RY.	RG.	WrY.	WrG.	
Data from Bateson, Saunders, and Punnet (1905)... ... From Mendel's data (1865, p 19).	4,926 315	1,621 108	1,656 101	478 32	8,681 556
Grand total................	5,241	1,729	1,757	510	9,237
Theoretically expected	5,195	1,731	1,731	577	9,237
Ratio......................	9	3	3	1	....

TABLE (20).—*Inheritance of factors I and (Tl)*

Cross.	F₂ classes and frequencies.				Total.
	TlY.	TlG.	AY.	AG.	
(P1−4×P39−3)−1............... (P39−1×P32−1)−1............. (P72−1×F41−5)−1, −2........	45 37 191	21 6 67	8 10 64	4 4 24	78 57 346
Grand total................	273	94	82	32	481
Theoretically expected	270.2	90.2	90.2	30.2	481
Ratio......................	9	3	3	1	....

7768°—17——7

TABLE 21.—*Inheritance of factors (Le) and (Tl)*

Cross.	F₂ classes and frequencies.				Total.
	TlT.	TlD.	AT.	AD.	
(P38—1×P35—7)—1 .. (P30—5×1—P38—20—1)—1, etc..	26 25	16 9	10 3	4 3	56 40
Grand total ...	51	25	13	7	96
Theoretically expected	54	18	18	6	96
Ratio....................	9	3	3	1	....

TABLE 22 —*Inheritance of factors (Le) and R*

Cross	F₂ classes and frequencies.				Total.
	RT.	RD.	WrT.	WrD.	
(P38—1×P35—7)—1............. (p67—6×P13—3—3)—1, −2, −3	26 47	17 17	10 18+1?	3 4	56 87
Grand total................	73	34	28+1?	7	143
Theoretically expected .	80.1	26.7	26.7	8.9	143
Ratio.....................	9	3	3	1	....

TABLE 23 —*Inheritance of factors G and (Tl)*

Cross.	F₂ classes and frequencies.				Total.
	GTl.	GA.	YTl.	YA.	
(P30—5—1×P38—20—1)—1...... (P30—5—4×P38—20—1)—1P.... (P30—5—4×P38—20—1)—1.....	16 15 Not recorded, but no indication of coupling.	12 1	8 Not recorded. 11	2 3	38 16 14
Grand total ..	31	13	19	5	68
Theoretically expected	37.8	12.6	12.6	4.2	68
Ratio...................	9	3	3	1	....

TABLE 24.—*Inheritance of factors (Bl) and S*

Cross.	F₂ etc. classes and frequencies.[a]				Total.
	GlFs.	GlAs.	EFs.	EAs.	
Mummy×Chenille F₂ generation[b].	138	2	6	31	177
F₃ progeny of F₂ double heterozygote...................	57	0	2	21	80
Do.......................	70	4	4	5	83
Grand total..............	265	6	12	57	340
Calculated on 8:1[c] basis.	226	17	17	64	324
Calculated on no-coupling basis...................	189. 9	63. 3	63. 3	21. 1	338

[a] Data from Vilmorin (1913).
[b] "Mummy" has glaucous foliage and free seeds, while "Chenille" has emerald foliage and adhering cause seeds.
[c] Ratio of noncrossover to crossover gametes approximates 8:1.

TABLE 25.—*Inheritance of factors A and (Lf)*[a]

Reference to cross. [b]	F₂, F₃, and F₄ classes and frequencies.				Total
	WE.	RpE.	WL.	RpL.	
P. 277 [c].......................	157	101	35	522	815
Calculated on 7:1 basis [d]	165. 9	137. 4	36. 3	469. 2	815
Ratio.......................	4. 6	3. 8	1	12. 9	....
Calculated on no coupling.......	51	153	153	459	815
Ratio.......................	1	3	3	9	....
P. 281 [c].......................	67	16	16	255	354
P. 281 [c].......................	59	21	26	197	303
P. 283 [c].......................	387	104	121	1, 361	1, 973
Grand total..............	513	141	163	1, 813	2, 630
Calculated on 7:1 basis [d]....	505	154	154	1, 823	2, 636

[a] Early White-Flowered Dwarf "(I. P.)" ×Late French Large-Podded "(G. P.)" and reciprocal.
[b] Data and interpretation from Hoshino (1915).
[c] First part of table represents character and frequency of F₂ population. For character of other populations, see Hoshino (1915).
[d] Ratio noncrossover to crossover gametes approximates 7:1.

TABLE 26.—*Inheritance of factors R and (Tl)*[a]

Reference to cross.	F₂ ,etc. classes and frequencies.				Total.
	RTl.	RA.	WrTl.	WrA.	
RTl×WrA (Pellew)	1,466	20	15	564	2,065
(P38—1×P35—7)—1	45	1	0	13	59
Grand total	1,511	21	15	577	2,124
Calculated on 63:1	1,568.7	16.3	16.3	512	2,113.3
Calculated expectancy on no-coupling	1,188	396	396	132	2,124
WrTl×RA (Pellew)	502	270	264	0	1,036
Calculated on 63:1	516.2	257.9	257.9	.063	1,036
Calculated expectancy on no-coupling	572.3	194.1	194.1	64.7	1,036

[a] Data from Pellew (1913) and the writer's own cultures. Interpretation from Vilmorin and Bateson (1911).

TABLE 27.—*Inheritance of factors G and O*[a]

Cross.	F₂ classes and frequencies.				Total.
	GCl.	GYl.	YGl.	YYl.	
(P21—15—1×P30—A—2)—1	11	0	0	3	14
(P21—15—1×P30—A—2)—2	18	2?	1?	2	23
(P30—5—1×P38—20—1)—1	25	4?	0?	10	39
(P30—5—4×P38—20—1)—1 (670) .	29	0?	3?	11	43
(P30—5—4×P38—20—1)—1 (671) .	23	0?	2?	5	30
(P30—5—4×P38—20—1)—2	15	0?	10?	5	30
(P57—2×P30—5—7)—1	4	0	0	3	7
(P57—2×P30—5—7)—2	9	0	0	1	10
Grand total	134	6?	16?	40	196
Calculated on no-coupling	109.8	36.6	36.6	12.2	196
Ratio	9	3	3	1	. . . .

[a] The question marks refer to the classifications of cotyledon color, the tendency to fade on more or less exposure making accurate determinations in the present case difficult.   The linkage in this case may possibly be absolute, but data from other crosses involving these foliage and cotyledon color characters show undoubted cases of plants with green cotyledons and yellow foliage.   Further, the writer has a race which breeds true to these characters.

FUNGUS FAIRY RINGS IN EASTERN COLORADO AND THEIR EFFECT ON VEGETATION

H. L. SHANTZ, *Plant Physiologist*, and R. L. PIEMEISEL, *Scientific Assistant, Alkali and Drought Resistant Plant Investigations, Bureau of Plant Industry, United States Department of Agriculture*

INTRODUCTION

The present paper deals with fairy rings caused by fleshy fungi. The rings due to other causes, such as the grass rings, are not considered. The fungus rings are marked either by the fruiting bodies of the fungus or by a stimulated or a depressed growth of the natural vegetation or the cultivated crop. Fairy rings may become so abundant locally as to affect materially the crop yields of fields. On lawns they cause unsightly bare spots and dark-green areas, and in small experimental plots cause either a total loss of crop or a greatly increased yield. They are undesirable in all cases, and their eradication is a matter of practical importance. The studies herein recorded were made on the High Plains at Akron, Colo., during the period 1907 to 1916, inclusive. The soil studies were made during the summers of 1914, 1915, and 1916. The native vegetation at Akron has already been described in some detail (Shantz, 1911).[1] It is typical short grass, composed largely of *Bouteloua gracilis* (*oligostachya*) and *Bulbilis dactyloides*.

The term "fairy ring," generally used to describe the arrangement of plants in an approximately circular form, originated in the belief that these circular growths marked the paths of dancing fairies. Early literature is filled with tales of superstition concerning these rings. The superstitions varied somewhat with the different countries. In Holland these circles often marked the places where the devil churned his butter. The presence of such a ring on a farm caused an inferior quality of butter if the cows ate the grass from a fairy ring. In France many people could not be induced to enter one of these rings, because enormous toads with bulging eyes abounded there; but no harm was experienced if the rings were unintentionally entered at night. In Sweden a person entering a fairy ring passed entirely under the control of the fairies. Treasures were marked by such rings in many places, but these riches could not be secured without the help of fairies or witches. In England it was regarded

---

[1] Bibliographic citations in parentheses refer to "Literature cited," p. 242-245.

Journal of Agricultural Research,
Washington, D. C.
ld

Vol. XI, No. 5
Oct. 29, 1917
Key No. G—124

(191)

eksenkringe.	Cercles mycogènes.	Hexent
olringe.	Cercles de sorcières.	Zauberr
ooverkringe.	*Danses de fées.*	Pilzring
huivels Karnpad.	Ronds de fées.	
uivelstjeinpad.	Ronds de sorcière.	
jenmolenpad.		Elfdans

CAUSE OF FAIRY RINGS

s indicated in the introduction, the cause of the f
,cure for a long time. At first these rings were ascri
ernatural influence. Later they were attributed to
ses.

radley (1717, pt. 3, p. 122–123) gave two probable c
s:

) Just under the turf where the mushrooms were growing, wa
e by pismires, which was not only hollow in many places at
covering of that passage was made of earth extremely fine whi
flung up: The fineness of the earth wrought by those laboriou
reasonably contribute to the extraordinary vegetation of the gr
d the hollowness of the ground underneath might produce that
hich afterwards might be formed into mushrooms.
) Garden-snails . . ., when they couple, always make choice
p upon . . . leaving upon the grass where they crept, a viscou
at it may be, that slime when it putrifies may produce the m
wing in circles upon Commons.

t is interesting to note in connection with Bradley's ob‹
.s have often been found working in the marasmius
ishington, D. C. They are probably attracted by the

I know that similar circles have been observed by naturalists, and by them ascribed to thunder; as we should certainly have done in this case, were it not for the regular annual progression, which, if the effect of thunder, must follow rules not yet investigated, either in electricity, vegetation, or the mineral system.

He examined the soil and found no clue. At times he also noticed the presence of fungi in the ring, but did not consider them significant.

Withering (1796, p. 222) definitely assigned the cause of fairy rings to *Agaricus oreades*.

I am satisfied that the bare and brown, or highly cloathed and verdant circles, in pasture fields, called *Fairy Rings*, are caused by the growth of this Agaric. . . . Where the ring is brown and almost bare, upon digging up the soil to the depth of about 2 inches, the spawn of the Fungus will be found, of a greyish white colour, but where the grass has again grown green and rank, I never found any of the spawn existing.

FUNGI CAUSING FAIRY RINGS

The tendency of all fungi to grow outward from the point of germination of the spore results in circular colonies in a widely varying group of fungi. Rings are often noted among the molds. This paper deals only with rings formed by Basidiomycetes.

The first fairy rings noted which were definitely assigned to fungi were caused by *Agaricus oreades*. Following this report many different species were associated with fairy rings.

Eleven years later Wollaston (1807) extended Withering's observations and reported rings formed by *Agaricus campestris, A. terreus, A. procerus,* and *Lycoperdon bovista*. Most of the writers following Wollaston have substantiated his results with the exception of the following: Persoon (1819, p. 4–5) called attention to the singular distribution of mushrooms in fairy rings, but stated that the cause was not well known. Lees (1869) attributed the cause of these rings to the action of moles. The burrows of these animals were supposed to be marked by the dead area in the fairy ring. Buckman (1870) believed that the fungi appeared as the result of the stimulation and death of the vegetation, and were not therefore the primary cause, although the ring was often continued and extended by them; in fact, he thought that rings to which fungi had not become attached soon broke up and disappeared. Gillet (1874, p. 22) states that it is difficult to explain the cause of the grouping of some species of mushrooms in fairy rings. Although it is understood that most of the writers thought that one species of fungus alone was the cause of a particular fairy ring, only a few writers stated so definitely. Greville (1828, p. 323) said that he never detected more than one species of fungus in the same ring. This statement was also made by McAlpine (1898). Williams (1901, p. 207) stated that fairy rings

showed a similar disposition to encourage the more luxuriant growth of other species of fungi that occurred elsewhere in the surrounding field

In eastern Colorado the fairy rings were never known to have more than one species in the fruiting zone, although a ring of *Calvatia cyathiformis* was seen in fruit that had smaller rings of *Agaricus campestris* also in fruit, in the inside (fig. 9).

Do....................	Psalliota campestris.......	
Agaricus sp................	Agaricus sp.............	
Amanita muscaria L.........	Amanita muscaria.........	Lu
Amanita phalloides Fr......	Amanita phalloides.......	
Boletopsis cavipes (Opat.) Henn.	Boletus cavipes..........	
Boletus bovinus L..........	Boletus bovinus..........	Ba
Boletus elegans Schum......	Boletus elegans..........	Lu
Boletus variegatus Swartz...	Boletus variegatus........	
Calvatia cyathiformis Bosc....	.....................:	Sh
Calvatia fragilis Vitt........		
Calvatia polygonia Lloyd....		
Cantharellus aurantiacus (Wulf.) Fr.	Cantharellus aurantiacus...	Lu
Cantharellus cibarius Fr.....	Cantharellus cibarius......	Le
Cantharellus cinerus Fr......	Cantharellus cinereus......	Lu
Catastoma subterraneum (Pk.) Morg.	....	Sh
Clavaria sp...............	Clavaria sp..............	Mü
Clitocybe geotropa Fr.......	Agaricus geotropus Bull...	Le
Do................	Agaricus (Clitocybe) geotropus Bull.	
Clitocybe gigantea Sow......	Clitocybe gigantea.........	Ba
Do................	Agaricus giganteus........	Jor
Do................	Agaricus (Clitocybe) giganteus Sow.	Le
Clitocybe infundibuliformis Fr.	Agaricus infundibuliformis Sch.	
Do................	Agaricus (Clitocybe) infundibuliformis Sch.	

[1] In the systematic portion of this work the writers are indebted to Mrs. Fl in charge of pathological collections, for many suggestions; to Miss Charles fo to the list of fungi causing fairy rings; to Prof. W. G. Farlow, of Harvard tion of *Agaricus tabularis*, and to Prof. C. G. Lloyd, of Cincinnati, Ohio (of *Calvatia polygonia*.

TABLE I —*List of fungi reported as causes of fairy rings and the names of the investigators*—Continued

IDENTIFIED FUNGI—continued

Species	Name used by author	Investigators
Clitocybe maxima (Gärtn and Meyer) Quel.	*Agaricus maximus* Gärtn and Meyer)	Münch (1914).
Do...................	*Agaricus (Clitocybe) maximus.*	Do.
Clitocybe nebularis Batsch..	*Clitocybe nebularis* Batsch	Stahl (1900).
Collybia confluens Fr.	*Agaricus confluens* Pers	Lees (1869).
Do	*Agaricus (Collybia) confluens* Pers.	Do.
Collybia sp.................	*Collybia* sp..............	Münch (1914).
Cortinarius amethystinus (Sch.) Sacc.	*Inoloma traganum*........	Ludwig (1906).
Cortinarius armillatus (Alb and Schw.) Fr	*Telamonia armillata* ...	Do.
Cortinarius traganus Fr .	*Inoloma traganum*	Do.
Hebeloma crustuliniforme Fr	*Agaricus crustuliniformis* Bull.	Lees (1869).
Do...................	*Agaricus (Hebeloma) crustuliniformis* Bull.	Do.
Hydnum compactum Pers....	*Hydnum compactum*	Ludwig (1906).
Hydnum repandum L........	*Hydnum repandum* L	Lees (1869); Ballion (1906)
Hydnum suaveolens Scop....	*Hydnum suaveolens* Scop	Thomas (1905); Coulter, Barnes, and Cowles (1911).
Hygrophorus virgineus (Wolf.) Fr.	*Hygrophorus virgineus* Fr.	Lees (1869).
Inocybe sp.................	*Inocybe* sp	Williams (1897).
Lactarius insulsus Fr.......	*Lactarius insulsus*	Ludwig (1906).
Lactarius piperatus (Scop.) Fr.	*Lactarius piperatus* Fr	Lees (1869).
Lactarius torminosus (Schaeff) Fr.	*Lactarius torminosus*	Ludwig (1906).
Lepiota morgani Peck	*Lepiota morgani*.........	Williams (1901), Shantz and Piemeisel.
Lepiota procera (Scop.) Quel.	*Agaricus procerus*.........	Wollaston (1807).
Lycoperdon bovista L..	*Lycoperdon bovista*	Do
Lycoperdon gemmatum Batsch.	*Lycoperdon pratense* Pers	Ballion (1906)
Lycoperdon cyclicum McAlpine.	*Lycoperdon cyclicum*, n. sp	McAlpine (1898)
Lycoperdon wrightii B and C	....	Shantz and Piemeisel.
Marasmius oreades (Fr.) Bolt.	*Marasmius oreades*........	Lawes, Gilbert and Warrington (1883), Williams (1897); Coville (1898); Massart (1910), Bayliss (1911).
Do.................	*Marasmius oreades* Fr....	Molliard (1910).
Do.................	*Marasmius oreades* Bolt....	Ballion (1906).
Do.................	*Agaricus oreades*.........	Withering (1796).
Do....~Marasmius~...	~~caryophylleus~~ (Sch.) Schroet.	Reed (1910).
Do.................	*Agaricus oreades*	Wollaston (1807); Jorden (1862), Buckman (1870), Van Tieghem (1884); Ritzema Bos(1901); Ludwig (1906);Lees (1869).
Marasmius urens Fr.......	*Marasmius urens* Fr.......	Lees (1869).
Morchella esculenta (L) Pers.	*Morchella esculenta* Bull.	Ballion (1906).
Morchella hybrida Pers......	*Mitrophora semilibera* DC	Do.

TABLE I.—*List of fungi reported as causes of fairy rings and names of the investigators*—continued

IDENTIFIED FUNGI—continued

Species.	Name used by author.	Investigators.
Paxillus involutus (Batsch.) Fr	*Paxillus involutus*	Ludwig (1906).
Pluteus cervinus Schaeff	....	Charles.
Tricholoma columbetta Fr ...	*Tricholoma columbella*	Massart (1910).
Tricholoma equestre L ...	....	Charles.
Tricholoma grammopodium Fr.	*Agaricus grammopodius* Bull.	Lees (1869).
Do.	*A g a r i c u s* (*Tricholoma*) *grammopodius* Bull	Do.
Tricholoma melaleuca (Pers.) Quel.	....	Shantz and Piemeisel.
Tricholoma personatum (Fr.) Quel.	*Agaricus bicolor*..........	Münch (1914).
	Agaricus personatus.......	Buckman (1870).
Do	*Agaricus* (*Tricoloma*) *personatus* Fr.	Lees (1869).
Do...................		
Tricholoma praemagnum ..	*Tricholoma praemagnum.*	Ramaley (1916).
Tricholoma sp	*Tricholoma* sp . .	Williams (1901).
Tricholoma terreum Schaeff	*Agaricus terreus.* . .	Wollaston (1807).
Tuber sp	*Tuber* sp.	Tulasne and Tulasne (1851).

FUNGI OF DOUBTFUL IDENTITY

Tricholoma graveolens (Pers) Quel.	*Agaricus graveolens*	Way (1847).
Tricholoma gambosum Fr	. do	Do
Hydnum cyathiforme Schaeff	*Hydnum tomentosum*	Ludwig (1906).
Hydnum candicans Fr ..	......do.................	Do.
Caldesiella ferruginosa Sacc..	do	Do
	Agaricus multifidus	Jorden (1862), Ritzema Bos (1901).

TYPES OF FAIRY RINGS

The present paper is concerned only with fairy rings produced as a result of the growth of fleshy fungi. When the fungi are in fruit, these rings are easily distinguished by the more or less regular arrangement of the fruiting bodies (Pl. 13, A). At other times the ring is easily distinguished by the appearance of the natural vegetation (Pl. 13, B). The differences in appearance of the vegetation in the ring as compared with that outside consist principally in a deeper color, due largely to greater chlorophyll content of the plants and in a more luxuriant growth, and in certain cases in a zone of bare ground or dead vegetation near the outer edge of the ring.

Before taking up a detailed description of the various types of rings found in eastern Colorado it seems desirable to review briefly the descriptions already published.

Hutton (1790) described a ring as consisting of three tracks, of which the first was a zone of dead or withered grass; the second lay just within and ran parallel to the first and appeared as a black zone of rotten grass; and the third, a dark-green zone of grass occurred partly on the black zone of rotten grass, but mostly on a smaller zone which had been formed the year preceding.

Practically every writer on this subject has included in his account a description of the rings. Few of these added materially to the description given by Hutton.

Westerhoff in 1859 (*in* Ritzema Bos, 1901) distinguished fairy rings formed by fungi and those formed by other causes. The fungus rings formed more or less complete circles, extending externally every year. They consisted of a zone of luxuriant grass of a dark-green color, surrounded by a circle of mushrooms (about August) and this in turn surrounded by a circle of withered grass. The inside of the ring appeared the same as the outside.

Jorden (1862) called attention to the green or brown rings and irregular dead patches, which were known as fairy rings.

Buckman (1870) and Kuperus (1876) described rings as formed of two bands, the inner one of fresh, green, luxuriant grass, and the outer

FIG. 1.—A sketch of a fairy ring produced by *Marasmius oreades*. The center of the circle is represented by O. The vegetation in the central part of the circle is represented at 1, the inner stimulated zone at 2, a bare zone at 3; and the outer stimulated zone at 4. The normal vegetation is shown at 5. The mycelium is represented by the dotted portion. The dotted line passing from zone 2 to 4 indicates the relative plant growth during the early spring when water is abundant (from Molliard, 1920).

one of more or less brown herbage or bare soil. Buckman stated further that the fungi appeared in this outer ring. He also described rings which consisted of only the fresh green ring and often produced no fungi.

Rings consisting of an outer dark-green zone, an intermediate yellow zone in which the grass was dead, and an inner green zone were described by Van Tieghem (1884, p. 1044–1045). In certain years the outer green band showed a large number of fungus fruits.

Sorauer (1886, p. 270–272) stated that the rings usually consisted of a green ring, and that it is only during an exceptional year that fungus fruiting bodies occur, while Stahl (1900, p. 666–667) described a ring distinguished by tall and robust plants of *Geranium robertianum*

Complete rings so bare of grass as to resemble footpaths, formed by *Lycoperdon cyclicum* on a bowling green, were described by McAlpine (1898).

Ballion (1906) gave detailed descriptions of rings formed by *Marasmius oreades, Tricholoma georgii, Psalliota arvensis, Lycoperdon pratense.* These were grouped on the basis of (1) the place in the ring occupied by the fruiting bodies; (2) whether the vegetation was killed.

A ring of *Marasmius oreades* in the month of September was described by Molliard (1910, p. 63, fig. 1) as follows: There were three distinct zones: Beginning at the center (o, fig. 1) the first zone was an internal one (2) where the phanerogamic vegetation contrasted sharply with the normal; the second zone (3) a middle one where the herbage was withered and where the fruits of *M. oreades* occurred; and finally an external zone (4) larger than (2) the internal zone, where the vegetation was greener and taller than the inside (1) or the outside (5).

Münch (1914) described and illustrated a ring formed by *Agaricus (Clitocybe) maximus* which had a stimulated zone outside as well as inside a well-defined dead zone. It differed from the rings described by other investigators in that the dead zone was lined with fruiting bodies both on the exterior and interior sides.

Fairy rings caused by fleshy fungi may be divided into types on the basis of their effect on the vegetation. It is apparent from a review of the literature that the effect of the various fungi on the vegetation varies greatly in different locations as the probable result of the different climatic and weather conditions.

In eastern Colorado the following types of fairy rings may be distinguished: (1) Those in which the vegetation is killed or badly damaged, caused by *Agaricus tabularis;* (2) those in which the vegetation is only stimulated, caused usually by species of Calvatia, Catastoma, Lycoperdon, marasmius, etc.; and (3) those in which no effect can be noted in the native vegetation, caused by *Lepiota* spp.

RINGS MARKED BY A ZONE OF DEAD VEGETATION

RINGS FORMED BY AGARICUS TABULARIS

The rings formed by *Agaricus tabularis* Peck vary in size from a few meters to 70 meters across. These rings are often complete and circular, the advance in all directions having been nearly uniform (Pl. 13, A; 18; 20; 30, A). More often, however, the ring is broken at certain points, and the larger rings are usually formed by a series of arcs which do not come into contact with each other but which show approximately equal radii (fig. 2).

In figure 3 is given a diagram and in figure 4 a bisect of a typical ring formed by *Agaricus tabularis*. At A the zones are shown as they are distinguished during a period favorable for the growth of the fungus and the production of fruiting bodies. (Pl. 13, A; 14, A; 19.) The ring consists of a series of three zones surrounding an area of normal short grass sod, which constitutes the inside of the ring (1). Next to this area occurs a broad zone differing from the natural sod in botanical composition, in the more luxuriant growth, and in the deeper green color of the vegetation. This is the inner stimulated zone (2). This wide green zone is the most prominent part of the rings in spring or in wet seasons. The bare zone (3) is not as broad as the inner stimulated zone and is somewhat more irregular. In this zone the vegetation is often entirely dead, but in many

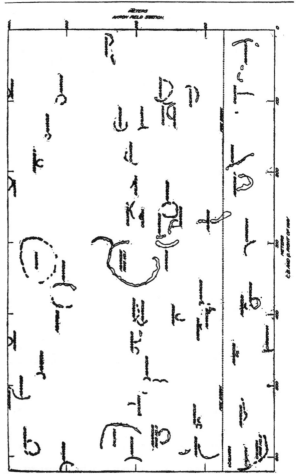

Fig. 1.—A map of an area of 400 by 600 meters, lying just west of the experiment station and north of the Chicago, Burlington & Quincy R. R. near Akron, Colo. The dotted line represents a fire break plowed each fall and of about 2 meters' width. This area includes a total of 62 rings or fragments of rings: 35 are *Agaricus tabularis*, 14 *Calvatia cyathiformis*, 3 *Catastoma subterraneum*, and 10 are unidentified rings. No rings of *Agaricus campestris* were noted in this section, although many grow at other places on the plains. From 0.5 to 1 per cent of the total land area here shown lies within the zone of influence of these fairy rings. The fruiting bodies at the time the map was made are represented by dots.

cases a few very poor perennials or short-lived annuals may be found. This zone is the distinguishing feature late in summer and fall or in dry seasons. If rains are frequent, scattered annuals succeed fairly well even on this area. Beyond the bare zone there occurs a rather narrow zone, the outer stimulated zone (4), resembling somewhat the inner stimulated zone, but being for the most part made up entirely of plants which characterize the native sod. The sporophores occur in this zone, near the outer edge. Outside this zone normal native short grass is found (5).

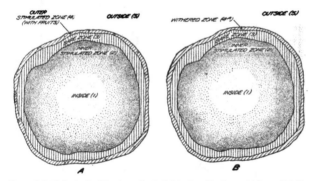

Fig. 3.—A sketch showing the different zones in a typical ring formed by *Agaricus tabularis*. At A the appearance of the ring is shown during a period of ample moisture supply. At B the appearance during a period of deficient moisture supply. The vegetation on the inside (1) is normal. In the inner stimulated zone (2) the greatest stimulation occurs near the bare zone. The bare zone (3) is usually devoid of vegetation, or contains only scattered plants. The outer stimulated zone (4) contains the fruiting bodies of the fungus and differs from the vegetation outside only in color and in more luxuriant growth; outside, the vegetation is normal (5). In B the same zones are shown as in A, except the outer stimulated zone which is here shown as the withered zone (4a) In this zone the plants not only wither but with continued droughts will die, and the area of this zone be added the following year to that of the bare zone.

At B the zones are shown as they appear during a period not favorable to fungus growth. During the late summer, when the moisture supply is deficient, or during a dry year following a wet year (Pl. 16) the ring presents a very different appearance from that described above. The inner stimulated zone (2) ripens or dries up as dry weather comes on and usually presents a brown, dead appearance, although the perennials remain alive in a dormant condition. If the season is dry, this zone (2), while noticeably different from the natural sod inside (1), does not show the luxuriant growth so characteristic of a wet year. During a dry year the bare zone (3) is unusually prominent, since not even the short-lived annuals appear, and the zone is named from its appearance during drouth periods. Such periods are characteristic of eastern Colorado. The withered zone (4a) is characterized by the dry vegetation, which during the more favorable period had marked the outer stimulated zone (4). During a dry year the short grass in this area is withered and often perfectly air-dry when the adjacent sod is still green.

RINGS MARKED ONLY BY A ZONE OF STIMULATED VEGETATION

Most of the fungi forming fairy rings in eastern Colorado produce only a temporary stimulating effect on the vegetation. In this group, therefore, the presence of fungi is indicated by an increase in the size, vigor, and chlorophyll content of the annuals and of the perennial grasses.

RINGS FORMED BY CALVATIA CYATHIFORMIS

A large number of the rings marked only by a zone of stimulated vegetation are produced by *Calvatia cyathiformis* Bosc (fig. 5, 6). They are usually much less conspicuous than those formed by *Agaricus tabularis*.

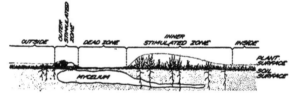

Fig. 4.—A bisect of the *Agaricus tabularis* ring shown in figure 3 at A. The vegetation on the inside and outside does not differ noticeably. The outer stimulated zone in which the fungus fruits are produced is separated from the inner stimulated zone by a bare zone in which plants are only occasionally found. The distribution of the mycelium in the soil is also indicated in the illustration.

In size they are often much larger. Several rings have been observed which exceed 200 meters in diameter (Pl. 24,B). In periods when the rings are not marked by fruiting bodies these rings can be distinguished from Agaricus by the sterile bases of the fruits of *Calvatia cyathiformis* which remain on the ground from one year to another, or by the natural vegetation which usually presents a stimulated appearance, but seldom, if ever, is damaged by the presence of the fungus. The annuals in the rings, which grow much taller than in the adjacent areas, are prominent both during their period of rapid growth and after they have ripened, at which time the rings appear as dark-yellow or brown circles on a uniformly light-green short-grass cover.

A sketch of a ring formed by this fungus is shown in figure 5. The first crop of fruiting bodies occurs at the outer edge of the stimulated zone. The vegetation in this zone consists of the same species as the native sod, but the growth is more luxuriant (Pl. 25,A). This is especially noticeable in the short-lived annual plants which stand up prominently above those in the surrounding sod   If the season is favorable and a second crop of puffballs is produced, they appear outside in advance of the stimulated zone (Pl. 26, A).

RINGS FORMED BY CALVATIA POLYGONIA

The giant *Calvatia polygonia* Lloyd is somewhat less abundant on the Great Plains than the one just discussed. It occurs frequently in eastern Colorado and forms large rings not differing essentially from those described for *Calvatia cyathiformis*. The rings formed by this fungus are usually intermediate in size, one ring having been noted with a diameter of 100 meters (Pl. 28, B).

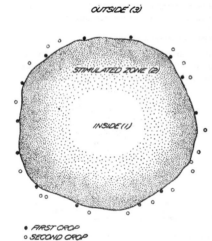

RINGS FORMED BY CATASTOMA SUBTERRANEUM

The interesting fungus *Catastoma subterraneum* (Pk.) Morg., the fruiting bodies of which develop underground and do not appear on the surface until after they have broken loose from their attachment, forms noticeable rings indicated

Fig. 5 —Sketch of a ring produced by *Calvatia cyathiformis*. The stimulated zone is most marked at the outer edge just inside the ring of fresh puffballs. If a second crop of puffballs are produced, they occur at the points marked o, In this ring there is a stimulated zone (2) lying between the normal vegetation inside (1) and the fruiting zone, which occurs at the inner edge of the normal vegetation outside (3).

by a stimulation of the native plant cover. The rings thus far noted by the writers have not been complete (Pl. 30, B). Fruiting bodies are formed near the convex side of the stimulated area. The effect on the native vegetation is the same as in the case of *Calvatia cyathiformis*, but the stimulated zone is even narrower.

RINGS FORMED BY OTHER FUNGI

Lycoperdon gemmatum and *L. wrightii*, which occur more commonly on cultivated and disturbed land, indicate little effect in stimulating the vegetation. *Calvatia fragilis* Vitt, *Marasmius oreades* (Fr.) Bolt, *Agaricus campestris* L. (Pl. 28, A), and *Tricholoma melaleuca* (Pers.) Quel. occur

rather commonly in the short-grass vegetation and form rings marked
by a stimulated area or by the occurrence of the fruiting bodies. The
rings are usually small, seldom exceeding 6 to 10 meters.

RINGS WHICH PRODUCE NO NOTICEABLE EFFECT ON THE NATURAL
VEGETATION

A single ring formed by *Lepiota morgani* Pk. was noted in the bunch-
grass vegetation near Yuma, Colo. (Pl. 26,B; 27,C). The ring, approximately 24 meters in diameter, was composed of about 63 fruits (fig. 7).
No effect could be noted on the vegetation. It is possible, however, that
if this fungus had developed on the hard land in the short-grass cover,
the effect would have been more noticeable.

Fig. 6.—A bisect of a ring sketched in figure 5. In the case of *Calvatia cyathiformis* the distribution of the
mycelium can scarcely be detected in the soil. The stimulated zone lies just inside the zone of fruiting
bodies.

DISTRIBUTION OF FAIRY RINGS

Fairy rings occur for the most part in grass lands, but have been
recorded in many cases in forests. Most of the investigators refer to
them as having been formed in grassy areas and marked by the more
luxuriant growth of grasses. The following authors discuss rings in this
type of vegetation: Bradley (1717, p. 122–123) Hutton (1790), Wollaston
(1807), Dutrochet (1834, 1837, p. 179–181), De Candolle (*in* Way, 1847),
Way (1847), Westerhoff in 1859 (*in* Ritzema Bos, 1901), Berkeley (1860),
p 41), Jorden (1862), Cooke (1866), Lees (1869), Buckman (1870), Gilbert (1875), Kuperus (1876), Lawes, Gilbert, and Warington (1883),
Van Tieghem (1884, p. 1044–1045), Sorauer (1886, p. 270–272), Treichel
(1889), Olivier (1891), McAlpine (1898), Coville (1897, 1898), Atkinson
(1900), Williams (1901), Ritzema Bos (1901), Beille (1904, p. 380–381),
Baillion (1906), Molliard (1910), Massart (1910), Bayliss (1911), Münch
(1914), and Ramaley (1916).

The following dealt with fairy rings in forests or about trees: Tulasne
and Tulasne (1851, p. 157–158), Stahl (1900, p. 666–667), Thomas (1905),
Ludwig (1906), Reed (1910), and Coulter, Barnes, and Cowles (1911,
p. 807).

Dutrochet (1837, p. 179–181) called attention to the fact that rings
develop most commonly on prairies that are not very fertile and where

the grass is short and yellow. Gilbert (1875) stated that it is known that fairy rings occur chiefly, though not exclusively, on poor pastures, and that they are discouraged by especially high nitrogenous manuring. Ritzema Bos (1901) stated that the best fairy rings are produced during rather dry seasons, and that mushrooms usually occur in meadows where the organic matter is not abundant.

On the Great Plains fairy rings have been noted in all sections from Texas to Montana. Their abundance in certain areas is illustrated in figure 2. This map covers an area of three-fourths of a mile by one-half a mile of unbroken native sod lying just west of the Akron experiment farm, near Akron, Colo. A total of 62 rings or fragments of rings is shown on this map. From 0.5 to 1 per cent of the total soil area lies within the zone of influence of these rings. This area contains 35 Agaricus rings, 14 Calvatia rings, 3 Catastoma rings, and 10 unidentified rings. Throughout the whole of eastern Colorado, especially on the "hard" land (characterized by the pure short-grass cover), fairy rings occur often in great abundance. Many other areas might have been chosen which would show an equal or possibly even greater number of rings.

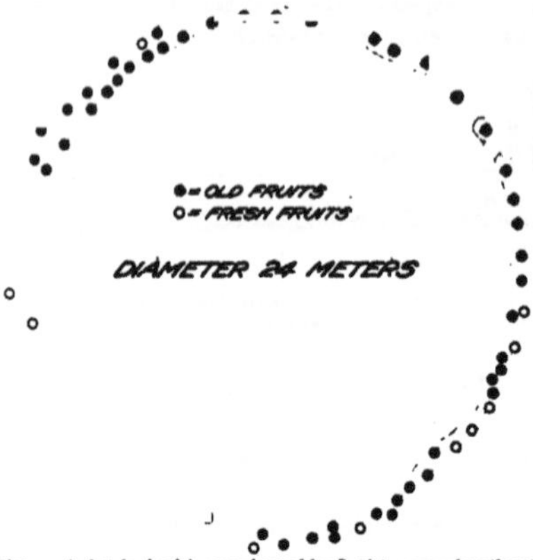

FIG. 7.—A sketch of a fairy ring formed by *Lepiota morgani* southeast of Yuma, Colo. No effect could be noted on the vegetation.

About one-half of the area shown in this map was plowed in the fall of 1915 and seeded to Turkey wheat. During 1916, an exceptionally dry year, the area was remapped. The results of this remapping are shown in figure 8.

CAUSE OF ADVANCE

Hutton (1790) noted the regular annual progression of fairy rings, but Wollaston (1807) was the first to discuss the cause of this progression. He came to the conclusion that the ring was formed by a progressive increase from a central point due to the exhaustion in the central area of some particular "pabulum" necessary for the further growth of the fungus; hence, the new growth of the fungus "roots" extended solely in the opposite direction—that is, outward. He confirmed his theory by the observation that, where two circles met, both were obliterated at the point of contact, and said:

The exhaustion occasioned by each obstructs the progress of the other, and both are starved.

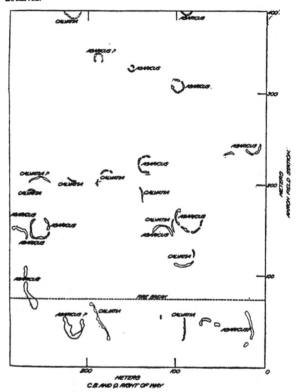

Fig. 8.—Remapped area shown in figure 2. During the fall of 1915 a portion of the area mapped in figure 2 was plowed and planted to fall wheat. During June, 1916, the area was remapped. This area is 400 by 290 meters and comprised the portion lying next to the experiment station grounds. A comparison with figure 2 will show that none of the Calvatia rings could be detected in the wheat crop the following year. Of 12 Agaricus rings mapped in 1915, 9 were easily detected in the wheat field in 1916

Way (1847, p. 43) thought that—

The theory of de Candolle, that these rings increase by the excretions of these fungi, being favorable for the growth of grass but injurious to their own subsequent development on the same spot, was insufficient to explain the phenomena.

tinually outward.

Reed (1910) stated that the tannin of the bark of
marking the center of a fairy ring had caused the deat
the downhill side. But Coville (1898) found the rings
on the downhill side, which suggests the harmful influen
position substances which are washed toward the lower

Investigators differ as to the exact cause of this outw
attributing it to the exhaustion of the nutrient mate
others attributing the cause partially to the fact that the
occurs only on the outside of the ring, and that th
mycelium on the inside is already old or dying. Th
grows outward from a central point and continues to
no more strange than that the horizontal roots grow o
and keep on growing out instead of back toward the
that the filaments grow into virgin soil by growing outw
the growth in that direction.

Where fungus fairy rings occur, the soil is usually rath
matter. The growth of the saprophytic fungus probab
available supply of organic matter. In the case of the
ring this is replenished by the death of the grass cover a
quent increase in growth of annuals and short-lived pere
and leave their root systems as available organic matter.
of organic matter do not therefore show a marked diff
inside the fairy ring. The active fungus filaments ar
outer edge, and a turning back into the central area woul
through from 1 to 5 meters of depleted soil or a contin
condition for a period of several years in the same spot
addition to these reasons why progression should be o
tions are numerous of the tendency of fungus filaments

(Fulton, 1906). This may be due to chemotropic stimulation or to the more obvious fact that food material or water is usually more abundant in the new soil. The high content of ammonia and nitrates in the soil

DIAMETER OF CALVATIA RING 65 METERS

○ FRESH CALVATIA CYATHIFORME FRUITS
● DRY CALVATIA CYATHIFORME FRUITS
× DRY AGARICUS CAMPESTRIS FRUITS

Fig. 5.—A ring formed by *Calvatia cyathiformis*, almost complete and 65 meters in diameter. The ring contained 90 fresh fruiting bodies and had apparently produced an earlier crop of 54 puffballs. This ring occurred in an area in which a great number of small rings produced by *Agaricus campestris* were found. These rings seem to have interrupted the Calvatia rings at all points except on the upper right-hand side. At no place is there evidence that the *C. cyathiformis* displaced *A. campestris*, although a possible condition of this kind is indicated on the upper right-hand portion of the figure. This ring was mapped northeast of Yuma, Colo., on June 29, 1916. It should be noted that, although Calvatia and Agaricus fruited abundantly in the region about Yuma and on the Wray Divide in 1916, no fruiting bodies were produced at Akron. The records in Table II are for the vicinity of Akron and are not general for the whole eastern portion of the State.

occupied by the older portion of the mycelium may also play a part. That rings do not continue when they come in contact with each other has been noted repeatedly. The inside of large rings of one species is often occupied with many smaller rings of another species (fig. 9), and in one case a ring of *Calvatia cyathiformis* had five large fruiting bodies in the area inside the ring. Two of these fruiting bodies were near the center of the ring, while three were situated about midway between the center and the stimulated zone. But the period of time which had elapsed since the soil mass was infected by the fairy-ring fungus may have been sufficient for the organic matter and the other conditions of the soil to have again become normal.

EFFECT OF SOIL AND WEATHER CONDITIONS ON THE PRODUCTION OF FRUITING BODIES

The production of fruiting bodies seems to be dependent largely on the soil and weather conditions. Jorden (1862) called attention to the influence of weather conditions, especially warmth and moisture, on the production of fruiting bodies in fairy rings. One ring of *Agaricus giganteus* fruited only once during a period of 50 years' observation.

In Table II are shown the dates on which the fungi fruited abundantly in the vicinity of Akron, Colo., during the years 1907–1916, inclusive. These records, taken from field notes, are probably complete, except for the occasional production of fruiting bodies in a small area under abnormally favorable conditions caused by local showers or drainage water.[1]

TABLE II.—*Dates on which fungi have been noted in fruit in the fairy rings at Akron, Colo.*

Year.	Date of occurrence of fruiting bodies.		
	Agaricus tabularis.	*Calvatia cyathiformis.*	*Calvatia polygonia.*
1907..........	June 21.............	July 26............	July 20.
1908..........			
1909..........	June 3–19........	{June 12–19......... {July 13–17........	July 6. August 10.
1910..........		August............	August.
1911..........			
1912..........		June..............	September 13.
1913..........			
1914..........		June 21............	
1915..........	{June 5–24......... {August 16–20.......	June 14–24........ August 16–20.......	}June 21.

Calvatia cyathiformis was noted in six of the nine years, *Calvatia polygonia* in five, and *Agaricus tabularis* in but three. *A. tabularis* fruited earlier in the season than the other fungi. During 1909 and 1915, when conditions were most favorable for the growth of these fungi, *A. tabularis* fruiting bodies appeared early in June, followed about 10 days later by fruiting bodies of *Calvatia cyathiformis*. A second and very light crop

[1] In 1911 the writers were absent during the latter part of May and the greater part of June, and no records were secured until June 19, after which date fungi were not noted. Mr. Robert D. Rands, who assisted in the alkali and drought resistant plant investigations at Akron, Colo., photographed a ring of *Calvatia cyathiformis* early in June. It seems improbable that *Agaricus* fruited during this year, since only a few *Calvatia* fruits and no fruits of *Agaricus* were noted by Mr. Rands.

of each of these fungi occurred on August 16, 1915, following a rain on August 14 of 0.43 inch.   Similarly *C. cyathiformis* produced on July 13, 1909, a second crop following a series of heavy rains (2.80 inches) on July 7–9.

Calvatia polygonia fruited later in the season.   The earliest record of its occurrence was June 21, 1915.   In 1909 it was found in fruit on July 6 and again on August 10.   It did not fruit as freely as did *Calvatia cyathiformis*, and often the rings produced only one or two fruiting bodies at a time.

In Table III is shown a summary of the weather conditions [1] for each of the years here considered, with the exception of 1907, during which year no determinations are available.   The data are summarized for the months of May, June, July, and August.

Soil moisture was the chief factor in controlling the fruiting of these fungi.   If during May and June the soil of the first and second foot was moist throughout, these fungi fruited rather abundantly.   If the soil was dry even during a part of this period, the fruiting bodies were not produced.   These data are not taken from moisture determinations in the fungus ring but from determinations in the typical short-grass sod.   As will be shown later, the moisture conditions in the Agaricus rings are not as favorable as those shown in this table.

In 1909 and 1915 the soil of the first and second foot was moist during the month of June, and the fungi fruited abundantly throughout this period.   In 1912 soil moisture conditions seem to have been favorable during June.   The failure of *Agaricus tabularis* to fruit at this time may have been due to the very extreme conditions of drouth during the preceding season.   In 1914 *Calvatia cyathiformis* fruited sparingly but *A. tabularis* produced no fruiting bodies.   The impervious nature of the soil in Agaricus rings probably explains the absence of fruiting bodies of this fungus in 1912 and 1914, following the dry years 1911 and 1913.   On the basis of soil moisture, 1909 and 1915 were the most favorable years.

The rainfall was heavier in 1909 and 1915 than in any other year.   The combined rainfall in May and June, the period during which it would be most important as a factor in fungus growth, was especially low in 1910, 1911, 1913, and 1916, the years during which fungi did not fruit.

The number of rainy days gives a better idea of the general conditions of humidity than does the amount of rainfall.   Rainy days were more numerous during May, June, July, and August in 1915 than during any other year, including 1909, when a good crop was produced.   In the period for 1915 more than half the days were rainy.   Evaporation was lowest in 1915.   A comparison of the evaporation rate in June, 1915, with that of the same month in 1909 shows conditions to have been

[1] For the weather data here recorded the writers are indebted to Dr L. J. Briggs, Biophysicist in Charge of the Office of Biophysical Investigations, Bureau of Plant Industry.

much more extreme in 1909 than in 1915. The years 1908, 1911, 1913, and 1916 were the most extreme, to judge from the evaporation rates. It was during these years that no fruiting bodies were observed.

TABLE III.—*Weather and soil conditions during the months of May to August for the years 1908 to 1916 at Akron, Colo.*[a]

Condition and month.	1908	1909	1910	1911	1912	1913	1914	1915	1916
Evaporation (inches):									
May	7.709	6.825	5.797	7.323	7.097	5.835	5.608	5.033	7.811
June	8.637	7.003ac	8.722	9.753	6.750c	8.178	7.309c	5.883ac	7.979
July	8.474	9.396c	9.763	9.774	7.618	9.359	8.654	6.660	11.116
August	7.806	8.538	7.142c	8.944	7.048	9.302	8.364	5.820ac	7.216
Mean	8.162	7.941	7.856	8.949	7.128	8.144	7.534	5.849	8.531
Rainfall (inches):									
May	3.30	1.87	2.06	1.15	3.86	1.44	1.46	4.13	2.24
June	2.37	3.32ac	1.38	1.48	3.39c	1.35	3.54c	3.75ac	2.09
July	2.42	4.61c	1.47	1.34	3.58	1.85	1.66	1.10	1.77
August	1.47	3.77	3.72c	1.30	1.58	1.14	1.05	3.51ac	2.82
Mean	2.30	3.39	2.16	1.38	2.60	1.45	1.93	3.12	2.23
Rainy days (trace or over):									
May	9	12	13	19	15	10	12	16	12
June	18	15ac	5	10	11c	12	11c	19ac	10
July	12	9c	5	13	17	10	17	21	13
August	12	6	12c	10	11	7	14	17ac	15
Mean	13	11	9	11	14	10	14	18	13
Mean temperature (°F.):									
May	55	52	53	58	55	57	57	52	55
June	64	64ac	67	70	63c	67	68c	50ac	64
July	70	71c	74	70	70	72	72	67	75
August	69	72	67c	69	69	75	71	64ac	69
Mean	65	65	65	67	64	68	67	61	66
Soil temperature (°F.).									
May	74+	57	59	62	60	63	63	56+	60
June	74	67ac	63	77	68c	75	76c	71+ac	73
July	77	64+c	85	80	76	80	81	74	81+
August	75	....	77c	77	76	83	79	71ac	77
Mean	75+	62+	71	74	70	75	75	71+	75+
Maximum air temperature (°F.):									
May	90	85	87	91	92	91	85	88	70
June	93	93ac	95	96	89c	97	93c	80ac	80
July	94	94c	100	95	96	103	96	95	93
August	98	95	95c	97	96	98	101	94ac	83
Minimum air temperature (°F.):									
May	24	24	29	28	28	31	32	27	41
June	41	41ac	32	43	37c	37	38c	32ac	40
July	42	51c	51	46	47	43	49	40	60
August	46	51	31c	41	48	53	42	39ac	56
Number of days during which available soil moisture was recorded in the first foot:									
May						19	31	31	8
June		30ac	17	10	30c	2	22c	30ac	23
July		18c	2	0	13	8	11	14	0
August	7	24	22c	6	12	0	0	31ac	2
Number of days during which available soil moisture was recorded in second foot:									
May						0	31	31	31
June		30ac	23	0	30c	0	30c	30ac	30
July		19c	0	0	23	0	14	30	23
August		0	0c	0	0	0	0	30ac	0

a += Record incomplete; a= *Agaricus tabularis;* c= *Calvatia cyathiformis* in fruit at sometime during the month.

The mean temperature of the soil during May and June, 1909, was 57° and 67° F., respectively. During 1915, when the fungi fruited abundantly, the soil temperature records were not complete, but indicate that conditions were nearly the same as in 1909, the values being 56° and 71°, respectively. The mean air temperatures did not vary markedly during the different years, except that 1915 was cooler than the other years. This is indicated in the maximum and minimum as well as in the mean temperatures.

WEATHER AND SOIL CONDITIONS DURING THE FRUITING PERIODS

The conditions during and preceding the principal fruiting periods in 1909 and 1915 are given in Table IV. The conditions during the period of fruiting are of interest in connection with the character of the fruiting bodies and the length of the fruiting period. The conditions immediately preceding determine whether or not fruiting bodies will be produced.

TABLE IV.—*Weather conditions for 10 days preceding and during the fruiting period of Agaricus tabularis and Calvatia cyathiformis*

Species and period.	Daily evaporation.	Daily rainfall.	Climatic conditions.					Soil moisture conditions.	
			Percentage of days with rain.	Temperature (°F.).				First foot.	Second foot.
				Soil. Mean.	Air. Mean.	Maximum.	Minimum.		
Agaricus tabularis:	*Inch.*	*Inch.*							
Fruiting period, June 9-19, 1909.	0.218	0.29	65	65	62	91	41	Moist	Moist.
Previous 10-day period, May 29-June 8, 1909.	.197	.08	60	57	55	85	36	...do.....	Do.
Fruiting period, June 5-24, 1915.	.208	.13	60	....	60	82	35	...do.....	Do.
Previous 10-day period, May 26-June 4, 1915.	.132	.25	80	....	54	71	37	...do.....	Do.
Fruiting period, Aug. 16-20, 1915.	.199	.06	80	71	63	82	49	...do.....	Do.
Previous 10-day period, Aug 6-15, 1915	.204	.12	80	71	57	82	35	...do.....	Do.
Calvatia cyathiformis:									
Fruiting period, June 12-19, 1909.	.232	.04	50	67	64	83	46	...do.....	Do.
Previous 10-day period, June 2-11, 1909.	.192	.28	80	63	60	91	41	...do.....	Do.
Fruiting period, July 13-17, 1909.	.294	0	0	59	71	91	54	do ..	Dry.
Previous 10-day period, July 3-12, 1909.	.293	.28	50	65	78	90	51	Dry to moist.	Moist to dry.
Fruiting period, June 14-24, 1915.	.186	.04	64	*70	62	82	46	Moist	Moist.
Previous 10-day period, June 4-13, 1915.	.234	.26	70	....	56	82	35	...do.....	Do.
Fruiting period, Aug. 16-20, 1915.	.199	.06	80	71	63	82	49	...do.....	Do.
Previous 10-day period, Aug. 6-15, 1915.	.204	.12	80	71	57	82	35	...do.....	Do.

* Record incomplete.

Agaricus tabularis fruited only once in 1909, the fruiting period extending over 17 days. At this period the soil-moisture determinations in short-grass land showed available moisture in both the first and second foot, and the mean soil temperature of $65°$ F. The mean air temperature was a little lower, $62°$, and the range in air temperature from 41 to 91°. The daily evaporation rate was relatively high, amounting to 0.218 inch from the free-water surface of a 6-foot tank. The rainfall averaged 0.29 inch per day, or a little greater than the evaporation.

In 1915 *Agaricus tabularis* fruited twice, the June period covering 20 days and the August period only five days. Soil-moisture conditions were favorable during both of these periods. Conditions of evaporation were similar. Although but a small amount of rain fell during the late period, the percentage of days with rain was very high. Soil temperature was high during the late period. During the early period continuous soil temperature records were not available, but occasional records showed the temperature to have differed but little from the temperature of the early period in 1909. The air temperatures did not differ markedly from those of 1909, except that the maximums were much lower for each crop. A comparison of the conditions in 1909 with those of 1915 shows only a slightly higher evaporation and higher maximum temperature in 1909. The effect of the more extreme conditions on the fruiting bodies of the fungi will be noted later.

Calvatia cyathiformis, which fruits later in the season, usually fruits during periods of higher evaporation and higher temperatures. In 1909 the first fruiting period extended over eight days. The mean soil temperature was a little higher then than during the Agaricus period and the evaporation a little more extreme, but the air temperature range not as great. Otherwise, conditions were similar. The second fruiting period came during July and lasted only five days. The soil temperature was unusually low, the air temperature reached its maximum on the last day of the period. Evaporation was high and there was no rain. The soil was rapidly drying off. This second fruiting period followed a rainfall of 2.39 inches on July 7. The soil of the first foot was moist, but that of the second foot dry during this period.

In 1915 the June fruiting period covered 11 days. The soil was moist at this time and the air temperature rather high, but the evaporation was unusually low. During this damp period with frequent rain and low evaporation the fruiting bodies were produced under conditions markedly different from those of 1909, in so far as aerial conditions are concerned. The second fruiting period covered only four days in August. The mean temperature of the soil was $71°$ F., although the air temperature was rather low. This period followed a period of rainy weather, as is shown by the 10-day period preceding. The evaporation was unusually low, and rain was recorded on four of the five days considered.

EFFECT OF WEATHER CONDITIONS ON THE CHARACTER OF THE FRUITING BODIES.

The effect of high temperature on the fruiting bodies is well known. Fruiting bodies which appear during damp, warm periods decay rapidly and are soon destroyed by maggots. On the other hand, those which appear during cool periods remain fresh for a much longer time.

On the Great Plains it often happens that a spell of wet weather will be followed by an exceedingly dry period in which the evaporation from the soil and transpiration from the plant surface is excessive. When the fruiting bodies of the fungus are produced during such a dry period, the shape and proportions of these bodies and the character of the top of the pileus are greatly modified. The fruiting bodies of *Agaricus tabularis* produced during a period of rapid evaporation have a very rough and scaly periderm, a thick, firm pileus, and a short, thick stipe. These present better herbarium material when dried in place than when dried in a drying oven. On the other hand, if the fruiting bodies are produced during a period of less excessive evaporation, the pileus is thinner and expanded, the stipe long, and the periderm often quite smooth. These decay rapidly, and are difficult to preserve. In 1909 the tops were mostly rough (Pl. 10, B, C). They were produced during a period of comparatively rapid evaporation. During the year 1915 the tops were either smooth or rough, depending upon the time of appearance and the exposure. The fruiting bodies produced among taller growing plants and grasses almost invariably had smooth tops (Pl. 11, A), while those which occurred in the open had rough tops (Pl. 11, B, C).

The *Agaricus tabularis* ring sketched in figure 10 was partly in a cornfield and partly in the open. The fruiting bodies developed in the cornfield were comparatively smooth on top, the pileus spread out very wide and thin, and the stipe comparatively long and slender, while the fruits produced in the short-grass sod had a thick, firm, pileus, rough on top, and a very short and stocky stipe. The differences were similar to those shown in Plates 11 and 12.

The amount of variation in the pileus is well illustrated by the fresh sporophores shown in Plates 10 and 11 and the dried sporophores shown in Plate 12. In Plate 12, A and B, the characteristic pileus of *Agaricus tabularis* is shown, while in Plate 12, C and D, this tabular structure is entirely lacking.

Duggar (1905), who has made an extensive study of wild and cultivated forms of Agaricus, calls attention to the great amount of variation produced in the different strains under different cultural conditions. This wide range of variation and the effect on morphology, color, etc., make doubtful the specific value of forms showing only slight morphological differences (Duggar, 1915).

It has been difficult to determine the number of distinct forms occurring as fairy rings in eastern Colorado. The first form to appear in the spring is *Agaricus tabularis*, ranging morphologically from typical *A. tabularis* to the smooth expanded type similar to *A. arvensis*. Later in the season, if moisture conditions are favorable, a series of smaller forms (*A. campestris*) appear which show almost as wide a range in morphological characters as does *A. tabularis*, but which are entirely distinct from the latter in time of appearance and in the character of the mycelium in the fairy ring. The mycelium is not dense in the soil. The whole plant is less vigorous, the fruiting bodies are smaller and of a more delicate texture. But in this group the fruiting bodies often become decidedly rough on top and present wide variability in the appearance either of the fresh or the dried forms. The appearance of the fairy rings produced by the *A. tabularis* group is entirely distinct from rings of the *A. campestris* group.

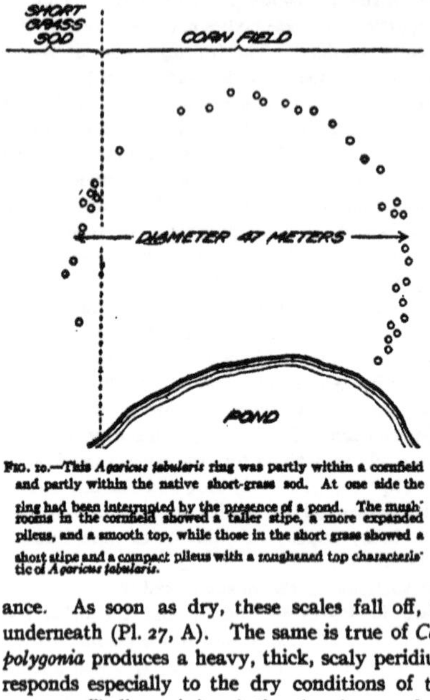

FIG. 10.—This *Agaricus tabularis* ring was partly within a cornfield and partly within the native short-grass sod. At one side the ring had been interrupted by the presence of a pond. The mushrooms in the cornfield showed a taller stipe, a more expanded pileus, and a smooth top, while those in the short grass showed a short stipe and a compact pileus with a roughened top characteristic of *Agaricus tabularis*.

The effect of the dry climate on *Calvatia cyathiformis* is not as marked as on *Agaricus tabularis*. The peridium, which is usually dried rapidly, has a rough tabular appearance. As soon as dry, these scales fall off, exposing the spore area underneath (Pl. 27, A). The same is true of *Calvatia fragilis*. *Calvatia polygonia* produces a heavy, thick, scaly peridium (Pl. 29, A, B), which responds especially to the dry conditions of the atmosphere. If the young puffballs are injured, they break open, forming abnormal fruiting bodies, such as shown in Plate 29, C. These fruiting bodies dry up before mature spores can be produced.

AGE AND RATE OF ADVANCE

Very little data are available on the rate of advance of fairy rings. Hutton (1790) observed this advance during a period of eight or nine years, but gave no data as to rate. Wollaston (1807) measured the

annual increase in the rings and found it to be from 8 inches to 2 feet
per year.

Thomas (1905) reported on a ring that had been studied especially as to
its outward increase for the period from 1896 to 1905. This ring was
formed by *Hydnum suaveolens*. In 1896 the ring was nearly complete,
but was never complete after that time and only during the years 1901,
1902, and 1905 was there any appreciable increase in outward growth.
The average increase of the radius for the period was found to be 23 cm.
From this the age of the ring was calculated to be about 45 years.

Ballion (1906) found an increase in the radius of a ring to be 12 cm.
during one year, but said the increase was very irregular. The advance
was said to be most rapid when the ring was young.

Coulter, Barnes, and Cowles (1911, p. 807) mention a colony of *Hydnum
suaveolens* which advanced from 9 to 11 meters in a period of nine
years. The age of the colony was estimated to be about 45 years.

Bayliss (1911) found the maximum increase of a ring of *Marasmius
oreades* to be 13½ inches per year, and the minimum increase 3 inches.
She thought rings might extend for 50 or even 100 years.

Very little data are available on the rate of advance of fairy rings in
eastern Colorado. Doubtless it is very unequal, and during dry years
little, if any, advance is made. During moist years, such as 1915, the rings
make a very decided advance.

The first crop of *Agaricus tabularis* usually occurred near the inner edge
of the outer stimulated zone (zone 4, fig. 3). The second crop, which in
1915 occurred two months later, had advanced an additional 8 to 30 cm.
(fig. 11).

It is often evident from the old fruiting bodies which have remained
in place that the advance is approximately the same for the first as for
the second crop. But the rings fruit only during the exceptionally wet
years. If it is assumed from these observations that the amount of
advance for each crop is approximately 30 cm., we may estimate the age
of the rings on the basis of the records of the last 10 years. Four crops
have been produced in 10 years, which would correspond to an advance
of 120 cm. in this period of time. From observations of the rings during
this period these estimates would seem to be approximately correct.
On this basis the average yearly advance would be about 12 cm. A
ring 60 meters in diameter would be approximately 250 years old. In
the southern portion of the area shown on the map (fig. 2) a large number of small or fragmentary rings are roughly arranged in the form of
two still larger rings. If these fragments have had a common origin,
which seems a correct assumption, the age of these large fragmented rings
would be approximately 600 years.

The rate of advance of the *Calvatia cyathiformis* ring is about the same
as that of the *Agaricus tabularis* ring during periods especially favorable

for growth. The second crop produced in 1915 showed an average
advance of from 8 to 30 cm. (fig. 5, 12, and Pl. 25, A, B; 26, A). Judging by the sterile bases of the old fruiting bodies the first crop shows a
similar advance over the crop the year before. If we assume this

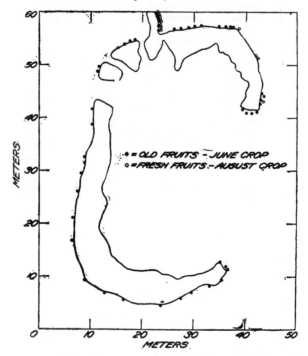

FIG. 11.—A sketch of ring 1 made in August, 1915. The fresh fruiting bodies are shown as circles and the
old fruiting bodies as dots. The ring has been interrupted at the upper left-hand side and on the right
has apparently come into contact with the second ring which produced a slight extension at the upper
part of the figure. The fungus was especially luxuriant in this portion and at this point produced a large
number of fruiting bodies in both the first and second crop. The distribution of the fruiting bodies near
the interrupted portion of the ring indicates a tendency on the part of these rings to grow around, producing an effect not dissimilar from the effect of tree growth in covering a wound or dead branch.

maximum rate to be 30 cm. per crop, the advance during the 10 years
would be eight times this amount, since the fungus fruited only eight
times, or a total advance of 240 cm. For the 10 years this would mean
an average yearly advance of 24 cm., or about twice as rapid as the esti-

mated yearly advance of *A. tabularis*. The age of the largest rings, which are 200 meters in diameter (Pl. 24, B), would be approximately 420 years.

DISTRIBUTION OF THE MYCELIUM IN THE SOIL

The distribution of the mycelium in the soil can not easily be determined, except for *Agaricus tabularis*. On the high plains the mycelium seldom penetrates deeper than the superficial soil layer, which is dark in color, owing to the presence of organic matter. In most cases the lower

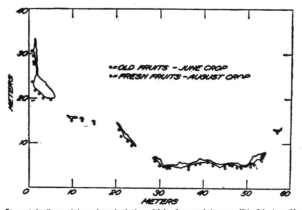

FIG. 12—A detail map of ring 5 shown in the lower left-hand corner of the map. This *Calvatia cyathi-formis* ring had produced two crops of puffballs, the first crop being indicated by dots and the second by circles. The zone of influence of the *C. cyathiformis* ring is much narrower than that of the *Agaricus tabularis* ring.

limit of penetration was reached near the inside of the area when the old mycelium had penetrated to a depth of from 5 to 10 cm. below the level of the young mycelium at the front edge of the fungus development.

In no case have the writers found the mycelium nearer to the surface than about 8 cm. This is probably due to the extremely dry condition of the surface soil during a large part of the year. Wollaston (1807), Molliard (1910, fig. 1), and Münch (1914) found the mycelium extending to the surface of the soil. This would be the case where the surface of the soil remained moist for a considerable time. In figures 4 and 13 and in Plates 14, C 15, and 17 the distribution of the mycelium of *Agaricus tabularis* is shown.

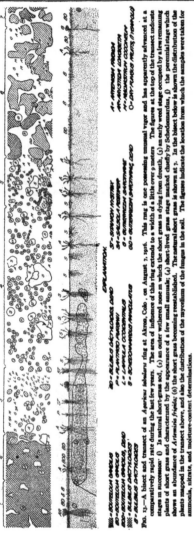

Fig. 13.—A bisect and transect of an *Agaricus tabularis* ring at Akron, Colo., on August 7, 1916. This ring is one possessing unusual vigor and has apparently advanced at a comparatively rapid rate during the last few years. The area of influence of this ring extends to a width of a little over 9 meters. The figures at the top of the transect indicate stages in succession: (1) In natural short-grass sod, (2) an outer withered zone in which the short grass is drying from drouth, (3) an early weed stage occupied by a few remaining plants of short grass and characterized by the appearance of a few small annuals; (4) short-lived grass stage marked chiefly by *Schedonnardus*, (5) the perennial stage which shows an abundance of *Artemisia frigida*; (6) the short grass becoming reëstablished. The natural short grass is shown at 7. In the bisect below is shown the distribution of the plants mapped in the transect above, and also the distribution of the mycelium of the fungus in the soil. The figures indicate the location from which the samples were taken for ammonia, nitrate, and moisture-content determinations.

EXPLANATION

A—*BOUTELOUA OLIGOS*
B—*BOUTELOUA GRACILIS*
BD—*BOUTELOUA GRACILIS, DEAD*
B'—*BULBUS DACTYLOIDES*

BD—*BULBUS DACTYLOIDES, DEAD*
P—*PLANTAGO PASMNI*
L—*LAPPULA OCCIDENTALIS*
S—*SCHEDONNARDUS PANICULATUS*

S'—*SOPHION VICIFER*
SH—*SALSOLI KENMER*
Q—*GUTIERREZIA SARATHRAE*
QD—*GUTIERREZIA SARATHRAE, DEAD*

A—*ARTEMISIA FRIGIDA*
M—*ARISTIDA LONGISETA*
N—*MUHLENBERGY COGGEN*
O—*DRY FUNGUS HOLES (TUNNELS)*

EFFECT OF THE MY-CELIUM ON SOIL CONDITIONS

The development of the mycelium brings about both chemical and physical changes in the soil.

CHEMICAL CHANGES

The chemical changes consist largely of the reduction of organic matter, brought about by the saprophytic nature of the fungus mycelium, and of the subsequent decay of the mycelium itself, owing to the action of molds or bacteria.

The protein portion of the dead organic matter of the soil is reduced to ammonia, which either unites with other compounds or is changed by bacteria to nitrites and these in turn by other bacteria into nitrates.

The reduction of the organic material thus furnishes a quantity of readily available nitrogenous material for the use of the green plants. Since on account of the dry subsoil and small amount of rainfall no leaching takes place in the soils in eastern Colorado, the chances for loss of nitrogen are greatly reduced.

As the mycelium passes on to new soil the old mycelium dies. The death of this mycelium is followed by decay produced by bacteria and molds which again liberates a supply of nitrogenous material for higher plants. The result is an increase in plant growth, the production of more roots in the soil, and consequently an increase in the organic matter contained in the soil.

Wollaston (1807) spoke of the exhaustion of the "pabulum" of the soil necessary for the growth of the fungus, but did not state of what this "pabulum" consisted.

According to Buckman (1870), anything which would kill a patch of grass and thus give the fungus something (organic matter) on which to live would produce a fairy ring.

A special study of the carbon content of the soil in fairy rings was made by Lawes, Gilbert, and Warington (1883). The percentage of carbon (Table V) was found to be uniformly higher outside the ring than either in the ring or within the ring. The determinations in and inside the ring differ only slightly, the smallest amount of carbon being recorded inside the ring.

TABLE V.—*Mean percentages of carbon a in the fine dry fairy-ring soils, according to Lawes, Gilbert, and Warington (1883)*

Description of ring.	Percentage of carbon.		
	Within the ring.	On the ring.	Outside the ring.
Grove Paddock (May, 1874)............................	3.06	2.72	3.34
Broadbalk (June, 1877).................................	2.38	3.36	3.29
Broadbalk (September, 1877).........................	2.48	2.60	3.12
Park (September, 1877)...............................	2.88	3.21	3.31
Park (April, 1878)...................................	3.12	3.04	3.44
Mean..	2.78	2.99	3.30
Carbon "outside" = 100............................	84.2	90.6	100.00

a Carbon determined "by combustion in oxygen."

The percentage of organic carbon in the soil in the different sections of *Agaricus tabularis* rings at Akron, Colo., is presented in Table VI. It is evident from this table that no significant differences are noted in the organic carbon content. The organic carbon content, with one exception, ranges from 2.15 to 3.79 per cent. Difficulties are encountered in determining the carbon content where the soil contains only a small amount of organic matter. It is evident in this case that the errors due to the inclusion of the roots of plants are so great as to obscure to a great extent the effect of the growing mycelium on the organic matter of the soil.

Sample No.	Location.	Ring B.	Ring 2.	
1	Outside............	2. 37	2. 76	3. 2
2	Youngest mycelium	2. 58	2. 87	2. 8
3	Dense mycelium....	2. 15	3. 17	3. 5
4	Below No. 3........	2. 23	2. 89	3. 7
5	Old dying my-celium..........	2. 23	2. 73	2. 8
6	Inside............	2. 34	2. 45	3. 4

a For the determination of organic carbon the writers are indebted to Dr. E
Soils. The determinations were made by the moist-combustion method with
bicromate.

One of the principal effects of the growth of the fung
soil is in changing the nitrate or ammonia content of t

Lawes, Gilbert, and Warington (1883) found mor
the ring than either in the ring or inside the ring (Tab
results were obtained for carbon (Table V) and the
nitrogen was found to range from 11.2 to 11.7. Th
the nitrates and found here a much greater amoun
either inside or outside (Table VIII). From the r
cluded that in fairy rings the nitrates of the soil wer
and that the source of the nitrates was the organic ma

TABLE VII.—*Mean percentages of nitrogen a in the fine, dry fai
to Lawes, Gilbert, and Warington (1883)*

TABLE VIII.—*Nitrogen as nitrates per million of dry fine soil, according to Lawes, Gilbert, and Warington (1883)*

Location.	Crum-Frank-land method.	Schlös-ing method.	Location.	Crum-Frank-land method.	Schlös-ing method.
Broadbalk fairy-ring soils, collected June 18, 1887:			Park fairy-ring soils collected Sept. 19, 1877:		
Within the ring	0. 23	. . . . . .	Within the ring	Trace.	. . . . . .
On the ring.	. 92	. . . . . .	On the ring (center). .	. 46	. . . . . .
Just outside the ring.	. 43	. . . . . .	On the ring (outer		
Outside the ring.	. 09	. . . . . .	edge).	1. 21	. . . . . .
Broadbalk fairy-ring soils, collected Sept. 15, 1877:			Just outside the ring.	Trace.	. . . . . .
Within the ring.	1. 31	1. 03	Park fairy-ring soils, collected Apr. 25, 1878:		
On the ring.	8. 07	11. 46	Within the ring.	. 17	. . . . . .
Outside the ring.	1. 10	2. 44	On the ring (inner		
			side).	1. 21	. . . . . .
			On the ring (outer		
			side).	None.	. . . . . .
			Outside the ring.	. 18	. . . . . .

TABLE IX.—*Quantity of ammonia (in milligrams) in 100 gm. of dry soil according to Molliard, 1910*

Zone.	Location.	Ammonia taken up by the water.	Ammonia retained by soil.	Total ammonia.
I.	Inside, no mycelium.	4	33	37
III.	Dead-grass zone, in mycelium.	14	66	80
IV (s). . . .	Outer stimulated zone (above mycelium area).	8	45	53
IV (p). . . .	Outer stimulated zone (in mycelium area). . .	17	56	73
V.	Outside (normal) no mycelium.	3	35	38

The determinations of [1] nitrogen in the form of ammonia and nitrates in rings formed by *Agaricus tabularis* in 1915 are given in Table X. The soil samples were taken with a soil tube and consisted of a core of soil extending through the first 6 inches, the first foot, or the second foot of soil. The results in the second foot are less significant than those in the first foot, since less of the mycelium lies in this area. It will be noted as a rule that the ammonia content is higher in the ring than inside, and higher inside the ring than outside. In the case of nitrates the results are not conclusive, although as a rule the nitrates are more abundant in the ring and inside than under the normal sod outside. In 1916, samples of soil were taken in six carefully selected places in a trench dug across a number of different rings formed by *Agaricus tabularis*.

---

[1] For the determination of nitrates and ammonia the writers are indebted to Mr. R. C. Wright, of the Office of Soil-Bacteriology Investigations, Bureau of Plant Industry. The determinations were made by a modified Ulsch method. Two hundred c. c. of soil extract were treated with one c. c. of concentrated sulphuric acid and five gms. of reduced iron powder, and the reduction was allowed to proceed from 12 to 15 hours. This solution was then made alkaline with magnesium oxid and distilled into standard acid.

ocation of these samples is indicated in the case of ring 2 in
results are shown in Table XIII. In these determinations
 indicates results on the comparatively fresh soil; the seco
 eterminatian made on a duplicate sample which had rema
 :ondition in the laboratory for about seven months. Nit
 abundant on the inside of the ring and most abundant in
 ied by the old dying mycelium. In the area occupied
g mycelium and just below it, and also in the mass conta:
 portion of the mycelium, the nitrate content was about
onia was least abundant inside the ring, and most abunda
 portion of the mycelium. In the case of the new mycel
rea just below the dense mycelium the ammonia content
ame. More ammonia was found outside in the native sod
 side of the ring. The total nitrogen was determined in
 dried samples. It is evident that the changes in the qu
tes or ammonia recorded in different portions of the soil
ccompanied by changes in the total nitrogen content of
her words, the nitrogen seems only to be changed from
other.

X.—*Quantity (in parts per million) of ammonia and nitrates in Agaric:
rings, Akron, Colo., 1915*

Ring No. and date.

s of nitrogen in the form of ammonia and nitrates
ormis rings for 1915 are summarized in Table XI.
of *C. cyathiformis* more ammonia was found in the
e inside or in the zone which had produced fruits.
tes the reverse was found to be the case, the nitrates
re abundant in the ring than either outside or inside.
from samples taken in 1916 are given in Table XII.

*(in parts per million) of ammonia and nitrates in Calvatia cyathi-
formis rings, Akron, Colo., 1915*

date.	Depth.	Nitrogen in the form of ammonia.			Nitrogen in the form of nitrates.		
		Out-side.	In ring.	In-side.	Out-side.	In ring.	In-side.
	Feet.						
....	1	19	21	19	3	0	8
....	2	11	11	11	3	5	0
....	1	......	......	......	10	8	6
....	2	......	......	......	Tr.	6	Tr.
....	3	......	......	......	Tr.	10	Tr.
....	½	13	37	34	8	11	11
....	½	......	......	......	Tr.	Tr.	Tr.
....	1	13	29	27	5	5	6
....	2	11	11	11	2	6	Tr.
....	3	......	......	......	Tr.	10	Tr.

transformed. In the old mycelium the ammonia content was very low and nitrate content relatively very high at the beginning of the experiment. After five days and two weeks the nitrate content was still high and the ammonia content relatively low.

TABLE XIII.—*Quantity (in parts per million) of nitrogen as ammonia and nitrates and N^3 in Agaricus tabularis rings, Akron, Colo., 1916*

Sample No.	Location.	Nitrogen in the form of ammonia.					Nitrogen in the form of nitrates.					Total nitrogen as N^3.[a]				
		Ring 6, June 21.	Ring 7, June 26.	Ring 8, June 28.	Ring 1, July 12.	Ring 2, July 22.	Ring 6, June 21.	Ring 7, June 26.	Ring 8, June 28.	Ring 1, July 12.	Ring 2, July 22.	Ring 6, June 21.	Ring 7, June 26.	Ring 8, June 28.	Ring 1, July 12.	Ring 2, July 22.
1	Outside (see fig. 13)	2.9	1.4	4.3	7.9	5.7	8.6	11.4	11.4	10.7	4.3	1,190	840	770	1,050	1,190
1	....do...........	5.0	4.0	9.5	7.6	3.8	4.0	5.0	3.8	1.9	.0	1,190	840	700	910	1,050
2	Youngest mycelium.	8.6	10.0	18.4	6.4	19.8	24.3	21.4	21.4	6.4	11.4	1,190	1,050	770	1,050	910
2	....do...........	17.1	.90	170.	5.0	11.0	.0	.0	.0	4.0	2.0	1,190	910	770	1,050	910
3	Mycelium densest, driest......	20.0	17.3	2.9	27.2	75.7	15.7	11.4	10.9	7.9	8.6	910	1,050	700	1,180	1,050
3	....do...........	17.0	21.0	21.6	27.0	22.9	2.0	4.0	9.5	11.4	5.7	770	980	840	1,050	1,050
4	Below No. 3......	11.4	2.9	4.3	13.0	11.4	20.0	19.9	22.9	11.4	8.6	1,050	840	770	980	1,330
4	....do...........	13.0	3.0	1.0	6.0	6.0	8.6	1.0	5.0	6.0	10.0	1,050	770	770	910	1,330
5	Old dying mycelium.	11.4	5.7	2.9	6.4	8.6	8.6	34.2	22.9	32.2	22.9	840	840	700	840	910
5	....do...........	6.0	17.1	3.0	5.7	14.0	.0	20.0	10.0	30.5	18.0	840	770	700	840	910
6	Inside...........	2.9	2.9	.0	3.6	2.8	10.0	8.6	1.4	7.2	7.1	1,050	770	700	910	980
6	....do...........	5.7	4.0	.0	6.6	9.5	.0	.0	4.0	1.9	980	700	700	910	840

[a] For these determinations the writers are indebted to Mr. H. W. Daudt, of the Bureau of Chemistry. The determinations were made by the Kjeldahl-Gunning-Arnold method.

Under field conditions the growth of the mycelium is accompanied with the production of ammonia. In the older mycelium, on the other hand, the nitrification is rapid and most of the nitrogen is found as nitrates and only a comparatively small amount as ammonia. The nitrification and ammonification determinations are in accord with the determinations made from field samples.

TABLE XIV —*Ammonification and nitrification (in parts per million) in Agaricus tabularis rings, Akron, Colo., May, 1917* [a]

Sample No.	Location.	Ammonification.			Nitrification.		
		Nitrogen as ammonia originally present in soil.	Nitrogen as ammonia gained after 5 days' incubation with 0.2 per cent of peptone.	Nitrogen as ammonia gained after 2 weeks' incubation with 0.2 per cent of peptone.	Nitrogen as nitrate originally present in soil.	Nitrogen as nitrate gained after 5 days' incubation with 0.2 per cent of peptone.	Nitrogen as nitrate gained after 2 weeks' incubation with 0.2 per cent of peptone.
1	Outside........	3.3	59.9	55.9	0.0	5.1	15.3
2	Youngest mycelium.........	1.5	34.7	39.5	9.0	−1.4	15.8
3	Dense mycelium	21.0	35.2	40.0	21.0	.5	22.8
5	Dying mycelium	1.5	29.0	17.1	34.5	28.5	117.5

[a] These determinations were made by Mr. R. C. Wright, of the Office of Soil Bacteriology, Bureau of Plant Industry.

From the results of previous investigations and from those here presented it is concluded that the progress of the fairy-ring fungus through the soil brings about the following chemical changes. The dead organic matter of the soil is utilized as a food supply for the saprophytic fungus. During the process the carbohydrates are consumed or reduced and parts of the protein material consumed by the fungus and reduced to ammonia. This combines readily to produce ammoniacal salts, or is changed by bacterial action to nitrites which are in turn converted into nitrates. The chief effect, in so far as soil chemistry is concerned, is to change the protein portion of the organic matter of the soil into compounds of nitrogen which are readily available to higher plants.[1]

PHYSICAL CHANGES

The physical differences in the soil in different portions of fairy rings are due to the growth of the fungus and to the effect of the amount of other vegetation developed on the different zones.

Waring in 1837 (Bayliss, 1911, p. 112–116) noticed the effect of the mycelium on the rate of absorption of water by the soil. In Table XV are presented the results of moisture determinations made by Lawes, Gilbert, and Warington (1883). Their results indicate less soil moisture in the ring than either outside or inside. Measurements made by Molliard (1910) showed that the soil was comparatively dry in the zone occupied by the mycelium (Table XVI).

TABLE XV.—*Percentage of water in fresh soil as collected, exclusive of stones, according to Lawes, Gilbert, and Warington (1883, p. 216).*

Location.	Surface soil.	Subsoil.
Grove Paddock fairy-ring soils, May 19, 1874:		
Within ring...	16. 03	15. 68
On ring..	12. 58	12. 30
Outside ring...	15. 71	16. 24
Park fairy-ring soils, Sept 19, 1877:		
Within ring..	22. 80	17. 04
On ring (center).....................................	19. 29	13. 13
On ring (outer edge).................................	18. 50	13. 23
Just outside ring....	23. 33	15. 03
Park fairy-ring soils, Apr. 25, 1878:		
Within the ring......................................	26. 34	19. 21
On ring (inside).....................................	26. 33	.
On ring (outside)....................................	21. 95	19. 14
Outside the ring.....................................	27. 96	19. 74

[1] While the details of the process of reduction of the proteid portion of the organic matter have formed no part of these studies, it is interesting to call attention here to the probability that some of the intermediate products, such as amino acids, are utilized directly not only by the fungi but by the higher plants. See Schreiner and Shorey (1910), Schreiner and Skinner (1912), Schreiner and Lathrop (1912), Schreiner (1913), and the review of this subject, with citations of the earlier literature in Lathrop (1917)

ss (1911) found soils which contained mycelium difficult
entioned a case where a rain which moistened the soil to
ches did not even penetrate the surface in the mycelium
conditions of moisture in the rings formed by *Agaricus t*
on, Colo., in 1914, 1915, and 1916 are indicated in Tables
, and XX. It will be seen from these tables that the soil
usually very dry in the middle and late summer. Dur
the soil in the ring differs but little from that outside o
g, provided the previous season was not too dry and a s
l had occurred during the winter or early spring to wet th
ass.

XVII.—*Moisture content above or below the wilting coefficient in two*
tabularis rings on May 16, 1914, Akron, Colo.

| epth. | Percentage of moisture. | | Depth. | Ring B. Percentage of mo Outside. |
	Outside.	In bare area, zone 3.		
nches.			*Inches.*	
....	+17. 8	+18. 6	1 to 3.....	+23. 3
....	+17. 3	+13. 7	3 to 6.....	+16. 4
....	+ 4. 4	− 3. 2	6 to 12.....	+13. 8
....	+10. 8	− .5	12 to 18....	+11. 5
....	+ 9. 1	. o	18 to 24....	+13. 0

difference in soil moisture may be due to several different
first place, if the soil is uniformly moist, the stimulated
g (zones 2 and 4, fig. 3) would be the first to become dry as

for two *A. tabularis* rings in May, 1915. These samples were taken in 3-inch sections. No differences were noted in the first 3 inches. In the second 3 inches the water content in zone 3 (fig. 3) was greatly reduced in the case of ring B, while in the second 6 inches no water at all was available in the ring, though the soil outside of the ring still contained available moisture. In Table XVIII the moisture content above or below the wilting point in three rings of *A. tabularis* on three different dates in 1915 is shown. While there was only a small amount of available water outside on June 16, there was no available water in the ring. On June 26 a similar condition obtained in the ring, while by July 13 the water content both inside and outside the ring had been reduced to below the wilting point. These differences are not so marked in the second foot. The year 1916 was a dry year and the fungi did not fruit. In figure 13 is shown a bisect in which the place of taking samples is shown. Samples take from four different Agaricus rings in 1916 are shown in Table XIX. In all cases the water content was low, the vegetation being in a dormant condition except outside the ring. During a period of heavy rain the samples at 2, 3, and 4 feet would remain dry, while others would readily become moist.

TABLE XVIII.—*Moisture content above or below the wilting coefficient in three Agaricus tabularis rings in 1915, Akron, Colo.*

Depth.	Ring No.	June 16.			June 26.			July 13.		
		Outside, zone 5.	Bare area, zone 3.	Inner stimulated, zone 2.	Outside, zone 5.	Bare area, zone 3.	Inner stimulated, zone 2.	Outside, zone 5.	Bare area, zone 3.	Inner stimulated, zone 2.
First foot	1	+ 9.3	+0.2	+ 6.7	+1.8	−1.7	+1.0	+0.1	−2.3	−2.9
Do......	2	+11.4	−2.1	+10.4	+3.0	−1.8	−1.4	−2.1	−4.2	−2.8
Do......	3	+ 6.7	−2.9	+ .1	+2.5	−3.5	−2.4	−2.5	−2.7	−3.9
Mean..		+ 9.1	−1.6	+ 5.7	+2.4	−2.3	− .9	−1.5	−3.1	−3.2
Second foot..	1	+10.4	+6.8	+ 5.8	+7.8	+1.4	+2.8	+4.4	+1.6	−3.7
Do	2	+ 9.2	+7.4	+ 9.0	+8.6	+2.0	+1.8	+3.5	−2.3	+2.7
Do ...	3	+ 8.5	−3.0	+ 5.4	+6.1	−2.8	+ .4	+2.5	−1.4	−1.4
Mean..		+ 9.4	+3.7	+ 6.7	+7.5	+ .2	+1.7	+3.5	− .7	−2.4
Third foot...	1	+ 9.5	+4.8	+ 5.9	+7.4	+2.9	+3.6	+4.2	+1.6	−2.1
Do.. ...	2	+ 9.2	+3.5	− .7	+8.2	+2.9	+ .6	+6.1	+ .6	+3.5
Do......	3	+ 9.5	−4.1	+ 4.3	+8.6	−2.3	+4.0	+6.5	+ .3	+1.3
Mean..		+ 9.4	+1.4	+ 3.2	+8.1	+1.2	+2.7	+5.6	+ .8	+ .9

In the case of rings of *Calvatia cyathiformis* the effect of stimulated growth is the same as in the case of *Agaricus tabularis*. The mycelium does not become dense enough to interfere with water penetration, and after each rain the soil is again moistened. Table XX indicates clearly the effect of the stimulated growth of vegetation on the reduction of soil moisture. The soil in the ring is first to be dried below the wilting point.

TABLE XIX.—*Soil-moisture content above or below the wilting coefficient in Agaricus tabularis rings in 1916, Akron, Colo.*

Sample No. and location.	Ring 2, June 12.	Ring 6, June 21.	Ring 7, June 26.	Ring B, June 28.	Average.
1, outside (see fig. 13)..............	−0.2	+0.3	+3.2	+4.8	+2.0
2, youngest mycelium.............	+.5	−2.3	−4.1	−1.8	−1.9
3, mycelium densest driest.......	−1.7	.0	−2.5	−2.3	−.6
4, below No. 3....................	−1.1	−1.5	−1.7	−.6	−1.2
5, old dying mycelium............	−.6	−2.2	−1.4	−2.5	−1.7
6, inside........................	−2.5	−.8	−4.1	−2.8	−2.6

TABLE XX.—*Moisture content above or below the wilting coefficient in a Calvatia cyathiformis ring in 1915, Akron, Colo.*

Ring No.	Depth of soil.	June 26.			July 13.		
		Outside, zone 5.	In stimulated area 2.	Inside, zone 1.	Outside, zone 5.	In stimulated area 2.	Inside, zone 1.
	Feet.						
Ring 5........	1	+2.2	−1.3	+0.5	−0.8	−2.6	−1.2
Do........	2	+6.4	+1.0	+5.5	+3.3	−2.9	+.7
Do........	3	+6.8	+1.7	+3.5	+4.1	−2.8	+2.7

For the purpose of studying the penetration of water both into the mycelium infected soil and into the natural sod outside, a strip about 9 meters long was selected on the edge of a ring of *Agaricus tabularis* where the zones were well defined. Holes were made 6 inches deep with a soil tube. They were made in rows 1 meter apart, consisting of three holes each. One row was outside in the natural sod (fig. 3, zone 5), the next row at the border between the withered (fig. 3, zone 4) and bare areas (fig. 3, zone 3), and the third row in the bare area. Two-liter flasks of water were then inverted into the holes to determine the rate of penetration (Pl. 23,A). The results are given in Table XXI.

In both records the number of cubic centimeters of water that penetrated into the soils was much greater in the natural sod outside the ring and least at the border line between the withered and bare areas where the mycelium was most dense. The individual records vary considerably. This is probably largely due to the varying density of the mycelium in the soil and to differences in the dryness of the soil. In the first measurements the penetration in the outside was three times as rapid outside as in the withered zone and twice as rapid as in the bare zone. The difference is a little less marked in the second experiment due partly to the fact that part of the soil had already been moistened by the first irrigation. After heavy rain the water stands on the soil over these rings of *Agaricus tabularis*, while on the adjacent land it soon penetrates and disappears. Several hours after a rain the surface soil will be muddy just above the dry hypha-filled portion. When these hyphae are dry, water is turned off just as from a dense mass of nonabsorbent cotton. The figures in Table XXI do not fully convey the difference in moisture penetration since under field conditions the moisture is free to run off on adjacent land and the relative amount of water which penetrates the soils of the *Agaricus tabularis* fairy ring is consequently smaller than would be indicated by the table. The effect of rain in moistening the soil of an *Agaricus tabularis* ring is shown in Plate 15. The mycelium-filled soil shows light in the soil trench. The water penetrated both in front of and back under the edge of the mycelium-filled soil without wetting the latter. The moist soil appears dark in the soil trench.

TABLE XXI.—*Rate of penetration (in cubic centimeters per hour) of water into the soil of an Agaricus tabularis ring, Akron, Colo., 1915*

Record No.	Outside.	At the border of withered and bare areas.	In the bare or nearly bare area.
First record, begun June 12..........................	12.5	20.9	22.7
	30.8	12.3	20.9
	41.7	6.4	9.3
	83.4	41.7	22.7
	41.7	20.9	22.7
	41.7	6.4	83.4
	83.4	7.6	60.0
	83.4	27.8	22.7
	83.4	10.8	22.7
Average......................................	55.7	17.2	31.9
Second record, same place, begun June 26.............	87.0	69.0	46.5
	46.5	44.5	69.0
	87.0	44.5	69.0
	105.0	69.0	46.5
	74.0	87.0	46.5
	87.0	27.4	69.0
	105.0	29.0	105.0
	95.3	69.0	27.4
	105.0	46.5	29.9
Average......................................	88.0	54.0	56.5

EFFECT OF FAIRY RINGS ON VEGETATION

RINGS OF AGARICUS TABULARIS

The first effect noted by the penetration of the filaments of this fungus into the soil occupied by short-grass vegetation is the slight stimulation of all the native plants. The stimulation (fig. 3, zone 4) is noted in the deeper color of the short grasses and in the greater and more rapid growth of the annuals (*Plantago purshii, Festuca octoflora, Hedeoma nana*, etc.)

and the perennial plants such as *Gutierrezia sarothrae* and *Artemisia frigida*, all of which grow in the native sod. The fruiting bodies of the fungus occur in this stimulated zone. The soil in the area just inside zone 3 (fig. 3) is permeated with filaments of the fungus from a depth of about 8 cm. to a depth of something over 30 cm. During years of normal or subnormal rainfall the soil of this zone is dry, and since water does not penetrate this soil, this zone is marked by dead or dying plants, or by the absence of a plant cover of any kind. Inside this bare zone, where the mycelium has died and partly disappeared and where moisture can penetrate more freely, the vegetation which is now composed largely of ruderals (fig. 13), is as luxuriant as is permitted by the soil-moisture supply and the increase in available nitrogenous material. The vegetation on the inside of the ring (zone 1) does not differ in composition and appearance from that outside (zone 5), although it has been destroyed and reestablished.

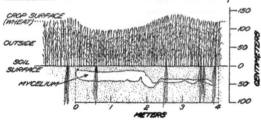

Fig. 14.—A bisect of a ring of *Agaricus tabularis* in a wheat field in 1915, a year of ample moisture supply.

The effect of the fairy ring caused by *Agaricus tabularis* on the native vegetation varies greatly with the moisture supply. The appearance of the ring is entirely different during a dry year (Pl. 16) and during a wet year (Pl. 19), and no description can be given which will apply at all times. The zones of greatest potential stimulation of plant growth due to available nitrogenous material may be brown and dry, owing to the lack of available soil moisture. The dead zone is only an expression of the extreme drouth produced by the combination of dry weather and unfavorable soil conditions. If rain is continuous and well distributed, the vegetation does not die, and a ring in which the bare zone has all but disappeared is the result. This condition was produced artificially in 1916 (Pl. 23, B). But since such a combination of favorable conditions seldom occurs in eastern Colorado the bare zone is well marked during normal years and can be distinguished even during wet years. The bare zone originates in the outer stimulated zone. During summer or late fall, or during dry years, this is changed to a withered condition, and if drouth is prolonged, to a bare condition due to the death of the grasses.

By gradual extension the mycelium develops another stimulated zone outside, and this in turn withers. In this way the dead zone is gradually increased on the outer edge. As the mycelium under this area becomes older it gradually dies out. Water penetration then becomes more normal and the ample supply of available nitrogenous material produces the inner stimulated zone.

The effect of rings of this type on cultivated plants is about the same as on the native plant cover. During a wet year the relative stimulation and depression of growth of the different areas is shown in the bisect (fig. 14; see also Pl. 21). The wheat just beyond the mass of mycelium showed a slight stimulation as compared with that entirely outside the ring. Over the area of dense mycelium the plants ripened prematurely and did not produce as much straw as in the normal crop. The grain production was very low. Back of the dense area of mycelium the fungus filaments were evident only at greater depths in the soil, and here the greatest stimulation of the wheat crop occurred. The results of croppings made at this time are shown in Table XXII.

TABLE XXII.—*Yield of wheat per square meter in a ring of Agaricus tabularis*

Item.	Inside (1).	Bare zone (3).	Outside (5).	Ratio 3:5.
During a wet year, Akron, Colo., Aug. 12, 1915:				
Wheat ring—				
Number of heads....................	....	108	190	57
Height............................cm..	....	93	210	44
Total dry matter.................gm..	....	401	500	80
Grain............................gm..	....	63	137	46
During a dry year, Akron, Colo., July 6, 1916:				
Wheat ring 8—				
Total dry matter.................gm..	91	0	108	....
Wheat ring 9—				
Number of heads....................	150	30	277	12
Height............................cm..	50	30	57	53
Total dry matter.................gm..	156	34	260	13
Wheat ring 10—				
Number of heads....................	54	0	160	....
Height............................cm..	35	0	50	....
Total dry matter.................gm..	57	0	132	....

In the same table the effect of the fairy ring on a wheat crop during a dry year is shown (Pl. 22). The zone just above the densest area of mycelium (bare zone 3) was then practically free of plant growth. The results in the table indicate clearly the harmful effect of this fungus on the crop during normal or dry seasons. Only in the case of ring 9 was any wheat produced in the bare zone, and the yield there was partly due to the width of the zone (only about 1 meter) and to the slight extension of the cropping into the adjacent zones.

....

It is evident that in this case drouth was not operat
_nt of the oat plants in such a wide portion of the are
pression of growth over the dead zone was not evid
imediate vicinity of the fungus fruits.   The land had
e spring, and the heavy and continuous rains had mair
celium-impregnated soil in a moist condition.

RINGS OF CALVATIA CYATHIFORMIS

In figures 5 and 6 the principal zones found in rings
gus are sketched.   The only effect produced on th
s fungus is to stimulate the growth on the area just in
e.   The short grasses are not injured in these rings,
nuals do not enter. *Festuca octoflora, Plantago pi
pida*, and the short grasses usually grow luxuriantly ir
e.   Their growth is more rapid, and the plants darke
e stimulated zone in these rings does not differ essen
tion from the outer stimulated zone of the ring of *A*
d since the perennial vegetation is not killed, the ba
st, nor is the inner stimulated zone (or weed zone)
g represented in the Calvatia ring.   The position of the
Calvatia rings is within the circle of fruits instead of o
case of the outer stimulated zone 4 in Agaricus rings
he vegetation inside does not differ from that outsid
ving years of more than normal moisture supply, whei
ver just inside becomes more dense than outside, due
the stimulated zone (Pl. 27, B).

figure 2 that only rings formed by *Agaricus tabularis* could be found in the wheat field during 1916. These were noticeable because of the depression of the growth of the wheat. Those formed by *C. cyathiformis* could not be distinguished, since they had produced no effect on the crop.

OTHER FUNGUS RINGS

Agaricus campestris, Calvatia polygonia, C. fragilis, Catastoma subterraneum (Pk.) Morg., *Marasmius oreades*, and *Tricholoma melaleuca* do not differ markedly from *C. cyathiformis* in their effect on the vegetation. *Lepiota morgani* occurred in a sandy soil in bunch-grass vegetation (Shantz, 1911, p. 54). No effect was noted on the native cover, although a great number of fruits had been produced.

GRAZING AND CULTIVATION

An especially interesting effect is seen in areas which are grazed. The grass in the stimulated zone appears to be more palatable than the less luxuriant grass in the adjacent sod and this zone is therefore often easily distinguished by the closely cropped grass.

Rings formed by *Calvatia cyathiformis* have not been found fruiting in cultivated fields. *C. polygonia* continues to fruit after land is cultivated, several rings having persisted in fields which have been under cultivation for a period of five years. *Agaricus tabularis* produces rings which seem to persist without injury under cultivation. The ring shown in the wheat field (Pl. 14, C; 21, A) was on land which has been under continuous cultivation for the past seven years. These rings often fruit as well as in cultivated fields as in the native soil.

CAUSE OF THE STIMULATED GROWTH

Practically every investigator of fairy rings has noted a stimulated growth of the natural vegetation on fairy rings. This stimulation was attributed to the decay of the fungus fruits or mycelium by Wollaston (1807), Way (1847), Westerhoff in 1859 (*in* Ritzema Bos, 1901), Berkeley (1860), Jorden (1862), Gilbert (1875), Van Tieghem (1884, p. 1044-1045), Olivier (1891), Stahl (1900), Ritzema Bos (1901), Beille (1904, p. 381), and Massart (1910). The principal constituents which contributed to this stimulation were potash, phosphated alkali, magnesia, and sulphate of lime, according to Way (1847), and nitrogenous and mineral manuring according to Gilbert (1875). Westerhoff in 1859 (*in* Ritzema Bos, 1901) actually produced the stimulated growth of the vegetation by scattering over the surface mushrooms cut up into small pieces. Jorden (1862) attributed the stimulation to the fertilizing effect of the ammonia which the fungus absorbed from the atmosphere. Lawes, Gilbert, and Warington (1883), and Ballion (1906), attributed the stimulation to the decay of fungus fruiting bodies and filaments and to the nitrates and residual products formed by the action of the mycelium on the organic matter of

dy pointed out, the growth of the mycelium produces
the organic matter of the soil, liberating ammonia,
up into nitrates or may form ammoniacal salts, both
mulate the vegetation. The principal stimulation is
se changes. Some indication of the effect of the dec
on the stimulated growth may be obtained by a co
s formed by *Agaricus tabularis* and those formed
the case of *A. tabularis* the ring is very broad, the s
extending back 2 to 5 meters or more behind the frui
vatia, Marasmius, or Catastoma rings, where only
mycelium is developed and where it is comparatively
ish the threads in the soil, the stimulated zone is alway
ending more than a meter behind the fruiting zone.
imulated area is the area in which the fungus filamen
organic matter of the soil while in the ring of *Agaricus*
nner stimulated zone is due largely to the decay of the m
son of the amount of ammonia and nitrates in the Cal
ing shows that the ammonia is comparatively more a
occupied by the active mycelium and that nitrates
more abundant in the area of dying mycelium.
the fungus mycelium reduces the protein of the organi
the end product ammonia but that during decay of
iich is accomplished by bacteria, and which probab
ly, this end product is rapidly built up into nitr
therefore that the greatest stimulation of vegetatio
the case of the fungus which had the greatest mas
his is undoubtedly true, for in the case of *Calvatia cyatl*

area consists of only a narrow zone near the edge of the ring, while in *A. tabularis* the zone extends far back toward the center of the ring.

From our results it seems likely that the greater amount of stimulation noted on the native sod and cultivated crop is due to the reduction of the nitrogenous organic matter to available nitrates and salts of ammonia, and to the subsequent decay of the fungus filaments which produce the same compounds. A stimulation of vegetation exactly similar to that in the fairy rings was produced artificially at Akron, Colo., by placing ammonium nitrate in shallow holes in the soil and by scattering it over the surface of the soil during the rainy period.

The investigations thus far conducted indicate that the stimulation of the grasses and other vegetation is due principally to the presence in the soil of nitrates and salts of ammonia which are derived from (1) the reduction of the organic matter of the soil; (2) the decay of the fungus fruits; and (3) the decay of the mycelium

TABLE XXIV.—*Effect of fairy rings on the chlorophyll content of plants, Akron, Colo., 1915.*[a]

RINGS OF CALVATIA CYATHIFORMIS, JUNE 28

Plant.	Percentage of alcoholic extracts of crude chlorophyll.		
	Outside.	In stimu-lated zone.	Inside.
	Per cent.	*Per cent.*	*Per cent.*
Bouteloua gracilis	100	340	125
Plantago purshii	100	280	120
Bouteloua gracilis	100	175	127
Plantago purshii	100	110	105
Mean	100	226	119

RINGS OF AGARICUS TABULARIS, JUNE 18

	Outside.	In stimu-lated zone.	Inside.
Gutierrezia sarothrae	100	110	90
Artemisia frigida	100	200	200
Malvastrum coccineum	100	225	300
Festuca octoflora	100	200	90
Hedeoma nana	100	200	135
Plantago Purshii	100	265	135
Bulbilis dactyloides	100	260	255
Mean	100	209	172

a The chlorophyll was extracted from equal areas of leaf lamina by cold 90 per cent methyl alcohol The extract was made in the dark. Colorimetric determination was made on the crude extract, the values outside being taken as 100.

CHLOROPHYLL CONTENT

One of the most characteristic features of fairy rings is the deep-green color of the vegetation in the stimulated zones. This effect was noted by many investigators, but no quantitative determinations of chlorophyll content have been made. The writers have determined by colorimetric

methods the relative density of the crude chlorophyll. Extracts were
made in the dark with cold 90 per cent methyl alcohol. The results
indicated in Table XXIV show that the chlorophyll content is about
twice as great in the stimulated zone as outside. The results are quite
uniform for the different plants and also for the Agaricus and Calvatia
rings. Similar measurements of the chlorophyll content in oat fields
showed the amount in the stimulated zone to be twice as great as normal.
The cause of the increase in vigor and chlorophyll content seems to be
amply explained on the basis of the available supply of nitrates and
salts of ammonia.

CAUSE OF THE DEATH OF THE VEGETATION

The death of the vegetation noted by many investigators has been
attributed to a number of different causes. The fungus filaments were
thought to absorb all nutrient material from the soil and thus cause the
death of the grasses, etc. This view was held by Wollaston (1807),
Van Tieghem (1884, p. 1044–1045), Olivier (1891), and Beille (1904,
p. 381).

Berkeley (1860, p. 41) attributed the death of the grasses to the effect
of the death of the spawn of the fungus, while Jorden (1862) attributed
the death of grasses to the entangling action of the filaments and their
strong "effluvia."

The cause of the death of the grasses was attributed to parasitism on
the part of the fungus by Ballion (1906) and Bayliss (1911), the latter
stating that the roots were directly attacked by the fungus filaments
and killed by toxic excretion.

The impenetrability of the soil to moisture, or the dry condition of
the soil, was believed by Ritzema Bos (1901) and Molliard (1910) to be
the chief cause of death of the grass. The latter attributed part of the
harmful effect to the high content of ammoniacal salts.

In the investigations of the writers the death of the vegetation noted
in rings formed by *Agaricus tabularis* was always accomplished by lack
of available soil moisture. No harmful effect could be noted on the
vegetation as long as the soil contained available water. During years
that were uniformly wet throughout the growing season the vegetation
was not noticeably damaged, while during dry years or periods of drouth
the plant cover would be partly or entirely destroyed just above the
dense mycelium. As stated elsewhere, the penetration of water into
the mass of soil filled with mycelium is very slow if the soil is once dried.
The stimulation of growth hastens the drying out of this soil mass; and,
when once dry, it remains dry even through heavy and continued rain.
During ordinary years, when the moisture supply was sufficient on the
natural sod, drouth was severe over the mycelium-impregnated soil.

If parasitism or toxic secretion were the cause of the death of grasses
in this area, the vegetation would be attacked while still growing and

ure remaine .   .. .. o main
grass roots in active condition.   In order to test the effect
m on the native vegetation, the following experiment wa.
ur.n the dry season in 1916.   A strip about 6 meters lo
wide was selected on the edge of a ring of *Agaricus tal*
he zones were well defined.   This strip extended 1 met
e border line between the withered zone (4) and the bare zo
neters inside of this line over the inner stimulated zone (2
inside (1).   The water was poured on slowly to prevent
ve all parts of the plot approximately the same amount
here was begun on June 12 and continued during the nex
equivalent of 2 inches of rainfall was added to the plo
the plot, in contrast to the dry sod, stood out sharply beca
growth and fresh green color of the vegetation.   A tren
he area just after the irrigation (Pl. 24, A) showed the soil to b
moistened under the bare zone (3) and outer stimulated zo
ch moister under the inner stimulated zone (2).   The upp
in the bare and outer stimulated zone were quite moist
the grass, which had grown up rapidly on the bare zon
ecome dormant, owing to drouth.   The plot was accor
a second time with an equivalent of 3 inches of rainfall
14 (Pl. 23, B) *Grindelia squarrosa, Psoralea tenniflora, Malvc
m, Gutierrezia sarothrae*, and *Artemisia frigida* were all
gorous here than in the natural sod.   Culms of grama gras.
inches high and abundant on the plot, but out in the norm.
he grass had fruited but rarely, they were only 4 to 6 inche
remaining plants in this bare area had revived and spreac
red close examination to distinguish the various zones, alt

total loss of the crop over the areas occupied by the ring. In some areas these rings are very numerous. It is estimated that they influence between 0.5 to 1 per cent of the area shown in the map (fig. 1). Even a 1 per cent reduction in gross yield is important when deducted from the net profit. Although there are areas where no fairy rings occur, other areas are badly infested with them. Often areas are found where the rings are more abundant than shown on the map.

In experimental plots the presence of fairy rings is a serious obstacle. A large ring of *Agaricus tabularis* may reduce the yield of a one-tenth acre plot to 50 per cent. Under a most favorable set of climatic conditions it may greatly increase the yield. It is therefore impossible to rely on results from plots occupied by fairy rings. Many of the inequalities of experimental plots are due to this cause, and the eradication of rings on such plots is as important as the experiments which are conducted on the plots.

Persoon (1819, p. 4) observed that if a ring was dug up successively for one or two years the same species reappeared eventually, but it was solitary and scattered.

Sorauer (1886 p. 270-272) stated that repeated digging up of the soil at the periphery of the ring would dry and kill the fungus.

McAlpine (1898) recommended soaking the ground thoroughly with sulphate of copper (or Bordeaux mixture), sulphate of iron 10 per cent, and strong lime water to kill the fungus.

Ritzema Bos (1901) stated that the rings could be destroyed by digging them up in dry, hot weather and exposing the soil to sunshine.

The principal methods suggested for the destruction of fairy rings have been the application of sulphates of copper and iron or repeated stirring of the soil, especially during dry periods. The latter method would be best when applied to experimental plots, since it would not introduce into the soil chemical substances which might possibly have a subsequent effect on crop production. Where the method of digging up the soil is not practicable, the application of fungicides, such as Bordeaux mixture, to the soil immediately over and a little in advance of the rings should be effective.

SUCCESSIONS INDUCED BY FAIRY RINGS

None of the previous investigators made a study of the revegetation following the destruction of the vegetation in the fairy ring. Tulasne and Tulasne (1851, p. 157) mentioned the fact that the production of weeds is due to the digging up of the soil by truffle hunters. Ballion (1906) noted the fact that at the first rains in the fall the perennials on the denuded zone, like "*ravenelles*" (wallflower) and "*oseille*" (sorrel), began to grow again and annuals sprang up from seeds fallen on the bare ground. None of the other investigators mentioned the effect on revegetation.

The effect of *Calvatia cyathiformis* and the other fungi which do not produce a bare zone is the temporary stimulation of the natural vegetation. This influence seldom covers a zone more than a meter wide. The effect on the short grasses is to thicken the sod during years of ample moisture supply (Pl. 27, B.).

Agaricus tabularis kills out the vegetation and initiates a succession (fig. 15) not essentially different from that occurring on abandoned tilled land (Shantz, 1911) or on abandoned roads (Shantz, 1917). The original vegetation is often entirely killed. On the inner part of the area of bare ground weeds develop to form the first stage in the succession. In this area, which lies above the dense mycelium, the weeds which start growth during the spring rainy season are usually killed by drouth while still very small.[1] This area of stunted annuals (fig. 15) constitutes the first or early-weed stage in the succession and consists chiefly of the an-

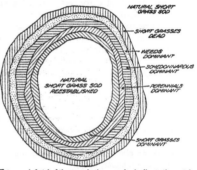

Fig. 15.—A sketch of the stages in the succession leading to the reestablishment of the short grass following its destruction by *Agaricus tabularis*.

nuals which are ordinarily abundant in the native sod, such as *Plantago purshii* R. and S., *Festuca octoflora* Walt., *Hedeoma hispida* Pursh., *Lepidium ramosissimum* A. Nels.

During a normal year these plants dry off about the middle of June and the area remains bare for the remainder of the season. Just inside this zone and not sharply separated from it is the inner stimulated zone. This zone is in the second or late weed stage and is dominated by a luxuriant growth of the plants listed in the first stage and the following more characteristic ruderals: *Chenopodium incanum* (S. Wats.) Heller, *Lappula occidentale* (S. Wats.) Greene, *Cryptanthe crassisepala* (T. and G.) Greene, *Amaranthus blitoides* (Wats.).

Following the late-weed stage several of the stages clearly recognized on abandoned roads or cultivated fields can be distinguished.

The third, or short-lived grass, stage is marked by *Schedonnardus paniculatus* (Nutt.) Trel., alone or with *Malvastrum coccineum* (Pursh.) Gray, and *Sitanion hystrix* (Nutt.) Smith.

---

[1] If deep-rooted perennials such as *Artemisia frigida*, *Gutierrezia sarothrae*, etc., occur in this zone, they occasionally continue to grow, due to a supply of available soil moisture in the deeper layers of the soil.

The fourth or perennial stage is shown by the entrance of _Gutierreza sarothrae_ (Pursh.) B. and R. and _Artemisia frigida_ Willd.

The fifth, or short-grass, stage is established gradually and _Bulbilis dactyloides_, which in this case must depend largely on reseeding, does not predominate over _Bouteloua gracilis_. As a result the two short grasses are established and the final stage reached without the development of the characteristic Bulbilis stage.

In figure 13 a bisect and transect of a ring of _Agaricus tabularis_ is shown. The zones here are unusually wide, and the mapping made in 1916 is characteristic of a dry year. Very few annuals appear. During 1915 the early weed stage to some extent, but especially the late weed stages, were marked by a very luxuriant growth and a dense stand. Such annuals as _Festuca octoflora_, and _Plantago purshii_ occurred abundantly in every zone and stage of revegetation. This ring is shown because it illustrates the succession stages somewhat better than the narrower rings in which the short-lived grass stage usually occupies a much narrower zone.

The zone of disturbed vegetation is usually about 3 meters broad in these rings of _Agaricus tabularis_. Estimates of the rate of advance of this fungus given earlier in the paper were 12 cm. per year. At this rate the time required for the vegetation to become reestablished would be 300 ÷ 12 or 25 years. This estimate is in accord with previous studies in succession on abandoned plowed fields (Shantz, 1911) and on abandoned roadways (Shantz, 1917).

GENERAL DISCUSSION

Under normal conditions in eastern Colorado _Agaricus tabularis_ produces an enormous superabundance of spores. From the number and age of the fairy rings which are shown in figure 2, new rings are only rarely formed. Most of the rings are fragments of old rings which have long since ceased to be true or complete circles. The conditions necessary for the germination of spores of _Agaricus_ spp. have been the subject of many investigations. More recent summaries and studies (Ferguson, 1902; and Duggar, 1905) show clearly that the conditions most favorable to germination would be seldom realized on the Great Plains. Nor would there be chances of reproduction by fragments of tissue. The effects of the presence of old fruits and mycelium on germination of spores would only tend to cause germination where the mycelium already existed. Only rarely do spores germinate and produce new rings. At first these would not present the appearance of older fairy rings, but would merely represent small more or less circular areas in which sporophores are produced and the native grasses are stimulated (Pl. 30, C.). The mycelium would spread out in all directions as practically all fungi do in culture media. During the first year or two there would be no differentiation into zones, but subsequent gradual outward growth would result in the separation of the zones shown in figure 3 (at A). Outward growth would be slow, approximately 12 cm. a year on an average. This progress would not be regular, but would be wavelike, being comparatively rapid, 30 to 60 cm. during favorable years and very slow or none at all during unfavorable years. The sod would first be stimulated by the increase of available nitrogen, owing to the reduction of the organic matter of the soil, then killed by insufficient soil moisture in the area of dense mycelium. As this mycelium began to decay, the native ruderals would again invade and develop luxuriantly, owing to the abundant supply of readily available nitrogenous material. These ruderals would give way to short-lived grasses, and these in turn to the short grasses of the natural sod To use a figure already employed by Ritzema Bos (1901), the spread of the mycelium would resemble the spread of a flame started by dropping a match into dry grass. Even if the grass were reestablished at a distance of a meter or so behind the flame, the chances of the flame's striking back would be very remote. We may think of the flame as representing the active mycelium and the grass as representing the organic matter of the soil on which the mycelium feeds.

The fairy ring starts from the point of germination of the fungus spore and spreads outward at approximately an equal rate in all directions. In case of an obstacle, such as an ant hill, or another fairy ring, growth stops at this point. In the case of an ant hill the ring will close in around it as soon as it is passed and tend to regain its original complete form. In case two rings meet they do not continue, but are exterminated at the line of contact. As the fungus filaments spread outward they consume a portion of the organic matter of the soil. Carbohydrates are consumed, and the proteid portion is consumed or changed into amino acids and then to ammonia. The fungus advances and never recedes, since the young growing filaments are always advancing into new soil and since only the old filaments lie on the inside of the ring where the available portion of the organic matter of the soil has already been consumed by the fungus.

The effect of the fungus filaments on the soil is to reduce a part of the organic matter to ammonia, which is combined to form ammoniacal salts or is converted by bacteria into nitrites and later into nitrates. When the mycelium dies, it is reduced by bacterial action to ammonia, which may later be built up into nitrates. The increase in available nitrogenous material in the soil occupied by the young mycelium stimulates the growth of the grasses or other plants, which consequently make greater demands on the soil moisture. When this is once exhausted (in *Agaricus tabularis*), the mass of fungus filaments prevents the penetration of rain water. The intense drouth to which plants are thus subjected kills off the buffalo and grama grasses and the other plants which may be associated with them. The area is thus left bare for the invasion of other plants. The mycelium after a few years dies, leaving the soil still more enriched and no longer impervious to water.

The first stage in the succession on this bare area is (1) an early-weed stage, followed by (2) a late-weed stage. This is followed by a (3) short-lived grass stage, and this by a (4) perennial stage, which gives way to the (5) original short-grass cover.

LITERATURE CITED

ATKINSON, G. F.
  1900. MUSHROOMS, EDIBLE, POISONOUS, ETC. 275 p., 223 fig. Ithaca, N. Y.
BALLION, Paul.
  1906. RECHERCHES SUR LES CERCLES MYCÉLIENS (RONDS DE FÉES). *In* Actes Soc. Linn. Bordeaux, v. 61 (s. 7, t. 1), p. lxii–lxxxviii
BAYLISS, Jessie S.
  1911. OBSERVATIONS ON MARASMIUS OREADES AND CLITOCYBE GIGANTEA, AS PARASITIC FUNGI CAUSING "FAIRY RINGS." *In* Jour. Econ. Biol., v. 6, no 4, p. 111–132, fig. A–G, pl 5–7.
    Cites Waring (p 112).
BEILLE, L.
  1904. PRÉCIS DE BOTANIQUE PHARMACEUTIQUE. t. 1. Paris
BERKELEY, M. J.
  1860. OUTLINES OF BRITISH FUNGOLOGY ... 442 p , 24 pl. (partly col) London
BRADLEY, Richard.
  1717. NEW IMPROVEMENTS OF PLANTING AND GARDENING ... pt. 1–3, 5 pl. London
BUCKMAN, James.
  1870. THE PRINCIPLES OF BOTANY. *In* Veterinarian, v 43, no. 509 (s 4, no. 185), p. 367–373, no 510 (s. 4, no. 186), p 439–444. (Continued article.)
COOKE, M. C.
  1866 FAIRY-RING CHAMPIGNON *In* Sci Gossip, 1866, no. 22, p. 225–227, fig 213–214.
COULTER, J. M., BARNES, C. R., and COWLES, H. C.
  1911. A TEXTBOOK OF BOTANY FOR COLLEGES AND UNIVERSITIES. v. 2, ECOLOGY New York, Cincinnati, Chicago.
COVILLE, F V.
  1897. OBSERVATIONS ON RECENT CASES OF MUSHROOM POISONING IN THE DISTRICT OF COLUMBIA U. S. Dept. Agr., Div. Bot. Circ. 13, 21 p., 21 fig.
  ———
  1898. THE FAIRY-RING MUSHROOM *In* Plant World, v. 2, no. 3, p. 39–41, 3 fig
DUGGAR, B. M.
  1905. PRINCIPLES OF MUSHROOM GROWING AND MUSHROOM SPAWN MAKING. U. S Dept. Agr. Bur. Plant Indus. Bul. 85, 60 p , 7 pl.
  ———
  1915. MUSHROOM GROWING. 250 p., 31 pl. New York, London,
DUTROCHET, Henri.
  1834. OBSERVATIONS SUR LES CHAMPIGNONS. *In* Nouv. Ann. Mus Hist. Nat. [Paris], t. 3, p. 59–79, pl. 4 (col)
  ———
  1837. MÉMOIRES POUR SERVIR A L'HISTOIRE ANATOMIQUE ET PHYSIOLOGIQUE DES VÉGÉTAUX ET DES ANIMAUX. t. 2. Paris.
FERGUSON, Margaret C.
  1902. A PRELIMINARY STUDY OF THE GERMINATION OF THE SPORES OF AGARICUS CAMPESTRIS AND OTHER BASIDIOMYCETOUS FUNGI. U. S. Dept. Agr. Bur. Plant Indus. Bul. 16, 43 p., 3 pl. Bibliography, p. 39–40.

FULTON, H. R.

  1906. CHEMOTROPISM OF FUNGI. *In* Bot. Gaz., v. 41, no. 2, p. 81-108. Literature cited, p. 107-108.

GILBERT, J. H.

  1875. NOTE ON THE OCCURRENCE OF "FAIRY-RINGS." *In* Jour. Linn. Soc. Bot. [London], v. 15, no. 81, p. 17-24.

GILLET, C. C.

  1874. LES HYMÉNOMYCÈTES, OU DESCRIPTION DE TOUS LES CHAMPIGNONS (FUNGI) QUI CROISSENT EN FRANCE . . . 828 p. Alençon.

GREVILLE, R. K.

  1828. SCOTTISH CRYPTOGAMIC FLORA. v. 6. Edinburgh, London.

HUTTON, James.

  1790. OF CERTAIN NATURAL APPEARANCES OF THE GROUND ON THE HILL OF ARTHUR'S SEAT. *In* Trans. Roy. Soc Edinb , v. 2, pt 2, p. 1-11.

JORDEN, George.

  1862. SOME OBSERVATIONS ON FAIRY RINGS, AND THE AGARICS THAT PRODUCE THEM. *In* Phytologist, [n s], v. 6, p 236-238

KUPERUS, D. J.

  1876. DUIVELS KARNPAD. *In* Landbouw Courant, jaarg. 30, no. 19, p. 91

LATHROP, E. C.

  1917. THE ORGANIC NITROGEN COMPOUNDS OF SOILS AND FERTILIZERS. *In* Jour. Franklin Inst., v. 183, no 2, p. 169-206, no. 3, p 303-321, no. 4, p. 465-498. Literature cited, p. 489-498.

LAWES, J. B., GILBERT, J. H , and WARINGTON, R

  1883. CONTRIBUTION TO THE CHEMISTRY OF "FAIRY RINGS " *In* Jour. Chem. Soc. [London], v. 43, p. 208-223

LEES, Edwin.

  1869. ON THE FORMATION OF FAIRY RINGS AND THE FUNGI THAT INHABIT THEM. *In* Trans. Woolhope Nat Field Club, 1868, p. 211-229

LLOYD, C. G.

  1916. CALVATIA POLYGONIA, FROM H L SHANTZ, EASTERN COLORADO. *In his* Letter no. 62, p. 8, note 444.

LUDWIG, Friedrich.

  1906. PILZRINGEL UND PILZWURZELN. *In* Prometheus, v. 17, no. 865, p 522-526 fig. 427-428.

McALPINE, Daniel.

  1898. "FAIRY RINGS," AND THE FAIRY-RING PUFF-BALL 12 p., 2 fold. pl. Melbourne.

MARCHAL, Émile.

  1902. IN BELGIEN IM JAHRE 1901 BEOBACHTETE PILZPARASITÄRE KRANKHEITEN. *In* Ztschr. Pflanzenkrank., Bd. 12, Heft 1/2, p. 47-49.

  ——

  1902. RAPPORT SUR LES ENNEMIS VÉGÉTAUX DES PLANTES PRÉSENTÉ AU CERCLE D'ÉTUDES AGRONOMIQUES PENDANT L'ANNÉE 1901. *In* Bul. Agr. [Brussels], t. 18, livr. 2, p. 228-230.

MASSART, Jean.

  1910. SUR LES RONDS DE SORCIÈRE DE MARASMIUS OREADES FRIES. *In* Ann Jard. Bot. Buitenzorg, sup. 3, pt. 2, p. 583-586, 1 fig.

MOLLIARD, Marin.

  1910. DE L'ACTION DU MARASMIUS OREADES FR. SUR LA VÉGÉTATION. *In* Bul Soc. Bot. France, t. 57 (s. 4, t. 10), no. 1, p. 62-69, 2 fig., pl. 5.

MÜNCH, Ernst.

  1914. ÜBER HEXENRINGE. *In* Naturw. Ztschr. Forst- und Landw., Jahrg. 12, Heft 3, p. 133-137, 2 fig.

OLIVIER, Ernest.
    1891. LES RONDS DES SORCIERS. *In* Rev. Sci. Bourbon., Ann. 4, p. 170.
PERSOON, C. H.
    1819. TRAITÉ SUR LES CHAMPIGNONS COMESTIBLES, CONTENANT L'INDICATION DES
        ESPÈCES NUISIBLES. 276 p , 4 col. pl. Paris.
RAMALEY, Francis.
    1916. MUSHROOM FAIRY RINGS OF TRICHOLOMA PRAEMAGNUM. *In* Torreya, v. 16,
        no. 9, p. 193–196, illus.
REED, H. S.
    1910. AN INTERESTING MARASMIUS FAIRY RING. *In* Plant World, v. 13, no. 1,
        p. 12–14, fig. 6.
RITZEMA Bos, Jan.
    1901. "HEKSENRINGEN," "KOL-" OF "TOOVERKRINGEN," "DUIVELS KARNPAD"
        OP WEILANDEN. *In* Tijdschr. Plantenziekten, jaarg. 7, afl. 4, p. 97–126.
        Cites Westerhoff.
SCHREINER, Oswald
    1913. THE ORGANIC CONSTITUENTS OF SOILS. U. S Dept. Agr. Bur. Soils Circ. 74,
        18 p.
────── and LATHROP, E. C.
    1912. THE CHEMISTRY OF STEAM-HEATED SOILS. U. S. Dept. Agr. Bur. Soils Bul.
        89, 37 p.
────── and SHOREY, E. C.
    1910. CHEMICAL NATURE OF SOIL ORGANIC MATTER. U. S. Dept. Agr. Bur. Soils
        Bul. 74, 48 p., 1 pl.
────── and SKINNER, J. J.
    1912. NITROGENOUS SOIL CONSTITUENTS AND THEIR BEARING ON SOIL FERTILITY
        U. S. Dept. Agr. Bur. Soils Bul. 87, 84 p., 11 pl. Reference list, p. 83–84.
SHANTZ, H. L.
    1911. NATURAL VEGETATION AS AN INDICATOR OF THE CAPABILITIES OF LAND FOR
        CROP PRODUCTION IN THE GREAT PLAINS AREA. U. S. Dept. Agr. Bur.
        Plant Indus., Bul. 201, 100 p., 23 fig., 6 pl.

    1917. PLANT SUCCESSION ON ABANDONED ROADS IN EASTERN COLORADO. *In* Jour.
        Ecol., v. 5, no. 1, p. 19–42, 23 fig.
SORAUER, Paul.
    1886. HANDBUCH DER PFLANZENKRANKHEITEN . . . Aufl. 2, T. 2. Berlin.
STAHL, E.
    1900. DER SINN DER MYCORHIZENBILDUNG. *In* Jahrb. Wiss. Bot. [Pringsheim], Bd.
        34, Heft 4, P. 539–668, 2 fig.
THOMAS, Friedrich.
    1905. DIE WACHSTUMSGESCHWINDIGKEIT EINES PILZKREISES VON HYDNUM SUAVE-
        OLENS SCOP. *In* Ber. Deut. Bot. Gesell., Bd 23, Heft 9, p. 476–478.
TREICHEL, A.
    1889. HEXENRINGE UND KÖRPERFÖRMIGE GRASFEHLE. *In* Verhandl. Berliner
        Gesell. Anthrop. Ethnol. und Urgesch., 1889, p. 352–355.
TULASNE, L. R., and TULASNE, Charles.
    1851. FUNGI HYPOGÆI. HISTOIRE ET MONOGRAPHIE DES CHAMPIGNONS HYPOGÉS.
        222 p , 21 pl. Parisiis.
VAN TIEGHEM, P. E. L.
    1884. TRAITÉ DE BOTANIQUE. 1656 p., 803 fig. Paris.
WAY, J. T.
    1847. ON THE FAIRY-RINGS OF PASTURES. (Abstract.) *In* Rpt. 16th Meeting Brit.
        Assoc. Adv. Sci., 1846, Notices and abst. Com., p. 43–44.

WILLIAMS, E. M.
  1901. FAIRY RINGS. *In* Plant World, v. 4, no. 11, p. 206–207, 1 fig.
WILLIAMS, Mabel E.
  1897. THE FAIRY RING AND ITS NEIGHBORS. *In* Asa Gray Bul., v. 5, no. 6, p. 94–98, 4 fig.
WITHERING, William.
  1796. AN ARRANGEMENT OF BRITISH PLANTS . . . Ed. 3, v. 4. London.
WOLLASTON, W. H.
  1807. ON FAIRY-RINGS. *In* Phil. Trans. Roy. Soc. London, 1807, pt. 2, p. 133–138.

PLATE 10

Agaricus tabularis:

A —Three fruiting bodies about 10 to 15 cm. in diameter showing variation in the pileus   Akron, Colo , June 14, 1915

B —Large, firm, fleshy fruiting bodies 10 to 15 cm. in diameter.   Akron, Colo., July 7, 1909.

C —Three typical fruiting bodies   From left to right these weighed respectively 207, 298, and 163 gms. each.   The largest was about 17 cm. in diameter.   Akron, Colo., July 12, 1909.

D.—Bottom view of fruits 7 to 15 cm. in diameter.   Akron, Colo., June 14, 1915.

(246)

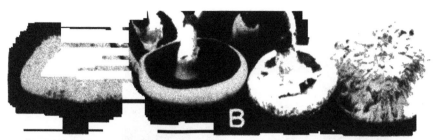

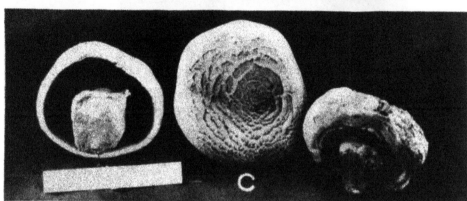

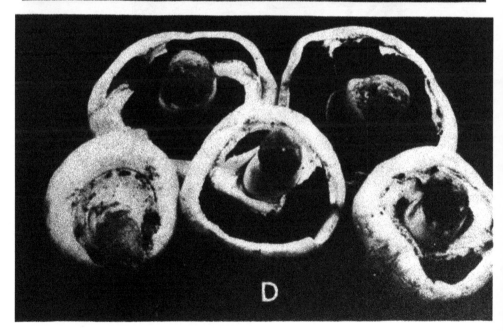

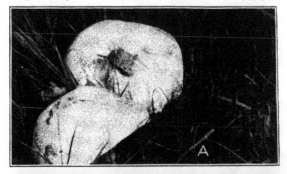

PLATE 11

Agaricus tabularis:

A.—Sporophores developed in a dense stand of *Agropyron smithi*, which protected them from rapid drying.   In this location the tops were smooth.   The fruiting body was 12 cm. in diameter.

B.—Sporophore from the same ring as figure A, but freely exposed and showing the rough scaly top.   The fruiting body was about 12 cm. in diameter.

C.—Typical of fruiting bodies exposed on dry hot days.   Fruiting bodies 10 cm. in diameter.   The figures of Plate 11 show variation in pileus.   Akron, Colo., June 15,

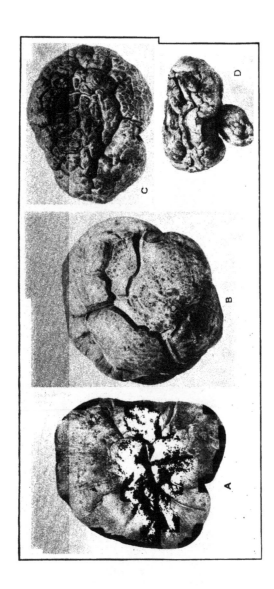

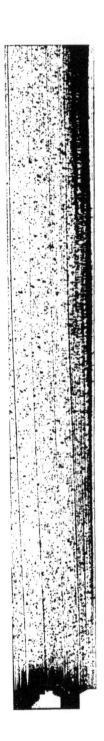

PLATE 13

Agaricus tabularis:

;s showing large fruiting bodies about 15 cm. in diameter and
:e.   Akron, Colo., June 7, 1909.

the fruiting period.   The area in the ring is distinguished by
1all plants of *Plantago purshii* and *Festuca octoflora* and unusu-
Artemisia frigida, the roots of which penetrate below the dry
The inside of the ring is marked by the abundance of annuals,
e most prominent and abundant.   Akron, Colo., June 25, 1915.

PLATE 14

Agaricus tabularis:

A.—A fruiting ring The vegetation outside of the ring is pure short grass with *Festuca octiflora* and *Plantago purshii*. At the outer edge of the ring 18 mushrooms occur in the portion shown in the photograph Just back of the fruiting area the short grass is dead or dying. In this area there are a few very small plants of *P. purshii*, *Lepidium ramossissimum*, and *Chenopodium incanum*. The inside shows a rank growth of *L. ramossissimum*, *Lappula occidentale*, *P. purshii*, and *Festuca octiflora*. Akron, Colo., June 14, 1915.

B —The mycelium as it appears on the soil mass. So dense is this fungus growth that all crevices in the soil are filled with white hyphæ. Water penetrates very slowly into a dry soil in this condition. Akron, Colo., August 14, 1915.

C.—A trench across a ring of *A. tabularis* in a Kubanka wheat plot, showing the mycelium in the soil A sketch of this trench is shown in figure 14 The mycelium in the ring fills the soil lying from 10 to 30 cm. in depth. The ring is about 4 meters wide, and on the inside of this area the mycelium occurs sparingly at a depth exceeding 30 cm. The mycelium may have grown down as it died above. The soil is very dry in the mycelium area, but moist where no mycelium occurs. The available water was 0 6 per cent in the mycelium and 11.6 per cent in the soil outside the mycelium area. After a rain the soil above the mycelium remained very wet, owing to the failure of moisture to penetrate the dry mycelium layer. Only that portion of the soil above the mycelium (3 or 4 inches) was moistened by the rain on August 6 and 7 of 1.89 inches, while the soil in the adjacent areas was wet to a depth of over 1 foot. Akron, Colo., August 10, 1915.

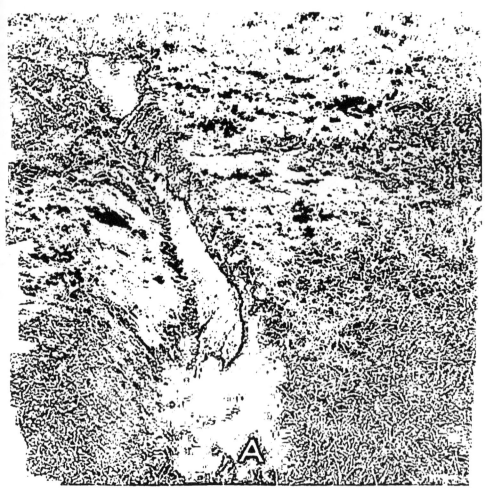

Agaricus tabularis:

A.—A trench across a fairy ring. This reproduction of a photograph which was taken after a rather heavy rain shows the dry soil area (light-colored) in the ring and extending back inside the ring. The darker soil above is freer from mycelium, while the light soil is filled with a dense growth of mycelium. In the foreground the moisture has penetrated around the end of the mycelium and back underneath, showing distinctly the difficulty with which the dry soil filled with mycelium is wetted. The mycelium comes to within 2 or 3 inches of the surface of the soil, and is usually about 1 foot in advance of the fruiting bodies. Akron, Colo., August 2, 1915.

B.—Another trench showing a different distribution of the mycelium in the soil. Akron, Colo., August 2, 1915.

PLATE 16

Agaricus tabularis:

A.—General appearance of a fairy ring formed by *A. tabularis* during a dry year. The bare zone shows clearly at the right. Scattered through this are a few remnants of the short grass and also a few plants of *Artemisia frigida*. The roots of the latter are partly below the soil occupied by the mycelium. Just at the left of this bare area the grass is closely cropped and withered This is the outer stimulated zone of wet periods. The illustration does not show the difference between this zone and the normal vegetation at the left. Akron, Colo., August 7, 1916.

B.—General view of a ring formed by *A. tabularis;* photographed during a dry year. The vegetation at the left is normal. Just at the right the withered zone extends to the lower right corner of the figure. The vegetation is dry and a little lighter colored in this zone. Several old sporophores produced in 1915 are still found in this zone. The darker area at the right is the bare zone and within this the inner stimulated zone, which, because of the dry period, is not luxuriant. A detailed transect and bisect of this ring are shown in figure 13. Akron, Colo., June 11, 1916.

A

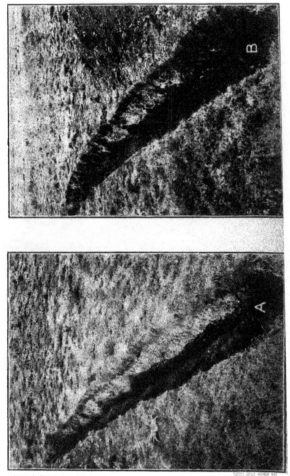

Agaricus tabularis:

A ring which produced only a few mushrooms in 1915   It is marked by the stimu-lated growth of *Festuca octoflora*.   In this stimulated area three distinct maxima can be distinguished which form concentric rings.   The first maximum occurs just inside the fruiting zone, the second about 2 feet inside, and the third about 4 feet inside the fruiting zone.

mis-bound
pp. 316 & 317.

RUN-OFF FROM THE DRAINED PRAIRIE LANDS OF SOUTHERN LOUISIANA

By Charles W. Okey,

Senior Drainage Engineer, Office of Public Roads and Rural Engineering, United States Department of Agriculture

INTRODUCTION

While the draining of low-lying lands by means of pumps has long been practiced in parts of England and Holland, the application of this means of securing drainage for lands in the United States is a comparatively recent development. A small amount of pumping of drainage water had been done on a number of sugar plantations in southern Louisiana before the middle of the last century, but these pumping plants were crude affairs. About 1908 the work of reclaiming the Gulf coast marsh lands by means of drainage pumping was actively started. Gradually, modern machinery was introduced until by the end of 1915 about 125,000 acres of land were under pump. Prior to the starting of this work there existed no very complete nor accurate data as regards the amount of water these plants would be called on to pump each year. Neither were there records as to what the maximum capacity of the plants should be. While there were available some more or less fragmentary records of the operation of drainage pumping plants in Europe, the conditions of climate, rainfall, and soil in the United States are so different that European records were not considered to be of practical value. The capacities of the earlier pumping plants were based upon very rough approximations. They were more often too small than too large. The need for information on this subject soon became so apparent that in 1909 Drainage Investigations, Office of Experiment Stations, United States Department of Agriculture, undertook the task of collecting such data. The work was actually begun in the month of April, 1909, by Messrs. W. B. Gregory and A. M. Shaw, who made the necessary measurements of the capacity of four drainage pumping plants and started a set of records of operation. Late in 1909 the work of carrying on these investigations was assigned to the writer and has been continued under his charge to the present time. The work of rating the capacity

Journal of Agricultural Research,
Washington, D. C.
kk

Vol. XI, No. 6
Nov. 5, 1917
Key No. D—14

(247)

of the pumps in the drainage plants has been extended along with records of operation of the same plants until the records now cover 10 drainage districts. At the end of each year progress reports have been issued giving the results collected. It is now proposed to summarize these reports in one publication.

OBJECT OF THE INVESTIGATIONS

The primary object of this work has been to establish the relation existing between the rainfall and the amount of water that it is necessary to pump from typical areas of land in order to secure a degree of drainage that will allow the growing of the ordinary field crops. When this relation is once determined, the rainfall data furnished by the Weather Bureau will enable the engineer to design adequate and economical drainage pumping plants for the swamp lands of the Gulf coast.

The records also disclose the total amount of pumping necessary in each year. In this paper the term "run-off" is used to denote the water that is removed by the pumps; in the tables the run-off is expressed as a uniform depth of water (in inches) over the drained area.

DESCRIPTION OF DISTRICTS

The districts on which the investigations are being carried on vary in area from 647 acres to 7,500 acres, and the various details of soil, crops, surface slope, and character of the drainage channels and levees, are different on each district. Therefore it is believed that it will be of advantage to describe each district, giving a brief summary of conditions prevailing during each year covered by the records

SMITHPORT PLANTING COMPANY TRACT—AREA, 647 ACRES

This tract adjoins the village of Lockport, Lafourche Parish. The land from which this plantation was reclaimed is typical of the open, grass-covered prairies that lie to the west of Bayou Lafourche. The surface slopes from the bayou to the low swamps in the rear. About half the area under consideration was originally covered with the water of Lake Fields to a depth of 1 to 3 feet. The humus or muck on this tract varied in depth from 18 inches on the higher portions to perhaps 8 inches in the lake bed. A considerable portion of the muck in the higher portions of the area is overlain by a layer of river silt about 10 inches thick. Silt of the same character forms the subsoil of the whole district. The total fall in the surface from the higher, or "front" portion to the lower side is about 4 feet. The ditch system is very complete, and the small field ditches go entirely across the area directly into the reservoir canal which skirts one side. The reservoir capacity up to the end of 1912 was 0.40 inch of water, and that of the pumping plant 1.10 inches, over the area drained. Thereafter the reservoir capacity was increased to 0.50 inch by further dredging. Except for a few short periods the entire area was well drained during the time covered by the records. During the early part of the period covered by the records the levees on this district were subject to a heavy seepage when the stage of the water was high on the outside. Later the levees were increased in size and the seepage was reduced greatly

CONDITIONS DURING 1909.—During this year about 25 per cent of the area was in cultivation. The remainder was covered with grass and weeds. The portion of the land cultivated was well drained throughout the year. Some of the lowest land near the reservoir was flooded for a short time during the heavy rainfall of September 20, but this was due to the fact that the tropical hurricane which accompanied the rainfall interfered with the proper operation of the machinery.

CONDITIONS DURING 1910.—About 65 per cent of the land was cultivated, and the entire area was well drained. The year was deficient in total rainfall and in heavy storms. The early part of the year was exceptionally dry.

CONDITIONS DURING 1911.—The cultivated area was increased to 75 per cent and drainage was maintained. Some quite heavy rainfalls occurred. Toward the end of the year fairly heavy rainfall coming at a time of low evaporation raised the water outside the district so that very heavy seepage came through the levees. The records for December (Table I) do not represent normal relations as between rainfall and amount pumped; they were not included in obtaining averages for the period.

CONDITIONS DURING 1912.—The cultivated area amounted to 85 per cent. A continuation during January of the high stage of water outside the district caused the amount of water pumped to be greatly above normal, and a recurrence of this stage of outside water in June and July, and again in December, caused the pumping for the year to be excessive. The averages for the tract exclude the above months.

CONDITIONS DURING 1913.—During this year the entire tract was brought under cultivation for the first time. The water level in the main canal was held unusually low, and drainage was complete at all times, with no flooding. During the year nearly the entire length of levee received further addition of material, and its ability to keep out seepage water was thereby much increased. It does not appear from the records that any great amount of seepage water entered the district except during the first part of January. Although the run-off during the month of October was greater than the rainfall, most of the rainfall came the last few days of September and was not pumped during the same month. By taking the two months together the percentage of rainfall in the run-off is about 75 per cent—not an excessive amount for such heavy rainfall at that time of year.

TABLE I.—*Rainfall and run-off on the Smithport Planting Company tract, June, 1909, to September, 1915, inclusive*

Month	Rainfall.	Runoff.	Rainfall in runoff.	Pump ran.	Rain fell.
	Ins.	*Ins.*	*P. ct.*	*Days.*	*Days.*
1909.					
June	10.63	4.20	39.5	12	12
July	2.19	.46	21.0	2	7
August	8.62	3.40	39.5	6	12
September	5.61	2.71	48.3	7	7
October	2.90	1.62	55.8	4	3
November	2.00	.00	.0	0	5
December	5.26	3.44	65.4	10	11
Total	37.21	15.83	42.5	41	57
1910.					
January	2.15	1.26	58.6	7	4
February	4.15	1.42	34.2	7	8
March	1.00	.47	47.0	3	2
April	.12	.00	.0	0	2
May	3.29	.12	3.6	1	7
June	5.43	.17	3.1	1	11
July	10.39	4.96	47.7	13	17
August	3.15	.28	8.9	4	9
September	3.84	.58	15.1	7	6
October	2.68	.37	13.9	3	8
November	2.52	.25	10.0	3	4
December	2.76	.95	34.4	6	6
Total	41.48	10.83	26.1	55	84
1911.					
January	1.50	.93	62.0	5	5
February	1.73	.66	38.2	5	5
March	2.05	.57	27.8	4	4
April	6.44	3.02	46.9	9	4
May	3.08	1.40	45.4	7	7
June	2.29	.00	.0	0	8
July	7.50	1.97	26.3	16	15
August	10.87	5.15	47.4	16	15
September	3.71	.82	22.1	7	12
October	4.07	1.97	48.4	8	6
November	4.81	2.14	44.5	6	5
December	6.51	7.21	110.7	19	10
Total	54.56	25.84	47.4	102	96
1912.					
January	3.85	6.83	177.4	16	8
February	3.78	2.61	69.1	7	5
March	3.66	2.10	57.4	7	7
April	3.62	1.87	51.6	7	10
May	6.09	4.65	76.3	13	6
June	7.73	10.74	132.0	24	9
July	5.39	6.90	128.0	19	11
August	5.50	1.53	27.8	5	10
September	1.99	.24	12.1	1	3
October	2.18	.75	34.4	2	4
November	3.72	1.64	44.1	8	3
December	11.35	10.98	96.7	22	15
Total	58.86	50.34	85.5	131	91

Month	Rainfall.	Runoff.	Rainfall in runoff.	Pump ran.	Rain fell.
	Ins.	*Ins.*	*P. ct.*	*Days.*	*Days.*
1913.					
January	4.31	4.11	95.3	12	6
February	2.60	1.76	67.7	7	7
March	4.57	3.27	71.6	10	7
April	2.68	1.05	39.2	3	4
May	3.49	1.69	48.6	3	4
June	2.77	.22	7.9	1	7
July	10.78	4.68	43.5	10	11
August	6.47	4.14	62.2	9	9
September	14.26	6.34	44.4	10	15
October	3.53	7.45	211.0	13	7
November	2.14	1.24	58.0	2	3
December	2.41	1.78	74.0	6	6
Total	60.01	37.73	62.9	86	86
1914.					
January	1.34	1.15	85.8	3	3
February	5.71	2.98	52.2	6	4
March	3.38	3.38	100.0	9	4
April	1.43	.75	52.4	3	3
May	1.43	.43	30.1	2	3
June	1.95	.49	25.1	2	2
July	10.01	4.81	48.5	10	15
August	6.06	3.71	61.2	12	9
September	4.22	2.04	48.3	7	5
October	2.31	2.98	98.7	4	2
November	10.01	7.73	77.2	10	7
December	4.25	6.39	150.4	18	5
Total	52.10	36.14	69.4	86	61
1915.					
January	5.03	4.68	93.0	12	5
February	5.56	3.35	60.3	10	6
March	2.23	2.41	108.0	6	4
April	.00	.00		0	0
May	2.58	.52	20.2	2	4
June	1.44	2.36	164.0	4	4
July	5.96	3.31	55.5	5	9
August	5.70	2.87	50.3	11	10
September	2.11	.31	14.7	1	3
October [a]	......	......	......	......	......
November [a]	......	......	......	......	......
December [a]	......	......	......	......	......
Total	10.61	19.81	64.7	51	45
Averages for period (5.91 years)	50.8	26.2	51.6	76.5	79.2

[a] Records omitted, as they do not represent normal conditions.

CONDITIONS DURING 1914.—During this year the whole of the tract was under cultivation. With the exception of one period in April, when the district was flooded with water from the adjacent highland, it was well drained throughout the year. The rainfall and amount pumped during the above-mentioned period are omitted from Table I, as they do not represent normal conditions. In November a very heavy rainfall occurred, which gave the highest rate of pumping, due to rainfall, so far experienced on this district (Table I). While the reservoir was full for

3 days, no flooding of the surface occurred. A small amount of seepage entered the district during the last day or two of November and the first 10 days of December. Continued drainage and cultivation had so compacted the humus or muck which originally covered the surface on this district that the top soil was no longer greatly different in appearance from that encountered on the average cultivated fields on near-by plantations. Careful examination would still show the origin of the top soil, but it had lost all of its turfy characteristics.

CONDITIONS DURING 1915.—The entire area was cultivated and well drained until the end of September. The early part of the year was very dry, but abnormally heavy rains occurred in the last few days of September and the early part of October and caused overflow from the adjoining highland. This water was pumped out, but it is not included in the records, as it did not represent typical conditions.

WILLSWOOD PLANTATION—AREA, 2,600 ACRES

This plantation, located about 10 miles above New Orleans, fronts on the Mississippi River. The soil along the river front is of a characteristic sandy nature, becoming finer grained as the distance from the river bank increases. A tract of about 800 acres which makes up the rear of the plantation was at one time a part of the adjacent wet prairie. It was formerly covered with a light growth of willows and the usual prairie grass. The humus or muck was originally 3 or 4 feet deep, although at present, after about 19 years of cultivation, it is well decayed and compacted, making a rather heavy soil. The total fall in the land surface from the front to the rear of the plantation is about 10 feet.

All the small field ditches run from the front to the rear of the plantation, down the greatest slope; but frequent larger cross ditches intercept these field ditches at intervals of about a quarter of a mile and carry the water to larger canals, which run in the same direction as the field ditches. The main reservoir along the rear of the plantation carries the water to the pumping plant. The capacity of the pumping plant was about 1.50 inches of water on the area drained per 24 hours. The capacity of the reservoir was about 0.40 inch of water over the area.

The levee was built around this district from material taken from immediately inside of the levee, and 'yet no appreciable amount of water seeps into the reservoir canals. All water is pumped out promptly from this district, and the reservoir canal is kept nearly empty, in readiness for heavy rains.

In the early spring of 1911 the area was increased by the addition of about 200 acres of river-front land, enlarging the total area to 2,600 acres. The pumping plant capacity was thus reduced to 1.40 inches per 24 hours, but, as the reservoir canals were later cleaned out, the reservoir capacity became about 0.42 inch. Except for high-water conditions in

1913.		*In.*	*P. ct.*		
January.........		5. 20	81. 8		
February.........		1. 51	61. 4		
March..........		2. 46	61. 2		
April...........		3. 23	57. 3		
May.............		2. 40	66. 2		
June............		. 10	5. 0		
July............		. 52	11. 8		
August..........		. 00	. 0		
September......		7. 37	62. 3		
October.........		2. 99	77. 5		
November.......		. 63	39. 9		
December........		. 67	58. 3		3
Total	51. 52	27. 08	52. 6	68	76
1914.					
January..........	1. 02	. 39	38. 2		2
February.........	5. 92	2. 19	37. 0		4
March...........	4. 51	3. 03	67. 2		5
April............	3. 73	1. 52	40. 8		4
May.............	. 88	. 14	15. 9		3
June............	2. 30	. 19	8. 3		6
July............	3. 84	......	......		13
August..........	3. 32	......	......		12
September.......	6. 70	1. 44	21. 5		6
October..........	2. 24	1. 08	48. 2		3
November.......	3. 71	1. 53	41. 2		3
December........	4. 08	3. 54	86. 8		5
Total......	42. 25	15. 05	35. 6	41	56
1915.					
January..........	6. 00	5. 76	96. 0		4
February.........	5. 95	4. 95	83. 1		6
March...........	1. 07	2. 78	260. 0		3
April............	. 00	. 24	......		0
May.............	1. 88	. 00	. 0		6
June............	2. 67	. 00	. 0	0	0
July............	7. 09	1. 54	21. 7	5	5
August..........	7. 69	1. 49	19. 4	5	10
September.......	8. 73	1. 76	20. 2	2	4
October..........	10. 33	14. 23	137. 3	12	5
November........	. 58	. 81	139. 5	4	3
December........	2. 06	. 68	33. 0	3	4
Total......	54. 05	34. 24	63. 4	63	56
Average for period (6.25 yrs.)			48. 9	65. 6	75. 4

LAFOURCHE DRAINAGE DISTRICT NO. 12, SUBDISTRICT NO. 1—AREA 835
ACRES

This district lies about 2 miles southwest of Raceland, mostly in typical
wet prairie land of the Lafourche section. It has a natural slope toward
the pumping plant. The highest portion of the district is perhaps 3
feet above mean low water in the surrounding waterways and the lowest
portion about 2 feet. About three-fourths of the district was originally
covered with a depth of from 1 to 2 feet of turfy humus or muck, while
the remainder was a cypress swamp which had been burned over and
the trees finally killed and leveled Pumping was started about 1908
and has since been continuous. In 1913 the inside reservoir canals were
relocated and made larger, the levees reinforced, and the old pumping
plant replaced by a new one. The present plant, when working at best
economy, has a capacity sufficient to remove a depth of 1.70 inches of
water from the entire district per 24 hours. The reservoir capacity is
about 0.40 inch of water over the district. The district is completely
ditched and practically all cultivated. As parts of the area have been
cultivated for five years, the muck has been greatly compacted and is at
present only a few inches thick; it is quite a heavy soil as compared with
the original condition.

CONDITIONS DURING 1915.—During the early part of the year there was
evidently considerable seepage entering the district. After the first
three months the weather was dry, and the evaporation took up practically the entire rainfall. During the month of October the district
experienced very heavy rains, and the pumping was greatly increased
because of seepage; therefore the records for the month are omitted from
the averages. (See Table III)

NEW ORLEANS LAND COMPANY TRACT—AREA 1,255 ACRES

This tract lies just within the northern city limits of New Orleans and
fronts on Lake Pontchartrain. It is bordered on the west by the New
Basin Navigation Canal and on the east by the outfall canal from the
city pumping station 7. The water on three sides is slightly above sea
level and is perhaps, on the average, about a foot above the average
land surface of the district. The tract was originally a part of the swamp
that bordered Lake Pontchartrain and was covered with a rather heavy
growth of trees, cypress being the predominating variety. This growth
had been cut down when the record started, but only about 100 acres
were cleared and cultivated The remainder was allowed to grow up in
weeds and brush. The soil is soft and spongy on the surface, with a high
percentage of vegetable matter. The subsoil is the usual Sharkey clay
to a depth of from 6 to 10 feet; this is underlain by almost pure sand
which undoubtedly allows a considerable seepage into the district. The
effect of the seepage on the run-off totals for each month is more noticeable during the winter months than at other times, especially if the

winter rainfall be light. The seepage is doubtless nearly constant so that during the months of small rainfall it forms a relatively large portion of the total run-off. The drainage water is collected by a few large canals (not many small ditches were cut when records were started) and discharged by gravity into the city drainage canal. The water was measured by means of a weir and a water register placed in the outfall canal.

CONDITIONS FROM 1909 TO 1912, INCLUSIVE.—During these years the area of cleared and cultivated land was increased so that by the end of 1912 the area well drained was about 400 acres. The remainder of the area was not completely ditched. Such water as entered the main canals was discharged promptly.

CONDITIONS DURING 1913.—The area ditched and cultivated or improved as residence property was increased during the year to about 500 acres. Owing to further ditching the area tributary to the main outfall canal was estimated at 1,180 acres after January 1, 1913. On October 25 the weir was placed farther down the main outfall canal, thus increasing the total area above the weir to 1,255 acres.

CONDITIONS DURING 1914 AND 1915.—During these years the area of drained and improved land was gradually increased. By the end of 1915 nearly all the area had been ditched and about 75 per cent had been cleared. About 300 acres had been improved with residences and sidewalks. A part of the remaining cleared land was under cultivation.

The amounts of water appearing in the column headed "Run-off" in Table IV represent the actual water that found its way into the main canals. These canals were of such size that water was discharged very quickly after it had once reached them. It was only during very extreme drouths that a small amount of water at least was not flowing over the weir. If the same area had been drained by a pumping plant, the run-off during these days of such small flow would have been combined in two or three days in the month, and the total quantity of water pumped would have been less, as the evaporation from the open canals would have been greater. This is shown by the fact that on the pumping districts the water in the canals is often lowered by evaporation alone during dry times

UNIT NO. 1, POYDRAS, LA.—AREA, 2,500 ACRES.

This district fronts on the Mississippi River about 20 miles below New Orleans, near the little town of Poydras, St. Bernard Parish. About 500 acres of the front portion were originally covered with a light growth of timber; the remainder was a part of the open grass-covered prairie typical of that section of the State. The top 3 or 4 feet of the soil in the prairie have a very high percentage of vegetable material, and would be classed as muck. This muck, however, is not as turfy as that originally on the Smithport tract or the districts around Raceland, and seems to have a higher percentage of river silt. The subsoil is Sharkey clay.

The district was first inclosed and the pumps started in March, 1912. At that time not many of the field ditches were cut. The work of cutting the ditches was carried on during the summer months and was finished in October or November. This district has a very complete system of drainage canals, spaced ¼ mile apart, although the usual spacing is ½ mile. The levee is of ample cross section and does not allow any seepage to enter the district. The pumping-plant capacity is approximately 1 inch per 24 hours, and the reservoir capacity is approximately 0.75 inch.

CONDITIONS DURING 1912.—As stated above, the land was not well drained until toward the end of the year. The results for 1912 are interesting chiefly because they show the amount of water to be removed from this kind of soil when it first drained. There were no ponds or other bodies of water within the district. The rainfall records for 1912 are those of the nearest station, New Orleans, distant about 20 miles. While in detail these records differ materially from those that would have been obtained on the tract, the total for the period covered is probably not greatly in error.

CONDITIONS DURING 1913.—During this year the entire area was drained by the pumping plant. The water was held about 4 feet below the surface and no flooding occurred. Owing to the soft nature of the soil, the small field ditches were quite badly choked with soft mud and did not carry the water very rapidly to the canals. During dry periods it was observed that the level of the ground water was considerably below the bottoms of these small field ditches; this indicated subdrainage of the soil by the canals which were spaced only ¼ mile apart. The rainfall records for 1913 were taken from the rain gage on the tract. About 400 acres of that part of the district bordering the river and about 100 acres of the prairie land were in cultivation in 1913.

CONDITIONS DURING 1914.—During the entire year the tract was well drained and the water held about 4 feet below the surface. In March two periods of heavy pumping occurred. In each period the plant operated at full capacity for a little over 24 hours without stopping. The water, however, did not approach nearer than 2 feet to the surface of the ground. About 400 acres of that part of the district bordering the river and about an equal amount of the prairie land were in cultivation. All the land that was cultivated was well drained by field ditches which had been cleared of soft mud and vegetation. The field ditches of the remainder of the tract were still in very poor condition.

CONDITIONS DURING 1915.—The amount of cultivated land in 1915 and the general conditions of field ditches and drainage canals were the same as in the previous year. While the first two months of the year were rather above the average in amount of rainfall, the remainder of the spring and the summer were very dry, except that one heavy rainfall occurred in June. At the end of September this district was flooded by a combination of excessive rainfall and water coming in from the adjacent land;

normal conditions were not restored until the end of the year. (See Table V)

GUEYDAN DRAINAGE DISTRICT, SUBDISTRICT NO. 1—AREA 7,500 ACRES

This district lies in the southwestern part of the State, and its natural features are quite different from those of the districts previously described. The land was taken from the typical wet prairie of southwestern Louisiana. While the surface was covered with a grass similar to that found in the aforementioned districts, the muck is very much shallower and more turfy in character. Before drainage the average depth was somewhat less than a foot. The muck is underlain by a clay subsoil that is very solid and impervious. The surface slope of the district is from north to south, toward the pumping plant, and totals about 2 feet in 3 miles. The area was first inclosed and the pumps started in March, 1912. As the subsoil was solid, the water was lowered very rapidly without damage to canal banks. The work of cutting the field ditches was begun in December, 1911, and was finished during the following summer months. The main drainage canals are spaced ½ mile apart.

The percentages of the rainfall in run-off during the first three months of operation were not calculated, as all the original surface water was not drained out until about the first of June. There were no bodies of open water inclosed in this district. On August 23, 1912, the area of the district was increased from 5,600 to 6,500 acres. For this latter area the capacity of the pumping plant was 1.25 inches per 24 hours and the reservoir capacity was 0.88 inch.

CONDITIONS DURING 1913.—During 1913 this tract experienced heavier precipitation than had any of the districts up to this time. It is apparent from the run-off records (Table VI) that considerable seepage entered the district during the early part of the year, and again in the last four months. While this was sufficient to affect the monthly totals, it does not make a relatively great difference in the maximum run-off for any particular storm. The amount of seepage per day depended on the height of the water on the outside of the district. Very little water appears to have seeped through the levees during a time when the water was not well over the surface of the surrounding undrained prairie. The entire seepage into the district was through a section of levee about 2,000 feet long which crossed a very soft area of land where the muck was from 5 to 15 feet deep and where the levee had not been built of good material nor to full height and cross section. As the drainage canal was immediately inside this section of the levee, the head against this soft material was often 4 to 7 feet. Cultivation was started on the higher parts of the area, and by the end of April 500 acres were plowed. Early in July the area of the district was increased to 7,500 acres by the addition of 1,000 acres of raw prairie land which had been ditched.

CONDITIONS DURING 1914 —Early in 1914 the area of this tract tributary to the pumping plant was decreased to 6,500 acres. Whenever the stage of water in the surrounding prairie was over the general surface there was considerable seepage through the weak section of levee. This is shown in the records for the first five months and the last two months of the year. During the other five months the water in the outside prairie was low, and seepage practically ceased. The total area cultivated was about 600 acres. Of this amount about 300 acres were in rice. Water was taken from the interior drainage channels to irrigate the rice land, and this further decreased the amount of water pumped out of the district during the months from June to August, inclusive.

CONDITIONS DURING 1915.—During the early months of 1915 the amount of water pumped, relative to the rainfall, was very heavy. However, by the middle of May the weak section of levee had been so much strengthened that very little seepage water entered. During the summer months a little over 1,000 acres of rice were grown in the district, and water to irrigate the rice was all taken from the inside drainage channels. It was owing to this practice that only a small amount of water was pumped during the summer months. On December 1 the area was again increased to 7,500 acres by the addition of the 1,000 acres which were drained during a part of the year 1913. As the two months previous to the time this land was again brought into the district had been very dry, no pumping was necessary to remove water from the new land. (See Table VI.)

LITTLE WOODS TRACT—AREA, 6,943 ACRES

This tract lies along the southern shore of Lake Pontchartrain, about 2 miles east of the New Orleans Land Co. tract. The frontage on the lake is about 7 miles. The width at the eastern end is about 2 miles and at the western end 1½ miles. Most of the surface was originally covered with a heavy growth of grass, only a small percentage of the area having been timbered. The surface was but a few inches above the mean lake level, and, except for a few small ridges and shell mounds, the tract was at one time subject to overflow whenever the lake was at high tide. A considerable depth of turfy humus or muck covered the entire area; the range in depth was from 1 foot along the lake shore to as much as 10 feet in the portion about a mile back from the lake. The average depth of muck was perhaps 5 feet.

Reclamation work was started in 1908, and in the fall of that year a pumping plant was erected with a maximum 24-hour capacity of 0.48 inch. Although a very small percentage of the canals necessary to drain the tract had at that time been constructed, the pumping plant was started in October, 1908. The construction of the canals and ditches was carried on during the time records of rainfall and pumping were kept. At times part of the tract was dammed off from the pumping plant, so that the relation between rainfall and amount pumped is not comparable to the results from the other districts already described. By the spring of 1912 about 600 acres of the tract had been drained with small field ditches, but in the remainder of the area only a few of the larger canals had been excavated. No rain gages were kept on this district during this period, so that the rainfall records from surrounding stations necessarily must be used. In view of the above features, it is not considered that the records of amount pumped during these earlier years would be of value for comparison with those of other districts. Therefore the records (Table VII) are given only for the period beginning January, 1913.

CONDITIONS DURING 1913.—During 1913 quite extensive changes were made. The construction of the main drainage channels was carried forward rapidly. The spacing of the main channels is ½ mile with the usual field ditches leading into them. A main canal carries the water to the pumping plants. A second pumping plant was constructed during the summer and placed in operation toward the end of the year. From the beginning of the year until May 1, the entire area was tributary to the old pumping plant; after that date the area was reduced to 2,400 acres, the remaining land being dammed off and not drained. From September 22 until October 29 water was allowed to run over these dams from the undrained portion into the canals of the drained portion. On October 29 these dams were closed, after which the old plant drained the 2,400 acres, and the new plant drained the remaining 4,543 acres. The records given in Table VII show the data from the old plant alone. The 2,400 acres drained by the old plant were almost completely ditched by the end of the year, and about 200 or 300 acres were in cultivation.

CONDITIONS DURING 1914 —Until the middle of May the district was still divided between the old and the new pumping plants. The 2,400 acres drained by the old plant were very completely drained and the canals were cleared of a great deal of soft mud and vegetable matter. The cutting of the field ditches in this portion was nearly finished. After the middle of May the two portions of the district were connected and all drainage water pumped by the new plant. The 24-hour capacity of the new plant is about 1.25 inches of water over the entire area of 6,943 acres. The canals in the newer portion of the district were deepened and widened, and the cutting of the field ditches was nearing completion. In the portion of the district near the old pumping plant, where cultivation was being started, the water was held very low for the entire year; while in the newer portions where ditching operations had not started, the drainage was not nearly so complete. The capacity of the completed and enlarged reservoir canals was now nearly an inch of water from the entire area. By the end of 1914 the area ditched had been increased to 5,400 acres, and 1,680 acres were in cultivation. (See Table VII.)

CONDITIONS DURING 1915.—The entire area was drained during this year by the new pumping plant. The early part of the year was marked by heavy rainfall, but after March the spring and summer months were unusually dry. The ditched area was gradually increased during the year to about 6,500 acres and the cultivated area to about 4,500 acres. The records were interrupted toward the end of September by the admission of some outside water. Normal conditions were restored in November and records resumed on December 1. (See Table VII.)

DES ALLEMANDS POLDER NO. 1—AREA 1,880 ACRES

This district includes a portion of the town of Des Allemands, in Lafourche Parish, and borders on Bayou Des Allemands. The land from which the district was taken is typical of that lying along Bayou Des Allemands. It has an elevation of from 1 to 2 feet above the mean water level in the bayou, and a large percentage of the land is made up of firm silt ridges with a very thin layer of muck on the surface. The muck was originally from 8 to 12 inches deep and was quite turfy. With the exception of a few scattering trees on the ridges, the tract was covered with a heavy growth of the natural prairie grass.

The reservoir canals extend across the district with a spacing of about 2,000 feet and have a storage capacity of about 0.35 inch. The pumping plant capacity under average conditions is about 0.75 inch per 24 hours. Pumping was started on this district in October, 1912, and was continuous thereafter. Ditching operations were commenced at the same time and carried on throughout the year 1913, so that by the end of the year about one-third of the area was completely ditched. Cultivation was started early in 1913, and by the end of the year 250 acres were cultivated.

During the months of January and February, 1913, a section of levee about 300 feet long allowed a considerable amount of seepage water to enter the district. This seepage had to be pumped out each day. In the latter part of February this defective spot was repaired. After that, except for a period of about a month in the latter part of September and the early part of October, it is considered that no appreciable amount of seepage entered the district.

CONDITIONS DURING 1914.—During 1914 about half the tract was completely ditched and drained. The remaining portion was partly drained with ditches having four times the usual spacing. The reservoir canals needed cleaning, and their poor condition made it necessary to pump much oftener than would otherwise have been the case. The water had to be lowered to near the bottom of the canal at the pumping plant before the effect was felt at the end of the district remote from the plant. If the canals had been in good condition, the number of days on which the pumps operated would have been much decreased, although the total amount pumped would have been little different. In the latter part of November a rainfall of 5.45 inches occurred in about three hours. This caused the pumping plant to operate at full capacity for about three days. The canal was full at the station for about 24 hours, after which it lowered very rapidly. If the canals had been in good condition, they would have delivered much more water to the pumping plant than it was able to handle, while with the canals in poor condition the pump was capable of taking the water away faster than it flowed to the pump. The area under cultivation during the year was increased to about 600 acres.

CONDITIONS DURING 1915.—During the first three months of the year
the rainfall was heavy and the water level in the surrounding canals outside the district was from 1 to 2 feet above normal. As a result considerable seepage entered the district, and the amount pumped during these
months was above normal.   The following three months were very dry
and no pumping occurred. The summer months were about normal as
to percentage of rainfall pumped.   At the end of September the records
were rendered abnormal by the entrance of outside water into the district,
and records were not resumed until the end of the year.   The amount of
cultivated land was increased to 800 acres. (See Table VIII.)

JEFFERSON DRAINAGE DISTRICT NO. 3—AREA, 5,000 ACRES

This district is located about 25 miles southwest of New Orleans on
Bayou Barataria   One entire side of the district is bounded by the above
bayou, and two other branch bayous border the ends of the district. The
remaining side is bordered by a canal dug for the purpose of obtaining
material for the levee. About three-fourths of the levee on this area
is located on the solid banks of the bayous; consequently little, if any,
seepage enters the district.   Except a very narrow strip of solid clay land
along the bayous, the soil of this district is the usual prairie muck. The
muck originally varied in thickness from 1 to 4 feet. Most of the surface
was covered with the ordinary prairie grass. Except for a small elevation of perhaps 2 feet along the bayous, the surface of the district is level
and has an elevation of 1 foot or less above average water in the surrounding bayous.   The inside canals are quite extensive and are so spaced that
the longest field ditch is but 2,000 feet in length. The storage capacity
of the canals is about 0.60 inch.

Ditching was started early in 1913, and by the end of the year nearly
the whole area had been drained with ditches at a spacing of 220 feet.
Pumping operations were started in February, 1913, and the record of
operations was started on March 12. Except for a small amount of land
along the bayou, no cultivation was done in 1913. Under average
conditions the capacity of the pumping plant is 1.25 inches.

CONDITIONS DURING 1914.—During the year the district was all drained.
The cutting of the field ditches was completed, and all canals were
cleaned and deepened. After the canals were cleaned it was not necessary to run the plant on as many days as was formerly the case, since
the water could be more completely taken out of all the canals and it was
not necessary to stop the plant and wait for the water to flow to it.
While three heavy rains occurred during the early part of the year, the
ground was unusually dry in each case before the rain, so that no very
heavy pumping was necessary; in fact, the plant was not operated at full
capacity during the entire year. By the end of the year 200 or 300 acres
of land were under cultivation.

CONDITIONS DURING 1915.—The entire area was well drained throughout the year.   Very heavy rainfalls occurred in September and October,

but the pumping plant removed all drainage water promptly. The cultivated area was about 250 acres at the beginning of the year and was gradually increased to about 600 acres by the end of the year. (See Table IX.)

LAFOURCHE DRAINAGE DISTRICT NO. 12, SUBDISTRICT NO. 3—AREA, 2,250 ACRES

This district is located about 3 miles southwest of Raceland in the typical prairie land of the Lafourche section. It has a natural slope toward the south to Lake Fields. The highest part of the district is perhaps 3 feet above mean low water in the lake and the lowest part is about at mean low water level. The whole of the area was originally covered with a layer of turfy muck from 1 to 3 feet thick. The subsoil is the usual Sharkey clay. The levees around the district were built from material taken from the outside of the drained area and have been reenforced by further work. Except at times of very high water in the surrounding waterways and canals outside the area, it is believed that very little seepage enters the district. The interior drainage system is quite extensive, but requires some smaller collecting ditches to take the water from the small field ditches. The storage capacity of these drainage canals is about 0.50 inch.

This area was first inclosed in March, 1910. Nearly all the field ditches were cut during that year, and the remaining ditch work was finished by the early part of 1912. Pumping operations were started in 1910, and, with an addition to the plant made in 1912, the capacity under average conditions is about 1.15 inches per 24 hours. Drainage has been maintained continuously since 1910. The records of operations were started in December, 1912.

Cultivation was started on this district early in 1910, and the area gradually increased until in 1913, the first year of records, about 800 acres were in cultivation. During 1913 the entire district was well drained, and conditions in general were practically normal.

CONDITIONS DURING 1914.—The entire district was well drained during the year. While some very heavy rains occurred, the plant was of ample capacity. The rain that gave the heaviest run-off occurred in the latter part of November on a very wet soil. In the first 24 hours of operation the pumps lowered the water 1.5 feet in the canals, and at the end of 40 hours they were stopped, as the water had been lowered 4 feet. The total pumping for the year was quite close to the amount to be expected normally.

During the year nearly all the field ditches were placed and maintained in good condition, and about 2,000 acres of the district were in cultivation.

CONDITIONS DURING 1915.—During the entire year, except October, the district was well drained. During that month some water entered the district from outside, and the records were interrupted. With the exception of the first two months of the year, the spring and summer

15752°—17——2

months were very dry and little pumping was done. All the field ditches were cleaned out, and the entire area was cultivated. (See Table X.)

TABLES OF RAINFALL AND RUN-OFF

In the tables (I–X) the monthly and yearly summaries for the various districts are given At the end of each of the tables is a statement of the average amounts of rainfall, run-off, etc , for the period covered by the records for that district, proper corrections having been made for such times as the records were temporarily suspended or the conditions as to rainfall or run-off were abnormal As each district has local characteristics, it is considered that any figure obtained by taking a general average of the values for all of the districts would be of no special significance. However, certain of the districts and years have been selected as being typical of conditions prevailing on lands reclaimed from the wet prairies of southern Louisiana These districts and years are as follows: Lafourche Drainage District No. 12, Subdistrict No. 3, 1913, 1914, 1915; Unit No. 1, Poydras, La., 1914, 1915; Des Allemands Polder No 1, 1914; Jefferson Drainage District No. 3, 1914, 1915; Gueydan Drainage District, Subdistrict No. 1, 1915. Where the period during 1915 did not cover an entire year, the yearly totals have been corrected proportionately before averaging. The results are as follows: Average rainfall, 48.95 inches; average run-off, 21.56 inches; average number of days on which pumps operated, 90.

The above annual rainfall is somewhat below the mean annual rainfall for New Orleans, but it is believed that the percentage of rainfall appearing as run-off (44 per cent) is not greatly in error for average conditions With this percentage and a mean annual rainfall of 56 inches, the amount of run-off to be expected per year under average conditions would be 24.6 inches.

TABLE III.—*Rainfall and run-off on Lafourche Drainage District No. 12, Subdistrict No. 1, April, 1914, to December, 1915, inclusive*

Month.	Rain-fall.	Run-off.	Rain-fall in run-off.	Pumps ran.	Rain fell.	Month.	Rain-fall.	Run-off.	Rain-fall in run-off.	Pumps ran.	Rain fell.
	In.	*In.*	*P. ct.*	*Days.*	*Days.*		*In.*	*In.*	*P. ct.*	*Days.*	*Days.*
1914.						1915—Continued.					
April	5. 20	3. 39	65. 2	3	2	April	0. 00	0. 00	......	0	0
May	2. 85	. 95	33. 3	5	4	May	1. 61	. 00	. 0	0	2
June	4. 63	. 64	13. 8	2	4	June	1. 66	. 00	. 0	0	3
July	8. 00	3. 14	39. 3	7	7	July	5. 09	2. 51	49. 4	3	3
August	7. 30	1. 94	25. 9	5	8	August	6. 99	. 00	. 0	0	12
September [a]	[b]7. 00	......	......	......	[b] 4	September	2. 10	. 00	. 0	0	3
October	. 00	. 44	......	2	0	October [c]	......	......	......	......	......
November	10. 90	8. 03	73. 6	10	3	November	1. 55	. 32	20. 6	1	3
December	3. 32	5. 37	161. 7	14	4	December	1. 30	. 00	. 0	0	1
Total	42. 40	23. 90	56. 4	48	32	Total	33. 23	20. 57	61. 75	42	36
1915.						Averages for period (1.58 years)	47. 7	28. 0	58. 7	56. 7	42. 9
January	4. 40	7. 06	160. 6	15	2						
February	6. 30	6. 65	105. 5	14	4						
March	2. 23	4. 03	180. 8	9	3						

[a] Averages do not include September, 1914.   [b] Not included in total.
[c] The record for October, 1915, is omitted.

TABLE IV—*Rainfall and run-off on New Orleans Land Company tract, June, 1909, to December, 1915, inclusive*

Month	Rainfall.	Run-off.	Rainfall in run-off.	Rain fell.	Month	Rainfall.	Run-off.	Rainfall in run-off.	Rain fell.
	In.	In.	P. ct.	Days.		In.	In.	P. ct.	Days.
1909.					**1913.**				
June	2.15	0.62	28.8	5	January	4.59	4.41	96.0	7
July	6.72	.88	12.3	13	February	1.76	2.78	158.0	6
August	5.34	1.21	22.7	14	March	4.07	2.70	66.5	8
September	1.40	.24	17.0	6	April	4.50	2.46	54.6	4
October	3.44	1.42	41.3	5	May	7.70	2.85	37.0	9
November	1.20	.66	52.4	5	June	2.40	.79	32.9	9
December	7.43	4.95	66.6	9	July	6.33	1.91	30.1	17
					August	6.61	3.25	49.2	11
Total	27.74	9.98	36.0	57	September	10.92	7.02	64.3	12
					October	5.41	3.41	63.0	7
1910.					November	3.02	.96	31.8	4
January	.90	.84	93.3	3	December	1.41	1.65	117.0	3
February	4.14	(a)	(a)	7					
March	3.21	2.22	69.2	2	Total	58.72	34.19	58.2	97
April	1.10	.76	69.0	2					
May	7.25	1.16	16.0	7	**1914.**				
June	6.39	1.07	16.7	12	January	1.33	.87	65.4	2
July	11.41	5.35	46.9	22	February	5.91	1.78	30.1	4
August	7.10	1.82	25.6	7	March	4.76	4.07	85.5	7
September	5.50	(a)	(a)	8	April	3.84	3.03	78.9	5
October	1.74	(a)	(a)	8	May	1.57	.78	49.7	3
November	2.06	.82	39.8	4	June	2.97	.22	7.4	5
December	3.12	1.73	55.4	9	July	7.70	2.70	35.1	9
					August	8.75	3.27	37.4	14
Total	53.92	15.77	29.3	91	September	7.89	4.92	62.4	10
					October	.88	1.92	218.2	4
1911.					November	4.30	1.71	39.8	6
January	1.90	2.43	127.9	3	December	4.19	2.84	67.8	5
February	1.49	.96	64.4	5					
March	5.19	3.73	71.9	3	Total	54.09	28.11	52.0	74
April	12.56	8.81	71.4	10					
May	2.39	3.27	135.8	7	**1915.**				
June	6.17	1.16	18.8	9	January	6.35	5.375	84.6	5
July	6.84	3.04	44.4	14	February	4.71	4.215	89.5	7
August	7.63	2.09	27.4	17	March	1.42	2.435	171.2	4
September	4.08	1.13	27.7	12	April	.04	.495	.....	1
October	2.41	.76	31.5	5	May	4.51	.385	8.5	6
November	4.18	.88	21.0	8	June	3.47	.805	24.9	3
December	7.38	4.43	60.1	15	July	8.21	3.360	40.9	8
					August	6.60	2.440	37.0	8
Total	62.22	32.69	52.5	108	September	2.70	.830	33.0	4
					October	10.25	9.520	92.9	5
1912.					November	1.64	1.150	70.2	3
January	4.48	5.60	125.0	10	December	4.12	2.095	50.9	4
February	5.70	2.34	63.2	6					
March	9.32	8.22	88.2	8	Total	54.02	33.225	61.5	58
April	8.42	7.59	90.2	12					
May	4.73	2.99	63.2	9	Averages for period (6.33 years)	59.5	30.78	51.7	96
June	2.58	1.25	49.6	10					
July	8.37	1.79	21.4	16					
August	6.99	.89	12.7	17					
September	2.88	.62	21.5	10					
October	1.72	.90	52.3	6					
November	2.17	1.21	55.8	3					
December	10.43	7.19	68.9	13					
Total	65.72	40.59	61.9	120					

Month.	Rain-fall.	Run-off.	Rainfall in run-off.	Days.	Rain fell Days.	Month.
1912.	*In.*	*In.*	*P.ct.*			**1914—Continued.**
March..........	10.81	3.72	......	20	10	March...........
April..........	8.62	1.96	......	12	16	April...........
May...........	16.80	5.29	......	24	12	May...........
June..........	4.11	1.34	......	8	13	June...........
July..........	7.58	6.20	......	14	18	July...........
August.........	4.93	1.10	......	5	15	August........
September.......	3.84	1.49	......	5	8	September.......
October.........	2.47	1.66	......	6	7	October........
November.......	2.50	.98	......	4	5	November.......
December........	11.21	6.93	......	21	18	December.......
Total......	72.67	30.67	......	119	122	Total......
1913.						
January..........	5.71	4.52	79.0	18	10	January.........
February........	2.82	4.81	170.0	16	8	February........
March..........	6.55	2.82	43.0	16	10	March.........
April..........	6.04	3.34	55.3	18	4	April...........
May...........	5.25	3.88	73.9	16	7	May...........
June..........	3.91	.50	12.8	4	3	June...........
July..........	9.58	4.58	47.8	15	11	July...........
August.........	4.74	2.34	49.3	12	7	August........
September.......	9.48	6.21	65.5	17	9	September.......
October.........	4.33	4.26	98.4	11	4	October a.......
November........	.57	.35	61.4	3	1	November a......
December	4.92	1.61	32.8	9	7	December a......
Total......	63.90	39.42	61.7	155	81	Total......
1914.						**Averages for period (2.75 years)**
January..........	0.89	0.76	85.4	5	1	
February........	6.35	2.63	41.4	8	6	

a Records omitted.

TABLE VI.—*Rainfall and run-off on Gueydan Drainage District, Su March, 1912, to December, 1915, inclusive* a

	Rain-fall.	Run-off.		fell.	Month.
1912.				Days.	**1914.**
h......					January..........

TABLE VII —*Rainfall and run-off on Little Woods tract, January, 1913, to December, 1915, inclusive*

Month.	Rain-fall.	Run-off.	Rain-fall in run-off.	Pumps ran.	Rain fell.	Month.	Rain-fall.	Run-off.	Rain-fall in run-off.	Pumps ran.	Rain fell.
1913.	In.	In.	P. ct.	Days.	Days.	1914.	In.	In.	P. ct.	Days.	Days.
January.........	4.40	5.36	122.0	31	8	June............	4.30	0.28	6.5	17	7
February........	2.10	3.33	158.5	28	8	July.............	7.20	1.18	16.4	27	16
March..........	5.30	3.40	64.2	26	8	August..........	8.40	2.66	31.7	25	13
April...........	3.50	1.58	45.2	25	3	September.......	2.80	1.20	42.9	21	4
May............	4.40	1.44	32.7	8	8	October.........	2.00	1.03	51.5	27	3
June............	3.45	1.13	32.6	6	10	November.......	3.90	1.40	35.9	26	7
July.............	2.87	.16	5.6	1	12	December........	4.10	2.76	67.3	31	6
August..........	3.95	.31	7.8	2	8						
September.......	9.90	2.82	28.5	15	17	Total....	30.60	20.46	40.0	299	78
October.........	3.77	2.83	36.6	18	6	1915.					
November.......	1.90	.94	49.5	6	3	January.........	7.60	6.16	81.1	31	7
December........	1.00	1.00	100.0	6	2	February........	4.60	4.89	106.2	28	5
						March..........	2.10	3.21	153.0	30	5
Total......	46.04	24.30	52.8	182	93	April...........	.00	.34	......	15	0
						May............	3.20	.28	8.8	5	6
314 (old plant).						June............	3.40	.43	12.6	9	6
January.........	1.20	0.91	82.5	5	5	July.............	7.30	3.39	46.5	27	6
February........	6.40	1.78	27.8	8	6	August..........	6.30	1.38	22.2	21	8
March..........	4.90	6.13	125.1	22	7	September.......	2.10	.85	12.9	6	3
April...........	4.80	3.88	80.8	18	4	October [b]......	......	......	......	......	......
May [a].........	.20	.43	215.0	6	1	November [b].....	......	......	......	......	......
						December........	6.00	4.06	67.7	99	7
Total......	17.40	13.13	75.5	59	20						
						Total....	42.80	24.59	57.00	202	54
314 (new plant).						[c] Averages					
January.........	1.20	1.20	100.0	24	2	for period					
February........	6.40	1.44	22.5	20	6	(1.83 years)	51.40	24.45	47.60	273	72.0
March..........	4.90	3.29	67.2	31	7						
April...........	4.80	3.27	68.1	26	4						
May............	.70	.85	121.4	24	3						

[a] 12 days in May.   [b] Records suspended.   [c] Includes only the time that conditions were typical.

TABLE VIII.—*Rainfall and run-off on Des Allemands Polder No. 1, January, 1913, to September, 1915, inclusive*

Month.	Rain-fall.	Run-off.	Rain-fall in run-off.	Pumps ran.	Rain fell.	Month.	Rain-fall.	Run-off.	Rain-fall in run-off.	Pumps ran.	Rain fell.
1913.	Ins.	Ins.	P. ct.	Days.	Days.	1914.	In.	In.	P. ct.	Days.	Days.
January.........	3.50	2.97	84.9	24	6	October.........	0.90	0.60	66.7	10	2
February........	2.16	3.18	147.0	22	6	November .	6.75	3.39	50.2	12	3
March..........	4.20	2.62	62.3	21	7	December........	3.70	4.85	131.1	27	5
April...........	2.15	1.86	86.5	13	3						
May............	3.58	1.31	36.6	11	5	Total....	45.70	23.83	56.5	177	39
June............	1.88	.34	18.1	3	2	1915.					
July.............	9.40	2.56	27.2	10	6	January.........	7.21	6.78	94.0	28	7
August..........	2.80	2.27	81.1	26	4	February........	6.15	8.26	134.2	28	5
September.......	9.60	5.15	53.7	23	9	March..........	2.15	3.84	178.5	17	3
October.........	4.46	5.57	125.0	26	5	April...........	.00	.00	.0	.0	1
November.......	1.64	1.02	62.3	11	3	May............	2.00	.00	.0	.0	1
December........	2.00	2.01	100.0	17	3	June............	.00	.00	.0	.0	0
						July.............	5.00	1.49	29.8	6	3
Total......	47.37	30.86	65.1	207	59	August..........	4.70	1.28	27.2	9	2
						September.......	2.20	.29	13.2	4	3
1914.						October [a]......	......	......	......	......	......
January.........	1.00	.92	92.0	17	1	November [a].....	......	......	......	......	......
February........	5.70	2.86	50.2	17	4	December [a].....	......	......	......	......	......
March..........	4.75	3.99	84.0	25	4						
April...........	4.60	3.52	76.5	22	2	Total ...	28.41	21.94	77.2	92	24
May............	.80	.49	61.3	11	2	Averages					
June............	2.40	.45	18.8	5	2	for period					
July.............	2.60	.43	16.5	5	3	(2.75 years)	44.2	28.6	64.7	173.0	44.4
August..........	8.20	2.49	30.4	12	7						
September.......	4.30	1.84	42.8	14	4						

[a] Averages do not include October, November, and December, 1915.

					Month.		
ber...	1.00				Mar 1............	..74	3.55 [a]
ecember.........	1.50		9	5	April............	.75	.44
					May.............	3.35	.00
Total......	49.60	57.5	120	74	June.............	2.44	.22
					July.............	6.07	1 24
1914.					August..........	3.10	37
nuary.........	1.06	112.3	8	1	September.......	6.76	57
bruary........	8.07	42.5	13	6	October.........	9.48	7 84
rch..........	6.00	129.7	19	2	November........	1.40	.36
ril...........	4.42	70.1	12	2	December........	4.11	1.80
y.............	2.58	35.7	5	4			
ne............	2.94	19.4	3	3	Total......	44.61	8.18
ly............	6.79	15.2	6	11			
ugust........	3.79	19.5	5	9	**Averages for**		
ptember.......	3.95	55.4	10	7	**period**		
tober..........	.43	128.0	4	3	**(2.75 years)**	51 8	26.6

a Averages do not include January, 1915.

TABLE X.—*Rainfall and run-off on Lafourche Drainage District No. No. 3, December, 1912, to December, 1915, inclusive*

Month.	Rain-fall.	Run-off.	Rain-fall in run-off.	Pump ran.	Rain fell.	Month.
	Ins.	*Ins.*	*P. ct.*	*Days.*	*Days.*	
1912.						**1914.**
cember........	11.56	6.79	58.7	24	14	September........
						October..........
1913.						November........
nuary.........	5.88	4.77	81.2	21	5	December........
bruary........	2.85	2.27	79.6	11	4	
arch.........	3.52	1.68	47.7	7	6	Total......
ril..........	2.07	.44	21.2	2	2	
y.............	3.40	.20	5.9	1	3	**1915.**
ne............	2.15	.00	.0	0	6	January..........
ly............	9.09	1.23	13.5	5	13	February........
gust.........	4.45	.77		3	5	March...........
ptember.......	9.09	2.42		9	15	April............
tober.........	4.20	4.02		17	5	May.............
vember........	3.14	.96		3	5	June............
cember........	2.20	1.72		9	7	July.............
						August..........
Total......	63.60	27.27		112		September........
						October [a]........
1914.						November........
nuary.........						December........
bruary........						
rch..........						Total......
ril...........						
y.............						**Averages for**
ne............						**period (3.0**
ly............						**years).....**
gust.........						

ACTUAL RATES OF RUN-OFF DUE TO HEAVY STORMS

In the foregoing tables of rainfall and run-off the figures under the columns headed "Run-off," with the exception of those shown for the New Orleans Land Company tract, are in fact the amounts pumped rather than the run-off of the water from the surface into the main drainage channels. In the case of the tract excepted, the water discharges by gravity as fast as it enters the main outfall channel. For almost the entire period covered by the record of each plant, an hourly reading has been made of the stage of water in the main reservoir canal during the operation of the pumping plant. The amounts of water stored in the main drainage channels at the different levels have also been calculated. By this means the amounts pumped per day during the heavy storms have been so adjusted for a number of typical storms on each district that the final results represent approximately the quantity of water which would have been removed had the pumps taken the water as fast as it entered the main reservoir channels, in other words, these results represent approximately the true run-off for the storms selected. With the exception of a very few heavy storms, only those were selected which occurred when the land in the district was saturated, or at least fairly wet.

Table XI gives the dates, rainfall, and revised run-off for these selected storms. In the columns headed "Rainfall" are indicated by the words "Moist," "Dry," etc., the condition of the land at the beginning of the storm periods. While the run-off is given as for consecutive dates after the beginning of pumping, this is not strictly correct, as the run-off was computed for 24-hour periods following the starting of the pumps, regardless of the calendar dates. Following each table are shown the number of heavy storms occurring when the land was dry and the number occurring when the land was wet. These figures include all of the 2- and 4-inch storms that occurred during the entire period of observation.

TABLE XI.—*Typical heavy storms and probable actual run-off of all districts*

SMITHPORT PLANTING COMPANY TRACT a

Date.	Rainfall.	Run-off.	Date.	Rainfall.	Run-off.
	Inches.	*Inches.*		*Inches.*	*Inches*
1909.			1913		
Aug. 10............	1.68	0.89	Sept 12 .	0.80	....
11............	2.80	1.15	13	3.00	1.27
12............	Moist.	.48	14............	.15	1.18
13............	....	.11	15............	Wet.	.03
			26............	.42	....
1911.			27............	2.80	1.20
Apr. 25............	1.02	....	28............	Wet.	.63
26............	2.02	.97	29............	....	.13
27............	Dry.	.50			
28............	....	.02	1914.		
Nov. 27	3.00	1.02	Feb. 27 .	2.00	....
28 .	Moist.	.48	28 ...	Wet.	1.10
29............	.	.03	Mar. 1............	....	.74

a From June, 1909, to December, 1915, of all of the storms of over 2 inches, 8 occurred on dry land and 21 on wet land, of those of over 4 inches, 3 occurred on dry land and 1 on wet land

			1912.	
....	4.10	1.29	Apr. 14............	3.93
....	1.00	1.36	15............	Wet.
....	Dry.	.44	May 10............	4.50
....	....	.18	11............	Wet.
....	4.14	1.23	12............	....
....	.30	.64	Dec. 5............	1.60
....	Wet.	.25	6............	2.73
....	3.12	....	7............	Wet.
....	.27	1.00	8............	....
....	Wet.	.37	9............	....
....	....	.27		
....	....	.36	1913.	
			Apr. 24............	3.09
			25............	Moist.
....	3.55	....	26............	....
....	Wet.	a 1.29	Sept. 13............	2.75
....	.30	.54	14............	Wet.
....	....	.42	15............	....
....	....	.40		
....	4.05	a 1.07	1915.	
....	2.21	1.15	Jan. 23............	2.97
....	Wet.	1.15	24............	Wet.
....	....	.64	Oct. 3............	1.38
			4............	1.90
			5............	1.46
....	2.18	....	6............	Satu-rated.
....	.60	1.27		
....	.00	.82	7............	....
....	Wet.	.25	8............	....
....	....	.42	12............	3.11
....	1.85	.20	13............	2.48
....	2.02	.98	14............	Satu-rated.
....	....	.72		
....	....	.40	15............	....
....	1.42	.73	16............	....

I.
.

for storage in canal.
909, to December, 1915, of all of the storms of over 2 inches, 14 occurred on dry land of those of over 4 inches, 2 occurred on dry land and 4 on wet land.

TABLE XI.—*Typical heavy storms and probable actual run-off of all districts*—Contd.

LAFOURCHE DRAINAGE DISTRICT NO. 12, SUBDISTRICT NO 1[a]

Date.	Rainfall.	Run-off.	Date.	Rainfall.	Run-off.
1914.	*Inches.*	*Inches.*	**1915.**	*Inches.*	*Inches.*
Nov. 13............	6.50	0.00	Jan. 23............	2.70	1.75
14............	Dry.	2.06	24............	Wet.	.46
15............		.69	25............		.36
27............	4.20	2.06	Feb. 4............	2.30	b.00
28............	Moist.	1.11	5............	Wet.	.96
29............		.65	6............		.49
30............		.29	July 3............	3.50	(b)
Dec. 1............		.29	4............	Dry.	1.23
			5............		.87

NEW ORLEANS LAND COMPANY TRACT [c]

Date.	Rainfall.	Run-off.	Date.	Rainfall.	Run-off.
1909.			**1912.**		
Dec. 12............	2.74	.34	Apr. 19............		.31
13............	Wet.	.33	Dec. 5............	1.00	
14............	.80	.32	6............	3.00	1.36
15............		.31	7............	Wet.	.45
1911.			8............		.32
Mar. 22............	3.80	.76	9............		.26
23............	Moist.	.41	**1913.**		
24............		.36	May 22............	2.95	.62
25............	1.15	.30	23............	1.63	.51
26............		.00	24............	Wet.	.30
Apr. 8............	4.03	1.19	25............		.22
9............	Saturated	.51	Aug. 17............	2.60	.38
10............		.27	Aug. 18............	Wet.	.27
11............		.00	19............		.22
26............	3.80	.95	Sept. 14............	4.32	1.51
27............	Saturated	.76	15............	{Water on	.85
28............		.58	16............	surface.}	.64
1912.			17............		.50
Jan. 8............	2.41	.64	18............		.38
9............	Saturated	.46	**1914.**		
10............		.33	Apr. 19............	3.14	.79
11............		.26	20............	Dry.	.36
12............	2.30	.42	21............		.24
13............	Wet.	.30	Aug. 26............	1.40	.00
14............		.23	27............	1.88	.48
15............		.20	28............	1.59	.39
22............	3.60		29............	Dry.	.29
23............	2.40	1.09	30............		.28
24............	Saturated	1.78	Sept. 24............	3.30	.78
25............		.84	25............	Saturated	.60
26............		.42	26............		.37
27............		.31	Dec. 29............	2.00	.34
Apr. 15............	2.10	1.42	30............	Moist.	.22
16............	1.60	1.08	31............		.16
17............	{Water on	.69			
18............	surface.}	.44			

[a] From November, 1913, to December, 1915, of all of the storms of over 2 inches, 4 occurred on dry land and 8 on wet land; of those of over 4 inches, 3 occurred on dry land and 3 on wet land.
[b] Not corrected for storage in canal.
[c] From June, 1909, to December, 1915, of all the storms of over 2 inches, 12 occurred on dry land and 24 on wet land; of those of over 4 inches, 1 occurred on dry land and 3 on wet land.

....	1.10	1.00		17............	1.03	
....	Wet.	.39		18............	Wet.	
				19............		
				20............		
....	2.30	....		23............	2.52	
....	.26	1.39		24............	Wet.	
....	Wet.	.71	June	13............	5.30	*b*
....	1.97	....		14............	.30	
....	.00	....		15............	Dry.	
....	2.75	....	July	3............	3.03	
....	Wet.	*b* 1.31		4............	Moist.	
....		.00		5............	.31	
....		.26		6............		

					1913.	Inches.
....	3.03	.82	June	17............		Dry.
....	.50	.87		18............		....
....	Wet.	.61	July	23............		5.00
....	3.25	1.27		24............		Wet.
....	2.60	1.28		25............		.10
....	Wet.	.69	Sept.	13............		2.20
....	....	.22		14............		Wet.
				15............		....
				26............		5.70
....	3.20	1.07		27............		.60
....	Wet.	.47		28............		1.60
....		.23		29............		.30
....	4.30	....		30............		3.20
....	1.10	.94	Oct.	1............		.40

, 1912, to Dec., 1915, of all of the storms of over 2 inches, 8 occurred on dry land
se of over 4 inches, 2 occurred on dry land and 4 on wet land.
d for storage in canal.
, 1912, to December, 1915, of all of the storms of over 2 inches, 11 occurred on dry land
of those of over 4 inches, 4 occurred on dry land and 3 on wet land.

TABLE XI —*Typical heavy storms and probable actual run-off of all districts*—Contd.

GUEYDON DRAINAGE DISTRICT, SUBDISTRICT NO 1—continued

Date.	Rainfall	Run-off		Date.	Rainfall.	Run-off.
	Inches.	*Inches.*			*Inches.*	*Inches.*
1909.				**1914.**		
Oct. 2	Wet.	0 96	Apr. 23 . . .			0. 39
3		. 92	24			. 02
4		. 84	28		1. 00	. 99
5		. 57	29 . . .		1. 35	. 53
22	5. 10	0. 96	30		Moist.	. 77
23	Wet.	1. 30	May 1			. 52
24		. 90	Nov. 13 .		. 75	1. 04
25		. 27	14		2. 95	. 80
1914.				15	Moist.	. 13
Feb. 12	3. 50	. 00	26		. 45	1. 45
13	Moist.	. 99	27		2. 10	. 55
14		47	28		Wet.	. 31
Apr. 18	5. 77	1. 77	**1915.**			
19	Dry.	. 82	July 4 .		3. 85	[a] 1. 19
20		. 60	5		. 12	1. 44
21		. 69	6		Wet.	. 41
22		. 94	7			. 16

LITTLE WOODS TRACT b

Date.	Rainfall	Run-off		Date.	Rainfall.	Run-off.
1915.				**1915.**		
Jan. 23	2. 70	1 06	July 6	1. 30	. 58
24	Wet.	. 60	7	Dry.	. 39
25		. 02	8		. 05
July 2	1. 00	...	Dec. 16	1. 30	. 61
3	3. 10	. 60	17	1. 80	. 66
4	1. 40	. 41	18	Wet.	. 08
5	. 10	. 19	19		. 18

DES ALLEMANDS POLDER NO. 1 c

Date.	Rainfall	Run-off		Date.	Rainfall.	Run-off.
1913.				**1914.**		
July 24	2. 00	. 63	Apr 18	4. 00	1. 09
25	Moist.	. 28	19	Dry.	. 80
Sept. 30	3. 30	1. 01	20		. 43
Oct. 1	Moist.	. 83	21		. 16
2		. 70	Nov. 27	5. 45	. 95
3		. 42	28	Dry.	. 65
4		. 28	29		. 66
5		. 11	30		. 51
1914.				Dec. 1		. 35
Feb. 28	2. 20	. 96	2		. 05
Mar. 1	Wet.	. 37	**1915.**		
2		. 12	Jan. 23 .	3. 00	1. 04
			24	Wet.	. 63
			25		. 43
			26		. 19

a Not corrected for storage in canal

b From October, 1908, to December, 1915, of all the storms of over 2 inches, 11 occurred on dry land and 11 on wet land, of those of over 4 inches, 2 occurred on dry land and 4 on wet land

c From December, 1912, to December, 1915, of all the storms of over 2 inches, 10 occurred on dry land and 11 on wet land; of those of over 4 inches, 2 occurred on dry land and 2 on wet land

1914.				1914.	
18	4. 68		Nov.	28	
19	Dry.	1. 35		29	
20		. 62		30	
21		. 10			
13	5. 25	. 00		1915.	
14	Moist-dry.	1. 30	Jan.	23	2. 5
				24	We
15		. 34		25	
16		. 19	Feb.	4	1. 8
26	4. 57			5	We
27	Wet.	1. 62		6	

'rom February, 1913, to December, 1914, of all of the storms of over 2 inches, 5 occurr
  on wet land; of those of over 4 inches, 3 occurred on dry land and 2 on wet land.
rom December, 1912, to December, 1915, of all of the storms of over 2 inches, 6 occurr
  1 on wet land; of those of over 4 inches, 3 occurred on dry land and 3 on wet land.

CONCLUSIONS

)wing to differences in the degree of wetness of the land, th
.he rainfall, the time of the year, the condition of the drai
, etc., the results for any particular district are not unifor
r, to make it possible to judge what run-off would occur und
ditions the results in the tables have been shown in curv
  amounts of the run-off (in inches of depth) for one, two,
r day periods have been plotted as ordinates and the r
es for each storm) as abscissas. Distinctive lines have b
ough the points representing the run-off for each length
od. Each curve has been drawn, so far as practicable,
se points and at the same time to conform to the other cur\

RT PLANTING COMPANY TRACT

Area_ _ _ _ _ _ _ _ _ _ 647 a.
Pumping Plant Capacity. 1.10"
Reservoir Capacity_ _ _ 0.50"

1 day Storms_ _ _ _ o
2 day Storms_ _ _ _ ●
3 day Storms _ _ _ _ ▲

WOOD PLANTATION

a_ _ _ _ _ _ _ _ _ _ _ 2600 a
mping Plant Capacity_ 1.40 "
ervoir Capacity_ _ _ 0.42"

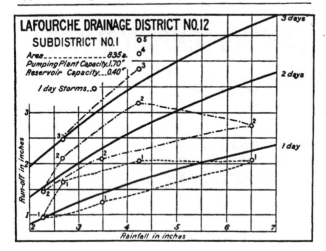

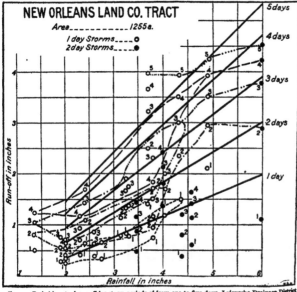

FIG. 2.—Probable actual run-off for storm periods of from one to five days, Lafourche Drainage District No. 12 and New Orleans Land Company tract.

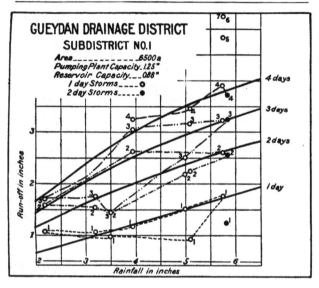

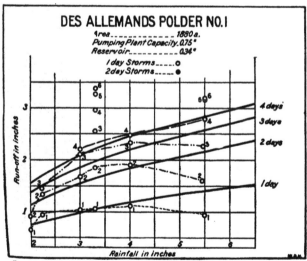

Fig. 3.—Probable actual run-off for storm periods of from one to four days, Gueydan Drainage District and Des Allemands Polder No 1

LAFOURCHE DRAINAGE DISTRICT NO.12
SUBDISTRICT NO.3

Area _ _ _ _ _ _ _ _ _ _ _ _ _ 2250 a.
Pumping Plant Capacity _ 1.15"
Reservoir Capacity _ _ _ 0.50"

1 day Storms _ _ _ _ o

Rainfall in inches

4.—Probable actual run-off for storm periods of from one to three days, Jefferson Dr
No. 3 and Lafourche Drainage District No. 12.

each set of curves is shown the area of the district, the average capacity of the pumping plant per 24 hours, and the storage capacity of the main drainage channels between the surface and a level 5 feet below. From these separate curves composite curves have been drawn which represent the average run-off from all the districts (fig. 5).

It is evident that the effect on the run-off of a change in the capacity of the pumping plant can not all be measured by the change of level of the water in the main reservoir channels, as greater slope and velocity

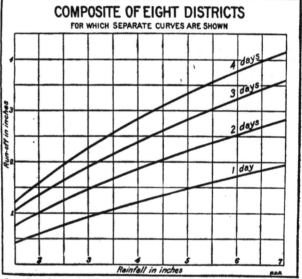

Fig. 5.—Composite curves for eight districts, showing probable run-off for storm periods of from one to four days

of water occur in the drainage channels when a larger pumping plant is installed. This effect extends even to the small field ditches.

In estimating the run-off likely to result from an assumed storm on a given district the curve used should be that for the district which resembles the given district in area, pumping-plant capacity, reservoir capacity, and general conditions. The curves may be considered only as representing general tendencies and not definite values. They should be of service, however, in making the proper adjustment between reservoir and pumping-plant capacity. A method for making this adjust-

15758°—17——3

ment has been outlined by Okey.[1] It is believed that, in general, after a capacity of about 0.50 inch of water over the area drained has been provided in the main reservoir channels between the surface and a level 5 feet below it will be cheaper to obtain increased capacity to handle storms by enlarging the pumping plant rather than the reservoir. This will depend, however, on local costs for excavation and pumping machinery.

TABLE XII.—*Average yearly number of storms of given intensities, based on daily rainfall records of the United States Weather Bureau for New Orleans, La., 1871 to 1915, inclusive*

Total rainfall.	Average number of storms per year.														
	1-day period.	2-day period.	3-day period.	4-day period.	5-day period.	6-day period.	7-day period.	8-day period.	9-day period.	10-day period.	11-day period.	12-day period.	13-day period.	14-day period.	15-day period.
Inches.															
3	1.622	3.020	4.060												
4	0.688	1.358	1.866	2.220											
5	.333	.644	.866	1.044											
6	.200	.233	.466	.644	.778	0.866									
7	.196	.222	.311	.356	.400	.466	0.600								
8	.111	.156	.156	.222	.289	.289	.333	0.376							
9	.022	.066	.089	.111	.156	.200	.200	.245	0.311						
10			.044	.066	.111	.156	.200	.200	.245	0.267					
11			.022	.044	.066	.089	.133	.133	.178	.178	0.200				
12					.044	.066	.089	.089	.089	.089	.089	0.111			
13						.022	.044	.044	.044	.044	.044	.044	.444	0.089	
14						.022	.022	.022	.022	.022	.022	.022	.022	.066	0.066
15						.022	.022	.022	.022	.022	.022	.022	.022	.024	.044	0.044

In determining the proper combined capacity of the pumping plant and reservoir the main factor to be considered is the amount of rainfall in an assumed period for which provision is to be made. Examinations of the daily rainfall records of the United States Weather Bureau at New Orleans from 1871 to 1915 have been made, and the results are shown in Table XII. As the accompanying curves were prepared with data from storms which occurred when the land was wet, a determination has been made of the proportion of storms which occur when the land is wet. Of all the storms of over 2 inches in 24 hours which have occurred on the districts during the time covered by the records, 64 per cent occurred on a wet and 36 per cent on a dry surface; of all storms over 4 inches the percentages are 54 per cent and 46 per cent, respectively. Of course the storage capacity of the land influences the run-off from the small storms relatively much more than it does that from the large ones. However, an examination of the daily rainfall and pumping records will show that heavy rains on a dry soil do not make very heavy demand on the pumping plant. It is believed, therefore, that a reduc-

[1] OKEY, C. W. THE WET LANDS OF SOUTHERN LOUISIANA AND THEIR DRAINAGE U. S. Dept. Agr. Bul. 71, p 71–75 1914.

tion of about 30 per cent in the average frequency of storms could be made safely and that the resulting figure would be the proper one for use. If the character of operations that are to be conducted on the land of a given district is known, a decision can then be made as to the heaviest storm for which provision must be made. On a district where staple crops are to be raised it would be economical to allow a certain amount of flooding oftener than would be advisable on land where high-priced truck crops are to be raised, while in residence districts it would be very desirable to prevent all surface flooding. In addition to the damage to crops due to flooding, there are other factors to be considered, such as inconvenience to residents and the possible depressing influence on land values of floodings occurring even at infrequent intervals.

DIAGNOSING WHITE-PINE BLISTER-RUST FROM ITS MYCELIUM

By REGINALD H. COLLEY,

Assistant Pathologist, Investigations in Forest Pathology, Bureau of Plant Industry, United States Department of Agriculture

INTRODUCTION

The mycelium of *Cronartium ribicola* Fischer, the causal organism of the white-pine blister-rust, can be found among the cells of *Pinus strobus* L. for some time before the spores of the fungus are produced. The examination of thousands of sections of diseased bark has convinced the writer that the morphological characters of the mycelium and its method of parasitism are sufficiently distinctive to warrant the assertion that the blister-rust can be recognized from the mycelium alone.

The stains generally employed in the preparation of slides for diagnosis are safranin and lichtgruen. The use of these two stains is, of course, already well established in both zoological and botanical work, and their employment in combination is by no means new. They are particularly valuable in the study of the blister-rust mycelium because they not only stain host and parasite at the same time but they also bring out extremely well the anatomical features of the host tissue. It has always seemed to the writer that more detailed accounts of the experiences of investigators in handling stains would be of very great value to students. Accordingly the method which has been employed is given below, step by step, with the hope that it may be useful and suggestive to investigators of both normal and pathological anatomy. Throughout, the writer has kept in mind the fact that there has been very little published in this country on the use of the stains in combination.

METHOD OF DOUBLE STAINING

(a) SECTIONS.—Free-hand sections of white-pine bark can be made to answer very well if no microtome is available, but many more sections of a better quality can be obtained with an ether or carbon-dioxid freezing attachment and a small microtome. Radial longitudinal sections are best, for in them the medullary rays are exposed, and in infected bark the mycelium of the blister-rust is generally closely associated with the cells of the rays. The sections should include, where practicable, the whole width of the bark, the cambium region, and three or four annual rings of wood. Such sections are, of course, larger than can be conveniently cut with a section razor by hand, but it is perfectly possible to

Journal of Agricultural Research,
Washington, D. C.
km

Vol. XI, No. 6
Nov. 5, 1917.
Key No. G—125

(281)

cut them with the freezing microtome if the operator is careful not to
cut them too thin. The advantage in having sections of such large size
lies in the fact that the observer can explore the whole of the susceptible regions for evidence of the mycelium of the parasite under conditions really favorable for its discovery.

No killing solution is used, but, so far as can be observed, no shrinkage
at all occurs in the hyphæ, and but little in the host cells of fresh specimens. The blocks of bark and wood should be cut into pieces approximately ⅛ to ¼ inch on a side. The prepared blocks are then soaked
for about 10 minutes in a 10 per cent aqueous solution of gum arabic
containing 0.5 per cent of phenol. Blocks cut from fresh specimens can
be put to soak at once. If the specimen is old and dry, the blocks should
be boiled before being put into the gum arabic. The same strength
gum-arabic solution is used as a freezing medium. If the material be
embedded in paraffin, the usual process of sectioning and mounting can
be followed. Excellent sections of young pine bark extending from
cambium to epidermis can be easily cut 5 μ thick from paraffin material.
However, for ordinary diagnosis or morphological study of the fungus
hyphæ sections 15 to 20 μ thick, cut on the freezing microtome, are very
satisfactory. Such sections should be removed from the knife with a moist
camel's-hair brush and rinsed in several changes of clean water. They
can be handled very conveniently in bevel-ground Syracuse watch glasses.

(b) STAINS.—The stains, safranin and lichtgruen, should be kept in
dropper bottles of 25-to-50-c. c. capacity and used from the dropper or
pipette. Similar bottles should be used for absolute alcohol, clove oil,
and xylol. The writer uses the alcoholic solution of safranin almost
exclusively. This method is described by Chamberlain [1] as follows:

The alcoholic solution is made by dissolving 1 g. of the alcoholic safranin in 100
c. c. of 95 per cent or absolute alcohol, and, after the safranin is completely dissolved,
adding 50 c. c. of distilled water.

This stain may be used over and over again, but it should preferably
be filtered back into the drop bottle after each staining. A convenient
stock solution of lichtgruen is a 1 per cent in 95 per cent alcohol. A convenient strength for use is one-half stock, made by diluting the stock
with an equal volume of 95 per cent alcohol. For thin sections, however, from paraffin material, better results can be obtained by the rapid
use of the stain at full strength. In addition to the regular drop bottle
pipettes, there should be two or three others of good capacity and small
tube opening for use in draining off the stains or excess reagents during
the steps of the process.

(c) STAINING PROCESS.—The sections should be rinsed several times
and then drained of all excess water and a generous quantity of safranin
added. This stain should remain for from two to four hours, after

---

[1] CHAMBERLAIN, C. J. METHODS IN PLANT HISTOLOGY. ed. 3, p. 51. Chicago, 1915.

which time it may be drawn off and kept to be filtered back into the drop bottle. By tilting the watch glass the sections can be grouped easily at one side of the watch glass. Then, after the safranin has been drawn off, any excess red in the dish may be wiped out with a cloth. Care should be taken that the sections do not dry out. The lichtgruen should be added as soon as the excess safranin is removed. A muddy mixture of red and green results. The lichtgruen replaces the safranin quite rapidly, so that a minute is usually the maximum time necessary to produce the proper green tint. The writer shakes the watch glass as soon as the lichtgruen is added, then draws off the red-green mixture, and adds a few drops of fresh lichtgruen. The green stain should not be allowed to act long enough to wash out too much of the safranin. The time required depends on (1) the thickness of the sections, (2) the time the sections were stained in safranin, and (3) the strength of the lichtgruen solution The used lichtgruen should be thrown into a waste jar.

(d) CLEARING AND MOUNTING.—After the removal of the green stain, absolute alcohol should be added immediately and the watch glass rocked or shaken. This absolute alcohol should be drawn off and saved in a waste jar, and the sections treated with fresh absolute alcohol for a minute or two, the total time depending on the number of sections handled at one time. This absolute alcohol should then be replaced by clove oil. While in clove oil the sections may be conveniently lifted with a small brush and arranged on the slide without danger of drying up, and for this reason clove oil is used between absolute alcohol and the more volatile xylol. After the sections are arranged, they should be rinsed with drops of xylol until the clove oil has been removed. A drop of balsam and a warm cover complete the mount. It is advisable to set the slide away for a day in the paraffin bath with a heavy lead weight on the cover to flatten out the sections.

(e) COLOR RESULTS.—The color results in the finished slide vary slightly, depending on the amount of safranin which was removed when the green stain was acting. A good diagnostic slide for *Cronartium ribicola* should have red host nucleoli and reddish nuclei, red wood cells, dead cells, and cork cells. Occasionally resin cells stain a deep red. The host cytoplasm and all cellulose walls should be green. The hyphæ of the parasite stain green, but they stand out well when they are colored a greenish pink or greenish red—that is, when the red is not removed too far. The nuclei of the parasite should be red. Although the stains employed do not result in the differential staining of the host and parasite, there is no difficulty in distinguishing them.

(f) ADVANTAGES OF THE METHOD.—The chief advantages of the method are its simplicity and rapidity. Permanent record slides from a suspected specimen can be completed easily in about two and one-half hours from the time the specimens reach the laboratory. Correctly prepared slides are sufficiently transparent for a thorough exploration with oil-immersion lenses. The practicability of the method has been demonstrated by constant use during the past year in the diagnosis of suspected cases of blister-rust and in the study of the anatomical features of parasite and host.

MYCELIUM OF THE PARASITE

The characters which distinguish the hyphæ of *Cronartium ribicola* from other hyphæ occasionally found in white-pine bark are their comparatively large size, fairly uniform diameter, and possession of haustoria. They are larger than any other parasitic hyphæ which the writer has observed in pine bark. While there is, of course, some difference in the width of individual hyphæ one rarely finds them deviating widely from the average width of 3 to 3.5 μ. The haustoria are particularly striking. They invade practically every cell in the infected region in young bark. In older bark they are most prominent in the cells of the medullary rays; and for this reason, as mentioned above, radial sections are particularly satisfactory for study. At the point where the haustoria pass through the cell wall, they are constricted. Once inside the cell their width increases until they average a little more than 4 μ in diameter (Pl. 31, B). Prominent bulges, which occur quite frequently, give them a characteristically irregular outline. When young and short they are straight; as they mature they usually develop pronounced bends, which may be as great as right angles. The host nuclei appear to attract the haustoria. Frequently their tips press dents in the sides of the host nuclei (fig. 1, a), recalling the occurrence of this well-known phenomenon in many other rusts; but, so far as the writer has been able to observe, the nucleus does not appear to suffer greatly from this close association with the parasite. In fact, the cells in which the haustoria are present seem to be quite as healthy as the cells which have remained free from attack. Usually the haustoria are unbranched; but in some instances, especially in the cells of the medullary rays, they branch (fig. 1, b). In no case have more than three tips been observed. All the cells of the mycelium, including the haustoria, are uninucleate until cell fusion occurs at the base of the æcium.

The Peronosporales and the Erysiphaceae are the only other groups of parasitic fungi beside the Uredinales which possess prominent haustoria. No members of the first two groups are known to parasitize the bark of *Pinus strobus*. Moreover, the second group is composed of leaf parasites whose mycelium is almost entirely exterior. Even if by any chance material containing mycelium from either of these two groups should be introduced into the mount along with the bark sections, there would be no danger of confusion. The fact that the members of the first group have cœnocytic hyphæ and that the members of the second group possess haustoria which are decidedly different in appearance 1 from those of the blister-rust, rules out any possibility of confusing the mycelium of members of either of the two groups with the mycelium which has been described.

METHOD OF PARASITISM

The manner in which the parasite attacks the host is also important as a distinguishing character. The hyphæ are intercellular in position. As stated above, practically every cell of the bark may be pierced by one or more haustoria. Parenchyma cells of the cortex and in the neighborhood of the resin ducts, sieve cells, medullary-ray cells, and even resin cells are all subject to attack. Only the cork and schlerenchyma

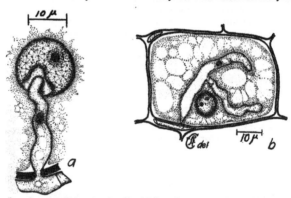

Fig. 1.—*Cromartium ribicola:* a, A moderately straight haustorium which has dented the side of the host nucleus. Note the manner in which the haustorium is constricted in passing through the cell wall. b, A branched haustorium in a young medullary-ray cell. The figures were drawn with the aid of a camera lucida.

cells of old bark and, of course, dead bark are apparently immune. The hyphæ may follow the medullary rays past the cambium and into the xylem for at least three annual rings, and in these medullary-ray cells the haustoria are quite as prominent as in the ray cells of the phloem. Haustoria do not enter the wood cells. The parenchyma cells surrounding the resin ducts in the last two annual rings of xylem are also heavily attacked. In tangential or radial sections one sees that the hyphæ are frequently massed in strands (Pl. 31, A). These strands are often formed in the region of the thin-walled sieve tubes, but they are also scattered through the parenchyma. Generally the medullary-ray cells are forced apart by single layers of hyphæ. Probably the swelling of the bark is due almost entirely to the spreading action of the hyphæ as they force themselves between the cells. This method of attack differs from that of other parasitic fungi observed on white-pine bark. In attacks made by species of Phoma and Fusicoccum the cells are killed, the hyphæ run through the cells, and the bark tends to dry up, turn brown, and crack; whereas in the case of the blister-rust attack the cells remain alive for a long time, the hyphæ run between the cells, the bark swells, turns a yellowish green color, and does not crack until the æcia are produced.

VALUE OF MYCELIAL CHARACTERS, HAUSTORIA, AND METHOD OF PARASITISM IN DIAGNOSIS

It is evident from the above discussion that the mycelium described must belong to the Uredinales. Furthermore, since there is only one rust known to attack the bark of *Pinus strobus*, the mycelium must be the mycelium of *Cronartium ribicola*. This obvious conclusion has been vigorously proved by the comparison of the mycelium in suspected cases with the mycelium in bark on which pycnia and æcia had developed. The presence in the bark of *P. strobus* of hyphæ and haustoria fitting the description which has been given above is sufficient evidence to warrant the conclusion that the pine is infected with *C. ribicola*.

SUMMARY

(1) A simple method for the use of safranin and lichtgruen in staining the mycelium of *Cronartium ribicola* and the cells of the host, *Pinus strobus*, is given in detail.

(2) The two stains employed give permanent preparations which are particularly favorable for the study of the relation of host to parasite.

(3) The mycelium of *C. ribicola* is distinguished from that of other fungi parasitic on white-pine bark by the size of the hyphæ and the possession of haustoria.

(4) The manner in which the parasite attacks the host cells is described in detail. This method of attack is characteristic of the blister-rust in white-pine bark.

(5) The presence in the bark of *P. strobus* of mycelium such as has been described is sufficient evidence to warrant the conclusion that the pine is infected with *C. ribicola*.

Cronartium ribicola:

A.—A tangential section showing strands of hyphæ among the white-pine host cells. Note the haustoria. In the upper left-hand corner a haustorium can be seen denting the host nucleus. × 325.

B.—A haustorium showing the characteristic bend and irregular outline. The section just missed the connection of the haustorium with the hypha outside the cell. The host nucleus was torn by the razor. × 1,250.

These figures illustrate excellently the relation of the parasite to the host in the phloem parenchyma Photomicrographs by the writer

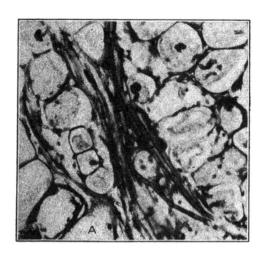

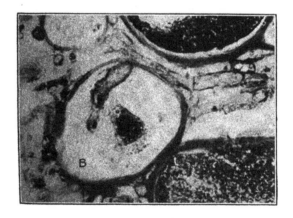

JOURNAL OF AGRICULTURAL RESEARCH

Vol. XI WASHINGTON, D. C., NOVEMBER 12, 1917 No. 7

EFFECT OF TEMPERATURE AERATION AND HUMIDITY ON JONATHAN-SPOT AND SCALD OF APPLES IN STORAGE [1]

By CHARLES BROOKS, *Pathologist*, and J. S. COOLEY, *Assistant Pathologist, Fruit Disease Investigations, Bureau of Plant Industry, United States Department of Agriculture*

INTRODUCTION

As the public becomes more critical in regard to foods, the troubles which affect the appearance of fruit but do not completely destroy it are brought into greater prominence, both from a pathological and an economic point of view. These troubles are of importance not only because of their damaging effects on the appearance of the fruit but also because of the part they may play in paving the way for the entrance of various rot-producing fungi. The present paper is a report on Jonathan-spot and scald of apples (*Malus sylvestris*), including a study of their relation to rot infection and the modifying effects of storage conditions and maturity of fruit.

EXPERIMENTAL APPARATUS

The apparatus used in the experiments has been fully described in an earlier publication (3).[2] The fruit was held in galvanized-iron boxes which had a 3-inch air space between them and the outer insulating walls of the chambers. The brine pipes were located outside of these metal boxes. This arrangement made it possible to keep the air in the inner boxes free from the drying effects of the brine pipes and thus to maintain higher humidities than could have been otherwise secured. The air in the inner boxes was circulated by means of fans. Where high humidity was desired, the air was fanned over pans containing water and saturated sponges; to obtain lower humidities it was circulated over calcium chlorid and lime. A record of the temperature and the humidity was kept by means of thermographs and hygrographs placed in the middle of the boxes and by wet and dry bulb thermometers in front of the fans.

---

[1] Studies in Fruit Rots and Spots. II.   [2] Reference is made by number to "Literature cited," p. 316-317.

Journal of Agricultural Research,
Washington, D. C.
kn

Vol. XI, No. 7
Nov. 12, 1917
Key No. G—1.6

(287)

Jonathan-spot (19) is a name applied to superficial
spots of the apple that are especially common on the J
In the early stages of the disease only the superficial co
are involved, and the spots are seldom more than 2 n
Later the spots may enlarge to a diameter of 3 to 5 mm.
sunken, and spread down into the tissue of the apples
depth. In the later stages of the disease rot fungi a
species of Alternaria being particurlarly common. T
occur at lenticels.

RELATION OF JONATHAN-SPOT TO APPLE ROT

Scott and Roberts (19) have pointed out that a apec
is often present in late stages of Jonathan-spot, and Coo
have found that a species of Alternaria is almost unive
with Jonathan-spot as it occurs in New Jersey. The wri
and observations indicate that fungi are not present in
of the disease, but that Jonathan-spots often serve as pc
for species of Alternaria and other apple-rot fungi.
repeated observations on this point, inoculation experin
to determine the influence of Jonathan-spot upon the sus
fruit to rot infection. A box of western Jonathan appl
the Washington market were separated into two lots,
Jonathan-spot and the other free from it. The apples
were practically alike in size and color. Part of the
were punctured with a platinum needle, two punctures
each apple. All of the apples were washed in water that
suspension of the spores of rot fungi, and the fruit wa
moist chambers at 15° C. The following results were

EFFECT OF TEMPERATURE ON JONATHAN-SPOT

Scott and Roberts (19) found that apples stored in a moderately cool basement developed much more Jonathan-spot than similar fruit held in commercial cold storage.

In 1915 and again in 1916 the writers carried out experiments to obtain further data in regard to the factors influencing the development of the disease. The fruit used in the experiments was shipped by express direct from the orchard to the laboratory, so that there was never a delay of more than four days after picking before the apples were placed under the various storage conditions. The fruit was always carefully sorted so that the apples placed at the various temperatures or humidities that were to be compared would be as nearly alike as possible in size, color, and maturity. This was accomplished by selecting a number of apples that were similar in all respects and distributing them one each in the containers that were to be placed under the different conditions and repeating the process until the desired number of apples was obtained. All the apples were Jonathan and all were free from Jonathan-spot at the beginning of the experiments. The apples used in the experiments reported in figure 1 were from West Virginia, those of figure 7 from New Jersey, and those used in the other experiments from Pennsylvania. The apples used in the experiments recorded in figure 6 averaged more than 3 inches in diameter, while those used in the other experiments averaged

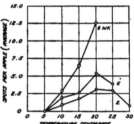

Fig. 1.—Graphs showing the effect of temperature on Jonathan-spot at the end of 2, 6, and 8 weeks

The apples were stored in moist chambers with moist filter paper added. The experiment was started on November 6, 1915.

about 2½ inches in diameter. In the former cases, 8 apples were used for each temperature or storage condition and in the latter, 10 to 15 apples for each condition.

The results obtained are shown in figures 1 to 8, inclusive. In all the figures the perpendiculars show the average number of spots per apple and the base lines the centigrade temperature.[1] In some of the experiments (as noted in the figures) a warm laboratory was used for the $25°$ C. temperature. This room temperature was not constant and averaged less than 25°. All the other temperatures were carefully controlled.

A study of the curves in the different figures shows that the amount of Jonathan-spot increased with a rise in temperature up to 20°, but that

---

[1] Temperature equivalents.

°C.	°F	°C.	°F.	°C.	°F.
0=	32	15=	59	25=	77
5=	41	20=	68	30=	86
10=	50				

the disease made practically no development at 30°. The increase in the disease between $_{10}$° and $_{1}$$_5$° was very striking, and it is of special interest to note that in most cases Jonathan-spot was held practically in check at $_5$° as well as at o°. It was thought possible that the spots might be present in an incipient form at the lower temperatures and that they

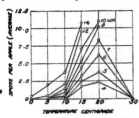

would make a rapid development upon removal to warmer temperatures, but experiments made to test this point gave negative results in all cases

The failure of the disease to develop at $_{30}$° indicates that its occurrence can not be primarily due to the presence of *Alternaria* sp., as has been suggested by Cook and Martin (6, 7), since it has been shown by the writers in an earlier publication (3) that *Alternaria* sp. makes a more vigorous growth at $_{30}$° than at any lower temperature

Fig. 2 —Graphs showing the effect of temperature on Jonathan-spot at the end of 4, 5, 6, 7, 8, 10, 12, and 14 weeks The apples were well matured and were stored in moist chambers with moist filter paper added The experiment was started on October 6, 1916.

The importance of uniform temperature in the keeping of fruit has often been emphasized. An experiment was made to determine the effect that fluctuations in temperature would have upon the development of Jonathan-spot. Apples that had been placed in moist chambers were moved back and forth from $_5$° to $_{25}$° at the end of each second day, and the results were compared with those obtained at constant temperatures. It was found

that slightly less disease developed on the fruit exposed to the different temperatures than on that kept at the average temperature of 15°. Moving the moist chamber to a different temperature naturally favored an exchange of air, and it is possible that this partial aeration of the container may have had some tendency to decrease the disease. However this may have been, the results indicated no harmful effects from fluctuating temperatures, so far as Jonathan-spot was concerned.

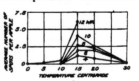

Fig. 3 —Graphs showing the effect of temperature on Jonathan-spot at the end of 4, 6, 8, 10, and 12 weeks. The apples were well matured and of the same lot as those of figure 2 but were stored in open containers in air that was gently stirred and that had a relative humidity of 85 to 95 per cent. The experiment was started on October 6, 1916.

The various experiments make it evident that temperature plays an extremely important part in the development of Jonathan-spot. It would be impossible to state just when physiological processes have been most concerned in the results, but it is evident that the effects are of a cumulative nature. The disease has not usually made much progress during the first month of storage, but after that time the number of spots have increased in a very gradual manner over a period of one or two months.

EFFECT OF AERATION AND HUMIDITY ON JONATHAN-SPOT

In most of the temperature experiments part of the fruit was placed in moist chambers and part in wire baskets or other open containers. The tops of the moist chambers fitted loosely, nearly always leaving cracks of .5 mm. or more, yet preventing a free circulation of air. The apples in the open containers were freely exposed to the air of the boxes, which was kept in circulation by fans. Anemometer tests showed that the movement of the air was in no case as rapid as ½ mile per hour. Gas analyses reported later under

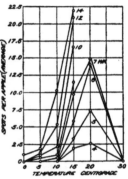

FIG. 4.—Graphs showing the effect of temperature on Jonathan-spot at the end of 4, 5, 6, 7, 8, 10, 12, and 14 weeks. The apples were of the same lot as those of figure 2 but were greener. They were stored in moist chambers with moist filter paper added. The experiment was started on October 6, 1916.

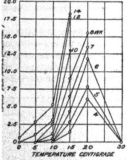

FIG. 5.—Graphs showing the effect of temperature on Jonathan-spot at the end of 4, 5, 6, 7, 8, 10, 12, and 14 weeks. The apples were of the same lot as those of figures 2 and 4 but were greener than the former and riper than the latter. They were stored in moist chambers with moist filter paper added. The experiment was started on October 6, 1916.

the head of "Apple-scald" (p. 300) showed that the amount of carbon dioxid in the free air of the boxes was little above normal, while that in the moist chambers was sufficiently high to have a distinctly depressive effect upon respiration.

The effects of aeration and humidity upon Jonathan-spot are brought out in figures 6 and 8. In figure 6 curve A shows the amount of the disease developed in moist chambers to which wet filter paper had been added, and curve B the amount developed in the open containers. The air in the former case was practically saturated, while in the latter case that at 10° had a relative humidity of 55 to 70 per cent and that at 25° was probably still drier. The experiments recorded in figure 8 were more carefully controlled. Curve C shows the amount of Jonathan-

spot developed at the various temperatures in moist chambers to which
wet filter paper had been added, and curve D the amount developed in
open containers in an atmosphere having a relative humidity that ranged

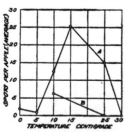

from 85 to 95 per cent, and usually
stood slightly above 90 per cent. The
weekly contrasts obtained under these
two conditions of storage are given in
figures 2 and 3. At 15° a second box
was maintained in which the relative
humidity ranged from 65 to 75 per
cent, and usually stood at 70 per cent.
Under this condition the fruit became
decidedly withered by the end of the
second month. The rate at which
Jonathan-spot developed on the apples
in the open containers in this dry box
was practically the same as that in the
more humid box at the same temperature (fig. 3).

Fig. 6.—Graphs showing the effect of aeration and humidity on Jonathan-spot.
Graph A shows the amount of disease
developed at the different temperatures in
moist chambers and graph B the amount
developed in open containers. The apples
were large. The experiment was started
on October 16, 1915, and the above records
were taken three weeks later.

A study of the results in figures 2, 3,
6, and 8 shows that the amount of Jonathan-spot in the moist chambers was
usually more than twice as great as that in the open containers, and
that this held not only for all the temperatures but for the various
periods of storage. The results seem
to leave no question that the storage in moist chambers favored the
development of the disease, but the
nature of the experiment makes it
difficult to estimate the relative importance of aeration and humidity.
In the one experiment reported above
in which the effect of a humidity of
70 per cent was compared with that
of a humidity of 90 per cent under
similar conditions of aeration, the
data furnished no evidence that the
amount of moisture in the air had any
influence upon the development of
Jonathan-spot. In so far as the results of a single experiment can be
relied upon, evidence is furnished that

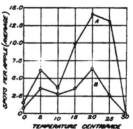

Fig. 7.—Graphs showing the effect o the maturity of fruit on Jonathan-spot. Graph A
shows the amount of disease on the green fruit
and graph B the amount on the ripe fruit. The
apples were stored in moist chambers with
moist filter paper. The experiment was
started on October 26, 1915, and the above
records were taken five weeks later.

the striking contrasts obtained between storage in moist chambers and
storage in the open containers are due to differences in aeration rather than
to differences in humidity. This interpretation of the results would be

entirely in harmony with the temperature responses and the cumulative nature of the disease. The question of aeration and humidity will be more fully discussed in connection with the results on apple-scald.

EFFECT OF MATURITY OF FRUIT ON JONATHAN-SPOT

Experiments were made to determine the effect of the maturity of fruit upon its susceptibility to Jonathan-spot. Jonathan apples of a particular picking and of a particular lot were divided into two classes that were as nearly alike as possible as to size of fruit, but one contained apples on which the ground color was decidedly more yellow than the other. There was usually also a larger blush area on the former than on the latter. The method of selection has been described on page 299. The apples of the two lots were placed in moist chambers and distributed at the different temperatures already mentioned. The results of the experiments are shown in figures 7 and 8. In figure 7 curve A shows the amount of disease developed on the green fruit and curve B the amount on the ripe fruit; in figure 8 curve A the disease on the green fruit and curve C the disease on the ripe fruit under like conditions. Further contrasts may be obtained by comparing the weekly development of Jonathan-spot on the ripe fruit, as shown in figure 3 with the weekly development on the green fruit, as shown in figure 6. In both the sets of experiments and at all

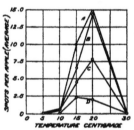

FIG. 8.—Graphs showing the effect of maturity of fruit and of aeration and humidity on Jonathan-spot. Graph A shows the amount of disease developed at the different temperatures on the greenest fruit of the particular lot, graph B the amount developed on the fruit that was somewhat riper, and graphs C and D the amount on the ripest fruit. The apples of graph D were stored in open containers and the others of the experiment in moist chambers. The experiment was started on October 6, 1916, and the above records were taken seven weeks later. See figures 2, 3, 4, and 5

the different temperatures the green fruit showed a greater susceptibility to Jonathan-spot than the ripe fruit.

The color of the spots that occurred on the lighter skin surfaces of the fruit was a lighter brown than that of the spots on the blush surfaces. The green fruit had less color than the ripe and consequently a relatively larger percentage of the light-brown spots. No clear-cut distinction could be drawn between the two kinds of spots, as the color of the one gradually shaded into that of the other, and they were exactly alike in other characteristics.

It is the general opinion of persons who have made orchard observations on Jonathan-spot that the disease is worst on the ripest fruit. The above experiments show that under certain storage conditions green fruit is more susceptible to the disease than ripe fruit, but it does not necessarily follow that the spots that develop under orchard conditions would obey the same law.

APPLE-SCALD

"Scald" is a term applied to a superficial browning that often occurs on apples in storage. Usually only the five or six surface layers of cells that form the color-bearing tissue of the apple are affected, but in aggravated cases the trouble may extend entirely through the firmer skin layer of the apple into the large isodiametric cells of the pulp. In such cases the flesh becomes soft, brown, and rotlike, and the trouble is sometimes spoken of as "deep-scald" Deep-scald has been of more common occurrence on York Imperial than on Grimes.

It will be noted that in the earlier and more typical cases of scald the tissue affected is the same as with Jonathan-spot. The skin color in the case of scald, however, is a light rather than a dark brown, the areas affected are always larger, and the demarcation between diseased and healthy tissue less definite than in the case of Jonathan-spot (Pl. 32 and 33).

SCALD AS A SOURCE OF ROT INFECTION

The fact that the protective skin layers of the apple are broken down in the case of scald naturally suggests the importance of the disease in paving the way for the attacks of fungi. The writers had often observed cases where apple rots appeared to follow scald, but it seemed desirable to make an experimental determination of the relation of scald to rot infection. The test was made on Grimes apples that had been used in the aeration and humidity experiments reported later. The scalded fruit had been picked at the same time and stored at the same temperature as the unscalded, but had received poorer aeration. The apples were washed in heavy spore suspensions of the fungus in the manner already described in the rot experiments on Jonathan-spot. Ten scalded and ten scald-free apples were used with each fungus. The experiment was started on November 7. Table II shows the results obtained with *Glomerella cingulata* after two weeks, and with *Penicillium expansum* and *Sclerotinia cinerea* after three weeks.

TABLE II.—*Relation of apple-scald to infection by apple rot fungi*

Fungus.	Number of infections on—	
	Scalded apples.	Scald-free apples.
Glomerella cingulata...	9	0
Penicillium expansum..	10	1
Sclerotinia cinerea..	8	0

The results indicate that the dead-skin areas produced by apple-scald furnish an excellent opportunity for attacks by the common rot fungi. According to the observations of the writers, *Alternaria* sp. is more likely to infect scalded tissue than any other rot-producing fungus. Unfortunately cultures of this fungus were not available for inoculation at the time of the above experiment. It will be recalled that Jonathan-spot tissue was found to be particularly susceptible to the attacks of *Alternaria* sp.

EFFECT OF TEMPERATURE ON APPLE-SCALD

Numerous experiments have been reported on the effect of temperature upon apple-scald. Powell and Fulton (17) found that prompt refrigeration was important in the prevention of apple-scald, but Greene (11) reported that holding the temperature high enough to permit the fruit to continue ripening delayed the appearance of scald and found that Grimes apples scalded much less when they were not stored immediately after picking. Some writers have reported that scald is worse in cold storage than in cellar storage, while others have found that the reverse condition apparently held.

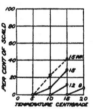

Fig 9 —Graphs showing the effect of temperature on apple-scald at the end of 9, 12, and 15 weeks. The dotted graph shows the amount of scald 3 days after removal from storage at the end of the 15 weeks. The apples were Jonathan and the same as referred to in figures 2, 4, and 5 of the Jonathan-spot experiments. The results we're similar for the different lots, and the average given is in the above graphs.

In the fall of 1915 and again in the fall of 1916 the writers carried out experiments to determine the effect of temperature on scald. The apparatus used in the experiments has already been described (p. 297). Great care was taken that the apples at the different temperatures should be alike in size and maturity (p. 299). Arkansas, Baldwin, Grimes, Jonathan, Northwestern, and York Imperial apples were used. The results obtained on the last three varieties are given in figures 9 to 16. The experiments made on the other varieties were not so complete nor the scald so bad, but the results were in agreement with those reported. In all of the figures the base line represents temperature centigrade and the perpendicular the per cent of scald. In obtaining records on the degree of scald the maximum scald that had been observed to occur on the variety was taken as 100 and the amount of scald in a particular case was measured by its relation to this standard. Both the area and the depth of the scald were considered. A record was made by two observers; and where differences occurred, the average was taken. The notes were made without reference to previous records. The amount of scald on the different apples at a particular temperature was remarkably uniform. In the early stages of scald at the higher temperature the apples took on a dry, russet appearance that was at first classed as distinct from the usual form of scald; but this soon passed over into the solid brown coloration typical of scald at lower temperatures, and therefore seemed to require no sepa-

rate classification. From 11 to 16 apples were used at each temperature in most of the experiments, but in the case of the results shown in figures 9 and 14, only 5 apples were used. Except where otherwise stated, the apples were in moist chambers.

The Jonathan showed less susceptibility to scald than the Grimes and York Imperial apples, but so far as temperature response is concerned, the results are so similar with the different varieties and in the different experiments that separate discussion is unnecessary.

Fig. 10.—Graphs showing the effect of temperature on apple-scald at the end of 6, 7, 9, 10, 13, and 15 weeks. The dotted graphs show the amount of disease 3 days after removal from storage at the end of the given week. The apples were York Imperial from Quincy, Pa. They were picked about October 8 and the experiment was started on October 11, 1916.

In all of the experiments apples were placed at 30° C., as well as at the temperatures indicated in the various figures. In most cases it was found necessary to discard the apples at this temperature on account of internal breakdown before there had been time for the development of scald. The only exception to this is shown in figure 12 and was obtained with apples that were quite immature. The results indicate that a temperature of 30° is unfavorable to the development of scald.

Scald always started sooner at 20° than at 15°, but in most cases the contrasts in the amount of disease at the two temperatures soon disappeared, and with Grimes apples of figure 15 the fruit finally showed a greater degree of injury at 15° than at 20°. The contrasts between the results at 10° and 15° were very striking with York Imperial and Grimes, especially during the second month of storage. During the third month scald developed rapidly at 10°, soon producing decided contrasts between the results at this temperature and those at 5°. During the fourth month scald began to appear at 5°, and in the following month traces of the disease were evident at 0° on the more susceptible apples.

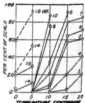

Fig. 11.—Graphs showing the effect of temperature on apple-scald at the end of 3, 4, 5, 6, 7, 8, 9, 10, 14, and 19 weeks. The dotted graphs show the amount of disease 3 days after removal from storage at the end of the given week. The apples were Grimes from Middletown, Va. They were picked on September 1 and the experiment was started on September 9, 1916.

With the Jonathan apples the disease developed more slowly, usually making its appearance on this variety about a month later than on the Grimes and York Imperial.

The amount of scald developed at the lower temperatures did not become fully evident until the fruit had been removed from storage.

This was particularly true of apples at 0°. The dotted lines in figures 9 to 15, inclusive, show the results at the end of the given week, but after the fruit had stood for three days at laboratory temperature. Half of the fruit under the particular condition was removed in each case. A comparison of this data with that obtained before removal, as shown by the solid lines having the same numbers, makes it evident that the scald on apples at 5° may appear several times as bad after a few days of free exposure to warm air, and that apples at 0° may be rather badly scalded without showing it while remaining at that temperature. In some cases the Grimes had developed a latent scald at 0° by the end of 10 weeks, but the York Imperial failed to show any scald upon removal from storage at that time, and the Jonathan had none at either 0° or 5° upon removal at the end of 15 weeks.

Fig. 12 —Graphs showing the effects of temperature on apple-scald at the end of 6, 7, 8, 9, 11, 14, 15, 18, 21, and 23 weeks. The dotted graphs show the amount 3 days after removal from storage at the end of the given week. The apples were Grimes from Vienna, Va. They were picked on August 10 and the experiment was started on August 11, 1916.

While the fruit at low temperatures showed a greater amount of scald upon removal, the nature of the temperature contrasts was not changed, the difference between the amount of disease at 0° and 5° usually being nearly as great after removal as before. The whole series of experiments has been consistent in showing that an increase in temperature is accompanied by an increase in rate of scald up to an optimum of 15° or 20°. The critical period for scald development appeared about a month earlier with each 5-degree rise in temperature.

Fig. 13 —Graphs showing the effect of temperature on apple-scald at the end of 4, 5, 6, 7, 8, 9, 12, 15, 16, 19, and 21 weeks. The dotted graphs show the amount 3 days after removal from storage at the end of the given week. The apples were Grimes from the same orchard as those of figure 12, but were picked on August 26 and the experiment was started on August 29, 1916.

Experiments were made to determine whether the increase in disease upon removal from storage was due to the sudden change in temperature. York Imperial apples that had been held in commercial cold storage till the first of April were removed and divided into various lots that were distributed at temperatures of 0°, 5°, 10°, 15°, and 20°. Some of the apples were allowed to remain at each of these temperatures, while others were gradually moved to higher temperatures. At the end of 10 days practically no scald was evident at 0° and 5°, and only about half as much at 10° as at 15° and 20°. At this time all of the apples from the lower temperatures were removed to 20°. After two days at this temperature all of the fruit was scalded practically alike, regardless of the gradation of temperature it had passed through in removal. It was evident that the method of removal had little or no effect upon the production of the disease, and further proof was obtained of the ability

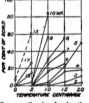

of the lower temperatures to inhibit the development of the scald phenomena after the diseased condition had actually been produced.

Experiments were also made to determine the effect of fluctuating temperatures upon the development of scald. Palladine (16) found that sudden changes in temperature caused an increase in the amount of carbon dioxid given off by seedlings, but Blanc (1), on the other hand, has reported that when seedlings and plant parts were moved from one temperature to another the rate of respiration was a mean between the normal rates at the two temperatures.

FIG. 14.—Graphs showing the effect of temperature on apple-scald at the end of 2, 3, 4, 5, 6, 7, 8, 9, 10, 15, and 17 weeks The dotted graphs show the amount 3 days after removal from storage at the end of the given week. The apples were Grimes from the same orchard as those of figures 12 and 13, but were picked on September 21 and the experiment was started on September 22

Grimes apples of the same lot as reported in figure 11 were placed in moist chambers and moved from one temperature to another at intervals of two days. In one case the apples were moved back and forth from 5° to 20°. The amount of scald developed at the different times of recording was found to be slightly less than on similar apples at a constant temperature of 15°, and the moved apples "stood up" approximately the same length of time as those at 15°, both being discarded on November 6.

In a similar experiment with apples of the same lot the moist chambers were moved back and forth from 0° to 30°. The higher temperature was secured in a warm laboratory and not in a closed incubator. The apples developed no scald and "stood up" till November 20—as long as similar apples at a constant temperature of 10°. Two reasons can be offered for the absence of scald, one that better aeration was secured by leaving the moist chambers in the open laboratory a part

FIG. 15.—Graphs showing the effect of temperature on apple-scald at the different temperatures at the end of 4, 5, 6, 7, 9, 10, 13, and 15 weeks. The dotted graph shows the amount of disease 3 days after removal from storage at the end of the given week. The apples were Grimes of the same lot as those of figure 11, but had been in commercial cold storage from September 9 to October 16.

of the time and the other that, as has already been pointed out, storage at a temperature of 30° is unfavorable to the development of scald. The fact that the apples "stood up" well in storage would emphasize the first hypothesis.

The results of the two experiments and of other similar experiments indicate that apples at a fluctuating temperature develop but little, if any, more scald than similar apples at an average but constant temperature.

EFFECT OF AERATION AND HUMIDITY ON APPLE-SCALD

There has been a great diversity of opinion as to the importance of ventilation in the development of scald. Powell and Fulton (17) found that paper wrappers did not reduce the amount of scald on apples, but that paraffin wrappers did. They further found that an atmosphere of moist oxygen favored the development of scald, while an atmosphere of nitrogen prevented it, at least during short periods of storage. Greene (11) reported that wrapping fruit retarded the ripening processes and delayed the appearance of scald.

Fulton (9) reported that extra wrappings on cartons of berries resulted in a decrease in the oxygen of the package and an increase in the carbon dioxid and caused the fruit to soften prematurely and take on a bad flavor. He found that these conditions were still more emphasized by storage in tightly stoppered glass bottles.

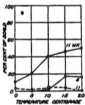

Hill (12) found that storing apples and peaches in either nitrogen or hydrogen resulted in a bleaching or browning of the skin and in the development of disagreeable flavors. Peaches stored in carbon dioxid developed similar conditions, but a more nauseating flavor. The peaches in hydrogen and those in nitrogen became soft, but those in carbon dioxid did not. He thought that the firmness of the fruit in the latter case was due to the fact that the gas decreased the hydrolysis of pectose.

Thatcher (21) stored apples in various gases. He found that in an atmosphere of air or of oxygen the apples became visibly overripe within

FIG. 16.—Graphs showing the effect of aeration and humidity on apple-scald. The solid graphs show the amount of scald developed in moist chambers at the end of 5 and 11 weeks and the dotted graph the amount in an open container. The apples were Grimes from cold storage. The experiment was started on November 6, 1915.

four weeks, that in hydrogen or in nitrogen they did not soften so rapidly, but soon became discolored and unhealthy in appearance, while in carbon dioxid they retained their color, flavor, and firmness for more than six months. He also found that berries softened much sooner in the air than in carbon dioxid. He concluded that the enzyms that participate in the changes of the carbohydrates during the ripening processes are oxidases and inhibited by carbon dioxid, an end product of oxidation.

The writers (2) have reported experiments indicating the importance of aeration and humidity in the prevention of apple-scald.

The experimental studies mentioned in the last reference (2) were continued in a much more extensive series of experiments carried out in connection with the temperature work already reported. In the temperature experiments, as has already been mentioned, the apples were placed in moist chambers of the usual laboratory style. The tops of the containers fitted loosely, nearly always leaving cracks of 0.5 mm. or more, yet preventing a free circulation of air. The moist chambers were usually opened once a week in note taking. With all of the different experiments and at all of the different temperatures similar apples were placed in the storage boxes by the side of the moist chambers, but held in baskets or other open containers that allowed a free circulation of air. The air in the boxes was gently circulated by fans, but its rate of movement was always less than ⅓ mile per hour.

Gas analysis tests were made to determine the amount of carbon dioxid to which the fruit was exposed under the various conditions. The apparatus used in the determinations was that of Bonnier and Mangin (with modifications by Dr. William Crocker and George T. Harrington), and the methods, those outlined by Thoday (22). The tests showed that the free air in the $_{20}°$ box had 0.42 per cent of carbon dioxid, that in the $_{15}°$ box 0.30 per cent, that in the $_5°$ box 0.24 per cent, and that in the $_0°$ box 0.02 per cent. The larger amounts of carbon dioxid in the warmer chambers is readily accounted for by the more rapid respiration of fruit at higher temperatures and by the better aeration that would naturally be secured in the colder boxes as a result of greater contrasts between the outside and inside temperatures. It was found impossible to obtain a satisfactory sample of air from the moist chambers, as the removal of the cover from such shallow vessels allowed too free an exchange of air. At the $_{15}°$ temperature it was possible to obtain samples from jars that had been handled as moist chambers, having small openings left at the stopper, and in which the percentage of scald had run parallel with that in the regular moist chambers. There was no contrast between the amount of scald on the fruit in the bottom of these jars and that on the fruit near the top. The air samples were taken from near the bottom of the jar. An analysis of these samples showed that the air contained 4.28 per cent of carbon dioxid. While complete data could not be obtained at the different temperatures, it seems evident that there was several times more carbon dioxid present in the air of the moist chambers than in the free air of the boxes.

The air in the moist chambers was kept in a saturated condition by the addition of wet filter paper, the fruit, upon removal to a warmer temperature, often being covered with a film of moisture or dotted with drops of water. The free air of the boxes was kept at a relative humidity of 85 to 95 per cent. Water always stood in the bottom of the metal boxes, but the fruit did not become perceptibly moist even upon removal for note taking; nor did it show any sign of withering even after several months of storage. In the experiments reported in figure 16, the relative humidity was lower, ranging from 55 to 70 per cent. In this case the apples became slightly withered toward the close of the experiment.

The results obtained in the moist chambers has already been reported in figures 9 to 16, inclusive. The apples scalded badly at all temperatures when the period of storage was sufficiently prolonged, but the amount of the scald decreased uniformly with the temperature. In the open containers, however, the results were very different. The apples scalded upon removal from 0°, while those at all higher temperatures remained free from scald even after removal. In the case of the apples in open containers corresponding with those of the moist-chamber experiment reported in figure 16, there was a slight trace of scald in the baskets at higher temperatures, but the fact that these apples had been in cold storage six weeks before the experiment started may furnish an explanation for this. In the experiments reported in figures 10, 14, and 15, no scald occurred in the open containers at 0°, while it was found in the moist chambers at that temperature in each case. The fruit used in these tests was riper than that used in the others and the duration of the experiments was several weeks shorter. The contrasts obtained on the less mature Grimes apples are brought out in figure 17. In all cases the fruit stood in the laboratory three days before the notes were taken. It will be seen that at $0°$ there was but little more scald in the moist chambers than in the open containers, while at $5°$ the contrast was extreme, all of the apples in the moist chambers being badly scalded, while those in the open were free from scald. The contrasts at $10°$, $15°$, and $20°$ were similar to those at $5°$, but appeared earlier in the experiment. It is evident that at $0°$ apples in open containers behaved entirely different with reference to scald from similar apples at any of the higher temperatures tested.

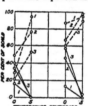

Fig. 17—Graphs showing the amount of scald on Grimes apples from 0° and 5° C, 3 days after removal from storage. The dotted graphs show the results obtained in moist chambers, the solid graphs the results in open containers. The graphs on the left show the amount of scald on December 22, those on the right the amount on January 23. The numbers near the graphs refer to the different experiments made. 1. Experiment started on August 11, 1916. (See fig. 12.) 2 Experiment started on August 29, 1916 (See fig 13.) 3 Experiment started on September 29, 1916. (See fig. 11.)

Careful notes were taken as to other differences between the fruit in the moist chambers and that in the open. It was found that the Grimes apples colored more rapidly in the latter case than in the former. At $15°$, $20°$, and $30°$ they reached a particular degree of color a week to 10 days earlier in the baskets than in the moist chambers, but as the lower temperatures were approached the color contrasts became less decided (fig. 18, 19, and 20). The results seem to justify the conclusion that the conditions which are favorable to the development of scald are unfavorable to the development of color in the fruit.

15753°—17——2

Other characteristics associated with ripening were also affected by the amount of ventilation. When the apples were fully ripe, those in the open were crisp, juicy, and agreeable to the taste, while those in the moist chambers, although less highly colored (in the case of Grimes) were insipid and mealy, often cracking open toward the close of the experiment. These contrasts were especially striking at the higher temperatures (including 30°). After several months of storage they became quite distinct at 5° and at the end of the experiment could usually be detected at 0°.

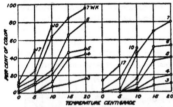

Fig. 18.—Graphs showing the development of color on Grimes apples in storage. The graphs on the left were obtained with the apples in baskets; those on the right with similar apples in moist chambers. The two series of graphs show the gradual development of the color at the different temperatures, the figures near the lines indicating the number of weeks between the starting of the experiment and the time of taking the notes. The apples were placed in storage on September 9, 1916. See figures 11 and 17 for the development of scald.

The differences in quality of the apples in the open and closed containers suggested the importance of sugar and acid determinations, but the apparatus for this work was not available till near the close of the experiment, which made it impossible to secure chemical data on the fruit that was at the higher temperatures.

The determinations reported in Table II were made the latter part of January. In sampling the fruit an 18-mm. cork borer was used, the cylinders of tissue being taken parallel to the core, but not including any part of it. All of the peel was also removed from the samples. Maceration was accomplished by the use of a sampling press devised by Clark (5). The pulp was reduced to extreme fineness, which made it easy to extract the acid and sugar.

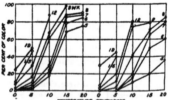

Fig. 19 —Graphs showing the development of color on Grimes apples in storage The graphs on the left were obtained with the apples in baskets; those on the right with similar apples in moist chambers. The two series of graphs show the gradual development of the color at the different temperatures, the figures near the lines indicating the number of weeks between the starting of the experiment and the time of taking notes. The apples were placed in storage on August 29, 1916. See figures 13 and 17 for the development of scald.

In the tests on acidity distilled water was first added to the weighed pulp at the rate of 5 c. c. to each gm. of apple tissue. The diluted mixture was then heated to 100° and held at that temperature for one hour in an Arnold sterilizer. The finely divided condition of the pulp may have made heating unnecessary for the extraction of the acid, but it was found that the hot extract could be filtered much more readily than the cold. The filtration was accomplished by means of ordinary filter paper. The pulp was allowed to drain and was then washed with distilled water until the total volume of the filtrate and washings equaled 12.5 c. c. for each gm. of original pulp. The filtrate was titrated against $N/20$ sodium hydroxid, using the color changes of the extract as an indicator described by Schley (18) and by Culpepper, Foster, and Caldwell (8). In the titrations, as well as in all the other steps of the procedure, the corresponding samples from the open and from the closed containers were carried through the tests side by side. The results of the analyses are given in Table III.

The method for sugar determination was essentially the same as that employed by Bryan, Given, and Straughn (4), with the exception that the extraction was made with water instead of alcohol.

Earlier tests on the efficency of the sampling press had shown that maceration was so thorough that

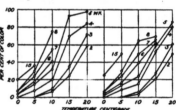

FIG. 20 —Graphs showing the development of color on Grimes apples in storage. The graphs on the left were obtained with the apples in baskets, those on the right with similar apples in moist chambers. The two series of graphs show the gradual development of color at the different temperatures, the figures near the lines indicating the number of weeks between the starting of the experiment and the time of taking notes. The apples were placed in storage on September 22, 1916. See figure 14 for the development of scald

the sugar of the pulp could be completely extracted with cold water in five minutes. The samples consisted of 20 gm. and, as already mentioned, the pulp was free from core, seeds, and peel. The proteins were precipitated with neutral lead acetate and the excess of lead precipitated with sodium oxalate. Inversion was accomplished by adding 5 c. c. of concentrated hydrochloric acid to 50 c. c. of lead-free solution and allowing it to stand overnight. The hydrochloric acid was neutralized with anhydrous sodium carbonate. The total sugars were determined by the Allihn[1] method, the cuprous oxid being determined by direct weighing. The results of the analyses are given in Table III.

In connection with the work on acids and sugars, dry-weight determinations of the pulp were also made, but only on the fruit that had been stored at 5°. In these tests the pulp was dried in an electric oven that stood at a constant temperature of 100°. The corresponding samples from the open and closed containers were always moved together, and care was taken to secure the lowest dry weight of the pulp. The

[1] WILEY, H. W., ed. OFFICIAL AND PROVISIONAL METHODS OF ANALYSIS, ASSOCIATION OF OFFICIAL AGRICULTURAL CHEMISTS. AS COMPILED BY THE COMMITTEE ON REVISION OF METHODS. U. S. Dept. Agr. Bur. Chem. Bul 107 (rev), 272 p , 13 fig 1908 Reprinted in 1912

results showed only fractional differences between the percentage of water in the fruit from the baskets and that in the fruit from the moist chambers; the contrasts being too slight in all cases to have any important significance in the interpretation of the data on acids and sugars. The average amount of water in the skin and core-free pulp of the fruit from the open containers was 87.30 per cent, while that from the fruit in the moist chambers was 87.67 per cent. Greater contrasts might have been expected, but it should be remembered that there was only a slight contrast between the relative humidity under the two conditions of storage.

TABLE III.—*Acid and sugar in Grimes apples, determinations made on January 24-29*

[The acidity is given in numbers of cubic centimeters of $N/20$ sodium hydroxid required to neutralize the filtrate from 100 gm of fresh pulp, the sugar as percentage of fresh pulp weight The apples of lot 1, 2, and 3 were the same as described in the legends of figures 12, 14, and 13, respectively]

Lot No	5° C.				0° C.			
	Acid.		Invert sugar.		Acid.		Invert sugar.	
	Basket.	Moist chamber.	Basket.	Moist chamber.	Basket.	Moist chamber.	Basket.	Moist chambe.
1.............	111.2	113.5	12.32	12.00	100.2	89.3	11.56	10.56
2.............	89.4	78.0	11.76	10.24	....	....	13.12	12.00
3.............	106.0	96.2	10.82	9.06	118.5	119.9	9.88	9.74

The results of the acid and sugar determinations as shown in Table III bring out some interesting contrasts. There was considerable variation in the different lots of apples, but in all cases the fruit from the open containers at 5° contained more sugar and more acid than that from the moist chambers at the same temperature. At 0° similar contrasts were obtained on the sugar content from all the lots and also on the acid content of samples from lot 1, but with lot 3 the acidity was slightly greater in the fruit from the moist chambers than in that from the baskets. Tests for acidity were also made on Grimes apples from 10° with results closely parallel to those obtained at 5°.

While the results indicate that the difference in quality of the fruit that accompanied the contrasts in aeration was at least partly due to differences in the sugar and acid content of the apples, they do not seem to explain the fact that the contrast in sweetness was much greater at 5° than at 0°. Since cane sugar is several times sweeter to the taste than reducing sugars, it is possible that much of the contrast in the taste of the apples from the two conditions of storage was due to differences in the relative proportions of these sugars. Unfortunately determinations were made only of total sugar. The conversion of starch into sugar does not need to be considered in this connection, as all of the starch had disappeared from the apples much earlier in the experiment.

While the analytical data will have to be regarded as of an incomplete and preliminary nature, the results as a whole are in agreement with the observations on the taste and quality of the fruit and when considered in connection with these would seem to indicate that the conditions of storage that favor the development of scald also favor the breaking down of the acids and sugars and possibly of other complex organic compounds of the apple.

Various other experiments were made to test the effects of aeration and humidity. In one case apples were stored in a sealed jar at $15°$ and the air renewed three times a week. The renewal was accomplished by connecting the jar with an exhaust that under the conditions of the experiment drew out a volume of air equal to that in the jar in approximately one minute. This air exchange was allowed to continue for about 10 minutes, the stale air being withdrawn from the bottom and the fresh air being freed at the top of the jar. The incoming air was from the closed storage box and therefore of the same temperature as the air removed from the jar. It was bubbled through a wash bottle containing water and must have gone into the jar in a practically saturated condition. The walls of the jar were always covered with a film of moisture. The method of renewing the air must have left but a fraction of a percentage of carbon dioxid in the jar, but gas analysis made after the jar had stood for two days showed that the carbon dioxid had then increased to 4.49 per cent, an amount slightly greater than that already reported for jars used as moist chambers. The apples of this experiment were of the same lot as those described in the legend of figure 12. Both scald and color developed much more slowly in the above experiment than on similar apples in moist chambers at the same temperature, a particular intensity of scald appearing about four weeks later in the former case, and a particular degree of color about six weeks later. (See figures 12 and 23.) After 15 weeks of storage as above described the apples were removed to the laboratory and allowed to stand in the open for several days. There was no increase in scald, and the fruit was still firm and in fairly good condition. Some of the apples seemed to have a very faint trace of the pungent alcoholic taste described later as characteristic of apples that had received poor aeration

A second experiment was carried out similar to the above, but the apples were of the lot described in the legend for figure 11, and the air was renewed six times a week instead of three. By reference to figure 11 it will be seen that similar apples in moist chambers at $15°$ had developed considerable scald at the end of four weeks and were badly scalded at the end of 5 weeks, but the apples receiving daily aeration had no scald at the end of 12 weeks and at that time had the same degree of color that those in the moist chambers had at the end of 6 weeks. (See figures 11 and 22.) There was no increase in scald when the apples were freely exposed to the air in the laboratory, and the fruit was still firm and entirely normal in taste several weeks after it had been found necessary to discard the corresponding fruit in moist chambers.

The results of the above experiments are extremely interesting, but the data are probably insufficient for a final interpretation. The delayed development of color on the fruit in the jars as compared with that in the moist chambers should probably be attributed to a difference in aeration rather than to a difference in humidity, but this hypothesis calls for an explanation of the fact that the suppression of color was not accompanied by an increase in scald. If we should assume that scald is due to the accumulation of products of incomplete oxidation in the apple, the above condition might be explained on the ground that the frequent aerations gave opportunity for the breaking down of these organic substances without furnishing the free and continuous supply of oxygen that was necessary for the development of the skin pigment.

Other experiments were made on the effects of aeration and humidity. In one case the air of the storage jar was renewed six times a week in a manner similar to that described above, but the oxygen of the ingoing air was increased to 2 per cent above the normal amount and the air stream was continued only long enough to carry in a volume of air four times as great as the air space in the jar. In another experiment the ingoing air had 4 per cent of oxygen added, but renewal was made only three times a week. In a third case the carbon dioxid of the air was increased to 2 per cent, and renewal was made six times a week. In another experiment small openings were left at the top of the jars, and with one jar a large quantity of calcium chlorid was placed in the bottom, with another a large amount of soda-lime, while with a third neither of these absorbents was added. All of these experiments were made at a constant temperature of 15°, and the apples were of the lots described in the legends of figures 10 and 15. Under all of the different conditions described scald developed at practically the same rate, but in no case quite as rapidly as on similar fruit in moist chambers. (See figures 10 and 15) As has already been mentioned, apples of these same lots did not scald when freely exposed in air having about 0.3 per cent carbon dioxid and a relative humidity of 85 to 95 per cent.

In an experiment at 15° with apples of the lot described in the legend to figure 11 the air was renewed three times a week, but the ingoing air had 5 per cent of carbon dioxid. The fruit remained green and did not show any sign of scald as long as it remained in the jar. Apples removed to the laboratory at the end of five weeks soon became slightly scalded, while those removed at the end of seven weeks were soon badly scalded. The fruit was found to have a pungent alcoholic flavor which it still retained after several weeks' exposure in the laboratory. It remained green, firm, and crisp under conditions that would soon have caused fresh apples to become yellow, withered, and overripe.

Apples were also stored in sealed jars at $0°$ and $15°$. As long as the jars remained unopened little or no scald developed and the fruit remained green. When the apples were removed to the laboratory after short periods of storage, their condition was found to be similar to that described above in the experiment with 5 per cent carbon dioxid, but upon removal after longer periods of storage they had a strong odor of alcohol and vinegar, were nauseating to the taste, and soon became brown and soft, taking on the appearance of frozen apples. Gas analyses made after several weeks' storage at $15°$ showed that the air of the jars had more than 50 per cent of carbon dioxid.

The results indicate that storage for a short time in an atmosphere in which the carbon dioxid has been greatly increased and the oxygen correspondingly decreased causes apples to become slightly alcoholic and to take on a rigor or an inactive condition from which they do not entirely recover and that longer storage under these conditions causes the intramolecular respiration to be carried to an extent that results in the death of all the pulp cells. The fact that alcohol can be readily detected in fruit that has been stored at $15°$ in air having more than 5 per cent of carbon dioxid and that scald develops on fruit stored in an atmosphere having slightly less carbon dioxid might be taken as further evidence that scald is due to the accumulation of products of incomplete oxidation in the surface layers of the apple.

Further evidence on the effects of aeration and humidity were obtained with rather green Arkansas and York Imperial apples placed in commercial cold storage. In two of the barrels the apples were packed in excelsior, while in the others they were packed in the usual manner. The apples were removed from cold storage on February 24, and after three days in the laboratory it was found that the former lot had about 10 per cent of scald, while the latter had 65 per cent.

Scald is sometimes of a somewhat local nature, and it can be produced on one part of an apple without the other parts showing any abnormal effects. In an experiment on aeration and humidity a small amount of water was placed in each of several tumblers and then an apple dropped into the top of each. The lines of contact between the apples and tumblers were sealed with a wax compound of beeswax, vaseline, and tallow. The lower halves of the apples were thus under moist-chamber conditions, while the upper halves were well aerated. The inclosed halves soon developed scald, while the exposed parts remained free from it (Pl. 33, C). It has been repeatedly observed that apples which were standing in films of water became scalded at the point of contact, while the rest of the skin remained normal. It has also been noticed that in the first stages of scald the brown discoloration sometimes appears in small spots instead of being evenly distributed over the skin. This was especially true where the conditions had been such as to condense the mois-

figure 12 were picked on August 10, those of figure
and those of figure 14 on September 21. By com
ponding curves in the different figures it will be seen
of the seventh or eighth week the ripe apples were
ing more rapidly than the green ones. This conditi
the fact that the apples of figure 12 were picked in mu
than the others, a circumstance that might be consi
the development of scald.

In taking notes on the various experiments alread
tions were also made on the comparative susceptibil
and ripest fruit of a particular picking. With the
was found that the ripest apples nearly always scal
the blush side of the fruit usually scalded before th
later pickings this condition did not hold and in so
confined almost entirely to the greener fruit surfaces

Similar results were obtained on rather green A
Imperial apples in commerical cold storage. The
on October 17, 1915, at Middletown, Va., and were pla
three days later. They were removed from storage o
was found that apples that were red on one side an
on the other were usually scalded only on the gree
apples that were a yellowish green or a streaked yell
a bright green on the other were scalded most on the

The results indicate that while the statement that
susceptible to scald than ripe fruit holds with usual cc
it does not hold for all degrees of maturity; nor does
fruit will scald more quickly than ripe. The data s

relationship between a particular degree of maturity and the development of scald.

In order to obtain further information on the relation of maturity of fruit to susceptibility to scald, careful notes were taken in the various experiments already reported on the development of the skin color in the apple. The degree of color was determined in a manner similar to that already given for the estimation of the amount of scald, the intensity of the color being compared with the maximum color that the fruit was likely to acquire. Records were always made by two observers and notes were taken without reference to earlier records.

The results given in figures 18, 19, and 20, were from the same lots of apples, respectively, as the scald notes of figures 11, 13, and 14. Temperature is indicated on the base line and percentage of color on the perpendicular. The conditions of the experiments have already been described; the apples in the open containers were exposed in an atmosphere that had less than 0.05 per cent of carbon dioxid and that had a relative humidity of 85 to 95 per cent, while those in the moist chambers were subjected to an atmosphere that was practically saturated and that sometimes had as much as 4 per cent of carbon dioxid.

It is interesting to note that the color changes were made more rapidly during certain periods of storage than in others and that there was often a correlation between the time of greatest color change and that of greatest scald development. If figures 11 and 18 are compared, it will be seen that at $15°$ and $20°$ there was a decided increase in color in the open containers in the fourth week and in the moist chambers in the fifth week and that at these temperatures scald made its most rapid development in the fifth week. At $10°$ the most rapid color changes came in the seventh week while the most rapid scald development came in the ninth week. A comparison of figures 13 and 19 brings out similar conditions. At $15°$ and $20°$ there was a big increase in color in the moist chambers during the sixth week and a big increase in scald during the seventh week. At $10°$ there was a big increase in color in the eighth week and a big increase in scald in the ninth week. In general the period of most rapid scald development has come about one week later at these higher temperatures than the time of most rapid color change. This correlation gives further evidence of the close relationship between the chemical changes in the apple and the development of scald.

A comparison of the results obtained in moist chambers with those in the open containers (as given in fig. 18, 19, and 20) shows that in the early stages of the experiment a particular degree of color was attained one or two weeks earlier in the latter case than in the former. The results indicate that the conditions that have already been reported as favoring scald have also tended to check the development of the skin color in the apple.

In the interpretation of the above contrasts in color development it becomes of interest to know whether the depressing effect of the moist chambers was due to the high humidity or the poor aeration. Some further experiments carried out at 15° throw light upon this subject.

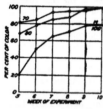

FIG. 21.—Graphs showing the influence of aeration and humidity upon the color development of Grimes apples in storage at 15° C. The perpendicular indicates the percentage of color, the base line the number of weeks of storage, and the curves give the comparative color development obtained with 70, 90, and 100 per cent relative humidity, respectively. With the graph marked "M100" the apples were held in moist chambers, while with the others they were kept in open containers. The apples were of the same lot as described in the legend of figure 13

As has already been stated, two boxes were maintained at this temperature, one in which the relative humidity ranged from 65 to 75 per cent but usually stood at 70 per cent, and another in which the relative humidity ranged from 85 to 95 per cent and usually stood slightly above 90 per cent. There was less fruit in the former box and the accumulation of carbon dioxid may therefore have been slightly less in this case than in the more humid box. In figures 21, 22, and 23 the color development in the open containers under these conditions is compared with that in moist chambers at the same temperature. In figures 22 and 23 other curves are given showing the color development on apples in jars where the air was kept saturated, but was renewed several times a week.

There was practically no contrast in color development between well-aerated apples exposed to a relative humidity of 70 per cent and similar apples exposed to a relative humidity of 90 per cent. It has already been pointed out that apples did not scald under either of these conditions. The color development in the moist chambers was much slower than in either of the above cases. The relative humidity was 10 per cent higher in this case than in either of the former, and, as pointed out

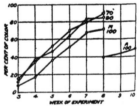

FIG. 22.—Graphs showing the influence of aeration and humidity upon the color development of Grimes apples in storage at 15° C. The perpendicular indicates the percentage of color, the base line the number of weeks of storage, and the curves give the comparative color development with 70, 90, and 100 per cent relative humidity, respectively. With the graph marked "M100" the apples were held in moist chambers, with the one marked "A100" they were in closed jars that had the air renewed six times a week, and in the other two cases they were kept in open containers. The apples were of the same lot as described in the legend of figure 11

earlier in the paper, the aeration was very poor. Since increasing the relative humidity from 70 to 90 per cent has caused no checking in color development, it does not seem probable that increasing it from 90 to 100 per cent would do so. In so far as this reasoning can be relied upon, evidence is furnished that the checking of color development in the moist chambers was due to poor aeration and not to high humidity. The contrast between the color development in the moist chambers and that in the sealed jars where the air was renewed but a few times a week was very striking. The air was apparently saturated in both cases, leaving no chance to attribute the contrast to a difference in humidity; but the carbon dioxid, although more intermittent, reached higher percentages in the latter case than in the former. The delayed color development in the latter case must therefore be attributed to poor aeration rather than to any difference in humidity.[1]

EFFECT OF DELAYED STORAGE ON APPLE-SCALD

Powell and Fulton (17) and Markell (14) have reported that a delay in storage is favorable to the development of apple-scald. Green (11) recently found that Grimes and Sheriff apples that were stored immediately after picking developed more scald than those that were held at a warmer temperature a few weeks before placing in storage. He attributed the difference to the fact that the fruit that was left out of storage for a time had opportunity to become more mature and therefore less susceptible to scald.

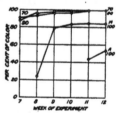

Fig. 23.—Graphs showing the influence of aeration and humidity upon the color development of Grimes apples in storage at 15° C. The perpendicular indicates the percentage of color, the base line the number of weeks of storage, and the curves give the comparative color development with 70, 90, and 100 per cent relative humidity, respectively. With the graph marked "M100" the apples were held in moist chambers, with the one marked "A100" they were in closed jars that had the air renewed six times a week and in the other two cases they were kept in open containers. The apples were of the same lot as described in the legend of figure 12.

The writers have obtained results similar to those of Green. Ten barrels of Grimes apples were picked at Middletown, Va., on September 7. Five barrels were shipped by express and were placed in cold storage on September 9, while the other five were shipped by freight and were stored on September 14. All of the apples were removed on December 19. After standing for three days at laboratory temperature, it was found that the scald on the apples that were delayed in reaching storage averaged 20 per cent, while that on the apples stored five days earlier was 54 per cent. The former lot of apples was found to have become decidedly riper than the latter, and it seems probable that the

[1] Since writing the above, the attention of the writers has been called to the recent work of Shamel (20) in which he reports that Bartlett pears were held for 30 days at an average temperature of 90° F. (32½° C) in an atmosphere having a relative humidity of approximately 90 per cent without any apparent change in the fruit and in which he expresses the belief that the ripening of the pears was prevented by the excessive humidity of the storage room. The room was partly filled with lemons undergoing the curing process and was ventilated for half an hour daily for a number of days about the end of the second week. The author does not discuss the significance of the ventilation; nor does he give any data as to the accumulation of carbon dioxid during the periods when the room was not ventilated.

The results of the various experiments point to a
conditions as the cause of scald. The disease has bee
by storing the fruit in moist chambers, in loosely stop
any condition that furnished a slight restriction to a
not developed on well-aerated apples (at temperatu
has it been found possible to produce typical sca
containers.

No evidence has been obtained that humidity has
development of scald except under conditions that h
osition of drops or films of water on the skin of the ap
it seems probable that the harmful effects were due p
tial exclusion of the air or to the retention of oxidati
water.

The rapidity of the development of scald has increa.
in temperature. In this fact it has followed the
activities. Morse (15) found that between 0° and 20°
in temperature caused a doubling of the amount of
off by Baldwin apples. Gore (10) reported that
including 40 different kinds of fruits it was found that r
from 1.89 to 3.01 times, an average of 2.376 times
rise in temperature. A study of figures 10 to 17 shov
rate at which the development of scald has increased
perature is approximately the same as the rate of incr
The question might be raised whether the more ra
scald at the higher temperatures might not be entirel

accumulation of carbon dioxid, but this does not seem probable. The tops of the moist chambers fitted loosely, and any increased accumulation in carbon dioxid would undoubtedly have been largely balanced by an increased air exchange. It is also probable that a higher percentage of carbon dioxid would be required to give a particular degree of inhibition at a high temperature than at a low one. Kidd (13) found that a rise of 10 degrees in temperature necessitated, roughly, the presence of three times as high a partial pressure of carbon dioxid to cause an inhibition of germination in the seeds of *Brassica alba*. It seems probable that the more rapid development of apple-scald at the higher temperatures should be attributed to the more rapid procedure of respiratory activities that have become abnormal rather than to any greater inhibitory action of the accumulated carbon dioxid.

Scald has not occurred at 30° either in moist chambers or in open containers. This may have been due to the fact that the carbon dioxid oxygen ratio did not become high enough to cause a sufficiently disturbing effect upon respiratory activities at such a high temperature. The skin color changes, however, were peculiar at this temperature, indicating the possibility of other causes for the absence of scald.

Scald occurred at 0° in the open containers, but not at 5° or at any of the higher temperatures. This contrast might be attributed to the greater inhibiting value of carbon dioxid at low temperatures, but the relatively small amount of this gas found in the 0° box has not furnished any support for this theory. It is the opinion of the writers that the development of scald at 0° under conditions that have not produced it at higher temperatures should be at least partly credited to the depressing effect of the low temperature itself. It is especially interesting in this connection to note that scald has not become evident while the apples were at 0°, even when it has been actually produced at that temperature. The skin sometimes took a slightly faded watery appearance, but it did not turn brown till the apples had been removed to a warmer temperature. The inhibition of the browning at 0° should probably be attributed to the suppressing effect of the low temperature upon the oxidizing enzyms of the apple skin, and it would certainly not be surprising if a temperature that has inhibited oxidation color changes should also have its peculiar effects upon the oxidation connected with respiration.

Scald apparently results from some effect that acts as a cumulative agent. Its rate of development is influenced by temperature, by the maturity of the fruit, and within certain limits by the degree of aeration, but it has not been found possible to produce it in a short period of time by the intensification of any favoring agency.

Apples stored under conditions favorable to the development of scald but removed to other conditions before a certain critical period is reached apparently do not retain their accumulated scald tendencies. A striking example of this is seen in the curves of the Grimes apples of figures 11 and 15. Both were of the same lot, but those of the latter figure had been in commercial cold storage five weeks before the particular experiment was started. These apples had a day's aeration in the laboratory and were then divided up and placed at the various temperatures of the experiment. Those in moist chambers at $0°$ scalded as quickly as those left in commercial cold storage or as those of figure 12 that were placed at $0°$ in the beginning, but those in moist chambers at $10°$, $15°$, and $20°$ scalded far more slowly than similar apples that had never been in storage. (See figures 11 and 15.) The cold-storage apples placed in open containers did not scald at any of the temperatures, while the apples from cold storage did not scald at $0°$ in open containers; those placed under this condition at first had 60 per cent of scald by the end of the experiment. (See curve 3, figure 17.) The checking effects of cold storage must not be overlooked, but the fact remains that apples stored at $0°$ for more than one-third of the time necessary to produce scald at that temperature have scalded much more slowly upon removal to other conditions than similar apples fresh from the tree.

Scald may be local in its effects; one side of the apple may be scalded and the other not; or scald spots may be scattered over the surface of the apple.

The production of apple-scald has been accompanied by a reduction in the sugar and acid content of the apple, by a retardation in skin color changes (in Grimes) and by an increase in mealiness, all of which might be considered as the expected result of respiration carried out under conditions of restricted aeration. When the aeration has been further decreased, scald has not occurred; but alcohol and various nauseating odors have been produced, suggesting a high degree of intramolecular respiration. The results seem to indicate that apple-scald is produced by the long-continued action of slightly abnormal respiratory conditions.

Much that has been said in regard to the nature of scald has been found true of Jonathan-spot. The two diseases are alike in the tissue affected; they have similar temperature responses and are similarly affected by aeration and humidity and by maturity of fruit. In all these cases the response has been much less striking with Jonathan-spot than with apple-scald, but the similarities are great enough to suggest some close relationship in the fundamental causes of the two diseases.

SUMMARY

(1) Jonathan-spot and apple-scald have shown many similarities. The initial stages of both diseases were found to be confined to the color-bearing cells of the skin. Both rendered the apples susceptible to rot infections. Both were decreased by good aeration and by a fair degree

of maturity in the fruit, and both increased with a rise in temperature, having an optimum at about $20°$ and a maximum at about $30°$ C.

(2) On Grimes and York Imperial apples scald developed in moist chambers at $0°$, $5°$, $10°$, $15°$, and $20°$ C., the rapidity of development increasing with the rise in temperature. In open containers, however, no scald developed at any of the above temperatures except $0°$. Grimes apples in closed jars but with the air renewed once a day have scalded much more slowly than similar apples in moist chambers

(3) Apples stored at $15°$ C. in an atmosphere having a high percentage of carbon dioxid (probably more than 5 per cent in all cases) have not developed typical scald, but have first passed into a sort of rigor, have later developed a pungent alcoholic taste, and have finally broken down throughout like a baked or frozen apple.

(4) The writers are of the opinion that apple-scald is largely due to abnormal respiratory conditions resulting from poor aeration. It has been pointed out that the occurrence of scald has been accompanied by a decrease in total acids and sugars, by a mealiness in the flesh of the apple, and by a checking in the color changes in the skin (Grimes). In looking for the final cause of scald, one should consider both the abnormal consumption of the organic food materials and the accumulation of the products of incomplete oxidation.

(5) The conditions within the apple tissue which finally result in scald are of a cumulative nature and apples removed from unfavorable storage conditions before a certain critical period was reached have shown little or no increased susceptibility to the disease.

(6) The rate of skin color development in Grimes apples has increased with a rise in temperature from $0°$ to $30°$ C. It has been checked by poor aeration, but apparently has been little affected by the degree of humidity.

(7) Apple-scald has been found more serious on green fruit than on ripe fruit, but it has developed more rapidly on the latter.

(8) Apples that have been delayed in storage have developed less scald than those stored immediately, but results of other experiments reported make it evident that the effect of delayed storage upon apple-scald will depend upon the initial maturity of the fruit and the degree of aeration given during the delay.

(9) Grimes apples held in commercial cold storage for five weeks, about one-third the time required to produce scald under cold-storage conditions, have shown much less susceptibility to the disease upon removal to other conditions than apples of the same picking tested five weeks earlier.

(10) The important rôle that aeration appears to play in the prevention of apple-scald may furnish an explanation for the small amount of this disease usually found in cellar and air-cooled storage.

LITERATURE CITED

(1) BLANC, L.
  1916. RECHERCHES EXPÉRIMENTALES SUR L'INFLUENC
        TEMPÉRATURE SUR LA RESPIRATION DES PLANTES
        t. 28, no. 237, p. 65–79.

(2) BROOKS, Charles, and COOLEY, J. S.
  1916. APPLE SCALD. (Abstract.) *In* Phytopathology, v

(3) ——— ———
  1917. TEMPERATURE RELATIONS OF APPLE-ROT FUNGI. *I*
        v. 8, no. 4, p. 139–164, 25 fig., 3 pl.

(4) BRYAN, A. H., GIVEN, A., and STRAUGHN, M. N.
  1911. EXTRACTION OF GRAINS AND CATTLE FOODS FOR TI
        SUGARS: A COMPARISON OF THE ALCOHOL AND TH
        DIGESTIONS. U. S. Dept. Agr. Bur. Chem. Circ

(5) CLARK, W. B.
  1917. A SAMPLING PRESS. *In* Jour. Indus. and Engin.
        788–790, 2 fig.

(6) COOK, M. T., and MARTIN, G. W.
  1913. THE JONATHAN SPOT ROT. *In* Phytopathology, v.

(7) ——— ———
  1914. THE JONATHAN SPOT ROT. *In* Phytopathology, v.

(8) CULPEPPER, C. W., FOSTER, A. C., and CALDWELL, J. S.
  1916. SOME EFFECTS OF THE BLACKROT FUNGUS, SPHAER
        THE CHEMICAL COMPOSITION OF THE APPLE. *In*
        v. 7, no. 1, p. 17–40. Literature cited, p. 39–40.

(9) FULTON, S. H.
  1907. THE COLD STORAGE OF SMALL FRUITS. U. S. Dept.
        Bul. 108, 28 p., 3 pl. (part col.)

(10) GORE, H. C.
  1911. STUDIES ON FRUIT RESPIRATION. U. S. Dept.
        142, 40 p., 17 fig.

(11) GREENE, Laurenz.
  1913. COLD STORAGE FOR IOWA GROWN APPLES. Iowa

Agaricus tabularis:

A.—Part of a large ring, showing 42 mushrooms The area just inside the fruiting zone is marked by dead or dying short grass. Inside this dead zone is a stimulated area chiefly characterized by *Festuca octoflora*. Akron, Colo., June 15, 1915.

B.—A large ring, showing 75 fresh mushrooms There are also 47 dry mushrooms. Outside the ring the vegetation is typical short grass. A narrow stimulated area, marked largely by *Festuca octoflora* extends several inches in advance of the fruiting zone. Just inside the fruiting zone the short grass is dead. Inside this dead zone the vegetation consists largely of annuals and *Malvastrum coccineum*, with short grass slowly becoming reestablished at a distance of about 4 meters inside the fruiting zone. Akron, Colo , June 14, 1915.

7709°—17——5

PLATE 20

Agaricus tabularis:

ing caused by *A. tabularis* and distinguished by a uniform growth
ii. This plant is almost as abundant outside of the ring as inside,
 ery small and therefore not as noticeable. This is a relatively smal
.imulation extends to the center. Akron, Colo., June 14, 1915.

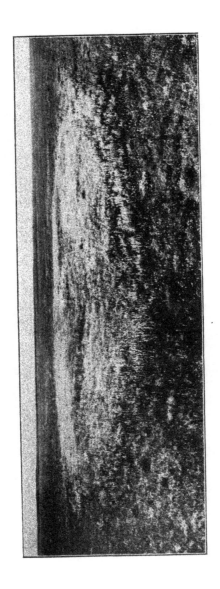

Agaricus tabularis:

A.—Effect of a ring on Kubanka wheat The wheat at the left shows normal production on the plot; at right, the straw is weak, the heads are poorly filled and prematurely ripened. The yield of dry matter at the left was 500 gm. per square meter, at the right, 400 gm. Even a greater difference was shown in the yield of grain, that in the ring being 63 gm. and that outside being 137 gm. per square meter. In this case the yield in dry matter inside was only 80 per cent of that outside and the yield of grain only 46 per cent. Akron, Colo , August 10, 1915.

B —The stubble showing the distribution of the ring of *A tabularis* The wheat ripened off early in the ring, and the straw turned dark The farther end of the trench shown in figure 2 ended at about the middle of the foreground of this picture. Akron, Colo., August 16, 1915.

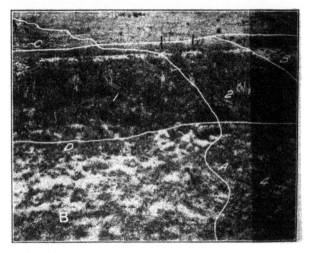

PLATE 23

A.—Method of measuring the penetration of water in three zones of a ring of *Agaricus tabularis*. The first three 2-liter flasks in the foreground are in the bare zone, the second three are in the withered zone, and the three in the background are outside. Those in the bare and withered zones are still almost full of water while those outside in the natural sod are empty. Akron, Colo , July 13, 1916.

B.—A ring of *Agaricus arvensis* showing the effect of irrigation during a dry year on the different zones. At this period the vegetation both in the zones of the ring and in the native sod was entirely dormant, due to drouth. On this account the withered zone could not be distinguished from the other vegetation. The effect of the surface flooding was marked between the lines C and D The outer edge of the bare zone is marked by the line A and the outside of the irrigated area by the line B. The area 3 represents the bare zone during a dry year and was in the same condition as that of area 1 at the beginning of the experiment. The effect of irrigation on the natural sod is shown by comparing area 2 with area 4. It is clear that, although the vegetation in area 3 was dying out owing to drouth, if water was continually supplied in quantities sufficient to keep the surface soil continually moist the grasses and other plants grew luxuriantly, indicating that the chief cause of the death of the vegetation was the deficiency in soil moisture. In the background of areas 1 and 2 is shown the trench illustrated in Plate 20, A. Akron, Colo., August 14, 1916.

PLATE 24

A.—A trench across the irrigated plot shown in Plate 19, B. Although water had been added continuously as rapidly as it was taken up by the soil, there was no penetration into the mycelium area, which still remained very dry. The growth of the native plants in this area was due entirely to the water available in the upper 3 or 4 inches of soil. Akron, Colo., June 28, 1916.

B.—A ring of *Calvatia cyathiformis* photographed August 21, 1907, 16 miles north of Cheyenne Wells, Colo. The ring is not complete; 16 of the 27 fresh puffbails should show in this illustration. Fifty-six sterile bases, the remnant of an earlier crop, were found in the same zone occupied by the fresh fruiting bodies. Although much of this section of country suffered severe drouth in 1907, recent rains had caused hundreds of these Calvatia rings to fruit. This ring was fully 200 meters across.

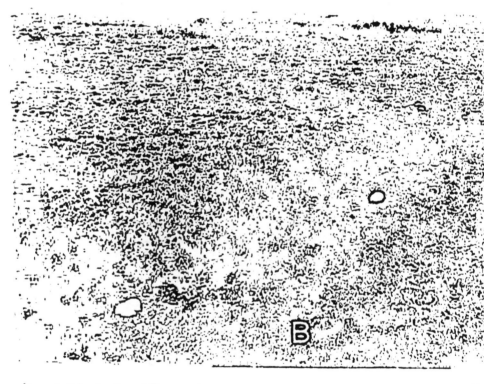

PLATE 25

Calvatia cyathiformis:

A.—A ring showing stimulation of native plants inside the fruiting zone. The plants showing increased growth are the annuals *Plantago purshu, Festuca octoflora, Lepidium ramosissimum,* etc., and the perennial short grasses and *Artemisia frigida.* The fruits are well in advance of this stimulated zone. The advance here is from left to right. The second crop of puffballs is seen. Akron, Colo., August 16, 1915.

B.—A trench through fruiting zone of *C. cyathiformis.* The soil is moist throughout, owing to recent rains. The mycelium can be distinguished with difficulty and does not render the soil impervious to water as does the mycelium of *Agaricus tabularis.* Fruiting bodies are seen about 30 cm ahead of the stimulated area in which *Plantago purshii* shows greatest response. Akron, Colo., August 16, 1915.

PLATE 26

A.—A ring of *Calvatia cyathiformis* producing a second crop of puffballs. The sterile bases of the June crop are still present and occur in the outer edge of the stimulated area. Eleven fruits and ten old stumps occur in the same portion of the ring. This ring is shown in figure 2 as ring 5 The stimulated effect is shown chiefly in the *Plantago purshii*, and is evident for a distance of 1 5 meters in the ring. The first crop appeared at the very edge of this area. The second crop shown in this illustration is about 30 cm. in advance of this area. Akron, Colo , August 16, 1915

B —A large ring formed by *Lepiota morganii*, showing a second crop of fruiting bodies. The sketch of this ring (fig 7) shows 10 fresh and about 50 dry fruiting bodies from the earlier crop. The vegetation is largely *Andropogon scoparius* and *Bouteloua hirsuta*, but no effect from the presence of the fungus is noticed on the native growth The ring is about 24 meters across and forms an almost perfect circle. The soil is sandy. The young fruiting bodies are about 15 cm. in advance of the old. Twelve miles southeast of Yuma, Colo., August 17, 1915.

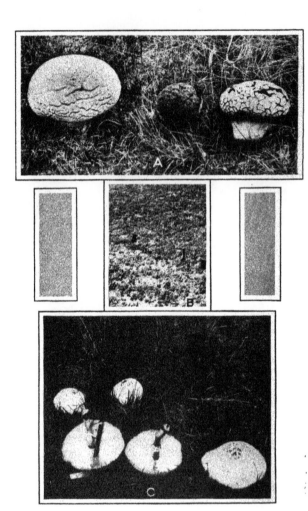

PLATE 27

A.—Three stages in the ripening of sporophores of *Calvatia cyathiformis*. The fruiting body at the left has just formed, and the flesh is still white and in good condition to eat. The peridium is already strongly marked. The fruiting body at the right is at least 2 days older, and the peridium is breaking off in rather thin plates. In the center is a still older fruiting body from the upper part of which the peridium has blown away, exposing the spore mass. The diameter of largest fruiting body is 15 cm.

B.—Effect of a ring of *C. cyathiformis* on the density of the short-grass sod. The area behind the short stakes is in the stimulated zone, while that in the foreground is in the natural sod. The effect here has been to thicken the sod cover. Akron, Colo., June 10, 1916.

C.—Three fruiting bodies of *Lepiota morganii* in place in the fairy ring shown in Plate 22 B, and sketched in figure 7, and two fruiting bodies inverted. These fruiting bodies are from 12 to 20 cm. in diameter.

PLATE 28

A.—A small ring formed by *Agaricus campestris*, showing a total of 29 mushrooms, the largest being about 10 cm. in diameter. The footpath in the upper right portion of the illustration has no connection with the ring. Akron, Colo., June 17, 1916.

B.—A large ring of *Calvatia polygonia*, showing 14 fruiting bodies, the largest of which are 33 cm. in diameter. The diameter of this incomplete ring was about 100 meters. The puffballs occurred at the edge of a very green area of short grass. No difference in composition of the native was noted, but the deep-green color and the somewhat more luxuriant growth of the short grasses just inside the fruiting zone was especially noticeable. Six miles southwest of Yuma, Colo., July 26, 1907.

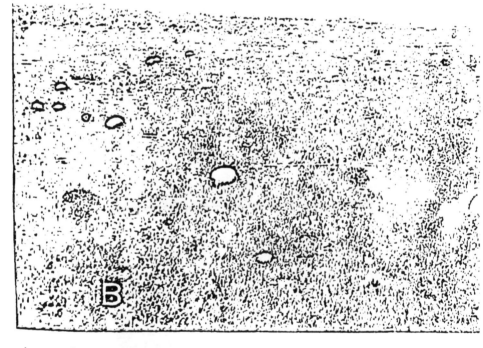

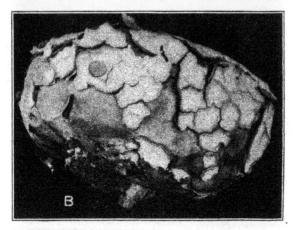

PLATE 30

A.—The effect of a ring of *Agaricus tabularis* on the native
ring the short grass is dying or dead, but *Gutierrezia sarothrae* is a li
greener in color than either outside or inside the ring.   Akron,

B.—A ring of *Catastoma subterraneum*.   The only outward evid
of this fungus is the stimulated growth of *Plantago purshii* and
green of the short grasses.   The mycelium is not easily distin
although the fruiting bodies are just breaking through the soil.
16, 1915.

C.—A very small ring apparently caused by a fungus but mar
luxuriant growth of *Festuca octoflora*.   Akron, Colo., June 15, 19

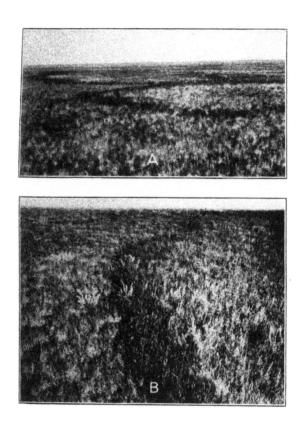

(17) POWELL, G. H., and FULTON, S. H.
    1903. THE APPLE IN COLD STORAGE.   U. S. Dept. Agr. Bur. Plant Indus. Bul.
        48, 66 p., 6 pl. (part col.)
(18) SCHLEY, Eva O.
    1913. CHEMICAL AND PHYSICAL CHANGES IN GEOTROPIC STIMULATION AND
        RESPONSE.   *In* Bot. Gaz., v. 56, no. 5, p. 480-489, 6 fig.   Literature
        cited, p. 489.
(19) SCOTT, W. M., and ROBERTS, J. W.
    1913. THE JONATHAN FRUIT-SPOT.   *In* U. S. Dept. Agr. Bur. Plant Indus.
        Circ. 112, p. 11-16, 2 fig.
(20) SHAMEL, A. D.
    1917. SOME OBSERVATIONS UPON THE RELATION OF HUMIDITY TO THE RIPENING
        AND STORAGE OF FRUITS.   *In* Mo. Bul. [Com. Hort. Cal.], v. 6, no. 2,
        p. 39-41.
(21) THATCHER, R. W.
    1915. ENZYMS OF APPLES AND THEIR RELATION TO THE RIPENING PROCESS.
        In Jour. Agr. Research, v. 5, no. 3, p. 103-116.   Literature cited,
        p. 116.
(22) THODAY, D.
    1913. ON THE CAPILLARY ENDIOMETRIC APPARATUS OF BONNIER AND MANGIN
        FOR THE ANALYSIS OF AIR IN INVESTIGATING THE GASEOUS EXCHANGES
        OF PLANTS.   *In* Ann. Bot., v. 27, no. 107, p. 565-573, 2 fig.

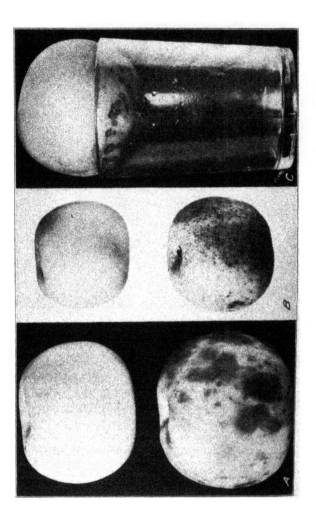

Apple-scald:

A.—Grimes apples, showing the effect of storing at 10° C. from September 9 to November 20. The upper apple was stored in an open container and remained free from scald. The lower apple was stored in a moist chamber and was badly scalded at the end of the period.

B.—Grimes apples, showing the effect of storing them at 10° C. from August 11 to October 17. The upper apple was stored in an open container and remained free from scald. The lower apple was stored in a moist chamber and became scalded.

C.—An apple with part of its surface inclosed in a tumbler. The inclosed portion shows an early stage of scald, while the part that is outside is free from scald.

STUDIES IN GREENHOUSE FUMIGATION WITH HYDRO-CYANIC ACID: PHYSIOLOGICAL EFFECTS ON THE PLANT [1]

By WILLIAM MOORE, *Head of Section of Research in Economic Zoology,* and J. J. WILLA-MAN, *Assistant Chemist, Minnesota Agricultural Experiment Station*

INTRODUCTION

In a previous paper (15)[2] the relationship between temperature and moisture to the degree of injury to plants fumigated with hydrocyanic acid was considered. The next step in the investigation has been to determine the action of the gas on the tissues of the plant. Little information on this point is to be found in the literature. In fact, some investigators have considered the injury to be not due to hydrocyanic acid, but to hydrochloric-acid fumes generated from sodium-chlorid impurities in the commercial cyanid (24). The reactions of hydrocyanic acid on isolated compounds have received considerable attention, but not from the standpoint of injury to plants during fumigation. The object of this paper is to show that hydrocyanic acid penetrates the tissues and is responsible for certain disturbances in the metabolism of the plant which may result in injury or even death.

EVIDENCE OF THE PENETRATION OF THE GAS

In order to demonstrate the actual presence of the hydrocyanic acid in the plant, chemical analyses were made at various intervals after fumigation.

From 60 to 130 gm. of tomato plants, usually 8 to 20 plants, were ground in a food chopper, suspended in a Kjeldahl flask in 300 c. c. of 5 per cent tartaric acid, and slowly distilled into a little 5 per cent sodium hydroxid until 100 c. c. of distillate were obtained. The cyanid in this distillate was then estimated by the method of Viehoever and Johns (22). A portion of the sample of tissue was used for the determination of dry matter.

The results are given in Table I. It is evident that there is an appreciable amount of cyanid absorbed, and that this amount is proportional to the injury produced. These figures are only comparative, since it is impossible to recover quantitatively the hydrocyanic acid contained in a plant tissue (1, 23). The acid disappears within a few hours after fumigation; and it is during this period, as will be shown later, that certain physiological disturbances are evident.

---

[1] Published, with the approval of the Director, as Paper No. 69 of the Journal Series of the Minnesota Experiment Station.
[2] Reference is made by number to "Literature cited," p. 336-338.

Journal of Agricultural Research,
Washington, D. C.

Vol XI, No. 7
Nov. 12, 1917
Key No. Minn —21

(319)

TABLE I.—*Amounts of hydrocyanic acid absorbed by the plants during fumigation*

[Results expressed as percentage of hydrocyanic acid in dry weight of plants]

Set No.	Time after fumigation.	Percentage of hydrocyanic acid to dry matter.	Injury to plants.	Dose.
	Hours.			
Set I.......	0. 0	0. 0037	Slight burning	1 gm. of potassium cyanid per cubic meter for 1 hour
Set II	. 0	. 0047	Severly injured	Do
	. 6	. 0041	do	Do.
	1. 2	. 0037	.....do..........	Do.
	2. 0	. 0027	.....do..........	Do.
Set III . .	. 0	. 0100	.....do..........	1 gm. of potassium cyanid per cubic meter for 1.5 hours.
	1. 5	. 0077	.....do..........	
	5. 6	. 0032	.....do..........	Do.
Set IV..	. 0	. 0006	Not injured	1 gm. of potassium cyanid per cubic meter for 1 hour.
	1. 5	. 0000	.....do..........	Do.
Set V...	. 0	. 0090	Severely injured	1.25 gm. of potassium cyanid per cubic meter for 1 hour.
	2. 0	. 0060	.....do..........	Do.

In the work reported in the previous paper (15) there was evidence that the absorbed hydrocyanic acid was given off from the leaves into the atmosphere after removal from the fumigation chamber, since plants were least injured when put under conditions favorable to rapid evaporation of the hydrogen cyanid. That the acid, however, may be partly or completely destroyed in the tissues is suggested by the work of Schmidt (18) and of Dezani (7), in which there was reported the destruction of a considerable portion of the absorbed or injected cyanid, by the sap of fruit, stem, and leaves. To determine, if possible, the fate in tomato plants of hydrocyanic acid absorbed during fumigation, resort was had to Mirande's (13) sodium-picrate test paper for the detection of the acid given off by leaves. After fumigation one plant was placed in a bell jar just large enough to contain it, with a strip of the moist test paper suspended among the leaves from the top of the jar. The leaves from another plant were packed loosely in a jar with strips of test paper. After two hours a slight orange tinge was observed on the latter paper, but none on the paper under the bell jar. The color represented an amount of evolved cyanid gas equal to only an extremely small fraction of the amount that must have been contained in the leaves. It seems, therefore, that most of the hydrocyanic acid is destroyed by the tissues, being either united with sugars or converted into ammonia. It is rather probable that the hydrocyanic acid which actually gains entrance to the cells is never evolved from them again and that the acid which is still within the intercellular spaces and in the cuticle at the close of the fumigation may be evaporated into the atmosphere if placed under the proper conditions.

EFFECT OF THE ENZYMIC ACTIVITIES OF THE PLANTS

Geppert's (10) work has proved conclusively that in the higher animals absorbed hydrocyanic acid limits oxygen transfer in the tissues. Loevenhart and Kastle (12) have shown that the gas is able to paralyze the catalytic activity of solutions of metallic colloids. Kastle and Loevenhart (11) showed that hydrogen cyanid inhibits the action of potato oxidases, while recently Shafer (19) has shown that it affects the activities of oxidases, catalase, and reductase in insect tissues. One would expect, therefore, to find that in plants which had absorbed hydrocyanic acid during fumigation the respiratory enzyms at least were affected. Enzyms other than those connected with respiration were tested for the effect of hydrocyanic acid as follows: Commercial pepsin acting on coagulated white of egg; proteases extracted from tomato leaves acting on coagulated fibrin; taka diastase and diastase from tomato leaves acting on soluble starch; zymase acting on dextrose; rennin. The result in each case was that there was no apparent effect on these enzyms, either in speed of action or in total change produced.

Among the respiratory enzyms studied were oxidases, catalase, and reductase. In this paper these terms should be construed as follows: Oxidases are the substances which will cause transfer of atmospheric oxygen to aromatic chromogens, such as hydroquinone, pyrogallol, and the cresols. Catalase liberates molecular oxygen from hydrogen peroxid. The reductase activity sought here was the decolorizing of methylene blue in the absence of oxygen. Two kinds of preparations for enzym work were used: First, the juice of the leaves, obtained by grinding them in a food chopper and squeezing through silk bolting cloth; second, the dried leaf powder, obtained by drying the leaves in a vacuum desiccator over quicklime, then powdering in a mortar and sifting through No. 12 bolting cloth.

OXIDASES.—For the quantitative determination of oxidase activity Bunzel's (6) simpler apparatus was employed. A temperature of $37°$ C. was used in all determinations; 0.1 gm. of leaf powder was allowed to act on 0.01 gm. of oxidizable chemical until the reading was constant, usually 1.5 to 2.0 hours. As a preliminary a number of oxidizable chemicals were tried. Since pyrocatechol and hydroquinone gave the best results, they were used in this work.

In order to demonstrate that the presence of hydrocyanic acid inhibits oxidase activity, powder from normal tomato leaves was placed in one arm of the Bunzel tubes and varying concentrations of hydrogen cyanid, together with hydroquinone, were placed in the other arm. In this way the cyanid and the leaf powder did not come in contact except during the reaction period. The results are shown in figure 1. It is evident that there is a rapid decline in activity in the presence of even very small amounts of the poison.

Next, in order to show whether the hydrocyanic acid destroys or combines with the enzym and exerts its effect even after the poison has been removed and when it is not present during the oxidase reaction, the following experiments were performed: 0.4 gm. of leaf powder on a watch glass was made into a paste with 1.0 c. c. of the hydrocyanic-acid solution to be tested, covered with another watch glass and allowed to stand for varying lengths of time; then placed in a vacuum desiccator for 24 hours.

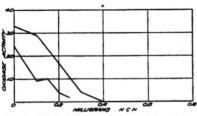

After powdering it in a mortar it was sieved through No. 12 silk bolting cloth and used for oxidase determination, as in the original powder. The results are shown in figure 2. The curves indicate that there is no permanent injury to the oxidase activity of these powders following treatment with hydrocyanic acid, and that if there is any change at all it is a slightly increased activity.

FIG. 1 —Curves showing in two samples of tomato leaves the effect on oxidase activity of the presence of hydrocyanic acid during the determination.

To show the effect of fumigation with hydrocyanic acid on the oxidases of tomato plants (*Lycopersicon esculentum*), a group of plants about 30 cm. high was divided into two sets. One set was fumigated with 1.0 gm. of potassium cyanid per cubic meter for one hour in the middle of the day; the other group was held as normal. At stated intervals before and after fumigation samples consisting of the upper three leaves of the plants (since the younger leaves always show effects of fumigations more

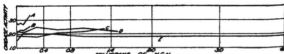

FIG. 2.—Curves showing the oxidase activity of leaf powders treated with varying concentrations of hydrocyanic-acid solution, the latter having been removed before the determination was made: *A*, treated 15 minutes before desiccation; *B, C, D, E,* treated 2 hours before desiccation.

and contain the more active oxidases) were taken and prepared in the powder form for analysis. The data are shown in the upper curves of figure 3. There is a sudden and striking drop in oxidase activity exhibited by the fumigated group at the close of the fumigation. This is followed by a rapid recovery to normal within 13 hours. The experiment was repeated with a second group of plants with similar results, as shown in the lower curves of figure 3. In this case the recovery was even more rapid, being complete in two hours.

The permanency of this recovery is shown by the following experiments. A uniform group of tomato plants about 17 cm. high was divided into four sets of 30 plants each. No. 6 was normal, unfumigated. No. 7 was fumigated from 7.20 to 8.50 p. m. with 1 gm. of potassium cyanid per cubic meter of air at a temperature of $13°$ C. Owing to the low temperature half the plants were killed, and experimental work could not be started on the other half until four weeks later. No. 8 was fumigated with 0.25 gm. of potassium cyanid per cubic meter overnight 3 times at

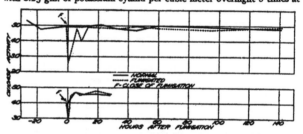

FIG. 3.—Curves showing the hourly activity of the oxidases of fumigated and of normal tomato plants.

2-week intervals. There was no visible injury. No. 9 was fumigated with 0.25 gm. of potassium cyanid per cubic meter overnight 10 times at 3-day intervals. Samples for oxidase analysis were taken at intervals and prepared in the powder form. Figure 4 gives the results. The curves indicate that there is in general no permanent decrease in oxidase activity except in the group very frequently fumigated.

Many other determinations on the fresh juice of tomatoes, working with both hydroquinone and pyrocatechol, confirmed the above results.

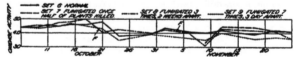

FIG. 4.—Curves showing the oxidase activity of fumigated and of unfumigated tomato plants over a period of eight weeks. *F*, close of fumigation.

CATALASE.—For the quantitative determination of catalase activity 0.025 gm. of the powder, or 0.13 c. c. of juice, was allowed to act on 2 0 c. c. of 3 per cent hydrogen peroxid in Bunzel tubes graduated to read positive pressures (2). A temperature of $37°$ C. was always used, and the reaction allowed to continue for 15 minutes. According to Appleman, catalases are rapidly destroyed by contact with acid plant juices, and calcium carbonate has to be added as a preventive. This was not found to be the case with tomato juices, however, and calcium carbonate was not used in these determinations.

The data in figure 5 show that with catalase, as well as with the oxidases, there is a temporary decrease activity at the close of fumigation, followed by a rapid recovery. The catalase differs from the oxidase activity, however, in that it not only recovers to the normal but that it actually exceeds the normal in considerable degree. The increase above normal at the end of the series of samples is much more marked than the decrease below normal at the beginning, which is the reverse of that found in the oxidase curves.

REDUCTASE.—Reductase activity could not be demonstrated in tomato leaves by the methylene-blue method of Shafer (19).

EFFECT ON RESPIRATION IN TOMATOES

Certain effects of hydrocyanic acid upon some of the individual enzyms connected with respiration in plants having been demonstrated, it was next deemed advisable to study the effects of the gas on the respiratory process itself.

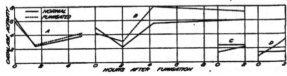

FIG. 5 —Curves showing the effect of hydrocyanic-acid fumigation on the activity of the catalase in tomato. *A* and *B*, juice used, *C* and *D*, leaf powder used.

The amount of carbon dioxid given off by a plant in the dark during a specified period was used as the index of comparative respiration between fumigated and unfumigated plants. The respiration chamber was a glass jar 30 cm. high and 13 cm. in diameter, fitted with a ground-glass stopper, through which were drilled two holes, one for intake and one for outlet of air. A black bag impervious to light inclosed the jar. The carbon dioxid was absorbed in a Truog tower of beads containing standard barium hydroxid (21). Another tower of beads containing saturated barium hydroxid was used to free the ingoing air of carbon dioxid. In order to avoid laboratory air, a long tube connected the apparatus with the outdoor air through a window. By means of a water pump air was aspirated vigorously through the apparatus for 40 minutes. It was found by testing against pure calcite that this length of time for this sized chamber always gave 95 to 98 per cent recovery of the contained carbon dioxid, which was sufficiently accurate for the comparative purposes at hand. Two jars and two plants, one normal and the other fumigated, were always run side by side, and all conditions were as nearly identical as possible. The temperature varied from $20°$ to $25°$ C. Five or six potted tomato plants from 15 to 20 cm. high were put in the dark for an hour. Three of them were then fumigated in the dark, the others meanwhile being kept in the dark. At the close of the fumigation one normal and one fumigated plant were selected for respiration measurements, the others being returned to the greenhouse to observe the extent of injury. The pots and soil surface of the two experimental plants were thoroughly paraffined to avoid the evolution of carbon dioxid. The plants were put into their respective jars, the latter sealed tight with paraffin, and the air freed from carbon dioxid. For the first day or two the respired carbon dioxid was measured every three to six hours, as is shown in figure 6. Since the atmosphere in the jars quickly became saturated with moisture, transpiration nearly ceased; hence the plants were not under normal conditions in this respect, and as a result there was a general decline in respiratory activity during the first 30 hours. As both

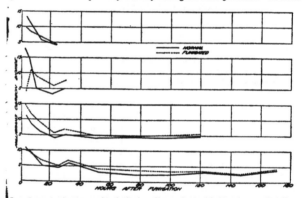

Fig 6.—Curves showing the rate of respiration (in milligrams of carbon dioxid per hour per 1 gm of dry matter in plant) of four pairs of tomato plants From top to bottom the groups are arranged in order of increasing injury from hydrocyanic-acid fumigation.

plants, however, were under the same conditions throughout, it was not deemed necessary to install a drying medium in the jars. At the end of the experiment the plants were removed and the dry weight of each determined, to form a basis for calculation. Although in all four of the experiments carried out the fumigation dose was the same, there was considerable difference in the gross effect on the plants, owing to other varying factors (15). The first showed no apparent injury; in the second the leaves curled, but did not wilt; in the last two the smaller leaves and stems wilted, and some did not recover. This selective injury produced different types of respiration curves.

As is shown in the two upper curves, a slight injury results in a material decrease in the amount of respired carbon dioxid at the end of the fumigation. This is followed by a rapid recovery not only to normal, but in excess of the normal. With greater injury (the two lower curves), however, the respiration is above normal even at the close of the fumigation. It remained above normal throughout the experiments, periods of five to seven days. A normal fumigation with hydrocyanic acid therefore results in a temporary decrease, followed by an increase in respiration. It is interesting to note at this point the correlation in the form of the curves of the catalase and respiratory activities, results similar to those found by Appleman in potato tubers. The curves of the oxidase activities, however, do not conform to either of these. Our results seem to confirm Appleman's, that catalase is more significant in respiration than are oxidases.

EFFECT ON PHOTOSYNTHETIC ACTIVITY

Inasmuch as the photosynthetic activity of a plant may be checked temporarily or permanently by heat, asphyxiation, or anesthetics (8, 9), and as hydrocyanic acid influences the respiration of the plant, it was naturally concluded that the cyanid might affect photosynthesis. With a view to ascertaining this point, the appearance and disappearance of starch in tobacco (*Nicotiana tabacum*) and geranium (*Pelargonium* spp.) was followed. The test for starch was made by steeping leaves in boiling water for a moment, then immersing in warm alcohol to extract the chlorophyll, after which they were placed in weak iodin solution. The plants to be tested were placed in the dark for 24 hours, or until a starch test on the leaves was negative. When free from starch, the plants were divided into two sets, one of which was fumigated in a tight box in the dark; the normal set was also kept in the dark during the fumigation. The dose was 1 gm. of potassium cyanid per cubic meter, for one hour in the case of the tobacco, and for one and a quarter hours in the case of the geranium. At the close of the fumigation both sets were brought into the light. A strong starch test was obtained in the normal plants within an hour, but the fumigated tobacco required 5.5 hours, and the geranium 5 hours before even a slight starch test was obtained. By evening the starch content was still much less in the fumigated than in the normal. Frequent tests showed that it required about three days for the fumigated plants to recover normal activity. Similar tests were made with potatoes, cucumbers, and tomatoes, with like results. It is noteworthy that the first evidence of recovered photosynthetic activity corresponds in general to the periods of recovery of the catalase and the respiratory activities.

EFFECT ON TRANSLOCATION OF FOOD MATERIALS

It was observed in the above experiments that the fumigated plants were not able to use up during the night even the small amount of starch that had been formed up to the previous evening. Evidently the translocation, as well as the manufacture of starch, is interferred with by hydro-cyanic-acid fumigation.

To confirm this point, tomato plants were placed in bright sunlight until in the afternoon they showed a high starch content. A number of these plants were fumigated in the sunlight with 1 gm. of potassium cyanid per cubic meter for 1.25 hours, the control also remaining in the sunlight. At the close of the fumigation both sets were placed in the dark. Eighteen hours later the unfumigated plants showed that about three-fourths of the starch had been removed. The fumigated, however, showed apparently no diminution in starch content. Six hours later the normal gave but a feeble starch test, while the fumigated set still gave a strong test. It required from 3 to 3.5 days for complete disappearance of starch in the fumigated plants. Similar trials were carried out with other sets of tomatoes and with geraniums, with confirmatory results

There are two possible explanations of the inability of the plant to remove its starch after submission to the influence of hydrocyanic acid. One is the inhibition of the diastases, the other is the inhibition of the oxidation of sugars in the growing parts of the plant, which would tend to maintain the concentration of sugar in the leaves and stems, and thus prevent the hydrolysis of starch.

In regard to the first possibility, it has already been mentioned under nonrespiratory enzyms that diastases are not affected by hydrocyanic acid. To test the point more accurately, leaf powders were prepared from fumigated and unfumigated leaves, and their diastatic activity tested quantitatively. One gm. of each of the powders liberated enough sugar from soluble starch to reduce 1.04 gm. of copper in Fehling's solution. Hence, there is no detectable effect of hydrocyanic acid on this enzym.

The inhibition of sugar oxidation was determined in the following way. A large geranium, plant containing two branches of about equal size was selected. About 5 inches of the growing shoot of one branch was inserted into a fumigation box through an opening, which was then sealed tight. This shoot was fumigated with 1 gm. of potassium cyanid per cubic meter for one hour. The plant was then placed in complete darkness, and from time to time leaves were removed for a starch test. The leaves in the normal branch lost all their starch within 24 hours. The basal leaves of the branch the tip of which had been fumigated required 48 hours for complete removal of starch. This shows that the translocation phenomenon brought about by hydrocyanic-acid fumigation centers in the growing parts, and thus is primarily a failure of the oxidation of food materials in those regions. A few days later it was noticed that the axillary buds of the leaves below the fumigated portion were awakened into growth just as if the growing shoot had been pruned. After attaining a length of about ½ inch, their further growth was prevented by the recovery of the terminal shoot.

EFFECT ON THE COMPOSITION OF THE PLANT

In view of the fact that the hydrocyanic acid absorbed by the plant is decomposed in the tissues, that the metabolism of the plant is changed because of the disturbed enzym equilibrium, and that several plants have been shown to utilize the nitrogen of the cyanid group (7, 14, 18), it was thought advisable to determine whether there was any change in the composition of tomato plants due to cyanid fumigation.

Six uniform groups of 30 tomato plants were selected and fumigated in different ways, as described in Table II. After six weeks of growth they were cut off at the ground and the fresh weight, dry weight, percentage of ash, and percentage of protein determined for each set. From an examination of the data, it will be seen that there is no apparent change in the gross composition of these plants due to any combination of fumigation doses.

TABLE II.—*Analysis of six sets of tomato plants fumigated with hydrocyanic acid, arranged in order of their total dry weight*

Set No.	Treatment.	Weight of plants.		Ash.	Protein.
		Fresh.	Dry.		
		Gm.	Gm.	Per cent.	Per cent.
11	1 fumigation, 0.25 gm. of potassium cyanid per cubic meter overnight...............	1,463	140	21. 22	14. 31
10	3 fumigations, 0.25 gm. of potassium cyanid per cubic meter overnight, 2 weeks apart...	1,295	123	16. 45	16. 84
12	1 fumigation, 1 gm. of potassium cyanid per cubic meter for 1.5 hours in daylight.......	1,174	120	19. 95	14. 28
14	1 fumigation, 0.5 gm. of potassium cyanid per cubic meter overnight.....................	1,089	106	20. 40	14. 18
15	Control, not fumigated........................	1, 101	103	21. 88	14. 37
13	9 fumigations, 0.25 gm. of potassium cyanid per cubic meter overnight, 3 to 4 days apart..	944	95	21. 00	14. 28

EFFECT ON THE GROWTH OF THE PLANT

Townsend (20) has shown that there is an increase in germination in seeds fumigated with hydrocyanic acid. Woodworth (25) found that scale insect eggs fumigated with a sublethal dose of hydrocyanic acid resulted in their earlier hatching. Many of the fruit growers of the West are convinced that whether their trees are infested with insects or not they will not receive a maximum yield unless they have been given cyanid fumigation. A similar stimulation in the growth of plants may be noted in Table II, where four fumigated sets at the close of the fumigation exceeded in dry weight the normal set. In this experiment very frequent fumigation, with a consequent constant disturbance of photosynthetic activity, resulted in a diminished dry weight.

To obtain more definite data on this point, a longer experiment with tomatoes was conducted, carrying them through their complete life cycle. Three sets (No. 16, 17, and 18) were selected according to uniformity in size and vigor. Wooden boxes 16 inches square and 12 inches deep were filled with rich soil and six plants in each allowed to attain a height of about 6 inches before the experiment was started. There were 10 boxes in each set. Set 16 was kept as a control. No. 17 was fumigated with 1.5 gm. of potassium cyanid per cubic meter for one hour during the night, with curtains drawn tightly. The effect of the fumigation was a severe curling of the leaves and a burning of the youngest leaves (Pl. 34, A). No. 18 was fumigated three times at two-week intervals with 0.25 gm. of potassium cyanid per cubic meter overnight. No injury resulted from any of these. Each week from three to six plants from each set were taken, and their dry weight was determined. The growth curves in figure 7 give these data. The plants were controlled to one growing stem, and at the end of seven weeks had reached the top of the greenhouse. The plants were now limited to one in a box and nine boxes to the set (Pl. 34, B).

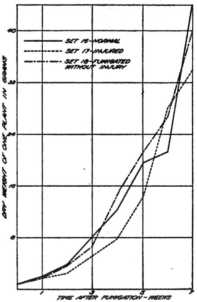

Fig. 7.—Growth curves of three sets of tomato plants as affected by hydrocyanic-acid fumigation.

As it was decided to follow the subsequent history of the sets by the quantity of ripe fruit produced, the growing point was cut off and all adventitious buds kept removed. Later, when the green fruit had begun to set, a number of large leaves from each plant were removed to admit more light. The daily production of ripe fruit is shown in figure 8. Attention should be called to the fact that set 18, fumigated without injury, produced fruit two weeks earlier than the normal, and that almost throughout the experiment both fumigated sets had a higher yield of fruit than the normal. At the close of the experiment the total weight of ripe fruit was less for the normal than for either of the fumigated sets.

It was thought desirable to obtain similar data on a quick-growing plant like lettuce. Small plants were set into boxes similar to the ones used in the above experiment with tomatoes, and divided into three sets. No. 19 was a normal set. No. 20 was fumigated with 0.25 gm. of potassium cyanid per cubic meter overnight, without apparent injury. No. 21 was fumigated with the same dose for one hour in bright sunlight. This set was severely injured, almost all of the mature leaves being burned. Samples for the determination of dry weight were taken every three to five days until the plants were fully grown. The data obtained are expressed graphically in figure 9. There appears to be practically no difference between the normal and the set fumigated without injury. The set which was severely injured was noticeably retarded, but soon recovered its power of growth and reached maturity about one week after the two other sets.

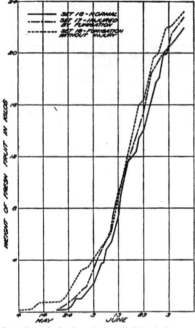

Fig. 8.—Curves showing the production of fruit by the three sets of tomato plants in figure 7.

From these results it appears that the growth factor in tomatoes at least is stimulated by hydrocyanic-acid fumigation. This is manifested not only in total dry weight of plants but in ripe fruit produced as well. These experiments further show that in greenhouse practice severe injury to plants may be not so serious as might appear, since the plants, after a temporary setback, grow with increasing vigor to maturity.

EFFECT ON PERMEABILITY OF THE LEAF

In the work of the Armstrongs (3) on hormones and anesthetics and their effect on permeability in plants, they distinguished between the action of hydrogen cyanid and that of the typical anesthetics. The latter disturbs the osmotic relations (and hence permeability) and brings enzyms and their substrates together more rapidly, while cyanid affects the enzyms directly, especially the ones concerned in pigment formation (the oxidases). Osterhout (16), using electrical conductivity as a measure of permeability, has demonstrated that both anesthetics and hydrogen cyanid cause at first a decrease in permeability, followed shortly by a rapid increase. This decrease is reversible if the anesthetic be removed. If the anesthesia be carried too far, however, it is followed by an increase in permeability which is irreversible. The change from one to the other probably represents the death point of the cells. When the Armstrongs (4) immersed leaves of *Prunus* sp. in water, the leaves

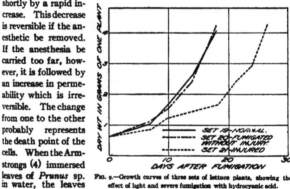

Fig. 9.—Growth curves of three sets of lettuce plants, showing the effect of light and severe fumigation with hydrocyanic acid.

gained slightly in weight, but lost sugar to the solution; if chloroform or ether was added to the water, the leaves gained still more in weight. When cyanid was used, there was no increase in weight; and the leaves lost no sugar to the solution.

To determine whether fumigation with cyanid affects permeability, and in which direction, the work of the Armstrongs was repeated with tomato leaves. The leaf blades were cut off, washed in distilled water, dried between blotting paper, and weighed. They were then immersed about two-thirds their length in a measured volume of the solution to be tested and allowed to stand the specified time. After drying and weighing, they were exposed freely to the air for an hour and weighed a third time. After the removal of the leaves, the solutions were tested with Fehling's solution for the approximate amount of reducing sugars lost by the leaves under various treatments. The results are exhibited in Table III.

TABLE III.—*Permeability of tomato leaves under the influence of ether, chloroform, potassium cyanid, and hydrocyanic acid*

No.	Material and treatment.	Original weight.	After treatment.		After exposure to air for one hour.		Sugar test in water after removal of leaves.
			Weight.	Change in weight.	Weight.	Change in weight.	
		Gm.	Gm.	Per ct.	Gm.	Per ct.	
1	Normal leaves in distilled water, 18 hours.	0.1953	0.1975	+ 1.1	0.1565	—19.0	Very faint positive.
2	.....do.................................	.2077	.2074	.0	.1330	36.0	Do.
3	.....do.................................	.1755	.1750	.0	.2312	16.0	Do.
4	.....do.................................	.1432	.1476	+ 2.9	.1074	25.0	Do.
	Average.................................	....	....	+ 1.25	....	—24.0	Do.
5	Normal leaves in ether water, 18 hours.	.1960	.1872	— 4.5	.1292	34.0	Faint positive.
6	.....do.................................	.1278	.1248	— 3.1	.0907	30.0	Do.
7	.....do.................................	.2913	.2730	— 5.6	.1810	37.0	Do.
	Average.................................	....	....	— 4.4	....	34.0	Do.
8	Normal leaves in a solution containing 2.4 mgm. of potassium cyanid per c. c.; 18 hours.	.2807	.2610	— 7.0	.1530	45.0	Negative.
9	.....do.................................	.2644	.2433	— 8.0	.1478	44.0	Do.
	Average.................................	....	....	— 7.5	....	44.0	Do.
10	Normal leaf in chloroform water, 18 hours.	.1408	.1200	—14.7	.0850	—39.0	Strong positive.
11	Normal leaves in water, 22 hours....	.1210	.1195	— 1.2	.1075	11.1	Faint positive.
12	.....do.................................	.0903	.0912	+ 1.0	.0820	9.2	Do.
13	.....do.................................	.2278	.2228	— 2.2	.2100	— 8.0	Do.
	Average.................................	....	....	— 0.8	....	— 9.4	Do.
14	Normal leaves in water containing 0.70 mgm. of hydrocyanic acid per c. c.; 22 hours.	.1708	.1358	—20.0	....	....	Negative.
15	.....do.................................	.0720	.0500	—30.0	.0280	61.0	Do.
16	.....do.................................	.1148	.0910	—20.0	.0735	36.0	Do.
	Average.................................	....	....	—24.0	....	—32.0	Do.
17	Normal leaves in water containing 0.35 mgm. of hydrocyanic acid per c. c.; 22 hours.	.1112	.0848	—23.0	....	....	Do.
18	.....do.................................	.1302	.0912	—30.0	....	....	Do.
19	.....do.................................	.1543	.1010	—34.0	.0658	57.0	Do.
	Average.................................	....	....	—29.0	....	—57.0	Do.
20	Leaves from plants fumigated one hour previously with 1.0 gm. of potassium cyanid per cubic meter, immersed in water 22 hours.	.2188	.2277	+ 4.0	.2065	5.6	Faint positive.
21	.....do.................................	.0700	.0571	—18.4	.0476	32.0	Do.
22	.....do.................................	.1012	.0928	— 8.3	.0842	16.8	Do.
	Average.................................	....	....	—10.0	....	—18.0	Do.
23	Normal leaves in water 24 hours....	.2255	.2133	— 5.3	....	....	Do.
24	.....do.................................	.1375	.1335	— 3.0	....	....	Do.
25	.....do.................................	.0631	.0621	— 1.2	....	....	Do.
26	.....do.................................	.0500	.0360	—28.0	....	....	Do.
	Average.................................	....	....	— 9.0	....	....	Do.
27	Normal leaves in water containing 0.10 mgm. of hydrocyanic acid per c. c.; 24 hours.	.0793	.0710	—10.5	....	....	Negative.
28	.....do.................................	.1250	.1141	— 8.7	....	....	Do.
29	.....do.................................	.0416	.0277	—33.4	....	....	Do.
	Average.................................	....	....	—17.0	....	....	Do.
30	Leaves from plants fumigated one hour previously with 1.0 gm. of potassium cyanid per cubic meter in water 22 hours.	.1853	.1603	—13.0	....	....	Faint positive.
31	.....do.................................	.1167	.1021	—12.0	....	....	Do.
32	.....do.................................	.0222	.0158	—29.0	....	....	Do.
	Average.................................	....	....	—18.0	....	....	Do.

Contrary to the findings of the Armstrongs, there was no gain in weight from the absorption of water. There was always a loss in weight, but this was less marked in normal leaves than in fumigated leaves, and when ether or chloroform, and more especially potassium or hydrogen cyanid were used, the loss was very pronounced. Fumigated leaves immersed in water showed less effect than normal leaves in a solution of hydrocyanic acid. Although the successive decrease and increase in permeability noticed by Osterhout could not be detected by this method, the very marked increase in the case of the leaves soaked in a solution of hydrocyanic acid no doubt indicates that the "irreversible increase" has been attained. Also, when leaves are "burned" in cyanid fumigation, it means that the "differential septa" have lost their power of differentiation, and they are disintegrated and the tissue is dead.

No explanation is offered for the failure of all the leaves under the influence of cyanid in Table III to lose sugar to the surrounding medium. One would expect the reverse from the fact, first, that respiration is below normal, and, hence, sugar is not used up; second, that diastases are not prevented by prussic acid from forming more sugar; and third, that the sugar if present can escape into the surrounding medium more readily because of the more permeable septa.

EFFECT ON TRANSPIRATION

Having shown that hydrocyanic-acid fumigation brings about an increased permeability in the leaf septa, the writers thought it desirable to ascertain how this is related to the wilting effects noticed in fumigation injury.

Water was withheld from a number of potted tomato plants until the wilting point was reached. On assuming that this gave all the pots about the same water content, 250 c. c. of water were added to each and the pot and soil surface thoroughly paraffined. After a preliminary test to show that all the plants transpired at about the same rate, half of them were fumigated. The dose and length of fumigation were varied to produce different degrees of injury. The controls were kept under similar conditions in each case. After fumigation the plants were removed to the greenhouse and weighed at intervals to follow the loss in water by transpiration.

The results are shown graphically in figure 10. Contrary to what was expected the fumigated plants in every case exhibited a diminished transpiration. The more severe the injury, the greater was the wilting effect and the less was the transpiration. In addition, stomatal measurements disclosed the fact that after fumigation, whether the plants were in the light or in the dark, the stomata rapidly closed. Recovery to normal transpiration required 24 to 36 hours in the case of plants lightly fumigated; severely injured plants, of course, never recovered.

The explanation of this phenomenon is, no doubt, that the inhibited photosynthetic and respiratory activities, together with the increased permeability of the cell walls, occasion the closing of the stomata. This reduces stomatal transpiration more or less completely. Cuticular transpiration, however, can continue as usual; in fact, with the increased permeability of the leaf septa, it might even be augmented. In order to demonstrate the hypothesis, two groups of tomato plants were kept in

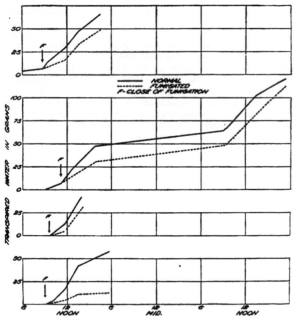

Fig. 10.—Transpiration curves of tomatoes as affected by hydrocyanic-acid fumigation. Arranged from top to bottom in order of increasing injury.

absolute darkness before, during, and after fumigation. The loss in water under these conditions represents loss through the cuticle of the leaves, since the stomata are closed. The curves in figure 11 show that there is a considerable increase in this phase of transpiration in the fumigated plants. Meanwhile the increased permeability has resulted in decreased osmotic pressure in the tissues, which would slacken, or stop altogether, the inflow of water from the stems. Consequently the water exchange in the leaf is in favor of the cuticular transpiration, and wilting results. If the disturbance in the osmotic relations does not reach the stage of "irreversible increase in permeability" described by Osterhout (16), the leaf will in time recover its normal turgescence. If it exceeds that point, however, the leaf is "burned" and dies, as was discussed above under permeability.

In order to demonstrate in another way that the septa of injured leaves are made equally permeable to solutes and to solvent, or, in other words, are "permeable" instead of "semipermeable," the experiment was tried of forcing water into wilted plants to cause them to recover their turgidity (17, p. 232). A plant was cut off at the base and the stem sealed into one arm of a U-tube. This arm was filled with water and the other with mercury. Normal tomato plants so wilted that they had bent completely over were forced into an upright turgid condition by 12 cm. of mercury within 10 minutes. Plants wilted by hydrocyanic acid, however, did not respond to the treatment. The less injured leaf blades would partly recover, and the petioles completely; the more severely injured blades would remain limp and grad-

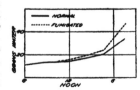

Fig. 11 —Curves showing the cuticular transpiration of fumigated and of normal tomato plants.

ually curl and shrivel, and the "burned" petioles also would fail to respond even slightly to the pressure.

CONCLUSIONS

From the data presented it may be stated that plants subjected to hydrocyanic-acid fumigation absorb more or less of the gas; that the immediate effect of the presence of this poison is a reduction in the activity of the oxidases and catalase, and, hence, in respiratory activity. Resulting from this is an inhibition of photosynthesis and translocation of carbohydrate, and a closing of the stomata. Another result is an increase in the permeability of the leaf septa, which causes less rapid intake of water from the stems and more rapid cuticular transpiration. In cases of mild fumigation this results in merely a temporary wilting; in more severe fumigations the wilting is followed by disintegration and death of the tissues. This increased permeability is no doubt due to the reduced respiratory activity. Budgett (5) has shown that changes in the permeability in protozoa treated with cyanid are similar to those produced by lack of oxygen. Within a few hours after fumigation the oxidase activity has returned to normal, while the catalase and the respiratory activities have exceeded the normal. By this time the recovery of photosynthetic action is first apparent; complete recovery, however, of this and of translocation of food material is not attained before from two to three days. Respiration remains above normal for several days.   If the increase in permeability is not so severe as to cause the death of the tissues, recovery is followed in many cases by a rate of growth and of fruit production (in the tomato) in excess of the normal. Hence, in greenhouse practice it is unwise to condemn injured plants too quickly.   The stimulation of growth may be due to at least two factors—namely, to the increased activity of the catalase which Zieger (26) found to be proportional to general metabolic activity in animals, and to the increased permeability of the cell walls, allowing readier exchange of food materials and of gases.   It is very improbable that the extra nitrogen of the cyanid has anything to do with increased nutrition, as is suggested by Woodworth (25) in the case of scale insect eggs.

In short, then, the primary effect of the presence of hydrocyanic acid in a plant is a disturbance of the oxidase and catalase activities.   All other physiological effects appear to be secondary to these.

LITERATURE CITED

(1) ALSBERG, C. L , and BLACK, O. F.
    1916. THE SEPARATION OF AUTOGENOUS AND ADDED HYDROCYANIC ACID FROM
        CERTAIN PLANT TISSUES AND ITS DISAPPEARANCE DURING MACERATION.
        In Jour. Biol. Chem., v. 25, no. 1, p. 133–140.
(2) APPLEMAN, C. O.
    1915. RELATION OF CATALASE AND OXIDASES TO RESPIRATION IN PLANTS. *In*
        Md. Agr. Exp. Sta. Bul. 191, 16 p., 2 fig. Literature cited, p. 16.
(3) ARMSTRONG, H. E., and ARMSTRONG, E. F.
    1910. THE ORIGIN OF OSMOTIC EFFECTS. III. THE FUNCTION OF HORMONES IN
        STIMULATING ENZYMIC CHANGE IN RELATION TO NARCOSIS AND THE
        PHENOMENA OF DEGENERATIVE AND REGENERATIVE CHANGE IN LIVING
        STRUCTURES. *In* Proc. Roy. Soc. London, s. B, v. 82, no. 559, p.
        588–602.
(4) ——
    1911. THE ORIGIN OF OSMOTIC EFFECTS. IV. NOTE ON THE DIFFERENTIAL SEPTA
        IN PLANTS WITH REFERENCE TO THE TRANSLOCATION OF NUTRITIVE
        MATERIALS. *In* Proc. Roy. Soc. London, s. B, v. 84, no. 571, p.
        226–229.
(5) BUDGETT, S. P.
    1898. ON THE SIMILARITY OF STRUCTURAL CHANGES PRODUCED BY LACK OF
        OXYGEN AND CERTAIN POISONS. *In* Amer. Jour. Physiol., v. 1, no. 2,
        p. 210–214, 9 fig.
(6) BUNZEL, H. H.
    1914. A SIMPLIFIED AND INEXPENSIVE OXIDASE APPARATUS. *In* Jour. Biol.
        Chem , v. 17, no. 3, p. 409–411, 1 fig.
(7) DEZANI, S
    1913. SUL COMPORTAMENTO DELL' ACIDO CIANIDRICO INIETTATO NELLE PLANTE.
        In Arch. Farmacol. Sper. e Sci. Aff., v. 16, fasc. 12, p. 539–546.
(8) EWART, A. J.
    1896. ON ASSIMILATORY INHIBITION IN PLANTS. *In* Jour. Linn. Soc. [London],
        Bot., v. 31, no. 217, p. 364–461.
(9) ——
    1897. FURTHER OBSERVATIONS UPON ASSIMILATORY INHIBITION. *In* Jour. Linn.
        Soc. [London], Bot v. 31, no. 219, p. 554–576, 1 fig.

(10) GEPPERT, J.
1888-89. UEBER DAS WESEN DER BLAUSÄUREVERGIFTUNG. *In* Ztschr. Klin.
Med., Bd. 15, Heft 3, p. 208-242, 1888; Heft 4, p. 307-369, 1889.
(11) KASTLE, J. H., and LOEVENHART, A. S.
1901. ON THE NATURE OF CERTAIN OF THE OXIDIZING FERMENTS. *In* Amer.
Chem. Jour., v. 26, no. 6, p. 539-566.
(12) LOEVENHART, A. S., and KASTLE, J. H.
1903. ON THE CATALYTIC DECOMPOSITION OF HYDROGEN PEROXIDE AND THE
MECHANISM OF INDUCED OXIDATIONS, TOGETHER WITH A NOTE ON
THE NATURE AND FUNCTION OF CATALASE. Pt. 1. *In* Amer. Chem.
Jour., v. 29, no. 5, p. 397-437, 1 fig.
(13) MIRANDE, Marcel.
1909. INFLUENCE EXERCÉE PAR CERTAINES VAPEURS SUR LA CYANOGENÈSE
VÉGÉTALE. PROCÉDÉ RAPIDE POUR LA RECHERCHE DES PLANTES À
ACIDE CYANHYDRIQUE. *In* Compt. Rend. Acad. Sci. [Paris], t. 149,
no. 2, p. 140-142.
(14) MOLLIARD, Marin.
1910. RECHERCHES SUR L'UTILISATION PAR LES PLANTES SUPÉRIEURES DE
DIVERSES SUBSTANCES ORGANIQUES AZOTÉES. *In* Bul. Soc. Bot.
France, t. 57, p. 541-547.
(15) MOORE, William.
1916. STUDIES IN GREENHOUSE FUMIGATION WITH HYDROCYANIC ACID. TEM-
PERATURE AND MOISTURE AS FACTORS INFLUENCING THE INJURY OF
PLANTS DURING FUMIGATION. *In* 16th Rpt. State Ent. Minn., 1915/16,
p. 93-108, fig. 23-27. Literature cited, p. 108.
(16) OSTERHOUT, W. J. V.
1917. SIMILARITY IN THE EFFECTS OF POTASSIUM CYANIDE AND OF ETHER. *In*
Bot. Gaz., v. 63, no. 1, p. 77-80, 1 fig.
(17) PFEFFER, Wilhelm.
1900. PHYSIOLOGY OF PLANTS. Ed. 2, v. 1. Oxford.
(18) SCHMIDT, H.
1902. UEBER DIE EINWIRKUNG GASFÖRMIGER BLAUSÄURE AUF FRISCHE FRÜCHTE.
In Arb. K. Biol. Anst. f. Land- u. Forstw., Bd. 18, Heft 3, p. 490-517.
(19) SHAFER, G. D.
1915. HOW CONTACT INSECTICIDES KILL. III. Mich. Agr. Exp. Sta. Tech.
Bul. 21, 63 p., 15 fig.
(20) TOWNSEND, C. O.
1901. THE EFFECT OF HYDROCYANIC ACID GAS UPON GRAINS AND OTHER SEEDS.
Md. Agr. Exp. Sta. Bul. 75, p. 183-198, 5 fig.
(21) TRUOG, E.
1915. METHODS FOR THE DETERMINATION OF CARBON DIOXIDE AND A NEW
FORM OF ABSORPTION TOWER ADAPTED TO THE TITRIMETRIC METHOD.
In Jour. Indus. and Engin. Chem., v. 7, no. 12, p. 1045-1049, 1 fig.
(22) VIEHOEVER, Arno, and JOHNS, C. O.
1915. ON THE DETERMINATION OF SMALL QUANTITIES OF HYDROCYANIC ACID
In Jour. Amer. Chem. Soc., v. 37, no. 3, p. 601-607.
(23) WILLAMAN, J. J.
1917. THE ESTIMATION OF HYDROCYANIC ACID AND THE PROBABLE FORM IN
WHICH IT OCCURS IN SORGHUM VULGARE. *In* Jour. Biol. Chem., v.
29, no. 1, p. 25-36.
(24) WOGLUM, R. S.
1911. HYDROCYANIC-ACID GAS FUMIGATION IN CALIFORNIA. THE VALUE OF
SODIUM CYANIDE FOR FUMIGATION PURPOSES. U. S. Dept. Agr. Bur.
Ent. Bul. 90, pt. 2, p. 83-90, pl. 9-10 (fold).

(25) WOODWORTH, C. W.
    1915. THE TOXICITY OF INSECTICIDES. *In* Science, n. s., v. 41, no. 1053, p. 367–369.
(26) ZIEGER, Rudolf.
    1915. ZUR KENNTNIS DER KATALASE DER NIEDEREN TIERE. *In* Biochem. Ztschr., Bd. 69, Heft 1/2, p. 39–110.

PLATE 34

A.—Tomato plants in the greenhouse at beginning of growth experiment. Note the curling of leaves in set 17.
B.—Tomato plants when in fruit, showing the method of handling.

EFFECT OF PUMPING FROM A SHALLOW WELL ON THE GROUND-WATER TABLE[1]

By Walter W. Weir,

Assistant Professor of Soil Technology, California Agricultural Experiment Station

INTRODUCTION

In many of the irrigated sections of California. the ground-water table has reached a point so near the surface that drainage is required. The usual methods of tile or underdrainage have in some instances proved expensive, especially where gravity outlets are not readily available. As a possible means of reducing the cost of drainage the suggestion has been made that equally satisfactory results might be secured by pumping from a shallow well into which the water from the surrounding area was allowed to seep without the use of tile. If adequate drainage could be secured by this method, it would be especially applicable in those sections where pumping is necessary to secure an outlet for tile drains and where the drainage water thus developed is needed for irrigation or alkali reclamation.

In order to obtain more definite information on the effect of pumping from a shallow well on the ground-water table and its relation to the drainage of irrigated lands, the experiment described in the following pages was undertaken at the Kearney Park Experiment Station, Kearney Park, California.

SOIL

The site for the pumping plant was chosen because it represented soil and culture conditions typical of a large area about Fresno. The land on three sides of the pumping plant was farmed and irrigated according to the usual practices in that vicinity, while on the fourth side it was more intensively cultivated and less frequently irrigated, owing to the experimental nature of the work carried on.

The soil is mapped by the Bureau of Soils[2] as the Madera fine sandy loam and is of fairly uniform texture from the surface to a depth of 7 feet. The excavation for the sump was made in December, 1913, this being the month when the water table is lowest. The soil was dry on the surface, but quite wet at 7 feet in depth. At from 7 to 8 feet a layer of grayish-colored hardpan was encountered, which, however, was seamy

---

[1] Based on work done under a cooperative agreement between the Office of Public Roads and Rural Engineering, United States Department of Agriculture, and the University of California Agricultural Experiment Station.

[2] Strahorn, A. T., Nelson, J. W., Holmes, L. C., and Eckmann, E. C. Soil survey of the Fresno area, California. *In* U. S. Dept. Agr. Bur. Soils Field Oper. 1912, [14th Rpt.], p. 2089-2166, fig. 56 pl. 27-31, A (col.) 1915

Journal of Agricultural Research,
Washington, D. C.
kj

Vol. XI, No. 7
Nov. 12, 1917
Key No Cal. —11

and easily excavated. Below the hardpan was found a layer of coarse sand 1 foot in thickness, which contained water. From 9 to 12½ feet a very dense, brown, iron-cemented hardpan was found, which, although somewhat seamy, was excavated with difficulty on account of the water. From 12½ to 13½ feet a second layer of coarse water-bearing sand was found. At 13½ feet a third layer of hardpan was encountered, but its nature and thickness were not determined. Later a 4-inch hole was bored through this hardpan to the loose material below.

METHOD OF PROCEDURE

The sump, which was approximately 6 feet square and 13½ feet deep, was cribbed with 3 by 6 inch redwood lumber and a ½-inch space was

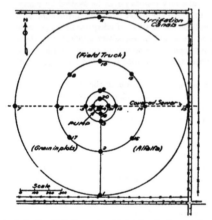

FIG 1.—Plan showing location of irrigation canals and test wells with reference to the pumping plant.

left between planks so that the lateral movement of the water would not be retarded.

The pumping equipment consisted of a 3-inch centrifugal pump of the vertical type, having the suction about 13 feet below the surface, and operated by a 5-horsepower belt-connected motor controlled by an automatic float switch.

Test wells of 3-inch galvanized-iron pipe, 8 feet deep, were located along lines extending in the four cardinal directions from the pump. The first wells were 100 feet, the second 300 feet, and the third 600 feet distant from the pump. On the diagonals wells were placed 300 feet from the pump, thus making eight wells on a radius of 300 feet and four at each of the other distances. One test well was located just outside the pump house and can be considered as being zero distance from the sump. Later, four more test wells were located 50 feet from the pump. Observations on these last four wells (No. 18, 19, 20, and 21) were not complete, as these wells were not cased and did not remain open during the entire season. Figure 1 shows the location of the test wells with reference to the pump and irrigation ditches, and the cultural features of the vicinity.

In March, 1914, observations were begun by noting the depth to water in the various test wells at intervals of a week. In April the pump was

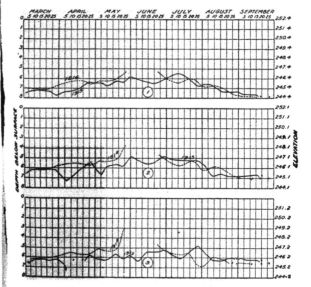

Fig. 2 —Ground-water curves for 1914 and 1915 (wells 1, 2, and 3).

started, but was kept in operation only a short time. On three or four occasions during the irrigation season the pump was run for a day or two. After about an hour's pumping the water would be entirely removed from the sump and the pump would stop for two or three minutes while the sump was being refilled to a height sufficient to start the motor. After the first lowering only two or three minutes were required to empty the sump. About 250 gallons were discharged during each pumping period.

It was soon found that an intermittent flow of such a small quantity of water was not sufficient to irrigate economically even the small demonstration plots of the Experiment Station; and, as no other provision was made for disposing of the discharge, very little pumping was done during the year. Observations were continued with reasonable regularity on the various test wells, the profiles of which are shown in figures 2 to 7. The pump was in operation for so short a time and the amount of water pumped was so small that there was no appreciable effect on the ground-water table, even in the immediate neighborhood of the pump.

TABLE I.—*Daily pump discharge (in gallons) from April to August, 1915, at Kearney Park Experiment Station*

Day.	April.	May.	June.	July.	August.
1.....................	....	125, 600	145, 800	133, 100	(a)
2.....................	....	125, 000	145, 800	125, 700	(a)
3.....................	113, 000	120, 800	142, 000	115, 200	(a)
4.....................	103, 300	122, 700	132, 400	136, 900	(a)
5.....................	103, 300	122, 700	(b)	112, 900	122, 700
6.....................	104, 000	125, 600	(b)	111, 400	118, 200
7.....................	100, 200	120, 800	(b)	125, 700	108, 500
8.....................	98, 500	129, 600	(b)	118, 200	105, 500
9.....................	101, 700	128, 800	(b)	112, 900	105, 500
10.....................	101, 700	128, 800	(b)	115, 200	103, 100
11.....................	c 81, 000	130, 000	(b)	119, 700	101, 800
12.....................	101, 700	124, 200	132, 300	115, 200	98, 000
13.....................	104, 000	122, 700	(d)	111, 400	95, 000
14.....................	101, 000	129, 600	(d)	109, 200	92, 800
15.....................	101, 700	128, 800	(d)	110, 000	98, 000
16.....................	100, 000	128, 100	(d)	112, 200	87, 500
17.....................	101, 000	125, 600	(d)	112, 200	85, 300
18.....................	102, 800	128, 800	(d)	(a)	c 41, 800
19.....................	103, 400	135, 500	(d)	(a)	c 40, 300
20.....................	121, 800	128, 800	(d)	c 98, 500	c 40, 000
21.....................	197, 000	128, 800	(d)	c 105, 600	(e)
22.....................	200, 000	128, 800	(d)	111, 500	(e)
23.....................	164, 200	130, 000	(d)	107, 000	(e)
24.....................	161, 800	132, 500	(d)	c 54, 200	(e)
25.....................	158, 600	128, 800	(d)	(a)	(e)
26.....................	144, 500	143, 000	(d)	c 80, 000	(e)
27.....................	156, 300	139, 300	(d)	115, 200	(e)
28.....................	c 74, 000	133, 300	(d)	c 70, 300	(e)
29.....................	138, 500	143, 000	c 71, 800	124, 200	(e)
30.....................	137, 800	140, 800	133, 000	125, 700	(e)
31.....................	....	140, 800	....	124, 200	(e)
Total.	3, 376, 800	4, 012, 600	903, 100	3, 119, 500	1, 550, 300
Total for season....	....	....	....	....	12, 791, 300

a Pump not running.
b Belt broken.
d Pump in operation only a portion of the day.

d Motor disabled.
e Pumping discontinued.

The profiles showing the water-table curves for 1914, therefore, represent very nearly normal conditions as they would have been had no pumping been done.

During 1915 much more complete data were obtained. The observation wells were measured weekly from the beginning of the irrigation season in March until they were all dry in September. The pump was run continuously, and daily records were made of the discharge from April 3 to August 21, except during about three weeks in June, when

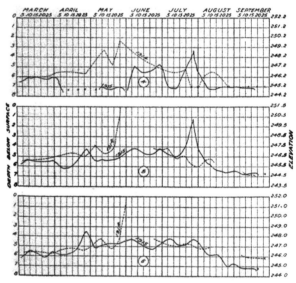

Fig. 3 —Ground-water curves for 1914 and 1915 (wells 4, 5, and 6).

the pump was dismantled for repairs. The weekly well readings for 1915 are plotted with those for the previous year in figures 2 to 8, and the daily pump discharge is shown in Table I. The data obtained during 1915 have been plotted separately for each month in figures 9 to 15, inclusive, showing the slope of the ground water along the different lines passing through the sump.

INTERPRETATION OF RESULTS

Observations taken at other places in the vicinity of Fresno covering a period of several years show that the ground water normally reaches its highest point during June. The rise up to this date is somewhat irregular, owing to irrigation and a somewhat fluctuating flow in the irrigation canals. The decline in the water-table curve after June is more regular and continuous until after the rainy season is well advanced in January or February. This condition is not so marked in the data shown in

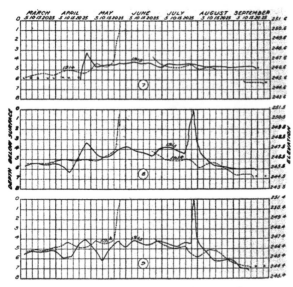

FIG. 4—Ground-water curves for 1914 and 1915 (wells 7, 8, and 9).

figures 2 to 8 as it probably would have been if the area covered by these observations had been larger. Local irrigations and variations in flow in the canals near by are responsible for the sudden rises and falls in the curves shown.

On April 3, 1915, when the pump was started, the water table averaged slightly lower than on the corresponding date in 1914. The pump ing for the first few days apparently affected the water table over the entire area, but showed a more marked effect in the wells nearest the pump. An irrigation late in April caused a sudden rise in the water table, which, however, had resumed a stable condition by May 5. Although the pump was in continuous operation during May, the water table maintained a rather constant level. In those wells which were 100 feet from the sump the water was about 1 foot lower during May, 1915, than during the same month in 1914. In the wells at a greater distance than 100 feet the difference is very slight.

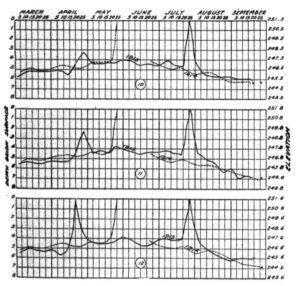

Fig. 5.—Ground-water curves for 1914 and 1915 (wells 10, 11, and 12).

During the week preceding and the week following the first of June, 1915, there was a perceptible rise in the water table over the entire area. It is apparent that this rise was well advanced before the pump stopped on June 4. Although the pump was in operation but one day between June 4 and June 28, there was no further rise in the water table until near the end of this period; in fact, there was a slight lower-

ing about the middle of the month. The rise recorded at the end of
June was apparently checked in the wells nearest the pump, but did
not reach the maximum in those farther away until the end of the first
week in July. An irrigation in July caused a sudden rise followed by
a decline which by August 10 had reached a point as low as any during
the season. The stopping of the pump from August 1 to August 4 did
not check the receding water table. During June, 1915, when the
pump was not in operation, the water table averaged lower than during

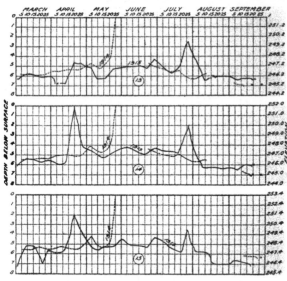

Fig. 6.—Ground-water curves for 1914 and 1915 (wells 13, 14, and 15).

June, 1914, while in July after the pumping was resumed, the water
table averaged higher than during the same month of the preceding
year. Well 4, which was very near the pump, was affected more than
any of the others by variations in the pumping. After August 10 the
curves for the two years coincide quite closely.

Figure 9 shows the water table for March, 1915, on four lines through
the sump. The pump was not yet started, and the water table was

practically stationary, the curve being nearly a straight line having a slope from north to south. Well 15 on the east-west line shows the effect of water being turned into the irrigation canal.

Figure 10 shows the effect of pumping on the east-west line as far as 600 feet from the sump, while on the north-south line the effect is very slight beyond 100 feet. This is particularly true south of the pump.

Figure 11 shows that during May, the month of maximum pumping, although the water table has a gradual slope from the north and east

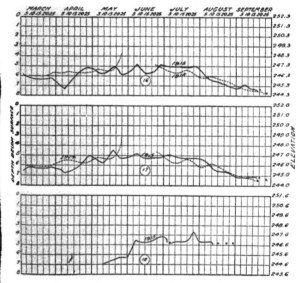

FIG. 7 —Ground-water curves for 1914 and 1915 (wells 16, 17, and 18).

toward the pump, it is approximately parallel to the ground surface. The pump apparently had no effect on the slope of the water table beyond 100 feet. During this month the wells at 50 feet distant from the pump were measured, which, no doubt, gives more accurate curves at these points.

During June the pump was in operation but a short time, and figure 12 indicates that the water table was approximately parallel to but about 2 feet higher than that during March. It will be noted from

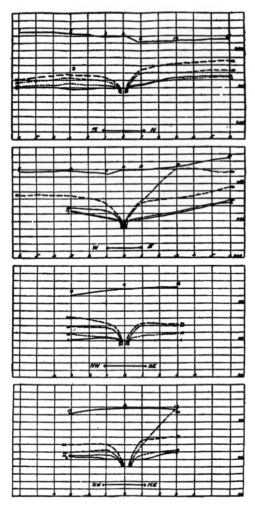

FIG. 10.—Profile of water table during April, 1915.

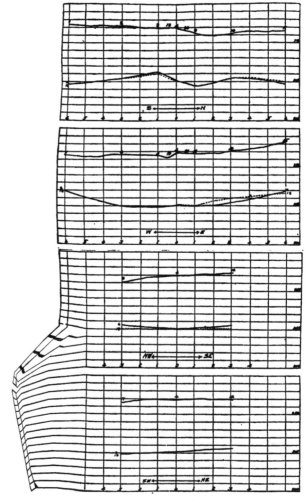

Fig. 15.—Profile of water table for September, 1915.

lower than that of March. After September 13 the water was everywhere more than 7 feet from the surface and could no longer be measured.

Figure 16 shows in composite form the same results as are shown in figures 2 to 8, each curve being a composite of the four well readings at the given distance from the pump. The wells on the diagonal at 300 feet distant from the pump are omitted from this diagram.

During May the average daily discharge from the pump was about 0.20 cubic feet per second, or slightly under 130,000 gallons per 24-hour day. This is sufficient to cover 1 acre 3¾ inches deep every 24 hours and without storage facilities is too small an amount for practical farm irrigation.

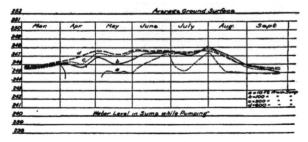

Fig. 16.—Composite curve showing water table from March to September, 1915, inclusive, at different distances from the pump.

CONCLUSIONS

The results of this experiment lead to the conclusion that under soil, irrigation, and farming conditions such as are found on the Kearney Park Experiment Station tract, pumping from a shallow well does not lower the ground-water table sufficiently to afford drainage to any considerable area. In this experiment, although the water table in the sump was maintained at a depth of about 12 feet below the ground surface and 5 to 7 feet below the normal ground water, the effect of the pumping was not appreciable beyond 100 feet from the pump. Except within a very short distance from the pump, the ground water rose to a point as near the ground surface in 1915 while the pump was in operation as it did in 1914, when no pumping was done. Seasonal variations are great enough to account for any differences observed.

Contrary to the results obtained here, it has been found that the water table can be materially lowered by the use of tile drains for greater distances away from the drain than is shown in this experiment. On the Kearney Vineyard Experimental Drain,[1] only 2 miles away, it was

---

[1] WEIR, W. W. PRELIMINARY REPORT ON KEARNEY VINEYARD EXPERIMENTAL DRAIN, FRESNO COUNTY, CALIFORNIA. Cal. Agr. Exp. Sta. Bul. 273, p. 101-123, 11 fig. 1916.

possible to keep the water table more than 2 feet lower at a distance of 160 feet from the drain than on surrounding untiled areas. This condition was obtained with the tile at a depth of less than 6 feet. On both the Dore tract[1] and the Toft-Hansen tract[2], about 6 miles from the Kearney Park experiment, the tile drainage systems had a noticeable effect on the ground-water table a mile away, although the tile were less than 5 feet deep on the Dore tract and only 3 feet deep on the Toft-Hansen tract. All of these drainage systems discharge into sumps from which the water is raised by means of pumps.

The fact that tile drains have proved more efficient than pumping from a well in lowering the ground-water table is due, no doubt, to the much larger area reached by the tile. No part of the 160 acres in the Kearney Vineyard Experimental Drain is more than 160 feet from a tile line, the Dore tract of 40 acres contains nearly 4,500 feet of tile, and there are 5,600 feet of tile on the 20-acre Toft-Hansen tract. Thus, with such a system any lateral movement of water is more readily intercepted, and any vertical pressure is relieved at more points than is possible where tile is not used, even though the water table is maintained at a greater depth in the well than is done by the tile lines. It would appear, therefore, from the results of these experiments that, although it has been proved feasible to reclaim water-logged land by means of tile drains, it would not be practicable to locate wells and pumping plants similar to the one described in this paper close enough together to lower the water table over any considerable area or develop enough water for practical farm irrigation without storage.

It is doubtful if the results of this experiment would have been materially different had it been located on the more poorly drained areas. Much the largest part of such areas about Fresno consists of soils mapped as belonging to the Fresno series. These soils contain hardpan layers nearer the surface, and the lateral movement of water is not so rapid as through soils of the Madera series such as are found on the experiment station tract.

In order to obtain the information given here, it was necessary to equip and operate a pumping plant for two years. The pump equipment, including sump, pump, motor, housing, power transmission, and discharge line, cost approximately $625. The power for the period covered by the experiment cost about $200, making a total of $825. There were some expenses incurred in installing test wells and making special investigations which would not be necessary under field conditions, and these have not been included in the total mentioned.

[1] FORTIER, Samuel, and CONE, V. M. DRAINAGE OF IRRIGATED LANDS IN THE SAN JOAQUIN VALLEY, CALIFORNIA. U S. Dept. Agr Office Exp Stas. Bul 217, 58 p, 8 fig, 2 pl (maps). 1909.
[2] MACKIE, W. W. RECLAMATION OF WHITE-ASH LANDS AFFECTED WITH ALKALI AT FRESNO, CALIFORNIA. U S. Dept. Agr. Bur. Soils Bul. 42, 47 p., 2 fig. 1907.

JOURNAL OF AGRICULTURAL RESEARCH

Vol. XI   WASHINGTON, D. C., NOVEMBER 19, 1917   No. 8

REACTIONS OF THE PHOSPHORUS OF THE THICK-ENED ROOT OF THE FLAT TURNIP [1]

By BURT L. HARTWELL, *Director, Agronomist,* and *Chemist;* F. S. HAMMETT, formerly *Assistant Chemist;* and P. H. WESSELS, *Associate Chemist, Agricultural Experiment Station of the Rhode Island State College*

INTRODUCTION

In earlier publications of the Rhode Island Agricultural Experiment Station (4, 7) [2] it was shown that the percentage of total phosphorus in flat-turnip roots (*Brassica rapa* L.) grown in different soils generally varied in the same direction as the variation in the amount of phosphorus which was available to the plant. The work herein recorded represents a study of the reactions, to various reagents and treatments, of different portions of the phosphorus in turnip roots, mainly for the purpose of ascertaining whether the variation in the percentage of phosphorus in roots from different soils is limited to certain of the groups of phosphorus compounds or is distributed proportionately among them all. This information was expected to throw light upon the question as to whether a determination of the total amount of phosphorus, or the amount in some particular group of phosphorus compounds, would be most useful in attempting to ascertain whether or not there existed sufficient phosphorus at the disposal of the plant.

ANALYSIS OF DRIED TURNIPS

To secure preliminary indications, the following work was done in 1907 with samples of turnips from different localities where there were found by soil tests to be wide differences in the amount of available phosphorus.

Dried turnips which had been ground to particles less than 1 mm. in diameter were first extracted with ether in a Soxhlet apparatus. As scarcely more than traces of phosphorus proved to be soluble in this solvent, the ether was removed, and the extraction continued with 95 per cent alcohol. These two extracts were united and evaporated and

---

[1] Contribution No. 135 from the Rhode Island Agricultural Experiment Station.
[2] Reference is made by number to "Literature cited," p. 370.

the phosphorus determined after ignition with magnesium nitrate, as has usually been done in connection with determining total phosphorus. The residue from the ether-alcohol extraction was treated successively at room temperature with five portions of 0.2 per cent hydrochloric acid. Each portion was allowed to stand at least 24 hours, with frequent stirring. In this extract was determined, in separate aliquots, the total phosphorus, that which was precipitated directly by molybdenum mixture according to the method used by Hart and Andrews (2) for determining "inorganic" phosphorus, as well as that precipitated by alcohol. The amount of phosphorus in the residue from the extractions was likewise determined. The results are given in Table I.

TABLE I —*Percentage of phosphorus pentoxid in dry flat turnips*

Flat turnips—	Percentage of phosphorus pentoxid.	Fractional solubilities expressed as percentages of total phosphorus pentoxid—				
		In ether and alcohol.	In 0.2 per cent hydrochloric acid.		Residue from preceding solvents.	Total.
			Precipitated by molybdenum mixture.	Precipitated by alcohol.		
From soil unmanured with phosphorus....	0.45	13	55	16	16	100
From the same soil manured with an insoluble phosphate........................	.45	11	48	20	21	100
From the same soil manured with bone......	.87	9	73	11	7	100
From soil containing very little available phosphorus	.27	10	53	20	17	100
From soil containing a medium amount of available phosphorus	.60	11	59	9	21	100
From soil containing much available phosphorus................................	1.14	10	66	8	16	100

It may be noticed, in case of the hydrochloric-acid extract of the turnips referred to, that larger proportions of phosphorus precipitated directly by molybdenum mixture are exhibited in the third column of the table (73 and 66 per cent) than in the others. These samples were grown in soils well supplied with available phosphorus, and the suggestion was thus received that possibly a plentiful amount of available phosphorus in the soil results in an increased proportion of that part of the phosphorus which has sometimes been classed as inorganic. The work of Koch and Reed (9), which included the growing of *Aspergillus niger* in nutrient solutions containing a very small as well as increasing amounts of phosphorus, was published at the time the preliminary work was going on and was of interest in the same connection. They found that the ratio of the protein phosphorus to the water-soluble or extractive phosphorus in the plant was 100 to 17 when the minimum amount of phosphorus was supplied and about 100 to 300 when there was a liberal amount. On account of the indications that possibly an increased percentage of phosphorus in the turnip root may be due in a considerable degree to an accumulation of inorganic phosphorus, attention was turned principally to the aqueous extractive portion rather than to the lipoid and residual portions, which constitute only minor parts of the phosphorus of the turnip roots.

It is shown by the following work that a more complete aqueous extraction of phosphorus can be made of the fresh pulp than of the dried turnip: 300-gm. lots from the same turnips were in two cases dried in slices at about 60° C., finely ground, and then digested for 36 hours, one with 800 gm. of distilled water and the other with 0.2 per cent hydrochloric acid; and in another case the undried pulp was digested similarly for 36 hours with about 475 gm. of distilled water, which with that in the turnips equaled the amount used to extract the dried turnips. The amount of extracted phosphorus pentoxid, in grams per 100 of dry turnip, was, with water, 0.561 from the fresh pulp and 0.500 from the dried material; whereas it was only 0.414 when the hydrochloric acid was used as a solvent with the dried material.

To avoid the changes which apparently take place during the drying of turnips, attention was next turned to a study of fresh turnips, including some of the transformations which occur in connection with the living cells.

MICROCHEMICAL EXAMINATIONS

A number of similar turnips which had been grown in soil in the greenhouse until the thickened roots were 1 to 2 inches in diameter were transferred to a nutrient solution in bottles. After the turnips had become adjusted to the new conditions some of them were deprived of phosphorus in the nutrient solution. Six days later it was found upon examination of one of these for phosphorus with a modified magnesium mixture containing only a small amount of ammonium hydroxid that practically no crystals of ammonium magnesium phosphate existed in microscopic sections, whereas similar turnips from which the phosphorus had not been withheld contained an abundance of them. An examination of the culture solution to which no phosphorus had been added showed that the phosphorus originally precipitable by the magnesium mixture had not passed out into the solution; therefore it probably had been used in the nutrition of the plant.

Some of the turnips which had been deprived of phosphorus were next supplied with liberal amounts in the nutrient solution, so that from time to time one of them might be examined for evidences of phosphorus absorption. Within 24 hours after phosphorus had been added in the nutrient solution crystals of ammonium magnesium phosphate were readily formed upon adding magnesium mixture to certain parts of the tissues.[1]

Repeated examinations for the relative abundance and for the first appearance of the crystals showed that they decreased in the order of the following locations: Taproot; bundles, and surrounding tissues near where the taproot merges into the thickened root; inside the cambium in the more active xylem tubes and surrounding parenchyma; outside of the cambium layer in the cortex; medullary rays, decreasing inward; and, finally, only traces in the pith and leaf petioles, from which it disappeared first upon again withholding phosphorus from the nutrient solution. The turnips which received maximum phosphorus nutrition increased to from 3 to 4 inches in diameter while growing in the solution.

It was noticed, while following the progressive decrease and increase of phosphorus directly precipitable by magnesium mixture that there was the opposite condition concerning the amount of starch. Simultaneous with an increase in the absorption of phosphorus it was observed that the leucoplasts containing the compound starch grains shrunk in size as the starch grains were corroded and dissolved, until finally when the maximum amount of phosphorus had been introduced into the nutrient solution, starch had practically all disappeared from the root tissues only to reappear when phosphorus was again withdrawn from the nutrient solution.

Similar observations have been made at this station (5) concerning starch, especially in potato vines growing with widely differing amounts of various nutrients, and it seems probable that a retardation in growth and in consequent carbohydrate requirement, owing to suboptimum nutrient conditions, tends toward an increased deposition of starch, although it is possible that the nutrients may serve as activators of amylases.

Examinations for precipitable phosphorus in the tissues of turnips grown under natural conditions in soil containing varying amounts of available phosphorus have been recorded previously (6). They furnished indications that the extent to which crystals were produced in the tissues upon treatment with magnesium mixture varied to a certain extent with the amount of available soil phosphorus.

Owing to the fact that the greater part of the phosphorus in turnips exists in the cell sap, the principal work now to be recorded deals with the soluble phosphorus. Some observations based upon the preliminary work have been made already in another connection (8).

---

[1] The principal features of the anatomical structure were found to be as follows. The taproot and, at the early stage, the thicker portion above it are diarch. As the plant develops the latter portion becomes collateral. The root portion finally consists of radial rows of bundles of vessels surrounded by fibrous, sharp-pointed cells between which the parenchymatous medullary rays are located. The remaining central area consists of thin-walled cells of irregular shape and size, which usually contain much starch and perform the functions of a storehouse. At the locality from which the leaves come the vascular xylem is limited to a narrow area surrounding a distinct pith which contains no differentiated vessels. This extreme upper part of the so-called turnip root must be considered as the stem and is differentiated by its voluminous pith from the enlarged and fleshy root portion with which it merges.

EXTRACTIONS OF THE FRESH PULP

To determine in fresh turnips the proportion of the total phosphorus which is present in solution, the following method was used: Trimmed and cleaned fresh turnips were weighed and halved. Wedge-shaped portions with the edges at the center were grated off and the remainder of the turnips weighed, so that the amount of the grated portion could be determined by difference.

The amount of water and phosphorus was determined in the ungrated portion and the following average procedure was also carried out: Two hundred gms. of pulp were taken, 80 c. c. of water added in transferring the pulp and juice from the grater to the receptacle in which all was allowed to remain for about 24 hours, with occasional stirring, 230 c. c. of extract were obtained by filtering under pressure. Fifteen samples taken from a given field at eight different times during a growing season yielded on an average, in the trimmed fresh turnips, 91.4 per cent of water and 0.107 per cent of phosphorus pentoxid, 80 per cent of the latter being present in the liquid portion. In the determination of the amount of phosphorus in solution it was assumed that the liquid remaining in connection with the pulp after filtering, contained the same percentage of phosphorus as the filtrate. It was found that when the extraction was made by 0.2 per cent hydrochloric acid less total phosphorus was extracted than when water was used, and furthermore that some analytical difficulty was experienced in this connection, especially with subsequent determinations involving ammonia, owing to the presence of pectin-like substances. Previous extraction of the fresh pulp with alcohol did not prevent 0.2 per cent hydrochloric acid from extracting the pectin-like substances from the pulp.

A 2 per cent acidity produced by acetic acid did not result in the same difficulties with the pectin substances; therefore the hydrochloric extraction of the pulp was abandoned. The following percentages of phosphorus pentoxid calculated to dry turnips show that when fresh turnips are grated without and with the acetic acid considerably less phosphorus is secured in the solution when the acetic acid is present.

Sample.	I	II	III	IV	V	VI	VII
Without acetic acid	0.54	0.76	0.38	0.54	0.33	0.37	0.13
With acetic acid...........	.22	.61	.31	.52	.26	.26	.06

It seemed to be immaterial whether the 2 per cent acidity was attained by adding acetic acid at the time of grating or to the turnip juice subsequently, for in different samples 0.260, 0.063, and 0.025 per cent were obtained in the filtered juice when the acetic acid was present at the time of grating, and 0.259, 0.055, and 0.026, respectively, when added to the juice itself before filtering—that is, the addition of acetic acid to the juice produced a precipitate which contained about the same amount of phosphorus as was left with the pulp when acetic acid was in contact with the same at the time of grating.

A number of measurements were obtained of th
total phosphorus in turnip extract, which failed to pas.
In different experiments the dialysis was carried out
an acidity of 0.2 per cent hydrochloric acid, or 2 per
in an aqueous solution. Thymol or chloroform was
growth of microorganisms. For a membrane parc
bladder, or chicken's crop was used. In some case.
replaced repeatedly with the fresh solvent until diffu
ceased, and the amount of phosphorus was then det
the dialyzate. Again, when fewer replacements we
allowed for equilibrium to be reached, and a deduc
the phosphorus in the dialyzate of the amount in
the diffusate, the remainder being considered as coll

In the various experiments there was always less
the phosphorus which seemed incapable of passing
and when the diffusates were replaced many times
and a long time was allowed for the final diffusions, al
of the phosphorus passed through the membrane.

If it be considered that dialysis separates crystall
phosphorus, it is evident that there is not much of
juice. It should be said, however, that enzymolys
place, and of course the possibility of hydrolytic cha
solvents should also be recognized.

Frequently, as will appear later, the main object i
nip extract was to obtain the dialyzates for the purp
to standard phosphate solutions, that an attempt

might still be considered as inorganic phosphorus which would combine readily with the usual precipitants of the same.

Realizing the indefiniteness of so-called inorganic phosphorus in such circumstances, the authors did not attempt to determine definitely the nature of the phosphorus precipitable by different reagents, because the main object was to find methods for determining differences in composition in turnips grown on soils containing different amounts of available phosphorus. In general the designation "inorganic phosphorus" has been used in the present paper more for convenience than because it is necessarily considered as representing a well-defined entity.

It was found that the Hart and Andrews method (2), as outlined for use with cereals, required with turnip extract so much nitric acid to prevent reduction in the molybdenum mixture and to enable a complete recovery of added phosphorus that the official strength of nitric acid was adopted in most cases with this reagent.

Many direct precipitations were made also with the official magnesium mixture. The comparative amounts of that portion of the phosphorus which was precipitated directly by the magnesium and by the molybdenum mixtures, from different lots of turnip juice, are shown by the following: If the amount precipitated with the magnesium mixture in each case is represented by 100, the amounts thrown down by the molybdenum mixture were 100, 93, 99, 95, 102, 95, and 98; average, 97. The two mixtures therefore gave practically the same results. The precipitates caused by the two reagents were so treated that the total phosphorus in them was determined.

Barium chlorid in neutral solutions was also used extensively as a precipitant; in fact, it was finally adopted as being in general the most suitable reagent.

The best criterion of the efficiency of the different methods for precipitating inorganic phosphorus in the presence of the colloidal organic constituents of turnip extract was considered to be the degree of recovery of phosphorus added in a standard solution to the dialyzates of the turnip extracts, from which practically all the phosphorus had diffused.

Evidence presented in regard to the correctness of a given method for determining inorganic phosphorus is of questionable value if it is based upon the extent to which phosphorus added in a standard solution is recovered from an extract in which inorganic phosphorus had been determined previously by the same method.

At first there was failure by any of the precipitants to precipitate completely the phosphorus added to the dialyzate, but as barium chlorid seemed the most promising the conditions for practically complete recovery were ascertained finally with this reagent. It was not determined whether or not conditions could have been imposed in connection with the molybdenum and magnesium mixtures which would lead to a complete recovery of added phosphorus, but such seemed probable from the results of the limited attempts which were made.

In preliminary work with undialyzed turnip juice and a standard phosphate solution containing phosphorus equivalent to 0 0200 gm of magnesium pyrophosphate, there was obtained by barium chlorid from a neutral solution in 4, 8, 16, 24, and 72 hours, respectively, 0.0120, 0 0186, 0.0194, 0.0200, and 0.0200 gm. more than was secured from the juice alone. This indicated that 24 hours was sufficient time for complete precipitation. It was also ascertained that a large excess of the reagent was of no advantage.

Based upon this preliminary work, added phosphorus was recovered completely a number of times from dialyzates of turnip juice from which practically all of the phosphorus had diffused. However, owing to the fact that in a few instances there was failure to recover all of the added phosphorus, it seems probable that a minus error is liable to occur when precipitating with barium chlorid from a strictly neutral solution, and it is probable that the method is more dependable if the solution is made slightly alkaline with ammonium hydroxid. A more detailed discussion of the method will occur in subsequent pages.

A limited amount of work was done with promising results by the method outlined by Collison (1), but the barium-chlorid method seemed well adapted for the work with turnip juice and was therefore mainly used.

ACTION OF OTHER REAGENTS AND CHANGES EFFECTED IN DIFFERENT WAYS

Although primarily interested in the determination of so-called inorganic phosphorus, the authors made certain examinations of turnip extract to ascertain the behavior of the phosphorus under different conditions, and these will be referred to briefly at this time.

Turnip juice is not of constant composition, but as an average less than one-tenth of the phosphorus in it is precipitated by acetic acid, more than seven-tenths by barium chlorid subsequently in the neutralized filtrate, and about two-tenths remain in the filtrate from the barium-chlorid precipitate.

The following data throw light upon the nature of the acetic-acid precipitate: Acetic acid was added to 1,500 c. c. of fresh juice until the acidity equalled 2 per cent; the mixture was allowed to stand for four days and was then filtered. The precipitate, after washing with 2 per cent acetic acid, was digested in a beaker for one hour at $37°$ C., with a solution of sodium hydroxid containing 0.5 per cent of free alkali. No reaction for inorganic phosphorus was secured when molybdenum mixture was added to the solution, thereby indicating that the phosphorus of the acetic-acid precipitate was not in compounds of a phosphoprotein nature (10). The alkaline digestion, however, so changed the acetic-acid precipitate that only two-thirds of its original phosphorus could be subsequently precipitated in a 2 per cent acetic-acid solution, the remainder passing into the filtrate but still not reacting as inorganic phosphorus. The two-thirds of the phosphorus mentioned above as being reprecipitated with acetic acid were subsequently again dissolved in 0.5 per cent sodium hydroxid and reprecipitated with acetic acid without any further change in the amount of phosphorus in the acetic-acid precipitate. The reprecipitated and washed material contained about 13 parts of nitrogen to 1 part of phosphorus.

An examination for phytin in turnip juice was made as follows: Two 400-c. c. lots of turnip juice, each containing phosphorus equivalent to 0.2272 gm. of magnesium pyrophosphate, were treated with enough acetic acid to make a 2 per cent solution. A precipitate containing 3 per cent of the total phosphorus was formed. The filtrate was examined for phosphorus in phytin according to the following method of Posternak (11): The solutions were made strongly alkaline with sodium hydroxid, and calcium chlorid was added. The white precipitate was filtered off and the filtrate found to contain only a trace of phosphorus. The precipitate was dissolved in hydrochloric acid, and sodium acetate and copper acetate were added. Phytin was found to be practically absent (3, p. 440).

By heating turnip juice, variable small amounts of phosphorus are thrown down with the not very voluminous precipitate which is formed. The filtrate may at once yield more or less inorganic phosphorus than an unheated aliquot. After the lapse of a week or more, chloroform having been added to the solutions, there was invariably an increase of inorganic phosphorus in the unheated juice and usually no increase in the heated juice. This indicated the possible presence of an enzym capable of increasing the amount of inorganic phosphorus, but efforts to prove that there was enzymic action were not universally successful. The amount of inorganic phosphorus precipitated by the immediate addition of the customary reagents to the juice usually constitutes so large a proportion of the total phosphorus that there is very little opportunity for further change.

PROPORTION OF INORGANIC PHOSPHORUS

The principle method which was used by the authors in the determination of the inorganic phosphorus in fresh turnips has been published elsewhere (6). It consists in brief in preparing the pulp in the presence of sufficient acetic acid to give about a 2 per cent acid extract, to which barium chlorid is then added as a precipitant, the solution being rendered neutral or barely alkaline with ammonium hydroxid. The inorganic phosphate is dissolved on the filter from the precipitate by treatment with hot water and acid, and the phosphorus brought into solution is determined in the usual manner with molybdenum and magnesium mixtures.

At first indications were afforded by this method that a larger proportion of the total phosphorus was secured from turnips grown on soil well supplied with available phosphorus than from turnips which had been grown on soil quite deficient in this ingredient. This was considered to be an important differentiation regardless of whether or not the determination could be considered strictly that of inorganic phosphorus, since it appeared to furnish an additional indirect method for securing information regarding the relative amount of available phosphorus at the disposal of the plant.

However, according to subsequent work with the method by other analysts on turnips grown in a number of different years, there were frequent failures to find any marked increase in the proportion of inorganic to total phosphorus in the turnips as a result of the application to the soil of liberal amounts of available phosphorus, although the crop and the percentage of total phosphorus in the same were much increased.

It was found in some instances that a small excess of ammonium hydroxid resulted in a considerable increase of inorganic phosphorus in the precipitate by barium chlorid. There was no certainty as to whether this increase represented a portion of the inorganic phosphorus not precipitable from a neutral solution, or whether the excess ammonia had caused a dissociation of phosphorus from organic combinations. By adding an excess of ammonia, on the assumption that all of the inorganic phosphorus was not precipitable in a neutral solution, practically the total phosphorus was frequently precipitated.

It has already been shown that phytin was absent from the juice, and it may well be questioned whether the small amount of phosphorus usually contained in the rather voluminous precipitate formed upon the addition of alcohol to the acidulated juice is an essential constituent of the same or has merely been dragged down by it.

If it is true that nearly all the phosphorus of the juice is inorganic, as seems likely to be the case, at most only the determination of total phosphorus in the juice could be of advantage for the purpose in mind; furthermore the determination of total phosphorus in the juice would be no more useful than that in the entire turnip unless differences in the amount of available soil phosphorus influence the composition of the juice more than of the rest of the turnip.

Such an influence seemed reasonable, and some evidence for that idea has been secured. At other times, however, an increase in the amount of available soil phosphorus seems to have increased the phosphorus of the juice no more than of the rest of the turnip.

Looking upon the inorganic phosphorus in the juice as being the excess above that required for the essential tissues, one would hardly expect to find much phosphorus in this form in turnips which had made only a small growth because insufficient soil phosphorus was available. Possibly, however, the turnip conserves a certain proportion of inorganic phosphorus in its thickened root, even if the supply is insufficient to produce a full-sized plant.   Inasmuch as the second year of this biennial is devoted to the production of seed, using as nourishment the root substance formed the previous year, a certain amount of inorganic phosphorus may be a normal requirement.   Turnips which have been stored still contain a large proportion of such phosphorus, but the authors have not ascertained whether this phosphorus supplies the early needs of the second-year's growth.

As a result of a critical consideration of all the data, the authors feel that the determination of the total phosphorus in turnips may prove as satisfactory for indicating the relative amount of available phosphorus in the soil upon which the turnips grew, as a determination of inorganic phosphorus in the juice by any method which they have tried, although in many instances the differences are more marked by the latter determination.

SUMMARY

In this paper is recorded work undertaken with the prime object of ascertaining whether the amount of any portion of the phosphorus of the turnip root is correlated more nearly than the total phosphorus with the relative amount available in soils.

Preliminary indications were derived from the successive extraction of dried turnips with ether, alcohol, and 0.2 per cent hydrochloric acid; but, since it was next found that larger amounts of phosphorus could be extracted from fresh than from dried turnips, all subsequent observations were made on fresh turnips.

Coincident with the introduction of phosphorus into a nutrient solution in which turnips were growing, the appearance of "inorganic" phosphorus and the disappearance of starch were traced microscopically in the different tissues; whereas upon withholding phosphorus the disappearance of inorganic phosphorus and the appearance of starch were similarly observed.   About four-fifths of the total phosphorus of fresh turnips was extracted with water.   When the latter was acidulated, somewhat less was secured because of partial precipitation.

Only a few per cent of the extracted phosphorus failed to pass through dialyzers.   Different precipitants of inorganic phosphorus were tested as to their ability to recover phosphate added in a standard solution to the dialyzates.

The phosphorus in the precipitate formed by adding acetic acid to turnip juice was not in phosphoprotein compounds.   There was no phytin in the juice.   The presence of a phosphatase was not shown.

Although the proportion of inorganic to total phosphorus in turnips was frequently made larger by phosphatic applications to the soil in which they were grown, this was not always shown to be the case by such methods as were used.

In most instances the phosphorus in the juice was so largely inorganic and constituted so large a proportion of the total that the determination of the latter seemed about as useful as of any portion for furnishing indications regarding the relative amount of soil phosphorus at the disposal of the turnip.

LITERATURE CITED

(1) COLLISON, R. C.
    1912. INORGANIC PHOSPHORUS IN PLANT SUBSTANCES. AN IMPROVED METHOD
        OF ESTIMATION. *In* Jour. Biol. Chem , v. 12, no 1, p. 65–72.
(2) HART, E. B., and ANDREWS, W. H.
    1903. THE STATUS OF PHOSPHORUS IN CERTAIN FOOD MATERIALS AND ANIMAL
        BY-PRODUCTS, WITH SPECIAL REFERENCE TO THE PRESENCE OF
        INORGANIC FORMS. *In* Amer. Chem. Jour., v. 30, no. 6, p. 470–485.
(3) ——— and TOTTINGHAM, W. E.
    1909. THE NATURE OF THE ACID SOLUBLE PHOSPHORUS COMPOUNDS OF
        SOME IMPORTANT FEEDING MATERIALS. *In* Jour. Biol. Chem., v. 6, no. 5,
        p. 431–444.
(4) HARTWELL, B. L.
    1913. THE PERCENTAGE OF TOTAL PHOSPHORUS IN FLAT TURNIPS AS INFLU-
        ENCED BY THE AMOUNT AVAILABLE IN SOILS. R. I. Agr. Exp. Sta.
        Bul. 154, p. 119–148.
(5) ———
    1916. STARCH CONGESTION ACCOMPANYING CERTAIN FACTORS WHICH RETARD
        PLANT GROWTH. R. I. Agr. Exp. Sta. Bul. 165, 23 p.
(6) ——— and HAMMETT, F. S.
    1911. THE EFFECT OF PHOSPHORUS MANURING ON THE AMOUNT OF INORGANIC
        PHOSPHORUS IN FLAT TURNIP ROOTS. *In* Jour. Indus. and Engin.
        Chem., v. 3, no. 11, p. 831–832.
(7) ——— and KELLOGG, J. W.
    1906. THE PHOSPHORIC ACID REMOVED BY CROPS, BY DILUTE NITRIC ACID
        AND BY AMMONIUM HYDROXID FROM A LIMED AND UNLIMED SOIL
        RECEIVING VARIOUS PHOSPHATES. *In* R. I. Agr. Exp. Sta., 18th Ann.
        Rpt , 1904/05, p. 253–285.
(8) ——— and QUANTZ, W. B.
    1910. THE PHOSPHORUS OF THE FLAT TURNIP. *In* Jour. Biol. Chem., v. 7, no 6,
        p. xxxviii.
(9) KOCH, WALDEMAR, and REED, H. S.
    1907. THE RELATION OF EXTRACTIVE TO PROTEIN PHOSPHORUS IN ASPER-
        GILLUS NIGER. *In* Jour. Biol. Chem., v. 3, no. 1, p. 49–52.
(10) PLIMMER, R H. A , and SCOTT, F. H.
    1908. A REACTION DISTINGUISHING PHOSPHOPROTEIN FROM NUCLEOPROTEIN
        AND THE DISTRIBUTION OF PHOSPHOPROTEINS IN TISSUES. *In* Jour.
        Chem. Soc. [London], Trans., v. 93, pt. 2, p. 1699–1721
(11) POSTERNAK, S.
    1903. SUR LA MATIÈRE PHOSPHO-ORGANIQUE DE RÉSERVE DES PLANTES À
        CHLOROPHYLLE. *In* Compt. Rend. Acad. Sci. [Paris], t. 137, no 3,
        p. 202–204.

THREE-LINED FIG-TREE BORER

By J. R. Horton,

Scientific Assistant, Tropical and Subtropical Fruit Insect Investigations, Bureau of Entomology, United States Department of Agriculture

INTRODUCTION

The adult three-lined fig-tree borer (*Ptychodes trilineatus* L.) is a large, longitudinally striped, long-horned, wood-boring beetle of the family Cerambycidae, which does considerable damage to fig trees (*Ficus carica*) in the Southern States by boring into the larger branches and trunks.[1]

This insect occurs throughout the southern United States from Florida to Houston, Texas, and from South Carolina to the Gulf. It has also been reported from parts of Mexico, British Honduras, Nicaragua, Costa Rica, Guatemala, Panama, and the West Indies, Colombia, and Venezuela, South America, and Tahiti, Oceania. The adult beetle causes some injury by feeding upon the fruit, leaves, and bark of fig trees and by ovipositing in the bark, but the greatest amount of damage is done by this insect while in the larval state. The larva is a white, flat-headed borer, or sawyer, which mines its way into the larger branches and trunks of the trees (Pl. 37), where it feeds upon the wood for from three months to more than one year, and reaches a length of nearly 2 inches, before changing to the pupa.

INJURIOUSNESS

The borers live in dry as well as in green wood, and specimens have lived for two or three weeks in other woods than the fig. Two larvae from one lot under observation ate their way out of their blocks of fig wood and into the top of a cypress table, where they tunneled for 2 and 3 inches, respectively, and completed the transformation to the adult stage. They appear to prefer wood which is partly dead and has lost some of its sap to healthy green wood, and therefore attack principally those trees or branches which are injured or diseased. Any injury, however, such as the breaking of a large limb, may invite the deposition of many eggs by the adult beetle, and the branch will then be killed by the borers. The first attack will, furthermore, generally be followed by others until the whole tree becomes involved and is finally killed by the insect.

Favorite points of attack are near wounds made by the breaking of large limbs, untreated saw cuts, splitting of the trunk, the knots formed

---

[1] Several other borers also attack fig trees in the Southern States—for example, *Leptostylus bnsttus* Lec., *Goes* sp., *Stephanoderes* sp., and *Atesss crypts* Say—and are found working in the same trees with *Ptychodes trilineatus*.

Journal of Agricultural Research,
Washington, D. C.
kp

Vol. XI, No. 8
Nov. 19, 1917
Key No. K—58

in the branches by fig canker,[1] injuries in the bark, etc. For example, a fig tree on the laboratory grounds at New Orleans, which through faulty pruning a comparatively light wind had caused to split down through the head, became so severely infested that it was estimated there were more than 1,700 borers in it at one time. Another fig tree which had been perfectly healthy and free from borers became severely infested in the trunk soon after the bark had been injured by a mercurial band applied to keep out ants. Still another healthy tree first became infested after a windstorm had broken off one of the larger branches, the eggs all being deposited in the branch from which it was broken, not far from the wound.

Every one of the six or eight injured or diseased fig trees on the laboratory grounds at New Orleans became heavily infested with borers, whereas not a single egg or borer could be found during three seasons in any of the 8 or 10 well-formed, healthy trees growing near them in the same yard. It may therefore be set down as a practically infallible rule that if the fig trees are kept in a thriving healthy condition, they will not be subjected to severe attacks, and may escape any injury whatever from the three-lined fig-tree borer.

BIOLOGY AND HABITS

THE EGG

DESCRIPTION AND PLACE WHERE FOUND

The egg (Pl. 35, A) is an elongate, nearly cylindrical, pure white to faintly yellowish or greenish object, ranging from 0.026 inch (0 66 mm) to 0.039 inch (1 mm.) broad, from 0.128 inch (3.25 mm.) to 0 144 inch (3.66 mm.) long, being on an average 0.036 inch (0.916 mm.) broad and 0.138 inch (3.52 mm) long. The shell is often more or less distinctly patterned like the grain of the fig wood in which the egg is deposited.

The eggs are deposited by insertion into the bark of the larger branches and trunk, thus being completely hidden from view and protected by the bark. They are usually deposited near wounds or decaying spots in the bark, or in limbs which have been cut or broken. They are even deposited in large branches which have been removed from the tree. They are seldom or never found in perfectly sound, healthy bark, a fact of importance in controlling the borer. The eggs usually occur singly, but sometimes two, three, or five are found together.

INCUBATION

The incubation period of eggs of the fig-tree borer does not vary greatly, at least in the period from May to September, when most eggs are developing. The maximum time required for the eggs to develop, as recorded in the season of 1914 and 1915, was 8 days, the minimum 3 days, the

[1] EDGERTON, C. W. DISEASES OF THE FIG TREE AND FRUIT. La. Agr. Exp. Sta. Bul. 126, P. 10-14, pl. 5. 1911.

average being approximately 5.6 days. The comparative regularity of the time required for incubation appears to be due to the relative uniformity of temperature and humidity prevailing in the bark tissues of the fig trees. The complete record on the incubation of eggs of the fig-tree borer at New Orleans, La., for the seasons of 1914 and 1915 is given in Table I.

TABLE I.—*Incubation period of eggs of the three-lined fig-tree borer, New Orleans, La., 1914 and 1915*

Number of eggs in the record.	Date of deposition.	Date of hatching.	Incubation time.	Number of eggs in the record.	Date of deposition.	Date of hatching.	Incubation time.
	1914.	1914.	*Days.*		1915.	1915.	*Days.*
9	July 2	July 5	3	2	June 1	June 7	6
17	July 4	July 9	5	2	..do..	June 8	7
5	..do..	July 11	7	1	June 2	June 7	5
6	July 27	Aug. 1	5	3	..do..	June 6	4
9	July 28	Aug. 3	6	1	..do..	June 8	6
6	July 31	Aug. 4	4	3	June 3	..do..	5
2	..do..	Aug. 5	5	4	June 7	June 11	4
4	..do..	Aug. 6	6	5	..do..	June 12	5
1	..do..	Aug. 8	8	7	..do..	June 13	6
6	Aug. 1	Aug. 5	4	8	June 9	..do..	4
1	..do..	Aug. 6	5	10	June 11	June 16	5
7	Aug. 3	Aug. 7	4	2	..do..	June 18	7
1	do	Aug 8	5	3	June 12	June 15	3
7	Aug 11	Aug 14	3	7	..do..	June 18	6
1	do	Aug 16	5	11	June 14	..do..	4
				6	June 15	June 20	5
	1915.	1915.		13	June 18	June 23	5
9	May 24	May 31	7	2	..do..	June 25	7
1	..do..	June 1	8	5	June 20	..do..	5
1	May 25	June 2	8	7	June 26	June 30	4
8	May 26	June 1	6				
1	do..	June 2	7	210			
3	May 29	June 5	7		Maximum................		8
1	May 30	June 6	7		Minimum................		3
2	..do..	June 5	6		Average................		5.6

HATCHING

When the development of the egg is completed and the larva ready to issue it chews its way out through the upper end of the eggshell. The larvæ generally feed at first upon the eggshell, sometimes devouring nearly all of it before tunneling into the bark layers.

LARVA OR BORER

DESCRIPTION

The larva, or borer proper (Pl. 35, B, C), is white to cream-colored, a legless grub, varying in length from about 0.125 inch (3 mm.) just after issuance from the egg to about 1.7 inches (43 mm.) when fully grown. It is broadest across the first thoracic segment, tapering gradually from the latter to the tip of the abdomen. The head is subrectangular, its sides converging posteriorly. The anterior border has a small 4-jointed appendage at each side of the base of the mandible. The anterior border and mandibles are dark brown to nearly black, and the posterior part of the head clear light amber. The anterior and lateral borders of the first body segment are of a shining, yellowish brown, and the posterior half velvety-brown; the remaining segments are creamy-white throughout. The dorsal and ventral surfaces of abdominal segments are tuberculate, the minute dorsal tubercles being arranged in four irregular rows, forming two irregular oval rings, one within the other. The last segment bears on its ventral surface a small group of amber-colored, chitinous spines.

Mr. F. C. Craighead, of the Bureau of Entomology, who has especially studied the classification of the larvae of the Cerambycidae, gives the following characters for distinguishing the larvae of *Ptychodes trilineatus* from others of that family:

Sides of head converging posteriorly; dorsal surface of prothorax vellured on posterior half, dorsal and ventral surfaces of abdominal segments tuberculate, tubercles on dorsal surface arranged in four irregular rows.

These characters will distinguish larvae of the subfamily Lamiinae, to which *Ptychodes trilineatus* belongs, from those of the other subfamilies. The presence of a small group of chitinous spines, on the ventral surface of the last abdominal segment, distinguishes the larva of the three-lined fig-tree borer from that of other species of the genus Ptychodes.

MOLTING AND GROWTH

The number of instars, or substages in the growth of the larva of the three-lined fig-tree borer, is variable and the occurrence of the molts irregular. A little more than half the specimens under observation molted only five or six times before transforming to the pupa, about one-fourth of them made the transformation after the eighth molt, and the remaining fourth after the fourth, seventh, ninth, or tenth molts. It is only rarely that as many as 10 molts occur in the larva. The insect will sometimes molt twice in close succession after being cut or otherwise injured, the extra molting being apparently a protective measure in such cases. In molting, the head cast splits slightly along the median ventral line, separates from the body integument, and is slipped off whole. The skin is slowly pushed back over the tip of the abdomen.

There is also a wide variation in the duration of the larval instars. Each of the first three instars may be completed in anywhere from 3 days to about 40 days, while each of the succeeding ones requires from about one week, as a minimum, to between 50 to 60 days, as a maximum, for those specimens which complete their growth in a single summer. With larvae which live through the winter from one season to the next, the later stages of growth will be much longer, requiring 5 or 6 months. The average duration of the different substages in the growth of the larva was as follows: Instar I, 8 days; II, 9 days; III, 12 days; IV, 16 days; V, 23 days; VI, 26 days; VII, 38 days for those specimens not passing and 5 months for those passing the winter in the larval stage;

VIII, 34 days for specimens completing their growth during the same summer in which they issued, and 6½ months for larvæ living over the winter; IX, 30 days for summer specimens, and 5½ months for those living through the winter as larvæ.

LENGTH OF LIFE AND HABITS

When the young borer issues from the egg, it mines its way along through the bark for several days. It then usually tunnels into the solid wood and often eats its way to the very heart of the branch. It lives and feeds in this manner, packing the burrow tightly behind it with "sawdust," for from about 2 to 15 months. About two-thirds of the borers observed completed this stage in the season in which the eggs from which they issued were deposited, while the remaining third lived through the winter, pupating the following season. Those borers completing the larval life in one season required approximately from 2 to 4½ months, the average larval life of the single season specimens being about 3 months. The overwintering borers required from 7½ to 15 months to complete the larval life, the average life in the wood being 11½ months. As the borers are feeding during most of this long period, and reach a size of 1.5 to 2 inches long and nearly 0.25 inch broad, it is seen that even a single insect can cause a great deal of injury. The records on the duration of the larval stage are given in Table II.

TABLE II.—*Duration of the larval stage of the three-lined fig-tree borer, New Orleans, La., 1915–16*

No.	Date of issuance.	Date of pupation.	Duration of larval stage.	No.	Date of issuance.	Date of pupation.	Duration of larval stage.
			Months.				*Months.*
1	May 21, 1915	Sept. 29, 1915	4.33	37	June 18, 1915	Sept. 7, 1915	2.75
2	...do...	Aug. 28, 1916	15	38	...do...	Sept. 18, 1915	3
3	June 1, 1915	Mar. 31, 1916	10	39	...do...	Sept. 7, 1915	2.75
4	June 2, 1915	Oct. 9, 1915	4.25	40	...do...	Sept. 29, 1915	3.5
5	June 1, 1915	Aug. 27, 1915	3	41	...do...	...do...	3.5
6	June 5, 1915	Sept. 29, 1915	4	42	do	Aug. 28, 1916	12
7	...do...	July 10, 1916	13	43	do	Aug. 31, 1915	2.5
8	June 7, 1915	Sept. 7, 1915	3	44	do	May 15, 1916	11
9	June 8, 1915	July 10, 1916	13	45	...do...	Mar. 31, 1916	9.5
10	June 7, 1915	Sept. 29, 1915	3.5	46	do	Sept. 18, 1915	3
11	June 6, 1915	Aug. 27, 1915	2.66	47	...do...	Aug. 31, 1915	2.5
12	June 8, 1915	Sept. 7, 1915	3	48	June 20, 1915	Mar. 31, 1916	9.5
13	...do...	Sept. 13, 1915	3.25	49	...do...	Sept. 18, 1915	3
14	June 11, 1915	Sept. 7, 1915	3	50	...do...	Mar. 10, 1916	8.6
15	...do...	Mar. 31, 1916	9.5	51	June 23, 1915	Oct. 9, 1915	3.5
16	...do...	do	9.5	52	do	Aug. 28, 1916	14
17	June 12, 1915	Aug. 17, 1915	2.25	53	...do...	Oct. 16, 1915	4
18	do	Aug. 28, 1916	14.5	54	...do...	Sept. 7, 1915	2.5
19	...do...	Aug. 17, 1915	2.5	55	...do...	Sept. 18, 1915	3
20	June 11, 1915	Aug. 27, 1915	2.5	56	do	Oct. 16, 1915	4
21	do	Sept. 18, 1915	3.25	57	...do...	June 12, 1916	11.6
22	...do...	Aug. 27, 1915	2.5	58	...do...	Oct. 9, 1915	3.5
23	June 13, 1915	Mar. 31, 1916	9.5	59	...do...	...do...	3.5
24	...do...	Aug. 27, 1915	2.5	60	June 25, 1915	Aug. 29, 1915	2.5
25	...do...	...do...	2.5	61	...do...	Sept. 29, 1915	3.25
26	...do...	Aug. 28, 1916	14.5	62	June 30, 1915	...do...	3
27	...do...	Aug. 27, 1915	2.5	63	do	Oct. 16, 1915	3.5
28	...do...	Oct. 25, 1915	4.5	64	do	Feb. 27, 1916	7.5
29	...do...	May 15, 1916	11	65	...do...	Sept. 29, 1915	3
30	June 16, 1915	Sept. 18, 1915	3.25	66	do	Aug. 31, 1915	2
31	...do...	Sept. 29, 1915	3.5	67	July 2, 1915	Mar. 31, 1916	9
32	...do...	May 15, 1916	11				
33	...do...	Aug. 19, 1915	2		Maximum		15
34	do	Aug. 28, 1916	14.5		Minimum		1.8
35	June 15, 1915	Aug. 19, 1915	2.25		Average		5.6
36	...do...	Sept. 7, 1915	2.75				

· When the larva is ready for pupation, it tunnels its way to a point near the surface, enlarges the cavity slightly, and completely surrounds itself with a sort of cell, or cocoon, composed of shreds and pellets of wood, tightly glued together. After making the cocoon, the larva ceases feeding, shrinks slightly, and molts, transforming to the pupal stage.

RESISTANCE OF THE LARVA TO INJURIES

While the larva of the three-lined fig-tree borer is soft bodied, and easily injured when removed from the protecting fig wood, its resistance to knife cuts and similar mechanical injury is great, as specimens cut so severely as to cause the loss of a considerable quantity of body fluid will often entirely recover and complete their development. Thus, one specimen which was cut nearly in two and had lost much of the body fluid recovered and reached the adult stage in the usual time, as did several others with injuries only slightly less severe. On one occasion, when a large-sized borer was chewed nearly in two and lost one of its mandibles as the result of meeting another borer in the wood, both wounds completely healed and the insect survived for more than a month. During this time the body cavity was closed by a thin, transparent membrane. The borer was, of course, unable to feed. A glossy black scab usually forms over a cut or abrasion in the larval skin. Infections often occur through abrasions, and this was the principal cause of the loss of injured specimens in the rearing work.

THE PUPA

DESCRIPTION

The pupa (Pl. 35, D, E) on first emerging is creamy-white, but soon takes on more yellow and develops a brown shading on the feet, the tips of the antennæ, and the mouth parts. As development progresses the eyes and mandibles turn amber-brown, then the prothorax, head, antennæ, sides of abdomen, and tips of wings take on similar coloring. The size varies slightly in the sexes and in different specimens of the same sex The length ranges from approximately 0.78 to 1.06 inches (20 to 27 mm.), with an average of approximately 0.94 inch (24.06 mm.); the breadth of the thorax from approximately 0.19 to 0.23 inch (5 to 6 mm.), with an average of approximately 0.21 inch (5.5 mm.); that of the abdomen from about 0.22 to 0.29 inch (5 75 to 7.50 mm.), with an average of approximately 0.23 inch (6.5 mm). The antennæ are 11-jointed but, owing to the fusion of the second and third joints, there appear to be only 10. They are directed backward along the sides of the body, the last seven joints resting curled together against the wing. The wings extend backward and downward under the abdomen, their tips reaching almost to the hind margin of the second abdominal segment. On each side of the dorsum of the prothorax are 16 small brown setæ or hairs arranged as follows: 1 long hair near the middorsal line near the front margin, 2 loose patches of 5 smaller hairs on the middle portion of each half of the pronotum, the rear patch being opposite its lateral marginal projection, the remaining 5 hairs set irregularly along the middorsal line and the hind margin of the pronotum. On the mesonotum are about 25 similar hairs so arranged as to form a V, the point directed toward the tip of the abdomen. A similar but larger V occurs on the mesothorax. On the dorsum of the first two abdominal segments is a thick brush of minute brown hairs, divided in the middle by the median dorsal line, consisting of the following numbers of hairs: On segment 3, about 44; segment 4, about 22; segment 5, about 19; segment 6, about 16; and segment 7, about 10. A group of similar setæ occurs at the base of each antenna and on the labrum and clypeus.

The life of the pupa is passed entirely within the sawdust cell constructed by the borer in its tunnel in the wood. When the pupal skin has been cast and the newly formed adult has hardened somewhat, it chews a circular hole through the bark and emerges into the outer air. The exit holes of the adult beetle are almost perfectly circular, and the occurrence of such holes in trunk and limbs indicates the escape of adults, not the entrance of larvæ.

DURATION OF THE PUPAL STAGE

The duration of the pupal stage, like that of the larval stage, is quite irregular. The extreme minimum duration in the specimens under observation was 5 days, but this occurred only in two cases. The extreme maximum was 73 days. The average duration of the pupal stage for all of the 70 specimens observed through that stage was 24 days The complete records on the duration of the pupal stage are given in Table III.

TABLE III.—*Duration of the pupal stage of the three-lined fig-tree borer, New Orleans, La., 1914, 1915, and 1916*

No.	Date of pupation.	Date of emergence.	Duration of pupal stage.	No.	Date of pupation.	Date of emergence.	Duration of pupal stage.
	1914.	1914.	*Days.*		1915.	1915	*Days.*
1	Sept. 1	Sept. 15	14	16	Aug 17	Sept. 6	20
2	...do....	Sept. 22	21	17.	do	Aug. 31	14
3	...do....	Sept. 14	13	18 ..	Aug. 19	do .	12
4	...do....	Sept. 22	21	19.	do	do	12
5	...do....	Sept. 15	14	20.	Aug. 27	Sept. 13	17
6	Sept. 9	Sept. 29	20	21..	...do..	do....	17
7	...do....	Sept. 22	13	22	...do....	...do....	17
8	...do....	...do.....	13	23..	do	do	17
9	...do....	Sept. 15	6	24.	...do.....	...do.....	17
10	...do....	Sept. 22	13	25	...do....	Sept. 16	20
11	Sept. 29	Oct. 16	17	26	Aug. 31	...do.....	16
				27	...do....	Sept. 12	12
	1915.	1915.		28	...do.....	...do.....	12
12	Apr. 24	May 15	21	29...	Sept. 7	Sept. 16	9
13	May 25	June 5	11	30.	do ..	Sept. 18	11
14	May 28	June 16	19	31..	do.....	...do ...	11
15	Aug. 17	Sept. 6	20	32	...do....	...do.....	11

TABLE III.—*Duration of the pupal stage of the three-lined fig-tree borer, New Orleans, La., 1914, 1915, and 1916*—Continued

No.	Date of pupation.	Date of emergence.	Duration of pupal stage.	No.	Date of pupation.	Date of emergence.	Duration of pupal stage.
	1915.	1915.	*Days.*		1915.	1915.	*Days.*
33	Sept. 7	Sept. 18	11	53	Oct. 16	Nov. 8	23
34	...do.	do....	11	54	...do ...	...do	23
35	..do..	do.. ..	11	55	...do....	Nov. 12	27
36	Sept. 18	Sept. 30	12	56	Oct. 25	Nov. 15	21
37	...do..	...do.....	12				
38	...do..	Sept. 23	5		1916.	1916.	
39	...do..	...do....	5	57	Feb. 17	Mar. 31	42
40	. do .	Oct. 4	16	58	Mar. 10	May 16	67
41	...do..	..do....	16	59	Mar. 31	June 12	73
42	Sept. 29	Oct. 13	14	60	...do....	...do....	73
43	...do..	Nov. 15	47	61	...do....	...do....	73
44	do..	Oct. 19	20	62	..do ...	. do....	73
45	...do..	Oct. 14	15	63	...do ...	.do....	73
46	.. do..	Oct. 18	19	64	.. do ...	May 15	45
47	..do .	Oct. 16	17	65	. do....	May 25	55
48	..do .	Oct. 24	25	66	Apr. 16	May 15	29
49	do.	Oct. 18	19	67	May 15	July 10	56
50	Oct. 9	Oct. 27	18	68	...do.....	...do....	56
51	do	Nov. 1	23	69	July 10	July 26	16
52	do.. ..	Nov. 3	25	70	...do.....	Aug. 28	49

Maximum... 73
Minimum... 5
Average... 24

THE ADULT

DESCRIPTION

The adult three-lined fig-tree borer (Pl. 36) is a long-horned beetle of the family Cerambycidae. The body is elongate, the females measuring from about 0.86 inch (22 mm.) to 1.1 inches (28 mm.) long from the vertex to the tip of the elytra, the average length being approximately 1 inch (25.5 mm.). The greatest width of the prothorax of the female is, on an average, about 0.2 inch (5.25 mm.), of the metathorax, across the base of the wing covers, about 0.27 inch (7 mm.).

The male varies from about 0.55 inch (14 mm.) to 0.82 inch (21 mm.) in length, being, on an average, about 0.75 inch (19.1 mm.) long. The prothorax is, on an average, about 0.16 inch (4.08 mm.), and the metathorax 0.21 inch (5.4 mm.) wide. The antennæ are 11-jointed, nearly 2½ times the length of the body (average 2.48 inches). The vertex is deeply and narrowly channeled, the channel extending in a distinct suture for the length of the head. The eyes are nearly divided, the lower lobe much broader than the upper. Prothorax cylindrical, narrowest in front. The legs are slender, the fore pair the longest. The body is broadest across base of the elytra, which taper toward their tips and end in a short sutural spine. The first and fifth ventral segments are longer than the intermediate ones. The general coloring of the dorsum, owing to a dense covering of fine appressed hairs, is brown and white in longitudinal stripes. Two broad brown stripes, punctuated with minute orange tufts in longitudinal rows, extend from antennæ to the tips of the elytra. A median scalloped white to yellowish stripe extends the length of the dorsum, ending abruptly at the anterior margin of the prothorax, and two similar lateral stripes extend from near the tips of the elytra along their margins to the bases of the antennæ. The integument beneath the tomentum is a dark amber-brown, finely punctate. The surface color of the antennæ, legs, and underside of the body is dark gray to brown. When first emerged, the general color is a very light brown to cream-yellow, with very faint white stripes.

FOOD OF THE ADULT

The adult beetles feed upon the tender bark of the smaller fig stems, and in captivity, when fed with such stems with the leaves attached, have eaten freely of the leaves. During the fruiting season they also feed to a considerable extent upon ripe and nearly ripe figs. They have been kept in captivity for several months, by feeding them leaves, fruit, and bark of the fig tree.

DURATION OF ADULT LIFE

The longevity of the adult three-lined fig-tree borer as determined upon 24 specimens in 1914_{-15} ranged from a minimum of about 2.5 months (75 days) to about 7.25 months (222 days). About one-fourth of the specimens lived 3 months or slightly longer, and the average length of life of all was 3.7 months. The complete records are given in Table IV.

TABLE IV.—*Longevity of adults of the three-lined fig-tree borer, New Orleans, La , 1914–15*

Number of specimens.	Date of emergence.	Date of death.	Duration of adult life.	
			Days.	Months.
	1914	1914.		
6	June 22	Sept. 5	75	2. 5
2	do	Sept. 14	84	2. 75
1	..do	Nov. 2	133	4. 33
5	..do .	Sept. 22	92	3
		1915		
3	..do.....	Jan. 21	213	7
1	..do.....	Jan. 30	222	7. 25
1	Sept. 22	...do.....	130	4. 33
	1915.			
1	Apr. 23	July 31	99	3. 25
1	..do.....	Sept. 14	144	4. 75
1	May 6	July 29	84	2. 75
1	May 15	Sept. 14	122	4
1	May 6	Sept. 1	118	3. 75
Total, 24.				
Maximum				7. 25
Minimum				2. 5
Average				3. 75

OVIPOSITION

In ovipositing, the insect usually stands with the body parallel to the axis of the trunk or branch, with the head toward its upper extremity, and makes a double transverse incision in the bark with the mandibles. The ovipositor is then inserted into one of the incisions, and the egg thrust downward and into the bark to a depth of from 0.125 to 0.25 inch. As a rule a cut is made in the bark for every egg, but sometimes two or three, and rarely as many as five eggs are deposited side by side. The "egg bite," consisting of two small, transverse slits made side by side in the bark by the mandibles, is usually the only external evidence of the presence of eggs; but on rare occasions the eggs are inserted so near the surface as to be visible as slight elevations in the bark.

The preoviposition period, or the time elapsing between emergence and oviposition, ordinarily varied from 8 to 16 days; but in a few specimens it was nearly a month. Most of the specimens required 8 days. The females also lived from 8 to 26 days after they had ceased to deposit eggs, but this was in the fall and winter when cool weather probably retarded oviposition.

NUMBER OF EGGS DEPOSITED

A complete record of oviposition was obtained from nine insects in 1914 and 1915. Three of the specimens deposited approximately 101 eggs each at the average rate of 1.5 eggs each per day; three more specimens deposited approximately 184 eggs each at the average rate of 1 egg each per day, and a seventh insect deposited 110 eggs at the rate of 1.4 eggs per day. The remaining two females deposited approximately 261 eggs each at the average rate of 2.4 eggs each per day. The oviposition record is given in Table V.

TABLE V.—*Oviposition record of the three-lined fig-tree borer, New Orleans, La., 1914 and 1915*

Number of specimens.	Date of beginning oviposition.	Date of ending oviposition.	Number of days of oviposition	Number of eggs deposited.
	1914	1914		
3	June 30	Sept. 4	66	302
3	June 30	Dec. 26	179	553
	1915	1915		
2	May 22	Sept. 6	107	523
1	Sept. 29	Dec. 15	77	110

From the foregoing it is seen that a single female three-lined fig-tree borer will deposit from 100 to 184 eggs in the course of her life at an average rate of from 1 to 2.4 eggs per day.

SEASONAL HISTORY

Adult beetles of the three-lined fig-tree borer first begin to emerge in March, but it is not until May that emergence is well under way, with the beetles appearing rapidly. The largest number of beetles appear in September, and the last specimens emerge in November or early December. These late-emerging beetles will deposit eggs on the warmer days of winter till at least as late as January 24. Fewest eggs are deposited in the months from February to April, inclusive, and the greatest number from May to September, inclusive. There is no true hibernation period in the latitude of New Orleans, but there is a period of comparative inactivity during the months from December to February, inclusive, during which, however, a slight amount of oviposition occurs, and eggs, larvæ, pupæ, and adults may be found in the fig wood.

The life cycle, from egg to adult, was passed by 54 specimens, or about five-sixths of the number observed through all stages, in the same season in which the eggs were deposited, their average life cycle being 3.5 months. Ten, or about one-sixth of them, lived through from May to June of one year to May and June of the following year, their average life cycle being 11.5 months. The generations are irregular, the hatch of the eggs deposited in the months from March to May reaching the adult state in the months from June to November, the hatch of a majority of those deposited in June and July becoming adults in the months from August to November, inclusive, and the remainder from March to June of the following year.

CONTROL OF THE BORER

Since the adult three-lined fig-tree borer does not as a rule oviposit in the perfectly sound limbs or trunk of the healthy trees, the most important measure of control is to keep the trees in the healthiest condition possible. The larger branches, 1 inch in diameter and upward, and the trunk are the parts particularly susceptible to attack.

Care should be taken to avoid bruising the bark or breaking limbs in cultivating and in picking the fruit, and the trees should be shaped to strengthen them as far as possible against breaking or splitting by heavy winds. The work of such diseases of the limbs as fig-canker and fig-limb blight should also be prevented as far as possible.

When a branch is accidentally broken, it should be immediately cut off smooth at its juncture with the larger branch or trunk and the wound painted with a mixture of five parts of coal tar and one part of creosote. At least a second and possibly a third coat should be applied when the preceding coat is dry.

It is important to prevent the first branch of a tree from becoming infested, as one infested and dying branch will invite further attacks upon the trees. When a branch has become infested, it should be removed and burned, as the borers will complete their development even in perfectly dead and dry wood and later infest other trees. Fig-wood prunings should always be destroyed, and never allowed to remain long in or near the orchard.

Freezing of the bark is likely to be followed by borer attack in the branches affected, and such branches should therefore be cut off and the surface of the cut painted as above recommended. When the bark becomes diseased or bruised, the affected area should be removed with a sharp knife, cutting square across to the sound wood, and coating with a protective paint so the bark will heal perfectly.

Trees already heavily infested with borers in the trunk, may as well be at once cut down and every scrap burned, as it will be practically impossible to save them and they will be a source of infestation and a menace to the healthier trees. Trees which are split through the head, those which are in a dying condition from any cause, and the volunteer trees which so persistently spring up from the roots of some varieties of older fig trees grown in southern Louisiana, should also be cut down and burned.

While it is best, where the infestation occurs only in certain branches, to remove the infested branches entire, the borers may be dug out of highly prized individual trees if the infestation has not progressed too far and its area is limited. Some good may also be accomplished, in such cases, by destroying the eggs with a knife or an awl. It would first be necessary, however, to become familiar with the appearance of the egg punctures and eggs. Yard trees should first of all, of course, be kept in a healthy condition in order to prevent attack by the borers. Oviposition may largely be prevented, in the case of a few yard trees, by insheathing the trunk and larger branches with wire netting. The screen would have to be kept in place practically throughout the year.

The borers in some cases may be killed by injecting carbon bisulphid into the tunnels and plugging the openings with putty, but this method is impracticable where the infestation is severe and well advanced.

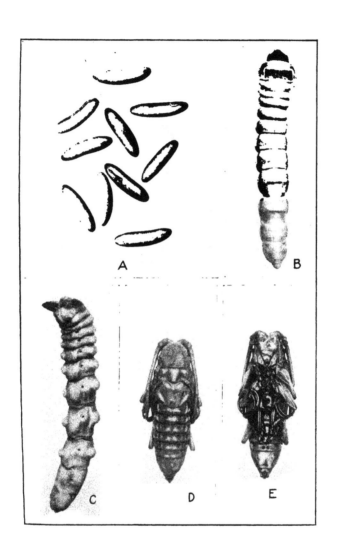

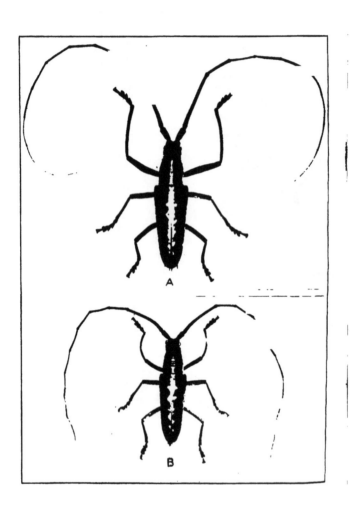

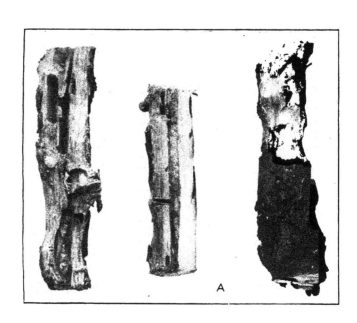

A STATISTICAL STUDY OF BODY WEIGHTS, GAINS, AND MEASUREMENTS OF STEERS DURING THE FATTENING PERIOD[1]

By B. O. SEVERSON, *Associate Professor*, and PAUL GERLAUGH, *Instructor, The Department of Animal Husbandry, Pennsylvania State Collge Agricultural Experiment Station*

INTRODUCTION

The necessity of judging beef cattle on a more scientific basis than is found in authoritative texts on the subject seems essential. Judgment in general is based on the merits of beef animals gathered from empirical results, and there exists little data on which scientific selection can be made. This necessity exists not only for the selection of breeding animals but for the selection of beef cattle purchased as "feeders." In the case of dairy cattle and the speed horse records of performance are available, and no one disputes their value as a means of selecting animals of merit. The feeder of beef cattle knows that breeding is important and gauges his selection by color markings, size, weight, form, sex, and condition. None of these are definite except sex; the rest are determined by observation, and judgment is made accordingly. Mitchell and Grindley,[2] of Illinois, have shown clearly that the selection of live stock for experimental purposes in many feeding experiments throughout the country has probably been the cause for the large experimental errors that these investigations show.

Is there any method of selection that can be used by experimenters and farmers to determine more accurately the probable gains in live weight and finish of beef cattle under a uniform system of feeding and management. At present the only real definite measurement used in the selection of steers for experimental purposes in feeding is live weight, while the judgment of the experimenter is employed in estimating uniformity of condition, general form, and quality. As shown in Table II of this paper, there is no correlation between live weight and the gains in live weight of steers fed during a fattening period of 140 days. What then can be used as a measure of determining probable gains and lowering the probable error caused by individuality of animals used?

OBJECTS OF THE INVESTIGATION

This paper does not attempt to answer these questions, but has as its object the presentation of information of a nature that may assist others

---

[1] Presented by permission of the Dean and Director of the School of Agriculture and Agricultural Experiment Station of The Pennsylvania State College.

[2] MITCHELL, H. H., and GRINDLEY, H. S. THE ELEMENT OF UNCERTAINTY IN THE INTERPRETATION OF FEEDING EXPERIMENTS. *In* Ill. Agr. Exp. Sta. Bul. 165, p 457-559, 8 fig. Bibliography. p. 578-579. 1913.

Journal of Agricultural Research
Washington, D. C.
kq

Vol. XI, No. 8
Nov. 19, 1917
Key No. Pa.—3

(383)

in making their solution possible. A series of body measurements were made of steers at the beginning and close of feeding experiments for three consecutive years. With these data a starting point for further statistical study is made possible. One object of this paper is to show the average body measurements of 2-year-old steers at the beginning and close of the fattening period caused by fattening and the relationship of other definite body measurements to each other, and to note those measurements that could be used in selection as a means of reducing the experimental error in feeding experiments and a study of variation in the measurements themselves. Another object is to show the correlation of gains to initial body measurements and to changes in body dimensions.

MATERIAL USED

During the winter months of 1914-15,[1] 1915-16,[2] and 1916-17,[3] 72 steers were divided into seven lots each year, a total of 216 animals. These steers were relatively uniform as feeders, varying in market grades from "fair" to "choice," the majority being "good" feeders. These market grades of "fair," "good," and "choice" feeders are illustrated respectively in figures A, B, and C of plate 38. In a total of 207 animals used in this study on which records were made of breeding, as shown by color of hair, 92 were Hereford grades, 84 were Shorthorn grades, 18 were Aberdeen-Angus grades, 7 were Shorthorn × Hereford crosses, 3 were Shorthorn × Aberdeen-Angus crosses, and 3 were Hereford × Aberdeen-Angus crosses. In no case did a steer fail to show some infusion of improved beef blood. The average initial weight of the 216 steers was 900.112 pounds with 700 and 1,300 pounds as extremes. The steers were as uniform in quality, weight, and condition as would ordinarily be obtained for feeding purposes. Each year 60 steers were divided into five lots of 12 each, selected with as much care for uniformity of weight, breeding, condition, and quality as possible.

The feeding of these various lots was done with rations affording very nearly the same opportunity for gains in live weight and condition of flesh for marketing.[4] From the entire group of 72 steers each year 12 "choice" feeders were selected (Pl. 38, C). These 12 steers were fed to "prime" condition and sold as "prime" steers. The 60 steers sold as "good" steers. The feeding period in all cases was 140 days, except those considered in Table II; and the measurements were made within two or three days of the beginning and the close of the feeding period. Table I includes all records for the three years, while the correlation tables (II-VIII) include the data collected during the first two years.

[1] TOMHAVE, W. H., and SEVERSON, B. O. BODY MEASUREMENT OF STEERS. *In* Pa. Agr. Exp. Sta. Ann. Rpt., 1914/15, p. 188-208. 1916.
[2] SEVERSON, B. O., and GERLAUGH, Paul. BODY MEASUREMENT OF STEERS. *In* Pa. Agr. Exp. Sta. Ann. Rpt., 1915/16. (In press.)
[3] ———— and BENTLEY, F. L. BODY MEASUREMENT OF STEERS. *In* Pa. Agr. Exp. Sta. Ann. Rpt. 1916/17. (Not yet published.)
[4] TOMHAVE, W. H., SEVERSON, B. O., and GERLAUGH, Paul. STEER-FEEDING EXPERIMENTS. Pa. Agr. Exp. Sta. Bul. 145, p. 17. 1917.

METHOD OF TAKING MEASUREMENTS

All measurements taken are indicated in Table I. The initial and final weights were averages of individual weights taken on three consecutive days at the beginning and close of the experiments. All body measurements except circumferences were made with the steel calipers shown in Plate 39, A. To the vertical standard, which was graduated in inches, were attached two movable horizontal crossbars. All circumferences were measured with a steel tape graduated in inches (Pl. 39, B).

The steers were tied and placed on a level wooden floor while being measured. Owing to their restlessness, some of the measurements were difficult to obtain, especially those of the head. The measurement of length of neck was discontinued after the second year because of the impossibility of getting correct position and dimensions. The distance from hock to ground was another measurement which possessed a large experimental error. The probability of error in measurements is a factor not considered, thus necessitating a larger number of measurements to reduce the probable error.

METHOD OF CALCULATION[1]

The method of calculation in the tables presented were as follows:[2]

The mean was obtained by the following formulæ: $A = \dfrac{\Sigma D^1 f}{n} + H$.

The standard deviation was calculated as follows: $\sigma = \sqrt{\dfrac{\Sigma (D^1)^2 f}{n} - \dfrac{(\Sigma D^1 f)^2}{n}}$

The coefficient of correlation coefficient was worked by the formulæ:

$r = \dfrac{\Sigma Dx Dy f}{n \sigma x \sigma y}$.

The coefficient of variation was obtained as follows: $C = \dfrac{\sigma}{a} \times 100$ per cent.

The probable error in each was determined as follows:

Probable error of mean: $E_A = \pm 0.6745 \dfrac{\sigma}{\sqrt{n}}$;

Probable error of standard deviation: $E\sigma = \pm 0.6745 \dfrac{\sigma}{\sqrt{2n}}$;

Probable error of coefficient of correlation: $Er = \pm 0.6745 \dfrac{(1 - r^2)}{\sqrt{n}}$;

Probable error of coefficient of variation: $E_c = \pm 0.6745 \dfrac{C}{\sqrt{2n}}$.

[1] DAVENPORT, C. B. STATISTICAL METHODS, WITH SPECIAL REFERENCE TO BIOLOGICAL VARIATION. ed. 3, p. 12–18, 44–45. New York and London, 1914.
[2] KEY TO FORMULÆ:

A = Mean;	H = Assumed mean;
σ = Standard deviation;	x = Character classified in vertical column;
r = Coefficient of correlation;	y = Character classified in horizontal row;
C = Coefficient of variation;	D^1 = Deviation from assumed mean.
n = Population;	f = Frequency.
Σ = Summation;	

TABLE I.—*Average initial and final measurements of 2-year-old steers fattened for market during a 140-day feeding period*

Measurement.	Initial measurement.		Final measurement.		Difference in measurements.	Percentage increase.
	Number of steers.	Average.	Number of steers.	Average.		
		Pounds.		*Pounds.*	*Pounds.*	
Weight....................	216	900. 112	216	1, 188. 398	288. 286	32.02
		Inches.		*Inches.*	*Inches.*	
Width of head....	214	8. 832	216	9. 112	. 280	3. 16
Length of head............	214	19. 411	216	19. 892	. 481	2. 48
Length of neck.............	103	19. 163	72	20. 990	1. 827	9. 53
Width of shoulders.........	214	16. 412	215	18. 459	2. 047	12. 42
Width of front flank.......	214	16. 378	216	18. 358	1. 980	12. 08
Width of paunch...........	214	23. 612	216	26. 101	2. 489	10. 54
Width of rear flank........	154	19. 527	216	22. 744	3. 217	16. 42
Width of loin..............	209	13. 984	216	15. 958	1. 974	14. 11
Width of hips..............	202	17. 662	216	19. 254	1. 592	9. 01
Width of thurls [a]........	209	17. 204	215	18. 533	1. 329	7. 72
Buttock to hip.............	214	18. 366	216	19. 622	1. 256	6. 83
Depth of chest.............	154	26. 730	215	27. 789	1. 059	3. 96
Shoulder point to ground ...	214	33. 033	216	34. 311	1. 278	3. 86
Chest to ground...........	214	22. 876	216	23. 013	. 237	1. 03
Hind flank to ground.......	214	29. 128	216	30. 202	1. 074	3. 68
Hock to ground............	142	20. 795	204	20. 914	. 119	. 57
Withers to ground.........	214	49. 224	216	53. 870	4. 646	9. 43
Hips to ground............	202	50. 855	214	52. 411	1. 556	3. 09
Shoulder to buttock........	214	53. 763	216	57. 988	4. 225	1. 85
Circumference of chest.....	214	73. 014	216	77. 694	4. 680	6. 41
Circumference of paunch	214	80. 256	216	88. 301	8. 045	10. 01
Circumference of hind flank.	214	71. 364	216	78. 685	7. 321	10. 25
Circumference of muzzle. ...	141	17. 198	143	17. 930	. 832	4. 85

a Hip joint.

These 216 steers used during the three years of 1914 and 1917, inclusive, were 2-year-olds, averaging in weight 900.112 pounds as "feeders" and 1,188.398 pounds as market animals, and gained at the rate of 2.058 pounds daily for 140 days. Although the steers were divided into six lots each year, they were in all cases finished into market steers varying from "medium" to "choice," the greater proportion being graded as "good" steers.

The following average initial measurements show a marked similarity: the length of head, length of neck, and width of rear flank, which vary from 19.163 to 19.527 inches; the width of shoulder and front flank differ by only 0.034 inch; the hips and thurls (hip joint) in width are 17.682 inches and 17.204 inches, respectively; and circumference of the body in the region of the chest and hind flank are 73.014 and 71.364 inches, respectively. The height at the withers of a "feeder" steer is 1.631 less than the height at the hips. The length of body from shoulder point to buttock is only 2.908 inches greater than the greatest height at the hips.

The average measurements at the conclusion of the fattening period show similarity as follows: The length of head, width of hips, and distance of buttock from hips varying within 0.638 inch of each other; the width of shoulder, front flank, and thurls are almost identical; and the circumference of the chest and hind flank are more alike than their initial measurements. The height has increased more at the withers than at the hips; thus, a 2-year-old steer changes his greatest height from the hips to the withers while receiving market condition   In circumference the increase was greater for the hind flank than for the chest; thus, the greater circumference of the chest at the initial measurement becomes less than the circumference of the hind flank in the finished steer.   In fattening, the greatest width at paunch and the greatest depth of body at the chest become more nearly alike, as shown by a difference of 3.118 inches at the initial measurement and 1.688 inches at the concluding measurement.

In all cases the difference between the initial measurement and the final measurement shows an increase in dimensions due to deposition of fat, muscular development, and growth.   The regions of the body covered by the greatest amount of muscular development show greater increased dimensions than those having less muscular covering.   In the regions where the growth would show the greater relative influence the least changes take place as shown in the width of head, length of head, distance from chest to ground and hock to ground.   The greatest increase in width took place in the hind flank rather than in the paunch, where it would seem natural to have the greatest increase because of feed capacity and condition.   The thick layer of flesh and fat deposits in the region of the hind flank, together with the distention of this region of the body in a fattened steer, are responsible for the greater width in this part of the body.   The width of loin, hips, thurls, shoulders, and front flank shows changes in dimensions caused mainly by increased condition of flesh.

The increase in height at the withers of 4.646 inches is not all due to growth alone, a larger portion of this increase being caused by the flesh covering over the withers and the deposition of fat in the muscular tissues of the shoulder region.   The fat deposit and muscular development causes the shoulder blade to be held more rigidly; thus, the body in the chest region rises between the shoulder blades, as indicated by the greater distance between the withers and the upper border of the shoulder blade. The greatest change in the body measurements was the circumference of the paunch.   This, however, was proportionately less than the increase of 7.321 inches in circumference of the hind flank.   The fact that the distance of chest to ground and hind flank to ground did not show greater difference was due to the lowering of the flank by deposition of fat in that region and the fat covering over the region of the chest.   The region of the body possessing the most valuable eatable parts on the whole are affected most in the fattening process.

The percentage of increase of the final measurements as compared with the initial measurements shows that increased weight was 32 02 per cent, the greatest increase of all measurements The measurements that showed more than 10 per cent increase were width of shoulder, front flank, paunch, rear flank, loin and circumferences of paunch and hind flank. Those which increased from 5 to 10 per cent were as follows: Length of neck, width of hips, width of thurls, distance from buttock to hip, height at withers, distance from shoulder point to buttock, and circumference of chest. Measurements that showed less than 5 per cent increase were: Width and length of head, depth of chest, distance from shoulder point to ground, from chest to ground, from hock to ground, from hip to ground, and the circumference to muzzle. Thus, again it is seen that the greatest relative changes in body dimensions occur in the regions affected most by deposition of fat and muscular development.

CORRELATION TABLES (II–VI)

The following correlation tables (II–VI) are presented as illustrations of the methods used in obtaining the facts presented in Tables VII and VIII.

TABLE II.—*Correlation of average daily gain of steers during a feeding period of 120 to 140 days and the weight at the beginning of the feeding period*

Daily gain (pounds).	Initial weight per steer (pounds).													
	1,300	1,250	1,200	1,150	1,100	1,050	1,000	950	900	850	800	750	700	To-tal.
3.2..............								1						1
3.0..............														0
2.8..............					1	1	3	1	1	2	1	1		11
2.6..............			1		1	4	2	2	4	1	3	3	1	20
2.4..............		1		1	2	1	5	8	11	7	6	3	2	47
2.2..............		1		2	2		4	7	4	7	4	5		36
2.0..............	1	1		3		1	4	7	9	16	9	12	2	65
1.8..............		1		3	2	2	5	7	11	10	9	7	6	63
1.6..............			1			1	5	3	5	4	4	5	4	32
1.4..............				1	2		4	4	10	3	7	1	1	33
1.2..............			2			2	2	4	2	6	2	1	1	22
1.0..............								1	2		1	1		5
.8..............						1			2					3
Total........	1	4	3	10	9	13	34	45	61	56	46	39	17	338

Mean live weight............... .pounds . 893. 93 ±3. 94
Mean daily gain ....do.... 1. 95 ± . 016
Standard deviation of live weight ....do. 106 88 ±2. 77
Standard deviation of daily gain .. do.. . . 446 ± . 011
Correlation................................... . 0364± . 0366

According to the literature cited,[1] there is some difference of opinion as to the average daily gain of steers of different ages which in reality is a difference in live weight, because, other things being equal, a steer when 2 years old will weigh more than a yearling. Table II shows a correlation of 0.0364, with a probable error of ±0.0366. The population includes 338 steers fed at The Pennsylvania State College during eight years. The correlation shows that there is no relation between the daily gain which may be expected from various weights of feeders of a fairly close range in age and condition.

TABLE III.—*Correlation of the initial circumference of chest of steers with gain in live weight during a 140-day feeding period*

Gain in weight (pounds).	Initial circumference of chest (inches).																			Total.
	78	77	76	75	74	73	72	71	70	69	68	67	66	65	64	63	62	61	60	
410						1			1											2
390	1					1		1	1											4
370					3	1	1	1	1			1								8
350				1	1	3	1	1	2			1								10
330			1		2	1	2	1	4	2		3	1	1						18
310			1			1	3	2	3	2	2	2		1	1					18
290				2	1	1		2	6	4	2	3	2		2					25
270			1				4	1	1		1	3	1		1					13
250				1		2	4	3	2	3		1	1		1					18
230					1	2			1		2		1	1	1	1	1			11
210						1			1	1	2		1							6
190			1							2	2									5
170									1		1								1	3
150														1						1
Total	1	0	4	4	9	13	15	11	23	16	12	13	9	3	4	3	1	0	1	142

Mean gain .... pounds.. 288.87 ±3.06
Mean circumference of chest... .... inches.. 69.824±0.221
Standard deviation of gain . ... pounds.. 54.180±2.168
Standard deviation of circumference of chest . inches.. 3.910±0.015
Correlation.. .238±0.053

The circumference of chest as an indication of the ability of a steer to make rapid gains is indicated here by the correlation 0.238±0.053. The spring of rib and depth of chest are regarded as indications of rapid gains by buyers. The circumference of the chest is of greater importance than the width or depth, as shown by these results.

---

[1] HENRY, W. A., and MORRISON, F. B. FEEDS AND FEEDING. ed. 15, p. 451. Madison, Wis., 1915.
WOLL, F. W. PRODUCTIVE FEEDING OF FARM ANIMALS. p. 258. Philadelphia and London, 1915.

TABLE IV.—*Correlation of increase of circumference of chest and gain in live weight of steers during a 140-day feeding period*

Gain in weight (pounds).	Increase in circumference of chest (inches)											
	12.5	11.5	10.5	9.5	8.5	7.5	6.5	5.5	4.5	3.5	2.5	Total.
412.5 .	1	.....	1	.....	2	1	1	.....	.....	.....	.....	2
387.5 .	.	.....	1	1	2	1	1	.....	.....	.....	.....	6
362.5 .	2	.....	2	2	2	1	.....	.....	.....	.....	.....	9
337.5 .	...	2	5	3	6	4	2	.....	1	.....	.....	23
312.5 .	...	1	2	5	3	4	5	.....	.....	.....	.....	20
287.5 .	.....	1	3	5	7	6	4	1	1	.....	.....	28
262.5 .	.....	.....	.....	2	4	4	3	2	.....	.....	1	16
237.5 .	.....	.....	1	3	4	4	8	.....	.....	1	.....	21
212.5 .	.....	.....	.....	.....	3	2	1	1	.....	1	.....	8
187.5 .	.....	.....	.....	.....	1	.....	.....	3	2	.....	.....	6
162.5 .	.....	1	.....	.....	1	.....	.....	1	.....	.....	.....	3
Total...	3	5	15	21	33	26	24	8	4	2	1	142

Mean gain....... . . pounds.. 288. 5 ±3. 101
Mean increase in chest circumference.. ... inches.. 8. 13 ±0. 105
Standard deviation in gain................ pounds.. 54. 77 ±2. 193
Standard deviation of gain in chest circumference,
  inches................................... 1. 87 ±0. 074
Correlation....................................... . 460±0. 044

The increase in the circumference of the chest varied from 2.5 to 12.5 inches, most of the steers showing an increase of 6.5 to 9.5 inches. Table II shows that the steers having a large increase in the circumference of the chest have been those steers that had made the greatest daily gains. (It is true that the more a steer gains the greater is the amount of covering of flesh over the ribs or around the chest circumference.)

TABLE V.—*Correlation of the initial width of the thurls with the gain in live weight of steers during a 140-day feeding period*

Gain in weight (pounds).	Initial width of thurls (inches)											
	20.25	19.75	19.25	18.75	18.25	17.75	17.25	16.75	16.25	15.75	15.25	Total.
412.5..... ..	.....	.....	.....	1	.....	.....	.....	.....	.....	.....	.....	1
387.5 ..	1	1	1	1	1	.....	.....	.....	.....	.....	.....	5
362.5	.....	.....	.....	.....	1	2	3	1	1	.....	.....	8
337.5	.....	.....	.....	.....	3	5	7	2	3	1	.....	21
312.5 ..	.....	.....	.....	6	1	2	2	3	4	1	.....	19
287.5	.....	.....	.....	.....	4	7	7	4	5	1	.....	28
262.5	.....	2	1	1	1	1	4	4	.....	2	.....	16
237.5... .	.....	.....	2	.....	1	5	6	2	2	2	1	21
212.5..........	.....	.....	.....	1	1	1	1	.....	4	.....	.....	8
187.5.........	.....	.....	2	.....	1	.....	.....	.....	2	.....	1	6
162.5.......	.....	.....	.....	.....	.....	.....	.....	1	.....	2	.....	3
Total...	1	3	6	10	14	23	30	17	21	9	2	136

Mean gain........................ ..pounds . 285. 48 ±3. 097
Mean width of thurls............... . inches . 17. 36 ±0. 059
Standard deviation of gain........... ..pounds . 53. 55 ±2. 185
Standard deviation of width of thurls . . . inches.. 1. 03 ±0. 041
Correlation......................................'.. . 224±0. 054

The width of the thurls is shown here to be a measurement which should be given considerable emphasis in foretelling the gaining capacity of steers. It is a measurement which usually is overlooked, but the correlation shows that it is more important than many measurements usually regarded as especially important.

TABLE VI.—*Correlation of the initial distance from chest to ground with gain of steers during a 140-day feeding period*

Gain in weight (pounds)	Distance from chest to ground (inches).								
	18. 5	19. 5	20. 5	21. 5	22. 5	23. 5	24. 5	25. 5	Total.
412.5					1	1			2
387.5				1	1	2	1	1	6
362.5				4	1	1	2	1	9
337.5				6	7	8	2		23
312.5	1		2	1	6	6	2	2	20
287.5			1	8	4	6	3	6	28
262.5		1		3	5	5	2		16
237.5			2	5	8	3	2	1	21
212.5				1	2	1	4		8
187.5				2	2	2			6
162.5				1	1	1			3
Total	1	1	5	32	38	36	18	11	142

Mean gain.................................... pounds.. 288. 55 ±3. 101
Mean distance from chest to ground.......... inches.. 22. 89 ±0. 066
Standard deviation of gains.................. pounds . 54. 77 ±2. 192
Standard deviation of distance from chest to ground, inches... 1. 18 ±0. 047
Correlation.................................... .077±0. 056

The distance from chest to ground does not show sufficient correlation with the gain to become an important factor in determining the choice of "feeder" steers, as shown by these results.

The results shown in Table VII are based on data collected on steers during the two winter periods of 1914–15 and 1915–16. All measurements considered in this table are initial measurements, except those that show increases of dimensions at the close of the fattening period as compared with initial measurements. The coefficient of variation is shown to be greatest on increases in circumferences of hind flank, paunch, and chest, and the increase of gains in live weight.

These measurements all relate to increased dimensions and occur in those parts of the body that show relatively high percentage increase over initial body measurements (Table I).

Of the initial measurements the greatest coefficient of variation is 11 9±0.04 per cent for initial live weight.

In all the chest measurements the coefficients of variation are relatively high: Width at point of shoulder 9.3±0.52, width of fore flank 10.4±0.6, depth of chest 6.1±0.34, and circumference of chest 6.9±

0.39. Likewise the mid and posterior regions of the body show variations, the rear flank with a coefficient of variation of 8.7±0.48, circumference of rear flank 6±0.34, width of loin 7.5±0.43, width of thurls 5.9±
0.34, circumference of paunch 11±0.62, and distance of hip to buttock
5.2±0.29.

The measurements affected most by growth show the least variations
and include the distance of·shoulder point, rear flank, and withers from
the ground, length of head, and distance of shoulder point to buttock.
In general, Table VII shows the greatest variation in those regions
of the body which change most in a fattening steer and those regions
affected most by deposition of fat and development of muscular tissue.

TABLE VII —*Means, standard deviations, and coefficients of variations shown in correlation tables*

Num- ber of steers.	Measurement.	Mean.	Standard deviations.	Coefficient of variation.
				Per cent.
338	Average daily gain.....pounds..	1. 93±0. 016	0. 446±0. 011	23. 1±0. 84
338	Initial live weight.........do....	803. 93±3. 94	106. 88 ±2. 77	11. 9±. 04
142	Total gain in live weight...do....	288. 55±3. 10	54. 77 ±2. 19	19. 1±1. 08
142	Increase in circumference of chest................inches	8. 13±. 10	1. 87 ± . 074	23. 0±1. 30
142	Width of rear flank.......do	19. 25±. 09	1. 69 ± . 06	8. 7±. 48
142	Circumference of rear flank, inches..................	72. 76±. 25	4. 40 ± . 17	6. 0±. 34
142	Height of shoulder point..inches..	34. 10±. 08	1. 57 ± . 06	4. 5±. 25
142	Height of rear flank........do....	30. 23±. 09	1. 67 ± . 06	5. 5±. 31
142	Distance, hip to buttock...do..	18. 69±. 05	. 99 ± . 03	5. 2±. 29
137	Width of loin. ... do..	13. 96±. 06	1. 05 ± . 22	7. 5±. 43
142	Circumference of chest... do .	69. 82±. 22	3. 91 ± . 01	6. 9±. 39
142	Depth of chest ......do....	25. 77±. 08	1. 59 ± . 063	6. 1±. 34
142	Width of fore flank ......do....	15. 02±. 08	1. 57 ± .	10. 4±. 60
142	Distance chest to ground . do .	22. 89±. 06	1. 18 ± . 86	5. 1±. 28
136	Width of thurls...........do....	17. 36±. 05	1. 03 ± . 84	5. 9±. 34
142	Length of head ......do....	19. 53±. 04	. 81 ± . 03	4. 1±. 23
142	Length of shoulder to buttock, inches..	55. 66±. 17	3. 02 ±2. 84	5. 4±. 30
142	Increase in circumference of paunch................inches..	8. 26±. 16	2. 96 ± . 11	35. 8±2. 02
142	Height at withers.........do....	49. 50±1. 07	1. 95 ± . 07	4. 9±. 27
142	Width at point of shoulder..do....	16. 42±. 08	1. 54 ± . 06	9. 3±. 52
142	Circumference of paunch...do....	81. 60±. 56	10. 04 ±4. 02	11. 0±. 62
138	Increase in circumference of rear flank.................inches..	7. 52±. 20	3. 51 ± . 14	46. 6±2. 67

TABLE VIII.—*Summary of correlations*

Correlating—	Correlation coefficient.	Correlating—	Correlation coefficient
Gain with initial live weight................	0.036±0.036	Gain with initial length of head..... .	0.182±0.037
Gain with increase in circumference of chest....	.460± .044	Gain with initial length from point of shoulder to buttock	−.020± .056
Gain with initial width of rear flank.	.079± 083	Gain with increase in circumference of paunch...	.306± .036
Gain with initial circumference of rear flank.....	.221± 053	Gain with initial height of withers..............	.163± .054
Gain with initial height of point of shoulder ..	.061± 056	Gain with initial width of shoulders..............	.144± .055
Gain with initial height of rear flank.............	053± .055	Gain with initial circumference of paunch	.124± .055
Gain with initial length of hip to buttock.	271± .053	Gain with increase in circumference of rear flank.	.203± .055
Gain with initial width of loin..................	.108± .085	Width of thurls with height of rear flank.....	.380± .049
Gain with initial circumference of chest........	238± .053	Circumference of chest with height of withers...	.621± .034
Gain with initial depth of chest .	.130± .046	Width of loin with height of chest from ground	.179± .057
Gain with initial width of fore flank.............	.164± .054	Width of chest with depth of chest............. ...	.365± .072
Gain with initial distance of chest to ground .	.077± .056		
Gain with initial width of thurls.....	.224± .054		

Table VIII, a summary of the coefficients of correlation, shows that increases in body measurements have a closer relationship with gains in live weight than the initial measurements. This would logically be expected, since gains are dependent upon the capacity of a steer to lay on fat and make growth in bone and body tissue. With the exception of initial weight, these measurements, as shown in Table VII, also had high coefficients of variation.

The correlation coefficients for the two body circumferences of chest and rear flank, the width of thurls (hip joint), and the distance of hip to buttock show the closest relationship of all the initial measurements with gains in live weight. This suggests the possibilities of using these measurements in the selection of feeding steers, at least for experimental purposes, as a means of reducing the experimental error caused by individuality of animals. The measurement of initial weight shows no relationship with gains, although this is usually considered one of the most important factors in selection of steers for experimental purposes.

Measurements that show intermediate relationship as indicated by coefficients of correlation are width of loin, depth of chest, width of fore flank, length of head, height at withers, circumference of paunch, and width at shoulders.

Measurements in which slight or no relationship exist as shown by correlation coefficients are initial live weight, width of rear flank, height at shoulder point, and the distance of rear flank and chest from the ground. One measurement, the distance of shoulder point to buttock, showed a negative correlation of -0.020 ± 0.056, but in which the probable error was greater than the correlation coefficient.

A close relationship of circumstance of chest with height of withers is indicated by $r = 0.621 \pm 0.034$. Likewise, the width of chest and depth of chest by $r = 0.365 \pm 0.072$ show a definite relationship.

The hind quarters of a steer are more important than the fore quarters in determining the gaining capacity of a steer, with the exception of the circumference of chest as shown by a correlation coefficient of 0.224 ± 0.054 for width of thurls, 0.271 ± 0.053 for distance of hip to buttock, and 0.221 ± 0.053 for circumference of rear flank.

The following points, held by authoritative judges of live stock to be important, are not substantiated by the results of this study thus far: Initial weight of steers, closeness to the ground of chest and hind flank, and the length of body from shoulder to buttock. The circumference of chest and rear flank are more important in ascertaining gains than feed capacity as indicated by the circumference of paunch.

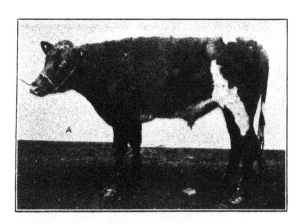

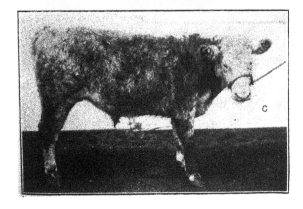

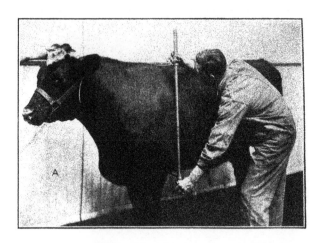

PLATE 39

A.—Measuring the depth of the chest of a steer with calipers.
B.—Measuring the circumference of the chest of a steer.

LIFE HISTORY OF ASCARIS LUMBRICOIDES AND RELATED FORMS

[PRELIMINARY NOTE]

By B. H. Ransom, *Chief*, and W. D. Foster, *Junior Zoologist, Zoological Division, Bureau of Animal Industry, United States Department of Agriculture*

Recently Capt. F. H. Stewart,[1] of the Indian Medical Service, in several publications has recorded the results of some experiments of great importance in the light which they throw upon the question of the life history of *Ascaris lumbricoides*, the common intestinal roundworm of man. The same parasite, or a form so closely related that it is morphologically indistinguishable so far as our present knowledge goes, is of very common occurrence in the intestine of pigs (*A. suum* or *A. suilla*). Stewart used both forms in his experiments. He failed in his attempts to infect pigs but found that if rats or mice were fed Ascaris eggs, the eggs hatched in the alimentary tract, and the embryos migrated to the liver, spleen, and lungs. In the course of their migrations they increased in size and passed through certain developmental changes, many of them finally reaching the alimentary tract again by way of the lungs, trachea, and esophagus. The young worms that succeeded in regaining the alimentary tract did not continue their development and soon passed out of the body in the feces, so that rats or mice surviving the pneumonia commonly caused by the invasion of the lungs became free of the parasites as early as the sixteenth day after infection.

As a result of his investigations Stewart was led to a conclusion contrary to the usually accepted opinion that the infection of man or pig with Ascaris results from the ingestion of the eggs of the parasite. He concluded that it is necessary in the life cycle for the eggs to be swallowed by rats or mice and that in these animals the embryos hatching from the eggs undergo certain migrations and developmental changes, after which they may be transferred in the feces or saliva of the rats or mice to food or other materials likely to be ingested by human beings or pigs, and thus ultimately reach their final hosts.

[1] STEWART, F. H. ON THE LIFE-HISTORY OF ASCARIS LUMBRICOIDES. *In* Brit. Med. Jour., 1916, v. 2, no. 2896, p. 5–7, 3 fig. 1916.

———. THE LIFE-HISTORY OF ASCARIS LUMBRICOIDES. *In* Brit Med. Jour , 1916, v. 2, no. 2909, p. 474. 1916.

———. FURTHER EXPERIMENTS ON ASCARIS INFECTION. *In* Brit. Med. Jour., 1916, v. 2, no. 2910, p. 486–488, 1916.

———. ON THE LIFE-HISTORY OF ASCARIS LUMBRICOIDES. *In* Brit. Med. Jour., 1916, v. 2, no. 2918, p. 753–754, 1916.

———. ON THE DEVELOPMENT OF ASCARIS LUMBRICOIDES LIN. AND ASCARIS SUILLA DUJ. IN THE RAT AND MOUSE. *In* Parasitology, v. 9, no. 2, p. 213–227, 9 fig., 1 pl. 1917.

Journal of Agricultural Research,
Washington, D. C.

ky

Vol. XI, No. 8
Nov. 19, 1917
Key No. A—32

In reviewing Stewart's work it appeared to us that, granting the correctness of his observations, the conclusions that he had drawn did not supply an adequate explanation of the mode of infection with Ascaris. On a number of occasions we had endeavored to infect pigs with Ascaris by feeding the and eggs, although the results of these attempts had been negative or uncertain and thus in harmony with the experience of Stewart and other investigators, we nevertheless did not feel justified in accepting these results as evidence against the hypothesis of a direct development without an intermediate host, nor did it seem that Stewart's experiments with rats and mice were sufficient to lead to the conclusion that these animals act.in any way as intermediate hosts in the life cycle of the parasite.

Repetition of Stewart's experiments in feeding rats and mice with Ascaris eggs gave results agreeing very closely with those which ·he has recorded. We have noted wider variations as to the time at which the larvæ may be found in various organs and have observed them in several locations in addition to those in which they were seen by Stewart, but the results of our experiments were essentially the same as his, and point to a migration of the larvæ to the liver, lungs, spleen, and other organs, and finally from the lungs to the alimentary tract by way of the air passages through the trachea and into the esophagus, during which migrations they undergo considerable development and structural change and increase to a size of 1.5 mm. or more in length.[1] In most of our experiments the eggs of the pig Ascaris were used, as Ascaris from pigs was more easily obtainable than parasites from human beings. In addition to our experiments on rats and mice we made further attempts, with negative results, to infect pigs.

Our unsuccessful attempts to infect pigs by feeding Ascaris eggs were made on animals several months old.[2] It is noteworthy that Epstein,[3] in his carefully controlled experiments with *A. lumbricoides* used very young subjects and that the positive results which he obtained can scarcely be explained upon any other assumption than that a direct development of the parasites occurred following feeding of the eggs—that is, development without an intermediate host. The experience of one of us (B. H. R.) in certain investigations[4] on the life history of *Syngamus trachealis*, our failures, and the failures of others to infect pigs with Ascaris, the fail-

[1] Since the preparation of this manuscript further investigations have shown that guinea pigs as well as rats and mice may be infected by feeding Ascaris eggs. In these animals the migration of the larvæ was, as far as observed, identical with that noted by Stewart for rats and mice. All of the six infected guinea pigs died from pneumonia between seven and eight days after feeding with Ascaris eggs, the lungs being heavily infested with Ascaris larvæ.

[2] The pigs which Stewart used with negative results were stated by him to have been 3 months old. The ages of the animals used by various other investigators who failed to obtain positive results in experiments to bring about infestations by feeding the eggs of Ascaris are generally not stated, but in many of the negative experiments with human beings it is clear that adults were used as subjects.

[3] EPSTEIN, Alois. UEBER DIE UEBERTRAGUNG DES MENSCHLICHEN SPULWURMS (ASCARIS LUMBRICOIDES). *In* Verhandl. Versamml. Gesell. Kinderh. Deut Naturf. u. Aerzte, v. 9, 1891, p. 1-16. 1892.

[4] Not yet published.

ures of various investigators to infect adult human beings, and Epstein's positive results in the case of young subjects, suggested the possibility that age is an important factor influencing the susceptibility of human beings and pigs to infection with Ascaris, and that many of the failures to bring about experimental infections would not have occurred if younger animals had been used as subjects. Our belief in this possibility was strengthened by the discovery of an Ascaris larva in a fragment of lung from a pig about 6 weeks old which had died from unknown causes in May, 1917, a finding which indicated a migration of larvæ like that which occurs in rats and mice. The intestine of this pig contained numerous immature ascarids, the largest about 5 cm. long. In order to test the possibility of infecting very young pigs, after several disappointments because certain sows reserved for the purpose of providing young pigs for experimental use either failed to farrow or devoured their newborn offspring, we finally succeeded in obtaining two young pigs from a sow which was found by fecal examination to be free from egg-producing ascarids. In the latter part of September, at the age of about 2 weeks, one of these pigs was given a large number of Ascaris eggs containing motile vermiform embryos. The number of eggs given was not determined, but there must have been at least several thousand. One week after feeding the pig which had been fed with Ascaris eggs was brought into the laboratory dead; death had occurred either the same day or late the day before; in any event, approximately one week had elapsed since the animal had been given a heavy dose of Ascaris eggs. The other pig continued in good health. Examination of the dead pig revealed a pneumonia, with numerous petechial hemorrhages in the lung tissue. Numerous ascarid larvæ, varying in length from 0.7 to 1.2 mm. in length, were found in the lungs, trachea, and pharynx; none in the liver, spleen, esophagus, small intestine, or large intestine.

It is of interest to note in this connection that when rats or mice are fed large numbers of Ascaris eggs they commonly die of pneumonia about a week later, at a time when numerous larvæ are present in the lungs, exactly as in the case of this pig. These findings are interpreted by us as clearly demonstrating that Ascaris larvæ in young pigs, presumably also in children, behave in much the same way as in rats or mice, and strongly support the hypothesis that the migrations and development of the parasites are very similar in the two cases, the only important difference being that in rats or mice the worms are unable to continue their development to maturity.

Stewart's very important discoveries concerning the behavior of Ascaris larvæ in rats and mice, the various contributions of other investigators toward the solution of the problem of the life history of *A. lumbricoides* and related parasites, and our own experiences outlined above, appear to justify certain conclusions, some of which in anticipation

of a more extended statement in a future paper, may be briefly given
as follows:

The development of *A. lumbricoides* and closely related forms is direct,
and no intermediate host is required.

The eggs, when swallowed, hatch out in the alimentary tract; the embryos, however, do not at once settle down in the intestine, but migrate
to various other organs, including the liver, spleen, and lungs.

Within a week, in the case of the pig Ascaris, the migrating larvæ may
be found in the lungs and have meanwhile undergone considerable
development and growth.

From the lungs the larvæ migrate up the trachea and into the esophagus
by way of the pharynx, and this migration up the trachea may already
become established in pigs, as well as in artificially infected rats and mice,
as early as a week after infection.

Upon reaching the alimentary tract a second time after their passage
through the lungs, the larvæ, if in a suitable host, presumably settle
down in the intestine and complete their development to maturity; if in
an unsuitable host, such as rats and mice, they soon pass out of the body
in the feces.

Heavy invasions of the lungs by the larvæ of Ascaris produce a serious
pneumonia which is frequently fatal in rats and mice and apparently
caused the death of a young pig one week after it had been fed with
numerous Ascaris eggs.

It is not improbable that ascarids are frequently responsible for lung
troubles in children, pigs, and other young animals. The fact that the
larvæ invade the lungs as well as other organs beyond the alimentary
tract and can cause a serious or even fatal pneumonia indicates that these
parasites are endowed with greater capacity for harm than has heretofore been supposed.

Age is a highly important factor in determining susceptibility to infection with Ascaris, and susceptibility to infection greatly decreases as the
host animal becomes older. This, of course, is in harmony with the
well-known fact that it is particularly children and young pigs among
which infestation with Ascaris is common, and that Ascaris is relatively
of rare occurrence in adult human beings and in old hogs.

JOURNAL OF AGRICULTURAL RESEARCH

Vol. XI   Washington, D. C., November 26, 1917   No. 9

EXPERIMENTS IN FIELD TECHNIC IN ROD ROW TESTS [1]

By H. K. Hayes, *Head of Section of Plant Breeding*, and A. C. Arny, *Head of Section of Farm Crops, Division of Agronomy and Farm Management, Department of Agriculturer University of Minnesota*

INTRODUCTION

Present field practices in Farm Crops breeding and variety testing show a lack of uniformity in methods of work. This is partly because some investigators have larger plots of land at their disposal than others. The principal reason for the lack of a standard practice, however, is the lack of sufficient experimental evidence as to the reliability of different methods.

The close relation between laboratory methods and reliability of experimental results is accepted without question by workers in such exact sciences as chemistry and physics. While it is not possible to develop as exact practices in field work, owing to the many uncontrolled environmental factors, there is a possibility of standardization.

In order to accomplish such an end, it is essential that workers situated in various sections of the country make a study of the reliability of their present field technic.

In a consideration of this question there are several methods of attack. Some of these are the use of check plots, the value of replications, the size and shape of plots, methods of correcting the error of any plot due to soil heterogeneity, and the effects of competition between adjacent rows or plots.

In recent years there has been a tendency to use the row method in small-grain breeding and variety testing. The present paper is the first of a series of studies in field practices in Minnesota. The experimental data are of two sorts: (1) A study of the effects of competition between small grains planted in rod rows when spaced at a distance of 1 foot apart. (2) The value of replications for rod row tests.

---

[1] Published, with the approval of the Director, as Paper No. 80, of the Journal Series of the Minnesota Agricultural Experiment Station.

Journal of Agricultural Research,   Vol. XI, No. 9
Washington, D C.   Nov. 26, 1917
kr   Key No. Minn. —22

(399)

PREVIOUS STUDIES IN FIELD METHODS

Before the development of the row method, a series of attempts were made to develop a standard practice for field plot variety tests. Regarding many of the points mentioned in the earlier papers, there is as yet insufficient experimental evidence.

Smith (*13*)[1] in discussing corn variety tests mentioned the disadvantages of growing a short variety beside a taller one. The probability that some varieties have stronger foraging powers than others is given as a further reason for discarding border rows. Plots five rows wide are recommended.

Cory (*2*) presents the plan of using row plantings to check the results of field plot experiments. The plan outlined was the use of two rows 12 inches apart and 50 links long. From result of these studies the author concluded that the row results were more reliable as an indication of yielding ability than the results obtained from plot tests.

The work of Marek is cited by Wheeler (*16*) as showing the possible errors due to greater growth on the edges of plots. The plan of removing a certain border part of each plot before obtaining yields is recommended

Piper and Stevenson (*10*) recommend certain practices for variety testing with the hope of standardizing methods. These recommendations are based on experimental studies of earlier investigators. Papers by Thorne (*15*) and Carleton (*1*) are especially mentioned as being valuable contributions to the idea of standardization. For corn variety work they adopted the method as outlined by Smith (*13*) with the addition of two to five replications for each variety and check plots every third or fourth plot. For small-grain testing the use of one-fortieth to one-tenth-acre plots in size with two to five replications and checks every third plot is recommended if one decides to use the plot method. For row tests a length of at least a rod with checks every fifth row and ten replications for each test is offered as a standard practice.

Mercer and Hall (*5*) reported a careful series of experimental tests in which they compared the relative reliability of results obtained from different-sized plots and gave results showing the reduction in error due to plot replication. The authors decided that one-fortieth acre plots were large enough for all practical purposes and recommended five replications for each variety under test. The conclusions of Lyon (*4*) in reference to plot size and replications are in substantial agreement with those of Mercer and Hall. Conclusive evidence is given to show that it is not possible to establish a schedule of relative yields for a series of plots even after several years' comparison.

Montgomery (*7*) presented an extensive study of the experimental error in nursery work with wheat for the characters nitrogen content

---

[1] Reference is made by number (italic) to "Literature cited," pp. 418-419.

and yield. He shows the value of a determination of the experimental error by the use of check plots. This error may then be used as á means of deciding which strains to eliminate. After determining the experimental error the rule given is

to double the error and subtract this sum from the highest variant. The remainder after the subtraction represents the figure below which all strains could be discarded without danger of discarding a high strain

for the character studied. This conclusion refers only to the seasonal conditions under which the test is made.

From a standpoint of a study of nitrogen content Montgomery (7) states that single plants should be replicated 40 or more times, 16-foot rows 5 to 10 times, and blocks 5.5 feet square 8 to 16 times. From a yield standpoint the author finds blocks 5.5 feet square repeated 8 to 10 times to be as accurate as 16-foot rows repeated 15 to 20 times. This he believes to be due to the competition between adjacent rows, although no accurate data on this question are presented.

In a previous paper Montgomery (6) shows the effects of competition on the yield of different varieties when grown in the same row and planted alternately in the row. Under such conditions some varieties seem to be much more vigorous than others.

Increasing the length of the row or size of block is said by the author (7) to be a means of securing accuracy. An excellent size for rows was found to be 2 to 4 rods in length and blocks 5 by 16 feet in area.

Montgomery believes the bulk method of planting in nursery trials to be more closely correlated with field results than when the nursery plots are planted in such a manner that each plant has individual room for development. Data given for 11 varieties of oats tend to substantiate this belief.

Von Rümker, Leidner, and Alexandrowitsch (12) followed the practice in trials of pure lines of wheat for yield of making the distance between rows of the same pure line 13 cm. (5.1187 inches) and between the plots of any two pure lines 24 cm. (9.448 inches) to obviate any influence one pure line might exert on another growing in an adjacent row. No data are presented of the reasons which led them to adopt the method.

A recent paper of Pritchard (11) gives a report of a study made in 1912 in sugar-beet breeding investigations. Two characters, total sugar production per row and percentage of sugar, were used in the study. These were the basis for a discussion of the value of check plots and the comparative effect of repeated plantings upon the experimental error. The author concludes that check rows as a standard for the comparison of individual progeny rows are more efficient than the use of the means of the progeny rows as a standard for comparison. Check rows 32 inches from the progeny rows varied less than checks at a distance of 64, 96, 128, and 160 inches. Check rows at a distance of 64 to 160 inches seemed to be about equally reliable as a comparison. In this work the employment of a check for every other row was said not to be sufficient to offset the variability due to soil irregularities; however, by means of 10 replications and alternate check rows the experimental error was reduced about 50 per cent. Reduction of the standard deviation was most rapid for the first 7 replications.

Numerous papers by Harris have called attention to the importance of the experimental error in variety testing. The more recent one (*3*) gives a method of determining the reliability of a field for experimental plot work. If irregularities in the field are large enough to affect areas larger than single plots, adjoining plots will be similarly affected. By the use of the correlation coefficient the correlation between the yield of the small plots and the yield of various groups of adjacent plots is determined. The more homogeneous the field the lower will be the correlation coefficient.

Surface and Pearl (*14*) have given a method of correcting plot yield which is partially based on Harris's correlation method. The calculated yield of any particular plot is determined by the contingency method. If this calculated yield is above or below the mean yield of the field, it is supposed that the soil of the plot is above or below the average of the field in productivity. The amount that the calculated yield is above or below the mean yield is then subtracted or added to the actual yield of the plot in question. The reliability of this method was then tested by the correlation plan as given by Harris. This test showed a considerable decrease in the heterogeneity of the field for the calculated yields as compared with the actual yields. The authors then show how this method may be used in variety tests.

MATERIAL AND METHODS OF WORK

In 1915 it was decided to make a test of the rod row as a preliminary method of determining performance. The method of handling the rows is about the same as is used at Cornell University in the Plant Breeding Department, although for the purposes of our work some modifications have been made.[1]

For our work 18-foot rows are used, spaced at a distance of 1 foot apart. At harvest 1 foot is discarded from each end of each row, thus leaving 16-foot rows. The use of the same length of row for wheat (*Triticum* spp.), oats (*Avena sativa*), and barley (*Hordeum vulgare*), increases the work of computing the yields per acre. The same length of row, however, for the various small grains allows a comparison of yield variability.

The rows were planted at the usual rate per acre for Minnesota, or 16.8 gm. per row for wheat, 15 gm. for oats, and 18.1 gm. for barley.

[1] Due acknowledgment is made to Dr. H. H. Love, of Cornell University, for descriptions of their row plans.

The work in 1915 was of a preliminary nature and does not necessitate an extensive discussion. It is only of interest as indicating the reason for the methods used in extensive tests made in 1916. In 1915 oat and wheat rows replicated nine times were grown for comparison with yields obtained in duplicated one-one hundred and tenth-acre plots. Correlation coefficients were calculated, giving the comparison between yields of 18 varieties of wheat and 21 varieties of oats. The coefficients are as follows:

$$\text{Wheat}\ldots\ldots r = +0.593 \pm 0.123$$
$$\text{Oats}\ldots\ldots r = +0.200 \pm 0.141$$

The same plot and series was used for these tests so that environmental conditions are relatively uniform. The fact that no closer relation in yield was obtained may be partially explained by the unreliability of the results as obtained in the duplicated one one-hundred-and-tenth-acre plot tests. (See following paragraph.) There is also the possibility of a greater competitive effect between adjacent rows of oats than for wheat.

A comparison of the relative reliability of single one one-hundred-and-tenth-acre plots and single rows of Bluestem wheat and Ligowa oats is given as a reason for further studies of the row method. The coefficient of variability is used as a measure of comparative reliability of the results of these tests (Table I).

TABLE I.—*Coefficients of variability for yield of wheat and oats as tested in one one-hundred-and-tenth-acre plots and rod rows, 1915*

Variety.	Type of plot	Coefficient of variability.
Haynes Bluestem wheat	Single rod row	13.73 ± 0.76
Do	Single one one-hundred-and-tenth-acre plot.	14.43 ± 2.16
Ligowa oats	Single rod row	18.87 ± 1.02
Do	Single one one-hundred-and-tenth-acre plot.	10.74 ± 1.29

These tests indicate that a single row is as reliable as a single one-one hundred and tenth-acre plot for Haynes Bluestem wheat, although for oats the plot test appears to be more reliable. The result of this test, together with experiments of other workers, has led to the tentative adoption of one-fortieth-acre plots at University Farm. Further studies will determine the number of replications necessary for the plot work. The rod row test appears very promising, although further studies seem necessary, particularly as to competition between adjacent rows and number of necessary replications.

The main part of the experimental results here reported consist of studies of competition between adjacent rows of different varieties and strains of wheat, oats, and barley and the decrease in variability due to replications.

The data obtained in 1916 is from two sources, namely, from rod row variety tests made by the Farm Crops Section and from rod row tests made in the Plant Breeding Nursery. The fields used for the work are at a considerable distance apart and have been handled in a different manner. In the presentation of results the tests will be spoken of as "Farm Crops test" and "Plant Breeding test," respectively.

In a considerable measure the uniformity of any particular field may be due to methods of handling. In the Farm Crops variety-test work the following rotations are used:

1. Oats, clover, corn.
2. Barley, hay, wheat, corn.

In the three-year rotation for the oats variety work there are two intervening crops. For the barley and wheat tests there is only one bulk crop intervening between the varietal tests.

In the Plant Breeding Nursery the following 7-year rotation is used·

Rye, 10 pounds of clover in spring, clover hay, small-grain nursery, ear-to-row corn, field peas for seed, soybean variety test, and small-grain nursery.

In the 1916 test, plots of three rows, each replicated three times, were used in the Plant Breeding Nursery if sufficient seed was available and replicated five times for each variety reported in the Farm Crops test. Check plots of a standard variety were used every third to fourth plot. No attempt was made to correct for yield by the use of check plots They were used, however, in a determination of the variability of the land and the value of replications.

It is realized that a single season's test does not furnish sufficient data upon which to draw a final conclusion as to a standard practice. The close relation between the development of reliable methods and the value of field tests, however, would seem to warrant the publication of results now available.[1]

COMPETITION BETWEEN ADJACENT ROWS

As each strain or variety tested was grown in at least one plot consisting of three rows of 16 feet in length, we may consider that each plot consists of one central row and two border rows which will be considered, respectively, as the right and left border rows. Each border row therefore grows near the same variety, the central row, on the one hand, and also beside some other variety or strain. In our work we have compared the yields of the border rows with each other, in relation to the height and yield of adjacent rows.

---

[1] Assistance in the computations was given by R W Garber and O A. Haenert, student assistants.

CORRELATIONS FOR HEIGHT COMPETITION

The effects of the height of adjacent rows on the yield of the border rows may be first considered. The method of determining such effect is as follows: The yield of the border rows in bushels per acre was computed and the yield of the right border compared with that of the left. If the left border yielded the most, the result is given as a positive, or plus (+) yield, the difference being expressed in bushels per acre. If the right border yielded the most, the yield difference is a negative, or minus (−) result.

In somewhat similar manner the height of adjacent rows was compared. If the row of the variety or strain near the left border row was taller than the variety near the right border row, the result is given as a positive, or plus (+) difference, in inches; if the row near the right border was taller, the result is given as a negative, or minus (−) difference, in inches.

It should be noted that with the plant breeding barley work the height was taken in centimeters.

Correlation tables are then made in which the comparative yield of the border rows of the same variety is correlated with the comparative height of adjacent rows. If there is an effect on yield of border rows of the same variety owing to the height of adjacent rows, a negative coefficient should be obtained. A representative table for barley is given (Table II).

TABLE II.—*Correlation between the differences in yield between the border rows of a variety of barley and the differences in height (in centimeters) between the neighboring border rows on the left and right, respectively. Barley Plant-Breeding Nursery, 1916*

[Left greater = +. Right greater = −]

Height, difference classes, adjacent border rows.	\	Yield, difference classes, border rows of same variety.													
		−9	−7.5	−6	−4.5	−3	−1.5	0	+1.5	+3	+4.5	+6	+7.5	+9	
	−25							1			1				2
	−20							1		1			1		3
	−15								1		1		1		3
	−10						1		1	2	1		1		6
	− 5					1	1	1	1	2	1			1	8
	0			1	1	2	1	3		3	2				13
	+ 5	2				2	3		1	1	1				10
	+10	1	1		1				1						4
	+15						1			1					2
	+20					1		1							2
	+25			1		1									2
	+30				1	1									2
	......	3	1	2	3	8	7	7	5	10	7	0	3	1	57

$r = -0.519 \pm 0.065.$

. . . .

In the results in Table III as given for barley (Plant
each test is an average of three replications as the va
strains are grown in the same order for each replicati
wheat results (Farm Crops Section) each test is an
results. With the exception of barley in the Plant
and winter wheat in Farm Crops work, each test rep
result.

In a consideration of these results certain enviro
are of interest. Until nearly the end of the crop sea
winter wheat in the Plant Breeding test stood up
The oats and spring wheat stood up well until after
nearly all oat plots lodged very badly, the spring w
siderably. A serious epidemic of black stemrust
tritici E. and H.), together with a very hot period sh
also caused serious reductions of yield in the Plant Bre
plots. In the Farm Crops work all cereals (with occ
stood up well.

The correlation coefficients for the Plant Breedin
siderable effect, owing to the height of adjacent rows
of the same variety in the barley tests and much sm
oats and winter-wheat tests.

In the Farm Crops test the correlation for the wint
±0.097. The odds against an occurrence of a devia
times the probable error is 1 to 82. There is only a
the effect of height in the barley tests, which is not
significant, and none for the oats or spring-wheat te

results are obtained in a similar way as for the effects of height. Table IV is given as an illustration of the method used. This result shows considerable effect due to yield of adjacent rows on the border rows of the same variety.

TABLE IV.—*Correlation between the differences in yield of the border rows of a variety and the differences in yield between the adjacent rows on the left and right, respectively. Barley Plant Breeding Nursery, 1916*

[Left greater = +. Right greater = −]

		Yield, difference classes, border rows of same variety.													
		−9	−7.5	−6	−4.5	−3	−1.5	0	+1.5	+3	+4.5	+6	+7.5	+9	
Yield, difference classes, adjacent border rows.	−21			1											1
	−18												1		1
	−15									1					1
	−12														0
	−9							3		1	3			1	8
	−6									2	2			1	5
	−3							1	1	4			1		7
	0				2	2	1	2	2						9
	+3	1	1			3	1		1	1	3				11
	+6	1				1	1		1	2					6
	+9	1		1		1	1								4
	+12						1		1		1				3
	+15														0
	+18				1										1
		3	1	2	3	7	5	6	6	11	9	0	2	2	57

$r = -0.394 \pm 0.075$

The correlation coefficients as obtained for the various crops are given in Table V.

TABLE V.—*Correlations between the differences in yield between the border rows of a variety and the differences in yield between adjacent rows of unlike varieties of cereals*

Cereal.	Designation of test.	Number of tests.	Correlation coefficients.
Barley	Plant Breeding Nursery	57	−0.394±0.075
Do.	Farm Crops variety test	48	−.040±.007
Oats	Plant Breeding Nursery	218	+.100±.045
Do.	Farm Crops variety test	121	+.017±.061
Spring wheat	Plant Breeding Nursery	193	+.007±.049
Do.	Farm Crops variety test	50	−.290±.087
Winter wheat	Plant Breeding Nursery	183	+.082±.050
Do.	Farm Crops variety test	38	+.315±.099

These results show less effect of competition due to the yield of adjacent rows than in the case of the height of adjacent rows. The coefficient obtained in the Plant Breeding results for the barley test was

−0.394±0.075. This is nearly six times the probable error, and thus it is statistically significant. The possibility that height and yield of adjacent rows was one and the same thing is dependent upon the question as to the relation of height and yield for the various crops. The correlation between height and yield for the 57 varieties of barley is +0.214± 0.085.

The coefficient as obtained for spring wheat, Farm Crops test, shows some competition between adjacent rows due to yield.

Perhaps the most striking correlation coefficient is that for winter wheat, which is +0.315±0.099. As has already been mentioned, minus (−) coefficients show competitive effect. Positive (+) coefficients might be explained as due to soil heterogeneity.

ACTUAL EFFECT OF HEIGHT OF ADJACENT ROWS ON BORDER ROWS OF
SAME VARIETY

The correlation coefficients as calculated show whether there is any competition between rod rows of small grains when the rows are spaced at a distance of 1 foot apart. The results as computed show that when varieties and strains of small grains are grown in rod rows there is an effect due to the height of adjacent rows in the case of barley, oats, and winter wheat in the Plant Breeding test and no effect for the spring-wheat test. The correlation coefficients for the oat and winter-wheat results show only a small effect, the coefficients being a little less than three and two times the probable errors, respectively.

In the Plant Breeding barley test the effect is a large one, and it seems worth while to demonstrate the actual effect of certain height differences. In considering these results certain facts will be repeated. The height of each plot was measured in centimeters, and there was no appreciable difference in height between the border and central rows in each plot. If there is any effect on border rows of the same variety due to differences in the height of adjacent rows one might expect, on the average, that the border row which grew near a short variety might yield more than a row of the same sort which happened to be near a tall variety. If this is so, the result is considered as a positive, or plus (+) difference. The average increase, in bushels per acre, is also given. Negative results are preceded by a minus (−) sign. The results are given in 3-cm. classes for the difference in height of the neighboring plot (Table VI).

The results in Table VI show that the actual effect on rows of the same variety due to the height of adjacent rows is of the utmost importance in rod row tests under conditions such as obtained for the barley in the Plant Breeding tests.

TABLE VI.—*Results on yield of border rows of the same variety of barley due to differences in height of adjacent rows*

Average differences in height of adjacent rows.	Number of tests.	Number positive.	Number negative.	Average differences in yield of border rows (in bushels per acre).
Cm.				
30.	2	2	0	4.1
27.	2	2	0	4.8
24.	2	2	0	1.7
21.	3	2	1	3.3
18.	2	2	0	1.5
15.	4	3	1	1.9
12.	7	7	0	4.9
9.	8	6	2	3.5
6.	10	8	2	2.1
3.	16	11	5	− .1
0.	9	7	2	+ .7

Similar results as presented for oats (Plant Breeding test) show a substantial agreement between the correlation coefficients calculated and results as obtained for the actual test of effects of certain specific height differences. These results are given in Table VII.

TABLE VII.—*Results on yield of border rows of the same variety of oats due to differences in the height of adjacent rows*

Average differences in height of adjacent rows	Number of tests.	Number positive.	Number negative.	Average differences in yield of border rows (in bushels per acre).
Inches.				
12.	2	2	0	9.8
10.	2	1	1	1.2
9.	2	1	1	−4.8
8.	2	0	2	−3.5
7.	4	2	2	3.3
6.	10	6	4	2.0
5.	13	9	4	1.7
4.	29	16	13	2.6
3.	34	21	13	3.4
2.	57	26	31	− .7
1.	67	31	36	−1.8
0.	53	23	30	− .3

The total minus (−) results for 1 to 12 inches difference in height are obtained by multiplying the average results by the number of tests and obtaining $\Sigma = 177.1$. The total plus (+) results are obtained in similar manner. $\Sigma = 268.3$.

lant Breeding Nursery.	51	11. 7 ±0. 190	2. 01±0. 134	17. 25±
...do.....................	51	11. 1 ± . 199	2. 11± 141	19. 08±
..do....................	51	12. 0 ± . 221	2. 34± . 156	19. 44±
...do..................	51	11. 6 ± . 157	1. 66± . 111	14. 30±
arm Crops variety test.	15	20. 6 ± . 261	1. 50± . 185	7. 28±
..do....	15	21. 1 ± . 317	1. 82± . 224	8. 63±
..do..................	15	20. 1 ± . 277	1. 59± . 196	7. 91±
..do..................	15	20. 7 ± . 240	1. 38± . 170	6. 67±
lant Breeding Nursery.	59	58. 3 ± . 598	6. 81± . 423	11. 68±
..do..................	59	55. 5 ± . 761	8. 67± . 538	15. 61±
..do..................	59	57. 7 ± . 731	8. 33± . 517	14. 43±
..do..................	59	57. 2 ± . 587	6. 68± . 415	11. 69±
arm Crops variety test.	20	61. 2 ± . 849	5. 63± . 600	9. 20±
..do..................	20	61. 8 ± . 908	6. 02± . 642	9. 74±
..do..................	20	61. 9 ±1. 106	7. 33± . 782	11. 84±
..do..................	20	61. 6 ± . 885	5. 87± . 626	9. 53±
lant Breeding Nursery.	9	36. 2 ± . 949	4. 22± . 672	11. 66±
..do..................	9	36. 6 ±1. 857	8. 26±1. 313	22. 57±
..do..................	9	38. 0 ± . 710	3. 16± . 902	8. 32±
..do..................	9	37. 0 ± . 924	4. 11± . 653	11. 11±
arm Crops variety test.	15	32. 0 ± . 381	1. 90± . 234	5. 94±
..do..................	15	31. 0 ± . 441	2. 53± . 312	8. 16±
..do..................	15	30. 5 ± . 509	2. 92± . 360	9. 57±
..do..................	15	31. 3 ± . 298	1. 71± . 211	5. 46±
lant Breeding Nursery.	72	27. 5 ± . 370	4. 65± . 261	16. 92±
..do..................	72	27. 4 ± . 419	5. 27± . 296	19. 21±
..do..................	72	26. 0 ± . 406	5. 11± . 287	39. 65±
..do..................	72	27. 0 ± . 327	4. 12± . 232	15. 26±
arm Crops variety test.	41	29. 1 ± . 504	4. 78± . 356	16. 43±
..do..................	41	28. 9 ± . 529	5. 02± . 374	17. 37±
..do..................	41	27. 6 ± . 473	4. 49± . 334	16. 27±
..do..................	41	28. 4 ± . 403	3. 83± . 285	13. 49±

en mentioned, the Farm Crops test is on a differ
ifferent system of rotation than the Plant Breed
, the check plots for the Farm Crops work occur
1 plot, while the Plant Breeding checks were pla
hese differences may partially explain the lack
efficients of variability as determined for the te:
allows for two main comparisons, both of which

It is entirely possible in a limited number of tests that if there is a competitive effect between adjacent rows that in some cases this might tend to lower the variability of the test. For example, if a plot happened to be situated in a particularly favorable soil location, the rows would be expected to yield more than the average for that particular variety. If, in addition, the border rows of this plot grew near a much taller variety or beside a stronger foraging variety, the effect of competition would be to lower the yield and thus reduce the calculated coefficient of variability for the variety in question. For this reason the coefficient of variability for the yield of the central rows is compared with the most variable border row test. Results of such comparisons are given in Table IX. The probable error of the difference between coefficients of variability being calculated by the following formula, the error of a difference is equal to the square root of the sum of the squares of the quantities entering into the difference (8).

TABLE IX.—*Comparison of the difference between the coefficient of variability of the border and central rows*

Cereal.	Designation of test.	Difference between the coefficient of variability of the border and central rows.	Difference ÷ probable error.
Wheat (Minnesota 169) . . .	Plant Breeding Nursery. . . .	2.09±1.788	1.2
Do.	Farm Crops variety test . . .	1.35±1.390	.9
Oats (Minnesota 281).	Plant Breeding Nursery . . .	3.93±1.230	3.2
Do.	Farm Crops variety test	2.64±1.613	1.6
Barley (Minnesota 105).	Plant Breeding Nursery. . . .	10.91±4.208	2.6
Do.	Farm Crops variety test ..	3.63±1.386	2.6
Winter wheat (Minnesota 529)	Plant Breeding Nursery.	2.74±1.509	1.9
Do.	Farm Crops variety test	.94±1.838	.5

The results show no significant difference for the spring- and winter-wheat tests between the coefficients of variability of border and central rows. The chances against an occurrence of a deviation as large as 1.9 times the probable error as obtained in the Plant Breeding winter-wheat test is 4.00 to 1. For barley the odds against an occurrence of a deviation as large as 2.6 times the probable error, as obtained for both Farm Crops and Plant Breeding tests, is 11.58 to 1 (9). The odds against an occurrence as great as 1.6 and 3.2 times the probable error as obtained for the Farm Crops and Plant Breeding tests for oats are, respectively, 2.57 to 1 and 31.36 to 1. It seems fair to conclude that border rows are more variable than central rows for barley and oat tests and that the differences in variability of border and central rows of spring- and winter-wheat tests here reported are of no great importance.

The comparison between the coefficients of variability for the 3-row plots and central rows is a means of determining the value of plot size. The results would seem to be more significant in the case of spring and

e results as reported for the comparison of central ro
(Table X) are in substantial agreement with the
parison of central and border rows (Table IX).   For
y the central rows are no more variable for yield
.   For these crops the border rows had a significantl
ent of variability for yield than was obtained for the
the winter- and spring-wheat tests there was lit
mpetitive effect of adjacent rows, with the exception
t Farm Crops test, as determined by the coefficient
e border and central rows.   Table X shows that wi
g wheat the use of all three rows for each check plot
was somewhat more accurate than the use of centra
ence between the coefficient of variability for yield f
and 3-row plots is 1.9, 0.5, 1.3, and 1.7 times the p
probability of the occurrence of deviations as large
the probable error are, respectively, 2.98 to 1 and 4.
t reduction in error in winter and spring wheat due
ree rows of the 3-row plots does not appear to be of
arrant the work involved in harvesting, threshing, an
order-row yields.

VALUE OF REPLICATIONS

e two methods of overcoming soil variations of ai
n's test are the determination of the sort and size o
ystematic plot replications.   Plots consisting of 3-rc
used for our test.   Each row was then harvested separ

for the central rows and for the 3-row plots, respectively. For these tests the standard varieties used in the check plots in the spring-wheat, oats, barley, and winter-wheat tests have been used.

The probable errors are often very large, owing to the small number of plots or rows in these tests in which many replications are studied. As the tests for the Farm Crops and Plant Breeding work are made in widely separated fields, except in the winter-wheat tests, it would seem that the conclusions for the particular crop year in question would be fairly sound.

The method of replication used for the work is as follows: If there were 50 check plots in a particular test and 2 replications were to be studied, an average of plot 1 and plot 26, plot 2 and plot 27, etc., was taken. Similarly for 3 replications and 51 plots the average of plot 1, plot 18, and plot 35, etc. was taken.

The value of replications in Haynes Bluestem wheat is given in Table XI.

TABLE XI.—*Value of replication of central rows and 3-row plots of Haynes Bluestem wheat (Minnesota 169) used as a check in the Farm Crops and Plant-Breeding tests*

Designation of test.	Method of test.	Number of tests.	Mean.	Standard deviation.	Coefficient of variability.
Farm Crops variety test...	Single central rows......	15	20.6±0.261	1.50±0.185	7.28±0.896
Do..................	2 replications............	7	20.6±.191	.75±.135	3.64±.656
Do..................	3 replications............	5	20.6±.323	1.07±.228	5.19±1.107
Do..................	4 replications............	3	20.7±.370	.95±.262	4.59±1.264
Do..................	6 replications............	2	20.8±.167	.35±.118	1.68±.566
Do..................	3-row plots........	15	20.7±.240	2.38±.170	6.67±.821
Do..................	2 replications............	7	20.7±.250	.98±.177	4.73±.853
Do..................	3 replications............	5	20.7±.314	1.04±.222	5.02±1.071
Do..................	4 replications............	3	20.7±.136	.35±.096	1.69±.465
Do..................	6 replications............	2	20.7±.119	.25±.084	1.21±.408
Plant Breeding Nursery..	Single central rows......	51	11.7±.100	2.01±.134	17.25±1.187
Do..................	2 replications............	25	11.6±.174	1.29±.123	11.08±1.067
Do..................	3 replications............	17	11.6±.239	1.46±.169	12.60±1.487
Do..................	4 replications............	12	11.8±.193	.99±.136	8.37±1.164
Do..................	6 replications............	8	11.6±.165	.69±.116	5.94±1.002
Do..................	9 replications............	5	11.4±.241	.80±.171	7.02±1.497
Do..................	3-row plots........	51	11.6±.157	1.66±.111	14.30±.974
Do..................	2 replications............	25	11.6±.158	1.17±.112	10.12±.975
Do..................	3 replications............	17	11.5±.128	.78±.090	6.76±.782
Do..................	4 replications............	12	11.8±.162	.83±.114	7.06±.972
Do..................	6 replications............	8	11.8±.157	.66±.111	5.62±.948
Do..................	9 replications............	5	11.6±.148	.49±.104	4.22±.900

For the Farm Crops test of Minnesota 169 wheat there were only 15 plots available. For this reason the Farm Crops results for the value of replications are less reliable than the Plant Breeding tests, where 51 plots were grown. In the Farm Crops work for the central-row replication studies the use of any number of replications from two to four seems about equally accurate. Six replications, as compared with two, gave a difference in the standard deviation of 0.40±0.179. This is slightly more than 2 times the probable error.

For the test of replications in the 3-row plots two or three replications appear to be about equally reliable. Four replications are much more accurate than three, the difference in the standard deviation of four

and three replications being 0.69±0.242.   This is 2.8
error.   The chances of a deviation as large as 2.8
error are 1 to 15.95.   Six replications appear to be
than four replications.

In comparing the test as made by the use of cent
plots it would seem that for one, two, or three repli
rows were about as accurate as the 3-row plots.
3-row plots are apparently as reliable as six repli
rows.

For the Plant Breeding test two or three replicatio
appear to be about equally reliable.   Four replicatio
as valuable as either six or nine.   The difference in
tion as given for three or four replications is 0.47
possible value.

For the test of replications in the 3-row plots there i
progressive reduction in variability as the numb
increases from one to nine.   The largest reductions a
and from two to three replications.   The progress
three and nine replications are only very small.

In like manner the value of replications has bee
oat check plots.   In a consideration of the results fo
3-row plots it is of interest to again call attention t
of competition in the Plant Breeding test between ad
than for the wheat tests.

Data for a comparison of the value of replications i
presented in Table XII.

Twenty plots were used for the Farm Crops test and 59 for the Plant Breeding work. There appears to be very little difference in variability as determined by the Farm Crops or Plant Breeding results, the Plant Breeding coefficients of variability for the respective replications being only slightly larger than those which were obtained for the Farm Crops test.

In the Farm Crops test there is a large reduction in the standard deviation for two replications as compared with a single replication. A slight reduction in variability for three as compared with two replications is apparent. Three or four replications appear to be about equally reliable. There is a large reduction in the standard deviation due to six replications as compared with three. For the central row tests the difference in the standard deviation for yield of six replications as compared with three is 1.80±0.884. This is slightly more than two times the probable error.

The comparison of 3-row plots and central rows shows little difference for any number of replications except nine, and such differences as are obtained for this number of replications are in favor of the use of central rows as compared with 3-row plots.

For the Plant Breeding test there is little difference between the use of 3-row plots and central rows for any number of replications below nine. There is an apparent advantage in favor of the use of 3-row plots for nine replications. As this result is exactly the reverse of the Farm Crops result for nine tests, it would indicate that these results are chance fluctuations.

For the central rows three replications appear to be as reliable as any number below nine. There is some advantage in the 3-row plot tests for six as compared with three replications. This is, however, of doubtful significance, the difference in the standard deviations being 0.99±0.541. Nine replications appear to be as reliable as nine replications for the Farm Crops work. It is of interest to note that three replications reduce the error about 50 per cent as compared with a single test.

The results for replications in barley are given in Table XIII.

TABLE XIII.—*Value of replication of central rows and 3-row plots of Manchuria barley (Minnesota 105) used as a check in the Farm Crops and Plant Breeding tests*

Designation of test.	Method of test.	Number of tests.	Mean.	Standard deviation.	Coefficient of variability.
Farm Crops variety test..	Single central rows......	15	32.0±0.331	1.90±0.234	5.94±0.731
Do...................	2 replications...........	7	32.0±.428	1.68±.303	5.25±.946
Do...................	3 replications...........	5	32.0±.317	1.05±.224	3.28±.670
Do...................	4 replications...........	3	32.0±.386	.99±.273	3.09±.851
Do...................	6 replications...........	2	32.0±.191	.40±.135	1.25±.422
Do...................	3-row plots.............	15	31.3±.098	1.71±.211	5.46±.672
Do...................	2 replications...........	7	31.2±.247	1.36±.245	4.36±.766
Do...................	3 replications...........	5	31.5±.284	.94±.200	3.01±.642
Do...................	4 replications...........	3	31.3±.167	.43±.118	1.37±.377
Do...................	6 replications...........	2	31.3±.095	.20±.067	.64±.216
Plant Breeding Nursery..	Single central rows......	9	36.2±.049	4.22±.672	11.66±1.872
Do...................	2 replications...........	4	36.6±.907	2.69±.642	7.35±1.753
Do...................	3 replications...........	3	36.2±1.050	2.70±.743	7.46±2.054
Do...................	3-row plots.............	9	37.0±.924	4.12±.653	11.11±1.764
Do...................	2 replications...........	4	37.6±.853	2.53±.603	6.73±1.605
Do...................	3 replications...........	3	36.9±1.102	2.83±.779	7.67±2.112

f test.	Method of test.	Number of tests.	Mean.	Standard deviation.	Co
ety test..	Single central rows......	41	29.1±0.504	4.78±0.356	1
....	2 replications...........	20	29.6± .593	3.93± .419	1
....	3 replications...........	13	29.1± .406	2.17± .287	
....	4 replications...........	10	29.4± .469	2.20± .332	
....	6 replications...........	6	28.5± .666	2.42± .471	
....	9 replications...........	4	28.8± .762	2.36± .539	
....	3-row plots.............	41	28.4± .403	3.83± .285	1
....	2 replications...........	20	28.4± .338	2.24± .239	
....	3 replications...........	13	28.5± .393	2.10± .278	
....	4 replications...........	10	28.4± .397	1.86± .281	
....	6 replications...........	6	28.1± .774	2.18± .547	
....	9 replications...........	4	28.0± .661	1.96± .467	
ursery..	Single central rows......	72	27.5± .370	4.65± .261	1
....	2 replications...........	36	27.6± .335	2.98± .237	1
:........	3 replications...........	24	27.4± .346	2.51± .244	
....	4 replications...........	18	27.6± .238	1.49± .169	
....	6 replications...........	12	27.3± .391	2.01± .277	
....	12 replications..........	6	27.3± .333	1.21± .236	
....	3-row plots.............	72	27.0± .327	4.12± .232	2
....	2 replications...........	36	27.1± .310	2.76± .219	1
....	3 replications...........	24	27.2± .304	2.21± .215	
....	4 replications...........	18	27.1± .234	1.47± .165	
....	6 replications...........	12	27.1± .323	1.66± .229	
....	12 replications..........	6	27.2± .333	1.21± .236	

very little difference for the Farm Crop tests betwe
as determined by central rows or 3-row plots with the
or two replications, which are lower in the 3-row pl
tral row tests. Three replications appear to be a
ny larger number.

Breeding test shows no significant difference in the
ermined by the use of 3-row plots or central rows
eplications studied. Four replications for both centra

and 3-row plots seem significantly better than three, although there is some doubt about the accuracy of this conclusion when one notes that four replications give a lower variability than six. Twelve replications as compared with three show a difference of 3.71 ± 1.041.

SUMMARY OF RESULTS

(1) In a study of competition between rod rows of small grains grown 1 foot apart there was some effect on the yield of border rows of the same variety due to height of adjacent rows of barley, winter wheat, and in one of two tests an indication of such effect in oats. The results were variable in different plots, such variation being due possibly to the environmental conditions. There was no apparent effect of height of adjacent rows on the yield of border rows of the same variety in spring wheat.

(2) The yield of adjacent rows appeared to be of some importance in barley tests and in the Farm Crops spring-wheat tests. Correlations obtained for other tests indicated considerable soil heterogeneity.

(3) The effects of the height of adjacent rows for the barley Plant-Breeding tests was unquestionable. The effects of the height of adjacent rows were sufficient to often cause differences of 4 or 5 bushels per acre in the yield of border rows of the same variety of barley.

(4) The comparison of yield variability of border and central rows of check plots of barley, oats, spring and winter wheat was further evidence of the competitive effect of rod rows of small grains when grown 1 foot apart. In nearly all tests the border rows proved to be more variable in yield than the central rows

(5) The check plots of oats, spring and winter wheat, and barley were used for a study of the value of replications in reducing error. In nearly all tests three replications as compared with a single plot reduced error by 25 to 50 per cent.

(a) In the spring-wheat Farm Crops tests for central rows four replications gave about the same coefficient of variability as nine replications for the Plant Breeding test. In the Farm Crops series six replications reduced the error by over 50 per cent as compared with three replications. The lack of agreement between the Plant Breeding and Farm Crops test may be explained by the serious black-stemrust epidemic which attacked the Plant Breeding Nursery.

(b) For the Farm Crops oat test six replications reduced the error by about 25 per cent as compared with three replications. For the Plant Breeding tests three replications were about as valuable as any higher number below nine.

(c) For the barley tests three replications seemed to be about as accurate as any number below six.

(d) For the winter-wheat results three or four replications reduced the error about 50 per cent and appeared to be about as accurate as any number below 12.

CONCLUSIONS

The studies here reported show the necessity of t field methods of technic in Plant Breeding and Farm

Conclusive evidence is given which shows that in considerable competition between rod rows of small a distance of 1 foot apart. This has led to the ado plots at University Farm. The results of a compa plot yields with central rows show that central row determination of yield as the use of all three rows.

In a study of replications for rod rows of small variability was shown for the different tests. In ge replications seem to be about as accurate a method number below nine. The indications are that 9 to 12 reduce error due to soil heterogeneity to a minimum

LITERATURE CITED

(1) CARLETON, M. A.

　　1909. LIMITATIONS IN FIELD EXPERIMENTS. *In* Proc. 3
　　　　Prom. Agr. Sci., 1909, p. 55–61.

(2) CORV, V. L.

　　1910. THE USE OF ROW PLANTINGS TO CHECK FIELD PL
　　　　Soc. Agron., v. 1, 1907/09, p. 68–70.

(3) HARRIS, J. A.

　　1915. ON A CRITERION OF SUBSTRATUM HOMOGENEITY
　　　　IN FIELD EXPERIMENTS. *In* Amer. Nat., v. 49,

(4) LYON, T. L.

　　1912. SOME EXPERIMENTS TO ESTIMATE ERRORS IN F
　　　　Proc. Amer. Soc. Agron., v. 3, 1911, p. 89–114,

(5) MERCER, W. B., and HALL, A. D.

　　1911. THE EXPERIMENTAL ERROR OF FIELD TRIALS. *In*

(12) RÜMKER, K. H. T. von, LEIDNER, R , and ALEXANDROWITSCH, J·
1914. DIE ANWENDUNG EINER NEUEN METHODE ZUR SORTEN- UND LINIEN-PRÜFUNG BEI GETREIDE. *In* Ztschr Pflanzenzücht , Bd. 2, Heft 2, p. 189–232, fig. 17–21

(13) SMITH, Louie H.
1910. PLOT ARRANGEMENT FOR VARIETY EXPERIMENTS WITH CORN *In* Proc Amer. Soc. Agron , v. 1, 1907/09, p 84–89.

(14) SURFACE, F. M., and PEARL, Raymond
1916. A METHOD OF CORRECTING FOR SOIL HETEROGENEITY IN VARIETY TESTS. Jour. Agr. Research, v 5, no 22, p 1039–1050 Literature cited, p. 1050.

(15) THORNE, C. E.
1905. SYLLABUS OF ILLUSTRATED LECTURE ON ESSENTIALS OF SUCCESSFUL FIELD EXPERIMENTATION U S. Dept. Agr Office Exp Stas Farmers Inst. Lecture 6, 24 p.

(16) WHEELER, H. J·
1910. SOME DESIRABLE PRECAUTIONS IN PLAT EXPERIMENTATION *In* Proc. Amer. Soc. Agron , v. 1, 1907/09, p. 39–44. Cites Marek (p 40).

HYDROCYANIC-ACID GAS AS A SOIL FUMIGANT

By E. Ralph de Ong,

Instructor in Entomology, Agricultural Experiment Station of the University of California

INTRODUCTION

Though much has been written on soil fumigation, it can not be said that we have a satisfactory method for treatment against soil insects. Most of such work has been done in France with carbon disulphid against phylloxera, but no reliable data on soil work exist regarding the availability of hydrocyanic-acid gas, the most efficient fumigant known in orchard and house work; hence, the following study was made of the action of hydrocyanic-acid gas in the soil and its effect on germinating seeds and plants.

Any extensive treatment of the soil must be considered in relation to crops grown thereon, especially those occupying the ground for a period of one year or more. Operations possible while the ground is occupied by a crop are much to be preferred, although knowledge of the treatment of unplanted areas is very desirable. The experiments to be described in this paper were made largely with the idea of the treatment of occupied ground, for it follows as a matter of course that treatment which is safe for planted areas is also safe for bare land.

Two distinct lines of work are necessary for the study of this subject:

(a) Establishing a definite ratio between the minimum point of toxicity to insects and the maximum dosage which is safe to germinating seeds and plants.

(b) A study of the physical and chemical action of the gas in the soil; the rate and extent of diffusion and absorption by soil water; the adsorption of the gas by soil particles and its decomposition by certain soil constituents.

A. RELATION OF POINT OF TOXICITY OF INSECTS AND PLANTS

Many data have been accumulated by the various Experiment Stations, and particularly by the California Station, regarding the toxicity of hydrocyanic-acid gas to insects. Likewise there are many data (*1-8*)[1] available on the effects of the gas on plants, but practically all of them are based on house experiments where the actual concentration of the gas is not accurately known. Therefore it was necessary to make a few tests to establish a basis of comparison.

A method elaborated by Prof. C. W. Woodworth, of the California Station, in the investigation of the action of hydrocyanic-acid gas on the eggs of certain insects, makes use of the vapor-tension laws of gases.

---

[1] Reference is made by number (italic) to "Literature cited," p. 436.

Journal of Agricultural Research,
Washington, D. C.
kn

Vol. XI, No. 9
Nov. 26, 1917
Key No. Cal.—12

It was found that the concentration of hydrocyanic-acid vapor in the air over a simple aqueous solution of the gas quickly becomes constant if a relatively large volume of the solution were held in a small closed container. Furthermore, if the temperature is constant, it was shown that the concentration of the hydrocyanic-acid vapor in air under the above conditions was dependent on the concentration of the solution.

A solution of hydrocyanic-acid gas was made by dissolving 1 gm. of sodium cyanid in 1 liter of water and adding 15 c. c. of 10 per cent sulphuric acid. (This is actually an aqueous solution of hydrocyanic-acid gas in a sodium-sulphate solution, probably containing a slight excess of sulphuric acid.) The air in a closed container above a relatively large volume of such a solution contained approximately 0.3 per cent of hydrocyanic-acid vapor,[1] which is within the range of strength used for fumigation work.

If air were slowly aspirated through this solution, it also would contain approximately 0.3 per cent hydrocyanic-acid vapor. Other solutions were made up similar to the above, but of different concentrations, the amount of sodium cyanid being reduced one-half in each succeeding culture until the minimum point of toxicity was reached.

Houseflies (*Musca domestica*) were exposed to the following range of dilutions in closed test tubes (Table I):

TABLE I.—*Concentration of gas used on houseflies*

Sodium cyanid per liter.	Approximate concentration of gas.	Sodium cyanid per liter.	Approximate concentration of gas.
Gm.	*Per cent.*	*Gm.*	*Per cent.*
1. 0	0. 3000	0. 0625	0. 0187
. 5	. 1500	. 0312	. 0093
. 25	. 0750	. 0156	. 0046
. 125	. 0375		

[1] The method of calculating the amount of hydrocyanic-acid gas in air:

A standard sodium-cyanid solution is made up as above, containing 1 gm. of sodium cyanid per liter and 15 c. c. of 10 per cent sulphuric acid. Air is aspirated first through this solution, then through N/100 iodin solution (containing a little sodium carbonate) until the latter is completely decolorized, the rate being 36.5 c. c. of air (the mean of a number of experiments) = 1 c. c. of N/100 iodin.

One equivalent of iodin is equal to two equivalents of HCN, according to the sense of the following equation:

$$HCN + I_2 = HI + ICN; \text{ therefore}$$
$$1 \text{ c. c. } N/100 \text{ I} = 0.0001351 \text{ gm. HCN}$$
$$1 \text{ c. c. } N/100 \text{ I} = 36.5 \text{ c. c. air}$$
$$36.5 \text{ c. c. air} = 0.0001351 \text{ gm. HCN}$$
$$1 \text{ c. c. air} = \frac{0.0001351}{36.5} \text{ gm. HCN}$$
$$1,000 \text{ c. c. air} = \frac{0.0001351}{36.5} \times 1,000 = \frac{0.1351}{36.5} = 0.0037 \text{ gm. HCN in 1 liter of air}$$
$$1 \text{ liter air} = 0.0037 \text{ gm. HCN}$$
$$1 \text{ liter air} = \text{weighs } 1.09 \text{ gm. at } 0° \text{ C. and } 760 \text{ mm.}$$

Barometric pressure 755.6
Thermometer 20° C.

$$1,000 \text{ c. c.} \times \frac{755.6}{760} \times \frac{273}{293} = 927.2 \text{ c. c. air}$$

927.2 c. c. weighs 1.2099 gm.

$$\frac{0.0037}{1.2099} \times 100 = 0.306 \text{ per cent HCN in air.}$$

In the concentration of 0.0156 gm. of sodium cyanid per liter (equivalent to 0.0046 per cent of hydrocyanic-acid gas over the solution) many of the flies were dead at the end of 1½ hours; the others, being apparently normal, were freed. A second series of experiments with hydrocyanic-acid gas was made on the following insects: Phylloxera (*Phylloxera vastatrix*), tree hoppers (*Stictocephala festina*), leaf hopper (*Agallia* sp.), houseflies (*Musca domestica*), beetles (*Diabrotica soror*), with the results given in Table II.

TABLE II —*Effect of hydrocyanic-acid gas on insects*

[Time of exposure one hour]

Sodium cyanid per liter.	Approximate concentration of gas above liquid.	Effect on insects experimented upon.			
		Phyllox-era.	Leaf hoppers.	House-flies.	Beetles.
Gm.	*Per cent.*				
1. 0	0. 3	Dead.	Dead.	Dead..	Dead.
. 5	. 15	...do.....	...do.....	...do....	Do.
. 25	. 075	...do.....	...do.....	...do....	Do.
. 125	. 0375	...do.....	...do.....	...do....	Do.
. 0625	. 0187	...do.....	...do.....	...do....	Alive.
. 0312	. 0093	...do.....	...do.....	...do....	Do.
. 0156	. 0046	...do.....	...do.....	...do....	Do.
. 0078	. 0023	do.....	.do....	10 per cent alive.	Do.
. 0039	. 0011	Alive.	Alive..	Alive..	Alive.

A concentration of approximately 0.0046 per cent hydrocyanic-acid gas in the atmosphere within the test tube was the minimum point of toxicity to flies in both experiments. For beetles a gas eight times as strong—that is, 0.037 per cent of hydrocyanic-acid gas—is necessary to cause death.

GERMINATION TESTS WITH HYDROCYANIC-ACID GAS SOLUTIONS [1]

Tests were made on alfalfa (*Medicago sativa*), turnip (*Brassica rapa*), corn (*Zea mays*), and lettuce (*Lactuca sativa*) seed to determine the effect of hydrocyanic-acid gas solutions on germination. Glass tumblers were used as containers for the solution, the seeds being placed in the glass on swings which just touched the surface of the water. The tumblers were sealed with paraffined paper held in place with a rubber band. The temperature of the room throughout the experiment ranged from 19.5° to 24° C. Fifteen dilutions of the hydrocyanic-acid gas solu-

---

[1] One point should be noted in this entire series of experiments—viz, the possibility of a very slight excess of sulphuric acid in the solution of hydrocyanic-acid gas. With these experiments it was the intention to use just enough sulphuric acid to combine with the sodium and free all the hydrocyanic-acid gas. In some cases all the acid might not have been neutralized and possibly would of itself cause some injury, or the sodium sulphate, one of the products resulting from the reaction of sulphuric acid with sodium cyanid, might also have some action on the plant.

re made, the amount of sodium cyanid being reduced
cceeding culture, No. 16 being a control of pure water
contained 2.3485 gm. of sodium cyanid to the liter, ap
per cent solution of hydrocyanic acid, or 0.704 per
acid gas in the air above the liquid. The seeds were pl
and allowed to remain for 48 hours before the first r
At this time most of the unsprouted seeds were ren
10 solutions and placed in swings similar to those abov
ure water. The record for these is found in Table II
"Removed from solution." In the first column wi
unt of sodium cyanid used in making the hydrocyar
s; the second column gives the record of the seeds
olution; and the third column the record of the see
maining 48 hours in the hydrocyanic-acid gas solutio

TABLE III.—*Summary of germination tests 11 days after planti*

ium cyanid in 1 liter.	Percentage of sprouted seeds—		Sodium cyanid in 1 liter.	
	Remaining in solution.	Removed from solution after 48 hours.		Remaining in solution.
Gm.	Per cent.	Per cent.	Gm.	Per cent.
2. 3485	0. 0	1. 25	0. 0091	98. 7
1. 1742	. 0	13. 00	. 0045	92. 0
. 5871	25. 0	27. 30	. 0022	97. 7 ....
. 2935	25. 0	57. 60	. 0011	100. 0 ....
. 1467	55. 1	80. 60	. 00056	98. 2 ....
. 0733	43. 3	79. 00	. 00028	97. 0 ....

effect (not shown in Table III) was the more rapid growth of the seedlings and their deeper color in the dilution immediately adjacent to where the definite retarding action ceased. A diagram of the heights is shown in figure 1.

The maximum height of the alfalfa seedlings was reached at a dilution of 0.0022 gm. of sodium cyanid per liter (equivalent to 0.0006 per cent of hydrocyanic acid over the solution).

EFFECT OF HYDROCYANIC-ACID GAS ON CUTTINGS AND SEEDLINGS

1. PLANTS IN SOIL WATERED WITH HYDROCYANIC-ACID GAS SOLUTIONS

Seedlings of lettuce and turnip and six weeks' old turnips and beets were planted in tumblers containing soil, and watered with hydrocyanic-acid gas solutions, 1 ounce of the solution being used to the tumbler.

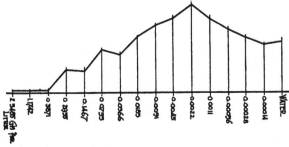

Fig. 1 —Diagram showing the relative heights of the alfalfa seedlings grown in hydrocyanic-acid gas solutions produced by adding sulphuric acid to sodium cyanid in solution.

The first solution contained 18.779 gm. of sodium cyanid per liter, with a proportionate amount of acid as in the germinating experiments, giving approximately a 5.6 per cent of hydrocyanic-acid gas over the solution, the amount of sodium cyanid being reduced one-half each time through a series of 12 graduations, the thirteenth culture being a control of pure water. One or more plants were placed in each tumbler and sealed with paraffined paper held in place with a rubber band. The covering paper was slit to the center, the plant inserted in the opening and the edges of the paper drawn together and sealed with melted paraffin close up to the stem of the plant. In this way the roots were exposed to the hydrocyanic-acid solution and to a sodium-sulphate solution, but the gas did not come in contact with the leaves.

All the plants in the first five containers were killed and only in the soil watered with the ninth solution which contained 0.0733 gm. of sodium cyanid per liter (equivalent to 0.0219 per cent of hydrocyanic acid over the solution) did the plants remain normal.

2. PLANTS WITH ROOTS IMMERSED IN AQUEOUS SOLUTIONS OF HYDROCYANIC-ACID GAS

Turnip and lettuce seedlings and three months' old turnip plants were removed from the soil in which grown and the roots were immersed in hydrocyanic-acid gas solutions and sealed in as above. In this experiment, as in the former, all the plants were killed down to and including dilution 1.1742 gm. of sodium cyanid per liter (equivalent to 0.3522 per cent of hydrocyanic-acid gas over the solution). The minimum point at which the plants were normal was very close to the first series; but there was this difference: In three dilutions, 0.0091 to 0.00225 gm. of sodium cyanid per liter (equivalent to 0.00273 to 0.000675 per cent of hydrocyanic-acid gas over the solutions) an acceleration in growth was noticed, while none was noted in the first series.

3 CUTTINGS IN AQUEOUS SOLUTION OF HYDROCYANIC-ACID GAS

The various solutions of hydrocyanic-acid gas were made up as before, and cuttings of chrysanthemum (*Chrysanthemum* sp.) and wandering jew (*Zebrina pendula*) were partially immersed in the solutions. The plants in this series were the most resistant of any of those tried. From Table IV it will be seen that only in solutions containing 9.394 gm. of sodium cyanid per liter (equivalent to 2.818 per cent of hydrocyanic-acid gas over the solution) were all of the cuttings killed, the point of normality being 0.0733 gm. of sodium cyanid per liter (equivalent to 0.0219 per cent of hydrocyanic acid gas over the solution).

The experiments on plants thus far have been with aqueous solutions of hydrocyanic-acid gas, in which some part of the plant was immersed. To make these experiments comparable with those on insects where the gas was used (instead of an aqueous solution of hydrocyanic-acid gas), it must be known whether or not the effect of hydrocyanic-acid gas on the leaves is comparable to the action of the gas solution on the roots.

4. PLANTS WITH LEAVES EXPOSED TO HYDROCYANIC-ACID GAS, WITH ROOTS PROTECTED

The ordinary-sized tumbler was partially filled with water, and inside the tumbler a whisky glass was placed containing ½ ounce of hydrocyanic-acid gas solution, the first one having 9.394 gm. of sodium cyanid per liter (equivalent to 2.818 per cent of hydrocyanic-acid gas over the solution) and being reduced one-half in each succeeding culture, as in previous experiments. Small seedling turnips were placed between the walls of the two glasses, with the roots extending into the pure water. Melted paraffin was then poured on the surface of the water, thus sealing in the plants and preventing the roots from coming in contact with the gas.

By comparing the results with those of plants in aqueous solutions of hydrocyanic-acid gas it will be seen that the action was similar. With the gas acting only upon the leaves, all were killed, up to and including a dilution of 0.5871 gm. of sodium cyanid per liter (equivalent to 0.1761 per cent of hydrocyanic-acid gas over the solution), while in the aqueous solutions all were killed, up to and including a dilution of 1.1742 gm. of sodium cyanid per liter (equivalent to 0.352 per cent of hydrocyanic-acid gas over the solution). Stimulation was shown in both series at about the same dilution. In the columns of Table IV showing plants in aqueous solutions of hydrocyanic-acid gas and soil watered with hydrocyanic-acid gas solutions, a very close similarity will be seen in the effect on the plant, whether the roots are entirely immersed or the plant is grown in a soil moistened with hydrocyanic-acid gas solutions.

TABLE IV —*Summary of effect of hydrocyanic-acid gas on plants*

Sodium cyanid per liter.	Soil watered with solutions of hydrocyanic-acid gas.	Roots in aqueous solutions of hydrocyanic-acid gas.	Cuttings in aqueous solutions of hydrocyanic-acid gas.	Leaves exposed to hydrocyanic-acid gas.
Gm.				
18.788	Dead..........			
9.394	.....do.........	Dead..........	Dead..........	Dead.
4.697	.....do.........	.....do.........	50 per cent shriveled. 50 per cent dead.	Do.
2.3485	.....do.........	.....do.........	Shriveled......	Do.
1.1742	.....do.........	.....do.........	Tips growing... Stems shriveled.	Do.
.5871	50 per cent shriveled. 50 per cent dead.	25 per cent alive. 75 per cent dead.	20 per cent alive. 80 per cent dead.	Do.
.2935	25 per cent normal. 75 per cent dead.	50 per cent normal. 50 per cent dead.	Tips growing... Stems wilty....	Partially dried.
.1467	50 per cent normal. 50 per cent dead	88 per cent normal. 12 per cent dead.	75 per cent normal. 25 per cent dead.	Do.
.0733	Normal.........	94 per cent normal. 6 per cent dead.	Normal.........	Tender leaves dead. Dead.
.0366	.....do.........	Normal..........	.....do.........	Acceleration.
.0183	.....do.........	.....do.........	.....do.........	Do.
.0091	.....do.........	20 per cent acceleration. 80 per cent normal.	.....do.........	Normal.
.0045	.....do.........	90 per cent normal. 10 per cent acceleration.	.....do.........	Do.
.00225	.....do.........	5 per cent acceleration. 95 per cent normal.	.....do.........	Do.
Water.	.....do.........	Normal..........	.....do.........	Do.

The minimum dosage for killing all the lettuce seedlings, the most susceptible plants experimented with, was, in the case of gas applied to the leaves only, a dilution 0.5871 gm. of sodium cyanid per liter (equivalent to 0.176 per cent of hydrocyanic acid gas over the solution), and at this concentration each of the plants tested showed a death rate of from 50 to 80 per cent when the solutions were applied to the roots. From this it will be seen that the gas alone is almost as toxic when applied to the leaves as the hydrocyanic-acid gas solution is when applied to the roots.

TABLE V.—*Minimum dosages toxic to insects experimented upon*[a]

Insect.	Sodium cyanid per liter.	Approximate concentration of gas.
	Gm.	*Per cent.*
Phylloxera...	0.0078	0.0023
Houseflies..	.0156	.0046
Beetles..	.125	.0375

[a] Maximum dosage harmless to all plants experimented with both in the gaseous and solution form of the acid was 0.0366 gm. of sodium cyanid per liter, approximately 0.0109 per cent of hydrocyanic-acid gas over the solution.

These experiments indicate that a dosage toxic to flies and phylloxera—viz, 0.0156 gm. of sodium cyanid per liter (equivalent to 0 0046 per cent of hydrocyanic-acid gas over the solution)—would be safe for all the plants experimented upon; but for the beetles used so much stronger dosage is required that it would be extremely injurious, if not fatal, to every plant experimented upon. Before planted areas could be treated it would be necessary to determine experimentally the minimum dosage fatal to the insects present and compare that with plants growing thereon, besides considering reactions of the gas and the soil. There would probably be a considerable margin of safety on some soils between the killing strengths necessary for an insect, such as the phylloxera, and injury to more hardy plants, especially in the dormant season.

B. PHYSICAL-CHEMICAL ACTION OF HYDROCYANIC-ACID GAS IN THE SOIL

If hydrocyanic-acid gas is used as a soil fumigant, it will be necessary to fill the soil with gas, either from sodium cyanid placed in the soil and decomposed into hydrocyanic-acid gas, or by generating the gas at the surface and forcing it into the soil. Tubes sunk into the soil to various depths make it possible to draw out the gas for measurement. The concentrations thus determined, compared with the data given above on the toxicity of the gas to insects, would indicate the probabilities of the presence of a killing strength of hydrocyanic-acid gas in the soil.

For drawing gas out of the soil ¼-inch gas pipe cut into 36-inch lengths was used. The tubes may be sunk to any depth, up to 30 inches, at which a determination is desired. To operate, the tube is sunk into the ground a short piece of rubber tubing is drawn over the end projecting out of the soil; into the open end of this tubing a rubber stopper (through which a short piece of glass tubing is inserted) is forced, effectually sealing the soil tube. The glass tubing in the stopper is the outlet through which the soil air is drawn. These tubes will be spoken of as "sampling tubes," in order to distinguish them from the "soil tube" used later on in the experiment for studying the diffusion of gas in the soil.

Experiments were begun in a clay-loam soil. In this the sampling tubes were sunk to a depth of 30 inches, and sodium cyanid in four different forms was introduced into the soil at the same depth

1. Lump sodium cyanid was placed in the ground, and double the amount of sulphuric acid necessary to free the gas was poured on it, but the acid soaked into the ground before the sodium cyanid was dissolved.

2. An aqueous solution of sodium cyanid without acid was poured down one of the tubes and air drawn off from a second tube 1 foot distant from the first and at the same depth, the air being aspirated through 25 c. c. *N/100* iodin; no trace of hydrocyanic-acid gas was shown in several trials.

3. An aqueous solution of sodium cyanid was poured into a pipe as before, followed with the usual amount of acid necessary to release all the gas, likewise without recovering the gas by aspiration through an adjacent tube.

4. Various forms of apparatus for generating gas on the surface and forcing it into the soil were used, the first one being known as the "Babo generator," in which only enough pressure was generated to force the gas out of the generator. A modified form of this apparatus was then prepared, in which large quantities of gas were generated, with slightly more pressure, but with no apparent advantage. Another style of generator was devised by Prof. Woodworth, giving a pressure of over 100 pounds per square inch; but, on trial, results similar to those in the other experiments were obtained, there being no apparent advantage in generating gas in quantity or at a considerable pressure. These were the most promising forms of apparatus used in generating gas on the surface, but all were so unsatisfactory that attention was turned to generating gas in the soil.

The actual generation of gas in the soil must be delayed during the operation long enough to allow for the covering of the generator with soil to a depth sufficient to prevent any loss of gas. Provision must also be made to prevent the filling of the generator with soil or the loss of acid before the reaction is complete.

A pasteboard box, paraffined on the inside to resist the action of the acid, or a tin can with a tight-fitting lid, may be used for the generator. In order to delay the action of the acid on the sodium cyanid until the preparations are completed, we wrapped the cyanid in paper or placed it in pasteboard boxes and then dropped it into the acid solution. Table VI gives the result of an experiment in a dry clay-loam soil.

At the top of the hole where the generator had
was no odor of gas at any time during or after the ger
trace of hydrocyanic-acid gas, except directly at the
6 inches from the generator, and even at this point
the point of toxicity to insects. This experiment
that heavy soils are very impervious to this gas.

Work was now begun on sandy soil.   A box 2 feet d
and 2 feet 4 inches long was filled with moist sand,
placed near the bottom and at one end of the box.
results.

TABLE VII.—*Hydrocyanic-acid gas generated in loose*

Depth of generator.	Charge of sodium.	Distance of test from generator.	Time elapsing after fumigating.	
Inches.	*Ounces.*	*Inches.*		
20	12	18	1 minute....	0.12 per acid gas.
20	12	18	16 hours......	0.3 per acid gas
20	12	18	64 hours......	Faint trac

By comparing the first and second reading it will
hours are required for the gas to diffuse even a short
When the gas was first generated the odor was very
generating box and also when the box was empti
last of the three readings. The gas, which is tempe
the water, is given off again as the partial pressure
phere decreases.

The water in the soil takes up large quantities of g

A series of experiments was therefore begun to determine the effect of water and of different soils on hydrocyanic-acid gas. Air from a jar containing 1.0 gm. of sodium cyanid and 15 c. c. of 10 per cent sulphuric acid in 1 liter of water was aspirated through 5 c. c. of $N/100$ iodin until the latter was decolorized, this requiring 468 c. c. of air. To determine the amount of gas absorbed by water, air from the same solution of hydrocyanic-acid gas as above was aspirated through 50 c. c. of water and then into iodin until the latter was decolorized, the result being that 3,456 c. c. of air decolorized 5 c. c. of $N/100$ iodin.

AIR ASPIRATED DIRECTLY INTO IODIN

It was found that 468 c. c. of air from the above solution decolorized 5 c. c. of $N/100$ iodin when introduced directly, while it was necessary to use 3,456 c. c. of air from the same solution to decolorize an equal amount of iodin when aspirated first through 50 c. c. of water, showing that the latter absorbed the gas from 2,988 c. c. of air, equivalent, according to the above data, to 31 c. c. of $N/100$ iodin.

To obtain an approximate idea of the amount of hydrocyanic-acid gas necessary to saturate the soil at a definite partial pressure of hydrocyanic-acid gas, air from the standard solution as above was aspirated through a tube containing 50 gm. of soil until no more gas was absorbed. At the first titration it required 2,740 c. c. of air when drawn through the soil to decolorize 5 c. c. of $N/100$ iodin, while 540 c. c. direct from the hydrocyanic-acid gas solution gave the same result. The difference between the amounts aspirated through the soil and that directly from the gas solution gave the total amount of hydrocyanic-acid gas absorbed, or adsorbed, or chemically decomposed by the soil, which was equivalent to 26 c. c. of $N/100$ iodin, in contrast with the equivalent of 31 c. c. of $N/100$ iodin absorbed by a similar amount of water. Table VIII indicates the gradual approach to the point of saturation.

TABLE VIII.—*Saturation of soil with hydrocyanic-acid gas*

Character of soil.	Titration.	Control.	Difference.
	C. c.	*C. c.*	*C. c.*
Sandy loam	2,740	540	2,200
Do	880	510	370
Do	620	510	110
Do	580	510	70
Do	460	415	45
Total quantity of air required to saturate soil			2,795

[1] The term "adsorption" will be used to designate the concentration of two or more heterogeneous phases at their contiguous surfaces in contrast with the volume concentrations of these same substances.
"Absorption" is the phenomenon of interpenetration of two or more phases until a certain concentration in the total volume is reached.

A large series of soils ranging from pure sand, and soils with a large percentage of gravel, to heavy adobe was then treated as before, sufficient air from the standard hydrocyanic-acid-gas solution being aspirated through 50 gm. of the different soils until enough hydrocyanic-acid gas came over to decolorize 5 c. c. of $N/100$ iodin.

Table IX gives the results of the experiments with the soil samples used. The column headed "Titration" gives the quantity of air from the hydrocyanic-acid-gas solution that was aspirated through 50 gm. of soil and then through 5 c. c. of $N/100$ iodin until the latter was decolorized. The "Control" column gives the quantity of air from the same hydrocyanic-acid-gas solution that was aspirated directly through a second 5 c. c. of $N/100$ iodin until it was decolorized, the difference between these two figures being an index of the amount of hydrocyanic-acid gas retained by the soil. The sixth column is a calculation of the number of pounds of sodium cyanid required to furnish as much hydrocyanic-acid gas for 1 acre, 3 feet deep, as was found to be adsorbed or decomposed chemically by the experimental sample, this amount being based on the difference between the results given in the columns under "Titration" and "Control," plus the amount required to saturate the soil at a definite partial pressure of hydrocyanic-acid gas. The seventh column gives the amount of sodium cyanid which it is estimated will give a killing strength of hydrocyanic-acid gas in the pore space of the soil and should be added to the amount which is adsorbed or decomposed chemically by the soil. The pore space in most soils was figured at 35 per cent by allowing 1 ounce of sodium cyanid per 100 cubic feet, it would require 28.5 pounds of cyanid to fumigate this space. This gives the effective dose for 1 acre, the amount being given in the eighth column. The estimated cost for fumigating 1 acre to a depth of 3 feet is given in the last column.

TABLE IX.—*Amounts of sodium cyanid required on different soils* a

Sample.	Locality.	Character of soil.	Titration.	Control.	Quantity of sodium cyanid per acre neutralised by soil.	Estimated amount per acre to give killing strength.	Effective dose per 3 acre-feet.	Cost per acre.b
			C. c.	C. c.	Pounds.	Pounds.	Pounds.	
1	Orland	Gravelly	1,742	404	831	25	856	$231.12
2	....do....	....do....	2,450	650	673	25	698	188.46
3	....do....	....do....	1,260	458	458	25	483	130.41
4	Marysville	Sandy loam	3,868	913	789	28	807	217.89
5	Chico	Clay loam	5,063	379	3,000	30	3,030	818.10
6	....do....	....do....	4,990	620	1,720	30	1,750	472.50
7	Oroville	....do....	2,106	446	940	28	968	261.36
8	Sacramento	....do....	3,540	515	1,430	28	1,458	393.66
9	....do....	....do....	2,840	560	730	28	758	204.66
10	Berkeley	....do....	3,632	728	975	30	1,005	271.35
11	Stockton	....do....	4,115	515	1,700	28	1,728	466.56
12	....do....	....do....	2,740	540	990	27	1,017	274.59

a Three feet, or even more, is the depth necessary to fumigate soil for phylloxera. For most insects one-half of this depth would be sufficient.
b The cost is figured on a basis of 27 cents a pound for sodium cyanid.

These requirements show the wide range of absorptive power of different soils, varying approximately from 3,030 pounds of sodium cyanid for a clay loam to 483 pounds for a gravelly soil. The figures given of the cost for the lighter soils indicate that the use of sodium cyanid, while comparable with fumigation with carbon disulphid, is not within the range of economy except in small areas, even when the value of the gas as a nitrogenous fertilizer is considered. The fumigation of heavy soils is very much more expensive.

From the difficulties encountered while working with the generators it seemed necessary to study more carefully the rate of diffusion and absorption of hydrocyanic-acid gas through the various soil media and the effect of soil moisture on this movement. Five experiments were made covering different types of soil and varying amounts of soil moisture. After introducing the hydrocyanic-acid gas into the soil, tube readings were taken at intervals ranging from a few minutes to 120 hours.

1. Heavy clay soil—Readings taken following the introduction of the hydrocyanic-acid gas and 48 hours later.

2. Wet clay-loam soil containing considerable humus—Readings taken immediately after charging the tube.

3. Very sandy, beach soil, containing about 5 per cent of broken shell slightly moist—First reading taken immediately, the next 16 hours later.

4. Clean river sand washed and dried at a temperature of $200°$ C. to remove the soil, organic matter, and moisture—Readings taken immediately after charging with gas and 5 days later.

5. Sand washed and dried as above and afterwards saturated with water—Readings taken immediately.

For this purpose a copper soil testing tube 36 inches long and 2 inches in diameter was made. One end of the testing tube is closed with the exception of a $\frac{1}{8}$-inch tube, the opposite end is closed with a rubber stopper through which is inserted a $\frac{1}{8}$-inch glass tube; at intervals of 6 inches along the sides $\frac{1}{8}$-inch tubes are inserted; to all of these small tubes a short piece of rubber hose is attached which may be closed with clamps. Air from a standard solution of hydrocyanic-acid gas was drawn in through the tube inserted in the stopper, by applying a vacuum at any of the tubes along the sides or at the opposite end of the testing tube, all the openings being closed except the ones where the air was admitted. Then by detaching the tube from the acid solution and drawing in pure air at the same point the amount of gas in that section of the tube could be measured, and also the amount lost could be calculated. On account of the small amounts of gas present in the different sections, qualitative tests only with silver nitrate ($AgNo_3$) was substituted for the iodin method. This overcomes the error from volatilization of

iodin caused by aspirating through it large quantitie:
with silver nitrate was reached with about 120 c.
standard hydrocyanic-acid gas solution when 450 c. c
concentration of gas would neutralize 1 c. c. of $N/$

These experiments, made in loose soil, emphasiz
diffusion of hydrocyanic-acid gas in the soil, which
sarily be very much slower, and perhaps would be
in field conditions. Diffusion seems to be accom
extent by at least a partial saturation of the soil b
it is moving (Table X). In experiment 1 the conce
seemed quite uniform throughout the tube, but in e.
standing 48 hours, a decided decrease is shown in th
tube, indicating that the outer part was not satu
extent as the sections through which the air was dri
the gas in water is indicated in experiment 2 (wet
of the tube not having a trace of gas, while the co
first three sections is very high. The chemical rea
of the dry gas by the clay soil over a period of tin
experiments 1 and 1a, where the quantity of air re
increases from 160 c. c. immediately after charging
c. c., 48 hours afterwards. The most striking differe:
of hydrocyanic-acid gas is shown between sand a
experiment 1, 108 liters of air from the hydrocyanic-ac:
aspirated through the tube and in experiment 3 o
from the hydrocyanic-acid gas solution were used, yet
hydrocyanic-acid gas was recovered from each.

SUMMARY

(1) The toxicity of hydrocyanic-acid-gas solutions varied with the insects experimented upon, from the minimum for houseflies of 0.0156 gm. of sodium cyanid per liter (equivalent to 0.0046 per cent of hydrocyanic-acid gas over the solution) to the maximum for beetles of 0.125 gm. sodium cyanid per liter (equivalent to 0.0365 per cent of hydrocyanic-acid gas over the solution), the latter being the most resistant of any insect experimented upon.

(2) Gas from a solution of approximately the same strength as that used above on houseflies retarded the germination total of lettuce seed 11.3 per cent.

(3) Lettuce seed is not killed by two days' exposure to hydrocyanic-acid gas as strong as 0.0366 gm. of sodium cyanid per liter (equivalent to 0.0109 per cent of hydrocyanic acid gas over the solution) and will give a good germination percentage if removed at the end of this time.

(4) Stimulation was greatest at a point one sixteenth of that causing retardation—viz, 0.0011 gm. of sodium cyanid per liter (approximately 0.00033 per cent of hydrocyanic-acid gas over the solution).

(5) Solutions of hydrocyanic-acid gas approximately 256 times as strong as that necessary to produce gas having the minimum killing strength for flies were fatal to all seedlings tested and to 50 per cent of the cuttings placed in the solution, while a solution approximately twice as strong as that required to produce a gas concentration fatal to flies had no effect even upon seedlings.

(6) Sodium-cyanid solutions introduced into the soil failed to give a trace of hydrocyanic-acid gas in air drawn from the soil.

(7) The use of pressure in forcing gas into the soil did not materially increase the rate at which it could be introduced.

(8) Soil and water are both strong absorbents of hydrocyanic-acid gas. Retention of hydrocyanic-acid gas by the soil is dependent upon the character of the soil, while that of water remains constant under uniform conditions of pressure and temperature.

(9) The variability of gas absorption by the soil makes it practically impossible in field work to estimate the dosage of sodium cyanid required to give a toxic effect on insects and at the same time to be within the margin of safety to plants.

(10) In small amounts of soil of a uniform character it is possible to determine experimentally the margin of safety between certain insects and plants.

(11) A heavy damp or a very wet sandy soil is almost impervious to hydrocyanic-acid gas.

(12) A pure sandy soil when wet will take up hydrocyanic-acid gas only in proportion to the amount of water present and this may again be given off, but gas in contact with a clay soil either enters into a chemical combination with some of the soil constituents or is adsorbed by the soil particles.

(13) Gas generated in a soil body diffuses with
clay soils or very wet sandy soils, but in sand with a
moisture, the diffusion of gas is much more rapid.

(14) The use of sodium cyanid offers a satisfact
gating masses of loose, porous soil, especially tho:
amounts of clay, or of seed beds and potting soil
allow of much wider range of concentrations when th
by a crop.

LITERATURE CITED

(1) BOURCART, E.
> 1913. INSECTICIDES, FUNGICIDES AND WEEDKILLERS.
> Grant. 431 p., 12 fig. London, New York.
> Cites Mouillefert, p. 136–137.

(2) FERNALD, H. T.
> 1910. FUMIGATION DOSAGE. *In* Mass. Agr. Exp. Sta.
> p. 214–247.

(3) FRANCE. MINISTÈRE DE L'AGRICULTURE ET DU COMMERC
> RIEURE DU PHYLLOXERA.
> 1880–1882. COMPTE RENDUE ET PIÈCES ANNEXES, 1879

(4) MAMMELLE, Th.
> 1910. SUR L'EMPLOI DU CYANURE DE POTASSIUM COMM
> RAIN. *In* Compt. Rend. Acad. Sci. [Paris], t.

(5) MORSE, F. W.
> 1888. EXPERIMENTS ON THE CAUSE AND AVOIDANCE OF
> THE HYDROCYANIC GAS TREATMENT OF TREES.
> Bul. 79, 3 p.

(6) SANDERSON, E. D.
> 1907. DIRECTIONS FOR TREATMENT OF INSECT PESTS AN
> N. H. State Bd. Agr. Rpt., 1905/07, p. 151–175

PLATE 40

Soil tube connected to air pump and jar containing hydrocyanic-acid-gas solution.

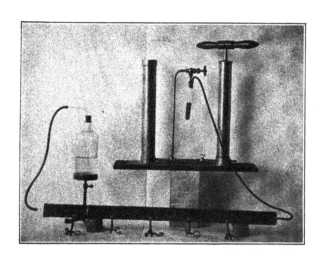

ENZYMS OF MILK AND BUTTER [1]

By R. W. THATCHER, *Director, Minnesota Agricultural Experiment Station,* and A. C. DAHLBERG, *Assistant Dairy Husbandman, Wisconsin Agricultural Experiment Station.*

INTRODUCTION

The deterioration of butter during storage is often attributed to the enzyms in the buttermilk which it contains. There has been some experimental study of the matter; but most of the published results of these investigations yield little information concerning the actual enzym content of butter, because of the fact that the studies dealt chiefly with the chemical changes in the butter itself during storage, and sufficient care was not used to prevent possible contamination by microorganisms. In fact, a survey of the literature dealing with the enzyms of milk and its products shows a lamentable lack of clearness on the part of the investigators in distinguishing between the enzyms of milk itself and those which are due to bacterial infections.

Rogers (*16*)[2] found lipase to be the cause of the increase in the acidity of canned butter on standing. Rahn, Brown, and Smith (*15*) stated that

all (storage) butter investigated showed an increase of "amid nitrogen," i. e., nitrogen not precipitated by copper sulphate, tannic acid, or phosphotungstic acid

Their methods were shown to be at fault by Rogers, Berg, Potteiger, and Davis (*17*), who could observe

no evidence of an increase in soluble nitrogen in butter on long standing at $_0$° F., even when the conditions of manufacture were most favorable to such changes.

They could detect no proteolysis in buttermilk held a long period of time in cold storage to which 18 per cent of sodium chlorid had been added, but active bacterial proteases, pepsin, and trypsin were not completely inhibited in their action by these adverse conditions. Lactose, they concluded, was oxidized only when a trace of iron and peroxid were added.

During the summer of 1915 one of us (Dahlberg), as a part of his duties in the Division of Dairy and Animal Husbandry of the Minnesota Agricultural Experiment Station, prepared several lots of butter under carefully controlled conditions of manufacture and placed them in stor-

[1] Published, with the permission of the Director, as Paper No 71 of the Journal Series of the Minnesota Agricultural Experiment Station.
[2] Reference is made by number (italic) to "Literature cited," pp. 448-450.

Journal of Agricultural Research,
Washington, D. C.
kt

Vol. XI, No. 9
Nov. 26, 1917
Key No. Minn.—23

(437)

age, in order to study the effect of varying methods of manufacture and storage upon the keeping qualities of the butter. From time to time studies of the bacterial development in the stored butter were made. The results of this work have recently been published elsewhere (25).

Since the conditions of manufacture and the bacterial development in these butters were known, they afford excellent material for a study of the enzym content of the butter after storage, and such a study was accordingly undertaken.

REVIEW OF PREVIOUS WORK

The following types of enzyms have been reported to be normally present in cow's milk: an amylase (27), a lipase (9), proteases (1–5, 7, 22) a peroxidase (19), salolase (22), catalase (12, 18, 20), reductase (20), and a lactose-fermenting enzym (21). Vandevelde (22), however, concludes that no lipase can be found in fresh milk and that salolase is present in only very slight amounts; while Grimmer (10), as a result of his study of the enzyms of the mammary glands, concludes that salolase, peroxidase, and catalase are the only enzyms which are normally secreted with milk. Further, Van de Velde and Landsheere (23) found no amylase in milk and criticize Spolverini's observations concerning peroxidase in milk, holding that its presence there is due to bacterial contamination.

These reports indicate something of the confusion which exists with reference to the normal enzyms of milk. The contradictory evidence which has been presented undoubtedly results, in part at least, from the rather crude methods of study of enzym action which were in use at the time when these investigations were in progress. Unfortunately, little attention has been given to these matters during recent years, when more refined methods of study might have given more concordant results.

The question as to whether any enzyms which may be present in milk will be carried down with the butter in churning has not been generally discussed. Kooper (13) states that during butter making catalase remains behind in the buttermilk and is therefore not a direct constituent of butter fat. It might be assumed, however, that any buttermilk which remains in the butter would contain the same enzyms that were present in the original milk. In fact, Hesse (12), who found small amounts of catalase in butter, calculated his results on the basis of its buttermilk content. Furthermore, since butter fat is undoubtedly an emulsion colloid, it might be assumed to carry with it into the butter, in the form of absorption complexes, large proportions of the enzyms originally present in the milk. But no definite information concerning these matters, other than that cited in the opening paragraphs of this paper, has been published.

EXPERIMENTAL WORK .

I. INVESTIGATIONS CONCERNING GALACTASE

The normal proteolytic enzym of milk, which aids in the slow decomposition of milk proteins into peptones, amino acids, and ammonia, was given the name "galactase" by Babcock and Russell (*1–5*). This name is unfortunately not in accord with the present system of nomenclature, and the enzym ought properly to be called "casease" or "lactalbuminase"; but, since the name suggested by Babcock and Russell has come into common use and its significance is properly understood, we decided to continue its use.

Before undertaking the proposed studies of the galactase content of milk, cream, skim milk, buttermilk, and butter it was necessary to determine upon certain details of technic which had not been satisfactorily worked out by former investigators. This preliminary work may be briefly summarized as follows:

(*a*) LENGTH OF PERIOD OF ACTION.—Previous experimental data give quantitative measures of proteolytic activity for periods varying from 1 hour for certain pepsin studies (*8*) up to the practically prohibitive time of nearly 3 years for galactase (*14*). The time in which most investigators have allowed galactase to act, in order to obtain comparative results, has varied from 30 days to 1 year. This appears to be unnecessarily long. The analytical data presented by Babcock and Russell give an average increase in the soluble-nitrogen content of a number of samples of skim milk, when expressed as percentage of the total nitrogen, of 13.49 per cent for the first 8 days and only 5.66 per cent for the following 12 days. One sample of skim milk increased 26 per cent in soluble nitrogen during the first 7 days, 2 per cent the following 7 days, and to increase the soluble nitrogen content another 26 per cent required a period of 3 days. There is very little difference in the comparative galactase content of various skim milks when analyses made at the end of 7 days or of 6 months are compared. In the present work the period of action was limited to 4 days, since it was found that the soluble nitrogen in skim milk increased 10 or 12 per cent during this time.

(*b*) MEASUREMENT OF HYDROLYSIS BY MEANS OF THE NINHYDRIN REACTION.—The ninhydrin method, as proposed by Harding and MacLean (*11*) for the determination of amino-acid or nitrogen, was compared with the official method of the Association of Official Agricultural Chemists (*26*) for determining the increase in soluble nitrogen in different samples of milk due to protease action (Table I). In the ninhydrin determination the sample was diluted 1 to 10, and 1 c. c. of this used in each analysis, but the results were calculated to the 10 c. c. basis. Bacterial action was inhibited by 0.5 per cent of chloroform. All samples in all the work reported in this paper were incubated at $40°$ C. for four days.

TABLE I —*Results with ninhydrin method of estimating the rate of hydrolysis of the proteins of milk*

Sample.	Total nitrogen.	Initial soluble nitrogen.	Initial ninhydrin nitrogen.	48-hour soluble nitrogen.	48-hour ninhydrin nitrogen.	Increase in soluble nitrogen.	Increase in ninhydrin nitrogen.
	Per cent.	*Per cent.*	*Per cent.*	*Per cent.*	*Per cent.*	*Per cent.*	*Per cent.*
Boiled skim milk . . .	0.0586	0.0033	0.0029	0.0039	0.0032	1.02	0.51
Skim milk + 5 per cent of sodium chlorid	.0586	.0108	.0029	.0137	.0031	4.95	.34
Skim milk................	.0586	.0108	.0029	.0177	.0032	11.77	.51

The ninhydrin method indicated no increase in nitrogen when there was an actual increase of 11.5 per cent in the soluble nitrogen. This was further substantiated by work on a pure casein solution. This gave only a trace of nitrogen by the ninhydrin method, with no increase after evident proteolysis had taken place. This method of estimating the rate of hydrolysis of proteins was therefore abandoned as not being applicable to investigations with milk proteins.

(c) THE INFLUENCE OF CHLOROFORM ON PROTEOLYSIS.—To prevent bacterial action in the samples, chloroform was added in amounts found ample by Harding and MacLean (*11*) and Van Slyke (*24*). Skim milk to be analyzed at varying intervals extending over a long period of time was treated with 1 per cent chloroform, which, according to the above investigators and Babcock and Russell (*5*), should have only slightly inhibited proteolysis. As shown in Table II, the soluble nitrogen of this skim milk did not increase on standing. In verifying this result the soluble nitrogen was obtained by a Kjeldahl determination of the nitrogen in the total filtrate from the casein precipitation. Closer agreement in the duplicates could be obtained from the filtrates than from the casein itself.

TABLE II —*Influence of chloroform on the proteolysis of milk as indicated by soluble nitrogen*

Sample.	Chloroform.	Total nitrogen.	Initial soluble nitrogen.	Soluble nitrogen after 4 days.	Increase in soluble nitrogen as per cent of total.
	Per cent.	*Per cent.*	*Per cent.*	*Per cent.*	
1. Skim milk..................	0.5	0.0583	0.0129	0.0183	9.26
	1.0	.0583	.0130	.0114	−2.74
2. Skim milk..................	.5	.0531	.0115	.0170	10.35
	1.0	.0531	.0115	.0120	.94
3. Skim milk..................	.5	.0406	.0119	.0136	4.18
	1.0	.0406	.0119	.0103	−3.94
4. Skim milk..................	1.0	.0547	.0106	.0079	−4.93
5. Buttermilk, neutral .	.5	.0551	.0125	.0201	13.60
6. Buttermilk, neutral	1.0	.0551	.0125	.0174	9.80
	1.0	.0463	.0064	.0059	−1.08

Sample 5 was buttermilk obtained from a small hand churning and probably contained 1 per cent of butter fat. Since fat destroys the anesthetic property of chloroform by combining with it, the results of No. 5 should not be considered. In every other case 1 per cent of chloroform in skim milk or buttermilk completely inhibited proteolysis.

(d) THE INFLUENCE OF SODIUM CHLORID ON PROTEOLYSIS.—The possibility that the sodium chlorid present in ordinary butter might inhibit proteolytic activity led us to make a study of the effect of varying percentages of sodium chlorid added to skim milk upon the rate of production of soluble nitrogen by galactase. After the period of incubation the unchanged casein was precipitated from the milk by means of dilute acetic acid and the soluble nitrogen in the filtrate determined by the Kjeldahl method. The samples were preserved during the incubation by 0.75 per cent chloroform. The following results were obtained (Table III).

TABLE III.—*Influence of sodium chlorid on the proteolysis of milk (with 0 75 per cent of chloroform)*

Sodium chlorid added.	Total nitrogen.	Initial soluble nitrogen.	Soluble nitrogen after 21 days.	Soluble nitrogen after 72 days.	Increase.
Per cent.	*Per cent.*	*Per cent.*	*Per cent.*	*Per cent.*	*Per cent.*
0.0 boiled......................	0. 9561	0. 0055	0. 0051	0. 0034	−3. 74
.0.............................	. 0561	. 0121	. 0188	. 0198	13. 72
.5.............................	. 0561	. 0121	....	. 0171	8. 91
1.0............................	. 0561	. 0121	. 0142	. 0131	1. 78
15.0...........................	. 0534	. 0121	....	. 0110	−2. 06
20.0...........................	. 0528	. 0121	. 0120	. 0103	−3. 40

The effect of sodium chlorid is marked; both 15 and 20 per cent, the concentration of the salt brine in butter, stopped all proteolysis. The very slight increase in the soluble nitrogen content of the skim milk containing 1 per cent of the salt is partially due to the excessive chloroform used, as the following tests, in which only 0.5 per cent of the anesthetic was used, will show (Table IV).

TABLE IV.—*Influence of sodium chlorid on the proteolysis of milk (with 0.5 per cent of chloroform)*

Sodium chlorid added.	Total nitrogen.	Initial soluble nitrogen.	Soluble nitrogen after 5 days.	Increase.
Per cent	*Per cent.*	*Per cent.*	*Per cent.*	*Per cent.*
0.0 boiled.................................	0. 0586	0. 0033	0. 0034	0. 51
.0...	. 0586	. 0108	. 0242	22. 86
.8...	. 0586	. 0108	. 0201	15. 87
3.0...	. 0586	. 0108	. 0143	5. 97

The actual checking of the increase in soluble nitrogen due to the addition of 1 per cent of sodium chlorid was approximately 10 per cent in either case, but 3 per cent of the salt did not entirely prevent proteolysis.

(e) OCCURRENCE OF GALACTASE IN MILK DURING PROCESS OF BUTTER MAKING.—In following the protease of the fresh milk to the finished butter, the reaction of the various products was in every case brought to that of the fresh milk—that is, 10 c. c. were exactly neutralized by 1.5 c. c. of *N/10* alkali when phenolphthalein was used as an indicator, by dipping a stick of sodium hydroxid into the sample. If the sample became too alkaline, it was neutralized by some of the original sample so that dilution was avoided. The various samples were treated with chloroform in the following proportions: The skim milk, buttermilk, and bowl contents, 0.5 per cent; the milk, 2.5 per cent; and the cream, which contained 23 per cent of butter fat, 5 per cent. The "bowl contents" was an emulsion of the slime in the wash water held in the bowl. An equal volume of boiled skim milk was added as a substrate, but the increase in the soluble nitrogen was calculated on the basis of the bowl contents alone. The "cream during ripening" refers to the proteolysis occurring during the 10 hours it was ripened at $85°$ F. and the 20 hours it was held at $54°$ previous to churning.

TABLE V —*Presence and concentration of galactase in milk and butter as indicated by the increase of soluble nitrogen*

Sample.	Total nitrogen.	Initial soluble nitrogen.	Soluble nitrogen after 4 days.	Increase.	Initial percentage of total nitrogen as casein.
	Per cent.	*Per cent.*	*Per cent.*	*Per cent.*	
Skim milk	0.0583	0.0129	0.0183	11.10	77.87
Whole milk	.0544	.0113	.0198	15.62	79.23
Cream..................	.0403	.0076	.0142	16.37	81.14
Bowl contents..................	.0189	.0046	.0262	114.28	89.93
Cream during ripening	.0403	.0076	.0081	1.66	....
Cream after ripening.... .	.0403	.0081	.0139	14.39	.
Buttermilk......................	.0551	.0125	.0201	13.60	....

It appears from these data that in separating milk, galactase is taken out of the skim milk part, slightly increased in the cream, and highly concentrated in the separator slime. While no relationship exists between the increase in soluble nitrogen and the total nitrogen, it is evident that the factors at work during milk separation which increase the percentage of casein in the total nitrogen also increase the galactase content. The cream underwent a slight proteolytic digestion during the ripening process, but the proteolysis after souring and neutralization was less than that of the sweet cream. This indicates that the chief proteolytic enzym of milk is not of bacterial origin, as Olson (*14*) recently reported.

(*f*) AMOUNT OF GALACTASE CONTAINED IN BUTTER.—The butter used in the experiments represented both good and bad qualities. The "fresh dairy" butter was made from the cream obtained from the milk of the University Farm herd, and was soured spontaneously without pasteurization. The "fresh creamery" butter was made in a cooperative creamery from sweet pasteurized cream ripened by a commercial starter. Both of these butters were of extra-fine quality. The "stored dairy" butter was the same as the "fresh dairy," except that it had been held in cold storage for eight months. The "fresh centralized" butter was made from the sour cream just as it was received by a central creamery. The last two butters mentioned were of poor quality. The "stored centralized" butter was made in the same way as the "fresh centralized," but had been held eight months in cold storage. It was of extremely poor quality.

No effort was made to obtain a pure enzym extract from the butter. A known weight, usually 400 gm., of butter was melted at 45° C. in two long glass tubes about an inch in diameter. After the separation was complete the clear fat was hardened by immersing the tubes in cold water, and the curd solution was then washed out. If an excess of fat remained in the extract, it was removed by rewarming and centrifuging. This extract was then dialyzed at a temperature never exceeding 13°C., in a parchment dialyzer, until free of sodium chlorid, the last six or eight hours of dialysis being with distilled water. This curd was made up to a given volume and used at once in the various tests.

The acetone-ether method of obtaining a fat-free, dry powder was also tried, but so high a percentage of sodium chlorid was left in the powder that dialysis was necessary. Consequently this method had no advantages over the other, and the enzyms probably would have been weakened by the precipitation.

The casein was precipitated by the optional official method (26, p. 118), rather than the official method because filtering was often more rapid, the volume to be filtered much less, and a clear filtrate more easily obtained. (Thirteen analyses of chloroformed-skim milk and butter-curd extracts gave an average of 0.00104 gm. more nitrogen in the form of casein by the use of alum as the precipitant; hence, the methods should not be used interchangeably.) A clear filtrate was more easily obtained and an excess of substrate assured by the addition of boiled skim milk to the curd solutions. In a few cases the filtering had to be done on a "Buchner funnel" through three filter papers and repeated 10 to 20 times to obtain a perfectly clear filtrate. Bacterial action was prevented by 1 per cent chloroform instead of 0.5 per cent because the extracts contained considerable fat (Table VI).

TABLE VI —*Presence of galactase in butter, as indicated by the increase in soluble nitrogen*

Kind of butter.	Concentration.		Initial soluble nitrogen.	Soluble nitrogen after 4 days.	Increase.	Increase for 10 gm. of butter.
	Ratio, butter to curd extract.	Ratio, curd extract to boiled skim milk added.				
			Gm.	Gm.	Gm.	Gm.
1. Fresh dairy.............	2:1	50:100	0. 0098	0. 0108	0. 0010	0. 0015
Fresh dairy, boiled......	2:1	50:100	. 0091	. 0091	. 0000	. 0000
2. Fresh dairy.............	2:1	50:100	. 0080	. 0090	. 0000	. 0015
Fresh dairy, boiled.....	2:1	50:100	. 0070	. 0045	. 0025	. 0000
3. Stored dairy.............	3:2	50:50	. 0043	. 0063	. 0020	. 0027
4. Stored centralized......	2:1	50:50	. 0042	. 0078	. 0036	. 0036
Stored centralized boiled	2:1	50:50	. 0039	. 0036	. 0003	. 0000
5. Fresh creamery.........	3:2	50:50	. 0049	. 0044	. 0005	. 0000

The two samples of fresh dairy butter gave uniform and measurable proteolytic action, while one sample of fresh pasteurized creamery butter showed no digestion of the casein. The increase in the soluble nitrogen of the stored butter was nearly twice that of the fresh butter.

II. LIPASE CONTENT OF BUTTER

Lipase hydrolyzes fats into alcohols and fatty acids, causing an increase in the acidity of the media acted upon. The increase in the acidity of butter on standing is supposed to be due to this enzym (16).

The curd solutions used in testing for lipase activity were just half the volume of the butter. The substrate was either 5 per cent of butter fat or olive oil, and the preservative was 2.5 per cent of chloroform. Ten c. c. aliquots of the extracts in 10 c. c. of neutral 95 per cent alcohol were titrated against $N/40$ sodium hydroxid, with phenolphthalein as indicator.

TABLE VII —*Presence of lipase in butter as shown by the increase of acidity*

Kind of butter.	Initial acidity.	Acidity after 4 days.	Increase over boiled.
	Per cent.	Per cent.	Per cent.
1. Fresh dairy, boiled.............................	0. 2	1. 3	0. 0
Fresh dairy......................................	. 2	1. 1	. 0
2. Fresh unsalted dairy, boiled......................	5. 3	5. 3	. 0
Fresh unsalted dairy	6. 3	6. 6	. 3
3. Stored unsalted dairy.............................	5. 5	5. 5	. 0
4. Fresh centralized.................................	2. 7	3. 0	. 3
5. Stored centralized	2. 4	3. 5	1. 1
6. Fresh creamery	1. 8	2. 2	. 4

The extracts from the unsalted butters were not dialyzed. This accounts for their higher initial acidity.

Lipase activity during four days at $_{40}$° C. was too small to be accurately measured, except in the extract from the stored centralized butter.

III. OXIDASE CONTENT OF MILK AND BUTTER

Since the investigations of oxidase activity in milk have been limited to a few easily oxidized substances and since oxidases show a specificity toward chemicals in their action, it was thought desirable to investigate by Bunzel's method the action of milk on several of the common chromogens (7). Four c. c. of milk were used in each test. Negative results were obtained with paraphenylene diamine, phenolphthalein, hydroquinon, pyrocatechin, phloroglucin, α-naphthol, and para-, meta-, and ortho-cresols. Positive results were obtained with metol and pyrogallol; but since the action of the latter was about 20 per cent the greater, pyrogallol alone was used for the final determinations. The results given in Table VIII were obtained in duplicate, the duplicates agreeing very closely.

TABLE VIII.—*Estimation of oxidase in milk as shown by the oxygen absorption of pyrogallol*

[Readings are expressed as millimeters of mercury, each millimeter representing an absorption of 0.005 c. c. of oxygen]

Time.	Skim milk 1.	Whole milk.		Skim milk 2.	
		Raw.	Boiled.	Raw.	Boiled.
Minutes.					
60		0.16	0.00	0.00	0.00
80		.40	.00	.00	.00
100		.56	.00	.17	.00
120	0.95	.70	.16	.17	.00
140	1.07	.85	.20	.35	.00
160	1.28	.97	.27	.40	.00
180	1.33	1.13	.40		
300		1.70	.91		
360		1.97	1.22		
420		2.25	1.48		
460		2.32	1.55		

There was no agreement in the results obtained from the skim milk from the same herd on different days; boiling failed to completely inhibit the activity of the whole milk, and the action was very slow in starting. It can not be caused by a normal milk oxidase. In every case a browning of the samples was quite marked before oxygen absorption started.

Seven curd solutions from different butters gave no action with pyrogallol.

15755°—17——4

IV. CATALASE CONTENT OF MILK AND BUTTER

Storch's method of measuring the catalase of milk, as given by Barthel (6), gave very closely agreeing duplicate results and was used in this work (Table IX). In milk the liberation of oxygen ceased in four hours. The curd extracts were equal in volume to the original butter. Five per cent of commercial hydrogen peroxid was added to the samples to be tested, and the concentration of peroxid in the boiled samples at the end of four hours were used as the basis for calculation.

TABLE IX.—*Catalase content of milk and butter as shown by reaction with hydrogen peroxid*

Sample.	Hydrogen peroxid content after 4 hours.		Reduction of hydrogen peroxid in unboiled milk or butter.
	Boiled.	Unboiled.	
	Gm.	Gm.	Gm.
1. Boiled milk............................	0. 0054	....	....
Milk.....................................	....	0. 0039	0. 0015
2. Boiled fresh dairy butter..............	. 0054	....	....
Fresh dairy butter.....................	....	. 0049	. 0005
3. Stored dairy butter...................	....	. 0049	. 0005
4. Fresh centralized butter...............	....	. 0026	. 0028
5. Stored centralized butter..............	....	. 0039	. 0015
6. Fresh pasteurized creamery butter......	....	. 0048	. 0006

The fresh centralized butter did not contain as much catalase as the results show, because a clear filtrate was not obtained. Every sample of butter contained catalase. That the centralized butter gave higher results than the other butter and as high as milk itself agrees with the belief that there is some relationship existing between catalase activity in milk and bacterial growth. Storage did not diminish its activity.

V. PEROXIDASE CONTENT OF MILK AND BUTTER

The common Storch test for heated milk was used on the butter extracts with the following results:

TABLE X.—*Color changes showing the peroxidase content of milk and butter*

Sample.	Time of heating (in minutes)				
	0	5	10	20	30
1. Fresh dairy milk boiled.	White...	White...	White......	White......	White.
Fresh dairy milk.........	...do....	Gray...	Gray......	Gray......	Gray.
2. Stored dairy milk.....	...do: ...	White...	White......	Faint gray.	Faint gray.
3. Fresh centralized......	...do....	Gray...	Gray......	Gray......	Gray.
4. Stored centralized......	...do....	White...	Faint gray.	Faint gray.	Faint gray.
5. Fresh pasteurized creamery.	...do....	...do...	White......	...do.......	Do.

All the samples gave some color change, but for the storage butters and the pasteurized creamery butter it was very slight. Milk diluted 1 to 160 with distilled water and to which 16 per cent of boiled milk was added gave a gray color similar to that of the fresh dairy butter; hence the peroxidase content of butter is very small.

COMPARISON OF MILK AND BUTTER ENZYMS ON THE BASIS OF TOTAL NITROGEN

The fat in butter acts as so great a diluent for the water-soluble constituents that a direct comparison of the enzyms of milk and butter is misleading. The more logical comparison is on the basis of the total nitrogen, since the proteins and enzyms exist in the same colloidal state and ought to be carried into the butter in the same proportions as they exist in the cream. The total protein of the fresh dairy butter was 0.55 per cent, that of the milk from which it was made 3.47 per cent, so the enzymic activities in the butter were multiplied by 6.31. On taking the milk as the standard and its enzymic content as 1, butter was found to contain the following amounts of the various enzyms which were studied:

Galactase	1. 100	Catalase	2. 000
Oxidase	0. 000	Peroxidase	0. 008

The galactase content of the butter was approximately equal to that of the milk, the catalase content was double, but the peroxidase content was very much less, and no oxidase activity could be detected.

RELATION OF ENZYMS IN BUTTER TO DETERIORATION DURING STORAGE

As mentioned in the introductory paragraphs of this paper, deterioration in quality of butter during storage has been considered by some investigators to be due to the action of enzyms contained in it. Fat-splitting (lipase) or protein-hydrolyzing ("galactase" or casease) enzyms have been suggested as possible agents in causing deterioration. Our work leads us to conclude that lipases are present in butter in very small amounts, if at all, and that they could not be conceived to be sufficiently active at the low temperature used in butter storage to cause any appreciable change in the quality of the butter. The protein-hydrolyzing enzym we found to be completely inhibited by sodium chlorid in the concentrations which are present in the water contained in all normally salted butters. This fact, together with the known inhibiting effect of low temperatures upon proteolysis by enzyms makes it impossible that the hydrolysis of proteins in the butter by enzyms plays any part in deterioration changes. We conclude, therefore, that enzyms are not to be considered as a factor in the deterioration of butter in cold storage.

SUMMARY

(1) Proteolysis in skim milk was completely inhibited by 1 per cent of chloroform and by 15 per cent of sodium chlorid. Galactase can not act in normal butter because of the high salt content.

(2) In the separation of milk the factors which increase the percentage of casein in the total nitrogen also increased the galactase content. The ripening of cream did not increase the rate of proteolysis.

(3) No oxidase was found in milk or butter.

(4) Only one sample of butter gave any evidence of lipase at the end of four days at $40°$ C.

(5) The enzym content of butter is very small, because of the high dilution in fat. Expressed on the basis of total nitrogen the butter examined contained as much galactase as fresh whole milk, twice as much catalase, but only one one hundred and sixtieth as much peroxidase.

(6) The cold storage of butter weakens the peroxidases, but has little effect on the catalase and galactase.

(7) Enzyms are present in butter in such small amounts and under such unfavorable conditions for enzym activity during cold storage that they need not be considered as a factor in the deterioration of butter during storage.

LITERATURE CITED

(1) BABCOCK, S. M , and RUSSELL, H L.
 1897. UNORGANIZED FERMENTS OF MILK. A NEW FACTOR IN THE RIPENING OF CHEESE. *In* Wis. Agr. Exp. Sta., 14th Ann. Rpt , [1896]/97, p. 161–193, fig. 14–15.

(2) —— —— and VIVIAN, Alfred.
 1898. DISTRIBUTION OF GALACTASE IN COW'S MILK. *In* Wis. Agr. Exp. Sta , 15th Ann. Rpt., [1897]/98, p. 87–92, fig. 13.

(3) —— ——
 1898. PROPERTIES OF GALACTASE A DIGESTIVE FERMENT OF MILK. *In* Wis. Agr. Exp. Sta., 15th Ann. Rpt , [1897]/98, p. 77–86.

(4) —— ——
 1899. INFLUENCE OF GALACTASE IN THE RIPENING OF COTTAGE CHEESE. *In* Wis. Agr. Exp. Sta., 16th Ann. Rpt., [1898]/99, p. 175–178.

(5) —— and others.
 1899. THE ACTION OF PROTEOLYTIC FERMENTS ON MILK, WITH SPECIAL REFERENCE TO GALACTASE, THE CHEESE-RIPENING ENZYME. *In* Wis. Agr. Exp. Sta , 16th Ann. Rpt , [1898]/99, p 157–174, fig. 31–42

(6) BARTHEL, Christian.
 1910. METHODS USED IN THE EXAMINATION OF MILK AND DAIRY PRODUCTS. Translated by W. Goodwin. 260 p., 65 fig. London.

(7) BUNZEL, H. H.
 1914. A SIMPLIFIED AND INEXPENSIVE OXIDASE APPARATUS. *In* Jour. Biol. Chem., v. 17, no. 3, p. 409–411, 1 fig

(8) EULER, Hans.
 1912. GENERAL CHEMISTRY OF THE ENZYMES. Translated by T. H. Pope. 323 p., 7 fig. New York, London.

(9) GILLET, Charles.
    1904. THE PRESENCE OF A LIPASE IN MILK. (Abstract.) *In* Exp. Sta. Rec.,
        v. 15, no. 10, p. 1002. (Original article in Jour. Physiol. et Pathol.
        Gén , 1903, no. 3. Not seen.)

(10) GRIMMER, W.
    1910. THE ENZYMS OF THE MAMMARY GLANDS. (Abstract) *In* Exp. Sta.
        Rec., v. 23, no. 3, p. 285–286. (Original article in Festschrift Otto
        Wallach, p. 452–466. Göttingen, 1909. Not seen.)

(11) HARDING, V. J., and MacLEAN, R. M.
    1915. A COLORIMETRIC METHOD FOR THE ESTIMATION OF AMINO-ACID α-NITRO-
        GEN. *In* Jour. Biol. Chem., v. 20, no. 3, p. 217–230, 3 fig.

(12) HESSE.
    1913. CATALASE IN BUTTER. (Abstract.) *In* Exp. Sta. Rec , v. 29, no. 6,
        p. 508. (Original article in Molk. Ztg. [Hildesheim], Bd 26, No. 6,
        p. 81–84, 1912. Not seen)

(13) KOOPER, W. D.
    1912. INVESTIGATIONS IN REGARD TO CATALASE. (Abstract.) *In* Exp. Sta.
        Rec., v. 26, no. 2, p. 112. (Original article in Milchw. Zentbl , Bd.
        7, No. 6, p. 264–271, 1 fig , 1911. Not seen.)

(14) OLSON, G. A.
    1908. MILK PROTEINS. *In* Jour. Biol. Chem., v. 5, No. 2/3, p. 261–281.

(15) RAHN, Otto, BROWN, C. W , and SMITH, L. M.
    1909. KEEPING QUALITIES OF BUTTER. Mich. Agr. Exp. Sta. Tech. Bul. 2,
        44 p., 6 fig. Literature cited, p. 44.

(16) ROGERS, L. A.
    1904. STUDIES UPON THE KEEPING QUALITY OF BUTTER. I. CANNED BUTTER
        U. S. Dept. Agr. Bur. Anim. Indus. Bul. 57, 24 p. Bibliography,
        p. 24.

(17) ——— and others.
    1913. FACTORS INFLUENCING THE CHANGE IN FLAVOR IN STORAGE BUTTER.
        U. S. Dept. Agr. Bur. Anim. Indus. Bul. 162, 69 p., 1 fig.

(18) SPINDLER, Franz.
    1911. BEITRÄGE ZUR KENNTNIS DER MILCHKATALASE. *In* Biochem. Ztschr.,
        Bd. 30, Heft 5, p. 384–412, 1 fig.

(19) SPOLVERINI, L. M.
    1904. THE OXIDIZING FERMENT IN MILK. (Abstract.) *In* Exp. Sta. Rec , v. 15,
        no. 10, p. 1002. (Original article in Rev. Hyg. et Med. Infantiles, t. 3,
        no. 2, p. 113–155, 1904. Not seen)

(20) STETTER, A.
    1915. CATALASE AND REDUCTASE DETERMINATION IN COW'S MILK IN PRACTISE,
        AND THE RELATION BETWEEN CATALASE AND REDUCTASE ON THE ONE
        HAND AND THE SPECIFIC GRAVITY, FAT, AND ACIDITY ON THE OTHER.
        (Abstract.) *In* Exp Sta. Rec , v. 33, no. 5, p. 414. (Original article
        in Milchw. Zentbl., Bd. 43, No. 14, p. 369–381, 2 fig., 1914. Not seen.)

(21) STOKLASA, J., and others.
    1905. ON THE ISOLATION OF FERMENTATIVE ENZYMS FROM COWS' MILK. (Ab-
        stract.) *In* Exp. Sta Rec , v. 16, no. 7, p. 700–701. (Original article
        in Ztschr. Landw. Versuchw. Oesterr., Bd. 7, No. 11, p. 755–774, 1 fig.,
        1904.)

(22) VANDEVELDE, A J. J.
    1907. NOUVELLES RECHERCHES SUR LES FERMENTS SOLUBLES DE LAIT. 85 p.
        Bruxelles.

(23) VAN DE VELDE, H., and LANDSHEERE, J DE
1904. THE FERMENTS OF MILK: AN EXPERIMENTAL AND CRITICAL STUDY. (Abstract.) *In* Exp. Sta. Rec., v. 15, no. 10, p. 1003. (Original article in Ann. Soc. Medico-Chirurg. Anvers, 1903. Not seen.)

(24) VAN SLYKE, D. D.
1910. EINE METHODE ZUR QUANTITATIVEN BESTIMMUNG DER ALIPHATISCHEN AMINOGRUPPEN; EINIGE ANWENDUNGEN DERSELBEN IN DER CHEMIE DER PROTEINE DES HARNS UND DER ENZYME. *In* Ber. Deut. Chem Gesell., Jahrg. 43, No 16, p. 3170–3187, 1 fig.

(25) WASHBURN, R. M., and DAHLBERG, A. C.
1917. THE INFLUENCE OF SALT ON THE CHANGES TAKING PLACE IN STORAGE BUTTER. *In* Jour. Dairy Sci., v. 2, no. 2, p. 114–126. References, p. 126

(26) WILEY, H W., ed
1908. OFFICIAL AND PROVISIONAL METHODS OF ANALYSIS, ASSOCIATION OF OFFICIAL AGRICULTURAL CHEMISTS. AS COMPILED BY THE COMMITTEE ON REVISION OF METHODS. U. S Dept. Agr. Bur. Chem. Bul. 107 (rev.), 272 p., 13 fig. Reprinted in 1912.

(27) ZAITSCHEK, A., and SZONTAGH, F VON.
1905. COMPARATIVE INVESTIGATIONS ON THE CONTENT OF PROTEOLYTIC AND AMYLOLYTIC ENZYMS IN DIFFERENT KINDS OF MILK. (Abstract.) *In* Exp. Sta. Rec., v. 16, no. 6, p 597. (Original article in Arch. Physiol. [Pflüger], Bd. 104, No. 9/12, P. 539–549, 1904. Not seen.)

JOURNAL OF AGRICULTURAL RESEARCH

VOL. XI   WASHINGTON, D. C., DECEMBER 3, 1917   NO. 10

INFLUENCE OF THE DEGREE OF FATNESS OF CATTLE UPON THEIR UTILIZATION OF FEED

By HENRY PRENTISS ARMSBY, *Director*, and J. AUGUST FRIES, *Assistant Director*,
Institute of Animal Nutrition of The Pennsylvania State College

COOPERATIVE INVESTIGATIONS BETWEEN THE BUREAU OF ANIMAL INDUSTRY
OF THE UNITED STATES DEPARTMENT OF AGRICULTURE AND THE INSTITUTE
OF ANIMAL NUTRITION OF THE PENNSYLVANIA STATE COLLEGE

INTRODUCTION

In the fattening of cattle it is a common experience that the gain in live weight secured per unit of feed consumed diminishes as the fattening progresses. Little reflection is required to make it evident that this phenomenon may be the combined result of a variety of causes. A supposed lower utilization of feed by the fattened as compared with the thin animal has been regarded not uncommonly as one such cause. It has been supposed that with the progress of the fattening the body cells become less efficient in the manufacture of fat from other nutrients or that the cells of the adipose tissue, as they become loaded with fat, offer, as it were, an increasing resistance to the deposition of added fat, to overcome which requires an expenditure of energy. In either case a unit of a resorbed nutrient, such as dextrose for example, would yield less fat than in the thin animal, while the heat production of the body would be correspondingly increased and the net energy value of the feed reduced.

The investigation here reported was undertaken to test this view by means of a direct comparison of the utilization of feed energy by the same steer in ordinary condition and when well fattened. Although it includes only a single comparison on one animal, the results appear of some interest in view of the paucity of experimental evidence on this point

OUTLINE OF EXPERIMENT

The subject of the experiment was a pure-bred Shorthorn steer about 2 years and 9 months old at the beginning of the experiment. He was a very quiet and docile animal. During the winter of 1912–13 he was the subject of a series of respiration calorimeter trials (unreported). During the following summer he was on pasture and gained some 240

Journal of Agricultural Research,
Washington, D. C.
kw

Vol. XI, No. 10
Dec. 3, 1917
Key No. Pa. —4

(451)

pounds in weight.   On October 21 he weighed 1,141 pounds and was put
on a preliminary ration of the same feed mixture used in the experiment
proper.   The latter began on November 2, 1913.[1]   At that time the
animal was in good condition but not fat.   A standard feed, consisting
of uniform proportions of alfalfa hay and a mixture of concentrates,
was used throughout the experiment.   During period 1 an approximate
maintenance ration of this standard feed was given, while during period
2 the amount was increased to the maximum which the animal would
consume.   A comparison of these two periods in the same manner as in
previous investigations served to show the utilization of the feed by the
animal.

At the close of period 2, on December 22, the steer was placed in the
hands of the Animal Husbandry Department of the College for fattening
and was fed by them until March 14, during which time he gained about
300 pounds and was brought into prime condition.   The appearance of
the steer before and after fattening is shown by the two figures of plate
41, taken November 3 and March 14, respectively.

In the second half of the experiment, beginning March 15, 1914, the
comparison of periods 1 and 2 was repeated in the reverse order.   In
period 3 the animal was given as heavy a ration of the standard feed
mixture as it was thought he would consume, although in fact a little was
left uneaten.   This was followed by period 4, in which an approximate
maintenance ration was fed.   A comparison of periods 3 and 4 served to
show the utilization of feed energy by the fattened animal.

RATIONS AND PERIODS

The mixture of concentrates employed consisted of 1 part by weight
(air-dry) of cottonseed meal, 2 parts of wheat bran, and 6 parts of maize
meal.   The standard feed consisted of 2 parts by weight (air-dry) of this
mixture and 1 part of alfalfa hay.   The composition of the dry matter
of the samples of these materials taken is shown in Table 1 of the Appendix.   The slight fluctuations observed in the moisture content of the
feeding stuffs from period to period was insufficient to cause more than
an entirely insignificant variation in the proportions of dry matter supplied by the several feeding stuffs.   As Table 2 of the Appendix shows,
the protein was ample for a fattening animal of this age according to the
accepted standards.

The dates of the several periods and the rations consumed are shown
in Table I.

TABLE I —*Periods and rations*

Period.	Preliminary period.	Digestion period.	Daily rations.	
			Hay.	Concentrates.
			Kgm.	*Kgm.*
Period 1............	November 2–12	November 13–22	1. 7	3. 4
Transition..........	November 23–29	....		
Period 2............	November 30–December 10	December 11–20. ..	3. 5	7. 0
Period 3............	March 15–25 ...	March 26–April 4...	3. 8	7. 6
Period 4............	April 5–15.	April 16–25..	2. 0	4. 0

DIGESTIBILITY

The digestibility of the several rations was determined in the usual manner, the results in detail being contained in Table 2 of the Appendix. The feces in period 3 were very watery, but it does not appear that this resulted in any lower digestibility. The percentage digestibility of the several nutrients in the four periods is shown in Table II.

TABLE II —*Digestibility of rations*

Constituent.	Before fattening.		After fattening.	
	Period 1 (light ration).	Period 2 (heavy ration).	Period 3 (heavy ration).	Period 4 (light ration).
	Per cent.	*Per cent.*	*Per cent.*	*Per cent.*
Dry matter................................	76. 4	74. 2	74. 8	76. 7
Ash......................................	45. 9	42. 6	46. 5	51. 7
Organic matter...........................	78. 1	75. 6	76. 4	78. 1
Protein..................................	75. 6	71. 6	75. 9	75. 7
Non-protein..............................	62. 4	59. 8	23. 2	67. 7
Crude fiber..............................	43. 2	45. 1	41. 4	42. 9
Nitrogen-free extract.....................	85. 7	83. 6	84. 0	85. 8
Ether extract............................	82. 2	79. 6	84. 3	81. 8
Total nitrogen...........................	75. 2	71. 4	69. 8	75. 8
Carbon..................................	76. 2	74. 0	74. 6	76. 1
Energy..................................	75. 6	73. 4	74. 1	75. 6

EFFECT OF AMOUNT OF FEED.—Both before and after fattening the heavier rations showed a distinctly lower digestibility than the lighter ones, the exception being the crude fiber in periods 1 and 2 and the protein and ether extract in periods 3 and 4. The averages for dry matter, nitrogen, carbon, and energy are contained in Table III.

TABLE III.—*Average digestibility of light and heavy rations*

Constituent.	Light ration.	Heavy ration.
Dry matter...per cent..	76. 5	74. 5
Nitrogen...do....	75. 5	70. 6
Carbon..do....	76. 1	74. 3
Energy..do....	75. 6	73. 8
Dry matter eaten per 1,000 pounds of live weight.....pounds....	8. 62	10. 20

Qualitatively the results correspond to those of other experiments both here and elsewhere (*1, p. 613–618*)[1], but quantitatively the difference is less than has usually been observed, a fact for which no obvious reason appears.

EFFECT OF FATTENING.—Fattening appears to have had no distinct influence on the digestibility, the exceptions being the protein, non-protein, crude fiber, and ether extract in period 3. The average percentages for the same ingredients as in Table III are shown in Table IV.

TABLE IV.—*Average digestibility before and after fattening*

Constituent.	Before.	After.
	Per cent.	Per cent.
Dry matter..	75. 3	75. 8
Nitrogen..	73. 3	72. 8
Carbon..	75. 1	75. 4
Energy..	74. 5	74. 8

URINARY EXCRETION

The results upon the urine are recorded in Table 3 of the Appendix and are summarized in Table V.

TABLE V.—*Average daily excretion in urine*

Period No.	Nitrogen.	Carbon.	Energy.	Ratio of nitrogen to carbon.	Energy per gram of nitrogen.	Energy per gram of carbon.
	Gm.	Gm.	Calories.		Calories.	Calories.
1......................	79. 5	107. 5	1,015. 9	1:1. 35	12. 78	9. 45
2......................	139. 4	191. 6	1,712. 1	1:1. 37	12. 28	8. 94
3......................	155. 6	218. 9	2,045. 8	1:1. 41	13. 15	9. 34
4......................	107. 0	136. 1	1,318. 8	1:1. 27	12. 32	9. 65

The percentages of the nitrogen, carbon, and energy of the feed which were excreted in the urine are shown in Table VI, from which it appears that the relative losses through this channel were less on the heavier

[1] Reference is made by number (italic) to "Literature cited," p. 464.

rations of periods 2 and 3 than on the lighter rations, and that they were on the average a little greater in the fattened than in the unfattened animal.

TABLE VI —*Percentage losses in urine*

Period No.	Percentage of total—			Percentage of digestible—		
	Nitrogen.	Carbon.	Energy.[a]	Nitrogen.	Carbon.	Energy.[a]
1	71.04	5.25	5.16	94.53	6.89	6.83
2	61.54	4.58	4.55	86.26	6.19	6.20
3	61.84	4.81	4.91	88.56	6.45	6.62
4	80.51	5.69	5.38	107.75	7.47	7.11

[a] Corrected to N equilibrium.

COMBUSTIBLE GASES

Fries (5) has shown that the large number of respiration experiments executed here during the last 15 years seem to demonstrate beyond question that the fermentation gases given off by cattle on normal rations have substantially the composition of methane. The results of the present trials (Appendix, Table 5) are entirely in accord with those of earlier ones and show a ratio of hydrogen to carbon corresponding substantially with that of methane (1 : 2.976). The slight deficiency of hydrogen in most cases we ascribe to the difficulty of securing complete oxidation of the last traces of this element. Table VII shows the ratio of hydrogen to carbon and the computed amounts of methane, both per head and per 100 parts of digestible carbohydrates. In accord with our earlier results (3, 4) the extent of the fermentation of the carbohydrates, as measured by the methane excretion, was decidedly less on the heavier than on the lighter rations. The range of the results is substantially the same as in previous experiments on similar rations, particularly those in which alfalfa hay was used.

TABLE VII.—*Production of combustible gases*

Period No.	Ratio of hydrogen to carbon	Methane computed from carbon.	
		Total per day and head.	Per 100 Gms. of digested carbohydrates.[a]
		Gm.	*Gm.*
1	1:2.970	137.3	5.2
2	1:3.063	206.0	3.8
3	1:3.010	215.8	3.7
4	1:3.048	154.0	5.0

[a] Crude fiber plus nitrogen-free extract.

METABOLIZABLE ENERGY

Tables 2 to 5 of the Appendix afford data for computing the percentage losses of energy in the several excreta and the percentage metabolizable. On the assumption that the corresponding values for the alfalfa hay were the same as those found for the same hay in an experiment upon the same animal during the previous year, the corresponding figures for the concentrate mixture may also be computed. The results are recorded in Table VIII.

TABLE VIII —*Percentage distribution of energy*

Item.	Percentage losses.			Percentage metabolizable.
	In feces.	In urine.	In methane.	
Total ration:				
Period 1	24. 44	5. 16	9. 15	61. 25
Period 2	26. 61	4. 55	6. 72	62. 12
Period 3	25. 88	4. 91	6. 48	62. 73
Period 4	24. 42	5. 38	8. 80	61. 40
Assumed for hay	42. 60	5. 92	6. 77	44. 71
Computed for concentrates:				
Period 1	15. 62	4. 79	10. 30	69. 29
Period 2	18. 81	3. 88	6. 70	70. 61
Period 3	17. 80	4. 42	6. 34	71. 44
Period 4	15. 64	5. 11	9. 78	69. 47

As in earlier experiments, the heavier rations suffered relatively greater losses of energy in the feces but smaller ones in the urine and methane, so that the percentage of the gross energy which was metabolizable was slightly greater in the heavier rations. It must be remembered also that the metabolizable energy as here computed includes that evolved as heat in the methane fermentation. Estimating this at 6.07 Calories per gram of methane (*2, p. 468*) the proportion of the gross energy of the rations which was available to sustain the tissue metabolism averaged 57.24 per cent for the lighter rations and 59.43 per cent for the heavier. No distinct difference in this respect is manifest between the unfattened and fattened condition.

The results recorded in Table IX show a close agreement with those of earlier experiments on similar rations.

TABLE IX.—*Average losses of chemical energy and metabolizable energy*

Item.	Gross energy per kilogram of dry matter.	Losses of chemical energy per kilogram of dry matter.	Metabolizable energy.	
			Per kilogram of dry matter.	Per kilogram of digestible organic matter.
	Calories.	*Calories.*	*Calories.*	*Calories.*
Total ration:				
Period 1	4,488	1,739	2,749	3,720
Period 2	4,470	1,693	2,995	3,856
Period 3	4,481	1,670	2,811	3,886
Period 4	4,478	1,728	2,749	3,719
Average	4,478	1,699	2,779	3,820
Assumed for hay	4,334	2,396	1,938	3,507
Computed for concentrates:				
Period 1	4,562	1,401	3,161	3,785
Period 2	4,537	1,334	3,203	3,980
Period 3	4,554	1,300	3,253	4,014
Period 4	4,548	1,389	3,159	3,788
Average	4,549	1,343	3,206	3,925

BODY INCREASE

From the income and outgo of nitrogen and carbon recorded in the tables of the Appendix the amounts of protein and fat stored in the body may be computed upon the usual assumption that the stock of carbohydrates in the body remained substantially unchanged. The average results for the four periods are contained in Table X.

TABLE X —*Daily gains of protein and fat*

Period No.	Protein.	Fat.	Total organic matter.
	Grams.	*Grams.*	*Grams.*
1	14.4	89.0	103.4
2	120.0	822.0	942.0
3	107.4	615.9	723.3
4	− 51.0	− 28.3	− 79.3

The rations of periods 1 and 4, as intended, were approximately maintenance rations. Those of periods 2 and 3 were sufficient to support a tolerably rapid fattening. As appears from Table 8 of the Appendix, the measured gain of energy differed from that computed from the gains of protein and fat, being greater in every instance. The conclusions of subsequent paragraphs are based on the energy balances corrected to 12 hours' each standing and lying, but the results for protein and fat may serve as the basis for some approximate comparisons. If the percentage of organic matter in the total gain in live weight is assumed to have been the same as that observed by Lawes and Gilbert for fattening cattle—viz, 73.89—the gains of organic matter recorded in Table X are equivalent to a gain in weight of 1.275 kgm. (2.8 pounds) per day in period 2 and 0.979 kgm. (2.2 pounds) in period 3. The experimental periods were too short to permit very satisfactory conclusions to be drawn from the live weights of the animals. Those weights, taken at about 7 a. m., after the morning feeding and before the daily watering, are shown in the accompanying graph (fig. 1). The straight lines in periods 2 and 3 represent the rates of gain, as computed above, which appear to correspond fairly well with the observed weights. The averages of the last six weighings for each period were as follows:

```
Period 1.................................................490 kgm.
Period 2.................................................536 kgm.
Period 3.................................................655 kgm.
Period 4.................................................642 kgm.
```

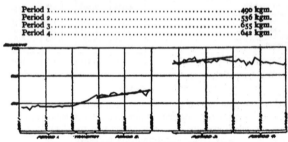

Fig. 1.—Graph of the fluctuations in the live weight of the steer.

A comparison of these estimated gains in weight with the feed consumption yields the results of Table XI. Obviously no great accuracy can be claimed for these computations, but nevertheless they show clearly a decreasing efficiency of the total feed.

TABLE XI.—*Feed consumed per unit of estimated gain in weight*

Period No.	Total feed.	Dry matter.	Digestible organic matter.
2	8.2	7.2	5.2
3	11.6	10.1	9.6

HEAT EMISSION AND PRODUCTION.

The measured emission of heat by the animal in the several periods and subperiods is recorded in Table 6 of the Appendix. In the periods on the heavier rations, both in the unfattened and fattened state, a notably larger proportion of the additional heat produced by the animal was disposed of by the evaporation of water, as appears from Table XII.

TABLE XII —*Paths of heat emission*

Period No	Radiation and conduction.	Latent heat of water vapor.
	Per cent.	Per cent.
1	75. 88	24. 12
2	52. 41	47. 59
3	54. 23	45. 77
4	68. 27	31. 73

Applying to the figures for heat emission the necessary corrections for the gain or loss of matter by the body and for the fluctuations of body temperature yields the results for heat production contained in Table 8 of the Appendix, while, in order to render the several periods comparable, a further correction to uniform standing and lying must be made, the corrected results being recorded in Table 9. The average daily heat production of the animal, corrected to a uniform period of 12 hours standing and 12 hours lying, as in our earlier experiments, was:

Period 1......................................10,905 Calories.
Period 2......................................16,511 Calories.
Period 3......................................19,992 Calories.
Period 4......................................14,095 Calories.

ENERGY EXPENDITURE CONSEQUENT ON FEED CONSUMPTION

The foregoing results show the same marked increase in heat production which has been uniformly found to follow an increase in the ration. From a quantitative comparison of the corresponding periods the energy expenditure per unit of total feed (hay and concentrates) consumed may be computed in the case of the unfattened and fattened animal, respectively. The results of this computation are contained in Table XIII. For the reasons stated in a previous paper (2, p. 471) no attempt has been made to correct these results for the differences in live weight.

TABLE XIII.—*Energy expenditure per kilogram of dry matter of total ration*

Condition and period No	Dry matter eaten.	Heat production.
Unfattened:	Grams.	Calories.
Period 2	9, 146. 3	16, 511
Period 1	4, 462. 9	10, 905
Difference	4, 683. 4	5, 606
Difference per kilogram of dry matter	....	1, 197
Fattened:		
Period 3	9, 911. 6	19, 992
Period 4	5, 215. 6	14, 095
Difference	4, 696. 0	5, 897
Difference per kilogram of dry matter	....	1, 256

The results are of quite the same order of magnitude as those obtained in previous experiments on similar mixed rations of hay and grain (*2, p. 477; 4, p. 385*). In the fattened animal there appears to have been an increase of about 5 per cent in the energy expended in the various processes intermediate between the prehension of the feed and the storage of protein and fat in the tissues. To this extent the experiment sustains the view outlined in the introductory paragraphs. The difference, however, is small and it is perhaps questionable whether it exceeds the experimental error. At any rate, it is far from accounting for the very marked difference in the economic utilization of the feed which is indicated by the approximate calculation of Table XI, and is of little significance in comparison with another factor to be considered immediately.

The relations shown in Table XIII may also be expressed in another way by comparing the percentages of the metabolizable energy which were recovered in the gain made in the unfattened and fattened states, respectively, as shown by the energy balances (Table XIV).

TABLE XIV.—*Percentage of metabolizable energy recovered in gain*

Condition and period No	Metaboliz- able energy.	Body gain.	Percentage recovered.
	Calories.	*Calories.*	
Unfattened:			
Period 2....................................	25, 398	8, 887	....
Period 1....................................	12, 269	1, 364	....
Difference................................	13, 129	7, 523	57. 3
Fattened:			
Period 3....................................	27, 865	7, 873	....
Period 4....................................	14, 338	243	....
Difference................................	13, 527	7, 630	56. 4

The foregoing results accord with those recently reported by Moulton(*8*), who computes from the results of comparative slaughter tests that in a fat and a very fat steer 53.39 per cent and 52.49 per cent, respectively, of the metabolizable energy supplied in excess of maintenance was recovered in the gain.

MAINTENANCE REQUIREMENT

The comparisons of Table XIII also afford data for computing the fasting katabolism of the steer in the manner described in an earlier publication (*3, p. 53*).

Each kilogram of dry matter consumed increased the katabolism by 1,197 Calories in the unfattened and by 1,256 Calories in the fattened state. It may be computed, therefore, by how much the katabolism would have been reduced had the feed been entirely withdrawn, while subtracting this amount from the observed heat production will give the basal katabolism. The calculations for periods 1 and 4 are:

Period 1 (unfattened), $10,905 - (1,197 \times 4.4629) = 5,563$ Calories.
Period 4 (fattened), $14,095 - (1,256 \times 5.2156) = 7,544$ Calories.

Similar computations for periods 2 and 3 would, of course, yield the same results.

The results per head and also those computed from the averages of the weights reported on p. 458 to a uniform live weight of 1,000 pounds for the sake of comparison with other data are shown in Table XV.

TABLE XV.—*Computed basal katabolism per day*

Condition.	Per head.	Per 1,000 pounds, live weight.	
		In proportion to weight.	In proportion to two-thirds power of weight.
	Calories.	*Calories.*	*Calories.*
Unfattened...	5,563	4,919	5,125
Fattened...	7,544	5,275	5,943

In the unfattened state the animal had a rather low basal katabolism, the average of 23 similar determinations by the writers (*1, p. 289*) being 5,906 Calories per 1,000 pounds, as compared with 5,125 Calories in this case.

It is evident that the major factor determining the lower economic efficiency in the fattened state was the very marked increase (36 per cent) in the basal katabolism. Doubtless this increase was due in part to the greater body weight to be supported while standing; but, as the table shows, the katabolism increased more rapidly than the weight or the body surface as estimated by the Meek formula. Apparently the accumulation of fat tended in some way to stimulate the general metabolism. These results are quite in harmony with those obtained by Kellner and Köhler (*6; 7, p. 14*) in experiments on the same subject. This greater maintenance requirement, together with the relatively somewhat smaller feed consumption, was chiefly responsible for the more expensive gains by the fattened animal.

NET ENERGY VALUES

From the foregoing the net energy values of the rations may also be computed as shown in Table XVI.

TABLE XVI —*Net energy values of total rations per kilogram of dry matter*

	Gross energy (Table IX).	Losses in excreta (Table IX).	Heat increment (Table XIII).	Net energy value.
	Calories.	*Calories.*	*Calories.*	*Calories.*
Period 1	4, 488	1, 739	} 1, 197 {	1, 552
Period 2	4, 470	1, 693		1, 580
Period 3	4, 481	1, 670	} 1, 256 {	1, 555
Period 4	4, 478	1, 728		1, 494
Average for light rations	4, 483	1, 733	1, 227	1, 523
Average for heavy rations.	4, 476	1, 682	1, 227	1, 567
Average in unfattened condition	4, 479	1, 716	1, 197	1, 566
Average in fattened condition	4, 480	1, 699	1, 256	1, 525

On the average of the experiments of the previous year on the same animal 1 kgm. of dry matter of the alfalfa hay increased the heat production of the animal by 1,018 Calories. If the values for the hay are assumed to be the same as those found in the previous year's experiment, the net energy values of the mixture of concentrates may also be computed as in Tables XVII and XVIII. Obviously this method of computation ascribes the entire difference between the results in the several periods to the concentrates and therefore tends to exaggerate it relatively.

TABLE XVII.—*Computed energy expenditure per kilogram of dry matter of concentrates*

Period No.	Dry matter eaten.		Total heat production.
	Hay.	Concentrates.	
	Gm.	*Gm.*	*Calories.*
Period 2	3, 088. 4	6, 057. 9	16, 511
Period 1	1, 507. 6	2, 955. 3	10, 905
Difference	1, 580. 8	3, 102. 6	5, 606
Difference due to hay			1, 609
Differences due to concentrates			3, 997
Difference per kilogram of concentrates			1, 288
Period 3	3, 335. 1	6, 576. 5	19, 992
Period 4	1, 754. 3	3, 461. 3	14, 095
Difference	1, 580. 8	3, 115. 2	5, 897
Difference due to hay			1, 609
Difference due to concentrates			4, 288
Difference per kilogram of concentrates			1, 376

TABLE XVIII.—*Computed net energy values of concentrates per kilogram of dry matter*

Period No	Gross energy (Table IX)	Losses in excreta (Table IX)	Heat increment (Table XVII).	Net energy value.
	Calories.	*Calories.*	*Calories.*	*Calories.*
Period 1....................................	4,562	1,401	1,288	1,873
Period 2....................................	4,537	1,334		1,915
Period 3....................................	4,554	1,300	1,376	1,878
Period 4....................................	4,548	1,389		1,783
Average for light rations..............	4,555	1,395	1,332	1,828
Average for heavy rations............	4,546	1,317	1,332	1,897
Average in unfattened condition......	4,550	1,368	1,288	1,894
Average in fattened condition .	4,551	1,345	1,376	1,830

SUMMARY

(1) A steer in medium condition received, in two successive periods, an approximate maintenance ration and a fattening ration of the same standard mixture of hay and concentrates. The animal was then fattened and the comparison of a maintenance ration and a fattening ration of the same standard feed mixture was repeated.

(2) The digestibility of the lighter ration was in both cases greater than that of the heavier one, although the difference was less than has usually been found in such comparisons.

(3) The corresponding rations were digested equally well by the fattened and unfattened animal.

(4) The relative losses of nitrogen, carbon, and energy in the urine were less on the heavy than on the light rations and a little greater in the fattened than in the unfattened state.

(5) The production of combustible gases (methane), both as compared with the total feed and with the digestible carbohydrates, was notably less on the heavier than on the lighter rations. There was no distinct difference in this respect between the fattened and the unfattened states.

(6) In consequence of the smaller losses in the urine, and especially in the combustible gases, the percentage of the gross energy of the feed which was metabolizable was greater in the heavier than in the lighter rations. No difference in this respect was observed between the fattened and the unfattened animal.

(7) By far the larger share of the additional heat produced on the heavier rations was eliminated by means of evaporation of water.

(8) The heat increment resulting from the consumption of a unit of feed was but little greater in the fattened than in the unfattened condition. Consequently the net energy values of the feed and the percentages of the metabolizable energy which were available for gain were but slightly less in the fattened than in the unfattened condition.

(9) The maintenance requirement of the steer was increased 36 per cent by a 3 months' fattening in which the live weight was increased by about 300 pounds. This increase was greater than corresponded to the increase in weight or in computed body surface.

(10) The lower economic efficiency of the fattened animal in this experiment was due chiefly to his higher maintenance requirement and only to a small extent if at all to a difference in the utilization of the surplus of feed over the maintenance requirement.

LITERATURE CITED

(1) ARMSBY, H. P.
    1917. THE NUTRITION OF FARM ANIMALS. 743 p., 44 fig. New York.
(2) ———— and FRIES, J. A.
    1915. NET ENERGY VALUES OF FEEDING STUFFS FOR CATTLE. In Jour. Agr. Research, v. 3, no. 6, p. 435–491, 2 fig. Literature cited, p. 489–491.
(3) ———— ————
    1911. THE INFLUENCE OF TYPE AND OF AGE UPON THE UTILIZATION OF FEED BY CATTLE. U. S. Dept. Agr. Bur. Anim. Indus. Bul. 128, 245 p., 17 fig., 3 pl.
(4) ———— ———— and BRAMAN, W. W.
    1916. ENERGY VALUES OF RED-CLOVER HAY AND MAIZE MEAL. In Jour. Agr. Research, v. 7, no. 9, p. 379–387.
(5) FRIES, J. A.
    [1912] THE COMBUSTIBLE GASES EXCRETED BY CATTLE. In Orig. Commun. 8th Internat. Cong. Appl. Chem., v. 15, p. 109–119.
(6) KELLNER, O., and KÖHLER, A.
    1898. UNTERSUCHUNGEN ÜBER DEN NAHRUNGS- UND ENERGIE-BEDARF VOLL-JÄHRIGER GEMÄSTETER OCHSEN. In Landw. Vers. Stat., Bd 50, Heft 3/4, p. 245–296.
(7) ———— ————
    1900. UNTERSUCHUNGEN ÜBER DEN STOFF- UND ENERGIEUMSATZ DES ERWACH-SENEN RINDES BEI ERHALTUNGS- UND PRODUKTIONSFUTTER. Landw. Vers. Stat., Bd. 53, 474 p.
(8) MOULTON, C. R.
    1917. THE AVAILABILITY OF THE ENERGY OF FOOD FOR GROWTH. In Jour. Biol. Chem., v. 31, no. 2, p. 389–394.

APPENDIX

The principal numerical data obtained in the experiments are recorded in the following tables. The computations involved have been carried out beyond the probable limit of accuracy of the experimental methods in order to guard against a possible accumulation of arithmetical errors.

COMPOSITION OF FEEDING STUFFS

The alfalfa hay used was taken from a stock of cut hay prepared in 1911–12 for an experiment of that year and again used in an experiment in 1912–13. The hay was old and rather hard, but was well eaten, except in period 3. Enough maize meal for two periods was freshly ground at the beginning of the first and third periods, respectively. The bran and cottonseed meal were purchased from the Dairy Husbandry Department. All the feeds were safely stored and protected from vermin.

The protein and nonprotein content of the feeding stuffs was computed from the nitrogen, using the following factors:

For protein:
Alfalfa hay.. 6.25
Wheat bran... 5.70
Maize meal... 6.0
Cottonseed meal 5.5
For nonprotein 4.7

TABLE 1 —*Composition of dry matter of feeding stuffs*

Feed and period.	Ash.	Pro-tein.	Non-protein.	Crude fiber.	Nitro-gen-free extract.	Ether extract.	Total nitro-gen.	Carbon.	Heat of combustion per kilo-gram.
	Per ct.	Per ct.	Per ct.	Per ct.	Per ct.	Per ct.	Per ct.	Per ct.	Calories.
Alfalfa hay:									
Period 1.............	9.66	12.13	3.01	29.60	43.59	2.01	2.58	45.04	4,344
Period 2.............	9.11	11.34	3.02	32.31	42.41	1.81	2.46	45.12	4,340
Periods 3 and 4......	9.49	11.85	3.05	30.30	43.04	2.27	2.54	45.16	4,339
Average............	9.42	11.77	3.03	30.74	43.01	2.03	2.53	45.11	4,341
Average of 3 samples, 1911-12....	9.06	12.39	2.86	30.10	43.63	1.96	2.59	44.98	4,368
Average of 7 samples, 1912-13....	9.26	11.29	3.14	30.46	44.09	1.76	2.47	44.90	4,350
Wheat bran:									
Period 1.............	6.54	14.91	1.48	9.70	62.64	4.73	2.93	45.85	4,597
Period 2.............	6.44	14.52	1.92	10.09	62.22	4.81	2.96	45.98	4,587
Periods 3 and 4......	6.50	14.56	1.80	9.77	62.23	5.14	2.94	45.93	4,573
Average............	6.49	14.67	1.73	9.85	62.37	4.89	2.94	45.92	4,586
Maize meal:									
Period 1.............	1.53	9.40	.22	2.36	82.16	4.33	1.61	46.22	4,488
Period 2.............	1.44	9.46	.25	2.27	82.35	4.23	1.63	45.74	4,457
Periods 3 and 4......	1.49	9.56	.22	2.01	82.50	4.22	1.64	45.97	4,465
Average............	1.49	9.47	.23	2.21	82.34	4.26	1.63	45.98	4,470
Cottonseed meal:									
Period 1.............	6.52	33.43	1.87	12.72	37.32	8.14	6.48	48.11	4,917
Period 2.............	6.37	33.30	2.22	12.31	37.70	8.10	6.53	47.99	4,904
Periods 3 and 4......	7.41	34.77	2.86	9.48	36.81	8.67	6.93	48.19	4,969
Average............	6.77	33.83	2.32	11.50	37.28	8.30	6.65	48.10	4,930

DIGESTIBILITY OF RATIONS

The digestibility of the rations was determined in the usual manner. In period 3 residues of mixed hay and grain were left uneaten. In the computation it is assumed that the proportions of the different feeding stuffs in the residues were the same as in the original ration, the actual composition of the residues agreeing closely with that computed on this assumption. The digestibility of the single feeding stuffs was not determined.

TABLE 2 —*Digestibility of rations*

Period and feed.	Dry matter.	Ash.	Organic matter.	Protein.	Non-protein.	Crude fiber.
	Gm.	*Gm.*	*Gm.*	*Gm.*	*Gm.*	*Gm.*
Period 1:						
Alfalfa hay............	1,507.6	145 7	1,361 9	182.8	45.3	446.2
Cottonseed meal ..	343 8	22 4	321 4	114.9	6.4	43.7
Wheat bran............	647 5	42 3	605 3	96.5	9.6	62 8
Maize meal............	1,964.0	30 0	1,934.0	184.7	4.2	46 3
Total fed	4,462.9	240.4	4,222.5	578.9	65.5	599.0
Feces.................	1,054.5	130.1	924.4	141.1	24.6	340.5
Digested............	3,408.4	110.3	3,298.1	437.8	40.9	258.5
Per cent digested....	76.37	45.88	78.11	75.63	62.44	43.16
Period 2:						
Alfalfa hay............	3,088.4	281.4	2,807.0	350.2	93.3	997.8
Cottonseed meal........	696.2	44.4	651.8	231.9	15.5	85.7
Wheat bran............	1,342.8	86.5	1,256.3	194.9	25.8	135.4
Maize meal............	4,018.9	57.7	3,961.2	380.3	10.0	91.1
Total fed............	9,146.3	470.0	8,676.3	1,157.3	144.6	1,310.0
Feces.................	2,358.7	269.6	2,089.1	328.2	58.1	719.9
Digested............	6,787.6	200.4	6,587.2	829.1	86.5	590.1
Per cent digested .	74.21	42.64	75.92	71.64	59.82	45.05
Period 3:						
Alfalfa hay............	3,508.1	333.0	3,175.1	415.7	106.9	1,063.1
Cottonseed meal........	813.3	60.3	753.0	282.8	23.2	77.1
Wheat bran............	1,540.0	100.0	1,440.0	224.2	27.7	150.5
Maize meal............	4,569.3	68.3	4,501.0	436.7	10.1	91.7
Total fed.	10,430.7	561.6	9,869.1	1,359.4	167.9	1,382.4
Refused................	519.1	37.2	481.9	67.7	8.4	77.3
Total eaten..........	9,911.6	524.4	9,387.2	1,291.7	159.5	1,305.1
Feces.................	2,497.4	280.4	2,217.0	311.3	122.5	764.9
Digested............	7,414.2	244.0	7,170.2	980.4	37.0	540.2
Per cent digested......	74.80	46.53	76.38	75.90	23.20	41.39
Period 4:						
Alfalfa hay............	1,754.3	166.5	1,587.8	207.9	53.4	531.6
Cottonseed meal........	406.6	30.1	376.5	141.4	11.6	38.5
Wheat bran............	770.0	50.0	720.0	112.1	13.9	75.3
Maize meal............	2,284.7	34.2	2,250.5	218.4	5.0	45.9
Total fed .	5,215.6	280.8	4,934.8	679.8	83.9	691.3
Feces.................	1,215.4	135.7	1,079.7	164.9	27.1	394.8
Digested............	4,000.2	145.1	3,855.1	514.9	56.8	296.5
Per cent digested	76.70	51.67	78.12	75.74	67.70	42.89

TABLE 2.—*Digestibility of rations*—Continued

Period and feed.	N.-free extract.	Ether extract.	Total nitrogen.	Carbon.	Energy.
Period 1:	*Gm.*	*Gm.*	*Gm.*	*Gm.*	*Calories.*
Alfalfa hay................	657. 2	30. 3	38. 9	679. 1	6, 549. 35
Cottonseed meal.............	128. 3	28. 0	22. 3	165. 4	1, 690. 39
Wheat bran.................	405. 6	30. 6	19. 0	296. 9	2, 976. 70
Maize meal.................	1, 613. 7	85. 1	31. 7	907. 7	8, 813. 63
Total fed.................	2, 804. 8	174. 0	111. 9	2, 049. 1	20, 030. 07
Feces...................	400. 1	30. 9	27. 8	488. 6	4, 895. 44
Digested................	2, 404. 7	143. 1	84. 1	1, 560. 5	15, 134. 63
Per cent digested..........	85. 74	82. 24	75. 16	76. 16	75. 56
Period 2:					
Alfalfa hay................	1, 309. 7	56. 0	75. 9	1, 393. 6	13, 402. 54
Cottonseed meal.............	262. 4	56. 4	45. 4	334. 1	3, 414. 42
Wheat bran.................	835. 6	64. 5	39. 7	617. 4	6, 159. 01
Maize meal.................	3, 309. 7	170. 1	65. 5	1, 838. 0	17, 910. 55
Total fed.................	5, 717. 4	347. 0	226. 5	4, 183. 1	40, 886. 52
Feces...................	940. 2	70. 8	64. 9	1, 086. 8	10, 879. 01
Digested................	4, 777. 2	276. 2	161. 6	3, 096. 3	30, 007. 51
Per cent digested..........	83. 56	79. 60	71. 35	74. 02	73. 39
Period 3:					
Alfalfa hay................	1, 509. 6	79. 8	89. 2	1, 584. 1	15, 220. 3
Cottonseed meal.............	299. 4	70. 5	56. 4	391. 9	4, 041. 3
Wheat bran.................	958. 3	79. 2	45. 2	707. 3	7, 042. 4
Maize meal.................	3, 769. 8	192. 6	74. 9	2, 100. 5	20, 399. 8
Total fed.................	6, 537. 1	422. 1	265. 7	4, 783. 8	46, 703. 8
Refused...................	315. 7	12. 8	14. 1	234. 3	2, 287. 0
Total eaten................	6, 221. 4	409. 3	251. 6	4, 549. 5	44, 416. 8
Feces...................	997. 3	64. 1	75. 9	1, 156. 2	11, 494. 1
Digested................	5, 224. 1	345. 2	175. 7	3, 393. 3	32, 922. 7
Per cent digested..........	83. 96	84. 34	69. 83	74. 59	74. 12
Period 4:					
Alfalfa hay................	754. 9	39. 9	44. 6	792. 2	7, 611. 2
Cottonseed meal.............	149. 7	35. 2	28. 2	195. 9	2, 020. 4
Wheat bran.................	479. 2	39. 6	22. 6	353. 7	3, 521. 2
Maize meal.................	1, 884. 9	96. 3	37. 5	1, 050. 3	10, 200. 1
Total fed.................	3, 268. 7	211. 0	132. 9	2, 392. 1	23, 352. 9
Feces...................	465. 8	38. 4	32. 2	571. 0	5, 704. 0
Digested................	2, 802. 9	172. 6	100. 7	1, 821. 1	17, 648. 9
Per cent digested..........	85. 75	81. 80	75. 77	76. 13	75. 57

URINARY EXCRETION

The collection was made with a urine funnel in the usual manner. There were occasional losses which were taken up with distilled water and weighed and analyzed separately. The weights given in the Table 3 include the water used to rinse out the urine funnel and tube. There is a

Total solids.　　Total nitrogen.　　Total carbon.

Total.

			Gm.	Per cent.	Gm.	Per cent.	Gm.	
		5.399	3,217.5	1.315	783.70	1.779	1,060.2	
			45.0	1.354	11.0	......	14.8	
......, ...	......		3,262.5	......	794.7	......	1,075.0	
6,040.4	......		326.3	......	79.5	......	107.5	
96,277.0	1.0364	5.597	5,388.6	1.434	1,380.3	1.970	1,896.7	
2,268.0	......	......	54.7	......	14.0	......	19.2	
98,545.0	......	......	5,443.3	......	1,394.3	......	1,915.9	......
9,854.5	......	......	544.3	......	139.4	......	191.6	......
75,120.0	1.034	6.560	5,379.5	1.549	1,163.8	1.996	1,636.8	186.55
5,506.0	......	......	375.8	......	81.3	......	114.3	......
80,626.0	......	......	5,755.3	......	1,245.1	......	1,751.1	......
10,078.3	......	......	719.4	......	155.6	......	218.9	......
0	1.031	4.950	4,106.5	1.289			1,361.4	97
0		..	410.7	......			136.1	......

EPIDERMAL TISSUE

was thoroughly brushed before entering the calorimete
er leaving it. The hair and dandruff obtained in the second
the small amounts swept up in the calorimeter were weighed
ontent of nitrogen, carbon, and energy determined. The
pped at the beginning of the experiment and at its close
r, and the same determinations made upon the hair obtained
clipping. The total of hair and brushings is regarded as
the production of epidermal tissue (Table 4).

TABLE 4.—*Average daily production of epidermal tissue*

GASEOUS EXCRETION

The excretion of carbon dioxid and methane, the latter being computed from the total carbon found in the combustible gases, is given in table 5:

TABLE 5.—*Gaseous excretion*

Period.	Water.	Carbon dioxid.	In combustible gases.		Methane computed from carbon.	Ratio, hydrogen to carbon.
			Hydrogen.	Carbon.		
Period 1:	*Gm.*	*Gm.*	*Gm.*	*Gm.*	*Gm.*	*1:*
Subperiod 1.............	2,320.90	2,305.76	16.58	47.39	63.32	2.858
Subperiod 2.............	2,173.04	2,310.16	17.27	51.93	69.39	3.007
First day...........	4,493.94	4,615.92	33.85	99.32	132.71	2.934
Subperiod 3.............	2,257.43	2,301.85	17.78	53.04	70.88	2.983
Subperiod 4.............	2,234.52	2,350.06	17.53	53.12	70.99	3.030
Second day..........	4,491.95	4,651.91	35.31	106.16	141.87	3.006
Average.............	4,492.95	4,633.92	34.58	102.74	137.29	2.970
Period 2:						
Subperiod 1.............	7,118.08	3,738.82	24.92	76.24	101.88	3.059
Subperiod 2.............	6,550.47	3,820.64	25.97	77.93	104.13	3.001
First day...........	13,668.55	7,559.46	50.89	154.17	206.01	3.029
Subperiod 3.............	6,712.28	3,730.76	25.90	78.60	105.03	3.035
Subperiod 4.............	6,965.47	3,727.53	23.88	75.54	100.94	3.163
Second day..........	13,677.75	7,458.29	49.78	154.14	205.97	3.096
Average.............	13,673.15	7,508.88	50.34	154.16	205.99	3.063
Period 3:						
Subperiod 1.............	5,480.06	4,319.04	24.97	75.05	100.28	3.005
Subperiod 2.............	7,662.40	4,534.97	26.93	81.01	108.25	3.008
First day...........	13,142.46	8,854.01	51.90	156.06	208.53	3.007
Subperiod 3.............	9,998.87	4,614.54	27.46	81.96	109.52	2.985
Subperiod 4[a]...........	10,011.85	4,678.52	27.94	84.92	113.48	3.039
Second day..........	20,010.72	9,293.06	55.40	166.88	223.00	3.012
Average.............	16,576.59	9,073.54	53.65	161.47	215.77	3.010
Period 4:						
Subperiod 1.............	3,707.71	2,937.49	19.96	61.04	81.57	3.058
Subperiod 2.............	3,972.70	2,928.18	18.58	55.95	74.76	3.012
First day...........	7,680.41	5,865.67	38.54	116.99	156.33	3.033
Subperiod 3.............	3,924.83	2,963.99	18.77	57.39	76.69	3.058
Subperiod 4.............	3,925.62	2,958.43	18.30	56.14	75.01	3.068
Second day..........	7,850.45	5,922.42	37.07	113.53	151.70	3.063
Average.............	7,765.43	5,894.05	37.81	115.26	154.02	3.048

a Computed from results for last 6 hours.

HEAT EMISSION

The amounts of heat given off by the animal by radiation and conduction and as latent heat of water vapor are given in Table 6.

TABLE 6.—*Heat emission*

Period.	By radiation and conduction.	As latent heat of water vapor.	Total.
Period 1:	*Calories.*	*Calories.*	*Calories.*
Subperiod 1.............................	4, 185. 0	1, 362. 4	5, 547. 4
Subperiod 2.............................	4, 182. 3	1, 275. 6	5, 457. 9
First day.............................	8, 367. 3	2, 638. 0	11, 005. 3
Subperiod 3.............................	4, 177. 7	1, 325. 1	5, 502. 8
Subperiod 4.............................	4, 048. 9	·1, 311. 7	5, 360. 6
Second day.........................	8, 226. 6	2, 636. 8	10, 863. 4
Average............................	8, 297. 0	2, 637. 4	10, 934 4
Period 2:			
Subperiod 1.............................	4, 418. 3	4, 178. 3	8, 596. 6
Subperiod 2.............................	4, 433. 2	3, 845. 1	8, 278. 3
First day.............................	8, 851. 5	8, 023. 4	16, 874. 9
Subperiod 3.............................	4, 795. 2	3, 940. 1	8, 735. 3
Subperiod 4.............................	4, 029. 0	4, 088. 7	8, 117. 7
Second day.........................	8, 824. 2	8, 028. 8	16, 853. 0
Average............................	8, 837. 9	8, 026. 1	16, 864. 0
Period 3:			
Subperiod 1............................:	7, 336. 8	3, 216. 8	10, 553. 6
Subperiod 2.............................	5, 765. 0	4, 497. 8	10, 262. 8
First day.............................	13, 101. 8	7, 714. 6	20, 816. 4
Subperiod 3.............................	5, 007. 1	5, 869. 3	10, 876. 4
Subperiod 4.............................	4, 953. 6	5, 827. 7	10, 831. 3
Second day.	9, 960. 7	11, 747. 0	21, 707. 7
Average............................	11, 531. 3	9, 730. 8	21, 262. 1
Period 4:			
Subperiod 1.............................	4, 725. 0	2, 176. 4	6, 901. 4
Subperiod 2.............................	4, 800. 9	2, 332. 0	7, 132. 9
First day..	9, 525 9	4, 508. 4	14, 034 3
Subperiod 3 . .	5, 194. 4	2, 303. 9	7, 498. 3
Subperiod 4	4, 890. 3	2, 304. 3	7, 194. 6
Second day.........................	10, 084. 7	4, 608. 2	14, 692. 9
Average............................	9, 805. 3	4, 558. 3	14, 363. 6

HEAT PRODUCTION

To ascertain the heat production, the measured heat emission as recorded in Table 6, must be corrected for any storage of heat in the body, due either to fluctuations of body temperature or to a gain or loss of matter by the body.

The rectal temperature of the animal was taken after it had stood in the calorimeter for about four hours—that is, about one hour before the beginning of the first subperiod—and again immediately after leaving the calorimeter. It would appear probable that the former temperature may be taken without serious error as the body temperature at the beginning of the experiment. Table 7 shows the body temperatures as taken and likewise the live weights of the animal when leaving the calorimeter.

TABLE 7 —*Body temperatures and live weights*

Period No	Temperature.			Live weight when leaving calorimeter.
	Entering.	Leaving.	Difference.	
	°C.	°C.	°C.	Kgm.
1	38. 8	38. 2	−0. 6	493
2	38. 7	38. 9	+0. 2	542
3	39. 0	39. 3	+0. 3	664
4	38. 7	38. 6	−0. 1	642

If the specific heat of the body is assumed to be 0.8, the corresponding correction to be applied to the heat emission, computed on the live weight when leaving the calorimeter, would be as shown in the second column of Table 8. The correction for the gain or loss of matter by the body is based on the observed gain or loss of protein, fat, and water and has been computed in the manner described in a previous paper (1).

Applying these corrections gives the results recorded in Table 8 for the heat production as compared with that computed in the usual way from the balance of nitrogen and carbon.

CORRECTION TO UNIFORM STANDING AND LYING

The proportion of time spent in the standing and lying positions, respectively, during the respiration calorimeter trials varied considerably, the percentage lying ranging, on the single days, from 26.7 to 58 4, equivalent, respectively, to 384.5 minutes and 840.5 minutes per 24 hours. The corresponding range for the entire 48-hour periods was 773 to 1,662 minutes per 48 hours, or from 26.8 to 57.7 per cent. In order to render the several periods comparable, the heat production has been computed to 12 hours standing and 12 hours lying, respectively, by the method described in a previous publication (2, *p. 454*).

	Heat emission.	Correction for body temperature.	Correction for body gain per day.	Observed heat production.	Computed heat production.	
	Calories.	*Calories.*	*Calories.*	*Calories.*	*Calories.*	*Per cent*
....	11,005.3	....	− 53.5	10,951.8	11,198.2	102.3
......	10,863.4	....	− 21.7	10,841.7	11,283.5	104.1
d for ire ...	10,934.4	−236.6	− 37.6	10,660.2	11,240.9	105.4
......	16,874.9	....	−165.2	16,709.7	17,106.7	102.4
......	16,853.0	....	−219.4	16,633.6	16,762.8	100.8
d for ire ...	16,864.0	+ 86.7	−192.3	16,758.4	16,934.8	101.1
......	20,816.4	....	+104.1	20,920.5	20,702.4	99.0
......	21,707.7	....	−218.3	21,489.4	22,129.6	103.0
d for re ...	21,262.1	+159.4	− 57.1	21,364.4	21,416.0	100.2
......	14,034.3	....	−156.5	13,877.8	14,611.2	105.3
......	14,692.9	....	−121.9	14,571.0	14,820.8	101.0
d for	14,363.6	− 51.4	−139.2	14,173.0	14,716.0	103.8

No.   **First day.**   **Second day.**

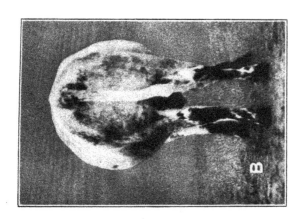

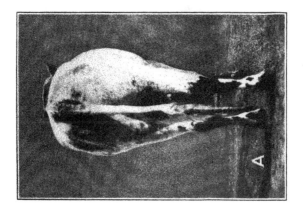

PODBLIGHT OF THE LIMA BEAN CAUSED BY DIAPORTHE PHASEOLORUM

By L. L. HARTER,

Pathologist, Cotton, Truck, and Forage Crop Disease Investigations, Bureau of Plant Industry, United States Department of Agriculture

INTRODUCTION

The Lima bean (*Phaseolus lunatus* L.), compared with some of the other truck crops, is relatively unimportant from a commercial standpoint. It is grown in a small way in the garden by nearly every farmer, and on a commercial scale in a few sections of the United States, particularly along the Atlantic seaboard and along the Pacific coast Like many other crops grown only for home use or on a commercial scale in a limited area, its several pests and diseases have either been overlooked or ignored. The Lima bean, however, has its share of diseases, some of which have been fairly well studied, while little is known about others. To the latter class belongs the podblight, a disease of considerable economic importance some seasons in commercial fields. It was first discovered by Halsted (16)[1] in 1891, in New Jersey. Further than this nothing is known as to the time and place of the origin of the disease.

There is no evidence from the search of the literature that the disease occurs outside of the United States, although it is impossible to state definitely that it does not, in view of the great number of species of Phoma, Phyllosticta, and closely related genera under which the casual fungus may have been described. A careful search through the descriptions of species of the genera Phoma, Phyllosticta, and Phomopsis on different species of Phaseolus has revealed none located in foreign countries identical with the organism causing the podblight. It therefore appears safe to assume at the present time that the disease is indigenous to the United States.

To judge from published accounts, Halsted (16) was the first to recognize this disease and found it causing much damage to the pods of pole Lima beans in New Jersey in the fall of 1891. He attributed it to a species of Phyllosticta, pointing out at the same time that large blotches were also produced on the leaves. That he studied the disease with some care and did not content himself with a few passing observations is evident from the fact that he found that the seeds from infected pods, when placed under a suitable environment, such as between moist cloths or paper, developed the fruiting bodies of the fungus.

---

[1] Reference is made by number (italic) to "i" Lterature cited." p. 502-504.

Journal of Agricultural Research,
Washington, D. C.
kx

Vol. XI, No. 10
Dec. 3, 1917
Key No G—128

In 1892 Ellis and Everhart (*15, p. 158*) evidently collected the same organism on the pods of *Phaseolus lunatus* at Newfield, N. J., and described it as a new species of Phoma, *Phoma subcircinata*, pointing out that it differed from *Phoma leguminum* West. in the subcircinate arrangement of the pycnidia and the rather larger binucleate sporules. Halsted later apparently concurs with Ellis and Everhart that this is a new species of Phoma, for in 1901 he (*17*) refers to the disease again by that name and enlarges his earlier description. Two illustrations are appended to his article, one showing the characteristic appearance of the disease on the pods and the other on the leaves, both typical of the trouble as we know it at the present time.

DISTRIBUTION, PREVALENCE, AND LOSS

For 14 years after this disease was found in New Jersey it had either not appeared in any other State or had not been discovered. In 1905, however, Clinton (*5, p. 265*) reports a leafspot of *Phaseolus lunatus* at New Haven, Conn., apparently the same as that described and illustrated by Halsted. He states that he did not observe the disease on the pods, but adds that it had not been looked for there. Although agreeing macroscopically with Halsted's description of the disease on the leaves, Clinton entertained some doubt of its identity with *Phoma subcircinata* E. and E. because of the fact that the spores averaged larger (5 to 12 by 2.5 to 3.5 μ) and were occasionally septate. He suggests the possibility that the fungus may be *Ascochyta phaseolorum* Sacc.

In 1912 Cook (*6, p. 517-518*) reports a severe outbreak of a disease of pole Lima beans in the State of New Jersey caused by *Phyllosticta* sp., which occurs on leaves and to some extent on the pods and stems. The following year it was less severe (*7, p. 801*), but in 1914 he (*8, p. 472*) reports the disease again as causing much damage, attributing it to *Phoma subcircinata* E. and E. A leafspot caused by *Phyllosticta phaseolina* Sacc. was also commonly met with.

In view of the relatively few times this disease has been reported during a period of 25 years, it is evident that it has generally been either overlooked or disregarded. This is probably due to the fact that the crop is relatively a minor one and that the disease appears usually after the low prices do not justify further pickings. It, however, is much more widely distributed than the published reports indicate. It was reported (unpublished) to the Plant Disease Survey from one locality in North Carolina by R. H. Fulton in 1913. The writer received and studied specimens of badly infected plants from Maryland in 1914 and 1915, and from Virginia in 1915. During the seasons of 1914 and 1915 the disease also caused heavy loss to the crop in New Jersey. In conclusion, it may be stated that the disease is rather widely distributed along the Atlantic seaboard and causes considerable damage to the crop.

DESCRIPTION OF PODBLIGHT

Although the ascogenous stage of this fungus has been connected by the writer with the pycnidial stage, the latter is alone concerned in the injury to the crop in the field. A macroscopic description of this disease therefore will be confined to the pycnidial or imperfect stage. In the field under suitable conditions the disease first appears on the leaves of plants from 1 to 2 feet high. Wet weather is conducive to the spread and warm weather promotes the development of this as well as many other diseases of this type. Serious outbreaks have often occurred immediately following a few days of rainy, warm weather in fields where previously the disease was scarcely noticeable. In such a case the loss is often serious; but if dry weather follows new growth of leaves and fruit will be produced comparatively free of the disease. If, however, the weather continues wet, a new outbreak is liable to occur.

The leaves function as host for the fungus during the earlier part of the season, from which it spreads later to the pods. Large, subcircular, brown, often bordered patches are produced on the leaves (Pl. 42, A). These patches, varying in size naturally with age, often attain a diameter of 1 to 3 cm. Infection is not restricted to any portion of the leaflets but the regions bordering the midrib are most frequently attacked. The fungus spreads more or less in all directions from the point of infection, but may often be delimited by a large vein or midrib. In this moribund tissue, or in tissue soon after it is killed, the fruiting bodies (pycnidia) of the fungus break through the epidermis. They are arranged more or less concentrically, and appear first as gray or grayish raised pimples, which later darken and become nearly black. They therefore may be seen as conspicuous, minute black specks on the upper surface of the leaves standing apart, sometimes confluent, on a brown background. The dead tissues finally become dry and fall out, leaving a ragged hole.

This disease must not be confused with a leafspot of Lima beans and other varieties of beans, as well as of cowpeas, caused by *Phyllosticta phaseolina* Sacc., the spots of which are smaller, more nearly round, and the pycnidia smaller and fewer in number.

The disease appears on the pods usually the latter part of July or early in August, at a time when the vines have nearly reached their full development. Under field conditions it is not certain at what stage in the development of the pod infection takes place. Young pods are rarely found diseased, and inoculation experiments have shown that they are infected less easily than ones nearly matured. The disease progresses rather slowly, requiring a week or 10 days after inoculation to produce a spot of any size or for the production of pycnidia. The growth of a bean pod is comparatively rapid, and it is likely that infection may take place when the pods are young, the fungus becoming visible as the pod approaches maturity. Infection occurs at any point on the surface of the pod, but more frequently at or near the ventral suture (Pl. 42, B). If infection occurs near or at the dorsal or ventral suture, the fungus spreads in all directions in a circular or semicircular manner, the hyphæ growing among the cells of the pod below the epidermis. The extent of the hyphal growth of the fungus among the cells is indicated by the darkening of the pod above the invaded area, evidently the first signs of the exhaustion of the food supply and approaching death of that part of the pod. This is soon followed by numerous, minute elevations over the surface, the points where the pycnidia are forming. These lifted points are at first gray, but as soon as the pycnidia break through the epidermis they become nearly black and protrude in a domelike manner from the surface. They are often, though not always, arranged concentrically and stand apart; or two or more may be confluent in a cluster or chainlike manner. Upon the death of the pod the fungus grows rapidly through the tissues, the pycnidia breaking out more or less over the entire surface (Pl. 42, C). The fruiting bodies which are formed subsequent to the death of the pod are usually not formed concentrically.

Under field conditions the fruiting bodies of the fungus are often found on the dead and more rarely on moribund stems. It is exceedingly uncommon to find typical infections of living stems, and the writer has never been able to produce infection of living stems artificially. The pycnidia, however, appear abundantly after the death of the vine on stems that have been sprayed with spores of the fungus.

The disease is much more common and destructive on the pole Lima beans than on the bush Limas.

ETIOLOGY

The podblight of *Phaseolus lunatus* is caused by a fungus pathogen to which various names have been given at different times. The use of different names is largely due to the lack of a knowledge of the life history of the fungus and to the practice of many mycologists in separating the genera Phyllosticta and Phoma, respectively, according to whether the fungus occurs on the leaf, or on the stem, or on other parts of the host.

Halstead (*16*) first reported this disease in 1891 as causing considerable damage to the pods and leaves, and attributed it to a species of Phyllosticta. He was evidently in doubt as to the species, for he does not refer it to *Phyllosticta phaseolina* Sacc. (*30, p. 149*), which had been described some years earlier and which occurs commonly on Lima beans, on *Phaseolus diversifolius*, Kansas (Kellerman), New Jersey (Ellis), Canada (Dearness) on *Phaseolus perennis*, Missouri (Galloway) and on cowpea, Kansas (D. B. Swingle). In 1905 Smith (*33*) proved the pathogenicity of this organism on the leaves of Lima beans and other varieties, such as kidney and wax, and on cowpeas. He found that the pods were not affected.

That the writer has been working with the same fungus which Ellis and Everhart described as *Phoma subcircinata* there is little doubt. Material, collected by Dr. J. B. Ellis at Newfield, N. J., and identified by him as *Phoma subcircinata*, has been examined and found to be identical with material collected at Vineland, N. J., only 4 miles distant from Newfield. It should be mentioned in this connection that the material from which Ellis and Everhart made their description was likewise collected at Newfield, in 1892. In 1890, three years before Ellis and Everhart's description of *Phoma subcircinata* appeared (*15, p. 158*), Ellis collected and identified as *Phyllosticta phaseolina* some material from Newfield which upon examination the writer has found to be identical with material later identified by him (Ellis) as *Phoma subcircinata* and also identical with the material the writer has been studying.

Phoma subcircinata E. and E. was described by Ellis and Everhart (*15, p. 158*) as follows:

On pods of Lima Beans, Newfield, New Jersey, October, 1892.

Perithecia subcuticular, 70–90 μ diam., sublenticular, subconfluent pierced above, membranous, black, subcircinately arranged in large (1 cm.), round, faintly zonate spots, finally spreading and occupying the entire surface of the pods. Sporules oblong-elliptical, hyaline, 2-nucleate, 5–6 by 2–2.5 μ, on simple basidia rather larger than the sporules

This description fits perfectly the fungus the writer has been studying, with the exception of the size of spores and pycnidia. An examination of the material collected and identified by Ellis as *Phoma subcircinata* shows the spores to be somewhat larger (6.4 to 8.0 by 2.4 to 3.6 μ, average 7.4 by 2.95 μ), and the pycnidia considerably larger (142 to 276 μ, average 185.8 μ). Material collected and identified by Ellis three years earlier as *Phyllosticta phaseolina* and which he probably later took for *Phoma subcircinata*, bore spores (6.0 to 8.0 by 2.8 to 3.2 μ, average 7.2 by 2.88 μ) and pycnidia (197.5 to 260.0 μ, average 219.0 μ) of about the same size. On material which the writer has collected and from which isolations have been made for inoculation work the spores varied from 6.0 to 8.6 by 2.4 to 4.1 μ, average 7.50 by 3.23 μ, and the pycnidia from 158.0 to 475.0 μ, average 245.86 μ. From the data at hand it appears evident that the fungus described by Ellis and Everhart as *Phoma subcircinata* is the same as the one the writer has been studying.

In view of the fact that the fungus causing podblight has been connected with a perfect stage, the genus to which the imperfect form should properly belong is of no great consequence, but it may be of interest to know that this fungus, like a number of others which have been classed as Phoma, belongs to the form genus Phomopsis. In morphological structures the podblight fungus is identical with similar structures of the genus Phomopsis as laid down in Diedicke's (*13*) revision of the group—

that is, chambering of the pycnidia, formation of sclerotial stroma, presence of pycnospores and stylospores, and structure of the pycnidial wall. No stroma is formed by this fungus on the leaves. If this fact should be disregarded and Saccardo's classification followed, it should be classed as a Dothiorella or a Fusicoccum. Stylospores have been found to be abundant in most of the material collected and studied, but none were found in the herbarium material collected and identified by Ellis as *Phoma subcircinata*. However, isolations have been made from material which did not bear stylospores but developed them in cultures later, which shows that, although not always present, the fungus may produce them. On the other hand, the organism isolated from specimens bearing stylospores do not always produce them in culture. This fungus is very similar to a large group of other organisms which have long been classed as Phoma. In a taxonomic study of the group Diedicke (*13*) transferred a number of species of Phoma to the genus Phomopsis. The podblight fungus does not differ essentially morphologically from the conidial stage of *Diaporthe batatatis* (E. and H.) Harter and Field, which was first described as *Phoma batatae* but found later by Harter and Field (*21*) to be a species of Phomopsis. It is similar also to the organism causing fruitrot, leafspot, and stemblight of eggplant, which was known as *Phoma solani* on the stem and fruit and as *Phyllosticta hortorum* on the leaves, both of which were found by Harter (*20*) to be caused by the same organism and to belong to the form genus Phomopsis.

Some years before *Phoma subcircinata* was described by Ellis and Everhart, Cooke and Ellis (*9, p. 93*) described *Sphaeria (Diaporthe) phaseolarum* occurring on bean stalks as follows:

Gregaria, tecta. Peritheciis globosis, immersis, minimis. Ostiolis spinaeformibus, atris, erumpentibus, Ascis clavatis. Sporidiis lanceolatis, quadrinucleatis. Sporidia 0.016 mm.

In the original description no type location was given; neither is it apparent in this or subsequent descriptions whether the fungus was found on Lima beans or other varieties. In the absence of these facts, together with the imperfect description, it would be difficult to identify correctly the fungus if it had not been more fully described later by Ellis and Everhart (*14, p. 460*) as follows:

Diaporthe phaseolorum (C. & E).
Sphaeria phaseolorum, C. & E. Grev. VI, p. 93.
Diaporthe phaseolorum, Sacc. Syll. 2635, Cke. Syn 2423.
Exsicc. Ell. N. A. F. 188.
Perithecia gregarious, buried, very small Ostiola spine-like, slender, projecting for ¼-½ mm. Asci clavate, 30-35 by 6-7 μ. Sporidia biseriate, oblong-lanceolate, 4-nucleate, scarcely or only slightly constricted, 10-12 by 3 μ (16 μ long, Cke)
On decaying bean vines left exposed through the winter, Newfield, N. J· Mostly around the nodes of the stem, the surface mostly blackened and the stroma limited within by a black line.

The writer collected a quantity of Lima beans (pods and stems) at Vineland, N. J., in October, 1914, and January, 1915, which bore an abundance of pycnidia of the podblight fungus, generally known as *Phoma subcircinata* E. and E. After isolating the causal fungus the material was wintered out at Washington, D. C., and covered with leaves and dirt. On June 2 perithecia of the genus Diaporthe were found to be abundant on stems and pods of this material. The perithecia were scattered and buried in the tissue of the host except the beaks which were long and projecting and somewhat bent as shown by Plate 43, D. A pycnidial fungus was isolated from single ascospores which was identical with the one isolated from the material before it was wintered out. Subcultures of this strain were used in inoculation experiments and produced infections identical with that produced by subcultures from isolations of single pycnospores. This strain (known as 598), as well as others isolated directly from pycnospores, produced stylospores. The perithecial stage has never developed in culture.

The perithecia produced on the wintered-out material varied in size from 158.0 to 355.5 μ, average 251.9 μ; the asci 28.0 to 46.2 μ by 5.2 to 8.0 μ, average 37.4 by 6.73 μ; the ascospores 6.4 to 12.0 μ by 2.3 to 4.0 μ, average 9.5 by 2.93 μ. The perithecia of type material varied in size from 158.0 to 237.0 μ, average 215.6 μ; the asci 28.0 to 44.0 μ by 4.8 to 8.0 μ, average 33.6 by 7.0 μ; the ascospores 8.0 to 12.0 by 2.4 to 4.0 μ, average 10.0 by 3.3 μ. From the fact that the ascospores, asci, and perithecia together with other morphological characters of the two organisms are practically identical, it is concluded that they are the same fungus and should be known as *Diaporthe phaseolorum* (C. and E) Sacc.

MORPHOLOGY

PYCNIDIAL STAGE

From the point of infection the hyphæ grow in all directions, invading practically all classes of cells except the bast, a layer of which (fig. 1, A, B, *d*) is found just below the epidermis. Figure 1, A and B, *c*, shows parenchyma cells traversed by the hyphæ. Soon after the death of the cells the pycnidia begin to form, and develop very rapidly thereafter. The hyphal growth increases in the cavities below the stomata or in the intercellular spaces. From here it passes up between the bast cells which are often somewhat separated beneath the stomata. As growth increases, gnarls of hyphæ accumulate under the stomata and between the epidermis and the bast cells (fig. 1, A, B, *b*). From this point the fungus spreads out just beneath the epidermis and develops a more or less circular plate, the base of which rests on the bast cells. Figure 1, A, shows an early stage in the development of the pycnidial plate. As the fungus growth continues, the epidermis is gradually elevated (fig. 1, B, *a*) as the

result of the pressure below. Definite cell layers
laid down about this time or a little earlier. Th
its full development soon thereafter (Pl. 43, B).

The inner layer of the pycnidium is compose
which the conidiophores arise. A slight cavity i
concomitant with the release of pressure by the ru
Inclosed in this cavity numerous conidiophores ar(
toward the center from which the spores are c(
increase in quantity, the cavity enlarges by the p
lining the cavity and probably by further lifting o
pycnidium. It has not been possible to determi(
opening from the pycnidial cavity to the outsid
formed, but it probably takes place only after the p
greatly increased by the continued production of

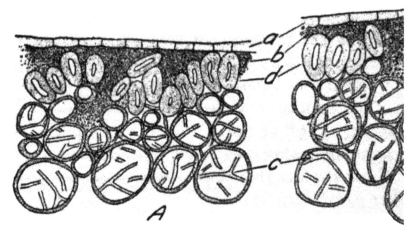

FIG. 1.—*Diaporthe phaseolorum: A*, An early stage in the formation of a
is not yet disturbed but the hyphæ (*b*) have accumulated directly b

The pycnidia vary greatly in size (158.0 to 475 μ, average 245.86 μ from host). They are irregular in shape, chambered with an inner hyalin layer from which the conidiophores arise, and a dark outer layer much thickened and darkened above (Pl. 43, B, C), all characters typical of the genus Phomopsis. Sometimes the contiguous sidewalls of two or more pycnidia are ruptured, forming a single cavity (Pl. 43, E). On the other hand, they may be pressed together but separated by a single dark wall and their respective hyalin inner layers, thereby forming a chain of pycnidia (Pl. 43, C). The pycnidia can hardly be said to be embedded in a true stroma.

Fic. 2.—*Diaporthe phaseolorum.* Long, slender, fragile sporophores with two pycnospores attached. (a) Sporophores; (b) pycnospores. ×500

The pycnidia on the leaves and stems are lenticular in shape, rarely, or not at all chambered (Pl. 43, A), fewer in number, and smaller than on the pods and usually isolated. They vary in size from 197.5 to 260.0 μ, average 219.0 μ.

The pycnospores are borne on slender, hyalin, simple conidiophores (fig. 2, a), 1½ to 3 times the length of the pycnospore. They are mostly oblong to fusoid (fig. 2, b; 3), hyalin, 1-celled, usually straight, seldom slightly curved, with two large oil droplets. These droplets, sometimes pressed together in the center of the spore, give the appearance of a septum, a characteristic common to the genus Phomopsis and so deceptive that at a casual glance one might believe that a true septum was laid down. That the spore is but 1-celled can readily be demonstrated by clearing it with a little salicylic acid. The variation in the size of the pycnospores from various sources is seen from the following measurements:

Fic. 3.—*Diaporthe phaseolorum.* A group of pycnospores. Note the oil droplets and variations in size and shape. ×500.

From host.—6.0 to 8.6 by 2.4 to 4.1 μ, average 7.5 by 3.23 μ.

Stems of *Melilotus alba.*—5.6 to 10.0 by 2.4 to 4.0 μ, average 7.82 by 3.11 μ.

Corn meal.—5.4 to 8.2 by 2.4 to 3.4 μ, average 6.49 by 2.79 μ.

Irish potato cylinders.—5.4 to 8.5 by 1.7 to 3.0 μ, average 6.36 by 2.59 μ.

Corn-meal agar.—5.1 to 8.3 by 2.4 to 3.4 μ, average 6.78 by 2.85 μ.

Rice.—5.4 to 8.5 by 2.0 to 3.4 μ, average 6.45 by 2.80 μ.

It will be noted that the spores are relatively large in artificial cultures on stems of *Melilotus alba*, and small on cooked Irish potato cylinders. This difference has been found rather constant, and is probably associated with vigor of growth. Of all media tried stems of *M. alba* were found exceedingly favorable for spore formation, while cooked Irish potato cylinders were rather unfavorable.

FIG. 4.—*Diaporthe phaseolorum:* Germinating pycnospores. ×500.

pycnospores in pycni have never been foun

They are long, slender, hyalin, nonseptate, straigh (fig. 5) borne on short, rather stout awl-shaped simple hyalin stylosporophores (fig. 6). Figure 5 shows some stylosporophores with stylospores attached. They are produced but sparingly the early part of the season or when the pycnospores are most abundant. As the season advances the stylospores increase in number and appear to be associated, though not always, with a saprophytic existence. Following the death of the host and the gradual disappearance of the pycnospores the stylospores increase and finally become very numerous. Specimens which bore practically no st

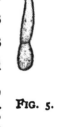

FIG. 5.

phore i:
(b) Sty
stylospt

bore them in increasing n the winter. There is a p data at hand that the s associated with a lack of f are produced in some cul but not until after the foo more or less exhausted. Al the pycnospores are usual beak in seven days, the sty

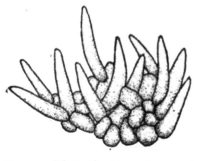

FIG. 6.—*Diaporthe phaseolorum:* A group of stylosporophores. ×500.

appeared sooner than 11 and generally not before 15 to lation. The average size of the winter crop of st under natural conditions is considerably larger than those produced in the fall or in culture. Measurements made of the fall stylospores from the host vary from 11.7 to 31.0 μ by 1.4 to 2.0 μ, average 22.83 by 1.73 μ, while those collected on February 7 varied from 20.6 to 54.4 μ by 1.38 to 2.4 μ, average 32.44 by 2.0 μ. They are produced in culture on rice, corn meal, Irish potato cylinders, and stems of *Melilotus alba* and vary in length from 11.7 to 54.4 μ and in width from 1.38 to 2.4 μ.

The function of these bodies is not known. Repeated attempts have been made to germinate them but without success. Reddick (*29*) has suggested the possibility that they were paraphyses and Saccardo (*31, p. 264*) regards them as conidiophores. Von Hohnel (*23, p. 526*), described a new genus, Myxolibertiella, to include those species having stylospores and pycnospores He first placed the genus in the order Melanconiales but later in the order Phomatales. Bubák (*2*) and Diedicke (*13*) both regard them as spores. The latter investigator for convenience designated the Phoma-like

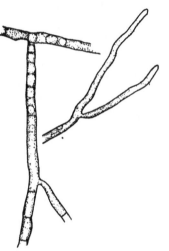

Fig. 7 —*Diaporthe phaseolorum*. Mature hyphæ Note the septations, oil droplets, and granules. ×500

spores by the Greek letter alpha (α) and the stylospores by beta (β). Shear (*32*), like Bubák, Diedicke, and others, regards these bodies as spores and designates them as scolecospores

There is nothing peculiarly distinctive about the mycelium. It is hyalin, frequently branched, and septate. It varies considerably in width and is often supplied with enlargements which bear a slight resemblance to chlamydospores. When young the cells are supplied with small protoplasmic granules. When the hyphæ are old, the cells are filled more or less with oil globules or droplets (fig. 7).

ASCOGENOUS STAGE

Specimens of Lima beans consisting of stems and pods collected in October, 1914, and January 1, 1915, at Vineland, N. J., were wintered out at Washington, D. C., and covered over with dirt and leaves. These

specimens were kept under observation through the remainder of the
winter and spring months.   Pycnidia and pycnospores were found from
time to time, as well as an increasing number of stylospores.   On June 2
pods brought into the laboratory bore a considerable number of mature
perithecia of an ascomycete, which was identified as *Diaporthe phaseolorum* (C. and E.) Sacc.   The writer was away for five weeks previous to
June 2 so it is not certain just when the development of perithecia began,
and for that reason it was not possible to follow the complete course of
development.   The perithecia are scattered and buried beneath the
surface, most of the beak protruding.   Plate 43, D and F, shows dia-
grammatically the arrangement of the perithecia and relatively the
depth they are sunken below the surface.   Whether or not the perithecia
are all formed in the cavities of old pycnidia has not been definitely determined.   The pycnidia are rather numerous
and often coalesce, and as a matter of chance a
perithecium might preempt a pycnidium and be
formed therein.   It is believed, however, that
this is not the rule.   The perithecia are not irregular in shape, as might be expected if they were
formed in old pycnidia; but they usually bear a
rather definite and constant form (Pl. 43, F)
They also have separate and distinct walls which
differ from the layers of the pycnidium.   In no
case have the spores or stylospores ever been
found associated with a perithecium, although
these bodies were present on the host at the time.

Fig. 8.—*Diaporthe phaseolorum* (a) Asci and (b) ascospores.   ×500.

The perithecia are slightly flattened, 158 to 355.5 μ, average 251.9 μ
in width and from 110.6 to 205 4 μ, average 166.5 μ in depth.   The beaks,
many times longer than wide, are usually curved or hooked (Pl. 43, D),
rarely straight, tapering gradually to the end.   The perithecium, circum-
scribed by a dark layer of cells, is lined within by a thin, hyalin layer
from which the asci arise.   The perithecium is evidently almost or
completely matured before the beak is formed.   Plate 43, D, shows
beaks just appearing through the epidermis.   Sections through perithecia with beaks in the incipient stage show the wall layers to be perfectly formed and many mature asci present.   The asci (fig. 8, a) are
sessile, clavate or fusoid, 28.0 to 46.2 μ by 5.2 to 8.0 μ, numerous,
and contain eight spores arranged mostly biserially.   The ascospores
are apparently bound together or embedded in a gelatinous mass, since
they are firmly held together even after the apparent disappearance of
the ascus wall.   The ascospores, (fig. 8, b) measuring 6.4 to 12.0 by 2.3
to 4.0 μ, are spindle-shaped, oblong-lanceolate, 1-septate at or near the
center, slightly or not at all constricted, and contain two to four oil
droplets.

The ascospores germinate readily in most any nutrient medium in about six to eight hours by the development of a single germ tube from one or both cells (fig. 9). The hyphæ branch is 12 to 24 hours, and septa are laid down at about the same time. Growth is very rapid. At ordinary room temperature a colony from a single spore becomes visible to the naked eye in about five days. The ascogenous stage has not been produced in culture. The transfer of a colony from a single ascospore resulted after 7 to 10 days in the production of the pycnidial stage on the same culture media on which pycnidia were developed by a transfer of pycnospores in the same length of time. In other words, there is no difference between the growth of the ascogenous strains and pycnidial strains on the same culture media.

INOCULATION EXPERIMENTS

The inoculation experiments on which the conclusions of the parasitism of the organism have been drawn were made in the greenhouse. It was found that Lima beans grow normally and produce fruit as well under greenhouse conditions as in the open field. Opportunities were therefore afforded to study the parasitism of this disease under controllable conditions and at any time of the year most suited to the needs of the work.

Fig 9.—*Diaporthe rum: Phaseolorum* germinating ascospores. ×500.

Controls were held with each experiment, the number and results of which are shown in Table I following a discussion of the inoculation experiments. The pods unless otherwise stated were inoculated by inserting spores and hyphæ into a wound made by a sterile needle. While later evidence will show that under suitable conditions infection may take place without wounding, the relative humidity of a greenhouse is usually so low that spores smeared on the surface often fail to germinate. It was necessary, therefore, to protect the spores from drying out by inserting them under the epidermis or by other means. In most cases it was impossible to protect the spores against drying, owing to the fact that when a Lima bean plant is fruiting abundantly it is generally too large to confine in an infection cage, though this was done with some smaller plants (see p. 486).

All inoculations were made from cultures grown on stems of *Melilotus alba*. The age of the culture was not considered or found important further than that the culture should be vigorous and fruiting abundantly. Spores were exuded from the pycnidium in about seven days after the culture was inoculated and continued to do so for 30 to 40 days thereafter or until the culture was dried. Spores were well suited for inoculation purposes until the culture medium was dried out.

Inoculations were made with six different strains. By a strain is here meant a difference in the source and not a difference in morphology or parasitology, since all were found to be identical organisms. For the sake of convenience in presentation each of the strains was given a separate number as follows: Strain 286 was isolated on October 26, 1914, from a Lima bean pod collected from a small patch at Vineland, N. J. Stylospores and pycnospores were abundant. Strain 420 was isolated from a Lima bean pod collected during the fall of 1914 by Mr. M. B. Waite in Maryland. Stylospores, as well as pycnospores, were present in abundance. Strain 447 was isolated from material collected on January 1, 1915, from the same field as No. 286. Strain 567 is a reisolation made on April 12, 1915, from a pod that had been inoculated in the greenhouse on March 13, 1915, with strain 447. Pycnospores were abundant, but there were no stylospores. Strain 598, as previously stated, resulted from the isolation on June 4, 1915, of a single ascospore from a pod that had been wintered out at Washington, D. C. Strain 601 was isolated on August 7, 1915, from leaves of Lima beans collected at Cape May, N. J., on August 5, 1915.

Other isolations of the same organism not used for inoculation purposes have also been made for the purpose of identification and comparison. Frequent reisolations were made from inoculated pods, one strain (567) of which was used to make inoculations. From specimens inoculated with this strain the causal fungus was again recovered, thus fulfilling the requirement of Koch's postulates, claimed necessary to prove parasitism.

The first infection experiments were made on January 15, 1915, when 20 nearly mature pods on six small plants were inoculated, 10 with organism 286 and 10 with organism 420. After the pods were inoculated by inserting spores and hyphæ, the plants were sprayed with spores suspended in water. They were then kept in a paper-covered infection cage for 24 hours. On February 20, pycnidia were abundant on the stems and on six pods inoculated with No. 286. Of those inoculated with No. 420, five pods were dead, and pycnidia were already on four and developed on one more after 10 days in a moist chamber. In this case there were no leaf infections, though the stems, particularly the lower parts, were covered with pycnidia, the pycnidia developing most abundantly on the moribund stems and pods.

On March 13, 21 small plants were sprayed with organism 447. There were only three small, immature pods in all and they were inoculated by inserting spores and hyphæ. These plants were inclosed for 48 hours in paper-covered infection cages. Twelve days later the three pods were drying up, and pycnidia were present on one and developed on the two others after 7 days in a moist chamber. Here again no leaves were infected, and the pods did not show typical infections. These results were disappointing at the time, but it was discovered later that young leaves and pods were difficult to infect, mature parts being much more susceptible to the disease

On April 9, 1915, twelve almost mature pods were inoculated by inserting spores and hyphæ in the usual way with organism 447. The plants were too large to place in the infection cage. Seven days later the tissue for half an inch or so around the point of infection was visibly darker. Three days later some of the pods were withering, and pycnidia began breaking through the epidermis ½ to ¾ inch from the point of inoculation. As long as the pods remain firm, the pycnidia form in more or less concentric rings; but concomitant with the withering of the pods the fungus rapidly invades all the tissue, and pycnidia break through the epidermis indiscriminately in a manner shown by Plate 42, C. Eight of these pods finally became infected

April 20, nine half-grown pods were sprayed with a suspension of organism 447 in sterile water and kept in an infection cage for 24 hours. These specimens were kept under observation until June 2, but no infection had taken place. It is believed that the failure here is again due to the relative immaturity of the pods. These results agree with those of Pool and McKay (*27*), who were able to infect only mature leaves of sugar beets with *Phoma betae* (Oud.) Fr. Apparently the writer has succeeded in infecting young pods by wounding, but there is reason for doubt, owing to the fact that a small wound frequently results in serious injury and death of the pod, in which case the fungus quickly invades it as a saprophyte. On April 23, spores of organism 447 were inserted in five pods, and by June 2 pycnidia and spores were abundant on three and a few days later on one more, one remaining sound.

On June 2, pods bearing perithecia which contained mature asci were brought into the laboratory and thoroughly washed and disinfected in mercuric chlorid. The perithecia were carefully picked out and macerated in sterile water in a watch glass Twenty-one nearly full-grown pods were inoculated by inserting ascospores and bits of broken tissue into wounds. Twelve days later thirteen pods showed unmistakable evidence of the disease, and by June 21 all but four of the pods were well covered with the fruiting bodies. The writer does not claim any proof of the connection of the conidial and perfect stage from this experiment. It is included here because it forms a link in the chain of experiments performed with this organism. The results are justly open to the criticism that spores and hyphæ of the imperfect stage may have been inserted, and it is not unlikely that such was the case.

On June 3, 1915, fifteen pods were inoculated in the usual way with organism 567. By June 14, pycnidia were abundantly produced over an area an inch or more in diameter on ten pods, and on four others by June 21. One pod alone remained sound.

On June 14, seventeen pods were inoculated with organism 598 (ascospore strain). In four days the tissue around the wound on some pods began to darken, showing unmistakable and characteristic symptoms of the disease. As the pods were killed by the fungus, they were gathered

and all but one bore numerous pycnidia by August 9.
of these infected pods produced a growth characteri.
stage. On the same date spores of organism 598 w
surface of five nearly full-grown pods. There were
on June 21, but by July 7 three showed evidence of inf
sound. Seven pods of various ages were sprayed wi
confined in an infection cage for 30 hours. Pycnidia
these pods on June 21 and on two others on Augus
stems and some of the dead branches. The writer wa
previous to August 2, and it is not definitely known
developed. It is probable, in view of the fact that
and studded all over with pycnidia, that they su
earlier.

On June 21 twenty-one pods were inoculated with
usual way. One diseased pod was removed on June
7, and one on August 17. One of the inoculated pods
and was therefore regarded as not infected. There v
for this experiment, one of which became infected an
the checks (66) held at various times.

On August 17, ten pods were inoculated with or
September 3 six were well covered with pycnidia. F
Of 19 pods inoculated with 601 on September 7, ten
fection on September 25. Nine remained sound.

A summary of the results is given in Table I.

TABLE I.—*Summary of inoculation experiments of*

PERPETUATION OF THE FUNGUS

The finding of mature asci of this fungus on June 2, 1915, and on June 11, 1916, and later on the dead vines and pods wintered out at Washington, D. C., and on similar material under field conditions at Vineland, N. J , at about the same season of the year shows one way the fungus may live through the winter. This method of perpetuation is, however, not the only one. At intervals from October until June pycnospores were frequently found, and the fungus was isolated at will from field material. It is interesting to note in this connection that the pycnospores were produced periodically during the winter season, depending upon weather conditions. Following a period of a few days of relatively warm, rainy weather, they could nearly always be found in more or less abundance. For example, pycnospores were found on material examined on October 24, January 2, 14, and 21, February 7, March 6, April 14, and June 2. Isolations were made on October 24, January 2, February 7, May 31, and June 2. Similar results were obtained with *Phomopsis vexans* (Sacc. and Syd.) Harter, the pycnospores of which were found in May and June on wintered-out material. However, the same material examined in December, January, February, and April bore no pycnospores but the stylospores appeared to be more numerous during the spring months. The appearance of pycnospores was roughly coordinated with rainfall and temperature, since on May 22, following a period of rainy weather and relatively high temperatures, they were found in great abundance.

It is likely also that the organism may live for a considerable time in the seed. The writer has isolated it in pure culture from the interior of both seeds and pods after surface disinfection in mercuric chlorid. From these results it is apparent that the fungus invades all parts and it is not unlikely that it is carried fron one season to the next on the seed. That the fungus is not readily killed by drying is evident from the fact that it was isolated from dried specimens kept in the laboratory for nine months. It would probably survive such conditions even longer. These results are supported by the earlier work of Halstead (*17*), who in 1901 called attention to the fact that the seeds were invaded by the fungus and suggested the possibility that it was carried through the season by that means.

MODE OF INFECTION

It is not always easy to explain how a parasite gains entrance into its host. It is safe to conclude that, if a fungus is parasitic, it may enter through wounds, it being merely necessary to show that the host is subject to wounding under natural conditions. Other types of penetration as, for example, through the unbroken epidermis or by way of the stomata are not always easy to demonstrate. There is no doubt that physical factors such as rainfall, temperature, humidity, and stomatal movement all play an important rôle in the penetration of the epidermis or stomata by certain parasites. In this connection it is interesting to note the results of Pool and McKay (*28*), who have studied in some detail the relation of stomatal movement to infection with *Cercospora beticola* Sacc. These investigators showed that the germ tubes entered only when the stomata were open, the movement of which was closely associated with a relatively high humidity, warm temperature, and light. Their results further showed that mature leaves were more readily infected and possessed a pore opening almost twice as large as young immature ones. The writer obtained similar results with the Lima bean fungus. It was found, as already pointed out, that young immature pods were not infected without wounding while mature pods could easily be. This difference may be due at least in part to the difference in the size of the stomatal aperture but more likely to the fact that immature stomata do not open, thus preventing the entrance of the hyphæ. Under humid conditions the spores germinate and produce a germ tube

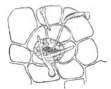 

Fig. 10.—*Diaporthe phaseolorum:* Some of the germ tubes of the pycnospores penetrating the stomata two days after the pods were sprayed, and other germinated spores lying alongside the stomata and growing toward or away from the aperture. X 260.

20 or 30 times the length of the spore in 24 to 48 hours. This is rather a feeble growth but the spores do not germinate readily in water, and then only a small percentage at that. On pods that had been sprayed with a spore suspension in water the germ tubes were found penetrating the stomata as shown by figure 10.

It is evidently a matter of chance when a germ tube finds its way into a stoma, since, if there was any attractive force, the germ tubes might be expected to head that way, which is apparently not the case. The germ tubes, as often observed by the writer and illustrated by figure 10, grow just as frequently alongside or away from the stomata as toward them, showing that there is no appreciable chemotactic influence. Entrance of the stoma would be easy so far as relative size of germ tube and stomatal aperature are concerned. In fact, it would not be impossible and it may actually happen that the spores fall directly through the pore into the intercellular space beneath. Measurements of the pores showed them to vary in length from 22 to 28 μ and in width from 5 to 7 μ when open, the germ tube of the spore in width from 2 to 3.5 μ, and the spore itself from 6 to 8.6 μ by 2.4 to 4.1 μ. Considering the relative size of the stomatal pore and the spore of the fungus, it is quite reasonable to conceive of an infection even during dry times, an event which actually does sometimes take place. It must of course be admitted, and the observed fact supports such a conclusion, that the number of infections by a fungus of this type during dry weather is relatively small compared with those during a period of rainfall or damp weather. Jones, Giddings, and Lutman (*25*) pointed out that moist, damp weather was conducive to the germination of the spores of *Phytophthora infestans* and made the interesting observation that the germ tube penetrated the leaf of an Irish potato through the epidermis as well as through the stomata. Similar observations were made by Jones (*24*), who found that the germ tube of *Macrosporium solani* E. and M. may penetrate the epidermis, as well as the stomata of Irish potatoes. The writer found no instance of penetration of Lima bean pods except through the stomata, and Pool and McKay (*28*) make a similar statement for *Cercospora beticola.*

The inoculation experiments already described have shown that infection takes place readily through wounds, perhaps made chiefly by wind, and by means of pickers. In ordinary farm practice the pole Lima bean is usually trained up several feet above the ground. Exposed thus to the wind, the vines are whipped around and the pods wounded by rubbing against the stems, poles, or wires on which they are trained. A careful examination will readily show the bruised spots on pods exposed to such conditions. Pickers also, in searching through the foliage, may likewise brush or press the immature pods against stems or poles, bruising them in a similar manner. The reddish, streaked appearance in localized spots often observed on bean pods is characteristic of wounds made by constant brushing or rubbing of the pod against a hard surface. It is in such areas that infections are frequently found.

The relation of insects to infection has not been studied

DISSEMINATION

The dissemination of the podblight of Lima beans must be considered from two standpoints: (1) Over short distances, and (2) over long distances, as from one district or State to another. As to the first of these, there are several ways in which the spores may be carried from plant to plant or from one field to another. It is not unlikely that insects, whether they feed on the pods or not, play an important part in the distribution of the spores. During wet periods the spores are exuded from the pycnidium, and any insect frequenting the diseased pods is likely to carry the spores about on its body. Pickers, too, in searching through the foliage for mature fruit, are likely also to distribute the spores to unpicked pods, where, under favorable conditions, new infections will start. It is interesting to note in this connection that this and certain diseases of other crops seldom become serious until about picking time. There are two possible explanations for this. One is that the host may be more susceptible at this age; and the other that picking of the fruit, especially if the foliage is wet with dew or from rain, results in a liberal distribution of the spores.

That strong winds may be responsible for severe outbreaks of this and other diseases is evident from the fact that Cook (*6, p. 517–518*) found a serious outbreak in a certain section of New Jersey previously visited by a severe windstorm. It is, of course, natural to conclude, as Cook assumed and the writer has frequently observed, that injury to pods and foliage during such a storm would add to the chances of infection. It is believed that the part played by the wind in disseminating fungus spores has been very greatly underestimated. It is a well-known fact that the pollen of pine trees is carried long distances. Dr. Albert Mann, of the Bureau of Plant Industry, found [1] a large variety of fungus spores collected on a gelatin plate exposed for about 15 minutes at a distance of 1 mile above the surface of the earth. This readily shows that spores are freely air-borne over long distances, and this fact alone may account for the serious outbreak of a disease in isolated sections.

Owing to the fact that the causal organism can readily be isolated from the seed, it is likely that this affords the principal means of distributing it over long distances. Fruiting bodies have also been found on infected seed. This shows the great necessity of saving seed only from fields where this disease does not occur.

CULTURAL CHARACTERISTICS OF THE FUNGUS

The fungus produces hyphæ on nearly all of the culture media in ordinary use. The pycnidia, on the other hand, are rarely produced on agars, but freely on media containing starch, such as rice or corn meal. A study of the growth and production of fruiting bodies was made on the following media: String-bean agar, Irish potato agar, beef agar, corn-meal agar, cooked rice, cooked corn meal, potato cylinders, and stems of *Melilotus alba*. These media were selected because they are in common use in the laboratory of Cotton, Truck, and Forage Crop disease investigations, and because some (corn meal and rice) were known to be particularly favorable for the growth of other very similar organisms, thus affording an opportunity for comparison. Other media might have served the same purpose.

The experiment was carried out by inoculating five tubes (100 c. c flasks, in case of corn meal) with spores from a 10-day-old culture on stems of *Melilotus alba*. After inoculation the tubes were kept in the light and exposed to the temperature of the laboratory room (21°–25°C).

STRING-BEAN AGAR.—This on the whole has been a poor medium for this fungus Growth of hyphæ started promptly and was visible in two days in slanted tubes of

---

[1] Unpublished data

the medium. At the end of 9 days three dark bodies, which were at that time thought to be pycnidia, and which probably were sterile ones, were developed. These cultures were kept under observation for 72 days, and no more pycnidia and no spores were found.

IRISH POTATO AGAR.—There was a visible vegetative growth in two days which continued to increase for about one week. A few scattered pycnidia appeared in all tubes but one. These were kept under observation for 72 days, and no spores or stylospores were ever produced.

BEEF AGAR.—A rapid vegetative growth during the first seven days of growth. Thereafter the hyphæ became fluffy and cottony. No pycnidia were formed in any of the tubes during the whole course of the experiment.

CORN-MEAL AGAR.—Growth was not visible in two days, but in four days the hyphæ had spread over an area ½ inch in diameter. In six days dark specks appeared which later developed into mature pycnidia. At the end of 13 days the slimy white, creamy masses began to exude from the pycnidia. These cultures were frequently examined thereafter for stylospores, but none were found until after 34 days, when they appeared in great abundance. When stylospores and pycnospores are both produced, the former always appear after, and never before, the production of the latter.

COOKED RICE.—Rice has proved to be a good medium for the growth of this organism. Hyphæ were conspicuous and pycnidia began to form in 4 days. From that time the pycnidia increased in number and spores began to exude in 7 days. A heavy, black stromatic mass, in which the pycnidia were embedded, was formed. As the medium dried out, these stromatic masses were left as conspicuous domelike projections over the surface of the medium. The pycnospores were exuded in slimy, yellowish, viscid masses in such quantities as to flow over the top of the stroma and mostly to cover the surface of the medium. Stylospores were found sparingly in 22 days and abundantly in 34 days.

COOKED CORN MEAL.—Corn meal was found to be one of the best media for this and other closely related fungi, such as *Phomopsis vexans*, *Plenodomus destruens*, and *Diaporthe batatatis*. In 5 days the hyphæ covered most of the surface of the corn meal in 100-c. c. flasks, and pycnidia began to form. The pycnidia were produced in great numbers in a more or less well-developed stroma which bore characteristics similar to those on rice. The small slimy droplets containing pycnospores were first noticed at the end of 13 days. The slimy exudate increased in amount until the droplets became quite large and flowed over the sides of the stroma, eventually covering nearly the whole surface of the medium. The stroma and pycnidia were dark on top and somewhat raised above the surface of the medium. As the corn meal dried out, the surface became more or less crusty and rugose. Stylospores were found for the first time at the end of 22 days.

POTATO CYLINDERS.—On this medium abundant mycelium was produced. A cottony, white growth covered one-third of the potato slant in two days and completely in four days. In five days incipient pycnidia were seen which later developed into a more or less stromatic mass. These masses were at first whitish on top, but later turned darker. In 22 days pycnospores were exuding slightly, and stylospores were present. At the end of 34 days the stylospores had become more abundant than on any other medium, the pycnospores, on the other hand, were relatively few.

STEMS OF MELILOTUS ALBA.—This medium, everything considered, is the best of any so far used. It gives a maximum amount of spores with a minimum of mycelium. There were no signs of growth in two days, but in five days numerous pycnidia began to form, and these soon became conspicuous as black specks studding the surface of the stems. The beaks were comparatively long and conspicuous. In one week under optimum conditions the viscid, creamy droplets in which the spores are embedded were noticeable on the end of the beak. These droplets gradually enlarged and finally overflowed the pycnidium, coalescing with similar droplets from other pycnidia and thereby forming an almost continuous liquid covering of the substratum. A stroma was not formed on stems of *Melilotus alba*. In this particular experiment no stylospores were formed. In other cultures, however, of the same organism on *M alba* stems a few stylospores were found.

EFFECT OF LIGHT ON GROWTH

The literature is full of the records of experiments on the effect of light on growth and development of chlorophyll-bearing plants, as well as fungi. It would be useless to review all or any considerable part of these records here. It is evident, however, from a perusal of some of these publications that no general, sweeping conclusions can be drawn that will apply to all fungi alike. Recent results, on the other hand, show that under similar conditions individual fungi may be expected to respond differently. In support of this last statement, reference may be made to the results of Coons (*10*), who found that light was a determining factor in the production of pycnidia with *Plenodomus fuscomaculans* (Sacc.) Coons. The writer (*19*), on the other hand, working with a different species of Plenodomus, *P. destruens* Harter, found that the absence of light did not inhibit the production of pycnidia. That the absence of light very greatly inhibited, but did not entirely prevent, the production of pycnidia by *Diaporthe batatatis*, was shown by Harter and Field (*21*) in another experiment. The writer has obtained similar results, as yet unpublished, with other fungi.

The influence of light on growth and production of fruiting bodies of the pycnidial stage of *Diaporthe phaseolorum* was determined by inoculating stems of *Melilotus alba* in test tubes and exposing one-half of the set to light and excluding the light from the other tubes by wrapping with black paper, which was found by previous test to prohibit the passage of light. The tubes were all tightly plugged with cotton. The cotton plugs alone were left exposed, so as to permit aeration. The two series were then placed on a table in the center of the laboratory room and kept under observation for 25 days. The cultures began to dry up about that time and development was accordingly arrested. The tubes kept in the dark were examined for a few minutes every few days for the purpose of taking notes and were for that length of time exposed to the light. While this brief period can not be regarded as having no influence on the results, exposure to the light for such a short time would certainly have no marked effect. The growths of mycelium in the cultures in the light and those in the dark were about the same at the end of three days, but from that time the mycelial growth in the dark was relatively more abundant and more cottony. In tubes in the light, on the other hand, there was, as has always been, a minimum of mycelium on stems of *Melilotus alba*. At the end of six days pycnidia were forming in large numbers in all tubes in the light and in smaller numbers in all those in the dark. The ratio between the number of pycnidia formed in the light and in the dark remained nearly constant, the fruiting bodies forming in all tubes, but they were better developed in the light. In one week spores began to exude from the pycnidia both in the light and in the dark. From this time spores continued to be exuded, but here again the amount of the exudate was relatively less in the dark. Toward the close of the experiment, however, there appeared to be a relative increase in the exudate from the pycnidia formed in tubes in the dark. But one conclusion can be drawn from these results, that while the dark retards the production of fruiting bodies and accelerates vegetative growth, it is not as marked as the results obtained with other fungi by the writer and others.

INFLUENCE OF TEMPERATURE ON GROWTH

To determine the influence of temperature on growth 12 (No. 2, 3, 5, 6, 8, 9, 10, 15, 17, 18, 19, and 20) of the 20 chambers in an Altmann incubator were used. Tubes of cooked rice were inoculated with spores of the fungus and immediately placed in the chambers, where they were kept exposed to the respective temperatures for a period of 27 days, at the end of which time growth had practically stopped. No growth took place in chambers 2 (4.1° C.), 3 (6.1° C.), 19 (35.4° C.), and 20 (38.8° C), and but a scant growth in 18 (32.9° C.). The optimum growth was found in chambers 15 (26.5° C.) and 17 (30.5° C.). The growth of hyphæ decreased relatively with the increase and decrease of the temperature from these limits of the optimum. In chambers 9 (17.4° C.), 10 (19.4° C.), 15 (26.5° C.), 17 (30.5° C.), and 18 (32.9° C.) stromatic masses of pseudoparenchymous tissue were produced which macroscopically looked like pycnidia. A careful examination of these bodies, however, showed that, while they bore some resemblance to pycnidia, no cavity had been produced within them and no pycnospores were found. The cause for this was later found to be due to a lack of sufficient aeration, resulting probably in a reduced oxygen supply. A subsequent experiment in which cultures of the organism on cooked rice were exposed in a chamber in which most of the oxygen had been removed by a mixture of pyrogallic acid and sodium hydroxid showed that while mycelial growth continued under such conditions the production of fruiting bodies was inhibited. In addition to the fact that the cultures in the Altmann chambers were kept in the dark, the absence of oxygen, which in itself was found, as already shown, to reduce somewhat the production of fruiting bodies, evidently accounted for the failure to produce pycnidia in the cited cases.

The growth of the podblight fungus in artificial cultures is schematically shown in figure 11.

INFLUENCE OF TOXIC AGENTS ON GERMINATION OF SPORES

Perhaps no chemical substance has been more generally studied in its relation to the growth of plants than copper salts. This is perhaps due to its well-known efficiency as a fungicide, following the discovery of which a large amount of work was undertaken to determine the concentration of copper salts necessary to prevent germination or bring about the death of the spores of a great variety of fungi. Much of this work was done with saprophytes, and consequently has little more than a scientific interest, while the practical side of the problem was evident from the fact that the toxicity to germinating spores of many well-known parasites was established. Because of the various uses that chemical substances may be put to, such as antiseptics, disinfectants, etc., the toxic action of many salts, metals, and acids, alone and in combination, on flowering plants and fungus spores have been extensively studied, and as a result an extensive literature on the subject has accumulated. As might be expected, the results have been very variable and often contradictory. In general, however, all investigations have shown that copper salts, among other chemical substances, are exceedingly toxic to all germinating spores and to the seedlings of flowering plants. Coupin (*11*) has shown by experiments that 1 part of copper sulphate to 700,000,000 parts of water is sufficient to retard the root growth of wheat seedlings, and the writer has proved (*18*) that a small amount of a single alkali salt, such as sodium chlorid, sodium carbonate, and magnesium sulphate in water cultures, inhibits the growth of the same plant. Heald (*22*) found 1/51,200 and 1/25,600 gram-molecule per liter of copper sulphate to be toxic to *Pisum sativum* and *Zea mays*, respectively. According to Moore and Kellerman (*26*) 1 part of copper sulphate to 500,000 parts of water killed *Closterium moniliferum* in 4 days and 1 part of copper sulphate to 3,000,000 parts of water killed *Anabaena flos-aquæ* in three days. *Uroglena americana* was even more sensitive and was killed in 16 hours in 1 part of copper sulphate to 10,000,000 parts of water.

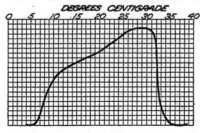

Fig. 11.—*Diaporthe phaseolorum*: Graphic representation of growth on cooked rice at different temperatures.

These few references show the extreme sensitiveness of some of the flowering plants and some algæ to copper salts, and at the same time the variability in their resistance. Although not strictly comparable with flowering plants, because of the different methods of experimentation and standards of measurement, the spores of fungi show equally great variability and sensibility to copper salts and other toxic substances. Clark (*3*) found that *Penicillium glaucum* was injured by copper sulphate at a concentration somewhere between $N/2,048$ and $N/1,024$, while *Aspergillus flavus* would endure a concentration of $N/512$ to $N/256$, two to four times as great. Crandall (*12*), on the other hand, showed that the apple scab fungus is quite resistant to copper sulphate, the spores being only slightly retarded in germination in a solution of 1 to 100,000, while a concentration of about 1 to 10,000 was necessary to entirely prevent germination. At this latter concentration some of the common molds, however, grew quite well. Temperature was found by Brooks (*1*) to influence the percentage of germination of the spores of certain fungi in the same concentration of a toxic substance. By an extensive series of experiments he found that *Penicillium glaucum* gave medium growth in $N/128$ copper sulphate at $_{20}$° C., *Monilia fructigena* did not germinate in $N/16$ copper sulphate at $_{25}$° to $_{30}$° C., but at $_{15}$° C. 12 per cent, at $_{10}$° C. 30 per cent, and at $_5$° C. 49 per cent.

No attempt has been made to review all the extensive literature on the effect of toxic substances on plant growth. The results of the work of the different investigators are very conflicting, due, perhaps in part at least, to the different methods of experimentation employed and the different standards of measurement.

During the time the podblight fungus of Lima bean has been under investigation many interesting observations have been made from time to time on the effect of certain chemical substances commonly used in the laboratory on the germination of the spores. The use of a solution of mercuric chlorid for disinfecting field specimens preparatory to isolating the fungus led to a study of the strength of this solution necessary to kill the spores in a given time. A similar study was made with formaldehyde because of its possible use in disinfecting infected seed, and with copper sulphate, a well-known constituent of Bordeaux mixture, sometime recommended as a spray in controlling the disease in the field.

METHOD OF EXPERIMENTATION

Preliminary experiments showed that the use of Van Tieghem cells to determine the toxic limit for the germination of the spores was exceedingly unreliable and was consequently abandoned. Although this method has been generally used, it is likely that many of the conflicting results may be attributed to it. In fact, the writer found that each test in Van Tieghem cells might be expected to give different results, probably owing to the variation in vapor pressure or to some slight accident overlooked, unexpected, and unaccounted for. After considering the different methods that might be used, the writer concluded

in o accoun ___ ___ ___ ___   e organism ra _
that inhibits growth. It seems that the same m
equally well to fungi and is easier of determination
since the length of time the organism is exposed
part also, and can be more accurately gauged. It i.
that a fungicide may merely inhibit growth when th
to it for a certain length of time but may kill ther
posure. In these experiments the time the spores v
fungicide was relatively short, the concentration of
proportionately strong to produce death in a compa
This method eliminated the necessity of taking in
other things, the temperature which Brooks (1) fou
important part in germination of the spores of some

The pycnospores of *Diaporthe phaseolorum* were fo
beef agar +15 on Fuller's scale; and therefore this
which to test germination. The spores from a cultu
were transferred by a small platinum loop to a solut
of the desired concentration and mixed by a wide
motion of the arm with the plugged end up, or by vig
the end of one minute, two minutes, and every two
up to the suspected limit of endurance, a loopful
transferred to a tube of liquefied agar and, after mixin
was poured. Control plates were made by transf
spores from the same test culture to a sterile water
from this tube to a tube of liquefied agar. This m
with each of the fungicides and repeated as often as n
the toxic limit. The plates were kept at room
laboratory. The colonies in the control plates appe

TABLE II.—*The results of exposing pycnospores of Diaporthe phaseolorum to different strengths of mercuric chlorid, copper sulphate, and formaldehyde for different lengths of time*

Time of exposure.	Copper sulphate			Formaldehyde			Mercuric chlorid.			
	N/12.5	N/31.25	N/62.5	1 100	1 250	1 500	N/135	N/1,350	N/2,700	N/4,050
Minutes.										
1	[a] 100	100	100	75	85	100	0	0	50	100
2	35	65	...	0	30	100	0	0	0	100
4	5	55	90	......	......	0	......	......	......	75
6	......	25	......	......	......	10	......	......	......	50
8	......	15	75	......	......	0	......	......	......	0
10	......	10	......	......	......	......	......	......	......	......
12	......	5	45	......	......	......	......	......	......	......
14	......	2	......	......	......	......	......	......	......	......
16	......	......	20	......	......	......	......	......	......	......
18										
20	......	......	2	......	......	......	......	......	......	......

[a] In this and corresponding columns are given the percentage of germination as estimated at the close of the experiment.   Water controls were made in every case in which the germination was arbitrarily placed at 100 per cent.   These plates, together with those poured from a 1 minute exposure, served as controls for comparison in estimating percentages of germination.

Reference to the table will show that the spores were nearly all killed in four minutes in a *N/12.5* copper-sulphate solution, and in 14 minutes in a *N/31.25*, and in 20 minutes in a *N/62.5* solution. It is interesting to compare these results with those of Clark (*3*), who found that the spores of several fungi exposed in a copper-sulphate solution were killed or injured as follows: *Aspergillus flavus* (?) Link *N/4* and greatly injured in *N/256*, *Sterigmatocystis niger* Von Tieghem *N/8* and greatly injured in *N/128*, *Oedocephalum albidum* Sacc. *N/64* and greatly injured at *N/256*, *Botrytis vulgaris* Fr. greatly injured at *N/128*, and *Penicillium glaucum* Link *N/1* and greatly injured at *N/256*. As will be seen, there is a wide difference between the strength of the solution necessary to kill and that required to injure greatly the spores. The writer found that none of the spores exposed for one minute to a solution of mercuric chlorid at a concentration of *N/135* and *N/1,350* would germinate, and that only 50 per cent germinated when exposed for the same length of time in a concentration of *N/3,700*. At the greater dilution, *N/4,050*, 50 per cent germinated after an exposure for six minutes. Clark (*3*) found that spores of *A. flavus*, *S. niger*, *P. glaucum* were killed in a *N/4,096* mercuric chlorid, *O. albidum* in a *N/16,384*, and *B. vulgaris* in a *N/65,536*. The spores of these fungi were killed in the solutions of the toxic substance more dilute than that required to kill the pycnospores of *Diaporthe phaseolorum*, but they were likewise subjected to the solution for a longer time. In another work Clark (*4*) has shown that *Rhizopus nigricans* was killed in 75 seconds in a solution of mercuric chlorid 1 to 3,700.

Formaldehyde was found to be very toxic to the spores, only 10 per cent germinating after a six-minute exposure in a 1-to-500 solution, and

Halsted (17) was the first, and probably the only
done any experimental work on the control of the
beans. In these experiments spraying the dwarf Li
mixture or soda-Bordeaux mixture yielded very
Although he evidently did not spray the pole Limas,
disease could be checked by so doing. It is now kno
attacks the leaves before the pods are set, and it is lik
infected from them. In view of this fact, the first spra
when the plants are 1 or 2 feet high, and should be re
to keep the foliage covered.

From badly infected pods the fungus invades the
live for a considerable time. Infected seed are usua
and immature. They probably would not grow, an
give weak, sickly plants. Only plump, bright seed
Sound seed may be disinfected to kill the adhering sp
for 5 or 10 minutes in a solution of mercuric chlorid (
hyde (1:100), or copper sulphate (1:100), after wh
rinsed in water.

A series of experiments has shown that germination
such a treatment. Healthy Lima bean seeds were
were treated for 5 minutes and a part for 10 minut
sterile water, they were laid between moistened fil
minated. The results showed that seeds exposed f
just as good germination as those soaked in sterile
length of time. It should be mentioned in this co
soaked in sterile water were all more or less covere

SUMMARY

(1) The podblight of Lima beans is probably indigenous to the United States.

(2) It was first recognized by Halsted in New Jersey in 1891, and has since been found in the States of Virginia, Maryland, Connecticut, and North Carolina.

(3) The disease forms circular brown spots on the leaves, and large unsightly spots on the nearly mature pods and on the stems. Numerous pycnidia are produced in the diseased areas.

(4) The loss, which in some fields is very large, is confined mostly to the pods.

(5) The Lima bean podblight fungus has been known as *Phoma subcircinata* E. and E. almost since its first discovery, although in a few cases it has been referred to as a species of Phyllosticta.

(6) The writer isolated single ascospores of a Diaporthe found on specimens wintered out and made inoculations with pure cultures of such strains which produced a disease identical with that produced by inoculations with pure cultures of a pycnospore strain. In 1892 *Diaporthe phaseolorum* was completely described, which agrees with the ascogenous fungus the writer has been studying. The fungus causing the disease is therefore referred to as *Diaporthe phaseolorum* (C. and E.) Sacc.

(7) The hyphæ course through and among the parenchymous cells of the pod, forming pycnidia just under the epidermis at points where the bast cells are more or less separated.

(8) Stylospores are abundantly found on field material and readily produced in culture on some media. The pycnospore stage belongs to the form genus Phomopsis and not to Phoma.

(9) The disease is readily produced by artificial inoculation with spores from both the pycnidial and ascogenous strains.

(10) Wounding of the pod is not necessary for infection, as the germ tubes may enter through the stomata.

(11) The fungus is carried through from one season to the next, either by periodic production of pycnospores and the persistence of hyphæ or by the production of an ascospore stage.

(12) The disease may be carried on the seed and spread by such agencies as wind, by the process of picking, and possibly by means of insects.

(13) The fungus fruits well on stems of *Melilotus alba*, on rice, corn meal, and other starch media, and but poorly or not at all on the agars tried with the exception of corn-meal agar.

(14) Cultures in the dark fruited more slowly than those in the light and had increased vegetative growth.

(15) No growth takes place at a temperature below 6.1° nor higher than 35.40 C. during a period of 27 days.

(16) Dilute solutions of formaldehyde, copper sulp
chlorid are toxic to the spores. The germination of t
retarded when exposed to a solution too dilute to k
time.

(17) To control the podblight it is recommended
selected. They should then be disinfected in mercur
or copper sulphate. The plants should be sprayed i
time when they are 1 to 2 feet tall, and often enoug
the foliage covered.

LITERATURE CITED.

(1) BROOKS, Charles.
   1906. TEMPERATURE AND TOXIC ACTION. *In* Bot. Gaz.,
      12 fig.   Literature cited, p. 375.
(2) BUBÁK, Franz, and KABÁT, J. E.
   1905. VIERTER BEITRAG ZUR PILZFLORA VON TIROL. *I*
      Jahrg. 55, no. 2, p. 73–79. (Continued article.)
(3) CLARK, J. F.
   1899. ON THE TOXIC EFFECT OF DELETERIOUS AGENTS
      AND DEVELOPMENT OF CERTAIN FILAMENTOUS
      v. 28, no. 5, p. 289–327; no. 6, p. 378–404, 10
      p. 402–404.
(4) ———
   1901. ON THE TOXIC VALUE OF MERCURIC CHLORIDE AND
      Jour. Phys. Chem., v. 5, no. 5, p. 289–316, 7 fi
(5) CLINTON, G. P.
   1906. REPORT OF THE BOTANIST. NOTES ON FUNGUS
      1905. *In* Conn. Agr. Exp. Sta., 29th Ann. Rpt.
(6) COOK, M. T.
   1913. REPORT OF THE PLANT PATHOLOGIST. *In* N. J

(14) ELLIS, J. B., and EVERHART, B. M.

　　1892. THE NORTH AMERICAN PYRENOMYCETES. 793 p., 41 pl. Newfield, N J·

(15) ——— ———

　　1894. NEW SPECIES OF NORTH AMERICAN FUNGI FROM VARIOUS LOCALITIES.
　　　　In Proc. Acad. Nat. Sci. Phila., 1893, p. 128–172.

(16) HALSTED, B. D.

　　1892. LIMA BEAN DISEASES. *In* N. J· Agr. Exp. Sta., 12th Ann. Rpt, 1891,
　　　　p. 287.

(17) ———

　　1901. BEAN DISEASES AND THEIR REMEDIES. *In* N. J· Agr. Exp. Sta Bul. 151,
　　　　28 p., 9 fig., 4 pl.

(18) HARTER, L. L.

　　1905. THE VARIABILITY OF WHEAT VARIETIES IN RESISTANCE TO TOXIC SALTS.
　　　　U. S. Dept. Agr. Bur. Plant Indus. Bul. 79, 48 p. Bibliography,
　　　　p. 47–48.

(19) ———

　　1913. THE FOOT-ROT OF THE SWEET POTATO. *In* Jour. Agr. Research, v. 1,
　　　　no. 3, p. 251–274, 1 fig., pl. 23–28.

(20) ———

　　1914. FRUIT-ROT, LEAF-SPOT, AND STEM BLIGHT OF THE EGG PLANT CAUSED BY
　　　　PHOMOPSIS VEXANS. *In* Jour. Agr. Research, v. 2, no. 5, p. 331–338,
　　　　1 fig, pl. 26–30. Literature cited, p. 338.

(21) ——— and FIELD, Ethel C.

　　1913. A DRY ROT OF SWEET POTATOES CAUSED BY DIAPORTHE BATATATIS.
　　　　U. S. Dept. Agr. Bur. Plant Indus. Bul. 281, 38 p, 4 fig., 4 pl.

(22) HEALD, F. D.

　　1896. ON THE TOXIC EFFECT OF DILUTE SOLUTIONS OF ACIDS AND SALTS UPON
　　　　PLANTS. *In* Bot. Gaz., v. 22, no. 2, p. 125–153, pl. 7.

(23) HÖHNEL, Franz von.

　　1903. MYCOLOGISCHE FRAGMENTE. *In* Ann. Mycol., Jahrg. 1, no. 5, p. 391–
　　　　414; no. 6, p. 522–534.

(24) JONES, L. R.

　　1896. REPORT OF THE BOTANIST. *In* Vt. Agr. Exp. Sta, 9th Ann Rpt., 1895,
　　　　p. 66–115, 15 fig.

(25) ——— GIDDINGS, N. J., and LUTMAN, B. F.

　　1912. INVESTIGATIONS OF THE POTATO FUNGUS, PHYTOPHTHORA INFESTANS. Vt.
　　　　Agr. Exp. Sta. Bul. 168, 100 p., 10 fig., 10 pl. Index to literature, p.
　　　　88–93.

(26) MOORE, G. T, and KELLERMAN, K. F.

　　1904. A METHOD OF DESTROYING OR PREVENTING THE GROWTH OF ALGAE AND
　　　　CERTAIN PATHOGENIC BACTERIA IN WATER SUPPLIES. U S. Dept. Agr
　　　　Bur Plant Indus. Bul. 64, 44 p.

(27) POOL, VENUS W., and McKAY, M. B.

　　1915. PHOMA BETAE ON THE LEAVES OF THE SUGAR BEET *In* Jour. Agr. Re-
　　　　search, v. 4, no. 2, p. 169–177, pl. 27.

(28) ——— ———

　　1916. RELATION OF STOMATAL MOVEMENT TO INFECTION BY CERCOSPORA BETI-
　　　　COLA. *In* Jour. Agr. Research, v. 5, no. 22, p 1011–1038, 6 fig, pl.
　　　　80–81. Literature cited, p. 1038.

(29) REDDICK, Donald

　　1909. NECROSIS OF THE GRAPEVINE. N. Y. (Cornell) Agr. Exp Sta Bul 263,

1879. SYLLOGE FUNGORUM. v. 18. Patavii.

SHEAR, C. L.

    1911. THE ASCOGENOUS FORM OF THE FUNGUS CAUSING DEAD
        In Phytopathology, v. 1, no. 4, p. 116–119, 5 fig.

SMITH, C. O.

    1905. THE STUDY OF THE DISEASES OF SOME TRUCK CROPS IN
        Agr. Exp. Sta. Bul. 70, 16 p., 6 fig., 2 pl.

STEWART, V. B.

    1916. THE LEAF BLOTCH DISEASE OF HORSECHESTNUT. *In* P
        6, no. 1, p. 5–19, 1 fig., pl. 2–4 (part. col.).

PLATE 42

Diaporthe phaseolorum:

.—A leaflet of *Phaseolus lunatus* L., showing a number of ragge
caused by the Lima bean podblight fungus. This specimer
May, N. J., August 5, 1915. The disease was produced on po
the organism (601) isolated from this leaf. Natural size.

.—A green Lima bean pod photographed 10 days after inoculati
strain. Pycnidia are abundantly formed over an area about ½
ide of this is another somewhat darkened area which has alre
e fungus, but the pycnidia have not yet broken through the ep

—A pod showing the characteristic manner in which the fungu
on after it is killed, pycnidia forming indiscriminately over
al size.

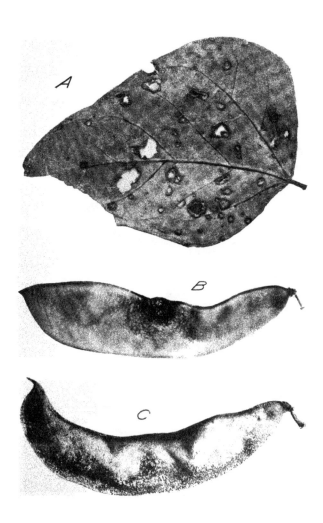

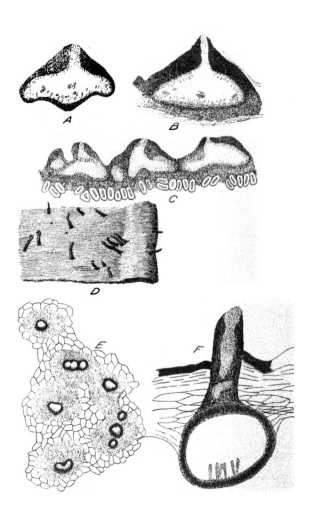

PLATE 43

Diaporthe phaseolorum:

A.—A vertical section through a pycnidium from the leaf. ×130.

B.—A vertical section through a pycnidium from a pod showing the beak and the thickened dark wall surrounding it.   Note also conidiophores arising from hyalin inner ayer. Some spores are attached. ×130.

C.—A vertical section through the epidermis and underlying host of a portion of a diseased pod showing the connection and chambering of the pycnidia. ×65.

D.—A somewhat diagrammatic drawing of a portion of the surface of a Lima bean pod, showing the distribution of the perithecia, the shape of the beaks and their projection from the surface. The beaks of two perithecia may be seen just breaking through the epidermis. ×16.

E.—A surface view of a portion of a bean pod showing arrangement of the pycnidia and fusion of the beaks. ×65.

F.—A vertical section through a perithecium from a pod. Note the dark wall surrounding the perithecium in which the asci are inclosed; also the portion of the beak buried below the surface of the pod. The outer end of the beak has been cut away. ×130.

TESTS OF A LARGE-SIZED REINFORCED-CONCRETE SLAB SUBJECTED TO ECCENTRIC CONCENTRATED LOADS

By A. T. Goldbeck, *Engineer of Tests*, and H. S. Fairbank, *Highway Engineer, Office of Public Roads and Rural Engineering, United States Department of Agriculture*

INTRODUCTION

During the past five years a series of tests to determine the distribution of stress in reinforced-concrete slabs carrying concentrated loads has been made by the Office of Public Roads and Rural Engineering. The slabs included in the series have varied from 3 to 16 feet in span, and 6 to 32 feet in width; and as a result of the tests much information has been collected, practically all of which has been published at various times.[1] (1, 2, 3, 4, 6.)[1]

In all of these tests the concentrated load has been applied to the center of the slab, and sufficient data have been gathered to indicate the manner and extent of stress distribution in slabs loaded in this manner. In general, it may be said that the stress varies from a maximum immediately under the load to a minimum at the extreme edges. The information as to the extent of the distribution has been made available for use in the rectangular-beam theory of design by the determination of a value known as the "effective width," which, when substituted for the width b in the formulas, will lead to the design of slabs of dimensions conforming to those shown by the tests to be necessary. This value is represented by that width of slab over which, if the stress were constant and equal to the maximum stress under actual conditions, the resisting moment would equal the resisting moment of a slab of the same depth and full width in which the stress is naturally distributed. It has been brought out in the previous papers that this effective width varies with the total width and span, the relation being approximately as expressed in the table following.

[1] Goldbeck, A. T. Tests of reinforced concrete slabs under concentrated loading. *In* Amer. Soc. Testing Materials, Proc., v. 13, p. 858–873, 10 fig. 1913

———— The influence of total width on the effective width of reinforced-concrete slabs subjected to central concentrated loading. *In* Amer. Concrete Inst., Proc. v. 13, p. 78–88, 13 fig. 1917.

———— and Smith, E. B. Tests of large reinforced concrete slabs. *In* Amer. Concrete Inst. Proc., v. 12, p. 324–333, 7 fig. 1916

———— ———— Tests of three large-sized reinforced-concrete slabs under concentrated loading. *In* Jour. Agr. Research, v. 6, no 6, p. 205–234, 28 fig., pl 26 1916

McCormick, E. B. Test of a reinforced concrete slab. *In* Amer. Concrete Inst. Proc., v. 11, p. 195–204, 8 fig. Discussion, p. 202–204. 1915.

Journal of Agricultural Research.
Washington, D. C.
ks

Vol XI, No. 10
Dec. 3, 1917
Key No. D—15

Total width÷span	Effective width÷span	Total width÷span	Effective width÷span
0.1	0.1	1.1	0 67
.2	.2	1.2	.68
.3	.28	1.3	.70
.4	.37	1.4	.71
.5	.44	1.5	.72
.6	.50	1.6	.72
.7	.55	1.7	.72
.8	.58	1.8	.72
.9	.62	1.9	.72
1.0	.65	2.0	.72

The tests indicate that the above values can be used for spans up to 16 feet, and they probably can be used for longer spans, although no longer spans were tested.

However, these conclusions were drawn from the tests of slabs under centrally applied concentrated loads only; and to make them properly applicable to the design of bridge slabs, which constantly are called upon to withstand concentrated loads applied near the parapet, it was recognized that the effect of eccentricity of loading must be determined.

With this end in view, another large-sized reinforced-concrete slab has been constructed and tested recently at the Arlington Experimental Farm of the United States Department of Agriculture, and it is the object of this paper to present the results of the test.

DESCRIPTION OF THE SPECIMEN

The specimen was made of machine-mixed concrete in the proportions of 1 part of Portland cement to 2 parts of Potomac River sand and 4 parts of Potomac River gravel. The sand was a good grade for use in concrete and the gravel was clean, well-graded, and free from weak pebbles. A rather wet mix was used, and the mixing and placing were done by experienced laborers at the Arlington Farm. As in the case of the other slabs tested, there was no attempt to make the concrete any better than it would be made in the field, but efforts were directed to secure work thoroughly representative of that which might be obtained under field conditions.

The slab, which was built in place and supported by concrete abutments, was 32 feet in width, and 16 feet in span. Its total thickness was 14 inches, and its effective thickness 13 inches. It was reinforced, in the longitudinal direction only, with ¾-inch plain square rods, spaced 5¾ inches apart, the sectional area of the steel being 0.75 per cent of that of the concrete above its center of gravity.

After the concrete was poured and while it still was soft, rows of bolts were set in the top surface of the slab, which, when withdrawn a few days later, left holes for the insertion of the plugs and feathers to be used for splitting off sections of the slab as the test progressed.

To provide definite points between which deformation readings might be taken, short brass plugs, drilled at one end with a No. 55 drill, were cemented in holes drilled for the purpose in the concrete, and similar No. 55 drill holes were made in bared sections of selected reinforcing rods. At points where it was planned to take deflection measurements steel plates were set in plaster of Paris on top of the slab. The locations of these deformation and deflection points are shown graphically in figure 1.

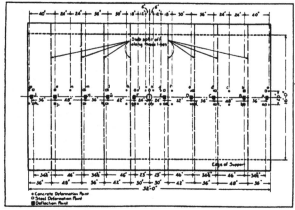

FIG. 1 —Diagram of slab A—25, showing location of deformation and deflection points

METHOD OF TESTING

Loads were applied over an 8-inch bearing block by means of a hydraulic jack mounted between the slab and a specially calibrated chromenickel beam, the deflection of which, observed with the aid of an Ames dial, provided a means of measuring the load. This apparatus has been described in detail in previously published papers.

Extensometer readings were made, between the points in both concrete and steel, with a 20-inch Berry strain gauge. Deflection measurements also were taken over the steel plates by means of a special apparatus designed for measuring the wear of concrete roads.[1]

The slab was tested first with the load applied in the center of the full width of 32 feet. Afterward successive sections were split off one side, making the width of the slab 29, 25, 22, and finally, 18.5 feet. As the actual position of the load remained unchanged, its position with respect to

[1] GOLDBECK, A. T. APPARATUS FOR MEASURING THE WEAR OF CONCRETE ROADS. *In* Jour. Agr. Research, v. 5, no. 20, p. 951–954, 1 fig., pl. 66. 1916.

the slab width was thus rendered increasingly eccentric. The position of greatest eccentricity was reached at the width of 18.5 feet, in which case the load was applied at a point 16 feet from one edge and only 2.5 feet from the other. Corresponding sections were then split off the other side of the slab, reducing the width to 15.5, 11.5, 8.5, and 5 feet, respectively, the point of application of the load remaining 2.5 feet from one edge. During this stage of the experiment the degree of eccentricity of the load was reduced with each section split off until finally in the 5-foot width the load was again applied in the center. In figure 1 the planes along which the slab was split are indicated by heavy dashed lines.

Complete sets of deformation and deflection readings were made on each width of slab, the repeated use of the slab cut to various widths being made possible by the fact that the load applied at no time stressed the specimen beyond its working stresses.

As will be noted by reference to figure 1, the concrete and steel deformation points and the deflection plates were spaced along the middle section of the slab parallel to the supports. As this is the dangerous section of a slab under such loading, the measurements taken represent the maximum deformation and deflection for the loads used at the various transverse distances from the point of application. The measurements have been plotted to scale, as the ordinates of curves, whose abscissæ are the distances between the points of measurements, and the results are shown in figures 2 to 10, inclusive.

RESULTS OF THE TESTS

As in the tests of centrally loaded slabs previously made, the curves of deformation of concrete and steel have been used to determine values of the "effective width" of the slab for the various total widths and positions of load. This is done by measuring the areas of the curves with a polar planimeter and dividing these areas by the maximum ordinates, the results in each case being the value of the effective width corresponding to the particular width of slab and position of load.

These values are shown on the respective curves. This method is based on the assumptions that the straight-line theory of fiber-stress distribution is applicable to slabs, and that the observed deformations are proportional to the extreme fiber stresses in the concrete and steel, though it is realized that for various reasons it is impossible to translate the deformations into fiber stresses.

The relation between the computed effective width and the total width for the several widths of the single slab tested is shown by the solid-line curve in figure 11. As will be noted, it is conservatively drawn through the lower edge of the belt of points derived from the concrete and steel deformation readings. It shows that the effective

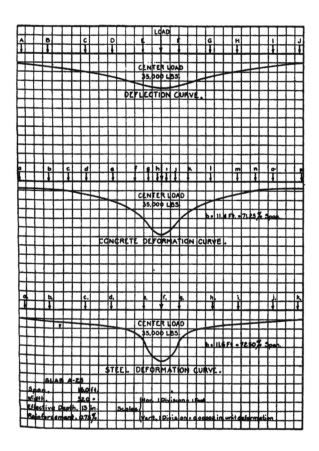

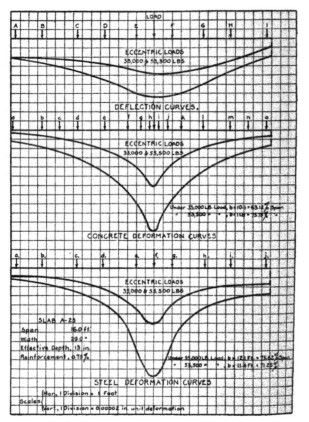

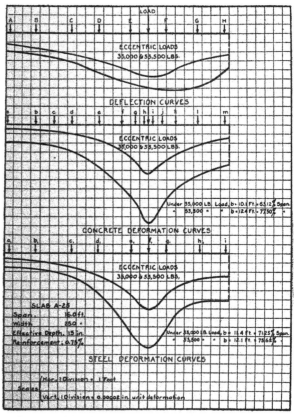

FIG. 4.—Eccentric load deformation and deflection curves of slab A-25. Slab width, 25 feet.

23716°—17——5

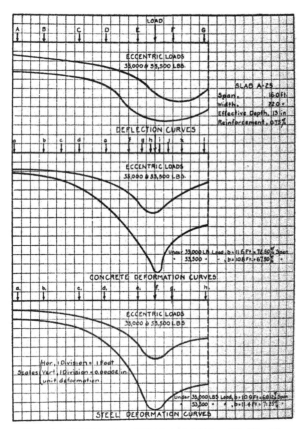

Fig 5.—Eccentric load deformation and deflection curves of slab A-25. Slab width, 22 feet.

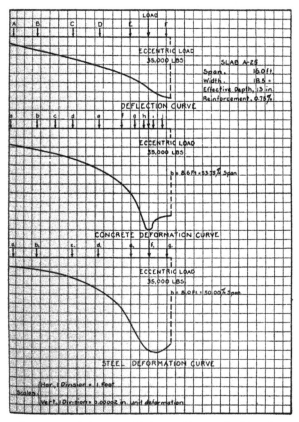

FIG. 6.—Eccentric load deformation and deflection curves of slab A-25.   Slab width, 18.5 feet.

Journal of Agricultural Research   Vol. XI, No. 11

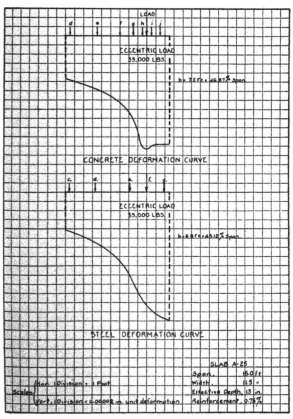

FIG. 8.—Eccentric load deformation and deflection curves of slab A-25. Slab width, 11.5 feet.

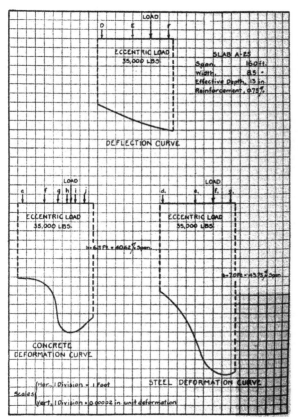

FIG. 9.—Eccentric load deformation and deflection curves of slab A-25.   Slab width, 8.5 feet.

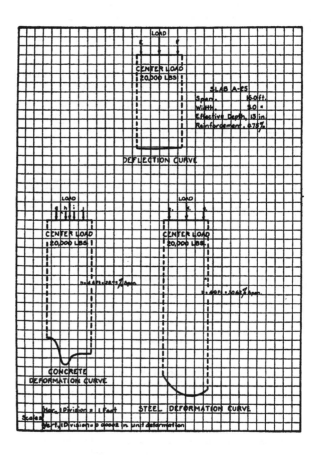

width of the slab 16 feet in span and 32 feet in width under the centrally
applied concentrated load was 11.2 feet, and that this value remained
unaltered when one edge was split off, reducing the total width successively to 29, 25, and 22 feet, although for these widths the load was
eccentrically applied. When, however, the total width was reduced to
18.5 feet, it will be observed that the effective width fell off abruptly to
a lower value, which was only slightly reduced by splitting off sections
from the other edge, so as to reduce the total width to 15.5, 11.5 and 8.5
feet, respectively. Finally, when the width was reduced to 5 feet, in

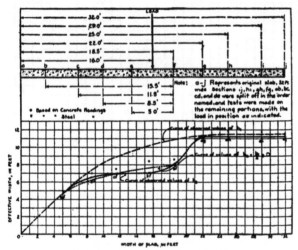

Fig. 11.—Curve showing effective width v. width of slab.

which case the load again was applied at the center, another falling off
in the value of the effective width is indicated.

The curves superimposed on the curve of observed values represent
what seems to be a logical interpretation of the somewhat peculiar
shape of the curve. The regular dash-line curve is a graph of the values
of the effective width for various total widths of a 16-foot slab subjected
to a concentrated central load. It is based upon the values given in
the foregoing table, showing the variation of effective width in slabs
loaded in that manner. The dotted-line curve is based upon the assumption that the effective width of an eccentrically loaded slab (b_e) is equal
to that (b_c) of a centrally loaded slab of the same total width when the
distance from the load to the nearer edge (D) is greater than one-half

the effective width of the centrally loaded slab, but that when D is reduced to less than $\frac{bc}{2}$ the corresponding value of $(be) = \frac{bc}{2} + D$. The approximate coincidence of this curve with that of the test values would seem to confirm the truth of the assumed relation. Unfortunately, however, the results can not be regarded as entirely conclusive, owing to the fact that the recent test involved only one span length and yielded only two points on the critical part of the curve between the total widths of 18 and 22 feet. It is hoped that this point will be cleared up by further tests.

CONCLUSIONS AFFECTING THE DESIGN OF SLABS

To provide against the frequently realized condition of a heavy concentrated load applied near the parapet, the test conclusively shows that the resisting moment required in the portion of a bridge slab near the outer edges is greater than that which is necessary in the central portion.

Further than this, if the relation indicated above be verified and shown to include other span lengths, it would seem that in designing a slab the necessary allowance for the concentrated load near the outer edge can be made very simple in the following manner:

(1) Use the formulas for narrow rectangular beams, substituting for the breadth (b) the value obtained from the foregoing table for central concentrated loads; (2) determine the loss in effective width due to the assumed eccentricity of the load; and (3) supply the deficiency by designing the curb of the parapet to provide a resisting moment equal to that of a slab of width equal to the loss in effective width due to eccentricity, making allowance for the greater stiffness of the section under the parapet. Thus, suppose a slab of 16-foot span and 20-foot width is to be designed to carry a concentrated load of 20,000 pounds applied at a point 4 feet from one edge, then

$$\frac{\text{Total width}}{\text{Span}} = \frac{20}{16} = 1.25$$

from the table for central concentrated loading, the effective width $= 0.69 \times$ 16 feet $= 11.04$ feet. Consider the load of 20,000 pounds to be carried by a width of 11.04 feet, use the ordinary formulæ for rectangular-beam design and determine the effective depth of the slab and the area of steel required. Now, by the relation indicated above, determine the effective width with the load in the critical position 4 feet from one edge, then

$$(be) = \frac{11.04}{2} + 4 = 5.52 + 4 = 9.52 \text{ feet}$$

the difference between the values of bc and be is $11.04 - 9.52 = 1.52$ feet.

Therefore, the curb of the parapet should be so designed that it will have a resisting moment equal to that of a width of 1.52 feet of the slab, making allowance for the greater stiffness of the parapet section.

In constructing slabs designed in this manner it is needless to say that the curb of the parapet must be added before the concrete of the slab has taken initial set.

TEMPERATURES OF THE CRANBERRY REGIONS OF THE UNITED STATES IN RELATION TO THE GROWTH OF CERTAIN FUNGI

By Neil E. Stevens,

Pathologist, Fruit-Disease Investigations, Bureau of Plant Industry, United States Department of Agriculture

INTRODUCTION

In his studies of the diseases of the cranberry (*Oxycoccus macrocarpus*) Shear (8)[1] pointed out that losses from fungus diseases are most serious in the southern sections of its area of cultivation, being greatest in New Jersey and decreasing northward through Long Island and Massachusetts. He further suggested that climatic conditions, particularly the longer hot summer, are chiefly responsible for the greater amount of disease in the southern localities.

As a basis for a more accurate study of this relation, the writer has undertaken to compare the temperature of these regions quantitatively. For this purpose a Weather Bureau observation station supposed to be fairly representative of the region has been chosen for each of the three chief cranberry-growing sections: Middleboro, Massachusetts (elevation 53 feet), Indian Mills, New Jersey (76 feet), and Grand Rapids, Wisconsin (1,021 feet). It is of course recognized that various localities in each region differ more or less widely, but these stations probably represent fairly well the difference in the three areas.

In addition to these representative stations, two others, North Head, Washington (elevation 211 feet), and Farmington, Maine (450 feet), are included, the former because it is close to the cranberry bogs now being developed in the Pacific coast region, the latter because it is the nearest station to a bog in Madrid, Maine, which is of interest as having perhaps the shortest and coolest summer of any commercial bog in the easter United States.[2]

RAINFALL IN CRANBERRY REGIONS

The relation of rainfall to the growth of the cultivated cranberry and its fungus parasites is probably not very direct. The cranberry grows on bog land where the water table is close to the surface, and, when cultivated, the land is drained and subjected to artificial flooding. Even for parasitic fungi the frequent heavy fogs, especially on the Pacific coast, and the abundant dews are probably as important as rainfall. The

---

[1] Reference is made by number (italic) to "Literature cited," p 529.

[2] The elevation of this bog is, according to Mr. Toothacher, roadmaster of the Sandy River and Rangeley Lakes R. R., about 750 feet.

Journal of Agricultural Research,
Washington, D. C.

kv

Vol XI, No 10
Dec. 3, 1917
Key No. G-127

curves of monthly rainfall[1] (fig. 1) indicate, however, that the precipitation of the four eastern localities is similar during the growing season, with the exception of the heavier August rainfall in New Jersey, while that of North Head, Wash., belongs to a quite different type.

LENGTH OF FROSTLESS SEASON

For most green plants the temperatures of greatest importance are those occurring between the last killing frost in spring and the first killing frost in autumn. For the cranberry under cultivation the length of the frostless season is to some extent artificially regulated, sometimes by holding the water late in the spring, often by flooding to avoid early frosts in autumn.

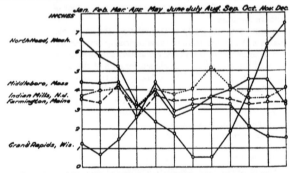

Fig. 1.—Curves of the monthly rainfall during 1916 at Middleboro, Mass., Farmington, Me., Indian Mills, N. J., North Head, Wash., and Grand Rapids, Wis.

The average length of frostless season probably furnishes, however, the best available indicator of the length of growing season for the cranberry and is used as a basis of calculation in the present study. As given by Day (4), the average length of the frostless season for these areas is as follows:

TABLE I.—*Length of frostless season in cranberry regions*

Locality.	Average date of last killing frost in spring	Average date of first killing frost in autumn.	Average length of crop growing season.
			Days.
Grand Rapids, Wis.	May 20	Sept. 20	123
Farmington, Me.	May 20	Sept. 20	123
Middleboro, Mass.	Apr. 20	Oct. 20	183
Indian Mills, N. J.	Apr. 20	Oct 20	183
North Head, Wash.	Feb. 9	Dec. 22	314

[1] The data on the monthly normal rainfall and temperature and the length of frostless season were kindly supplied by Mr. P. C. Day, Chief of the Climatological Division of the Weather Bureau.

DAILY NORMAL TEMPERATURES

The daily normal temperature and precipitation for a considerable number of stations in the United States have been made available by the work of Bigelow (2). Of the five stations used in the present study, however, only North Head is included. In computing the normal daily temperature for the remaining stations, the method used was that outlined by Bigelow (2, *p. 3*).

Figure 2 shows the monthly normal mean temperatures for the five stations. North Head differs markedly from the others in having a much more even temperature throughout the year, while Grand Rapids, Wis., and Farmington, Me., show a greater difference between the summer and winter temperatures than do Middleboro, Mass., or Indian Mills, N. J.

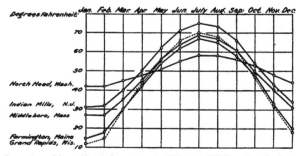

FIG. 2.—Curves of the monthly normal mean temperature during 1916 at Middleboro, Mass., Farmington, Me., Indian Mills, N. J., North Head, Wash., and Grand Rapids, Wis.

TEMPERATURE EFFICIENCY

In order to evaluate the temperature of a region with regard to its effectiveness for plant growth, both the height and duration of favorable temperatures must be considered. In the absence of exact data as to the temperature relations of a given species, and where daily mean temperatures are available, a simple summation of daily means probably gives as satisfactory an index of temperature efficiency in our climate as our present knowledge permits. Briefly, the method is as follows: A certain minimum temperature is assumed as a starting point and the amount added to the summation each day is the difference between this assumed minimum and the number which represents the mean temperature for that day. The minimum is sometimes the freezing point, but often a somewhat higher temperature.

Obviously this method of estimating temperature efficiency is far from ideal. The daily mean temperature obtained by averaging the observed maximum and minimum probably does not exactly represent the heat available on that day. Neither are the daily normal temperatures obtained by Bigelow's method (2) accurate indexes of temperature. They are, however, the best approximations available, and, as the writer has elsewhere stated, if progress is soon to be made in understanding the climatology of plant diseases, a serious effort should be made to utilize the climatic data now at hand.

A summation of remainders as a method of estimating temperature efficiency is open to serious criticism on theoretical grounds, and other methods of integrating temperature have been suggested by Livingston (6). For temperate climates it is not apparent that either of these is greatly superior to a summation of remainder indexes. In a recent study the writer (9) has compared the growth of *Endothia parasitica* in various localities with the temperature computed by the methods of "Physiological summation indexes," "Exponential summation indexes," and "Remainder summation indexes." In that case at least, the last method gave results fully as satisfactory as the others. McLean (7) compared the two last-named methods of temperature summation in connection with a study of climatic conditions in Maryland and found that the two methods agreed in showing a clear relation between temperature and plant growth at the stations used, but he states that—

no evidence was wrought out as distinctly in favor of either of the two methods

TEMPERATURE RELATIONS OF CRANBERRY FUNGI

Two fungi known to be important causes of fruitrots of the cranberry have been selected as types: *Glomerella cingulata* (Stonem.) S. and v. S, abundant during certain seasons in New Jersey and on early varieties in Massachusetts, and *Fusicoccum putrefaciens* Shear, which occurs commonly in Wisconsin and Massachusetts, especially on late varieties, but which is comparatively rare in New Jersey. The temperature relations of these fungi as indicated by their growth on corn-meal-agar plates have been tested in the constant-temperature apparatus described by Brooks and Cooley (3) and are given in Table II. The figures given are in all cases averages of the results of several tests.

TABLE II.—*Average growth of fungi on corn-meal-agar plates for 15-day periods*

Temperature.	Growth of *Glomerella cingulata*.	Growth of *Fusicoccum putrefaciens*.
°C.	Cm.	Cm.
25	8. o	4. o
20	5. 8	3. 3
15	2. 9	2. 3
10	1. o	1. 3
5	o	. 8
o	o	. 4

The rate of growth of both fungi increases with rise of temperature up to at least $25°$ C., but their minimum temperatures are very different, *Fusicoccum putrefaciens* growing somewhat even at $0°$ C., while the minimum temperature for the growth of *Glomerella cingulata* is apparently not far from $8°$ C. The results obtained with *G. cingulata* from cranberry agree fairly well with those obtained by Ames (*1*) for this fungus on bean agar, and by Brooks and Cooley (*3*) for the same organism on apple. Edgerton (*5*) in his studies does not give minimum temperatures, but the maximum for most of the strains of Glomerella used by him is well above $25°$ C.

The temperature relations of a number of other fungi causing fruitrots of cranberries[1] has been determined by their growth on corn-meal-agar plates in an ice thermostat While the temperatures of this apparatus were not absolutely constant and the results obtained are therefore not strictly comparable to those given in Table II, they furnish a satisfactory indication of the temperature relations of these fungi.

TABLE III.—*Average growth (in centimeters) of fungi on corn-meal-agar plates for 15-day periods*

Approximate temperature.	Guianardia.	Acanthorynchus.	Phomopsis.	Pestalozzia.	Sporonema.	Pythium sp.
°C.						
25	3.2	1.7	5.2	(15.0) 7 cm. in 7 days	6.5+	(22.5) 9 cm. in 6 days.
20	2.3	1.0	4.5	9+	3.7	(17) 9 in 8 days.
15	1.6	.5	2.1	7	2.5	9+
10	.8	0.0	.1	2.5	1.5	5.5
5	0.0	0.0	0.0	.2	.5	.6
1	0.0	0.0	0.0	0.0	?	0.0

None of these rot-producing fungi grew at $1°$ C.; Pestalozzia and Sporonema gave some growth at $5°$; and all except Acanthorynchus grew somewhat at $10°$. This last-named fungus, which has the highest minimum temperature of any cranberry fungus tested, has also the highest maximum, growing somewhat even at $34°$ C.

TEMPERATURE INDEXES

The temperature efficiency of the five localities was calculated separately for the two fungi, with the minimum temperature for growth on corn-meal agar as zero points—that is, $47°$ F. for *Glomerella cingulata* and $32°$ for *Fusicoccum putrefaciens*. On this basis the remainder summation indexes are as follows (Table IV).

---

[1] All fungi used in these temperature tests were isolated from rotten cranberries during the fall and winter of 1916 by Mr. Bert A. Rudolph, of the Bureau of Plant Industry. In many cases cultures of a species from several different localities were compared without finding any differences in temperature relations.

Locality.	Glomerella cingulata.		Fusicoccum putrefaciens.		Ratio of index for G. cingulata to that for F. putrefaciens.
	Index.	Per cent.	Index.	Per cent.	
Grand Rapids, Wis..............	2,371	51	4,216	66	0.56
Farmington, Me.................	2,027	44	3,872	61	.52
Middleboro, Mass..............	2,555	56	5,315	83	.48
Indian Mills, N. J..............	4,600	100	6,360	100	.72
North Head, Wash..............	1,896	41	6,067	95	.31

In calculating percentages for purposes of comparison the largest indexes, those at Indian Mills, have been considered 100 in all cases.

COMPARISON OF EASTERN LOCALITIES

On comparing the three eastern stations it is apparent that the effective temperature for the growth of *Glomerella cingulata* in New Jersey is nearly twice that at Middleboro, and more than twice that at Farmington. Temperature differences alone would therefore indicate a very much greater growth of fungi having the temperature relations of *G. cingulata* in New Jersey. This difference might not result disadvantageously if the greater amount of heat could be utilized by the cranberry. This, however, does not appear to be the case. In fact, it is probable that, as suggested by Shear (8), the higher summer temperatures in New Jersey are an actual disadvantage to the cranberry. Certainly the summer temperatures at Middleboro are sufficient for the production of good crops of cranberries and the abundant growth of vines. Good crops of cranberries of excellent quality are regularly produced in New Jersey. This, however, may be due to the fact that careful cultural methods and thorough spraying more than overcome the disadvantage of a slightly unfavorable climate.

For *Fusicoccum putrefaciens* the difference in the three localities is not so great, the temperature efficiency in New Jersey being less than 20 per cent greater than that at Middleboro. On the basis of temperature alone one might expect endrot (caused by *F. putrefaciens*) to be slightly more abundant in New Jersey. Apparently, however, it is much more abundant in Massachusetts. The writer believes that this is due to a greater freedom from competition.

That these two fungi and doubtless others do compete, even within a single berry, is clearly shown by observations made in the fall of 1916. Cranberries of the Early Black variety taken from a bog in Wareham, Mass., were used in temperature experiments. The few berries which rotted at temperatures of 5° and 0° C. yielded chiefly *Fusicoccum putrefaciens*, while those at higher temperatures, 15° to 25°, yielded chiefly *Glomerella cingulata*. The crop of Early Blacks from this bog showed during the first half of September, a considerable portion of rotten berries which had yielded chiefly *G. cingulata*. Apparently at temperatures favoring its growth the more rapidly growing *G. cingulata* had obscured the *F. putrefaciens*. Cranberries of the same variety but of superior keeping quality from another bog showed a somewhat similar relation. Of the berries which rotted at $_0$° and $_{10}$° C. over 90 per cent yielded only *F. putrefaciens*. Of those rotting at $_{20}$° C. about 30 per cent yielded only *F. putrefaciens*, while the remainder showed fungi belonging to five different genera.

That this may happen on a large scale seems to the writer to be indicated by the relations of the two fungi in Massachusetts, where *Fusicoccum putrefaciens*, though known to occur on the early varieties, is especially common as a storage-rot of late berries, making the greater part of its growth at temperatures well below the optimum for the growth of such fungi as *Glomerella cingulata*. It must not be supposed that the competition is solely or even perhaps chiefly between these two fungi. The relations must be very much more complex than this and may be fur-

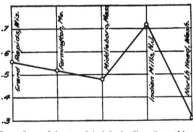

FIG. 3.—Curves of the ratios of the index for *Glomerella cingulata* to that for *Fusicoccum putrefaciens*.

ther confused by the susceptibility of some varieties of cranberries to certain fungi. These fungi are, however, taken as representative of different classes, within each of which there is great variation and undoubtedly competition. The fifth column in Table IV gives the ratio between the index for *G. cingulata* and that for *F. putrefaciens* at the different stations, the index for the latter being considered unity in each case. These ratios, which are shown graphically in fig. 3, may be taken to represent in a rough way the relative amounts of heat available for the two fungi in each locality and indicate to some extent the relative amount of competition to which *F. putrefaciens* is subjected. Where this ratio is high, as in New Jersey, fungi having temperature relations similar to *G. cingulata* predominate. Where it is low, as in Wisconsin and Massachusetts, *F. putrefaciens* apparently becomes more abundant.

WISCONSIN AREA

The temperature efficiency index of Grand Rapids, Wis., is between those of the two New England stations, and during the growing season the rainfall of the three localities is similar. One would then expect the fungus diseases common in the Massachusetts cranberry region to be found also in Wisconsin, and this is generally the case.

23716°—17——6

PACIFIC COAST REGION

As indicated by Table IV, the effective temperature for *Fusicoccum putrefaciens* at North Head, Wash., is very much greater than that at Middleboro, Mass., and only slightly less than that at Indian Mills, N. J., while the effective temperature for *Glomerella cingulata* is less even than than that in Maine. The ratio of temperature indices of the two fungi is accordingly much lower than that at any of the other stations.

As still further emphasizing the difference in this region as regards temperature, it should be noted that daily normal temperatures above 47° F. occur at North Head from April 15 to November 15, while the lowest normal temperature for the year is 41°, and this is reached only for six days. So far as temperature is concerned, then, fungi having the temperature relations of *Fusicoccum putrefaciens* could grow throughout the year.

The absence of high temperatures, combined with the long period of moderate or low temperatures, would indicate the suppression of fungi with relatively high temperature requirements and the correspondingly greater abundance of fungi with lower temperature requirements. On this basis one would be tempted to predict that when the cranberry industry becomes widely developed in the Puget Sound region there will be an increasing abundance of fungi, known chiefly from the northern portion of the present area of cranberry cultivation, possibly even the occurrence in epidemic amounts of fungi occasional or even rare in northern New England. Whether such predictions as to the development of fungi in the coastal region are justified or not, and even though the relation may be affected by growing different varieties in the several localities, it seems certain that the problem of fungus control will be quite different on the Pacific coast than in the eastern United States.

CONCLUSIONS

(1) The principal areas in the United States within which the cranberry is grown commercially have more or less widely different climates.

. (2) This difference appears not only in the rainfall, length of growing season, and daily normal temperatures of these regions, but in their temperature efficiency, as indicated by a summation of daily mean temperatures.

(3) The fungi known to cause decay of cranberry fruits vary greatly in their temperature relations, as shown by their growth in pure culture on artificial media.

(4) In general, the rate of growth of these fungi is reduced by low temperature, and most of them are unable to grow below 5° C., but *Fusicoccum putrefaciens*, the cause of endrot, is able to grow even at 0° C.

(5) The fungus diseases of the cranberry have, in general, proved the most serious and destructive in regions with long growing seasons, combined with high summer temperatures. This applies particularly to the diseases caused by *Glomerella cingulata, Guignardia vaccinii,* and other fungi having similar temperature relations.

(6) Certain other cranberry fungi, notably *Fusicoccum putrefaciens,* are, on the contrary, more abundant in regions which have only short periods of high temperature—that is, periods having daily mean temperatures of $60°$ F., or above.

LITERATURE CITED

(1) AMES, Adeline.
   1915. THE TEMPERATURE RELATIONS OF SOME FUNGI CAUSING STORAGE ROTS. *In* Phytopathology, v. 5, no. 1, p. 11–19.

(2) BIGELOW, F. H.
   1908. THE DAILY NORMAL TEMPERATURE AND THE DAILY NORMAL PRECIPITATION OF THE UNITED STATES. U. S. Dept. Agr. Weather Bur. Bul. R. 186 p.

(3) BROOKS, Charles, and COOLEY, J. S.
   1917. TEMPERATURE RELATIONS OF APPLE-ROT FUNGI. *In* Jour. Agr. Research, v. 8, no. 4, p. 139–164, 25 fig., 3 pl.

(4) DAY, P. C.
   1911. FROST DATA OF THE UNITED STATES; AND LENGTH OF THE CROP-GROWING SEASON, AS DETERMINED FROM THE AVERAGE OF THE LATEST AND EARLIEST DATES OF KILLING FROST. U. S. Dept. Agr. Weather Bur. Bul. V, 5 p., 5 fold. maps.

(5) EDGERTON, C. W.
   1915. EFFECT OF TEMPERATURE ON GLOMERELLA. *In* Phytopathology, v. 5, no. 5, p. 247–259, 4 fig.

(6) LIVINGSTON, B. E.
   1916. PHYSIOLOGICAL TEMPERATURE INDICES FOR THE STUDY OF PLANT GROWTH IN RELATION TO CLIMATIC CONDITIONS. *In* Physiol. Researches, v. 1, no. 8, p. 399–420, 4 fig.

(7) McLEAN, F. T.
   1917. A PRELIMINARY STUDY OF CLIMATIC CONDITIONS IN MARYLAND, AS RELATED TO PLANT GROWTH. *In* Physiol. Researches, v. 2, no. 14, p. 129–208, 14 fig. Literature cited, p. 207–208.

(8) SHEAR, C. L.
   1907. CRANBERRY DISEASES. U. S. Dept. Agr. Bur. Plant Indus. Bul. 110, 64 p., 7 pl. (part col.). Bibliography, p. 55–57.

(9) STEVENS, N. E.
   1917. THE INFLUENCE OF TEMPERATURE ON THE GROWTH OF ENDOTHIA PARASITICA. *In* Amer. Jour. Bot., v. 4, no. 2, p. 112–118, 1 fig. Literature cited, p. 118.

JOURNAL OF AGRICULTURAL RESEARCH

Vol. XI   Washington, D. C., December 10, 1917   No. 11

MOVEMENT OF SOLUBLE SALTS THROUGH SOILS

By M. M. McCool, *Professor of Soils*, and L. C. Wheeting, *Instructor in Soils, Michigan Agricultural Experiment Station*

INTRODUCTION

That there is a tendency for soluble materials to be distributed from regions of high to those of lower concentration in moist soils is to be expected from our knowledge of the phenomenon of diffusion. Moreover, it is generally recognized that the translocation is affected by moisture. One of us (McCool), while conducting soil cultural studies, has repeatedly observed that certain soluble salts when added to soils soon accumulate at or near the surface unless loss of water by evaporation is prevented. Moreover, it has recently been shown in our laboratories that the actual concentration of the soil solution induced by the addition of single salts required to inhibit the growth of higher plants, as well as to retard certain bacteriological activities in the soil, varies appreciably with different soils, which, of course, is to be predicted when we take cognizance of the exchange of bases that may result when salts are added to the soil, as reported by numerous investigators. Thus, during the progress of our researches on the concentration of the soil solution and related subjects several questions with respect to the movement of soluble salts in soils, as well as changes in the composition of the soil solution resulting from the addition of various substances, have arisen. Although reports concerning certain phases of the subject have been obtained, a critical examination of available literature has led to the conclusion that additional well-controlled experiments under both laboratory and field conditions are desirable. On account of the fundamental importance of this subject, the present status of the researches bearing upon it, and certain improved methods, it is being exhaustively investigated in our laboratories, as well as under field conditions. In this report, however, are presented results of some of our laboratory studies of the translocation of certain salts when added in varying amounts to soils of different texture and water contents, together with changes induced in the composition of the soil solution.

Journal of Agricultural Research,
Washington, D C.
b

Vol. XI, No. 11
Dec. 10, 1917
Key No. Mich.—6

(531)

REVIEW OF LITERATURE

Müntz and Gaudechon,[1] several years ago conducted extensive experiments to determine the extent of the movement of sodium nitrate and potassium chlorid in soils, losses of water by evaporation being prevented. When crystals of sodium nitrate or potassium chlorid were placed in a light silicious soil of low water content, the soil containing the salt crystals gradually became moist at the expense of the surrounding soil. After eight days the changes in the water content of the soil receiving sodium nitrate were found to be as follows:

Effect of salt on water movement in the soil

	Per cent
Water content of soil when placed in container	3. 2
Water content of soil containing a salt after 8 days.......	7. 3
Water content of surrounding soil after 8 days.........	2. 6

Moreover, seed were placed in the salted layer of soil and also in the surrounding mass. Germination was retarded in the one region by the high concentration of the soil solution, and in the other by lack of moisture. These authors maintain that when salt is added to a homogeneous mass of soil two systems result: One in which the vapor pressure is high and the other in which it is lower. As a result distillation of water takes place from the one to the other.

Analyses of samples of the soil taken at different distances from the point of application of the salts revealed that the movement of the same was negligible, even after 15 days. In another experiment a garden soil which contained 17½ per cent of water, spoken of as being quite moist, was also used as the medium. Under these conditions the distillation of water vapor did not take place; neither was the movement of the salts detected, 20 mm. from the point in which they were placed in the soil mass.

The extent of the vertical movement was determined by half filling a box 40 cm. high and 11 cm. thick with light soil containing 16 per cent of water, depositing 5 gm. of sodium nitrate, and then filling the box with the soil. The loss of water by evaporation was prevented. After three days samples were taken and the nitrate content was determined. Movement of the salt had not proceeded through a distance of 40 mm. Similar results were obtained with a soil somewhat more moist where determinations of the nitrate content were made six days after the experiment was set up.

Two glazed pots were filled with a sandy clay soil; one was gently, and the other heavily, compacted. Two gm. of powdered potassium chlorid were placed in the center of each, 15 mm. below the surface. Movement of the salt was found to have taken place to a slightly greater

[1] MÜNTZ, and GAUDECHON, H. DE LA DIFUSSION DES ENGRAIS SALINS DANS LA TERRE. *In* Ann. Sci. Agron., s. 3, anné 4, t. 1, no. 5, p. 379–411, 30 fig. 1909.

s were expose   o 15.5 mm. o   , _ o

lied water equal in amount to a rather heavy precipitatio
authors conclude, as a result of their extensive studie.
fertilizer salts when placed in rather dry soils take up
e surrounding soil and long remain localized in the humi
:sult of this condition, seeds may not germinate; but thi
ded by fertilizing and seeding at different intervals.   Wh
·e somewhat moist, the water movement does not take
the other hand, even in moist soils, the salts diffuse exeee
on account of the discontinuity of the soil mass.   Heav
aids this movement, inasmuch as more particles are broug
e contact; yet in a very compacted soil the movement is
is in free liquid.   When salts are applied to soils under fiel
they remain localized for long periods, although the
appreciable.

eaux and Lefort [1] report results of investigations of the
nitrates in a sandy and loam soil, respectively, under lab
ns.   The soils were placed in tightly compacted wooden
o cm. deep.   In some cases the nitrates were placed in f
soil, while in others the salt was deposited in holes made
the center of the boxes.   It was found necessary to spr
h water several times during the experiment in order to
t by evaporation.   Samples of soil were taken at 5, 10,
tances both horizontally and vertically downward from th
ication of the nitrates.   The authors bring out that the
salt by diffusion, either horizontally or vertically downw
slow.   It seems that the translocation of the salts mus
ected by the movement of the water originally present

days. Between August 25 and October 25, 5.45 inches of rain fell. On
October 25 samples of soil were taken at 10 cm. intervals from the
surface to 100 cm. in depth. The nitrates were found to have been
returned to lower levels by the rain water, but on the whole they were
nearer the surface at the end of the period than they were at the
beginning, the total rainfall amounting to 7.92 inches. These authors
point out that nitrates would bring better results in many cases if they
were turned under or applied very early in the season.

They [1] later continued the field experiments and studied the movements of nitrates in the soil by sampling and determining the nitrates
present, and also by employing the sugar beet as the indicator. The
previous field experiments were confirmed. The deep application of the
nitrates afforded higher yields of beets than if they were placed nearer
the surface.

Demolon and Brouet [2] consider that the mode of application of fertilizers should vary with the nature of the soil and with the cultural methods
in vogue. In some
cases part of the
material should
be applied in the
autumn and the
remainder in the
spring when the
seed bed is being
prepared, and in others surface applications are desirable, and in still others
plowing under is best. They maintain that salts have a general tendency to remain at the surface of the soil, rather large amounts of rain
being required to carry nitrates into the subsoil.

FIG. 1.—Apparatus used in studying the movement of salts through soils.

There are on record results of numerous studies of the translocation
of salts in soils induced by percolating water, Sharp [3] having made recent contributions. Jensen [4] has also shown that the solubility of certain inorganic constituents of the soil mass are increased by the presence
of decaying organic matter; but, inasmuch as these do not bear directly
upon our preliminary investigations, they are not considered at this time.

EXPERIMENTAL RESULTS

Usually brass tubes 1½ inches in diameter were employed as the
containers in studying the movement of soluble salts in the soil, but
in some of the early series glass cylinders 8 inches in length and 2 inches

[1] MALPEAUX, L., and LEFORT, G. LA CIRCULATION DES NITRATES DANS LE SOL; LEUR MODE D'EMPLOI. *In* Ann. Sci. Agron., s. 4, ann. 2, t. 1, no 6, p. 705–726, 2 fig. 1913.
[2] DEMOLON, A., and BROUET, G. SUR LA PÉNÉTRATION DES ENGRAIS SOLUBLES DANS LES SOLS. *In* Ann. Sci. Agron., s. 3, ann. 6, t. 2, no. 6, p. 401–418, 2 fig. 1911.
[3] SHARP, L. T. FUNDAMENTAL INTERRELATIONSHIPS BETWEEN CERTAIN SOLUBLE SALTS AND SOIL COL-LOIDS. *In* Univ. Cal. Pub. Agr. Sci., v. 1, no. 10, p. 191–339, 3 fig. 1916.
[4] JENSEN, C. A. EFFECT OF DECOMPOSING ORGANIC MATTER ON THE SOLUBILITY OF CERTAIN INORGANIC CONSTITUENTS OF THE SOIL. *In* Jour. Agr. Research, v. 9, no 8, p. 253–268. 1917.

in diameter were used. As illustrated by figure 1, each of the brass containers was made up of three sections, the short central portions being 2 inches and each of the longer ones 8 inches in length. In setting up the series the short tubes were filled with the soil which had been treated with a definite amount of salt, the longer tubes joined to them, and then carefully filled with the moist untreated soil. The water content of the entire soil column was as nearly uniform throughout as could be made by careful mixing and screening. After the tubes were filled they were carefully sealed, unless otherwise specified, and then placed in a closed chamber, the temperature of which was maintained constant at 18° C. At the close of each period the tubes were unsealed; cross sections of the soil were removed by means of a spatula, dried, cooled in a desiccator, and carefully mixed; then the freezing-point lowering was determined in the usual manner.

MOVEMENT OF SOLUBLE SALTS THROUGH FINE-TEXTURED SOILS

In the first series of experiments the changes induced in the concentration of a rather heavy silt-loam-soil solution by the addition of sodium chlorid were studied. Water to the extent of 50 per cent of the oven-dry weight of the soil was added when the depressions of the freezing-point lowerings were determined. The results presented in Table I are quite striking, changes in the concentration of the soil solution being measurable 3 inches from the soil mass treated with 1 per cent sodium chlorid in solution, after seven days, and 2 inches from that receiving 0.1 of 1 per cent of the salt solution. Fifteen days after the treatment it is worthy to note that the concentration of the soil solution showed an increase throughout the tubes which had been treated with the larger amounts of sodium chlorid; but, on the other hand, changes induced in the soil which had received a more dilute salt solution were less striking. Under the above conditions the rate of salt movement becomes more rapid with an increase in the mass of salt added to the soil.

TABLE I.—*Changes in the concentration of the soil solution induced by the addition of sodium chlorid to a heavy silt loam*

Distance from salt layer.	Freezing-point lowerings.				Distance from salt layer.	Freezing-point lowerings.			
	After 7 days.		After 15 days.			After 5 days.		After 15 days.	
	1 per cent NaCl.	0.1 per cent NaCl.	1 per cent NaCl.	0.1 per cent NaCl.		1 per cent NaCl.	0.1 per cent NaCl.	1 per cent NaCl.	0.1 per cent NaCl.
Inches.					*Inches.*				
4.	0.000	0.000	0.025	0.000	1.	0.100	0.012	0.075	0.013
3.	.003	.000	.040	.000	2.	.045	......	.050	.005
2.	.045	.006	.080	.010	3.	.003	.000	.035	.000
1.	.105	.010	.090	.020	4.	.000	.000	.020	.000
0.	.110	.015	.095	.010					

In the next series of experiments another but lighter silt loam was used as the medium for studying the changes brought about in the concentration of the soil solution in the soil by the addition of sodium chlorid, 0.5 per cent of the salt was added in solution to 100 gm. of the soil, glass tubes being employed as the containers. An additional feature, notably the effect of different water contents upon the rate of change in the concentration of the soil solution, was included in this series.

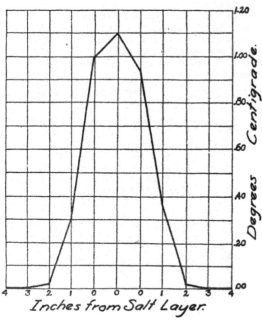

FIG. 2 —Graph of the freezing-point-lowerings after 5 days, induced by the addition of 2 per cent of potassium chlorid to silt loam with 20 per cent of moisture.

According to data presented in Table II, an increase in the concentration of the soil solution was measurable 2 inches from the salt layer after 5 days, becoming more apparent after 10 days; and at the close of the 20-day period the movement of the soluble material had taken place throughout the soil column. An examination of these data also reveals that the rate of movement through the soil containing 20 per cent of water was somewhat more rapid than it was through the soil which contained 10

	After 20 days.			tance from salt	Freezing-point lowerings.					
					After 5 days.		After 10 days.		After 20 days.	
	20 per cent of water.	10 per cent of water.	20 per cent of water.		10 per cent of water.	20 per cent of water.	10 per cent of watre.	20 per cent of water.	10 per cent of water.	20 per cent of water.
				Inches.						
o	0. 000	0. 022	0. 100	1......	0. 430	0. 380	0. 390	0. 350	0. 480	0. 260
o	. 130	. 168	. 160	2......	. 010	. 110	. 040	. 130	. 185	. 180
o	. 160	. 422	. 230		. 000	. 020	. 000	. 010	. 046	. 070
.....	.....	.....	.....	4......	. 000	. 000	. 000	. 000	. 030	. 020

5 Centigrade.

centration of the soil solution in the soil were measurable in the second layers of soil after 5 days, and at the end of 10 days movement had taken place throughout the soil mass in one case and to the third section in another.

TABLE III.—*Changes in concentration of the soil solution induced by the addition of 2 per cent of potassium chlorid to a silt loam with 20 per cent of water*

Distance from salt layer.	Freezing-point lowerings.		Distance from salt layer.	Freezing-point lowerings.	
	5 days.	10 days.		5 days.	10 days.
Inches.			*Inches.*		
4	0.000	0.000	0	0.930	0.840
3	.000	.018	1	.360	.435
2	.015	.070	2	.012	.135
1	.290	.425	3	.000	.050
0	.980	.720	4	.000	.010
0	1.100	.825			

MOVEMENT OF SOLUBLE SALTS THROUGH COARSE-TEXTURED SOILS

After having proved conclusively that an appreciable movement of salts from regions of greater to those of lesser concentration take place in fine-textured soils, it was considered advisable to continue the studies with coarse-textured materials. Accordingly a coarse grade (No. 1) of quartz sand was treated with 0.45 per cent of sodium chlorid, and the changes in the concentration in the various layers resulting from the treatment after 5-, 10-, and 20-day periods, respectively, were determined in the usual manner. In this series glass containers were employed, but instead of being placed in a horizontal position they were stood upright in the chamber. The results given in Table IV reveal that the upper movement after 5 days reached the first inch of soil and had progressed downward into the second inch. After 10 days the concentration of the solution was slightly greater in these sections than at the end of the previous period, and after 10 days' movement had progressed 1 inch farther in each direction. The more rapid downward movement, it seems, may be accounted for by the downward movement of water, inasmuch as 3 per cent approaches the maximum film capacity of this material.

TABLE IV.—*Changes in the concentration of the soil solution induced by the addition of 0.45 per cent of sodium chlorid to a coarse quarts sand with 3 per cent of water*

Distance from salt layer.	Freezing-point lowerings.			Distance from salt layer.	Freezing-point lowerings.		
	5 days.	10 days.	20 days.		5 days.	10 days.	20 days.
Inches.				*Inches.*			
4	0.000	0.000	0.000	1	0.140	0.130	0.015
3	.000	.000	.000	2	.060	.070	.020
2	.000	.000	.005	3	.000	.000	.010
1	.020	.040	.015	4	.000	.000	.000
0	......	......	......				

t and elaborate series a medium sandy soil was employe
tor.   The effect of the mass of sodium chlorid added, as w
content upon the movement of salts in the soil, was studie
mmarized in Table VI reveal that even within five days t
n of the soil solution in the soil, which contained 3 per ce
s measureable in the second inch from the treated layer, an
in the third inch, and at the end of 20 days' movement w
e progressed 4 inches.   Where the water content of the so
ent, the distance of movement was consistently greater
ach period, and it is notable that after 20 days the concer
e soil solution in the last, or fifth, inch of soil, was compar
or there seems to be a tendency under these and simil
ward a uniform concentration of the soil solution througho
f soil.   Indeed, several preliminary experiments, which we
out 75 days, showed this to be the case.   Inasmuch as th
subject is to be reported upon in detail later, these data a

*hanges in the concentration of the soil solution induced by the addition
I per cent of sodium chlorid to a medium sand*

9 per cent of	3 per cent of w
	5 10
	days. days.
0. 005	0 400

In another series, reported in Table VII, 0.1 per cent of sodium chlorid in solution was added to a portion of the medium sand, in order to determine if the mass of salt present affects the rate of salt movement through sandy soil. It is obvious, from a comparison of the data with those in Table VI, that the movement becomes somewhat less rapid with the decrease in the amount of salt added to the soil. After 5 days the concentration of the soil solution in the first inch of soil was far less than it was in the previous series, and at the end of 10 days the increase in concentration of the soil solution was measurable in the second inch, whereas in the previous series the concentration had increased 5 inches from the salt layer.

TABLE VII.—*Changes in the concentration of the soil solution induced by the addition of 0.1 per cent of sodium chlorid to a medium sand*

Distance from salt layer.	Freezing-point lowerings.		Distance from salt layer.	Freezing-point lowerings.	
	5 days.	10 days.		5 days.	10 days.
Inches.			*Inches.*		
4	0.000	0.000	0	0.130	0.080
3	.000	.000	1	.018	.025
2	.000	.018	2	.000	.010
1	.025	.020	3	.000	.000
0	.085	.060	4	.000	.000
0	.140	.080			

Potassium chlorid was added to the medium sand, and the changes in the concentration of the soil solution in the soil were likewise determined after 5 and 10 days, respectively. The middle section of the brass tubes were filled with soil which had previously been treated with 0.9 per cent of the salt in solution, the moisture content of the entire soil column being 9 per cent. It is notable from Table VIII, and graphically represented in figures 4 and 5, that the concentration of the soil solution had increased in the first inch of soil after 5 days, and that the material in solution had passed into the third inch of soil after 10 days.

TABLE VIII.—*Changes in the concentration of the soil solution induced by the addition of 0.9 per cent of potassium chlorid to a medium sand containing 9 per cent of water*

Distance from salt layer.	Freezing-point lowerings.		Distance from salt layer.	Freezing-point lowerings.	
	5 days	10 days.		5 days.	10 days.
Inches.			*Inches.*		
4	0.000	0.000	0	0.240	0.210
3	.000	.009	1	.090	.095
2	.000	.014	2	.000	.015
1	.075	.120	3	.000	.005
0	.265	.255	4	.000	.000
0	.260	.265			

In order to approach more nearly field-soil fertilization, 2 gm. of solid potassium chlorid and sodium carbonate, respectively, were placed in the middle of the tubes containing medium sand; otherwise the studies were conducted in the usual manner. The results presented in Table IX show that the concentration of the soil solution in the soil changes soon after solid salts are applied to the moist soils. The increase in the concentration within 5 days was appreciable 2 inches from the layer of potassium chlorid in the tubes containing 3 per cent of water, and 3 inches from it in those containing 9 per cent of water.

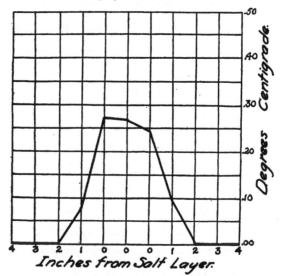

Fig. 4.—Graph of the freezing-point lowerings after 5 days, induced by the addition of 0.9 per cent of potassium chlorid to sand with 9 per cent of moisture.

TABLE IX.—*Changes in the concentration of the soil solution induced by the addition of 2 gm. of powdered potassium chlorid to a medium sand*

Distance from salt layer.	Freezing-point lowerings.				Distance from salt layer.	Freezing-point lowerings.			
	9 per cent of water.		3 per cent of water.			9 per cent of water.		3 per cent of water.	
	5 days.	15 days.	5 days.	15 days.		5 days.	15 days	5 days.	15 days.
Inches.					*Inches.*				
4...........	0.000	0.085	0.000	0.003	1...........	0.650	0.368	0.255	0.282
3...........	.030	.251	.000	.096	2...........	.260	.335	.050	.141
2...........	.180	.493	.132	.214	3...........	.030	.117	.000	.113
1...........	.418	.691	.300	.277	4...........	.000	.067	.000	.005

It is notable that the changes induced by the addition of sodium carbonate were measurable after 5 days 1 inch from the salt in the soil containing 3 per cent and 2 inches from it where the moisture content was 9 per cent, and at the close of the 15-day period the soluble material had moved into the next layer of soil. The results obtained from the addition of this salt are given in Table X.

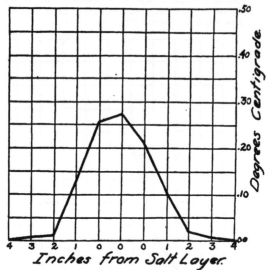

Fig. 5.—Graph of the freezing-point lowerings after 10 days, induced by the addition of 0.9 per cent of potassium chlorid to sand with 9 per cent of moisture.

TABLE X.—*Changes induced in the concentration of the soil solution by the addition of 2 gm. of powdered sodium carbonate to a medium sand*

Distance from salt layer.	Freezing-point lowerings.				Distance from salt layer.	Freezing-point lowerings.			
	9 per cent of water.		3 per cent of water.			9 per cent of water.		3 per cent of water.	
	5 days.	15 days.	5 days.	15 days.		5 days.	15 days.	5 days.	15 days.
Inches.					*Inches.*				
4............	0. 000	0. 000	0. 000	0. 000	1............	0. 332	0. 415	0. 225	0. 279
3............	. 000	. 013	. 000	. 000	2............	. 025	. 115	. 005	. 029
2............	. 029	. 005	. 000	. 007	3............	. 000	. 015	. 000	. 000
1............	. 349	. 220	. 303	. 255	4............	. 000	. 000	. 000	. 000

In another series the tubes were stood upright, the upper ends being open in order to permit loss of water by evaporation and to thus determine if the downward movement of the soluble material is appreciable when water is being lost in this manner. The results obtained and presented in Table XI show that the downward movement took place, the concentration of the solution being appreciably increased in the bottom layers of silt-loam soil, where 1 per cent of potassium chlorid was added, although the water content of the various layers was found to have decreased. It is possible that some of the soluble material in the lower layers would be translocated later on to the upper ones by the film water movement, this of course, could be determined by carrying on the experiments over a longer period. The greater concentration of the soil solution in the upper layers indicates an upward film movement of the water in this particular soil, of these water contents. We have experiments in progress which should show at what water content film movement ceases in different classes of soils.

TABLE XI.—*Movement of 1 per cent of potassium chlorid through silt-loam soil in open tubes, showing the concentration after 10 days*

Distance from salt layer.	Low moisture.			High moisture.			Distance from salt layer.	Low moisture.			High moisture.		
	Moisture content.	Moisture loss.	Freezing point lowerings.	Moisture content.	Moisture loss.	Freezing point lowerings.		Moisture content.	Moisture loss.	Freezing point lowerings.	Moisture content.	Moisture loss.	Freezing point lowerings.
Inches.							*Inches.*						
4.....	2.93	7.88	0.080	11.75	6.21	0.165	1......	9.88	0.93	0.125	12.74	5.22	0.062
3 ...	5.68	5.18	.065	11.95	6.01	.175	2......	9.78	1.03	.038	12.64	5.32	.040
2.....	8.37	2.44	.110	12.21	5.75	.*172	3......	9.93	.88	.045	12.95	5.01	.058
1.....	8.85	1.96	.227	12.30	5.66	.130	4......	10.02	.79	.060	13.00	4.96	.048
0.....	7.52	2.51	.270	12.00	5.96	.098							

Where sand was employed, rather strikingly different conditions have ensued. It may be cited that, for example, according to the data in Table XII, the amount of soluble material that moved downward was indeed slight in comparison with the upper translocation, or with the downward movement in the silt-loam soil. It is evident from these results that soluble salts such as sodium nitrate when applied to sandy soils, or present in them, are more likely to be lost to the crop by their accumulation in the upper layers of soil during a drouth than they are in case of the finer-textured classes of soils. Field experiments now under way may afford additional evidence bearing on this question.

TABLE XII.—*Movement of 1 per cent of potassium chlorid through medium sand in open tubes, showing the concentration after 10 days*

Distance from salt layer.	Low moisture.			High moisture.			Distance from salt layer.	Low moisture.			High moisture.		
	Moisture content.	Moisture loss.	Freezing-point lowerings.	Moisture content.	Moisture loss.	Freezing-point lowerings.		Moisture content.	Moisture loss.	Freezing-point lowerings.	Moisture content.	Moisture loss.	Freezing-point lowerings.
Inches.							*Inches.*						
4.....	0.46	2.73	0.048	3.68	4.69	0.285	1......	2.25	0.94	0.020	4.73	2.95	0.018
3.....	1.23	1.96	.082	3.98	4.39	.045	2......	2.37	.82	.007	4.82	3.55	.014
2.....	1.54	1.65	.090	4.11	4.26	.085	3......	2.40	.79	.000	5.11	3.26	.002
1.....	2.00	1.19	.070	4.22	4.15	.050	4......	2.45	.74	.000	5.30	3.07	.000
0.....	2.20	.99	.030	5.45	2.92	.013							

CHEMICAL STUDIES

Chemical studies of the solution obtained by extracting 1 part of the various layers of soil with 1 of distilled water and passing it through the Chamberland filter have been made. In the first series 300 gm. of medium sand and silt-loam soil, respectively, were treated with 1 per cent of sodium chlorid in solution, and were placed in the bottom of 3-gallon jars and the filling completed with untreated moist soils. One set was unsealed, in order to permit loss of water by evaporation. At the close of the experiment each 1-inch layer was air-dried, extracts obtained as above indicated, and certain bases determined. The depression of the freezing point of the soil and the amount of the bases found in the extraction are given in Table XIII.

TABLE XIII.—*Changes in the composition of the soil solution induced by the addition of 1 per cent of sodium chlorid to a medium sand and a silt loam. Duration, 15 days*

Distance from salt layer.	Medium sand, containing 9 per cent of water.				Silt loam, containing 20 per cent of water.			
	Freezing-point lowerings of soil.	Iron and aluminium.	Calcium.	Magnesium.	Freezing-point lowerings of soil.	Iron and aluminium.	Calcium.	Magnesium.
Inches.	*P. p. m.*	*P. p. m.*	*P. p. m.*	*P. p. m.*	*P. p. m.*	*P. p. m.*	*P. p. m.*	*P. p. m.*
4.................	0.042	No appreciable changes	+124.6	+6.90	0.002	No appreciable changes	+14.6	+5.58
3.................	.077		+74.3	+6.59	.050		+45.8	+11.50
2.................	.117		+51.7	+4.28	.322		+148.9	+19.95
1.................	.140		+26.7	+3.27	.522		+103.4	+12.70
0.................	.277		+39.4	+1.31	.687		+89.8	+9.47

It is notable that the addition of sodium chlorid resulted in an increase in the amount of the bases calcium and magnesium released to the solution upon extraction, the amount increasing from the salt-treated layer of soil upward in the case of the sand in the open containers. The magnesium and calcium were at the maximum in the second layer in case of the silt loam. The bases iron and aluminium were not measureably affected by the treatment.

The addition of the sodium carbonate, as may be seen from Table XIV, resulted differently, inasmuch as the amount of iron and aluminium in the extract from the treated and adjacent layers of sandy soil were strikingly increased, while the parts per million of calcium were appreciably less, and the magnesium decreased slightly in the two lower layers. In case of the silt loam the amount of calcium in the extraction from the treated and the adjacent layer of the soil was decreased, while the changes in the amount of magnesium were negligible.

TABLE XIV —*Changes in the composition of the soil solution induced by the addition of 1 per cent of sodium carbonate to a medium sand and a silt loam. Duration of experiment, 15 days*

Distance from salt layer.	Medium sand, containing 9 per cent of water.				Silt loam, containing 20 per cent of water.			
	Freez-ing-point low-erings of soil.	Iron and aluminium.	Calcium.	Magne-sium.	Freez-ing-point low-erings of soil.	Iron and aluminium.	Calcium.	Magne-sium.
Inches.		*P. p. m.*	*P. p. m*	*P. p. m.*		*P. p. m.*	*P. p. m.*	*P. p.*
4...................	0.004	0.00	+26.70	+2.75	0.000	No ap-pre-ciable changes	+ 3.9	+1.74
3...................	.005	.00	+ 6.50	+1.14	.000		+ .8	.00
2...................	.053	+ 7.2	— 8.50	+ .61	.000		.00	.00
1...................	.257	+84.6	—10.00	— .15	.012		—14.5	.00
0...................	.287	+60.3	— 8.00	— .22	.092		—14.31	— .97

Where the containers were sealed, the presence of the sodium chlorid in medium-sand soil resulted in similar but somewhat less striking conditions than in the previous series. According to the data presented in Table XV, the amount of calcium and magnesium increased upward from the treated to the top layer, or these bases obviously were liberated to the soil solution and then passed onward to regions of lower concentration.

TABLE XV.—*Changes in the composition of the soil solution induced by the addition of 1 per cent of sodium chlorid and 1 per cent of sodium carbonate to medium sand. Duration of experiment, 25 days*

Distance from salt layer.	Sodium chlorid added.				Sodium carbonate added.			
	Freezing-point lowerings of soil.	Iron and alumi-nium.	Calcium.	Magne-sium.	Freezing-point lowerings of soil.	Iron and alumi-nium.	Calcium.	Magne-sium.
Inches.		*P. p. m.*	*P. p. m.*	*P. p. m.*		*P. p. m.*	*P. p. m.*	*P. p. m.*
3..........	0.002	+3.60	+ 3.60	+0.48	0.000	Trace.	+16.17	+0.92
2..........	.015	+3.40	+27.08	+4.21	.020	Trace.	+ 6.46	+ .39
1.	.085	+1.20	+12.01	+1.79	.025	Trace.	— 4.65	— .44
0..........	.225	+4.60	+ 2.89	+ .26	.095	+63.00	— 3.40	+ .70

One per cent of sodium carbonate was added to a portion of the third layer of silt-loam soil taken from the closed containers and the second layer of sand, respectively. They were then moistened and let stand for five days in closed containers. At the end of this period the parts per million of the bases calcium and magnesium in the extracts were determined, the results obtained being given in Table XVI. It is notable that the addition of the sodium carbonate resulted in an appreciable decrease in the amount of these readily soluble bases, probably changing them from the chlorids to the carbonates

TABLE XVI.—*Changes in the composition of the soil solution induced by the addition of sodium carbonate*

Soils.	Iron and aluminium		Calcium.		Magnesium.	
	Before adding sodium carbonate.	After adding sodium carbonate.	Before adding sodium carbonate.	After adding sodium carbonate.	Before adding sodium carbonate.	After adding sodium carbonate.
		P. p. m.	*P. p. m.*	*P. p. m.*	*P. p. m.*	*P. p. m.*
Silt loam.	Trace.	70. 00	60. 50	16. 66	14. 43	6. 55
Sand	..do...	102. 00	61. 90	27. 76	6. 20	2. 79

DISCUSSION OF RESULTS

We have presented experimental data which show that soluble salts are translocated from regions of high to those of lower concentration in moist soils when inclosed in sealed containers; and in case of silt loam in the open containers upward movement is very rapid and the downward translocation is marked, the water movement evidently decreasing the downward translocation. We are at loss to account for the contradictory report of Müntz and Gaudechon, unless the methods employed failed to detect the changes that may have taken place in the nitrate content of the soil. In case of the potassium chlorid, the lack of movement reported may have been and probably was due to the retention of the potassium by the soil, other bases being forced into the solution.

As stated in the introduction, such movements are to be expected, especially if the moisture coats the soil particles in the form of films, in view of the fact that diffusion of salts take place in solution, but on the other hand the movement may not be and probably is not due wholly to diffusion. It does not seem untenable to assume that the reactions which take place when salts are added to the soil play their rôle. A given base coming in contact with a particle or a group of particles may be held and others liberated, adjacent particles may not be satisfied, so far as one or more of these bases are concerned, and by removing them from solution may aid in the translocation of soluble material in the soil.

mical studies show that the addition of soluble salts to a gi
he soil results in changes of the composition of the soil solu
not be confined to the soil mass receiving the applicat
hat such conditions are of far-reaching importance in com
the results obtained from the use of soluble-fertilizer salts
empts to bring about a so-called balanced soil solution. I
ils undoubtedly vary with respect to the action that takes p
are treated with various soluble substances, it does not s(
work out a balanced soil solution by studying a few soil

CONCLUSIONS

results presented show that soluble salts move from regi
lower concentration in moist soils, the rate being rather ra
ore do not long remain localized, as reported by earlier inv
oreover, the rate of movement is affected by the water con
and the mass of salt present.
ere moisture is being lost by evaporation, the upward m
ore rapid than the downward translocation in heavier s
e of sands the downward translocation is indeed slight, t
that soluble salts, such as sodium nitrate, are more likel
7 upward movement to the surface of sandy soils durin
an in case of heavier soils.
emical studies show that as the salts move through the
solution in the various layers of soil changes in composi
ges, it seems, have an important bearing upon the result
d from the use of fertilizer salts, especially upon attempt

BREEDING SWEET CORN RESISTANT TO THE CORN EARWORM

By G. N. Collins, *Botanist*, and J. H. Kempton, *Scientific Assistant,*
Office of Acclimatisation and Adaptation of Crop Plants and Cotton Breeding,
Bureau of Plant Industry, United States Department of Agriculture

INTRODUCTION

The production of sweet corn (*Zea mays*) in the southern part of the United States and throughout the American Tropics is seriously interfered with by the ravages of the corn earworm (*Chloridea obsoleta* Fab.).

The geographic range of this insect is practically coextensive with maize culture, extending to the northern boundaries of the United States. But Quaintance and Brues[1] state that nowhere in the transition zone, comprising in the main the New England States, New York, Pennsylvania, Michigan, Wisconsin, and Minnesota, is the pest of regular occurrence or a cause of any considerable damage. For some distance south of this region the injury is also comparatively slight, but in many sections near the southern border of the country sweet corn is not grown at all, its place on the table being taken by field varieties.

The exclusion of sweet varieties from these regions may not be entirely due to the corn earworm, but it is probably safe to consider this insect the major factor.

The corn earworm does not confine its depredations to sweet corn, but also attacks field varieties. From the fact that northern varieties of field corn, when grown in the South suffer much more than do the local sorts, it would appear that the especial susceptibility of sweet varieties is not due to the character of the seeds alone, and that the southern varieties of field corn must possess some additional peculiarity that renders them at least partially immune.

A comparison of the general characteristics of northern and southern varieties at once suggests that the greater immunity of southern varieties may be due to the greater development of husks in the southern varieties. Attention was early called to this possibility by Mr. O. F. Cook,[2] from observations made on a variety of corn growing near Brownsville, Tex. This variety produces small ears inclosed in very long husks. Mr. Cook noticed that, while many larvæ were found inside of the projecting husks, few had reached the ears.

---

[1] Quaintance, A. L., and Brues, C. T. THE COTTON BOLLWORM. U. S. Dept. Agr. Bur. Ent. Bul. 50, 155 p., 27 fig., 25 pl., 1905.
[2] Bionomist in Charge, Office of Acclimatization and Adaptation of Crop Plants and Cotton Breeding, Bureau of Plant Industry.

Journal of Agricultural Research,
Washington, D. C.
k

Vol. XI, No. 11
Dec. 10, 1917
Key No. G—199

With the idea that southern field varieties owe their relative immunity to the thick covering of long husks which protect their ears, it seemed worth while to endeavor to breed varieties of sweet corn possessing numerous long husks. Since the distinctive character of the seeds of sweet varieties behaves in hybrids as a Mendelian unit, strains that would breed true to the sweet character might be expected in the second generation of a cross between field and sweet varieties. It was hoped that from the plants producing sweet seeds, strains possessing the desired husk characters could be isolated.

The present paper is an account of an attempt to secure this result, with a discussion of some of the factors of worm resistance on which light has been thrown in the course of the experiment.

Even in the worst worm-infested regions it is largely a matter of chance whether any particular ear is injured or escapes. To select intelligently, it is therefore highly important to know something of the plant characters that minimize injury and to use these characters as a basis for selection. The study of the characters associated with worm resistance was carried on simultaneously with the breeding work, and the value of the results is believed to lie in the analysis of the characters and the method of breeding quite as much as in the material results. These material results comprise two strains of sweet corn possessing marked resistance to the corn earworm.

PROTECTIVE CHARACTERS

Four protective characters were in mind at the beginning of the experiment: (1) The distance which the husks extend beyond the tip of the ear, with the idea that larvæ frequently gain access to the ear by entering at the tip of the shoot and eating their way down the silks. It would obviously be advantageous to increase the distance they must travel. (2) The thickness of the husks' covering. Many ears are found with perforations through the husks, and a thicker covering might be expected to hinder the invasion of the larvæ from this direction. (3) The texture of the husks. In most sweet varieties the husks are relatively soft and smooth, while in field varieties, especially those from the Tropics, the husks are firm and harsh. The outer husks of some varieties are covered with firm spicules, providing a surface almost as silicious as sandpaper. This character might be expected to deter the insects from eating their way to the ear through the husks. (4) Husk leaves. It was thought that ears without husk leaves might be less attractive to moths.

PLAN OF EXPERIMENTS

The experiments were begun in 1912. Mr. John H. Kinsler, at Victoria, Tex., made various crosses between three commercial varieties of sweet corn, Stowell's Evergreen, Early Evergreen, and Early Cory, and two varieties of field corn, Brownsville and Marrainto. Brownsville is a selected strain of the variety in which Mr. O. F. Cook had first observed the worm resistance; Marrainto is a variety from northern Mexico, with rather thicker and harsher husks than those of Brownsville. These also extend well beyond the ear.

Isolated plantings of the first generation of these crosses were made at Victoria in 1913. The four lines which have been continued are Ph75 Brownsville × Early Cory; Ph77 Early Evergreen × Brownsville; Ph79, Stowell × Brownsville; and Ph80, Marrainto × Evergreen. The ears from the first-generation hybrid plants contained a mixture of sweet and horny seeds. The sweet seeds from selected ears of each of the four hybrids were planted in separate rows at Lanham, Md., in 1914, one row from each ear.

The procedure in 1914 was to make pollinations between plants of a similar appearance, and usually within the same row. The distance the husks extended beyond the ear, the thickness and number of husks, and the extent to which husk leaves were developed were recorded for all plants used in making pollinations. An attempt was also made to grade the plant with respect to the texture of the husks.[1] In selecting plants for pollination preference was given to those with long husks and few husk leaves; but other types were also pollinated, including a few that were distinctly inferior with respect to the characters thought to denote worm resistance.

Fourteen ears were selected for ear-to-row tests in 1915. The designations of these ears with their ancestry are given in Table I.

TABLE I.—*History of the corn plantings made at Chula Vista, Cal., in 1915*

1912. Victoria, Tex. Original cross.	1913. Victoria, Tex. Treatment.	1914. Lanham, Md. (Sweet seeds planted.) Plant combinations.	1915. Chula Vista, Cal. Planting designation.
Brownsville × Early Cory ...	Isolated block..	121×122	Ph121
Do....................	....do...............	124×122	Ph123
Do....................	....do...............	1×9	Ph127
Do....................	....do...............	8×111	Ph128
Evergreen × Brownsville..	Hand-pollinated.......	10×12	Ph129
Do....................	....do...............	12×10	Ph130
Do....................	....do...............	12×15	Ph131
Stowells × Brownsville	Isolated block	159×158	Ph120
Do....................	....do...............	137×75	Ph122
Do....................	....do...............	140×138	Ph124
Do....................	....do...............	89×30	Ph125
Do....................	....do...............	162×160	Ph126
Marrainto × Evergreen........	Hand-pollinated.......	182×179	Ph118
Evergreen × Brownsville..	....do............:....	32	
		×	
Stowells × Brownsville........	Isolated block..........	21	Ph119

[1] The surface of the husks in the hybrid progenies is distinctly harsher than in commercial varieties of sweet corn. It was not found possible, however, to distinguish the different progenies in this particular with serviceable accuracy and the notation of the character has been discontinued.

The 1915 planting was made at Chula Vista, near San Diego, Cal., on March 15. The corn earworm is a more serious pest in this region than at Lanham, Md., and a more uniform infestation resulted.

In comparison with other plantings of sweet varieties in the same neighborhood, there seemed little doubt that the field of hybrids as a whole was less injured than other varieties. An effort was made, however, to secure definite quantitative data on the effect of the selection, and to determine the characters most closely associated with immunity. The characters of the individual plants were recorded in a series of measurements described below:

(1) DAMAGE.—The portion of the ear rendered inedible was estimated on a scale of 10—that is, an ear in which the larvæ had eaten completely to the base, rendering it worthless, was classed as 10. The slightest damage was recorded as 1 and an ear one-half of which was destroyed as 5. With intermediate stages estimated on the same scale, the grading was all done by the junior author and experiments showed that the maximum uncertainty regarding the class to which any particular ear should be referred was not greater than one grade.

(2) NUMBER OF LARVÆ.—The number of larvæ found inside the husks at harvest, together with any which it could be seen had escaped. When the infestation is very severe, the fact that the larvæ are cannibalistic would doubtless cause the recorded number to be lower than the true number.

(3) DAMAGE PER LARVA.—The figure indicating the total damage of each progeny divided by the total number of larvæ in the same progeny.

(4) PROLONGATION.—The distance from the tip of the ear to the tip of the husks, recorded in centimeters.

(5) LENGTH OF HUSKS.—This measurement was obtained by adding the prolongation and length of ear.

(6) LENGTH OF EAR.—The length of the ear in centimeters, including any damaged portion.

(7) NUMBER OF HUSKS.—The total number of husks surrounding the ear.

(8) NUMBER OF LAYERS.—A small hole was cut through the husks at the side of the ear at a point about midway between the tip and the base, and the number of layers of husks at this point were recorded.

(9) DAYS TO SILKING.—The number of days that elapsed from planting to the first appearance of silk.

(10) MATURITY.[1]—The degree of maturity was judged by slicing off the tops of the grains, and estimating the proportion of opaque to transparent endosperm. Prime eating condition was designated "grade 10." The lowest grade that would be marketable on our scale would be about 6 and the highest about 15.

(11) SILKING TO HARVEST.[1]—The number of days that elapsed between silking and harvesting of the ear.

(12) HUSK LEAVES.—The extent to which husk leaves were developed was graded in accordance with an arbitrary scale ranging from 0 to 10. The same system of grading husk leaves has been used for a number of years in recording the behavior of all varieties grown and has proved to be a reliable measurement.

(13) NUMBER OF ROWS —The number of rows of grains on the ear was included in the notes largely as an indication of the circumference of the ear.

Table II gives the mean value of each of the characters for each of the progenies grown in 1915.

---

[1] Not taken in 1915.

TABLE II.—*Measurement of plant characters at Chula Vista, Cal., in 1915*

Progeny.	Damage.		Number of larvæ.		Damage per larva.		Prolongation.	
	First ears.	Second ears.	First ears.	Second ears.	First ears.	Second ears.	First ears.	Second ears.
Ph118............	2.1±0.3	3.1±0.3	1.3±0.1	1.4±0.1	1.7±0.3	2 1±0.2	9.8±0.6	9.9±0.4
Ph119............	2.7± .3	1.6± .2	1.6± .2	1.2± .1	1.7± .3	1.3± .2	9.7± .5	10.9± .5
Ph120............	3.3± .6	1.9± .4	1.7± .2	1.2± .1	2 0± .5	1.5± .3	9.8± .6	9.8± .8
Ph121............	2 5± .3	1 6± .2	1.8± .2	1.5± .1	1.4± .2	1.1± .2	9.7± .2	9.8± .5
Ph122............	2.2± .2	1.4± .3	1.4± .1	1.3± .1	1 5± .2	1.1± .2	11.2± .3	11.6± .8
Ph123............	1.8± .3	1.5± .3	1 3± .1	1.3± .1	1 4± .9	1 2± .2	8.3± .6	8.7± .5
Ph124............	.8± .2	1.9± .5	1.1± .1	.8± .1	.7± .2	2.3± .8	13 1± .6	11 9± .8
Ph125............	5.9± .8	4.6± .8	2 2± .1	2.0± .2	2.8± .4	2.3± .4	9 7± .5	8.1± .8
Ph126............	3 1± .3	4.4± .7	2 1± .1	1.8± .3	1.5± .2	2.4± .5	6.5± .4	7 8± .9
Ph127............	3.1± .5	4 3± .4	1.9± .2	2.1± .1	1 8± .3	2.1± .2	6.0± .7	5 2± .4
Ph128.	5.0± .8	3 1± .3	2.1± .1	1.8± .2	2.4± .4	1.7± .3	12.5± .6	11 9± .7
Ph129............	5.2±1.1	4 4± .8	2 3± .2	1 7± .2	2 3± .5	2 6± .5	7.7± .1	10.3± .7
Ph130............	3.1± .4	2.7± .3	1.7± .2	1.9± .2	1 9± .4	1.4± .2	8.9± .5	8.4± .4
Ph131............	3.8± .4	2 4± .3	1.9± .1	1.5± .1	2.0± .2	1.6± .2	7.3± .5	10.1± .6

Progeny.	Length of ear, first ears.	Number of husks, first ears.	Number of layers.		Days to silking, first ears.	Husk leaves.		Number of rows, first ears.
			First ears.	Second ears.		First ears.	Second ears.	
Ph118............	13.1±0.3	12.7±0.3	10.7±0.3	10.0±0.3	98±1	2.8±0.3	3.3±0.2	14.0±0.2
Ph119............	14.7± .3	13 3± .4	10.0± .3	10.4± .3	108±1	.6± .1	1.5± .2	15 0± .3
Ph120............	14.3± .5	16 4± .7	11.6± .4	12.2± .4	113±2	2.2± .4	2.7± .3	16 7± .4
Ph121............	14.3± .3	12.3± .4	8.8± .4	7 9± .3	107±1	4.3± .3	4.8± .4	12 4± .2
Ph122............	15.8± .3	14.1± .3	10.9± .2	12.2± .3	107±1	.4± .1	.9± .2	13.0± .2
Ph123............	15.4± .3	12.4± .3	10.6± .4	10.3± .2	105±2	1.8± .3	2.3± .2	13.2± .3
Ph124............	15 3± .4	13.7± .4	11.6± .2	10.9± .3	111±2	.3± .2	1 1± .3	17 1± .4
Ph125	15 6± .6	10.1± .4	8 7± .3	9.2± .9	96±2	1.1± .3	1 0± .3	14.8± .3
Ph126........	15.1± .5	12 8± .4	10.2± .4	12.1± .7	115±1	1 8± .3	1 6± .4	17.5± .5
Ph127............	11.9± .6	10.8± .6	7.3± .2	7 1± .5	104±3	1 4± .4	3 7± .5	13.6± .4
Ph128............	12.3± .4	11.7± .6	7.8± .5	8.4± .3	107±2	1.4± .4	2 8± .4	14.5± .2
Ph129............	14.7± .8	15.0± .4	10.0± .4	9.3± .5	92±4	.7± .4	1.6± .2	15.3± .5
Ph130............	14.3± .3	14.7± .6	10.0± .2	9.4± .2	90±1	1.2± .2	3.9± .3	14 7± .3
Ph131............	14.0± 5	12.8± .6	9 1± .3	9.4± .3	103±2	1.2± .1	3.0± .3	13.1± .2

ANALYSIS OF RESULTS IN 1915

The first step in analyzing the results was to determine whether the several rows showed real differences in the amount of damage inflicted by the corn earworm. If significant differences were not developed, it would hardly be worth while to proceed with selection.

The average damage for the different rows is shown in columns 2 and 3, Table II, together with the probable errors. It will be seen that some rows are damaged much more than others, and a consideration of the probable error shows that many of these differences may not be ascribed to chance. An analysis of the first, or upper ears, for which the data are more complete, shows that the row with the greatest amount of damage, Ph125, was damaged seven and one-half times as much as Ph124, which was the least affected.

It was thought possible that the degree of damage of the different rows might be influenced by their position in the field. The infestation might come from one side of the field and the rows nearest its source thus be more severely damaged. The arrangement of the progenies in the table corresponds with that in the field, and no general trend is apparent. It happens that the least and most damaged rows stood side by side near the middle of the field.   If location was a factor, adjacent rows should, on the average, show a closer agreement in the extent of damage than pairs taken at random.   On eliminating Ph129, which is the reciprocal of Ph130, the correlation between adjacent rows for the upper ears alone is 0.161 ±0.189 and for upper and second ears combined 0.173±0.189. There is, thus, little or no tendency for adjacent rows to be damaged to a similar extent, and position in the field seems not to be an important factor in causing the observed differences.

On the other hand, if the immunity which some of the rows enjoyed is due to plant characteristics, there should be some agreement between the degree of immunity of upper and second ears of the same row.   A comparison of the value in columns 2 and 3, Table II, shows such an agreement.   The row in which the upper ears were most damaged is also the row in which the second ears were most damaged.   It can also be seen that the row with the least damage to the first ears has a very low damage in the second ears.   Beyond these rather outstanding cases, the agreement or lack of agreement is not obvious from inspection.   The correlation of damage between upper and second ears is believed to be a fair measure of this agreement; it was found to be 0.718±0.087.   This correlation alone would seem to establish the fact that the individuality of progenies is an important factor in determining the extent of damage.   From the 1915 results we may safely conclude that there is something about the plants descended from certain ears which affords them an appreciable measure of protection.   The next step was to determine, if possible, whether this protection could be referred to any of the recorded plant characters.

As soon as a detailed analysis of the 1915 data was attempted it became evident that there were sources of error that had not been adequately guarded against.   The ears were not harvested at a uniform stage of development, and those left longer were more severely damaged.   There were also many ears bagged to secure pure seed, and since the bagging was in a measure selective, it introduced another source of possible error. These disturbing factors made it appear unwise to place confidence in any detailed analysis of the 1915 data.   These results will therefore be considered only in connection with the results of the following season.

EXPERIMENTS IN 1916

The 1916 plantings were made at Lanham, Md., on May 14 and consisted of 35 rows, as follows: A repetition of the 14 progenies grown in 1915 (ancestry described on p. 551), 9 progenies from ears secured by hand pollinations within the rows of the progenies grown in 1915, 8 progenies from ears obtained by crosses between the rows in 1915, and 2 first-generation crosses between 1915 progenies and Hopi maize.   There were

....	3330	Ph119	3601	Ph125
....	3334	Ph119	3286	Ph118
....	3368	Ph120	3328	Ph119
....	3369	Ph120	3328	Ph119
....	3373	Ph120	21	Hopi.
....	3403	Ph121	3328	Ph119
....	3489	Ph122	3320	Ph119
....	3557	Ph124	3523	Ph123
....	3602	Ph125	21	Hopi.

ɔf taking notes has already been described (p. 552).   The
was followed as in 1915, except that the number of days
harvest and the degree of maturity were added.

ɔssible to secure ears at approximately the same degree
exact dates of the first pollen and the first appearance
·orded on a tag attached to each plant.   After a few pre-
ments it was found that from 16 to 18 days after silking
prime eating condition.   The attempt was made, there-
all ears within these dates.   There was some deviation
but the results showed that these departures did not
t the damage.

ᵉs the mean values of the characters (first and second ears
each of the progenies, with the probable errors of the
As in 1915, the order of the progenies in the table is
t in which they were planted.

al progenies, which are repetitions of the 1915 series, are
asterisk.   The 1915 ancestry of the new progenies is
II, but to trace the ancestry back of 1915 reference may
le I.

TABLE IV.—*Means of different characters of corn, first and second years combined, in 1916*

Progeny.	Number of plants.	Damage.	Number of larvæ.	Damage per larva.	Prolongation.	Length of husks.	Length of ear.	Number of husks.	Number of layers.	Days to silking.	Maturity.	Days silking to harvest.	Husk leaves.	Number of rows.

* 1915 progeny.

results of the 1916 experiments will first be examined to det
r they corroborate the 1915 data in showing that there
differences among the progenies in the extent of damag
xamination of the average damage of the different pro
IV, column 2) shows the results to be in fair accord wit
revious year and shows even more clearly that the differer
ent of damage are not the result of accident.
4, again the least damaged of all the progenies, is with one
parated from all other progenies by significant differences
n, Ph140, is a descendant of Ph124, produced in 1915, th
being a plant of Ph123. From the remaining progenies P
ed by differences that range from 2.4 to 16.5 times the probabl
ay be seen also that Oregon Evergreen, P129, one of th
varieties included in the experiment, was damaged mor
the hybrid progenies except Ph127, Ph127C1, and Ph
these, Ph127C1, is the only progeny in which the difference
. Oregon Evergreen was chosen for comparison as one
orm-resistant of the commercial varieties of sweet corn.
variety most generally grown in the worm-infested reg
thwest. The relatively high damage in the two Hopi
and Ph141 should also be noted. The following notes wer
Eastern variety of sweet corn grown by the side of the
m experiment and maturing at the same time.

longation. . . . . 2. 0±0.
ber of larvæ . . . . 2. 4± .
mage. . . . . 3. 8± .
gree of maturity . . . . 8. 9± .

the days from silking to harvest were not recorded

in opposite parts of the continent and under widely different environmental conditions. It may well be that the dry, cool climate of the coast of California, where corn is grown under irrigation, would bring into prominence a somewhat different complex of protective factors than the relatively hot, moist climate of Maryland.

Table V gives the general average of the 14 progenies that were planted both seasons with respect to the different characters based on the upper ears and the interannual correlations for the different characters. It will be seen that the infestation was much more severe at Chula Vista, Cal., than at Lanham, Md. This is best indicated by the average number of larvæ. The damage per larva and the total damage were both increased at Chula Vista by the fact that in 1915 many of the ears were allowed to mature before the notes were taken. The pronounced environmental differences between the two localities is indicated by the fact that the time from planting to silking was, on the average, 23 days longer at Chula Vista than at Lanham the following year. In view of the much slower growth of the plants at Chula Vista, it is reasonable to assume that the grains were also slower in maturing. This would expose the ears to the attacks of the larvæ for a longer time and would tend to increase the damage per larva and total damage.

TABLE V.—*General mean of the different characters of 14 progenies in 1915 and 1916 and the interannual correlation*

Factor.	Damage.	Number of larvæ.	Damage per larva.	Prolongation.	Length of husks.	Length of ear.
			Grades.	Cm.	Cm.	Cm.
General mean, 1915.............	3.22	1.75	1.76	9.32	23.6	14.3
General mean, 1916.............	.65	.84	.76	5.88	26.6	20.7
Interannual correlation........	.37±0.16	.73±0.09	.38±0.16	.79±0.07	.76±0.08	.02±0.19

Factor.	Number of husks.	Number of layers.	Days to silking.	Husk leaves.	Number of rows.
				Grades.	
General mean, 1915.............	12.9	9.8	105.2	1.55	14.6
General mean, 1916.............	12.7	9.9	82.0	3.24	15.3
Interannual correlation........	.53±0.13	.62±0.11	.51±0.13	.39±0.16	.72±0.09

Seventeen of the progenies grown in 1916 were descended from the fourteen progenies grown in 1915. The behavior of these progenies affords some evidence regarding the intensity of the inheritance of the measured characters. To reduce the agreement between parent and offspring to a quantitative basis, the mean values of the seventeen progenies and the mean values of their parent progenies, both grown in 1916, were correlated. Eight of the new progenies grown in 1916 were not descended from single 1915 progenies, but resulted from crosses between different 1915 progenies. In these cases a midparental value was taken by averaging the mean values of the two parents. The correlation coefficients are given in Table VI.

All of the correlations are positive and all are apparently significant. The average of the coefficients for the 11 measured characters was 0.60.

With such small numbers little confidence can be placed in differences in the correlations found for the different characters. It is interesting, however, that damage and number of larvæ have coefficients as large as those of morphological characters.

TABLE VI.—*Correlations between parents and offspring in corn experiments in 1916*

Damage.	Number of larvæ.	Damage per larva.	Prolonga- tion.	Length of husks.	Length of ear.
		Grades.	*Cm.*	*Cm.*	*Cm.*
0.66±0.09	0.72±0.09	0.40±0.14	0.68±0.09	0.72±0.09	0.56±0.11

Number of husks.	Number of layers.	Days to silking.	Husk leaves.	Number of rows.	Average.
			Grades.		
0.46±0.13	0.43±0.13	0.61±0.10	0.53±0.12	0.77±0.12	0.60

MEASURES OF INJURY

Of the characters recorded, three were measures of injury. These are given under the headings "Damage," "Damage per larva," and "Number of larvæ." As might be expected from the nature of the characters, the three measures of injury constitute a closely correlated group. The damage per larva was calculated by dividing the total amount of damage in each progeny by the total number of larvæ. The damage per larva is thus, of course, definitely associated with both damage and number of larvæ. There is, however, a factor in the degree of damage not covered by the two other characters—that is, the number of ears that escape without infestation. These uninfested ears reduce the average damage but do not affect the damage per larva.

Since to lessen the amount of damage is the practical object sought, primary consideration will be given to the relations existing between this character and possible protective characters. The correlations with number of larvæ and damage per larva will be considered only as they may help to elucidate the correlations with damage.

CHARACTERS CORRELATED WITH INJURY

In Table VII are given the interprogeny correlations of the characters measured.

With 31 progenies little significance may be attached to any correlation that is less than 0.35, since to exceed 3 times the probable error the correlation must be at least 0.33. Of the plant characters measured, the following showed a correlation with damage of 0.35 or closer: prolongation, length of husks, number of layers, and days to silking. In atddition to these, maturity and husk leaves are significantly correlated with number of larvæ. Among the plant characters the following significant correlations appear: Prolongation with length of husks; length of husks with length of ear; number of husks with number of layers; number of layers with days to silking; days to silking with maturity, days silking to harvest, and husk leaves; maturity with days to harvest and husk leaves.

Table VII.—*Interprogeny correlations*

Factor.	Damage.	Number of larvæ.	Damage per larva.	Prolongation.	Length of husks.	Length of ear.	Number of husks.	Number of layers.	Days to silking.	Maturity.	Days silking to harvest.	Husk leaves.	Number of rows.
Damage			*Grades.*	*Cm.*	*Cm.*	*Cm.*				*Grades.*		*Grades.*	
Number of larvæ													
Damage per larva													
Prolongation													
Length of husks													
Length of ear													
Number of husks													
Number of layers													
Days to silking													
Maturity													
Days silking to harvest													
Husk leaves													
Number of rows													

correla _ being nega v .   erprogeny correla
longation and damage was −0.71 ±0.06 and the regr
on prolongation was 1.02—that is, with an average in
the prolongation an additional 1 per cent of the cro
meaning of this regression of 1.02 may be further
paring the damage in the progenies with the great
longation. There were 14 progenies whose mean
less than 5 cm., the average prolongation of this gro
The average damage in this group of progenies was o
cent of the total crop. The 12 progenies with the
prolongation, all of which were 6 cm. or over, had an
tion of 7.7 cm. and a damage of 0.5 grades, or 5 pe
average increase of 4.8 cm. in prolongation was ac
average reduction of 5 per cent in damage.

The correlation of 0.71 between prolongation an
ciently close to justify the hope that the method follo
and that by increasing the prolongation through hyb
tion substantial reductions in the damage can be s
study, however, indicates that the relatively close co
prolongation and damage is probably not to be con
on the basis of a simple physical protection. The
doubting the apparent direct relation between prolong
is that prolongation appears to have nearly as much aff
of larvæ as on the amount of damage, the correlation b
Since all the larvæ found inside the husks were coun
had gained access to the ear or not, prolongation can
have reduced the number of larvæ in any such man
expected to reduce the damage.

If, to gain access to the ear, the larva must eat i
silks, prolongation would seem necessarily to be a ver

small larvæ are frequently found among the silks inside the husks with no indication of having eaten their way there, it is inferred that these small worms can crawl down the silks to the tip of the ear. Larvæ hatched on other parts of the plant would on the average be larger when they reach the silks and must eat their way to the ear. Against these larger larvæ a greater prolongation should prove an adequate protection, but they are a comparatively small factor in the total damage, as their depredations are largely confined to the tip of the ear.

In both 1915 and 1916 an attempt was made to record for each ear whether the larvæ entered the ear by traveling down the silks or by eating through the husks. Of the 1,449 earworms found in the damaged ears 1,384 were recorded as having entered from the end and only 65, or 4.5 per cent, by penetrating the husks. The holes so frequently observed in the husks, the presence of which led to the belief that thick, harsh husks would afford protection, are explained as being made by the larvæ when emerging. The few larvæ which gain access to the ear by eating through the husks do not leave a continuous track, for the husks of a young ear are elongating at different rates and the continuity of the hole left by the larva is soon broken.

From the fact that in the compacted silks the larvæ would be close together and the débris left by one larva might conceal other small larvæ, it was thought that there would perhaps be a greater tendency to overlook larvæ in the ears with great prolongation with the result that the number of larvæ in such ears would be underestimated. If errors of this kind were of sufficient magnitude, they might account for the correlation between prolongation and number of larvæ.

As a check against errors in counting the number of larvæ, the percentage of ears that had no larvæ was correlated with prolongation. Whatever may be the difficulties in counting the larvæ, whether an ear contains larvæ or not is a fact easily observed, and in dividing the ears into those with and those without larvæ there would be no tendency to overlook the presence of larvæ in ears with the greater prolongation. The correlation between prolongation and the percentage of ears with larvæ was found to be −0.59, a very close agreement with the −0.60 correlation between number of larvæ and prolongation.

If the relation between prolongation and damage as measured by the interprogeny regression is one of cause and effect—that is, if each increase of a centimeter actually reduces the damage by 1 per cent, the same relation should hold among the individual plants of a progeny. In other words, the intraprogeny regression should be the same as the interprogeny regression. The average intraprogeny correlation of damage and prolongation, which seems the best expression we have for the relations existing among individuals, is -0.254 ± 0.024 and the intraprogeny regression of damage on prolongation is 0.72.[1]

The apparent effect of prolongation on damage within the progeny is thus found to be only about 70 per cent of the effect indicated by the interprogeny regression.

The closer relation found to exist among the means of the progenies might come about through the interprogeny correlation of prolongation with other protective factors. The general absence of genetic correlations in maize characters would render this explanation improbable, but any explanation of the relation of prolongation to damage should also apply to the relation between prolongation and number of larvæ. It is not clear how prolongation can directly affect the number of larvæ, and the coherence of prolongation with other protective characters is the only explanation that suggests itself. For example, the progenies with the greatest prolongation might be later in maturing. If this were the case and larvæ became less numerous as the season advanced, the closeness of the interprogeny correlation between prolongation and damage would appear to be greater than it really is. To approximate the true effect of prolongation on damage, an attempt must be made to eliminate, as far as possible, the effects of other correlated characters. To do this, resort may be had to "partial correlations."

In the present example the partial correlation between prolongation and damage with respect to days to silking will give, so far as the data permit, the degree of relationship between prolongation and damage

---

[1] There are many difficulties in the way of securing a satisfactory expression for the intraprogeny correlations of damage and prolongation.

To combine the crude determinations of all the individuals into a single population is to confuse the inter- and intra-progeny correlations. To avoid this it seems better to calculate the intraprogeny correlation for each of the progenies.

There is a further difficulty in the choice of method. The customary product moment method, which is perfectly applicable to the means of the progenies, can not properly be used with the individuals of a single progeny owing to the pronounced skewness of the distribution of damage. In a great many of the progenies approximately one-half of the individuals have zero damage. This division of the plants into two groups, those that were damaged and those that were not, would seem to indicate that the biserial correlation may properly be used.

Differences between the mean prolongation of first and second ears prevent the combining of first and second ears in a single correlation table, but the independent calculation of the coefficient for first and second ears in the separate progenies provides an added check on the reliability of the method.

The method followed has been to calculate the biserial correlation in each of the progenies for both first and second ears. In most of the progenies the division was made between zero damage and a damage of one or more. In a few progenies a more equal division was secured by making the division between palants with a damage of one or less and two or more. No correlation was calculated where the smallest class fell below 10 individuals.

A weighted average of all the coefficients is taken as the best single expression of the intraprogeny correlation.

The mean intraprogeny regression was calculated by the formula. Regression of damage on prolongation=

$$Rdp\frac{\sigma d}{\sigma p}$$

where Rdp=average intraprogeny biserial correlation and σd and σp=the square root of the mean of the weighted squares of the standard deviations.

- - , - - g o. --- e ween p
other measurements of injury are all in the same directi
ect which days to silking may have had is to make tl
nship less close than the true one. The partial corr
nt days to silking change the direct correlation be
ion and number of larvæ from −0.60 to −0.73, tha
gation and damage per larva from −0.75 to −0.76.
like manner the elimination of difference in "mat
g to harvest" fails to reduce the correlation between p
e measures of injury. The partial correlation of prolo
e for constant maturity raises the direct correlation
.79; when constant for "silking to harvest," the direct
anged.
ther character that might be suspected of affecting
.n prolongation and damage is husk leaves. The
eaves may make the ears either more or less attracti
they are depositing their eggs and thus change the
gaining access to the ear.
partial correlation between prolongation and dama;
husk leaves indicates that the net result is negligib
:s the direct correlation of −0.71 to only −0.70.
pplying the formula for partial correlations a second
time an expression may be obtained for the correlati
gation and damage with season, maturity, "silking t
sk leaves all constant. This was found to be −0.83.
re conclude that if the relatively close interprogeny
n prolongation and damage is due to the associa
on with other protective characters, these character

The difference between the intra- and inter-progeny regression remains unexplained, and in the light of this disparity it should be kept in mind, no asurance can be given that an increase in the prolongation in other stocks will be followed by the same rapid increase in immunity found in the course of these experiments.

LENGTH OF HUSKS

Length of husks and prolongation are measurements of nearly the same thing. Prolongation, however, may increase in either of two ways. The husks may be longer or the ear may be shorter. The first selections were made for prolongation without special regard to the length of the ear, and it was feared that in so doing there might have been a loss in the length of the ear. Fortunately prolongation is more closely correlated with length of husk than with length of ear. The loss in length of ear has not been material, and the worm-resistant strains have a satisfactory ear length. The average for the different progenies ranges from 15.2 to 26.7 cm. Since little is to be gained by reducing the damage at the expense of the length of ear, it would probably be safer in future work to use length of husks as a basis of selection than to rely on the prolongation.

HUSK LEAVES

The correlation between damage and husk leaves is 0.31. Since husk leaves afford additional surface on which moths can deposit eggs the larvæ of which may gain access to the ear, a positive correlation would be expected. On the other hand, eggs so deposited are to some extent at the expense of eggs which in the absence of husk leaves would be deposited on the silks. Larvæ hatching on the husk leaves would be somewhat delayed in reaching the ear. These larvæ might be expected to do less damage than those hatching on the silks, and for this reason the damage per larva should be negatively correlated with husk leaves.

That the husk leaves do attract the moths or at least afford a location for the eggs is indicated by the positive correlation of husk leaves with number of larvæ, 0.52. The second assumption of an opposite relation with damage per larva does not appear in the direct correlation, which is also positive though only 0.12. It will be shown, however, that in general as the number of larvæ increase the damage per larva also increases, and the partial correlation of husk leaves with damage per larva for constant number of larvæ is, in fact, negative, − 0.32. This makes it appear that there is also support for the view that husk leaves tend on the average to reduce the damage done by each larva.

The final results are therefore in accord with the supposition that the manner in which husk leaves increase the damage is through providing additional opportunities for the moth to deposit eggs near the tip of the

s a pro u. _ arac er num _ ayers appears
tage.   The direct correlations between number of hus
three measurements of damage are negative, but too l
cance; and since the correlations of number of layers
of injury are in every case higher, the correlation
husks and damage is doubtless largely a secondary re
relatively close relation between husks and layers.
tions between number of husks and the measures of 1
number of layers are, in fact, all positive instead of n

On the other hand, a large number of layers appea
character second only to prolongation in importance.
lation between number of layers and damage is $-0.$
must be largely independent of prolongation, for the
between layers and damage for constant prolongatio
With days to silking constant, the correlation is red
for both days to silking and prolongation constant
$-0.40.$

Since the records show that only 4.5 per cent of
access to ears by penetrating the husks, it is diffic
correlation between number of layers and damage
direct protection.   A large number of layers, which,
wide husks, might bring about a closer wrapping of t
to some extent impede the progress of the larvæ.   It
able, however, that the true relation is that suggested a.
tion for the relation between prolongation and damag
of layers is positively correlated with some protec
considered in these experiments.

Since the correlation between layers and number of l
that between layers and damage per larva, a large

would appear to reduce damage more by reducing the number of larvæ than by reducing the damage per larva, and it is difficult to imagine how a number of layers can have any direct effect on the number of larvæ—unless the idea is entertained that a large number of closely wrapped layers causes the larvæ to desert the ear.

The relation between layers and number of larvæ is not the result of any relation existing between layers and prolongation, since the partial correlation for constant prolongation is −0.47. The interprogeny regression of number of larvæ on layers is 0.12—that is, the number of larvæ is reduced on the average by 0.12 of a larva with the addition of each layer.

The average intraprogeny correlation for these characters is −0.04, and the regression of larvæ on layers is 0.01.

In Table VIII the inter- and intra-progeny standard deviations, correlations, and regressions of the measures of injury and the more important protective characters are brought together for comparison. The uniformly lower values of the intraprogeny regressions as compared with the interprogeny regressions support the idea that there are other import, ant characters not included among those recorded.

TABLE VIII —*Inter- and intra-progeny standard deviations, correlations, and regressions of the measures of injury and the more important protective characters of corn*

Factor.	Standard deviations		Correlations.						Regressions					
	Interprogeny.	Mean intraprogeny.	Prolongation.		Number of layers.		Husk leaves.		Prolongation.		Number of layers.		Husk leaves.	
			Interprogeny.	Mean intraprogeny.	Interprogeny.	Mean intraprogeny.	Interprogeny.	Mean intraprogeny.	Interprogeny.	Mean intraprogeny.	Interprogeny.	Mean intraprogeny.	Interprogeny.	Mean intraprogeny.
Damage...........per cent..	3.74	9.77	−0.71	−0.26	−0.52	−0.10	0.31	0.01	−1.02	−0.72	−1.72	−0.51	0.97	0.05
Number of larvæ...........	.23	.58	−.60	−.13	−.51	−.04	.52	.03	−.54	−.02	−.12	−.01	.10	.02
Prolongation.........cm..	1.58	3.48	......	......	......	......	......	......	......	......	......	......	......	......
Number of layers.........	1.13	1.90	......	......	......	......	......	......	......	......	......	......	......	......
Husk leaves.......grades..	1.20	2.01	......	......	......	......	......	......	......	......	......	......	......	......

MATURITY

A character often ascribed to sweet varieties is that they remain in an edible condition for a longer time than do field varieties. Comparatively little damage is done after the ears have passed the edible stage, and, if the hardening of the grains was hastened, it seems not unreasonable that the injury should be diminished, and this might be a partial explanation of comparative immunity of field varieties. In the 1916 experiments the majority of the ears were harvested between 16 and

regressions than
difference is 1.3 grades of maturity.

It might have been expected that the progenies re
season to reach the silking stage would also have req
ter silking before reaching the edible condition.

the case. The correlation between the degree of
to silking was positive and close, 0.924. This was
due to the fact that silking to harvest averaged sli
ate season progenies, but with silking to harvest cc
correlation of maturity and days to silking was 0.88.
on this determination the ratio of "maturity" to "
was correlated with days to silking and found to be 0.

Two possible explanations are suggested for this un

(1) The field varieties used as parents in the c
season than the sweet varieties, and field varietie
mature the seed more rapidly after fertilization. If t
between these two characteristics, the later maturi
mature their seeds more rapidly. It should be reca
with the same number of days from silking to harves
to detect significant difference in the degree of ma
different progenies or the nonsweet varieties included

(2) The climatic conditions following the flowering
enies may have been more conducive to rapid mat
than earlier in the season. It is difficult to find su
in the meteorological conditions of the latter part of tl
days are shorter and the temperatures no higher.
observed, however, that varieties planted late in
with greater rapidity than the meteorological conditio

◡   arvæ.   ,   ˅   ,

relation between number of larvæ and damage per la₁
   a of negative—that is, on the average the more l
an ear the greater is the damage done by the indivi
light of our present knowledge this would seem to in
tincts of the moth are in accord with the requirem
   In other words, the moth deposits more eggs on the
ible to the larvæ. The only alternative explanation th
   that the larvæ desert ears that are distasteful to then
rther indication that the moths exercise choice in attac
_loseness of the correlation between the extent of damæ
cond ears of the same rows. Since this interprogeny
r than that between damage and prolongation, the mo
˙ve character measured, it follows that the close corr
the damage of first and second ears does not result
   t that both have a similar prolongation.
˙ght be urged that, since the two ears on a plant frequ
k simultaneously, there would be a tendency toward si
   To test this point, the interprogeny correlation b
r of days that elapsed between the silking of first and s
e difference in the number of larvæ found in first and
ilculated. The correlation was 0.24+0.10. This
n the right direction, is too low to explain the similarity
   first and second ears, and may be the result of chance
   is correct to assume that a great part of the immuni
iecured has come about through a correlation bet
ters and some protective character or characters no
ild not be surprising if there were similar correlatio
usks and characters of field varieties that are unde
orn.

delayed in sweet varieties. It is possible that in these particulars the immune strains might tend to resemble the field parent. So far as has been observed, this has not been the case. Both the percentage of sugar and the retardation of transformation are very difficult to measure accurately, for, in addition to the labor of chemical analysis, the problems are seriously complicated by variations due to the time when the analysis is made and by individual variation. It would seem that, to compare two strains with respect to these characters, it would be necessary to analyze a sufficient number of samples of each variety to secure a reliable average and to repeat this entire process at short intervals, beginning soon after fertilization and continuing until the sugar content was practically constant.

The only evidence obtained on these points is that when gathered at the proper time the immune strains were pronounced by a number of different observers to be fully as sweet as the parent sweet varieties, and that in the regression of maturity on days silking to harvest no consistent differences were found between the immune strains and commercial sweet varieties. That sweet segregates from a cross between sweet and field varieties are not deficient in sugar is shown by the work of Pearl and Bartlett,[1] who found the percentage of sugar in the F_2 segregates of a sweet with dent cross to be higher than in the sweet parent.

CONCLUSIONS

In the southern part of the United States and throughout the Tropics very little sweet corn is grown. The chief reason for this is believed to be the ravages of corn earworm (*Chloridea obsoleta* Fab.).

Attempts to grow sweet varieties in the South usually result in an almost complete destruction of the crop by corn earworms. The native field varieties, on the other hand, escape with relatively slight injury, and are largely used as a substitute for sweet corn.

The most obvious difference between sweet and field varieties that might be expected to affect the activities of the corn earworm is the extent to which the ears are protected by husks. Sweet varieties generally have the husks poorly developed. A possible reason for this may lie in the fact that in the northern part of the Corn Belt one of the most desired characteristics in sweet corn is an early season. Generally speaking, early varieties produce few leaves and few leaves are associated with few husks. There is, therefore, a simple explanation of why commercial varieties of sweet corn have poorly protected ears and the poorly protected ears of sweet varieties afford at least a theory as to why they are especially susceptible to the ravages of the corn earworm.

With these facts in mind the problem was to combine the well-protected character of the ears of southern varieties of field corn with

1 Pearl, Raymond, and Bartlett, J M. MENDELIAN INHERITANCE OF CERTAIN CHEMICAL CHARACTERS IN MAIZE. *In* Ztschr. Indukt. Abstam. u. Vererbungslehre, Bd. 6, Heft 1, 2, p. 1–28, 1 fig. 1911.

the table qualities of sweet varieties in the hope that a sweet variety with some degree of immunity would thus be secured.

Crosses were made between commercial varieties of sweet corn and southern varieties of field corn. Sweet seeds were selected from the first-generation ears, and in the second generation plants with well-covered ears were chosen and propagated. The descendants of these plants have been found to be much less subject to injury from the corn earworm than commercial sweet varieties.

The earworm resistance was tested in 1915 near San Diego, Cal., and in 1916 near Washington, D. C. In both seasons the series of hybrids as a whole was found to be less damaged by the corn earworm than commercial sweet varieties. There were also pronounced differences in the immunity of the progenies derived from the different F_2 generation ears. The close agreement between the extent of damage of first and second ears of the different progenies is taken to indicate that the constitution of the plant is an important factor in the immunity. An effort was made to determine the plant characters which give rise to this comparative immunity.

That the factors concerned in immunity are inherited, and are, thus, capable of improvement is indicated by the correlation between the extent of damage in related progenies. The correlation between the average damage of parent and offspring was 0.66 ± 0.09.

Low damage was found to be significantly correlated with a number of morphological characters. For the most part these morphological characters were also correlated with one another. Of the characters measured, prolongation, or the extent to which the husks exceed the ear, was found to be the most closely correlated with low damage. The interprogeny correlation between prolongation and damage was -0.71, with a regression of 1.02 per cent in damage for each centimeter of prolongation. Within the progenies the average correlation was -0.26 and the regression 0.72. This difference between the inter- and intra-progeny regression is believed to indicate that the protection is in part due to other characters correlated with prolongation and not included in those measured.

The thickness of the covering provided the ear by the husks was also found to be associated with low damage, but since only 5 per cent of the larvæ that reach the ear do so by penetrating the husks here again the relation can hardly be a direct one of cause and effect. From the standpoint of worm resistance husk leaves are also shown to be an undesirable character.

By recording the number of larvæ in each ear it is possible to resolve the total damage into (1) number of larvæ and (2) the average amount of injury done by each larva.

It was found that in the more immune progenies
larvæ and the damage per larva were low.   Since th
must be determined largely by the choice exercise
depositing eggs, it follows that the plants avoided by
the plants which the larvæ find most distasteful
between the instincts of the adult insect and those of
to explain as the result of morphological characters of
argue that at least a part of the immunity is the
differences, perhaps the presence of some volatile s
alike to the moth and the larva.   Both in Califo
during the period of the experiments the injury from
found to decrease slightly as the season advanced.

From the experiments here reported it appears
the length and thickness of the husk covering and
leaves varieties of sweet corn can be produced in v
the corn earworm is materially lessened.   No difficul
in securing by hybridization and selection the desir
in combination with the seed characters of sweet

ease and the causal organism.

flaxwilt problem has been of great importance in this c
also a serious problem in some of the flax-growing countries
, *12, 19, p. 211–217*). In America the flax industry ha
rapidly westward, owing to the loss from wilt. The dis
of the wilt diseases which are produced by various spe
im. Plants at any age from germination to maturity
·d and killed by the parasite. The disease is manifeste
wilting of young seedlings and a yellowing of the foliage o
followed by a wilting which may involve the entire plant
e of it, thus causing a bending or twisting of the plant
lted side. The disease is highly destructive to comm
usitatissimum) when grown on thoroughly infected so
ten the entire crop is destroyed.

SOURCE OF MATERIAL

seed and "flax-sick soil" were kindly supplied for this w
. L. Bolley,[2] of the North Dakota Agricultural Expe
. From plants grown from these seeds in the "sick
of Fusarium was isolated which agreed in cultural charact
thogenicity with *Fusarium lini* Bolley (*4*). There was
)f spore size than Bolley gave in his original description, b
be accounted for by the difference in environmental con
which the spores were produced, as environment was fo

nce is made by number (italic) to "Literature cited." p. 604–605.
riter wishes to express his sincere appreciation to Prof. H. L. Bolley for his hearty
invaluable suggestions and in supplying material for the work. He is also indebt
es, of the Plant Pathology Department, and to Prof. L. J. Cole, of the Experimental
t, of the University of Wisconsin, for their aid in defining the problems at the outs
ective supervision and kindly criticisms of the pathological and breeding phases as

In undertaking a study of the inheritance of wilt r(
most important considerations was to obtain for th
flax which were highly resistant and strains which w
ible to the disease.   A number of both resistant and
were obtained from Prof. Bolley.   Several varietie
were also obtained from various places in North Dak
Before the crossing work was begun all of these st
were thoroughly tested as to resistant and susceptibl
sick soil" from North Dakota.   Plants for these test
infection experiments, were grown in the greenho
strains proved satisfactory under greenhouse conditio
not.   All of the resistant strains except North Dakota
discarded, that variety being used exclusively, as it
rior in resistance to all other strains tested (Pl. 44, B
designated as No. 4 throughout this work.   Prof. B
seed of No. 4 says:

This flax has been growing on the North Dakota State gro
years and ought to be highly resistant to wilt.

The most satisfactory strain of susceptible flax usec
Dakota Pure-Seed Laboratory No. 14654; Pl. 44, B,
that this strain died out completely on his seed plots a
station and should be well suited for work of this
strains of common flax have been used to some ex
been so satisfactory as No. 3 in giving uniform res
No. 5—a white-flowered variety(Pl. 45, B, b)—wa.
linseed mill at Red Wing, Minn.   This was the only wh

naturally arisen the question as to the nature of individual variation and the cause of disease resistance. Although there has been considerable theorizing concerning the possible cause, or nature of resistance, very little intensive research has been directed toward a positive solution of the problem. Biffen's (*2, 3*) hybridization experiments with wheat were the first to throw light on disease resistance as being due to inheritable factors which behave according to the laws of Mendel. It hardly seems possible, however, to explain the inheritance of such a character as resistance on so simple a basis as that stated by Biffen.

Having in mind these questions concerning the nature and inheritance of wilt resistance in flax, the writer undertook the present investigations with the Departments of Plant Pathology and Experimental Breeding at the University of Wisconsin in February, 1915, his objects being (1) to study the mode of penetration of flax plants by *F. lini*, (2) to make comparative studies on the penetration of cabbage seedlings by *F. conglutinans* (3) to determine whether or not *F. lini* enters the resistant flax plant, (4) to study the relation of the fungus to the tissues of susceptible and resistant flax plants, and (5) to study the inheritance of wilt resistance through hybridization. Flax was found to be a most suitable plant for the hybridization work for a number of reasons—namely, it has a short growing season, can be grown to maturity in the greenhouse, is easily cross-pollinated, and highly resistant and susceptible strains are available. Even with these advantages it was not expected that more than a clue as to the nature of the inheritance of wilt resistance might be obtained in the time allotted for the work, since breeding is a slow process and it was necessary to develop methods as the work progressed.

NATURE OF WILT RESISTANCE

MODE OF PENETRATION

Before making a detailed study of the relation of the fungus to the various host tissues it was considered of fundamental importance to know how it enters the host, and to know whether or not it penetrates the resistant plant. There seems to be very little definite information as to the exact mode of entrance of the parasitic soil fungi into their hosts. It was hoped that by using pure-culture methods penetrations of flax and cabbage seedlings by species of Fusarium might be obtained so that they could be detected by aid of the microscope. Bolley (*4*) states that *F. lini* penetrates the young flax plants at any point, through the seed, leaves, stem, or roots. His illustration shows very clearly that the fungus is able to penetrate the cell walls at any point, but he does not show conclusively the initial points of entrance. However, he was dealing with the subject mainly from a practical rather than from a standpoint of detailed microscopical study.

In order to make a careful study of penetration, special culture methods were necessary. Test tubes were prepared by placing in the bottom rolls of filter paper (Pl. 44, A) which were moistened and rendered suitable as a medium for fungus growth by pouring a small amount of melted potato agar over them. These tubes were autoclaved and then planted in equal numbers with seeds of both resistant and susceptible varieties of flax. These seeds had been previously treated for five minutes with a 1-to-1,000 solution of mercuric chlorid. At the time of planting, or in some cases just after the seeds had germinated, some of the tubes containing each strain were inoculated with *F. lini* from a pure culture, while others were left uninoculated to serve as controls, and in most cases these remained free from fungus growth (Pl. 44, A, *b*, *c*). Cabbage seedlings were also grown for penetration studies in tubes prepared as above and inoculated with *F. conglutinans* from a pure culture. The potato agar served as an excellent medium for the growth of the species of Fusarium, which had immediate access to the young seedlings growing in the tubes. As soon as any signs of wilting could be seen, the seedlings were carefully removed from the tubes, mounted on slides in a 20 per cent glycerin solution and examined carefully under the microscope. In this way the root hairs can be easily observed, and almost the entire young root system is so transparent that penetrating hyphæ can be detected. In some cases it was necessary to separate the cortical layer from the inner part of the root tissue in order to be able to examine it closely. This was done by carefully splitting the root on one side with a sharp scalpel and removing the cortical layer, which was then spread on the slide and mounted in glycerin, as mentioned above. Penetration studies were also made with young flax seedlings grown in very loose, infected soil. It is very difficult to obtain clean root hairs from plants grown in soil. The best results were secured by taking the plant up with a large lump of soil which was then dissolved away without destroying all of the root hairs by placing it in a vessel of still water and agitating gently. These roots were then mounted and examined in the same manner as those taken from the tubes. A small amount of eosin placed under the cover slip was often of considerable aid in observing root hairs and hyphæ.

A careful study of slides prepared as above described revealed root-hair (fig. 1, G, H, I), epidermal (fig. 2, 3), and stomatal (fig. 4), penetration of flax seedlings taken from tube cultures; root-hair (fig. 1) and epidermal (fig. 3), penetration of flax seedlings grown in infected soil; and root hair penetration (fig. 1, A, B) of cabbage seedlings taken from tube cultures. The cabbage seedlings produce a more abundant supply of root hairs, and are, for this reason, more desirable for the study of root hair penetration than are flax seedlings. By cross inoculation in tube culture it was found that *Fusarium conglutinans* could penetrate the root hairs of flax seedlings (fig. 1, K). Likewise, *F. lini*

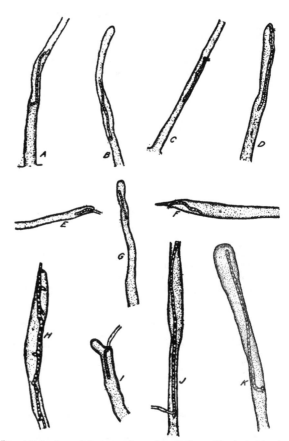

Fig. 1.—A-F, *Fusarium conglutinans* penetrating root hairs of cabbage seedlings in pure culture in test tubes. G, H, and I, *F. lini* penetrating root hairs of flax seedlings grown in test tube cultures. J, *F. lini* penetrating root hair of flax plant grown in loose, infected soil. K, *F. conglutinans* penetrating root hair of flax seedling in pure culture in test tube. Camera-lucida drawings.

was evidently able to penetrate cabbage seedlings as they were killed
by it in tube cultures. However, no penetration was observed in the
latter as examination was limited. When flax was planted on "cabbage-
sick soil" and the reverse, no wilting or yellowing occurred. The fungus,
very likely, enters the root hairs to some extent, but probably is unable
to invade the tissues of the plant for the same reason that *F. lini* is unable
to invade the tissues of the resistant flax plant. It was shown by the

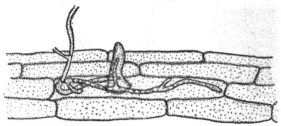

Fig. 2.—*Fusarium lini* penetrating epidermis of young flax root grown in loose, infected soil.

tube-culture method (Pl. 44, A, *a*, *d*) that *F. lini* can penetrate the
young seedlings of the resistant strain of flax as readily as it can penetrate the seedlings of the susceptible strain under those conditions.

By what exact means the fungus is able to penetrate the cell walls
of the host is not known. Perhaps the most feasible explanation is that
given by Ward (*26*) that the fungus protoplasm overcomes the resistance

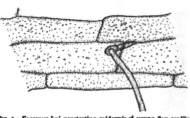

Fig. 3.—*Fusarium lini* penetrating epidermis of young flax seedling
in test-tube culture.

of the cells of the host
by means of enzyms
or toxins. This might
be interpreted as
meaning that the fun-
gus secretes an enzym
which has a solvent
action on the cell wall,
or that it may secrete
a toxin which prevents
any reaction on the part
of the host-cell proto-
plasm by killing or weakening it, thus making possible the invasion of the
cell. Perhaps both of these phenomena occur simultaneously. In the case
of root-hair penetration there is a slight depression at the point of entrance
of the fungus, and the diameter of the opening made by a hypha is somewhat less than the regular diameter of the hypha in question. Stomatal
penetration was found on the stem of a young flax seedling near the
point where the root began branching. Seedlings taken from the soil

showed stomata on the parts which were below the soil surface; therefore it is possible that stomatal penetration is a source of infection as well as root-hair penetration, and the penetration of the epidermis of young roots. The fungus is also capable of infecting through wounds, as was shown by artificial inoculations (Table I). This discovery of root-hair, epidermal, and stomatal penetration bears out Bolley's assumption that the fungus is able to penetrate the young plant at any point.

RELATION OF THE FUNGUS TO THE SUSCEPTIBLE PLANT

The relation of fungi to their host tissues is a very complicated one which varies greatly, according to the fungus and host under consideration. The object of this work was to study the relation of the fungus to the various tissues of the host and to seek any evidence that might be of value in explaining the possible cause for certain disease phenomena.

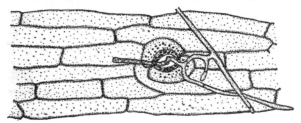

FIG. 4.—*Fusarium lini* entering stoma of young flax seedling in test tube culture.

After entering the susceptible plant, the fungus passes directly through the cell walls of the parenchyma tissues to the vascular system, which it invades to its limits. Bolley (4) states that sections through the stems and roots of wilted plants show that the parasite is able to penetrate the cell walls at any point and pass directly through any of the tissues, not excepting the woody parts. This statement holds true especially for plants in the later stages of wilt. Very few fungus hyphæ can be found in the cortical tissues of the stems of plants which have just wilted, although these tissues become thoroughly ramified with hyphæ as the plant begins to decay. The cortical parenchyma of the roots, however, is the first tissue to be invaded by the fungus. In the early stages of the disease the hyphæ are confined largely to the woody tissues in the stem.

Eight newly wilted plants ranging from half-grown to the late-flowering stage were stripped of their foliage, treated for five minutes in a 1-to-1,000 solution of mercuric chlorid, washed in sterile water, cut in pieces with sterile instruments, and plated out on potato agar. Each piece was noted carefully, in order to be sure just what part of the plant it came from. Pure cultures of *F. lini* were obtained from all parts of

he stems of six of these plants up to the terminal bu
rowth was obtained from the seed capsules. The
bowed growth up to within 1 inch of the terminal bu
rom near the top of the stem of a plant which had just
vith "Pianeze IIIb stain" (24) showed the fungus hy
ar system but not in the cortex. The method use
varied somewhat from that described by Vaughan (24)
passed from xylol through absolute and 95 per cen
tain, where they were allowed to remain overnight.
ashed rapidly in water to remove loose stain and d
ent alcohol until the desired point was reached. Aft
ere passed through absolute alcohol and xylol and

growth of the fungus and which increases the transpiration of the plant. This seems reasonable, so far as it goes, but does not seem sufficient for a complete explanation. If it were a case of the water supply being cut off by the clogging of the vessels, we should not expect so much of the one-sided wilting of plants which is so common with flax. The leaves on one side of the stem may become yellow, while those on the other side remain perfectly normal. Stems of plants which are wilted on one side only become characteristically twisted or curved, owing perhaps to unequal growth and shrinkage of tissues. If this wilting were due to the mere cutting off of the water supply at some point, it would be reasonable to except that when a normal plant has its stem cut half-through it would turn yellow and wilt on the cut side. Five flax plants were cut in this way, but none of them showed any yellowing or wilting from the wound. It is certain that by the time the foliage of the plant begins to wilt the root system has been invaded rather severely

Fig. 6.—Longitudinal section of the woody tissue of the susceptible flax plant showing the invading hyphæ of *Fusarium lini*. Notice the microspores of the fungus in the host cells.

by the fungus, and the root hairs are largely destroyed. This is especially true in case of the flax, and would likely account to a considerable extent for any lack in the water supply, and for the general weakening of the plant. Furthermore, there must be a protoplasmic disturbance in cells of the invaded tissues, which helps to produce the local symptoms. Phenonema of this kind might be due to toxic substances produced by the fungus, which interferes with the normal functions of the host protoplasm. The fungus also consumes a part of the food and water supply of the plant. There are very likely, a number of factors which aid in the production of wilt symptoms due to the invasion of flax by *F. lini*— namely (1) Partial destruction of the root system which limits the food and water supply of the plant; (2) use of part of the food and water supply of the plant by the fungus; (3) an increase in transpiration and an increase in the growth of the fungus due to a rise in temperatures; and (4) the possible production of toxic substances by the fungus, which interferes with the normal functions of the host protoplasm.

FIG. 7.—Cross section of the vascular tissues of a susceptible flax plant, sho'
Fusarium lini. Camera-lucida drawing.

into needle wounds in the stems of plants. Some w
above and others just below the soil surface. Ste
ground were wrapped with moist cotton, while tho
the surface of the ground were covered by replacin
gives the results of these inoculations. It will be s
that the inoculations made in the field were less su
made in the greenhouse. The plants in the field w
stage, which is rather late for infection, and were growi
where the moisture could not be satisfactorily contro
house young plants were inoculated and kept unde
moisture conditions.

TABLE I.—*Results of artificial inoculations of resistant and susceptible flax plants with Fusarium lini*

Date.	Strain of flax.	Number of plants inoculated.	Number of field plants infected.	Number of greenhouse plants infected.
1915.				
August 7......	Resistant............................	7	0	....
Do........	Susceptible.........................	6	0	....
August 17.'....	Resistant............................	4	....	0
Do........	Susceptible.........................	2	....	2
Do........	.....do.............................	3	....	3
August 26.....	Resistant............................	5	0	....
Do........	Susceptible.........................	5	0	....
Do........	.....do.............................	5	0	....
November 29...	Resistant............................	11	....	0
Do........	Susceptible.........................	1	....	1
Do........	.....do.............................	10	....	3
1916.				
November 26...	Resistant............................	15	....	0
Do........	Susceptible.........................	15	....	13

Since no infection was obtained by inoculating plants of the resistant strain, while those of the susceptible strain did become infected, it seems that there must be some immediate reaction on the part of the protoplasm of the resistant plant to check invasion by the fungus.

Whatever the nature of resistance, it is not manifested to the highest degree unless the plant is kept under perfectly normal conditions. This fact was shown by the way in which seedlings of the resistant strain of flax were killed by the fungus in tube cultures (Pl. 44, A, d), and was further tested by planting disinfected seeds of both the resistant and susceptible strains in flasks of soil which had been sterilized and inoculated with *F. lini*. When these flasks were plugged with cotton and kept in the laboratory, the resistant strain of plants showed slightly more resistance at first, but later succumbed to the attack. However, when flasks were prepared as above and placed in the greenhouse without the cotton plugs, some of the plants of the resistant strain lived to the flowering stage, while plants of the susceptible strain immediately died of wilt. Even under greenhouse conditions, where the temperature runs above normal, some of the plants of the resistant strain wilt.

Careful examination of the root system of resistant plants grown in infected soil showed that some of the smaller roots were decaying and there were brownish spots on the larger roots. A number of these plants which showed no signs of wilt above ground were taken and the root system disinfected thoroughly on the surface with a 1-to-1,000 solution of mercuric chlorid for 2½ to 5 minutes. They were then washed thoroughly and plated out on potato agar. In a large percentage of cases pure cultures of *F. lini* were obtained from these roots. In some cases other

Oct. 4, 1915..

Oct. 7, 1915..

Feb. 21, 1916.. 5

Sept. 20, 1916... 5

Sept. 26, 1916... 5

This table shows that the fungus is at least able enough beneath the surface of the resistant plant to mercuric chlorid used in disinfecting the surface, bu evidence as to the extent of invasion.

Parts of the roots of resistant flax plants which shov were fixed in Flemming's medium fixative, embed stained with "Pianeze IIIb stain" (24) as previous stain gives a pink color with the parenchyma cell wa with the fungus tissues. With lignified, cutinized, ar it gives a light green. A careful study of these se the fungus entered the parenchyma tissues of the resi but seldom, if ever, penetrated so far as the xyl limited invasion by the fungus is accompanied by a changes on the part of the host tissues which are vicinity of the invading hyphæ.

(1) There is a slight breaking down of the invade ever, is not sufficient within itself to disconnect th rounding cells. Marryat (17) and Ward (25) state wheat which resists yellow-rust (caused by *Puccini* is a sudden breaking down of the first cells to be inva

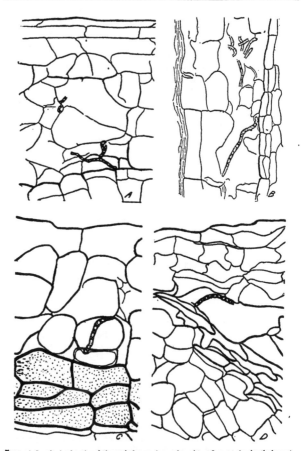

FIG. 8.—A, Longitudinal section of the cortical parenchyma of a resistant flax root showing the formation of cork walls around the point of invasion by *Fusarium lini*. B, Longitudinal section of the cortical parenchyma of a resistant flax root showing a cork layer formed between the point of invasion and the vascular system. Notice the increased cell division beneath the cork layer. C, Longitudinal section of the cortical parenchyma of a resistant flax root showing cell-wall penetration by *F. lini* and the formation of cork walls between the invading hypha and the vascular system of the root. The protoplasm in the cork cells was granular as indicated by stipling. D, Cross section of the cortical parenchyma of a resistant flax root showing the heavy cork walls formed around the point of invasion by *F. lini*. Camera-lucida drawings.

the fungus off from connection with other cells of the host. The fungus then dries up and dies or remains dormant in the dead host cells. The resistant flax plant behaves differently toward *F. lini* than does wheat toward *P. glumarum*. The rust fungus is an obligate parasite and is not capable of growing in dead tissues; *F. lini* may grow either as a parasite or as a saprophyte, and for this reason its development would not be checked by the death of the host cells. Furthermore, the breaking down of the cells in the flax plant is not so complete as that stated for wheat by Marryat and Ward, and would doubtless play very little part in decreasing further invasion by the fungus.

(2) In some cases the protoplasm of the host cells immediately surrounding the point of invasion becomes granular in appearance and stains green with the Pianeze stain, whereas the protoplasm of the normal cell fails to take the green at all. The writer is unable to offer any definite explanation for this change other than to say that it is possibly a chemical change in the host protoplasm excited by the presence of the fungus. There is doubtless a certain amount of injury to the host protoplasm, which may account in part for the coarse granular condition. Furthermore, as will be mentioned later, there may be some substance produced which is injurious to the fungus.

(3) Surrounding the area in which the cells show the granular appearance, there is a stimulation to cell division. This cell division is more abundant toward the vascular system from the point of invasion. In some cases the dividing walls are formed more or less irregularly, while in other cases a typical cork cambium seems to be formed. The newly-formed cells are to all appearance cork cells.

(4) Accompanying this cell division and other phenomena is a thickening of cell walls, which is much more noticeable toward the vascular system from the invaded point. This thickening of walls may extend three or four cell layers beyond the point of invasion and is more pronounced with newly formed cells. However, the walls of cells which were formed previous to invasion may become thickened. The process of thickening seems to be a laying down of additional material which is produced by the protoplasm of the affected cells. The modification of old walls is more noticeable below and above the point of invasion and toward the epidermis, where cell division is not abundant. These thickened walls stain green with Pianeze stain, which fact indicates that they are either lignified, cutinized, or suberized. In parenchyma tissues of this kind we should hardly expect to find lignin or cutin. When treated with concentrated potassium hydroxid, these walls gave the typical yellow reaction for suberin, which confirms the conclusion that they are of a corky nature.

Taking into consideration the above-mentioned phenomena with other possibilities, it seems that a combined explanation might be offered for the resistance of flax to extensive invasion by *F. lini*. In the first place the protoplasm of the resistant plant may naturally contain a substance or substances injurious to the fungus. We know that the resistant plant differs from the susceptible plant in respect to its physiological nature. This difference might be due to some permanent chemical composition of the protoplasm as suggested above, or it might possibly be due to a hypersensitiveness of the protoplasm of the resistant plant which causes it to react much more readily than does the protoplasm of the susceptible plant in producing the phenomena which cause resistance. The fungus seems to be less vigorous in the invaded cells of the resistant plant than in the invaded cells of the susceptible plant; in other words, it is less abundant in the cells of the resistant plant. It seems possible, therefore, that some toxic or other chemical substance is produced by the protoplasm of the host which has a deleterious effect on the fungus. The coarse, granular appearance and staining reaction of the protoplasm of the invaded cells indicate that considerable change has taken place. Apparently this change is accompanied by an injury to both the host cells and the fungus hyphæ. Perhaps some substance is produced by the host protoplasm during the change which has an injurious effect on the fungus. When the hyphæ of the fungus come in contact with the modified or corky walls of the cells they fail to penetrate, and further invasion is prevented. Possibly these thickened walls would not be sufficient within themselves to prevent invasion, but they serve as a barrier to the fungus after it has been weakened by protoplasmic reaction on the part of the invaded host cells. These phenomena seem to indicate that resistance is due either directly or indirectly to the chemical nature of the host protoplasm. Appel (*1*) believes resistance in plants to be of a chemical nature and makes the following statement:

Efforts must be made to find the causes of immunity, and after solving this question to determine without infection the disease-resistant qualities in different varieties and individuals in order to be able to establish the desired resistance and at the same time eliminate undesirable qualities.

Such a theory might at first seem entirely feasible; but, when the multiplicity of constitutional and environmental factors influencing the production of the resistant character is considered, it appears more improbable that any such analysis will ever be satisfactorily made. Ward (*25, 26*) also speaks of the chemical nature of resistance. He (*26*, p. *21*) says:

Infection, and resistance to infection, depend on the power of the Fungus-protoplasm to overcome the resistance of the cells of the host by means of enzymes or toxins; and, reciprocally, on that of the protoplasm of the cells of the host to form anti-bodies which destroy such enzymes or toxins, or to excrete chemotactic substances which repel or attack the Fungus-protoplasm.

This theory might be offered as a partial explanation for the resistance of *F. lini* by flax plants.

Finally, if we take into consideration the apparently weakened condition of the fungus in resistant host cells, the change in the nature of the protoplasm of the invaded cells, the new cell division, and the formation of cork walls around the point of invasion, all of which seem to play a part in the prevention of further invasion by the fungus, and all of which are due more or less to the chemical reaction of the host protoplasm, it seems safe to conclude that the resistance of flax to *F. lini* is essentially of a chemical nature.

INHERITANCE OF WILT RESISTANCE THROUGH HYBRIDIZATION

In undertaking a study of the inheritance of wilt resistance though hybridization it was very necessary that highly resistant and susceptible strains of flax be secured and thoroughly tested on infected soil before making crosses. The object of this chapter is to deal with methods of procedure in the work and to give such results as have been obtained from crosses up to date.

METHODS OF SOIL INOCULATION

Before the progeny from crosses could be tested it was highly important that the soil on which the plants were to be grown should be thoroughly infected with the wilt-producing organism, *F. lini*. The soil sent by Prof. Bolley from North Dakota was found by the preliminary tests to be satisfactory (Pl. 44, B), but the quantity was not sufficient. An attempt was therefore made to inoculate soil with pure cultures of the organism. In order to try this out, a small flat of greenhouse soil was sterilized in an autoclave one half being planted to flax No. 3 (susceptible), and the other half to flax No. 4 (resistant). After planting, about half a dozen tube cultures of *F. lini*, which were fruiting abundantly, were mixed thoroughly in a small pot of water and poured over the flat. Wilting of the susceptible plants did not begin until they were of considerable size, but they were completely killed in a short time after the disease started (Pl. 44, C). A large bench of soil was then inoculated. Water suspensions of the organism from pure culture were poured over the soil and worked into the surface. Seeds of the susceptible strain of flax were planted in abundance in the soil. As more fruiting cultures of the organism were obtained, the inoculation was repeated. When the plants from this seed showed considerable wilt they were turned under the soil and more seed planted. Only three or four plantings of this kind were necessary with the pure-culture inoculations to put the soil in suitable condition for the growth of hybrid plants (Pl. 46, A, B). It was also found that from 1½ to 2 inches of the North Dakota soil spread over Madison soil was sufficient to produce thorough wilting of susceptible plants (Pl. 46, C, b).

tion to a greater extent, and for this reason it was more desirable to remove them as soon as possible after the seed began to set.

For selfing plants the paper bags were used very little. The whole plant was covered with cloth. The most satisfactory method was to use wire cylinders about 3 inches in diameter and about 12 inches long, made of screening and covered with slips of finely woven white cloth made to fit. These cloth slips should be considerably longer than the cylinders, so that they can be tied at both ends. A piece of stout wire is pushed into the ground beside the plant, so that it extends a few inches above the top of the plant. The cylinder is then placed over the wire and the plant, and the cloth is brought together at the upper end and tied tightly around the large wire above the top of the plant. The lower end of the cloth is then tied around the wire and the plant above ground. The fruiting part of the plant is thus protected within the cylinder, where it produces seed in a fairly normal manner. These cylinders should be removed as soon as the flowering period is over and the fruit has set.

METHODS USED IN GROWING THE PROGENY FROM CROSSES

Plants of the first and second generations which were to be tested for resistance were grown in flats and on benches of North Dakota "flax-sick soil" and Madison soil which was inoculated as previously stated. These experiments were conducted in the greenhouse throughout the year. There is a slight variation in temperature in the greenhouse with change of seasons, a condition that can not be prevented. It was shown by temperature studies (23) that a difference of a few degrees might greatly influence the rate and amount of attack of flax by *F. lini*. By comparing the controls grown in winter and summer this difference will be observable. The high summer temperatures increased the severity of the wilt, even a few plants of the resistant strain wilting.

The soil was well pulverized and the seed planted about 1 inch apart in rows about 4 inches apart. This made it possible to grow a large number of plants in a comparatively small area. In every case rows of both parent strains were planted in every flat or bench in sufficient numbers to serve as controls, and in some experiments selfed seed from the parent plants of the crossed seed were planted as controls. The number of seeds planted was recorded in order to ascertain the percentage of germination. In the summer, when the greenhouse temperature ran high, the percentage of germination was low. In some cases practically none of the seeds germinated. This perhaps was not due to temperature alone but to the increased activity of other physical and biological agents in the soil. As soon as the seed germinated and the seedlings appeared above the ground, they were counted and recorded as plants. Any plant that made its appearance above the ground was counted, even though it died in this stage from wilt. The results are given with this stage of the plant as a starting point. It would be impossible to determine in every case the exact reason for the failure of the seed to germinate; and as some seeds of both strains did not germinate, it was not thought wise to attribute any of the failure directly to *F. lini*, although it is quite likely that this organism was partly responsible. If there was any doubt as to the cause of the wilting of the seedling after it appeared above the ground, it was determined by isolation methods in the laboratory. After the number of plants was recorded they were kept under almost daily observation. Notes were taken every week where possible and the number of healthy, wilted, and dead plants recorded. Any plant which showed undoubted wilt symptoms was recorded as wilted. This method of note-taking made it possible to compare the rate of wilting of the hybrid plants with that of the susceptible strain. As time and space were very limited it was found to be undesirable to grow all plants to complete maturity. It was then necessary to select some stage in the development of the plant as the end point for observation and note-taking. At this stage all plants which were not to be kept for seed could be removed and other seed planted. The flowering stage was selected as being the most favorable to cease note-taking, for at this stage the plant has reached its maximum activity and very little noticeable infection takes place after this time.

Part of the hybrid seed was grown in clean soil in a different greenhouse where there was no chance for infection by *F. lini*. These plants were self-fertilized as previously described in order to obtain seed for the next generation. By this means seed was obtained from plants that might have been destroyed by wilt if they had been grown in infected soil.

RESULTS OBTAINED FROM THE CROSSES

The parent strains, as previously stated, were thoroughly tested on "flax-sick soil" before the crosses were made and were found to be uniformly resistant or susceptible, as the case might be. However, some of the plants of the resistant strain wilted in later experiments. Results obtained from the progeny of crosses show that there is a great difference in individuality among plants of any strain with respect to its resistance to wilt. This difference was shown very strongly in the two generations grown, the results being so widely different from individual crosses that it will be necessary to discuss the different crosses or groups of crosses separately. The first generation from certain crosses proved to be entirely, or almost entirely, resistant to wilt. Others were intermediate with respect to resistance, while still others were entirely susceptible. In cases where there was entire or partial susceptibility a difference in the time and rate of infection was noticed as compared with the time and rate of infection of the common, susceptible flax. In some experiments the common flax would be almost entirely

dead before the first-generation plants began wilti
Later all of these F_1 plants would perhaps suceu
In some of the experiments there was segregation in
offspring, part of the plants wilting, while others wer
resistant parent.   In this case there was an interm
to the time of infection of the F_1 plants also. ·This
not due to excessive vigor of the plants, since vigor se
in the resistance to wilt by the flax plánt.   The plar
seem to be even more vigorous than resistant plant
cumb very readily to the attacks of the fungus.

No such simple ratios were obtained from these fl
(2, 3) reported from his wheat crosses.   Biffen was
of the inheritance of resistance by wheat to yellow-
of experiments (2) he crossed Rivet wheat, which is
rust, with Michigan Bronze, which he states is pro
tible to yellow-rust than any other wheat in existence
tion from this cross was entirely susceptible to attac]
seed was obtained for a sesond generation, owing to
attack.   Red King, a very susceptible variety, wa.
and the first generation was susceptible in this case
eration from these plants gave practically a 1-to-3
interprèted as indicating that resistance and suscepti
acters, the latter being dominant to the former.   Biff
where he drew the line between resistance and su
important point.   He says the plants which he pla
group were "relatively" or "almost" free from ru
that the third generation gave results which confirm

generation according to Mendelian laws. Unless there is some possibility that these fruits were not subjected to uniformly favorable conditions for blossom endrot, this is apparently an unusual case of inheritance.

Orton (*20, p. 463*) says, in writing on the resistance of farm crops to disease:

When a disease-resistant variety is crossed with a nonresistant variety, the resulting offspring inherit resistance to a limited and varying extent.

The writer found that in crosses between resistant and susceptible flax plants there was considerable difference in the results obtained from the different progenies. The results seemed to depend largely on the individual plants crossed. One of the most promising crosses made was one between the resistant strain No. 4 and the most susceptible strain, No. 3. The resistant plant was used as female parent, and the progeny was designated as $4D20$, "$_4$" referring to the strain, "D" to the plot where the female parent was grown, and "$_{20}$" to the number of the female plant. With these data on the female parent it was easy to refer to the original records of the cross for the strain of the male parent. This system of recording was used throughout the work. From crosses between the 2 plants mentioned above 5 capsuls were obtained, and 26 first-generation plants were grown on soil thoroughly infected with *F. lini* (Pl. 45, D, *b*). These plants were distributed among three flats, one containing 8 and the two others containing 9 plants each. The plants in one flat were growing at a lower temperature than those in the two other flats, the temperature of the former ranging from $_{14}°$ to $_{19}°$ C., while that of the latter ranged from $_{18}°$ to $21°$. None of these F_1 plants, however, were infected, although the controls of susceptible flax No. 3 wilted completely (*23*). They were grown to maturity and seed was obtained for a second generation by selfing. There was a segregation in the second generation into resistant and nonresistant plants. Table III gives the results obtained from the first and second generations and their controls. In this case selfed seed from the parent strains were grown as controls. In order to show the comparison between hybrid plants and plants of the susceptible strain, the number of plants wilted at the end of three weeks and the number killed by wilt at the end of the experiment were recorded.

The second-generation plants, as indicated by the controls given in the table, were grown under somewhat more severe conditions than the first generation, as is shown by the fact that a number of the plants of the resistant parent strain were killed or infected by wilt. The F_1 plants were grown in pure North Dakota soil, while the plants of the second generation were grown in artificially infected soil. Bolley (*9*) says that a change in the type of soil may cause a weakening of the resistant character. There was also a difference of temperature under which these F_1 and F_2 plants were grown, which may have played some part. The F_2 plants were grown in the summer and autumn, while the F_1 plants were grown in winter and early spring, when the temperature was considerably lower. A slight rise in temperature serves to accelerate the growth of the fungus, and the disease becomes more severe.

TABLE III.—*Resistance to flaxwilt obtained from the progeny of a cross between resistant flax No. 4♀, and susceptible No. 3♂* [a]

Parent strain.	Date of planting.	Number of plants grown.	Ratio at end of three weeks.		Ratio at end of experiment.		Number of plants killed by wilt.
			Wilted.	Not wilted.	Wilted.	Resistant.	
	1916.						
4D20(F₁)	Feb. 2.......	26	0	26	0	26	0
Resistant No 4. ...	.do......	39	0	39	0	39	0
Susceptible No. 3............	....do.......	52	34	18	52	0	52
F₂ generation:							
4D20–1.................	July 22......	68	37	31	45	23	40
4D20–2.................	Sept. 23.....	98	42	56	62	36	53
4D20–3.................	....do.......	115	40	75	68	47	47
4D20–4.................	....do.......	28	9	19	22	6	19
4D20–5.................	....do.......	101	59	42	76	25	64
4D20–6.................	....do.......	50	29	21	34	16	30
4D20–8.................	....do.......	70	54	16	61	9	53
Total	....	530	270	260	368	162	306
Resistant No. 4.. .	Sept 23 .	56	6	50	10	46	6
Susceptible No. 3.......	do . .	82	76	6	82	0	82

a ♀=female; ♂=male.

Since the first generation from this particular cross was entirely resistant, it was hoped that some reasonable explanation might be given for the results obtained in the second generation, although it was apparent at once that they could not be explained on a unit-factor basis. As the number of susceptible F₂ plants was very large, it was thought that they might be explained by Little's (*16*) hypothesis, which is an explanation of cases that appear to be a reversal of dominance. The individuals showing the character in question decrease in number in the F₂ generation, as there is an increase in factors which produce the particular character. The general principle is that with the addition of each factor involved the number of F₂ individuals possessing the character in question is multiplied by 3, while the total number of F₂ individuals is multiplied by 4. The difference between the number of individuals with the character and those lacking it grows progressively greater with each factor added. With the flax cross under consideration, the first generation was entirely resistant to wilt. In the second generation, from a total of 530 plants 162 were resistant and 368 susceptible. These figures approach very closely the expectation, if four factors are concerned in producing resistance. The actual expectation would be 81 resistant to 175 susceptible, which is a ratio of 1 to 2.16. The actual proportion obtained was 81 resistant to 184 susceptible, which is a ratio of 1 to 2.27. This ratio is fairly close to the expectation. The number of susceptible plants ran rather high, although this was to be expected, since the F_3 plants were grown under slightly abnormal conditions, as we have already seen. Some of these plants showed only slight signs of infection and would no doubt have resisted entirely had they been under less severe conditions. In this case the discrepancy is in the direction expected, the number of susceptible plants running high. Since this is the only cross that gave definite ratios, too much emphasis should not be placed on these results until further experimental evidence is obtained. It is furthermore very desirable that experiments of this kind be conducted in environmental conditions which are kept fairly constant.

TABLE IV.—*Resistance to flaxwilt obtained with reciprocal crosses of resistant flax No. 4 with susceptible No. 3 and their controls*

Parent strain.	Date of planting.	Number of plants grown.	Ratio at end of three weeks.		Ratio at end of experiment.		Number of plants killed by wilt.
			Wilted.	Not wilted.	Wilted.	Resistant.	
	1915.						
4E1 (F₁)	Oct. 15	9	2	7	9	0	9
6E1	do	9	1	8	9	0	9
4E1 (self)	do	17	0	17	0	17	0
6E1 (self)	do	17	13	4	17	0	17
F₂ generation (6E1 ♀ and 4E1 ♂):	1916.						
6E1-1	June 27	76	56	20	67	9	57
6E1-2	do	133	100	33	128	5	115
6E1-3	do	125	56	79	123	2	118
6E1-4	do	104	49	55	101	3	97
6E1-5	do	66	22	44	66	0	65
6E1-6	do	34	15	19	34	0	33
Total		538	298	240	519	19	485
Controls, F₂ generation:							
Resistant No. 4	do	84	12	72	30	54	22
Susceptible No. 6	do	8	8	0	8	0	8
Susceptible No. 3	do	24	24	10	24	0	24
F₂ generation (4E1 ♀, 6E1 ♂):							
4E1-1	do	93	61	32	92	1	84
4E1-2	do	74	52	22	73	1	68
4E1-3	do	36	33	3	36	0	36
4E1-4	do	86	63	23	86	0	84
4E1-5	do	56	31	25	56	0	55
4E1-6	do	84	55	19	80	4	78
Total		429	295	134	423	6	405

Another cross which was followed up with great care was a cross of resistant flax No. 4, with susceptible No. 6. Reciprocal crosses were made, and selfed seed was obtained from both parent plants. The resistant parent is designated as 4E1 and the susceptible parent as 6E1. The first generation (Pl. 45, A, b, c) proved to be entirely susceptible, although it might be said that there was an intermediate condition, owing to the fact that there was considerable difference in the time of wilting of the first-generation plants as compared with plants grown from selfed seed from the susceptible parent. The second generation was almost entirely susceptible. These plants, however, were grown under more severe conditions than were the plants of the first generation. The severeness of these conditions can be seen from results of the resistant control given in Table IV.

In this case, as is shown by the table, the second-generation plants were grown under fairly severe conditions. Thirty of the eighty-four plants of the resistant control were infected. The second generation was grown on artificially infected soil. Considering the behavior of the first generation and the severe conditions under which the second generation was grown, together with the results obtained, it seems impossible at present to give any direct explanation on a Mendelian basis.

In order to see what the result would be if the seed from the crosses on a number of plants were mixed, seed from five different plants were thrown together and planted. In this case sight was lost of the individuality of parent plants, due to the mixing; consequently, no definite ratios could be expected as individuals behave very differently. Table V gives the results obtained from mixed progeny.

TABLE V.—*Resistance to flaxwilt of the mixed progeny of crosses between resistant flax No. 4♂ and susceptible No. 3♀*

Parent strain.	Date of planting.	Number of plants grown.	Ratio at end of three weeks.		Ratio at end of experiment.		Number of plants killed by wilt.
			Wilted.	Not wilted.	Wilted.	Resistant.	
3 E–Mix. (F₁)................	1915. Oct. 2	153	53	100	106	47	76
Resistant No. 4..............	...do....	20	0	20	0	20	0
Susceptible No. 3............	...do....	46	43	3	46	0	46
F₂ generation:	1916.						
3 E–Mix–1................	Apr. 22	89	41	48	49	40	46
3 E–Mix–2................	...do....	81	63	18	78	3	72
3 E–Mix–3................	...do....	88	55	33	84	4	67
Total......................		258	159	99	211	47	185
Resistant No. 4...........	Apr. 22	45	1	44	a 3	42	0
Susceptible No. 3..........	...do....	43	33	10	43	0	43

a Slight.

As is shown by the controls in Table V, these plants were grown under very good conditions for the best test. The resistant parent strains stood up almost perfectly, while every plant of the susceptible parent strain died out completely. Since there are all gradations between an entirely resistant and an entirely susceptible first generation from individual crosses, we would expect the intermediate condition when seed from a number of individuals are mixed. This was what actually happened. In the second generation there was great variation in the progeny from individuals of this first generation (Pl. 46, B, *c, h*). This is what might be expected, since the chances are that each group of second-generation plants came from a different cross. This method of growing the progeny was found to be very unsatisfactory as it gave no direct line on individual crosses which have been found to behave so differently.

TABLE VI.—*Resistance to flaxwilt of the progeny from crosses between resistant and susceptible strains of flax in which about an equal number of resistant and susceptible plants occurred in the first generation*

RESISTANT NO. 4 ♀ × SUSCEPTIBLE NO. 6 ♂

Parent strain.	Date of planting.	Number of plants grown.	Ratio at end of three weeks.		Ratio at end of experiment.		Number of plants killed by wilt.
			Wilted.	Not wilted.	Wilted.	Resistant.	
	1916						
4E2 (F₁)	Apr. 25	6	1	5	3	3	1
Resistant No. 4	do	6	0	6	0	6	0
Susceptible No. 6	do	9	9	0	9	0	9
F₂ generation:							
4E2-1	Sept. 23	26	8	18	15	11	14
4E2-2	do	37	17	20	24	13	21
Total		63	25	38	39	24	35
Resistant No. 4	Sept. 23	26	4	22	7	19	5
Susceptible No. 3	do	24	14	10	24	0	24

RESISTANT NO. 4 ♀ × SUSCEPTIBLE NO. 3 ♂

	1916						
4D3 (F₁)	Feb. 2	18	0	18	10	8	2
Resistant No. 4	do	17	0	17	0	17	0
Susceptible No. 3	do	16	2	14	16	0	11
F₂ generation:							
4D3-1	July 21	74	47	27	73	1	53
4D3-2	do	63	43	20	61	2	46
D3-3	do	105	69	36	103	2	82
Total		242	159	83	237	5	181
Resistant No. 4	July 21	37	6	31	14	23	8
Susceptible No. 3	do	26	25	1	26	0	26

RESISTANT NO. 4 ♂ AND SUSCEPTIBLE NO. 3 ♀

	1916						
3E22 (F₁)	July 21	9	1	8	4	5	3
Resistant No. 4	do	15	0	15	0	15	0
Susceptible No. 3	do	18	16	2	18	0	18
F₂ generation:							
3E22-2	Sept. 23	24	20	4	23	1	22
Resistant No. 4	do	26	4	22	7	19	5
Susceptible No. 3	do	24	14	10	24	0	24

cases were discovered in which the first generation split
equal numbers of resistant and susceptible plants, but
be no definite order of segregation in the second genera-
this kind.   The severeness of conditions under which
e grown can be judged from the controls in each case.
esults of crosses of this kind.

ing of tables has been made to show the results of
ere was a larger number of resistant than susceptible
rst generation.   There seems to be but little relation
aavior of the first and second generations from these
II gives the results of crosses of this kind.

ance to flaxwilt of crosses between resistant and susceptible strains of
reater number of resistant than susceptible plants occurred in the firs

RESISTANT NO. 4 ♀ × SUSCEPTIBLE NO. 3 ♂

in.	Date of planting.	Ratio at end of	Ratio at end of
	1916.		
....	Mar. 29		
....	...do.....	95	
....	...do.....	85	
....	July 22		
....	...do.....		
....	...do.....		

TABLE VII.—*Resistance to flaxwilt of crosses between resistant and susceptible strains of flax in which a greater number of resistant than susceptible plants occurred in the first generation*—Continued

RESISTANT NO. 4 ♂ × SUSCEPTIBLE NO. 5 ♀

Parent strain.	Date of planting.	Number of plants grown.	Ratio at end of three weeks.		Ratio at end of experiment.		Number of plants killed by wilt.
			Wilted.	Not wilted.	Wilted.	Resistant.	
	1916.						
5E22(F₁).....................	Apr. 26	14	0	14	5	9	1
Resistant No. 4..............	...do.....	18	0	18	0	18	0
Susceptible No. 5............	...do.....	16	16	0	16	0	16
F₂ generation:							
5E22-1...................	Sept. 23	53	21	32	40	13	32
Resistant No. 4...........	...do	26	4	22	7	19	5
Susceptible No. 5..........	...do.....	2	2	0	2	0	2

RESISTANT NO. 4 ♂ × SUSCEPTIBLE NO. 3 ♀

Parent strain.	Date of planting.	Number of plants grown.	Ratio at end of three weeks.		Ratio at end of experiment.		Number of plants killed by wilt.
			Wilted.	Not wilted.	Wilted.	Resistant.	
3E14(F₁).....................	Feb. 2	9	0	9	2	7	0
Resistant No. 4..............	...do.....	14	0	14	0	14	0
Susceptible No. 3............	...do.....	21	7	14	21	0	21
F₂ generation:							
3E14-1...................	July 21	80	62	18	79	1	67
Resistant No. 4...........	...do.....	26	4	22	7	19	5
Susceptible No. 3..........	...do.....	24	14	10	24	0	24

RESISTANT NO. 4 ♀ × SUSCEPTIBLE NO. 5 ♂

Parent strain.	Date of planting.	Number of plants grown.	Ratio at end of three weeks.		Ratio at end of experiment.		Number of plants killed by wilt.
			Wilted.	Not wilted.	Wilted.	Resistant.	
4E11(F₁),	Oct. 18	5	0	5	2	3	0
Resistant No. 4..............	...do.....	79	3	76	6	73	3
Susceptible No 5. .	...do.....	51	36	15	51	0	50
F₂ generation a:							
4E11-1.....................	...do.....	202	55	147	137	65	86
4E11-2.....................	...do.....	128	63	65	116	12	86
4E11-3.....................	...do.....	141	30	111	101	40	56
Total........................	....	471	148	323	354	117	228

RESISTANT NO. 4 ♀ × SUSCEPTIBLE NO. 5 ♂

Parent strain.	Date of planting.	Number of plants grown.	Ratio at end of three weeks.		Ratio at end of experiment.		Number of plants killed by wilt.
			Wilted.	Not wilted.	Wilted.	Resistant.	
4E7(F₁).....................	Oct. 18	5	0	5	2	3	0
Resistant No. 4..............	...do.....	79	3	76	6	73	3
Susceptible No. 5............	...do.....	51	36	15	51	0	50
F₂ generation a:							
4E7-1.....................	...do.....	44	10	34	18	26	10
4E7-2.....................	...do.....	52	7	45	16	36	11
Total........................	....	96	17	79	34	62	21

Table VIII shows the results of crosses in which the first generation was entirely or almost entirely susceptible. The second generation in these cases behaved about as did the second generation from crosses which have been given above. There seems to be no common explanation for segregation in this case, which, of course, is to be expected when the first generation segregates irregularly, as was true with these crosses.

TABLE VIII —*Resistance to flax wilt of crosses between resistant and susceptible strains of flax in which the first generation plants were entirely or almost entirely susceptible*

RESISTANT NO. 4 ♀ × SUSCEPTIBLE NO. 5 ♂

Parent strain.	Date of planting.	Number of plants grown.	Ratio at end of three weeks.		Ratio at end of experiment.		Number of plants killed by wilt.
			Wilted.	Not wilted.	Wilted.	Resistant.	
	1916.						
4E13 (F₁)	Feb. 2	14	4	10	12	2	3
Resistant No. 4	...do....	18	0	18	0	18	0
Susceptible No. 5................	...do....	12	6	6	12	0	12
F₂ generation:							
4E13–1....................	Sept. 23	127	82	45	126	1	125
Resistant No. 4...............	...do....	26	4	22	7	19	5
Susceptible No 5. .	....do....	2	2	0	2	0	2

RESISTANT NO 4 ♀ × SUSCEPTIBLE NO. 5 ♂

Parent strain.	Date of planting.	Number of plants grown.	Wilted.	Not wilted.	Wilted.	Resistant.	Number of plants killed by wilt.
4E12 (F₁)	Oct. 18	11	5	6	11	0	8
Resistant No 4 . ....do....	79	3	76	6	73	3
Susceptible No 5 .. ....do....	51	36	15	51	0	50
F₂ generation:[a] 4E12–1 ....do....	123	62	61	106	17	83

[a] Control same as for F₁.

A large number of first-generation plants have been grown from different crosses from which no second-generation plants have been grown. The first generations from these crosses vary from those which are entirely resistant to those which are entirely susceptible, as have been given in the previous tables. Table IX will show these results.

DISCUSSION OF RESULTS

The inspection of Tables III to IX reveals no common explanation for the results obtained. Except for one case (Table III), no definite ratios could be detected. However, there are a number of possible reasons for these results: (1) The plants were grown under abnormal greenhouse conditions. (2) The temperature was too high in some cases, as previously emphasized. (3) A different type of soil was used for some of the tests, which, according to Bolley (9), is likely to cause a breaking down of the resistant character. (4) There is a great variation among individual plants within a strain.

1	5	3	3	1
3	5	4	4	2
4	1	5	0	3
7	3	9	1	4
0	10	1	9	0
4	1	5	0	5

xplanation for the results
een bred by selection to
tions which we might call
istance is probably due to
ed that under the normal
nplc, are required to proof equal or unequal value
us, as is evidenced by the
vhat is termed a normal
s made more severe, there
its of the resistant strain,
g in the greenhouse in the
is one or more additional
equired to cause the plant
ted to resist under these
has been made to make
ay be absent entirely, or
be either homozygous or

heterozygous. Then, a certain amount of segregation would be expected when the plants are subjected to these conditions. Segregation would likewise be expected to take place in the first-generation offspring from crosses of these plants with plants of the susceptible strain, provided the plants are grown in the abnormal environment. Differences of the individual plants of the susceptible strain with regard to the resistant character might account for considerable variation in the amount of segregation in the first generation from such crosses. A difference of this kind has been the basis for the selection of resistant strains. It is possible that some of these plants have one or more of the factors for resistance, which may be either homozygous or heterozygous, but do not have a sufficient number of them to cause the plant to resist where the disease develops normally. In crossing susceptible plants of this kind with plants of the resistant strain various factor combinations would be obtained, some of which would produce resistance in the first generation under the severe conditions while others would not. Results obtained in these experiments could easily be correlated with a theory of this kind.

The perfecting of methods rather than the development of any definite proof of laws governing the inheritance of wilt resistance in flax has been the main accomplishment in this work. Therefore a few helpful suggestions might be offered by way of conclusion of this chapter.

(1) The resistant plant should be tested under severe disease conditions before making the cross. (2) Reciprocal crosses should be made in each case, and selfed seed should be obtained from both parent plants, this to be planted along with the crossed seed. (3) First-generation plants should be grown in disease-free soil to obtain seed for a second generation. (4) First and second-generation plants which are to be tested should be grown under similar conditions with selfed seed from the parents planted as controls. (5) The experiments should be conducted under uniform environmental conditions in order to obtain conclusive results.

CONCLUSIONS

(1) The flax plant is most suitable for a study of the nature and inheritance of wilt resistance, since it grows very well in the greenhouse, has a short growing season, is easily crossed, resistant and susceptible strains are available, and conditions for infection can be produced with certainty.

(2) *Fusarium lini* penetrates the flax plant through root hairs, young epidermal cells, stomata of seedlings, and perhaps through wounds.

(3) *Fusarium conglutinans* is able to penetrate the root hairs of young cabbage seedlings when the seedlings are grown in tube cultures.

(4) *F. lini* invades the various tissues of the susceptible flax plant causing the disease known as flaxwilt. No considerable clogging of

vessels can be seen. Wilting may be due to the combined action of several factors: (a) Destruction of the young active root system by the fungus, which cuts off a part of the food and water supply of the plant. (b) Use of the food and water supply of the plant by the fungus. (c) More vigorous growth of the fungus and increased transpiration of the host plant due to a rise in the temperature. (d) The possible production of toxins by the fungus which injure the host protoplasm.

(5) *F. lini* penetrates the resistant flax plant and stimulates division and cork wall formation in cells adjacent to those attacked, but is not able to invade the tissues to any considerable extent owing to a number of possible reasons: (a) The permanent chemical composition of the resistant plant may be of such nature as to be injurious to the fungus. (b) The protoplasm of the resistant plant may be more highly sensitive than that of the susceptible plant, thus reacting more readily in the production of those phenomena which cause wilt resistance. (3) The stimulation to new cell division and the laying down of cork walls which seem to serve as a barrier to further invasion by the already weakened hyphae.

(6) Wilt resistance in flax is an inheritable character which is apparently determined by multiple factors.

(7) There is a great difference in the individuality of plants of a strain with respect to the resistant character, as shown by their offspring. The first generation from some crosses is entirely resistant, from some intermediate, and from others entirely susceptible.

(8) The degree of resistance shown by a strain of flax depends to a considerable extent on the environmental conditions under which the plants are grown. A strain which was bred to resist under certain conditions may break down under a more severe environment. Plants of North Dakota Resistant No. 114, the best strain employed in this work, was not entirely resistant with the high summer temperatures in the greenhouse.

(9) All parent strains to be used in crossing should be thoroughly tested on infected soil under favorable disease conditions before making the crosses. The resistant parent to be used in the cross should be grown on infected soil.

(10) Hydridization experiments should be conducted under uniform environmental conditions in order to obtain conclusive results. The fact that such varied results were obtained in this work is probably due to the different environments under which the plants were grown.

LITERATURE CITED

(1) APPEL, Otto.
   1915. DISEASE RESISTANCE IN PLANTS. *In* Science, n. s., v. 41, no. 1065,
       P. 773–782.
(2) BIFFEN, R. H.
   1905. MENDEL'S LAWS OF INHERITANCE AND WHEAT BREEDING. *In* Jour.
       Agr. Sci., v. 1, pt. 1, p. 4–48, 2 pl.
(3) ——
   1907. STUDIES IN THE INHERITANCE OF DISEASE RESISTANCE. *In* Jour. Agr.
       Sci., v. 2, pt. 2, p. 109–128.
(4) BOLLEY, H. L.
   1901. FLAX WILT AND FLAX-SICK SOIL. N. Dak. Agr. Exp. Sta. Bul. 50, p. 27–58,
       16 fig.
(5) ——
   1906. FLAX CULTURE. N. Dak. Agr. Exp. Sta. Bul. 71, p. 141–216, 22 pl.
(6) ——
   1907. PLANS FOR PROCURING DISEASE RESISTANT CROPS. *In* Proc. Soc. Prom.
       Agr. Sci., v. 28, p. 107–114.
(7) ——
   1908. THE CONSTANCY OF MUTANTS: THE ORIGIN OF DISEASE RESISTANCE
       IN PLANTS. *In* Amer. Breeders' Assoc. Rpt., v. 4, [1907]/08, p. 121–
       129.
(8) ——
   1908. [REPORT OF THE] DEPARTMENT OF BOTANY. *In* N. Dak. Agr. Exp. Sta.,
       18th Ann. Rpt., [1907]/08, pt. 1, p. 45–82.
(9) ——
   1909. SOME RESULTS AND OBSERVATIONS NOTED IN BREEDING CEREALS IN
       A SPECIALLY PREPARED DISEASE GARDEN: *In* Amer. Breeders' Assoc.
       Rpt., v. 5, [1908]/09, p. 177–182.
(10) ——
   1911. [REPORT OF THE] DEPARTMENT OF BOTANY. *In* N. Dak. Agr. Exp. Sta.,
       21st Ann. Rpt., [1910]/11, pt. 1, p. 43–47.
(11) ——
   1912. [REPORT OF THE] DEPARTMENT OF BOTANY AND PLANT PATHOLOGY. *In*
       N. Dak. Agr. Exp. Sta., 22d Ann. Rpt., [1911]/12, pt. 1, p. 23–60.
(12) BROKEMA, L.
   1893. EENIGE WAARNEMINGEN EN DENKBEELDEN OVER DEN VLASBRAND.
       In Landbouwk. Tijdschr., 1893, p. 59–71, 105–128.
(13) EYRE, J. V., and SMITH, G.
   1916. SOME NOTES ON THE LINACEAE. THE CROSS POLLINATION OF FLAX.
       In Jour. Genetics, v. 5, no. 3, p. 189–197.
(14) GILMAN, J. C.
   1916. CABBAGE YELLOWS AND THE RELATION OF TEMPERATURE TO ITS OCCUR-
       RENCE. *In* Ann. Mo. Bot. Gard., v. 2, no. 1, p. 25–84, 21 fig., 2 pl.
       Literature cited, p. 78–81.
(15) JONES, L. R., and GILMAN, J. C.
   1915. THE CONTROL OF CABBAGE YELLOWS THROUGH DISEASE RESISTANCE
       Wis. Agr. Exp. Sta. Research Bul. 38, 70 p., 23 fig. Literature cited,
       p. 69–70.
(16) LITTLE, C. C.
   1914. A POSSIBLE MENDELIAN EXPLANATION FOR A TYPE OF INHERITANCE
       APPARENTLY NON-MENDELIAN IN NATURE. *In* Science, n. s., v. 40,
       no. 1042, p. 904–906.

(17) MARRYAT, Dorothea C. E.
　　　1907. NOTES ON THE INFECTION AND HISTOLOGY OF TWO WHEATS IMMUNE TO
　　　　　　THE ATTACKS OF PUCCINIA GLUMARUM, YELLOW RUST. *In* Jour. Agr
　　　　　　Sci., v. 2, pt. 2, p. 129–138, 2 pl.
(18) NORTON, J. B.
　　　1913. METHODS USED IN BREEDING ASPARAGUS FOR RUST RESISTANCE. U. S.
　　　　　　Dept. Agr. Bur. Plant Indus. Bul. 263, 60 p., 4 fig., 18 pl.
(19) NYPELS, Paul.
　　　1897. NOTES PATHOLOGIQUES. *In* Bul. Soc. Roy. Bot. Belgique, t. 36, pt. 2,
　　　　　　p. 183–276, 18 fig.
(20) ORTON, W. A.
　　　1909. THE DEVELOPMENT OF FARM CROPS RESISTANT TO DISEASE. *In* U. S.
　　　　　　Dept. Agr. Yearbook, 1908, p. 453–464, pl. 39–40.
(21) ———
　　　1913. THE DEVELOPMENT OF DISEASE RESISTANT VARITIES OF PLANTS. *In*
　　　　　　Compt. Rend. 4th Internat. Conf. Génétique, 1911, p. 247–265, 9 fig.
(22) STUCKEY, H. P.
　　　1916. TRANSMISSION OF RESISTANCE AND SUSCEPTIBILITY TO BLOSSOM-END
　　　　　　ROT IN TOMATOES. Ga. Agr. Exp. Sta. Bul. 121, p. 83–91, 3 fig.
(23) TISDALE, W. H.
　　　1916. RELATION OF SOIL TEMPERATURE TO INFECTION OF FLAX BY FUSARIUM
　　　　　　LINI. *In* Phytopathology, v. 6, no. 5, p. 412–413.
(24) VAUGHAN, R. E.
　　　1914. A METHOD FOR THE DIFFERENTIAL STAINING OF FUNGUS AND HOST CELLS.
　　　　　　In Ann. Mo. Bot. Gard., v. 1, no. 2, p. 241–242.
(25) WARD, H. M.
　　　1902. ON THE RELATIONS BETWEEN HOST AND PARASITE IN THE BROMES AND
　　　　　　THEIR BROWN RUST, PUCCINIA DISPERSA (ERIKSS.). *In* Ann. Bot.,
　　　　　　v. 16, no. 62, p. 233–315, 3 pl.
(26) ———
　　　1905. RECENT RESEARCHES ON THE PARASITISM OF FUNGI. *In* Ann. Bot., v.
　　　　　　19, no. 73, p. 1–54. Literature cited, p. 50–54.

PLATE 44

A.—Flax seedlings as grown in the tube culture for penetration studies: *a*, Seedlings of susceptible flax in tubes inoculated with *Fusarium lini*. *b*, Susceptible flax seedlings growing in uninoculated tube. *c*, Resistant flax seedlings growing in uninoculated tube. *d*, Resistant flax seedlings in tubes inoculated with *F. lini*. These seedlings were killed as readily as susceptible seedlings under these conditions.

B.—Flax plants growing in North Dakota "flax-sick soil." *a*, Resistant flax No. 4. *b*, Susceptible No. 3.

C.—Flax plants grown on soil from Madison, Wis. This soil was sterilized and inoculated with pure cultures of *F. lini*. *a*, Susceptible No. 3. *b*, Resistant No. 4.

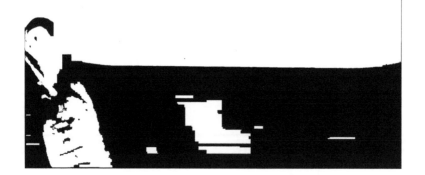

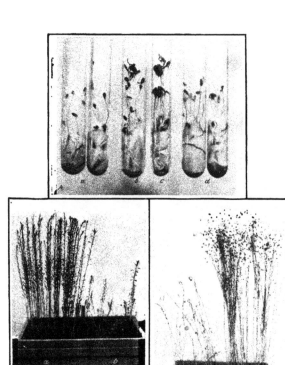

PLATE 45

A.—Flax plants growing on North Dakota "flax-sick soil." *a*, Resistant No. 4. *b*, *c*, Reciprocal crosses between resistant No. 4 and susceptible No. 6. All of these F_1 plants died later of wilt. *d*, Susceptible No. 6.

B.—Flax plants growing in North Dakota "flax-sick soil." *a*, Resistant No. 4. *b*, Cross between resistant No. 4 and susceptible No. 5. *c*, Susceptible No. 5.

C.—Flax plants growing in North Dakota "flax-sick soil." *a*, Resistant No. 4. *b*, *c*, Crosses between resistant No. 4 and susceptible No. 5. *d*, Susceptible No. 5.

D.—Flax plants growing in North Dakota "flax-sick soil." *a*, Resistant No. 4. *b*, Cross 4D20 between resistant No. 4 and susceptible No. 3. *c*, Cross 3E14 between plants of the same two strains as *b*. *d*, Susceptible No. 3 (all dead).

PLATE 46

A.—Flax plants growing in soil inoculated artificially with *Fusarium lini*. *a*, Resistant No. 4. *b*, Susceptible No. 3. *c-h*, First-generation plants from crosses between these two strains. Notice that some of the crosses were almost entirely resistant to the disease while others wilted almost as badly as plants of the susceptible strain.

B.—Second-generation flax plants growing on artificially infected soil. *a*, Susceptible No. 3. *b*, Resistant No. 4. *c-h*, Second-generation plants from crosses between these two strains. These plants show the difference in individual plants of the first generation. *c* and *d*, *e* and *f*, *g* and *h*, respectively, came from individual plants of the first generation.

C.—Second-generation flax plants growing in North Dakota "flax-sick soil." *b*, Susceptible No. 3. *c*, Resistant No. 4. *a*, *d-h*, Second-generation plants from the cross 4D20 in which the first generation were all resistant. Of the 530 plants of this F_2 generation 162 were resistant and 368 susceptible.

NATURE AND RATE OF GROWTH IN LAMBS DURING THE FIRST YEAR

By E. G. Ritzman,[1]

Animal Husbandman, New Hampshire Agricultural Experiment Station

THE PROBLEM

The data presented herewith furnished material for an analysis of the nature and rate of the development of lambs during their first year of growth.

The primary object in view in obtaining these data was the establishment of the normal rate of growth of definite somatic characters for specific periods during this first year of development. The data secured for this purpose, however, were used also in a study of correlation.

In planning this experiment it was decided to take the first record at 2 weeks of age after the lambs had straightened out somewhat, the second at 4 weeks, thereafter monthly up to 3 months of age, and from this time on quarterly up to a year of age. Consequently each lamb was weighed and measured on the dates when it attained these respective periods of age for the first three periods. After the third record was taken, which was at 8 weeks of age, it was found impractical to proceed on this basis, because of the time required to attend to each individual lamb on the date it would attain the age for measurement, so they were divided into groups each of which contained lambs which were within a week of the same age. By taking records of a whole group at one time considerable time was saved. The fourth period covers only 3 weeks instead of 4, as it was found necessary to turn the flock out on pasture at this time when the oldest group of lambs reached the age of 11 weeks. This record was therefore taken before going to pasture, which was one week ahead of the intended schedule because the first week on pasture might result in a temporary reaction to growth owing to the ewes' adjustment to change of feed. The growth periods for which measurements were taken beginning with the second week as shown in Table II are:

1st period	2 weeks	4th period	13 weeks
2d period	4 weeks	5th period	16 weeks
3d period	3 weeks	7th period	12 weeks

[1] Acknowledgment for courteous cooperation is due to members of the Station for Experimental Evolution of the Carnegie Institute, notably to Dr. C B. Davenport for valuable suggestions and criticism in the preparation of the manuscript and to Dr. J A. Harris for recalculating and checking the tables of data.

Journal of Agriculture Research,
Washington, D. C.
k2

Vol. XI, No. 11
Dec. 10, 1917
Key No. N.H.—4

(607)

The characters selected for measurement include the following, which it was thought would best serve for the interpretation of size and conformation:

1. Weight (in pounds).
2. Height at shoulder.
3. Head length (sagittal length to horn ridge).
4. Head width (transverse diameter at poll).
5. Neck length (horn ridge to second dorsal vertebra).
6. Trunk length (second dorsal vertebra to tail head).
7. Chest depth (vertical diameter behind foreleg).
8. Chest width (transverse diameter behind foreleg).
9. Loin width (between fourth and fifth lumbar vertebra).
10. Croup length (anterior point of sacrum to tail head).
11. Foreleg length (elbow to ground).
12. Hindleg length (stifle joint to ground).

As nearly all of these measurements of conformation depend on skeletal dimensions, they are not materially influenced by variation in degree of flesh; so that, with the exception of depth and width of chest, the variation in measurements caused by variation in state of flesh is a negligible factor. As the whole flock was maintained under identical conditions of feed and management and was never kept in a high state of flesh, the chest measures give a uniform comparison and vary closely with variations of the skeleton of the chest. Measurements of purely quantitative flesh traits, such as body circumference at chest, neck circumference, and hindleg circumference, were taken but are not included here because they represent condition of flesh rather than size. Body size is determined by the quantitative development of structural characters which are measured as height, length of vertebral axis, depth and spread of frame.

SIZE

Because of a lack of sufficient distinction between size as expressed by weight (which is variable according to condition of flesh) and by structural development (as a framework for carrying flesh) the term when applied to an animal as a whole may become confusing.

Size, in the empirical sense, is expressed in various ways. In meat animals size is commonly expressed by weight; in draft oxen by chest circumference; and in horses by height at withers or weight, or both.

From the geneticists' and breeders' point of view these terms are not only inapplicable, but are often misleading, as no single dimension is adequate to express size.

Structurally, size is expressed in the summation of those dimensions that determine the limits of bulk—dimensions accordingly that are representative of the body type as regards length, depth, and width.

Bulk, or volume, which is probably the best single criterion of size, is here for convenience of comparison measured by the form of two types of planes: 1 One formed by length with vertical measurements and the other by length with transverse measurements. Length. includes four units: (1) Head, (2) neck,[2] (3) trunk, and (4) croup. Height includes four: (1) Total height, (2) chest depth, (3) foreleg length, and (4) hindleg length. These transverse measurements were taken: (1) Head width, (2) chest width, and (3) loin width. The uniformity of the correlations shown in Table III suggests that loin and chest widths reflect either directly or by correlation the relation in variability between length and width.

CONFORMATION

Accepting the length of the vertebral axis as the best linear dimension with which to compare other linear dimensions, we can separate it into three distinct divisions: (1) The head, (2) the neck, and (3) the trunk.

From the economic point of view the first two of these divisions are unimportant, since they contribute no material value to the carcass; nor would any variation in their conformation within normal limits produce any material change in the quantity of fleece.

Conformation is essentially expressed by proportions, in which size as a whole plays no part. It is a term expressing only dimensional relations, of which the ratio of length to breadth and depth are the most important. In a general way conformation may be divided into four types.[3] Assuming that we have four animals of equal trunk length, we may designate these classes as (1) those with relatively deep but narrow bodies; (2) those with relatively broad but shallow bodies; (3) those with bodies that are relatively both deep and broad; and (4) those with bodies that are both shallow and narrow. By these types as a basis we are able to express the general traits of conformation of an animal.

What the specific effect of any variation in ratio of depth to length or of width to length has on the development of two of the major traits (constitution and mutton form) of an animal remains still to be established empirically. Experiments under way here along this line suggest that relative depth of body is the most important if not a perfectly satisfactory index of constitutional development.[4] On the other hand, it is also found that thick flesh, while always associated with relative

---

[1] These planes are also used to contrast differences in conformation.

[2] Of all dimensions this is the most difficult to obtain with accuracy.

[3] These of course represent extremes between which we find all possible degrees of intermediates.

[4] This is based on the assumption that a chest cavity which is deep in relation to body circumference is associated with relatively larger internal capacity for performing its vital functions and a consequent greater constitutional vigor. In explanation of this point it may be added that cross sections of a number of different types of animals show entirely different results from internal measurements of cross sections than might be expected from external measurements. While external measurements are a good index to internal conditions, they may easily be misinterpreted in animals with a heavy covering of flesh. Furthermore, an animal may be capable of developing a high degree of mutton (with good care and feed) and not possess the desirable degree of constitutional vigor, or the reverse condition may be found as well.

external width,[1] is not necessarily accompanied by a proportional increase in depth. If we view these conditions in relation to length of body, considered in sheep of approximately equal length, we can describe the above classes as (1) strong constitutionally, but of inferior mutton type; (2) good mutton type, but weak constitutionally; (3) strong constitutionally and of good mutton type; and (4) weak constitutionally and of poor mutton type. However, this can be only a generalization applicable within breed types, as there is a marked racial difference in the ratio of depth to width between some breeds, as, for example, the South, down and Rambouillet.

CHANGES THAT OCCUR IN PROPORTIONS DURING GROWTH

In all of our domestic animals the conformation at birth differs more or less from that of the mature state. Young foals, calves, and lambs seem to have disproportionately long legs. In man the reverse is true. Young animals are in many respects undesirably proportioned. While nature exacts these conditions to meet the immediate physiological exigencies attendant on birth, she later makes the correction by differences in the rate and degree of development of different somatic characters; in other words, the growth of the various somatic characters of an animal organism is by no means proportional. These type differences between new-born and adult are shown in Table I. This table shows measurements and weights of 62 animals taken at 14 days of age and again at an average age of 280 days.[2]

Since these measurements are intended to compare conformations rather than to compare size, the 14-day measure has been enlarged, first on the basis of shoulder height at 280 days to contrast comparisons of depth and length, and, second, on the basis of total length[3] to contrast relationship of width and length.

---

[1] In addition to width (transverse diameter) the hereditary tendency to develop thickness of flesh must also be present.

[2] Only 40 animals are included in the later data, the remainder having been sold or killed by dogs.

[3] Total length is the sum of three other dimensions—head length, neck length, and trunk length.

TABLE I.—*Measurements and weights of 62 sheep when 14 and when 280 days old*

Item.	1 H × R and Sn × R crosses.	2 Pure R.	3 Average at 14 days.	4 Average at 280 days.	5 Enlarged scale height.	6 Enlarged scale length.	7 Increase.
							Percent. (a)
Number of animals	56	6	62	40	40	40	.....
Age...............days	14	14	14	280	14	14	.....
Weight...........pounds	15.5	14	15.2	68.7	.....	.....	.....
Height, shoulder......mm	385	405	390	575	575	640	47.5
Head length..........mm	110	120	115	175	170	185	52.2
Head width...........mm	80	80	80	115	120	130	43.8
Neck length..........mm	175	170	175	285	260	280	62.9
Trunk length.........mm	320	330	320	525	470	520	64.1
Chest depth..........mm	145	135	140	260	205	230	85.7
Chest width..........mm	100	95	100	170	150	165	70.0
Loin width...........mm	75	60	70	120	105	115	71.4
Croup length.........mm	85	75	80	135	120	130	68.8
Foreleg length.......mm	265	280	270	395	400	440	46.3
Hindleg length ......mm	300	315	305	445	450	500	45.9
Ratios:							
$\frac{\text{Head width}}{\text{Head length}}$	.....	.....	.....	0.657	0.70	.....	.....
$\frac{\text{Foreleg length}}{\text{Trunk length}}$	.....	.....	.....	.753	.846	.....	.....
$\frac{\text{Chest width}}{\text{Chest depth}}$	.....	.....	.....	.654	.717	.....	.....
$\frac{\text{Chest width}}{\text{Trunk length}}$	.....	.....	.....	.324	.317	.....	.....
$\frac{\text{Loin width}}{\text{Trunk length}}$	.....	.....	.....	.229	.221	.....	.....
$\frac{\text{Croup length}}{\text{Trunk length}}$	.....	.....	.....	.257	.250	.....	.....
$\frac{\text{Chest depth}}{\text{Trunk length}}$	.....	.....	.....	.495	.442	.....	.....
$\frac{\text{Hind leg length}}{\text{Trunk length}}$	.....	.....	.....	.404	.962	.....	.....

ª Column 7 shows the percentage of increase from 14 to 280 days. It is therefore equal to (column 4 — column 3) ÷ column 3.

Columns 5 and 6 of Table I show the average conformation of all lambs measured at exactly 14 days of age[1] and again at an average of 280 days. Owing to losses and sales, the number represented in the older lot is smaller than it was at the beginning.

A comparison of these figures shows the proportions relative to height that exist in conformation at these respective stages of growth. These may be summarized as follows: In developing from 14 to 280 days (comparing columns 5 and 4).

1. The head becomes relatively longer and narrower.
2. The neck becomes relatively longer.
3. The trunk becomes relatively longer.

[1] Measured at 14 days magnified to scale of height at 280 days, column 5, and to scale of length at 280 days, column 6.

4. The chest becomes relatively deeper.
5. The chest becomes relatively wider.
6. The loin becomes relatively wider.
7. The croup becomes relatively longer.
8. The leg length remains nearly the same.

In relation to total length of body in developing from 14 days to 280 days (comparing columns 6 and 4).

1. Head narrows and becomes shorter.
2. Neck remains nearly similar in length.
3. Fore and hind legs become much shorter.
4. Chest depth increases.
5. Chest width does not change significantly.
6. Chest depth consequently increases in proportion to chest width.
7. Loin width does not change significantly.
8. Croup length does not change significantly.

These conditions are quantitatively expressed by percentage of increase in column 7. The general deduction is that total height, head width, and leg length increase to about the same relative degree and show least change; chest and loin width, and croup and trunk length show a more marked increase; chest depth increases relatively much more than any other dimension, and that dimensions of width increase relatively slightly in excess of dimensions of length, the latter (length) being similar in neck and trunk but less in the head.[1]

These conditions are further illustrated in figure 1, A–D, in which A and B represent the horizontal plane limited by the total length of vertebral axis and the structurally important transverse dimensions of head width, chest width, and loin width. B represents size and proportions at 280 days, and A, the proportions at 14 days enlarged to equal length of B.

Figure 1, C and D, represents two vertical planes of limits at 14 and 280 days, respectively, where total height, chest depth, leg length, and trunk length are shown in relation to each other. C represents the proportions at 14 days enlarged to equal to the total height of D.

RATE OF GROWTH BY PERIODS

Table II shows measurements and weights by periods for each dimension recorded. A comparison of the quantitative values of dimensions in the upper portion of the table gives a general idea of increase made by each character during each period. The ratios below show some of the changes that occur in the relative proportion between different characters.

---

[1] The necessity for an unequal rate of development of the various characters becomes obvious when we consider the differences in conformation between the young and the adult stage.

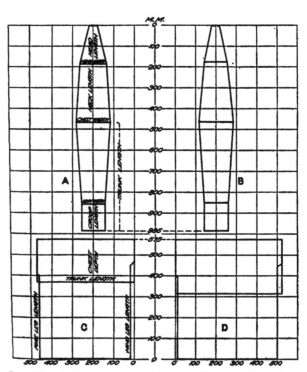

Fig. 1.—Diagrams showing the proportionate growth of lambs at 14 and 180 days: *A*, width of head, chest, and loin when 14 days old; *B*, same when 180 days old; *C*, height, depth of chest, length of legs and trunk when 14 days old; *D*, same when 180 days old.

TABLE II.—*Average measurements and ratios of measurements of lambs by periods (not including pure Rambouillet sheep)*

Item	Age 2 weeks.	Age 4 weeks.	Age 8 weeks.	Age 11 weeks.	Age 24 weeks.	Age 40 weeks.	Age 52 weeks.
Interval since last measurement, days..........................	....	14	28	21	91	112	84
Weight...................pounds..	15.5	21.5	37.5	450	53.5	660	67
Height.......................mm..	385	415	475	505	545	575	585
Head length..................mm..	110	120	135	145	160	175	185
Head width...................mm..	80	85	95	100	105	115	115
Neck length..................mm..	175	190	205	220	265	280	280
Trunk length.................mm..	320	365	440	475	500	520	540
Chest depth..................mm..	145	160	195	210	235	260	265
Chest width..................mm..	100	115	130	140	155	170	175
Loin width...................mm..	75	80	95	100	110	120	120
Croup length.................mm..	85	95	110	120	125	135	135
Foreleg length................mm..	265	280	310	320	360	380	395
Hindleg length...............mm..	300	320	360	375	400	435	445
Ratios:							
$\frac{\text{Head width}}{\text{Head length}}$	0.73	0.71	0.70	0.69	0.66	0.66	0.67
$\frac{\text{Chest width}}{\text{Trunk length}}$	.31	.31	.30	.30	.31	.33	.33
$\frac{\text{Loin width}}{\text{Trunk length}}$	.23	.22	.22	.21	.22	.23	.22
$\frac{\text{Croup length}}{\text{Trunk length}}$	.26	.26	.25	.25	.25	.26	.25
$\frac{\text{Foreleg}}{\text{Trunk length}}$	.83	.77	.70	.67	.72	.73	.73
$\frac{\text{Chest depth}}{\text{Trunk length}}$	.45	.44	.44	.44	.47	.50	.49
$\frac{\text{Trunk length}}{\text{Height}}$	.83	.88	.93	.94	.92	.90	.92
$\frac{\text{Chest width}}{\text{Chest depth}}$	.69	.72	.67	.67	.66	.65	.66

The curves in figure 2, of which Table II forms the basis, illustrate this point more forcefully. The most rapid development occurs during the earlier stages of life, decreasing with advancing age, as would be expected. The marked feature of these curves, however, is the fact that, allowing for some variations between different dimensions, they show such an exceedingly rapid growth for approximately the first three months. This period produces at least 50 per cent of the whole year's growth in dimensions and over 60 per cent of the total weight increase for the year under normal care. In the following three months' period there is a considerable decline, probably due in part to two severe temporary checks that occur at this time in this locality, the first occurring when the lambs go out on pasture and the second occurring when they are weaned. Approximately 20 per cent of the whole year's growth is the limit that can be expected for this second quarter. There is about an equal increase in the third quarter which comes during the fall of the year and not over 5 per cent in the last quarter, or winter period.

Economically this is a matter of vital importance in modern methods of sheep husbandry. It emphasizes the fact that under ordinary farm conditions the possibilities of profits are greater when all surplus is sold off at 3 or 4 months of age.[1]

Full structural growth (maturity) is probably attained near the end of the second year under normal conditions. The date must naturally vary somewhat with different breeds, and also with care and environment, as exceptionally favorable conditions of feed, climate, and health accelerate growth, resulting in early maturity with a regular growth curve, whereas

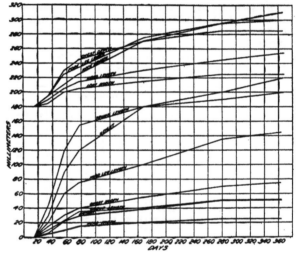

FIG. 2.—Graphs of the development of lambs.

unfavorable conditions of health, feed, and climate, which are incompatible with a normal growth curve earlier in life, may be compensated by growth prolonged beyond the normal time for its logical termination.[2]

If the potential limit of size of these animals is intermediate between the size characters of their adult parentage, then the measurement at 280 days shown indicates that 75 per cent of the total growth has been com-

[1] RITZMAN, E. G. EWES' MILK: ITS FAT CONTENT AND RELATION TO THE GROWTH OF LAMBS. *In* Jour. Agr Research, v 8, no. 2, p. 29-36, 1 fig. 1917 Literature cited, p. 35-36

WATERS, H. J. THE CAPACITY OF ANIMALS TO GROW UNDER ADVERSE CONDITIONS. *In* Proc. 29th Ann. Meeting Soc. Prom Agr Sci , 1908, p. 71-76, 5 fig. 1908

———. THE INFLUENCE OF NUTRITION UPON ANIMAL FORM. *In* Proc. 30th Ann. Meeting Soc. Prom. Agr Sci , p. 70-98, 6 fig. 1909.

MENDEL. L B. NUTRITION AND GROWTH *In* Harvey Lectures delivered under the auspices of the Harvey Society of New York, s. 10, 1914/15, p. 101-131. 1915. Bibliography, p. 128-131.

pleted at this period[1] in all characters except chest width and loin width, which have attained, respectively, only 55 and 65 per cent of their adult state. This exception is quite in accord with general observed facts that widening continues after height or length stature has been completed. All the other curves shown in figure 2 would therefore flatten during the second year even more than during any previous stage, which is quite in harmony with other investigations on growth.

CORRELATION BETWEEN CHARACTERS

Three types of correlations are shown in Table III: (1) Those between characters representing length, (2) those between characters which represent length and depth, and (3) those between characters which represent width. The degree of correlation is shown at the ages of 14 and 280 days, respectively.

TABLE III.—*Correlations between the body measurements of sheep*

HEAD LENGTH RELATIVE TO FORELEG LENGTH AT 14 DAYS OF AGE

[r=0.443±0.068]

Head length.	Foreleg length.										
	205	215	225	235	245	255	265	275	285	295	Total.
95								1			1
100		1			1	1					3
105	2		1	1	3	3	5				15
110			1	2	1	5	3	2			14
115						2		6	2	1	11
120						2	4	5	2		13
125						1		3			4
130								1			1
Total	2	1	1	2	6	10	14	19	6	1	62

HEAD LENGTH RELATIVE TO FORELEG LENGTH AT 280 DAYS OF AGE

[r=0.693±0.063]

Head length.	Foreleg length.									
	355	365	375	385	395	405	415	425	435	Total.
150	1		1							2
155										0
160		1	1		1					3
165		2								2
170	2	3	1							6
175	1		5		3	1				10
180			1	3	3		2			9
185					1	1				2
190					1					1
195					1					1
200				1			1			2
205							1		1	2
Total	4	6	9	4	10	2	4	0	1	40

[1] This is based on a large number of adult measurements and is quite uniform for all dimensions shown, excepting chest width and loin width.

TABLE III.—*Correlations between the body measurements of sheep*—Continued

HEAD LENGTH RELATIVE TO TRUNK LENGTH AT 14 DAYS OF AGE

[r=0.554±0.059]

Head length.	Trunk length.																Total.
	235	245	255	265	275	285	295	305	315	325	335	345	355	365	375	385	
95									1								1
100						1	2										3
105					1	3	3	5	1	1	1						15
110		1			1	1	1	2	4	1	2	1					14
115								3	2	3	2	1					11
120							2	4	1	5	1						13
125									2		1				1		4
130														1			1
Total		1			2	5	6	9	14	6	12	4	1		1	1	62

HEAD LENGTH RELATIVE TO TRUNK LENGTH AT 280 DAYS OF AGE

[r=0.537±0.076]

Head length.	Trunk length.																	Total.
	465	475	485	495	505	515	525	535	545	555	565	575	585	595	605	615	625	
150					1		1											2
155																		0
160					1	2												3
165					1	1												2
170	1				2	1	1		1									6
175		1		1	2	2	1	3										10
180			1	1	1	1		3	1	1								9
185									1	1								2
190					1													1
195				1														1
200				1									1					2
205				1									1				1	2
Total	1	1		2	10	8	4	6	3	2			2				1	40

FORELEG LENGTH RELATIVE TO TRUNK LENGTH AT 14 DAYS OF AGE

[r=0.498±0.064]

Foreleg length.	Trunk length.															Total.
	245	255	265	275	285	295	305	315	325	335	345	355	365	375	385	
205				2												2
215						1										1
225					1											1
235					1	1										2
245					2	2		1		1						6
255					1	1	1	1	2	3	1					10
265	1					1	2	5	1	3	1					14
275							5	5	2	3	1			1	1	19
285							1	2		2	1					6
295									1							1
Total	1			2	5	6	9	14	6	12	4	1		1	1	62

TABLE III.—*Correlations between the body measurements of sheep*—Continued

FORELEG LENGTH RELATIVE TO TRUNK LENGTH AT 280 DAYS OF AGE

[r=0.700±0.054]

Foreleg length.	Trunk length.																	Total.
	465	475	485	495	505	515	525	535	545	555	565	575	585	595	605	615	625	
355	1	1	...	...	...	1	1	...	...	...	...	...	...	...	...	...	...	4
365	...	...	...	...	4	1	1	...	...	...	...	...	...	...	...	...	...	6
375	...	...	...	1	2	2	2	1	1	...	...	...	...	...	...	...	...	9
385	...	...	...	1	2	...	...	...	1	...	...	...	...	...	...	...	...	4
395	...	...	...	...	1	4	...	4	...	1	...	...	...	...	...	...	...	10
405	...	...	...	...	...	...	...	1	1	...	...	...	...	...	...	...	...	2
415	...	...	...	...	1	...	...	...	...	1	...	...	2	...	...	...	...	4
425	...	...	...	...	...	...	...	...	...	...	...	...	...	...	...	...	...	...
435	...	...	...	...	...	...	...	...	...	...	...	...	...	...	...	1	...	1
445	...	...	...	...	...	...	...	...	...	...	...	...	...	...	...	...	...	...
Total	1	1	...	2	10	8	4	6	3	2	...	...	2	...	...	...	1	40

HINDLEG LENGTH RELATIVE TO TRUNK LENGTH AT 14 DAYS OF AGE

[r=0.557±0.074]

Hindleg length	Trunk length.																Total.
	235	245	255	265	275	285	295	305	315	325	335	345	355	365	375	385	
235	...	...	...	...	2	...	1	...	1	...	...	...	...	...	...	...	4
245	...	...	...	...	...	...	...	...	...	...	...	...	...	...	...	...	0
255	...	...	...	...	1	...	...	...	...	...	...	...	...	...	...	...	1
265	...	...	...	...	...	1	...	...	...	...	...	...	...	...	...	...	1
275	...	...	...	...	2	1	...	1	...	...	...	...	...	...	...	...	4
285	...	...	...	...	2	2	1	...	1	1	1	...	...	...	...	...	8
295	...	1	...	...	...	1	2	4	2	3	1	...	...	...	...	...	14
305	...	...	...	...	...	...	3	2	2	1	...	1	...	...	...	...	9
315	...	...	...	...	...	...	2	4	1	7	1	...	...	...	...	...	15
325	...	...	...	...	...	...	...	2	...	...	1	...	...	...	1	...	4
335	...	...	...	...	...	...	1	...	...	...	...	...	...	...	...	...	1
345	...	...	...	...	...	...	...	...	...	...	...	...	...	1	...	...	1
Total	...	1	...	...	2	5	6	9	14	6	12	4	1	...	1	1	62

TABLE III —*Correlations between the body measurements of sheep*—Continued

HINDLEG LENGTH RELATIVE TO TRUNK LENGTH AT 280 DAYS OF AGE

[r=0. 474±0.066]

Hindleg length.	Trunk length																	Total.
	465	475	485	495	505	515	525	535	545	555	565	575	585	595	605	615	625	
385							2											2
395																		0
405	1						1											2
415		1			1													2
425				1	2	2												5
435				1	2	3	1	3										10
445						1	1	1		1			1					5
455						2	2		1	1								6
465						2							1					3
475						1				1								2
485							1		1									2
495																		0
505																	1	1
Total	1	1		2	10	8	4	6	3	2			2				1	40

CHEST DEPTH RELATIVE TO TRUNK LENGTH AT 14 DAYS OF AGE

[r=0. 596±0.062]

Chest depth.	Trunk length.																Total.
	235	245	255	265	275	285	295	305	315	325	335	345	355	365	375	385	
100																	0
105									1								1
110																	0
115						1											1
120						1			1								2
125						1	3	1	1								6
130					1	1	1	1			3						7
135						1	1		2	1	1						6
140		1					1		1	4	2	1					10
145							1	2	3	1	3	1					11
150									4	2	3	2	1				12
155										1						1	2
160													1				1
165							1				1				1		3
Total		1				2	5	6	9	14	6	12	4	1		1	1	62

TABLE III.—*Correlations between the body measurements of sheep*—Continued

CHEST DEPTH RELATIVE TO TRUNK LENGTH AT 280 DAYS OF AGE

[r=0.544±0.075]

Chest depth	Trunk length																	Total
	465	475	485	495	505	515	525	535	545	555	565	575	585	595	605	615	625	
225					1													1
230																		0
235							1											1
240				1	2	1												4
245					1													1
250		1			3	1		1										6
255	1					2	1	1										5
260				1	1	1	1	2										6
265						2	1		1	1								5
270					1				1							1		3
275							2			1								3
280					1				1									2
285						1												1
290													2					2
Total	1	1	...	2	10	8	4	6	3	2	...	...	2	...	...	...	1	40

HEIGHT RELATIVE TO TRUNK LENGTH AT 14 DAYS OF AGE

[r=0.964±0.058]

Height	Trunk length																Total
	235	245	255	265	275	285	295	305	315	325	335	345	355	365	375	385	
335							2										2
345						2		1		1	1						5
355							2		2								4
365						2	2	1	1								6
375		1			2			2	2	2							9
385						1		2	1	1	1	1					7
395								2	3		3	2					10
405								2	3	1	1						7
415								1	2		2	1					6
425										2		1					3
435										1					1		2
445																1	1
Total	...	1	...	...	2	5	6	9	14	6	12	4	1	...	1	1	62

TABLE III.—*Correlations between the body measurements of sheep*—Continued

HEIGHT RELATIVE TO TRUNK LENGTH AT 280 DAYS OF AGE

[r=0.609±0.067]

Height.	Trunk length.																	
	465	475	485	495	505	515	525	535	545	555	565	575	585	595	605	615	625	Total.
515				1														1
525					2													2
535	1	1			1	1	1											5
545							1											1
555					1	1	1			1								4
565					2	1	1											4
575						1	1		1		1							4
585					1	1	2	1	2									7
595							1		1	1	1							4
605						1			2				1					4
615							1		1				1					3
625																	1	1
Total	1	1		2	10	8	4	6	3	2			2				1	40

HEAD WIDTH RELATIVE TO CHEST WIDTH AT 14 DAYS OF AGE

[r=0.477±0.066]

Head width.	Chest width.									
	85	90	95	100	105	110	115	120	125	Total.
70	2	5	3	1						11
75	1	3	4	3	1	1				13
80	2	3	5	6	4	8				28
85			2	5	1				1	9
90				1						1
Total	5	11	14	16	6	9			1	62

HEAD WIDTH RELATIVE TO CHEST WIDTH AT 280 DAYS OF AGE

[r=0.63±0.064]

Head width.	Chest width.														
	135	140	145	150	155	160	165	170	175	180	185	190	195	200	Total.
90						1									1
95															0
100					1										1
105			1	1		2									4
110			1	1		1	2	2	3						10
115	1		1		1		1	1	4		1	3	1		14
120					1				4	1				2	8
125									4		1				5
130											1				1
Total	1		3	2	3	4	3	3	11	2	2	3	1	2	40

TABLE III.—*Correlations between the body measurements of sheep*—Continued

LOIN WIDTH RELATIVE TO CHEST WIDTH AT 14 DAYS OF AGE.

[$r = 0.699 \pm 0.044$]

Loin width.	Chest width.									
	85	90	95	100	105	110	115	120	125	Total.
60...........................	4	2	3	....	....	....	....	....	....	9
65...........................	1	7	7	3	....	....	....	....	....	18
70...........................	....	2	2	7	2	....	....	....	....	13
75...........................	....	....	2	2	4	7	....	....	....	15
80...........................	....	....	....	3	....	2	....	....	1	6
85...........................	....	....	....	1	....	....	....	....	....	1
Total....................	5	11	14	16	6	9	....	....	1	62

LOIN WIDTH RELATIVE TO CHEST WIDTH AT 280 DAYS OF AGE

[$r = 0.62 \pm 0.066$]

Loin width.	Chest width.														
	135	140	145	150	155	160	165	170	175	180	185	190	195	200	Total.
95...........................	..	..	1	..	..	..	..	..	..	..	..	..	..	..	1
100..........................	..	..	2	..	..	..	..	..	..	..	..	..	..	..	2
105..........................	..	..	..	..	..	..	..	..	..	..	..	..	..	..	0
110..........................	..	..	..	..	2	2	..	..	..	..	..	..	..	..	4
115..........................	1	..	..	2	..	1	2	..	3	..	..	..	..	..	8
120..........................	1	..	..	..	..	1	1	1	5	..	..	1	1	..	11
125..........................	..	..	..	..	..	..	1	1	2	2	1	..	..	1	8
130..........................	..	..	..	1	..	..	1	1	..	..	1	..	..	1	5
135..........................	..	..	..	..	..	..	..	..	..	..	..	..	..	..	0
140..........................	..	..	..	..	..	..	..	..	1	..	..	..	..	..	1
Total..............	1	..	3	2	3	4	3	3	11	2	2	3	1	2	40

HEAD WIDTH RELATIVE TO LOIN WIDTH AT 14 DAYS OF AGE

[$r = 0.363 \pm 0.074$]

Head width.	Loin width.						
	60	65	70	75	80	85	Total.
70...	2	7	2	....	....	....	11
75...	3	2	3	3	2	....	13
80...	3	8	5	9	3	....	28
85...	1	1	2	3	1	1	9
90...	....	....	1	....	....	....	1
Total.......................	9	18	13	15	6	1	62

TABLE III.—*Correlations between the body measurements of sheep*—Continued

HEAD WIDTH RELATIVE TO LOIN WIDTH AT 280 DAYS

[r=0.58±0.071]

Head width.	Loin width.										
	95	100	105	110	115	120	125	130	135	140	Total.
90.................................	....	....	....	....	1	....	....	....	....	....	1
95.................................	....	....	....	....	....	....	....	....	....	....	0
100................................	....	....	....	1	....	....	....	....	....	....	1
105................................	....	1	....	1	1	1	....	....	....	....	4
110................................	1	....	....	1	4	3	1	....	....	....	10
115................................	....	1	....	....	1	5	2	4	....	1	14
120................................	....	....	....	1	1	2	3	1	....	....	8
125................................	....	....	....	....	....	....	1	....	....	....	1
130................................	....	....	....	....	....	....	....	1	....	....	1
Total..................	1	2	....	4	8	11	8	5	....	1	40

SUMMARY OF DATA ON CORRELATIONS BETWEEN BODY MEASUREMENTS

Age	Dimension.	Mean.	Standard deviation.	Coefficient of variability.	Coefficient of correlation	
Days. 14	Trunk length.	316.29	23.72	0.075	Head length—foreleg length.	0.443±0.068
280	do	523.75	28.91	.055	....do..............	.693±.063
14	Head length.	112.34	7.44	.066	Head length—trunk length.	.554±.059
280	....do.......	176.50	12.51	.071	....do..............	.537±.076
14	Foreleg length	262.58	18.55	.071	Foreleg length—trunk length.	.498±.064
280	....do........	384.50	19.23	.050	....do.......	.700±.054
14	Hindleg length.	297.10	23.29	.078	Hindleg length—trunk length.	.557±.074
280	....do.........	441.50	25.06	.057	....do..............	.474±.066
14	Chest depth.	140.16	12.08	.087	Chest depth—trunk length.	.526±.062
280	....do.........	259.63	14.97	.058	....do.......	.544±.075
14	Height shoulder.	386.13	26.16	.066	Height—trunk length.	.564±.058
280	...do.........	573.25	28.54	.050	....do..............	.609±.067
14	Head width	78.065	4.947	.063	Head width—chest width.	.477±.066
280	....do....	113.375	6.920	.061	....do..............	.630±.064
14	Chest width ...	101.855	7.418	.072	Loin width—chest width.	.699±.044
280	....do.........	170.000	15.49	.091	....do..............	.620±.066
14	Loin width......	69.348	6.380	.092	Head width—loin width.	.363±.074
280	....do.........	119.125	8.795	.074	....do..............	.580±.071

All correlation coefficients shown are positive, though this association in variation is never very close, only three of the correlations being as great as the smallest that Castle [1] found between skeletal dimensions in rabbits. The correlations at 14 days are both smaller and less uniform than at 280 days, indicating that animal proportions are more unstable at that immature age.

The smallest correlation at 280 days is between hindleg length and trunk length. It is questionable whether the hindleg length is a satisfactory character for this purpose because of the unavoidable erorrs in measurement, owing to the tendency of sheep to hunch down when handled, thereby producing fluctuations in measurement of material value. Except this case, the 280-day correlations show a range between extremes which is approximately equal to the range in correlations between the skeletal dimensions in rabbits shown by Castle, the range shown by Castle being from 0.658 to 0.858, and the range shown here being from 0.537 to 0.700. It is very probable that the higher degree of association between the variations in dimensions shown by castle is due to the greater accuracy that can be obtained in measuring prepared single bones than by measuring characters in live animals, such as are represented in this work, most of which involves dimensions of several bones.

The similarity in the ranges of correlations, however, indicates that it is true of sheep as Castle finds it of rabbits that

to a large extent the factors which determine size are general factors affecting all parts of the skeleton simultaneously.

The foreleg length shows the highest correlation with other measurements. Its coefficient with head length is 0.693 and with trunk length it is 0.700.

The correlations between dimensions representing width are somewhat lower than this but fairly uniform. Since the growth curves (fig. 2) show that length dimensions attain full development somewhat earlier than dimensions of width, it is probable that, as width approaches maturity, its coefficients will more nearly approximate those of length, especially as the correlation does not seem to be as definite in the immature as in those approaching maturity.

---

[1] MacDowell, E. C. size inheritance in rabbits. With a prefatory note and appendix by W. E. Castle. Carnegie Inst. Washington. Pub. 196, p. 51. 1924.

JOURNAL OF AGRICULTURAL RESEARCH

| VOL. XI | WASHINGTON, D. C., DECEMBER 17, 1917 | NO. 12 |

BACTERIAL-BLIGHT OF BARLEY

By L. R. JONES, *Professor of Plant Pathology*, A. G JOHNSON, *Assistant Professor of Plant Pathology*, and C. S. REDDY, *Assistant in Plant Pathology, University of Wisconsin.*

INTRODUCTION

For several years especial attention has been given to diseases of barley (*Hordeum* spp.) by the Department of Plant Pathology of the University of Wisconsin. In 1912 a peculiar leafblight was observed at Madison, Wis., at first causing considerable damage in a plot of two-row Montana barley, later appearing on the adjacent common six-row varieties. The general symptoms indicated that it was a bacterial disease, and laboratory studies strengthened this judgment. The disease has recurred in the departmental plots each year since and has been observed elsewhere as well, but it was not until the summer of 1915 that sufficient critical study could be given to justify publication. These field observations and laboratory studies show that this is a widespread disease, capable of producing economic loss, and is caused by a heretofore undescribed bacterial parasite.

APPEARANCE OF THE DISEASE

ON THE LEAVES.—The first evidence of the disease which has been noticed is on the young, green leaves in the form of small water-soaked areas which later enlarge. In nearly all cases the progress is chiefly longitudinal, the invasion being retarded or limited by the veins. Thus, the more prominent lesions may assume the form of somewhat irregular stripes which may extend in certain cases the full length of the blade and sheath, but are apparently terminated at the node. Only in rare cases has a lesion been found on the culm. The most vigorous development results when the invasion occurs along the midvein, owing apparently to the greater thickness and succulence of these parts. In such cases the striped development becomes most pronounced (Pl. B, 2). Sometimes two or more stripes may develop on a leaf, giving an appearance somewhat resembling the so-called "stripe disease" of barley caused by *Helminthosporium gramineum* (8, 9).[1] In addition to these stripe-like lesions, which are the more common, blotchlike lesions may also occur. At times these may involve so much of adjacent leaf tissue as to cause the entire blade, or considerable portions of it, to shrivel and turn

---

[1] Reference is made by number (italic) to "Literature cited," p. 643.

Journal of Agricultural Research,
Washington, D C
M

(625)

Vol. XI, No. 12
Dec. 17, 1917
Key No. Wis.—9

light brown. Following infection, the invaded tissues sooner or later lose their normal green, passing through translucent yellowish to brownish tints. Translucency is indeed so constant and distinctive a character of the lesions, even in the later brown stage, as to suggest the specific name chosen for the parasite.

If a leaf showing a strong, freshly developed invasion along the mid-vein and extending down the sheath (Pl. 47, B) is cut sharply across at the ligule and either portion squeezed between thumb and finger, a milky-gray droplet appears on the cut surface. Microscopic examination of this droplet proves it to be teeming with bacteria. If such or similar lesions are cut across and mounted in water, the bacterial extrusion may be readily demonstrated under the microscope. In a rather advanced stage of the lesion the slime expressed between thumb and finger as referred to above may be of the consistency of thick cream, and, when exposed to the air, quickly hardens to a brittle resinous mass.

These facts are the more interesting since one of the distinctive characters of the disease is the occurrence in nature, under favorable conditions, of a bacterial exudate hardening thus upon exposure. Under humid conditions, especially in the early morning, such exudate may appear on either surface of recently invaded leaf tissues as tiny, clouded droplets. These differ from the ordinary droplets of guttation water in their distribution on the lesions, their smaller size, and in the fact that they are distinctly grayish or milky, whereas the guttation water is crystalline clear. If such exudate droplets remain undisturbed they soon harden into yellowish, resinous granules studding the surface of the lesions (Pl. 47, D), but are easily detachable. Frequently these droplets coalesce with each other or with the guttation water and form conspicuous drops (Pl. 47, A) which later may spread over the leaf surface and dry down as a thin, grayish, almost transparent film (Pl. B, 3; 47, C, b). This exudate may occasionally be sufficiently abundant and sticky to entrap small insects (Pl. 47, C, a). It is noteworthy also that this dry exudate is quickly redissolved in water (Pl. 48, A); and doubtless this characteristic aids materially in the dissemination of the organism as discussed later.

Since the appearance of the leaves striped by this bacterial blight may sometimes resemble the Helminthosporium stripe disease, distinctions are worth noting as follows: (a) the Helminthosporium disease is systemic, striping usually all the leaves of the attacked host plant, whereas the bacterial attack is local; (b) the water-soaked and translucent appearance, and (c) the exudate are peculiar to the bacterial blight.

ON THE HEADS.—In case the flag leaf is badly attacked before the head escapes, the exudate may so seal up the sheath that the head can not escape (Pl. 49, A). In some cases the heads may emerge, but may be variously bent and distorted, and a portion of the grains blighted·

Ordinarily the effects of the disease are not serious enough to be very conspicuous, causing only water-soaked lesions on the glumes like those on the leaves, but usually with less exudate. Such blighting of the glumes does not destroy the grain as a rule, although the kernels may become discolored (brown) and more or less stunted or shrunken. The most important aspect of such attacks, which will be discussed in a later chapter, relates to the possible overwintering and distribution of the disease with such attacked kernels.

COMPARISON WITH SIMILAR PREVIOUSLY DESCRIBED BACTERIAL DISEASES

From the preceding description it will be noted that this bacterial-blight of barley bears some similarities to certain previously described bacterial diseases. There are also distinct differences.

Ráthay's disease of orchard-grass, caused by *Aplanobacter rathayi* E. F. S., (*10, v. 1, p. 171; v. 3, p. 155–160*) and O'Gara's disease of western wheat-grass, caused by *Aplanobacter agropyri* O'Gara (*6, 7*), both have a characteristic exudate which is yellow in color and is produced in much greater abundance than in the case of the disease on barley. While those diseases are most conspicuous on the inflorescences and upper leaf sheaths the disease on barley occurs chiefly on the leaf blades, with distinct lesions, first on the lower leaves, then progressing upward, and usually less prominently on the heads. Culm and head distortions occasionally occur with barley in a manner somewhat similar to those with Ráthay's and with O'Gara's diseases. Futhermore, while the diseases of orchard-grass and western wheat-grass are caused by nonmotile organisms, the disease of barley, as will be brought out later, is caused by a motile organism.

Manns's bladeblight of oats (*4*) is characteristically different, both as originally described and as observed by the writers, from the bacterial-blight of barley treated in this paper. While both diseases produce local infections in a somewhat similar manner, the bladeblight of oats, as far as noted by the writers, is commonly characterized by a rather wide, conspicuous, light-colored, halo-like margin about the lesions on the leaf blades, well shown by Manns (*4, Pl. XIII, fig. 1 and 3*). Characteristic lesions of this type have been produced on oats by inoculating with pure cultures of what the writers take to be *Pseudomonas avenae* Manns, isolated from oats. Thus, the characteristic lesions on oats are distinctly different from those on barley, as previously described, the light-colored, halo-like margin not occurring on the barley. The causal organisms are also different. *P. avenae*, as described by Manns and also as isolated by the writers, is white in culture, monotrichous, and pathogenic on oats. The writers find it nonpathogenic on barley. The causal organism of the disease on barley studied by the writers is yellow in culture, monotrichous, and pathogenic on barley, but not on oats. The writers conclude, therefore, that the bladeblight of oats as described by Manns is different from the bacterial-blight of barley treated in this paper.

While Manns (*4, p. 104*) supposes that barley is attacked by his blade-blight of oats, this is regarded by the writers as doubtful. He states that:

> Among the barleys grown at the Station, one variety, known as Primus, showed a susceptibility to the disease even more marked than did the Wideawake variety among the oats. The Oderbrucker variety of barley also showed an occasional blade infected. These observations were verified by cultures.

In this statement it is possible that Manns refers to the same bacterial-blight which is now under consideration, since the present writers have observed this blight on both the Primus and Oderbrucker barleys. And, furthermore, his statement that "observations were verified by cultures" does not seem to imply any carefully conducted cross-inoculation experiments.

DEVELOPMENT OF THE DISEASE

The disease will not ordinarily have blighted enough leaves to attract attention until the plants are two-thirds grown. The earlier stages of its development were, however, followed with care in the experimental fields in 1915 and 1916. In 1915 it was first detected in a field of Beldi barley, a common six-row variety. The seed of this was from a crop badly infected in 1914 and saved for this trial planting. A good, vigorous stand was secured, and the plants were 8 to 10 inches high before the first lesions were observed. These were well scattered over the field, and doubtless a few earlier primary lesions had been overlooked, since no trouble has been met in infecting younger plants in the experimental inoculations. From this time on the disease made steady progress, and two weeks later from 2 to 5 per cent of the plants showed dead, blighted leaves, so that even a casual observer would notice it. In 1916 earlier lesions were noted, but the subsequent development was very similar to that in 1915. This probably represents extreme developments in Wisconsin, but, as will be explained later, the disease is capable of producing greater injury under certain conditions.

OCCURRENCE AND HOST RANGE

GEOGRAPHICAL DISTRIBUTION

The bacterial-blight of barley has been noted and specimens collected by one of the writers (A. G. Johnson) in eight different States as follows: Akron and Greeley, Colo.; Ames, Iowa; St. Paul, Minn.; Amsterdam, Bozeman, and Moccasin, Mont.; Williston, N. Dak.; Corvallis, Oreg. Aberdeen, Brookings, and Highmore, S. Dak.; Janesville and Madison, Wis. If the reference to the disease by Manns (*4, p. 104*) from Wooster, Ohio, is included, the bacterial-blight of barley has been found in nine States. This distribution is indicated in figure 1.

HOSTS

This disease has been noted and collected on a large number of strains of barley including some 52 named varieties, in addition to a large number of unnamed hybrids and unclassified imported stocks. It has been definitely observed on the three main groups of barley: namely, the *Hordeum distichum* group (two-row barleys), the *H. vulgare* group (common six-row barleys), and *H. hexastichum* group (erect six-row barleys). These observations are summarized in Table I.

In connection with these studies the question has been constantly kept in mind as to the occurrence of this or like diseases on other grains. As a result similar bacterial diseases have been found on wheat (*Triticum* spp.), spelt (*Triticum spelta*), rye (*Secale cereale*), timothy (*Phleum pratense*),

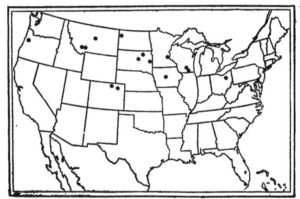

FIG. 1.—Outline map showing known distribution of bacterial-blight of barley in the United States.

and oats (*Avena sativa*). In each case, except that of timothy, the bacterial parasite has been isolated and the disease reproduced by inoculation and the parasite recovered.

With the exception of the bladeblight of oats the symptoms of these diseases are so similar as to be easily considered identical with the barley blight. Detailed study of the organisms, however, shows them to be different from that of barley. Work in connection with these is still in progress, and the results will be presented in subsequent publications.

VARIETAL SUSCEPTIBILITY

Observations have been made in the trial plots of the Office of Cereal Investigations,[1] United States Department of Agriculture, at Highmore,

[1] The writers wish in this connection to express their appreciation for courtesies extended by the various persons concerned at these places.

S. Dak., St. Paul, Minn., and to some extent at Moccasin, Mont., and Williston, N. Dak. (see Table I). In all cases special attention was given to the virulence of the disease on the different varieties of barley upon which it occurred. The differences noted are indicated in the table as follows: "0" is reserved for absence of the disease; "+" indicates very meager occurrence, yet definitely observed; a figure in the numerical range from 1 to 10 is used to represent the degree of infection in any particular plot as compared to that in other plots. As standards "1" is used to represent the condition where infection was very slight—that is, no special search being necessary to find the disease; it being present in small amounts on a considerable number of plants in whatever part of plot noted; "10" is reserved to represent a very abundant infection, where every leaf on every plant was very severely attacked. The intervening numbers between 1 and 10 are used to represent the respective intervening conditions.

TABLE I —*Bacterial-blight of barley: Varietal range and susceptibility*

Variety.	Highmore, S. Dak.	St. Paul, Minn	Moccasin, Mont	Williston, N. Dak.	Madison, Wis.	Akron, Colo.
Hordeum distichum group						
Benzin....................	... ..	6	....	....	....	....
Black Egyptian	5	....	....	....	....	....
Bohemian	5	....	....	....	....	....
Chevalier................	4	....	....	....	....	....
Chevalier, Scotch.........	....	....	....	....	+	....
Frankstein...............	6	....	....	....	....	....
Golden Melon.............	5	....	....	....	....	....
Gold Foil	6	....	....	....	....	....
Hanna...................	6	8	....	4	....	....
Hannchen................	4	....	....	4	....	+
Highland Chief...........	....	2	....	....	....	....
Holland ｡	6	....	....	....	....	....
Horn....................	6	....	....	....	....	....
Moravian................	7	....	....	....	....	....
Ouchac	4	....	....	....	....	....
Primus..................	5	....	....	5	....	....
Princess................	4	....	....	4	....	....
Proskowetz..............	4	....	....	....	....	....
Scholeys	5	....	....	....	....	....
Smyrna	....	2	....	....	....	....
Streigum................	5	2	....	....	....	....
Svanhals................	5	....	....	....	....	....
White Austrian...........	6	....	....	....	....	....
Hordeum vulgare group:						
Abyssinian...............	....	4	....	....	....	....
Albacete................	....	....	6	....	....	....
Bay Brewing.............	....	9	....	....	....	....
Beldi...................	6	....	8	....	6	....
Benzin..................	....	8	....	....	....	....
Bernard	....	....	....	4	....	....
Blue....................	....	2	....	....	....	....
California...............	....	....	....	....	....	+
California Prolific........	....	....	....	....	4	....
Caucasian...............	4	....	....	....	....	....
Chinese Turkestan........	....	9	....	....	....	....
Cuzco..................	6	....	....	....	....	....
Daniel's................	4	....	....	....	....	....

TABLE I.—*Bacterial-blight of barley: Varietal range and susceptibility*—Continued

Variety.	Highmore, S Dak.	St. Paul, Minn.	Moccasin, Mont.	Williston, N. Dak	Madison, Wis.	Akron, Colo.
Hordeum vulgare group—Con.						
Eagle............................		8	....	....	....	....
Featherston...................	6	2	....	....	....	....
Gatami..........................	....	....	....	5	....	....
Luth.............................	4	2	....	....	....	....
Manchurian....................	....	2	....	....	....	....
Minnesota 105................	....	....	....	....	....	....
Oderbrucker..................	4	....	....	4	....	....
Odessa..........................	4	6	....	2	+	....
Peru.............................	4	....	8	....	....	....
Poda.............................	4	....	....	....	....	....
Red River......................	4	....	....	....	....	....
Summit..........................	+	....	....	....	....	....
Williston C. I. 882...........	....	....	....	4	....	....
Wisconsin 5....................	....	....	....	....	+	....
Wisconsin 14..................	4	....	....	....	....	....
Hordeum hexastichum group:						
Mariout.........................	6	....	3	....	....	....
Williston Selection 17.......	....	....	....	+	....	....

While, at each of the above-mentioned places, a number of other varieties of barley were observed to be free from bacterial blight, a possible indication of some degree of resistance, no mention is made of these in Table I, since no experimental tests have been made on them as yet in this regard.

These observations indicate that the bacterial blight is widespread in its occurrence, and is probably to be found in all barley-growing sections of the United States. It seems to be considerably more prevalent and destructive in the Dakotas than farther east. It attacks all three groups of barley without evident preference. Within a group there is evidently a difference in relative susceptibility. For example, in the *vulgare* group, Oderbrucker seems to be less susceptible than Beldi, and in the *distichum* group, Hanna seems somewhat more susceptible than Chevalier. The observations here recorded were, however, not sufficiently extended to justify any close final comparisons as to such varietal rankings in regard to relative disease resistance.

THE ORGANISM

ISOLATION AND REINFECTION

The freshly invaded tissues as well as the exudate are teeming with bacteria and, after acquaintance with the specific organism, little difficulty has been experienced in isolating it from either source. As illustrating method and results, the following may be cited: A strong, succulent, lower leaf was selected showing natural midvein infection extending down past the ligule. This lesion was in the earlier stages of rapid development and showed the characteristic water-soaked appearance. The leaf was washed carefully in sterile water, dipped into 95 per cent alcohol for a few seconds, and rinsed well in sterile water. The leaf was then cut across with a sterile scalpel at the ligule where the midvein is thickest. This thickened portion, when cut thus, exposed two comparatively large, interior surfaces practically free from external contaminations. By gentle pressure on the midvein a small drop of the milky gray bacterial slime was forced from the interior to the cut surface. This was touched with a sterile platinum needle and agar strokes made. Eight such cultures were made and, within three days at room temperature, these gave similar growths, all apparently dominated by one organism. Plates poured from one of these cultures gave abundant colonies of one type, and several pure cultures were obtained from these. All proved to be the same organism with the characteristic cultural characters and all proved to be pathogenic when sprayed on young barley plants. Infections similar to the natural ones resulted in all cases from such inoculations, and the organism was reisolated. Control plants remained healthy.

The organism has also been obtained, although with more admixture of other species, by each of the following methods:

(1) Similar surface washing and disinfection have been made of invaded leaf tissues. These were then cut into pieces and laid upon the surface of nutrient agar. Abundant bacterial growth spread promptly from the cut edges of such leaf sections. This proved to contain a mixture of species, but from it the parasitic organism has been isolated by making further streak and dilution cultures.

(2) The surface exudate from diseased tissues is teeming with bacteria. Isolations have been made from both fresh and dried exudate at various times. While other organisms may occur, the pathogenic species is readily secured from the exudate in either condition. Isolations have been made from fresh exudate taken directly from diseased plants in the field, also from such plants when the exudate has been forced out in the moist chamber and from plants artificially infected in the greenhouse.

In the same way the organism has been obtained from the hardened granules of the exudate. In one case the granules were picked off the leaves, placed in sterile water blanks, in which they quickly soften and diffuse, and from these successful platings were made. In another case small exudate granules were planted directly on agar plates and the organism recovered from the resulting growth.

(3) The organism has been secured also from the invaded leaf tissues which had been kept as dry herbarium specimens for eight months. Small pieces cut from the lesions were immersed directly in bouillon for one-half hour and platings then made from this. Pure cultures of the organisms were thus obtained and pathogenicity proved by inoculations.

(4) The organism has also been isolated from old barley kernels. On July 18, 1916, pieces of glumes from infected barley 2 years old were

incubated for two hours in bouillon. The organism was subsequently obtained from this by platings and its pathogenicity proved by inoculations (Pl. B, 4; 49, B).

Further details regarding inoculation experiments are given in a later chapter.

MORPHOLOGY

Both in the host plant and in culture media the organism is a short, actively motile rod, usually single or in pairs except in nutrient peptonized beef broth containing 2 per cent of sodium chlorid, in which it forms long nonmotile chains. Growing on peptone beef agar (3-day-old cultures), the organism measures 0.5 to 0.8 by 1 to 2.5 μ when stained with gentian-violet or Ziehl's carbol-fuchsin. It stains less readily with Loeffler's methylene-blue.

No spores have been definitely seen. Capsules are formed on agar and in milk. The organism is motile by means of a single polar flagellum. Repeated stainings for flagella both by V. A. Moore's modification of Loeffler's method and by Zettnow's method agree in showing the monotrichous character of the organism (fig. 2). It is Gram-negative and not acid-fast.

FIG. 2.—*Bacterium translucens:* from 96-hour growth on pot*to agar, stained by Zettnow's method to show flagella. X 2,000.

CULTURAL CHARACTERS

Unless otherwise specified, these have been studied at approximately 25° C., a temperature favorable for the organism. Reference to reaction of media is made in terms of Fuller's scale. Unless otherwise noted all reactions were determined by titration with phenolphthalein as indicator, following boiling of solution tested. Notations of color have been made in comparison with Ridgway's Color Standards.[1]

AGAR POURED PLATES.—On peptone-beef agar, reaction +10, colonies appear in 48 to 60 hours and at end of four days are 1 to 5 mm. in diameter, circular, smooth, shining, and amorphous except for inconspicuous somewhat irregular concentric striations within the colonies 2 Margin entire. Colonies are wax-yellow tinged with old-gold in color, soft but not viscid in consistency. Buried colonies are lenticular or granular (Pl. 48, C).

AGAR STABS.—Stabs in +10 peptone-beef agar when 3 days old show a raised, smooth, shining, wax-yellow, opaque surface growth 5 mm. in diameter. Growth is only at the surface

---

[1] RIDWAY, Robert. COLOR STANDARDS AND COLOR NOMENCLATURE. 43 p., 53 col. pl. Bibliography, p. 42-43. 1912.

2 These striations within the colonies are evident only under slight magnification, about X10, and under favorable lighting; semidirect lighting best. These striations are so inconspicuous as to have been overlooked by the writers until Dr Erwin F. Smith, of the Bureau of Plant Industry, called attention to similar colony structure in a closely related organism.

AGAR SLANTS.—On peptone-beef agar slants, stroke cultures make a moderate, filiform, wax-yellow, opaque, smooth, shining, convex growth.

GELATIN PLATES.—Surface colonies show a characteristic margin. This may be due to liquefaction, which takes place slowly at 18° C. In about 10 days liquefaction is complete and the colonies float.

GELATIN STABS.—At 20° C. in +10 peptone gelatin the surface growth is abundant after several days with a medium pit of liquefaction. At the end of four days the growth becomes stratiform and slowly liquefies its way toward the bottom of the tube. It progresses somewhat more than halfway in 40 days.

BEEF BOUILLON.—In +10 peptone-beef bouillon uniform clouding occurs within 48 hours. This clouding subsequently becomes rather strong; a membranous pellicle forms at about the fourth day and then slow clearing begins, becoming first evident just under the pellicle. Sediment is flocculent. Crystals have been noted.

POTATO CYLINDERS.—Growth on steamed potato cylinders in two days is abundant, spreading, slimy, flat, glistening, smooth, without odor, and in color wax-yellow with a tinge of old-gold. Later the growth becomes very abundant and the elevation convex, but no change in medium is noticed at the end of two weeks.

MILK —Inoculated milk coagulates slowly and then clears, beginning at the top. In a month there is partial digestion, some soft coagulum, and a heavy, watery, yellow growth at the top. Reaction showed increase in acidity.

LITMUS MILK.—No marked color change is noticeable during the first two weeks, although a heavy yellow growth develops at the top. Then reduction begins, and in a month the medium is amber-colored.

FERMENTATION TUBES.—The tests were made in 2 per cent Witte's peptone water and 2 per cent, respectively, of each of the following carbon compounds· Dextrose, lactose, saccharose. maltose, glycerin, and mannit. Clouding was first noted at end of two days. At end of six days nearly all were clouded in the open end. Mannit and glycerin were the slowest to cloud In all cases there was a definite line of demarcation across the inner part of the U and no visible growth in the closed arm No gas was formed in the closed arm in any case

When 20 days old, they were titrated with phenolphthalein as indicator, boiling to drive off carbon dioxid. In each case there was slight acid production. They were similarly tested again when 30 days old, and found to have produced more acid, When the test was made without boiling, they were found to be almost neutral to phenolphthalein. Both ammonia and acids are produced, the ammonia apparently slightly in excess.

A strong test for ammonia was found by Nessler's reagent and also by testing the gas evolved with red litmus paper when a portion of the culture was heated with potassium hydroxid.

No gas was formed in fermentation tubes containing sterile milk. The milk in the open end cleared gradually, followed later by the slow clearing in the closed end. The reaction was acid to litmus.

NITRATE BOUILLON in fermentation tubes gave a good clouding in open end; none in closed end. No gas was produced. There was a strong test for ammonia and nitrates Since the organism produces ammonia from peptone, the test does not show whether nitrates were reduced or not. There was a slight production of acid when tested after boiling. No nitrites found.

TOLERATION OF SODIUM CHLORID.—Neutral peptone-beef bouillons containing, respectively, 2, 5, 6, and 7 per cent of chemically pure sodium chlorid were inoculated from young bouillon cultures. There was uniform clouding after four days in the 2 per cent strength. All other strengths inhibited growth. The experiment was repeated, using 2, 3, and 4 per cent of sodium chlorid. Growth occurred in the 2 per cent strength only, and in this the organism developed long chains and soon died.

OPTIMUM REACTION AND TOLERATION LIMITS.—Peptonized beef bouillon was used with sodium hydrate as the alkali under test and hydrochloric as the acid. Bouillons were prepared titrating +25, +20, +10, +5, 0, −20, and −30, and uniformly inoculated from a 4-day-old bouillon culture. At the end of 48 hours, growth was evident in +10, +5, and 0. Clouding was weak in the 0 bouillon at this time, but the +5 and +10 were heavily clouded. At the end of four days a pellicle was evident on the +10 tubes. The +5 later formed a weak pellicle, but none formed on the 0. Gradual clearing began on the fifth day, but the cultures never cleared entirely. The sediment in the +10 was somewhat flocculent, but in the +5 and 0 it had more of a slimy, stringy consistency. None of the other reactions showed any growth.

It seems evident, therefore, that +10 Fuller's scale represents the optimum reaction and that the organism is highly sensitive to both alkali and acid.

Another test was made of acid toleration, using +10 bouillon as the basic medium, to which was added 0 1 and 0.2 per cent, respectively, of malic, tartaric, and citric acids. The cultures were kept under observation for 20 days following inoculation, but no growth was observed in any of the acidulated tubes.

USCHINSKY'S SOLUTION.—A moderate, uniform clouding was noticeable on the fifth to eighth day. The clouding never became heavy and no pellicle was formed. There was slight change in the medium. Only a slight yellow tinge was given to it and a small amount of yellow sediment. The bacteria formed long chains in this medium.

FERMI'S SÓLUTION.—No growth.

COHN'S SOLUTION.—No growth.

STARCH AGAR.—No evidence of diastasic action on potato starch suspended in peptone-beef agar, tests being made with potassium iodid-iodin.

REDUCTION OF NITRATES.—Cultures in nitrate bouillon were tested with Trommsdorf's reagent, first after 24 hours and repeatedly thereafter, and gave negative tests for nitrites. At the end of 30 days the cultures gave strong tests for ammonia and nitrates but negative test for nitrites. It is doubtful if nitrates are reduced, since good tests for ammonia are obtained from bouillon which does not contain potassium nitrate.

INDOL.—Tests in peptonized Uschinsky's solution and in peptone water at the end of two weeks and four weeks showed slight indol production. B. coli run in comparison showed much stronger indol production.

METHYLENE-BLUE IN MILK.—Reduction is slow, first evident at end of three weeks. At end of two months the cultures are white except for a surface layer and some sediment which is blue.

BLOOD SERUM —Stroke cultures on Loeffler's blood serum gave a moderate, waxyellow, glistening, smooth, convex, filiform growth. Medium not liquefied.

AEROBISM.—The organism appears to be strictly aerobic. It does not grow in the closed end of fermentation tubes with any of the carbon foods tested. No growth occurs in the lower end of stab cultures of agar or gelatin. Shake cultures of agar showed growth at or near the surface only.

LITMUS AGAR WITH SUGARS.—On litmus-lactose agar stroke cultures there is abundant growth and the medium becomes somewhat bluer, probably due to production of acids and ammonia, the ammonia a little in excess. The same may be said about the growth on litmus-maltose agar.

On litmus-dextrose agar stroke cultures there is very abundant growth, slight bluing along the stroke, reduction to an amber color in two weeks.

Optimum reaction for growth is +10 Fuller's scale

TEMPERATURE RELATIONS

In +10 peptone beef bouillon the optimum temperature was 26° C. or slightly above. No growth was secured at 36°, and only part of the tubes clouded at 35°. Slow growth occurred at 10°; none at 6°. Numerous trials show the thermal death point of 2-hour cultures made from 48-hour cultures to be apparently about 50°. In the case of old cultures growth has been obtained and pathogenicity of organism proved after a 10-minute exposure to 55°, indicating the development of a slightly more resistant type of cell under these conditions.

DESICCATION

The organism is highly resistant to drying as it occurs normally in the host tissues, since, as previously explained, it has been isolated from dried herbarium specimens of leaves 8 months old and it may live on the seed at least two years. It also survives long on culture media. When dried on sterile cover glasses it seems more sensitive. Thus, a young well-clouded bouillon culture was diluted with equal parts of sterile water and a full 2-mm. loop transferred to each of a series of sterile cover glasses and these dried in a sterile chamber. When tested by plunging these in broth at intervals, it was found that most perished within 24 hours, and all within two days.

SUSCEPTIBILITY TO MERCURIC CHLORID

Various tests have shown the organism to be very sensitive to mercuric chlorid.

TECHNICAL DESCRIPTION

On the basis of the foregoing studies the parasite in briefly characterized as follows:

Bacterium translucens, n. sp.[1],[2]

Cylindrical rods rounded at ends, solitary or in pairs; individual rods 0.5 to 0 8 by 1 to 2.5 μ, motile by a single polar flagellum; aerobic, no spores.

Superficial colonies in peptone-beef agar plates round, smooth, shining, amorphous except for inconspicuous somewhat irregular concentric striations within, wax-yellow tinged with old-gold; margin entire.

Liquefies gelatin slowly; produces slight acidity in milk; digests casein, nitrates not reduced; acid produced in cultures with various sugars. No gas produced. Gram-negative. Group number 211.2222532.

Pathogenic in leaves of *Hordeum vulgare, H. distichum, H. hexastichum,* forming translucent elliptical to striaform lesions.

Type locality: Madison, Wis., on *Hordeum vulgare.*

Distribution: Northern Mississippi Valley and westward to Pacific coast.

---

[1] According to Migula's classification, the combination would be *Pseudomonas translucens,* n. sp.

[2] *Bacterium translucens,* sp. nov., aerobium, asporum; coloniis cum striis inconspicuis internis aliquid irregularibus concentricisque; baculis cylindricis apicibus rotundatis, solitariis vel binis; bacillis singulis 0.5—0.8×1.0—2.5 μ, flagello uno polare mobilibus.

Coloniae superficiales in agar-agar rotundae, leves, nitentes, mellese. Liquefacit gelatinam lente. Lac sterile acidum facit; casein segregat. Nitrum non redigit. Culturae in mediis cum saccharo sacchari, saccharo uvae, saccharo lactis acidae fiunt. Gas non facitur. Methodo Grami non coloratur.

Habitat in foliis vivis *Hordei vulgaris, H. distichi, H. hexastichi* in lineis vel maculis translucentibus. Madison, Wisconsin. Amer. bor.

INOCULATION EXPERIMENTS

ON BARLEY.—The disease has been reproduced on barley with the characteristic symptoms by artificial inoculation and the original organism recovered.   While a considerable percentage of infections have resulted where wound inoculations were made, the best results and those leading to the typical lesions have followed the spraying of the organisms in watery suspension upon young uninjured leaves.

The following details of one of the earliest series of inoculations illustrates this:

Beldi barley was sown in 6-inch pots, and on June 28, 1915, when the vigorous young growth was about 4 inches high, four uniform pots, each containing about 25 plants, were selected for experimental uses.   Of these pot I was inoculated by atomizer spray with a water suspension of the organism obtained as follows: A pure culture of the organism was plated out from the milky ooze squeezed from a typical leaf midvein lesion.   Sterile water was poured into an agar tube streak culture and the bacterial growth allowed to diffuse into this, which was then used to spray upon the young barley plants.   In pot II a wound inoculation was made on the first green leaf of each plant.   This was done by touching with the needle the bacterial growth of a young agar streak culture of the same source as described above.   With this needle a slight wound was made about midway the leaf, in the midvein or near it, the attempt being to puncture but one epidermal layer and to avoid a perforation through the entire leaf.   In pot III wound inoculations were made as in II except that the inoculation needle was dipped directly into the milky ooze squeezed from a typical barley leaf midvein lesion.   Pot IV was held as a control, the plants being wounded as in II and III, but with sterile needle.

These pots of barley plants had all been grown in the greenhouse previous to inoculation and were kept there in a damp chamber for 24 hours following this, then transferred to the open garden.   The pots were here sunken to the soil level and thus given normal outdoor exposure, the weather being warm and rather dry.

Infection was noted in pots I, II, and III on July 7, an incubation period of nine days.   There was no infection in IV.   On July 9 pot I showed approximately 100 per cent of plants infected, II approximately 40 per cent, III about 20 per cent, and IV no plants infected.   After the occurrence of splashing rains, secondary infections appeared, and at the end of a month, when these were at their height, pots I, II, and III showed a like condition, with abundant lesions on practically every plant, whereas the control, pot IV, showed only four or five recently developed lesions.   Since pot IV was only 18 inches removed from the nearest infected pot and fully exposed for the month, splashing rains were held responsible for these few late infections.

On July 7 wound inoculations were repeated (Beldi barley, 4 inches high), making a single prick with a very fine needle into a single marked leaf on each plant in the pot; plants were transferred to the open garden immediately following inoculation, no artificial means being used to increase aerial moisture. About 90 per cent of infection resulted.

In another series of inoculations, in which pot-grown greenhouse plants were used, a small drop from a young broth culture of the organism was placed on each leaf and the epidermis underneath punctured in one place with a fine-pointed needle. The plants were kept in the greenhouse in a rather humid atmosphere. Water-soaked infection areas appeared just above the pricked spot in four days, one of the shortest incubation periods we have noted. This series gave 100 per cent of infection.

Barley was planted in rows in the open garden on July 15. On July 22, when the seedlings were about 1 inch high, they were inoculated by spraying with a water suspension of pure culture. No shading or cover was used, but the weather was moist. On July 28 the first infections were noted, an incubation period of six days under normal outdoor conditions. On July 30 the lesions were numerous and typical, with an abundance of liquid exudate (Pl. 47, A).

In other cases abundant infections have been obtained by similarly spraying older plants (8 to 25 days) in the open without subsequent covering.

These results show that infection is easily secured under normal environmental conditions without artificially rupturing the epidermis. It is noteworthy that the application of the organisms with the atomizer insures better results than wound inoculation. Probably this is due to the rapid drying out of the wounded tissues when the puncture method is employed. Another interesting outcome is the lesser percentage of success where the inoculations were made by direct transfer of the bacterial ooze from infected tissues as compared with pure-culture inoculations. This is possibly due in part to the low degree of vitality of the organisms in such direct transfer. It may be in part attributed to the fact that such ooze hardens very quickly upon exposure, thus imprisoning the organisms temporarily at least; and meanwhile the tissues in the wounded spot dry out to a degree unfavorable to infection.

ON OTHER GRAINS.—Several trials were made to infect other cereals with this organism. Two series of inoculations were run on barley, wheat, rye, oats, spelt, emmer (*Triticum dicoccum*) einkorn (*Triticum monococcum*), and timothy, and a number of different series on barley, wheat, rye, and oats. All gave uniformly negative results except in the case of barley which was in every instance heavily infected. Cross-inoculations with this and other cereal-blight organisms have been briefly reported previously (1), and work in connection with these is still in progress.

RELATIONS TO HOST TISSUE

A study of razor sections of the lesions as they occur in fresh leaves show the general facts to be as follows: The bacterial invasion is in the parenchyma. The parenchyma cell walls become disintegrated and in the interior tissues long rifts result, which extend lengthwise of the leaf blades. In such lesions the bacterial slime fills the interior cavities, expelling the air; hence, the translucency of the invaded areas. Later this bacterial slime hardens to a resinous texture, making the translucent condition a permanent characteristic of the older lesions.

For detailed histological studies material was secured from recently infected young barley leaves. These had been inoculated by spraying with a water suspension of a pure culture and showed numerous small translucent infection areas. Pieces of these were fixed in boiling absolute alcohol saturated with mercuric cholorid, embedded in paraffin, and sectioned. Such sections stained with carbol-fuchsin showed infections to be stomatal. In some cases the invasion was as yet confined to the stomatal chamber which might be only partially occupied or crowded full of the organisms. Numerous cases showed intercellular invasions in the parenchyma in the early development of the lesions. In other cases later stages were in evidence where the bacteria had developed in large masses, evidently spreading from the stomatal chambers through the intercellular spaces, crowding apart and crushing the parenchyma cells. Instances of advanced invasion were found where the cell walls had partially disappeared, but the details of wall solution have not received a critical study as yet. In the older lesions, as noted above, there occurred long cavities filled with the bacterial growths. It is from such cavities in the advanced stages of the lesion that the bacterial slime may be squeezed out which has been the source of material for the isolation and inoculation work previously described.

OVERWINTERING AND DISSEMINATION

The fact that the disease may attack the heads early in their development has already been pointed out. The glumes may show water-soaked areas like those on the leaves and, upon these, drops of the bacterial exudate may appear and harden into tiny resinous granules or spread as a glistening film. Observation of these facts at once suggested two questions, each of much importance. (1) How much is the grain checked or injured by such attack? (2) Does the organism overwinter on such grains or other barley tissues?

As to the first, the writers' observations have failed to detect any case where the grains are killed, although in certain cases there may be considerable injury. Although they may be more or less shrunken, they generally mature and are viable. This seemed somewhat surprising, in view of the extent of the damage to the leaves, but is apparently explained in part by the fact that the heads are susceptible for so brief a period. The only cases observed where the heads were destroyed have occurred when the flag leaf and its sheath were badly attacked while the head was still inclosed. The bacterial exudate may then so seal the developing sheath that the normal escape of the head is impossible and these culms can not "head out" (Pl. 49, A).

As to the overwintering of the organism, evidence has been obtained in two ways: First, blighted leaves showing the well-developed translucent lesions, which had lain in the laboratory as dry herbarium specimens for eight months (July-March), were used as material from which to make poured-plate isolations. From these the characteristic organism was secured, its identity determined by cultural characters and its pathogenicity by inoculations on young barley leaves. Apparently the organism lives longer in the interior of these desiccated leaves than it does in the dried exudate upon the surface, since poured plates from the exudate granules from similarly stored leaves failed to show the organism. Only the smaller granules were available for such trial, and further tests may show that in the larger exudate granules the organism may persist longer.

The second and more direct evidence relates to the persistence of the organism with the seed grains which may be stored over the winter or disseminated widely. Field evidence indicated early that in certain cases the disease was introduced with seed from certain western sources. In following up this matter seed was collected in the summer of 1914 from a field of Beldi barley in Montana severely attacked by the blight. Lesions were evident on the glumes of these plants before maturity and showed, although less clearly, upon the ripe grain.

Some of this seed was planted in the trial grounds at Madison in the spring of 1915. The bacterial-blight was detected on the leaves in this plot when the plants were 8 inches high, scattered infections then occurring throughout the plot. At the same time no disease could be detected on the other barley plots in series which were planted with Wisconsin-grown seed supposed to be free from the disease. Later in the summer, however, the blight was produced by inoculation upon certain of these others, indicating that the difference was not primarily in varietal susceptibility.

As previously reported (3), isolation cultures were made in July, 1916, from the glumes of some of this same infected grain, which was collected in Montana in 1914. The characteristic barley-blight organism was found to be present, was secured in pure culture, and its continued pathogenicity was proved by inoculation experiments (Pl. 49, B). It is thus apparent that the organism may persist with the stored grain and remain pathogenic after at least two years of dormancy.

The conclusion that the disease is introduced with the seed seems further justified by the writers' observations in preceding years. Thus, in 1912, it first appeared in a plot of Chevalier, a two-row barley, the seed of which was brought from Montana. Later it developed on the adjacent six-row varieties, having evidently spread from the Chevalier plot. The following season this same plot of soil was replanted with clean barley, and none of the disease reappeared. Our conclusion is, therefore, that probably under Wisconsin conditions the disease usually comes from organisms overwintering with the seed rather than in the soil, although the organism is capable of living over the winter in the dry, blighted straw.

It is doubtless carried with the seed to new localities, and not only its overwintering but its long-range dissemination is thus accounted for. For local dissemination in the field water is apparently the chief agent, although thrips and aphids may possibly play a considerable part. As already shown, surface application of the organism by spray or otherwise without wounding suffices for infection. The abundant bacterial exudate from blighting tissues is readily distributed by beating rain or trickling dew and may doubtless be carried farther, though in lesser amounts, by visiting insects. The prompt hardening of this exudate into resinous masses preserves the vitality of the organism for a long time without hindering this mode of water dissemination, since the dry exudate again quickly softens and diffuses upon the application of water. (Pl. 48, A.)

CONTROL MEASURES

It is not practicable to exercise any control measures aiming to check the spread of the disease after it appears in the barley field. On the other hand, the conclusion of the writers that its overwintering and long-range dissemination occur chiefly, if not wholly, with the seed encourages the hope that practical control measures may be developed.

Ordinarily it may suffice to avoid the use of seed from badly infected fields or localities when these conditions are known. In case of doubt it seems probable that seed disinfection may greatly reduce the danger of introducing the disease, if not eliminate it altogether. The writers have not as yet had opportunity for a convincing trial of this matter. An experiment bearing upon it was, however, made in March, 1916, as follows: Barley grain harvested from an infected field in 1914 had meanwhile lain dry in the laboratory. Isolations from the glumes then (1916) showed the organism still alive. After soaking this grain for two hours in formaldehyde solution (1 part of Merck's 40 per cent formalin in 320 parts of water) attempts to isolate the organism from the material failed. These results should not, however, be given too great weight, since negative results had also been obtained with some samples of untreated seed suspected of harboring the organism. Owing to the high degree of sensitiveness of the organism to mercuric chlorid, it is probable that this will be even more reliable for use as a seed disinfectant. The hot-water method may also prove useful, especially if later studies show that the organisms penetrate the kernel.

SUMMARY

(1) Bacterial blight of barley is a widespread disease capable of producing serious loss and is caused by a heretofore undescribed parasite.

(2) The lesions on the leaves begin as small water-soaked areas which enlarge to yellowish or brownish somewhat translucent blotches or irregular stripes. Similar lesions may appear later on the glumes. The chief injury, however, is apparently to the foliage.

(3) A bacterial exudate may appear scattered over the lesions as tiny clouded droplets which harden into yellowish resinous granules or spread to form a grayish, flaky surface film. This exudate and the translucency of the invaded parts are characters distinguishing the bacterial blight from the Helminthosporium diseases.

(4) This disease bears some resemblance to Ráthay's disease of orchard grass, to O'Gara's disease of wheat grass, and to Mann's blade blight of oats. It is, however, distinct from any of these.

(5) Primary lesions may appear on seedling plants very early in their development and secondary lesions when the plants are 8 to 10 inches high. Later the disease spreads with increasing rapidity.

(6) The disease has been found widely distributed from the eastern Mississippi Valley to the Pacific coast. It attacks the barleys of all three of the main groups: viz, two-row, common six-row, and erect six-row. Within each group there is a considerable range of varietal susceptibility and, so far as observed, some are free from attack, but further work is necessary for final comparisons.

(7) The invaded tissues and exudate are teeming with the bacterial parasite which has been isolated not only from these sources but also from dry overwintered leaves and from grain 2 years old. The organism is a monotrichous rod, yellow in culture, with the group number 211.2222532 and described as *Bacterium translucens*, n. sp.

(8) Inoculation experiments have shown that the disease may be readily induced on barley by spraying with water suspensions of the organism. Negative results were obtained from inoculations on oats, rye, wheat, spelt, emmer, einkorn, and timothy.

(9) The channels of invasion are stomatal and intercellular.

(10) Although the organism may overwinter in infected leaves, diseased kernels doubtless constitute the chief means of dissemination and source of spring infection.

(11) While control measures are not as yet fully worked out, the most promising methods consist in the avoidance of infected seed and in seed disinfection.

LITERATURE CITED

(1) JONES, L. R., JOHNSON, A. G., and REDDY, C. S.
    1916. BACTERIAL BLIGHTS OF BARLEY AND CERTAIN OTHER CEREALS. *In* Science, n.s., v. 44, no. 1134, p. 432-433.

(2) ———— ————
    1916. A BACTERIAL DISEASE OF BARLEY. (Abstract) *In* Phytopathology, v. 6, no. 1, p. 98.

(3) ———— ————
    1917. BACTERIA OF BARLEY BLIGHT SEED-BORNE. (Abstract.) *In* Phytopathology, v. 7, no. 1, p. 69.

(4) MANNS, T. F.
    1909. THE BLADE BLIGHT OF OATS—A BACTERIAL DISEASE. Ohio Agr. Exp. Sta. Bul. 210, p. 91-167, 1 fig., 15 pl. (partly col.) Literature cited, p. 166.

(5) MIGULA, W.
    1900. SYSTEM DER BAKTERIEN. Bd. 2. Jena.

(6) O'GARA, P. J.
    1915 A BACTERIAL DISEASE OF WESTERN WHEAT-GRASS. FIRST ACCOUNT OF THE OCCURRENCE OF A NEW TYPE OF BACTERIAL DISEASE IN AMERICA. *In* Science, n. s., v. 42, no. 1087, p 616-617.

(7) ————
    1915. A BACTERIAL DISEASE OF WESTERN WHEAT-GRASS, AGROPYRON SMITHII. *In* Phytopathology, v. 6, no. 4, p. 341-350, pl. 9-13 (partly col.).

(8) RAVN, F. K.
    1900. NOGLE HELMINTHOSPORIUM-ARTER OG DE AF DEM FREMKALDTE SYGDOMME HOS BYG OG HAVRE. *In* Bot. Tidsskr., Bd. 23, Heft 2, p. 101-321, 26 fig. 2 col. pl. Litteratur, p. 317-321.

(9) ————
    1901. ÜBER EINIGE HELMINTHOSPORIUM-ARTEN UND DIE VON DENSELBEN HERVORGERUFENEN KRANKHEITEN BEI GERSTE UND HAFTER. *In* Ztschr. Pflanzenkrank., Bd. 11, Heft 1, p. 1-26, 8 fig., 2 col. pl.

(10) SMITH, E. F.
    1905-14. BACTERIA IN RELATION TO PLANT DISEASES. 3 v, illus., pl. Carnegie Inst. Washington Pub. 27. Bibliography, v. 1, p. 203-266.

1.—Healthy barley leaf for comparison.

2.—Early stage of bacterial blight (natural infection) showing a translucent, water-soaked stripe along the midrib, yellowing toward the apex. Often such lesions as this will be studded with exudate granules (see Pl. 47, D) or show a drop of bacterial ooze at the base (see Pl. 47, B).

3.—More advanced stage of bacterial blight of barley (natural infection) where numerous local invasions have developed. Note the many water-soaked spots; the glistening grayish exudate scales below, which are shown in sharper color contrast in this painting than in nature. The tissues, as in the more advanced lesions above, commonly retain their translucence as they yellow and die because of the hardened bacterial ooze retained within.

4.—Barley seedling showing abundant infection spots on the lower leaf from atomizer inoculation with strain isolated from old infected seed. The individual greenish water-soaked spots represent stomatal infections. The general dying and yellowing above resulted from such invasions.

5.—Barley seedling similarly infected with strain isolated from blighting leaf.

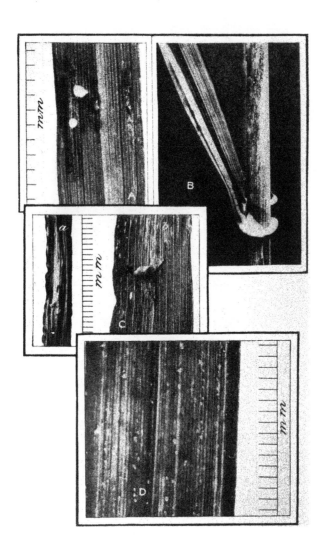

PLATE 47

A.—Barley leafblight, $\times$ 5½, natural infection, showing exudate gathered in three grayish glistening drops on the surface of the water-soaked blighting area.

B.—Blighting barley leaf, $\times$ 1½. Early stage of mid leaf stripe invasion, this stripe showing the water-soaked translucent appearance; the grayish exudate so abundant as to have formed a glistening droplet just above the ligule.

C.—Blighting barley leaves natural infection, $\times$ 2. a, A small insect (midge) entrapped in the exudate. b, The dry thin grayish exudate film (X) at the lower part of the blighted area is partly detached and turned back. The exudate film is so nearly transparent that the leaf veins are slightly visible through it.

D.—Barley leaf, $\times$ 4, showing general blighting, the surface studded with numerous exudate droplets hardened into yellowish resinous granules, readily dislodged. These granules tend to appear in longitudinal rows upon the stripes.

A.—Photomicrograph of a bead of hardened exudate dissolving in water under microscope. The granular effect is due to the bacteria which quickly diffuse at the margins.

B.—Five-day-old agar slant cultures of *Bact. translucens* incubated at $24°$ C. *a*, On peptone beef agar. *b*, On potato agar. Upon the latter medium the organism grows profusely and in a characteristic manner—viz, growth heaps up abundantly, is slightly more watery, and a little lighter in color than on peptone beef agar.

C.—Colonies of *Bact. translucens* on peptone-beef agar plate, dilution culture from bouillon, photographed by semidirect lighting to show characteristic internal somewhat irregular concentric striations. $\times 5\frac{1}{2}$.

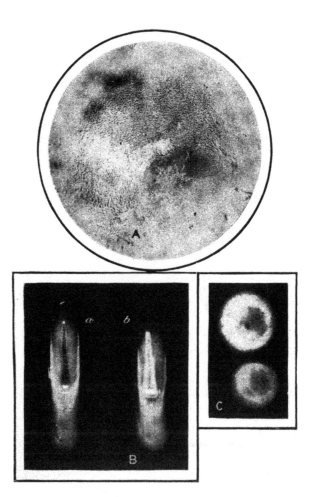

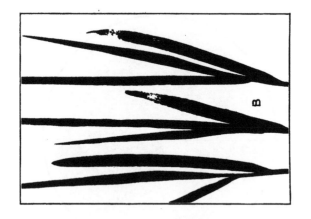

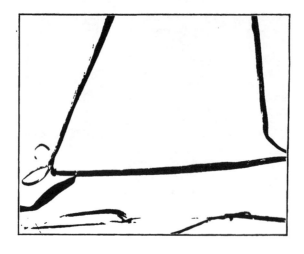

PLATE 49

A.—Blighting barley culms at heading stage. The sheaths so sealed by the bacterial exudate as to prevent the escape of the inclosed heads which thus perish undeveloped. × ⅔.

B.—Barley seedlings showing abundant infections on first leaves three days after atomizer inoculation with strain of organism isolated from 2-year-old seed. The photograph was made, using chiefly transmitted light to show the translucent character of the invaded areas. × ⅗.

INFLUENCE OF THE AGE OF THE COW ON THE COMPOSITION AND PROPERTIES OF MILK AND MILK FAT

By C. H. ECKLES, *Dairy Husbandman*, and L. S. PALMER, *Chemist, Department of Dairy Husbandry, Missouri Agricultural Experiment Station*

INTRODUCTION

The question of the changes in the composition of milk with successive lactation periods has been one which has from time to time attracted the attention of chemists and physiologists. That phase of the question relative to the percentage of fat in the milk has been of great practical importance to the dairyman, who is interested in knowing whether the percentage of fat in the milk of the heifer can be used as an index of what may be expected of the mature animal. Data bearing on this question have been compiled by several investigators, but the conclusions drawn from these data have not been entirely uniform. These data have been gathered in most cases from dairy herds located at various experiment stations, both abroad and in this country.

It is the purpose of the present paper to offer data on this question taken from the records of the University of Missouri dairy herd. It is believed that the data have an added value because of the fact that this herd has always been composed of pure-bred animals only, comprising the Jersey, Holstein, Ayrshire, and Shorthorn breeds, and also because of the fact that all of the animals were born and raised at the Station, with the exception, of course, of the individuals of each breed which were purchased when the herd was started.

Another phase of the question of the influence of age on the composition of milk is that of the effect of old age. Whether or not very old cows, which are not infrequently found in long-established herds or in private families, continue to give normal milk has never, to our knowledge, been studied. A very thorough search of the literature has failed to disclose any investigations of this character. The authors have had an opportunity to study the composition of the milk of one aged cow in the University herd and also of two other cows in the same herd which were advanced in years. The data secured from these animals will also form a part of the present paper.

HISTORICAL REVIEW

The principal investigations of the influence of age upon the percentage of fat in the milk have been by La Cour and by Haas and Högström in Europe and by Wing and by Hills and Hooper in this country.

La Cour (7)[1] reported the results of five years' observations of the percentage of cream in the milk of a large number of cows, using the Fjord

---

[1] Reference is made by number (italic) to "Literature cited," p. 657–658.

Journal of Agricultural Research,
Washington, D. C.
le   (645)

Vol. XI, No. 12
Dec. 17, 1917
Key No. Mo.—4

control centrifuge. When the results were compiled with regard to the
age of the cows, it appeared that the percentage of fat in the milk was
highest during the first three years of the cow's lactation life and de-
clined at the rate of about 0.1 per cent of cream at intervals of two to
three years. A few very old animals—that is, 15 to 18 years old—
showed cream percentage 0.6 to 0.9 per cent of cream lower than normal,
which was that observed during the fifth and sixth years of the cow's
life.

Haas(2) studied the data gathered at Allgäu from 1894 to 1898 from the
standpoint of the influence of the age of the cow upon the percentage of
fat in the milk. The conclusion drawn from the data was that the fat
content increases up to the fifth calf, after which there is generally a
gradual decline. Individual animals, however, showed exceptions to
this rule. It is stated also that the specific gravity of the milk is highest
with the first calf and lowest with the sixth.

Högström (4) compared the percentage of fat in the milk of 393
Ayrshire cows of different ages with the average percentage of fat of 799
lactation periods of these cows. The conclusion drawn was that the fat
content of the milk is highest at 3 years of age, is high at 4 years, and
remains close to the average for the family during the full activity of the
cow—that is, through the tenth year—after which the fat percentage
again rises slightly. The difference between the highest fat percentage
(at 3 years of age) and the average was 0.14 per cent of fat.

Wing (8) reported the average fat percentage of individuals in a herd
of 25 grade Jersey and Holstein cows from the second to the sixth year
with the following results: 3.71, 3.71, 3.68, 3.60, and 3.49 per cent of
fat. The data were based on the first 40 weeks' production of each
cow's lactation period.

Hills (3) has made a very careful study of the records of 99 cows at the
Vermont Experiment Station with regard to the effect of advancing age
on the quality of the milk. The records of this herd for a period of 10
years were studied with this question in view both from the standpoint of
separately following the same animals through their lives in the herd,
making comparisons of groups composed strictly of the same individuals,
and also from the standpoint of making comparison of groups according
to age, regardless of the changing individuality of the animals. On
account of the large number of animals considered, practically the same
conclusion was reached in each of the methods of study.

When the same animals were studied throughout their lives in the
herd in only 3 out of 46 cases was the percentage of fat higher after the
first lactation. Of the remaining 43 cases 29 showed a downward trend
of the fat percentage from the first lactation and the remaining 14
showed a stationary fat percentage throughout the period of greatest
activity, after which a gradual decline occurred, owing to old age. Hills
concludes that—

heifers practically "strike their gait," so far as the quality of the milk flow is concerned, in their first lactation; and whatever the effect of advancing years upon milk quality may be, it is not profound enough to be of importance until old age becomes imminent.

Hooper (5) has recently studied data showing the records of 634 Holstein cows on official test. When the data were classified according to the age of the cows, the percentage of fat in the milk was practically constant from the second through the seventh year. The author concluded that the heifers gave richer milk than the mature cows, but the differences do not appear to be sufficiently great to justify this conclusion. The percentage of fat beginning at 2 years 2 months of age and ending with 7 years 2 months of age was as follows: 3.51, 3.55, 3.54, 3.78, 351, 3.50, and 3.52.

With regard to the question of whether aged cows give normal milk, Klein and Kirsten (6) studied the composition of the milk fat of cows of different ages, the oldest of which was 13 years of age. All the younger cows gave milk fat of normal composition throughout the lactation period, the characteristic changes due to advanced lactation, as manifested in the decline in the saponification value and Reichert-Meissl number and the increase in the iodin value, occurring in each case. The old cow in the group, however, failed to show these changes at the close of the lactation period. The milk fat was also abnormal at the beginning of lactation, showing a Reichert-Meissl number of 43.5 and an iodin value of 31.7. The authors attributed the abnormalities in the case of this cow to individuality rather than to the age of the animal. At the same time they express the view, based on practical experience, that very old cows no longer produce good butter.

INFLUENCE OF AGE ON PERCENTAGE OF FAT IN MILK

The records from which the data presented in this paper are drawn consisted of the true average percentage of fat for the entire lactation period of each cow. This figure has been calculated from the total milk and fat production for the period, the milk production being based on the actual amount of milk produced at each milking for the entire period, and the fat production being based on the percentage of fat in a composite sample of five days' duration taken at the middle of each month. It should also be stated in connection with the lactation periods considered for each animal that whenever this exceeded 12 months the milk and fat production for the 12 months were considered that of the lactation period. This, coupled with the not infrequent failure for periods of parturition to be 12 months apart, results in it being impossible to make the number of lactation periods of an individual animal a criterion of its age. This means, for example, that the 20 Jerseys who completed six lactation periods were not necessarily of the same age at the end of the sixth period. Our data should, therefore, be interpreted more

correctly as representing the influence of the number of lactation periods
upon the average percentage of fat for each period. The differences
between the results presented in this form and the actual influence of
age upon the percentage of fat did not present the probability of being

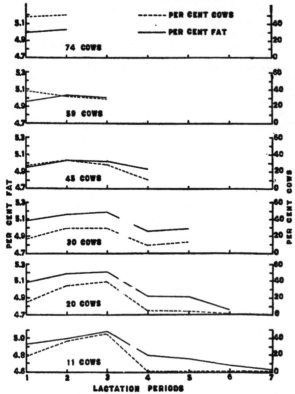

FIG. 1.—Graphs of the variation in percentage of fat of Jersey cows in successive lactation periods, and proportion of total number of cows in each group showing highest test in each period.

of sufficient magnitude to justify the calculation of the data on the basis
of the actual age of the animals.

Although the dairy herd from whose records our data are drawn
comprised animals of the four breeds already mentioned, not enough
animals of the Shorthorn breed were included to warrant the use of

their records from this source. The data, therefore, include the records of animals of the Jersey, Holstein, and Ayrshire breeds only.

The data showing the average percentages of fat for the successive lactation periods of all animals of each breed and similar data for the

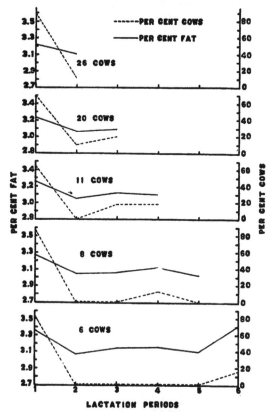

Fig. 2.—Graphs of the variation in percentage of fat of Holstein cows in successive lactation periods, and proportion of total number of cows in each group showing highest test in each period.

entire herd are given in Table I. An examination of the records of the individual animals of each breed, which it would be impracticable to publish on account of their extensiveness, showed rather striking differences among the breeds with respect to the frequency with which the individuals of each breed showed the highest fat test in each lactation period. This frequency, calculated as the percentage of cows of each group showing the highest test for each period, is presented for each breed in Table II. The combined data for each breed are shown graphically for a portion of the cows in figures 1, 2, and 3.

TABLE I.—*Percentage of fat in milk in successive lactation periods*

JERSEY COWS

Number of cows.	Period—											
	1	2	3	4	5	6	7	8	9	10	11	12
74	4.99	5.04										
59	4.96	5.04	5.01									
45	4.95	5.04	5.03	4.84								
30	5.09	5.17	5.20	4.97	5.00							
20	5.09	5.20	5.22	4.92	4.92	4.76						
11	4.93	5.00	5.09	4.80	4.76	4.68	4.63					
4	4.89	4.90	5.05	4.68	4.54	4.68	4.61	4.58				
3	4.86	4.83	4.83	4.47	4.56	4.64	4.62	4.48	4.47			
3	4.86	4.83	4.83	4.47	4.56	4.64	4.62	4.48	4.47	4.22		
2	4.99	4.90	4.75	4.50	4.44	4.76	4.70	4.48	4.49	4.33	4.22	
2	4.99	4.90	4.75	4.50	4.44	4.76	4.70	4.48	4.49	4.33	4.22	4.01

HOLSTEIN COWS

Number of cows.	Period—											
	1	2	3	4	5	6	7	8	9	10	11	12
26	3.22	3.10										
20	3.23	3.06	3.09									
11	3.26	3.05	3.12	3.10								
8	3.27	3.04	3.06	3.12	3.03							
6	3.36	3.07	3.15	3.16	3.10	3.41						
5	3.33	3.06	3.11	3.11	3.06	3.18	3.11					
1	3.29	2.99	3.03	3.08	3.09	3.05	3.00	2.88				

AYRSHIRE COWS

Number of cows.	Period—											
	1	2	3	4	5	6	7	8	9	10	11	12
9	3.90	3.88										
9	3.90	3.88	3.79									
6	4.04	3.92	3.81	3.64								
4	4.01	3.97	3.78	3.68	3.82							
2	4.06	4.08	3.79	3.83	3.84	3.79						
1	4.10	3.90	3.87	3.99	3.87	3.76	3.65					
1	4.10	3.90	3.87	3.99	3.87	3.76	3.65	3.38				
1	4.10	3.90	3.87	3.99	3.87	3.76	3.65	3.38	3.58			

ALL COWS

Number of cows.	Period—											
	1	2	3	4	5	6	7	8	9	10	11	12
109	4.48	4.48										
88	4.45	4.47	4.45									
62	4.56	4.58	4.58	4.41								
42	4.64	4.65	4.66	4.49	4.51							
28	4.65	4.66	4.67	4.47	4.45	4.36						
17	4.42	4.36	4.44	4.25	4.21	4.18	4.13					
6	4.46	4.42	4.52	4.30	4.19	4.26	4.14	4.10				
4	4.67	4.60	4.59	4.35	4.44	4.42	4.38	4.21	4.25			
3	4.86	4.83	4.83	4.47	4.56	4.64	4.62	4.48	4.47	4.22		
2	4.99	4.90	4.75	4.50	4.44	4.76	4.70	4.48	4.49	4.33	4.22	
2	4.99	4.90	4.75	4.50	4.44	4.76	4.70	4.48	4.49	4.33	4.22	4.01

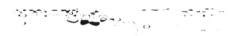

In calculating the data in Table II, which may be referred to as a frequency table, it was found in certain cases that a cow showed the same average test in two lactation periods, which usually followed in succession. Those cases in which the average percentage of fat in two periods did not differ more than 0.05 per cent were assigned to this class. In calculating the frequency tables for these cases the assumption was made that the period showing the highest average test meant that no other period showed a higher test, and the frequency calculation was made in favor of the first period which showed the high test.

There are several outstanding features of the tables and figures so far cited, of which mention may be made. In the first place one is impressed by the fact that the differences in the average fat percentage from lactation period to lactation period among the different breeds are not large. Indeed, it is necessary to plot the fat percentages on a rather large scale in order to make apparent the differences which occurred. Inasmuch as the figures do not show all the data given in the tables, it is necessary to turn to them to see that the differences in the average percentage of fat from lactation period to lactation period accumulate to rather large figures as the number of lactation periods of a cow's life becomes greater.

A second striking feature of the data is the marked similarity between the fat percentage and distribution curves. This fact indicates that the distribution curves and the data from which they are drawn are in reality of much greater significance than the curves of percentage of fat. They show clearly that a high percentage of fat in any one period among a large number of cows is due in most cases to the fact that a greater number of cows in that group showed a higher test in that period. This similarity between the curves of fat percentage and distribution of the highest test is especially striking for the 30 Jersey cows shown in figure 1, and for the Holstein cows shown in figure 2.

A third feature of the data is the striking breed characteristic with respect to the frequency of distribution of the highest test in the different lactation periods. In the case of the Jersey cows the data in Table II show that, when only the first two lactation periods are involved, the tendency is for the second period to give a slightly higher average test than the first period, but that this tendency is greater among a few cows than among a large number. In fact, the data show that with the 74 cows having only two lactation periods the distribution was practically equal between the two periods. For cows of the Jersey breed in the herd the tendency for the second period to show the highest average test diminished greatly as the number of lactation periods increased, and in many cases passed to the third period. In general the data show that Jersey cows may be expected to show their highest average test in either the first, second, or third periods, but the chances favor the second and third periods over the first to a rather marked degree. From the third period on, a gradual diminution of the average test may be expected, and this may accumulate to an appreciable decrease in advanced age. Jersey cows, then, do not necessarily "strike their gait" in the first lactation period with respect to the average percentage of fat in their milk.

TABLE II.—*Relative distribution of highest average test for the lactation period*

JERSEY COWS

Number of cows.	Lactation periods.	Proportion of cows showing highest test in period—						
		1	2	3	4	5	6	7
74	2	48.7	51.3					
59	2	47.5	52.3					
59	3	39.0	32.2	28.8				
45	2	44.5	55.5					
45	3	33.3	33.3	33.3				
45	4	26.7	33.3	28.9	11.1			
30	2	40.0	60.0					
30	3	30.0	36.7	33.3				
30	4	23.3	36.7	30.0	10.0			
30	5	16.7	30.0	30.0	10.0	13.3		
20	2	30.0	70.0					
20	3	20.0	40.0	40.0				
20	4	15.0	40.0	40.0	5.0			
20	5	15.0	35.0	40.0	5.0	5.0		
20	6	15.0	35.0	40.0	5.0	5.0	0.0	
11	2	27.3	72.7					
11	3	18.2	36.4	45.5				
11	4	18.2	36.4	45.5	0.0			
11	5	18.2	36.4	45.5	0.0	0.0		
11	6	18.2	36.4	45.5	0.0	0.0	0.0	
11	7	18.2	36.4	45.5	0.0	0.0	0.0	0.0

HOLSTEIN COWS

Number of cows.	Lactation periods.	1	2	3	4	5	6	7
26	2	88.5	11.5					
20	2	90.0	10.0					
20	3	70.0	10.0	20.0				
11	2	100.0	0.0					
11	3	81.8	0.0	18.2				
11	4	63.7	0.0	18.2	18.2			
8	2	100.0	0.0					
8	3	100.0	0.0					
8	4	87.5	0.0	0.0	12.5			
8	5	87.5	0.0	0.0	12.5	0.0		
6	2	100.0	0.0					
6	3	100.0	0.0	0.0				
6	4	100.0	0.0	0.0	0.0			
6	5	100.0	0.0	0.0	0.0	0.0		
6	6	83.3	0.0	0.0	0.0	0.0	16.7	
5	2	100.0	0.0					
5	3	100.0	0.0	0.0				
5	4	100.0	0.0	0.0	0.0			
5	5	100.0	0.0	0.0	0.0	0.0		
5	6	83.3	0.0	0.0	0.0	0.0	16.7	
5	7	83.3	0.0	0.0	0.0	0.0	16.7	0.0

TABLE II.—*Relative distribution of highest average test for the lactation period*—Con.

AYRSHIRE COWS

Number of cows.	Lactation periods.	Proportion of cows showing highest test in period—						
		1	2	3	4	5	6	7
9............	2	55. 5	44. 5					
9............	2	66. 7	33. 3					
9............	3	55. 5	22. 2	22. 3				
6............	2	83. 3	16. 7					
6............	3	83. 3	16. 7	0. 0				
6............	4	83. 3	16. 7	0. 0	0. 0			
4............	2	75. 0	25. 0					
4............	3	75. 0	25. 0	0. 0				
4............	4	75. 0	25. 0	0. 0	0. 0			
4............	5	50. 0	25. 0	0. 0	25. 0			

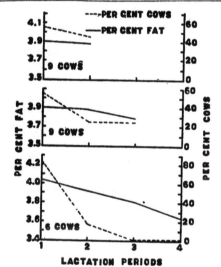

Fig. 3 —Graphs of the variation in percentage of fat of Ayrshire cows in successive lactation periods, and proportion of total number of cows in each group showing highest test in each period.

The data for the cows of the Holstein breed shown in Table II and figure 2 are in marked contrast to those for the Jersey breed. The chances for the highest fat test are so overwhelmingly in favor of the first lactation period that the conclusion seems justified that this is a breed characteristic. A similar tendency is shown by the data from the Ayrshire cows, shown in Table II and figure 3. The tendency is not so

great as for the Holstein cows, but
but the data are less conclusive bec
animals involved.

The fact that certain of the anin
whose records are included in the d
mine the influences of the plane o
first calving upon the dairy qualitie
factors may have contributed to :
data from the two breeds. If a :
should have a depressing effect up
during the first lactation period, :
explain to a certain extent why th
ber of lactation periods showed th
test in the second than in the first
oldest animals in the herd; and it
calves were not reared on nearly s
earlier days of the herd than has be
be stated in this connection that th
kept continuously for a period of 2

The data for the first two lactati
cows used in the above-mentioned
Tables III and IV with respect to t
tion and (2) plane of nutrition dur

The data bearing on the age of
indicate that a difference of 16 me
the Jerseys had little influence up
for the higher test to accompany
case of the Holsteins, however, ca
to decrease the breed tendency for
higher average test.

TABLE III —*Influence of age at first part
lactati*

Gr., up No	Num- ber of cows.	Breed of cows.	A\ a l pe t
			M
1.....	22	Jersey...............	:
2.....	21	.....do...............	:
3 .	16	.....do...............	.
4 .	14	.....do...............	.
5.. .	4	Holstein............	:
6. .	12	.....do...............	.
7	9	.....do...............	.

TABLE IV.—*Influence of plane of nutrition during growth on percentage of fat in milk of first two lactation periods*

Group No.	Number of cows.	Breed.	Plane of nutrition during growth.	Fat in milk.		Cows showing highest test in—		
				First Period.	Second Period.	First Period.	Second Period.	Stationary.
				Per cent.	*Per cent.*	*Per cent.*	*Per cent.*	*Per cent.*
1.....	8	Jersey	Heavy fed.....	5. 32	5. 07	62. 5	25. 0	12. 5
2.....	10	....do......	Light fed......	5. 13	5. 21	20. 0	60. 0	20. 0
3.....	7	Holstein.....	Heavy fed.....	3. 35	3. 10	100. 0	0. 0	0. 0
4.....	7	....do......	Light fed......	3. 05	3. 00	42. 8	14. 4	42. 8

The influence of the plane of nutrition during growth, shown in Table IV, is more striking. The data indicate that this factor may be of importance in connection with the average fat test for the first lactation period. The light-fed Jerseys showed both a lower average test in the first lactation period and also a less frequent tendency for this period to show the highest test, while the heavy fed Jerseys showed exactly opposite results. These data seem to justify the tentative explanation already advanced as to why the Jersey group as a whole showed different results from the Holstein and Aryshire groups. The data in Table IV show, in addition, that the strong breed tendency for the Holsteins to show a higher average test in the first lactation period than in any subsequent period was materially diminished by light feeding during the growth of the animals. It may also be pointed out that the light feeding of the animals of both breeds was by no means extreme, although it was widely different from the heavy feeding carried out with the other animals in the experiment.

COMPOSITION AND PROPERTIES OF MILK OF AGED COWS

The fact that certain animals in the Jersey herd of the Missouri Station had the breed characteristic of persistent milking developed to a marked degree has made it possible to examine the composition and properties of the milk of two cows which had attained an advanced age. Complete analyses of the milk fat of cow 124 were made on two occasions in her sixteenth year and during her twelfth lactation period, and again on two occasions in her nineteenth year, when she was in her thirteenth lactation period. Similar analyses were made of the milk of cow 16 when she was 14 years of age, and during her ninth lactation period. Two analyses have also been made of the milk of cow 403, of the Dairy Shorthorn breed, when she was 11 years of age, and during her fourth lactation period. These data are presented in Table V.

TABLE V.—*Influence of extreme age on composition and properties of milk and milk fat*

Cow No.	Age of cow	Stage of lactation	Date of sample.	Average daily milk.	Composition of milk.							Composition of fat.			
					Total protein	Casein	Albumin and globulin	Residual proteo-nitrogen	Fat	Lactose	Ash	Saponification value	Reichert-Meissl number	Iodin value	Melting point.
	Yrs.	Days.		Lbs.	P. ct.	P. ct.	P. ct.	P. ct.	P. ct.	P. ct.	P. ct.			Habl.	°C.
a124	16	282	June 17-19 1913....	26.3	3.20	2.61	0.59	0.038	4.20	4.65	0.72	232.7	30.82	27.74	33.1
124	16	188	June 24-26, 1913....	22.7	3.02	2.60	.59	.039	4.30	5.00	.71	234.2	24.55	29.17	33.1
124	19	848	May 16-18, 1916....	7.3	3.11	2.64	.47	.034	4.80	.....	.....	226.0	28.39	37.23	33.3
124	19	864	June 11-13, 1916...	12.0	3.64	2.96	.68	.042	4.50	4.32	.51	224.0	26.85	35.64	36.5
a16	14	207	June 17-19, 1913....	17.2	3.37	2.83	.54	.031	4.15	5.55	.75	231.1	32.74	29.73	35.6
16	14	233	June 24-26, 1913....	15.1	3.48	2.95	.53	.039	4.08	5.31	.76	228.9	31.89	31.16	37.0
b403	11	155	June 17-19, 1913....	23.5	2.88	2.27	.61	.034	3.63	4.97	.70	224.5	24.90	35.12	33.3
403	11	161	June 24-26, 1913....	21.5	2.89	2.99	.60	.038	3.63	4.96	.70	225.1	32.38	35.58	33.0

a Jersey.   b Shorthorn.   c Cow was farrow.   Duration of lactation was 1,000 days.

These data fail to reveal any significant features of the composition of either the milk or the milk fat which can be attributed to advanced age. The milk of cow 124 in the nineteenth year of her age failed to show any abnormalities even after she had been in milk for the astonishingly long period of 864 days. It is true that the first analyses of the milk fat made in 1913 in the case of cows 124 and 16 showed a higher Reichert-Meissl than iodin value, as was observed by Klein and Kirsten (6) for the milk fat of a cow 13 years of age. Inasmuch as this relation was not borne out in the other analyses no significance can be attached to the two isolated cases.

In order to obtain support for the statement of Klein and Kirsten that old cows no longer give normal butter, a semicommercial churning of butter from cow 124 was carried out with the milk obtained in May, 1916. The cow was on pasture at the time and the butter churned out exceedingly rapidly and was very soft when first made. After hardening, however, it was judged by several persons, including the authors, to be of excellent quality. One-pound portions of the butter were packed in glass jars and placed in the refrigerator. After three months the butter was still of good quality, although it had developed a slightly old taste. It is evident that the statement of Klein and Kirsten does not hold true for all cases, for it was not supported by the case of cow 124, which was much more advanced in years than any of the animals studied by these investigators.

SUMMARY

The percentage of fat in the milk of Jersey cows attains its maximum with respect to the average for the entire lactation period during any one of the first three periods, but the chances appear to be greater that this will be attained in the second or third period rather than the first.

Holstein cows almost invariably show the highest average percentage of fat for the lactation period during the first period.

Ayrshire cows more frequently show a higher average lactation test during the first than during subsequent periods, but less frequently than in the case of Holstein cows.

The variations in the average percentage of fat among the first few lactation periods are not sufficiently great to be of much practical importance, but the gradual decline in average test accumulates to a figure of considerable importance as the number of periods of lactation becomes greater.

A low plane of nutrition during growth and prior to the first lactation period probably contributes materially to a decrease in the average percentage of fat for the first lactation period from that which it would be if the period of growth is supported by a more liberal plane of nutrition.

Neither the percentage composition of the milk nor the physical and chemical constants of the milk fat of aged cows show any abnormalities attributable to old age.

Butter made from the milk of a cow 19 years old and in her thirteenth lactation period was pronounced to be of excellent quality, and kept for a period of three months at a temperature of $8°$ to $10°$ C., without showing any marked deterioration.

LITERATURE CITED

(1) ECKLES, C. H.
1915. THE RATION AND AGE OF CALVING AS FACTORS INFLUENCING THE GROWTH AND DAIRY QUALITIES OF COWS. Mo. Agr. Exp. Sta. Bul. 135, 91 p., 26 fig.

(2) HAAS.
1898. EINFLUSS DES ALTERS DER KÜHE AUF DIE MENGE UND GÜTE DER MILCH. *In* Milch Ztg., Jahrg. 27, No. 53, p. 841.

(3) HILLS, J. L.
1907. THE EFFECT OF AGE ON THE YIELD AND QUALITY OF MILK. *In* Vt. Agr. Exp. Sta., 19th Ann. Rpt., 1905/06, p. 339-350.

(4) HÖGSTRÖM, K. A.
1906. KOMJÖLKENS FETTHALT, DESS NORMALA VÄXLINGAR OCH ÄRFTLIGHET. *In* K. Landtbr. Akad. Handl. och Tidskr., Årg. 45, No. 3/4, p. 137-176, 18 diagr.

(5) HOOPER, J. J.
1915. FAT CONTENT OF MILK FROM HEIFERS AND COWS. *In* Breeder's Gaz., v. 68, no. 19, p. 808.

(6) KLEIN, J., and KIRSTEN, A.
    1902. UNTERSUCHUNG ÜBER DIE CHEMISCHE ZUSAMMENSETZUNG DES MILCH-
        FETTES EINZELNER KÜHE VON VERSCHIEDENEM ALTER IM LAUFE EINER
        LAKTATION. *In* Milch Ztg., Jahrg. 31, No. 37, p. 577-578; No. 38, p.
        594-596, No. 39, p. 611-613.
(7) LA COUR, A.
    1894. ON THE IMPORTANCE OF FAT DETERMINATION AS AN ACCESSORY IN THE
        IMPROVEMENT OF DAIRY CATTLE. (Abstract.) *In* Exp. Sta. Rec., v.
        6, no. 5, p. 455-457. (Original article in Tidskr. Landökon., Bd. 13,
        p. 303-336. 1894. Not seen.)
(8) WING, H. H., and ANDERSON, Leroy.
    1899. STUDIES IN MILK SECRETION DRAWN FROM THE RECORDS OF THE UNIVER-
        SITY HERD, 1891-1898. N Y. (Cornell) Agr. Exp. Sta. Bul. 169, p.
        517-552, fig. 98-100, 3 diagr.

SOIL ACIDITY AND THE HYDROLYTIC RATIO IN SOILS[1]

By C. H. Spurway,

Assistant Professor of Soils Physics, Michigan Agricultural College

HISTORICAL REVIEW

Recent investigations pertaining to soil acidity indicate a relationship between iron and aluminium compounds in soils and their reaction to litmus-paper-phenolphthalein indicator. Abbot et al. (*1*)[2] found that soil acidity in certain marsh soils may be due mainly to the presence of aluminum nitrate, and the quantity of aluminum found in an unproductive marsh soil was directly proportional to the amount of alkali required to neutralize the soil solution. Veitch (*9*) and Harris (*6*) found that when acid soils were treated with certain salt solutions iron and aluminium salts came from these soils and the iron and aluminium could be precipitated from the solutions as hydroxids by the addition of alkali. Daikuhara (*4*), working on soils of Japan found a relationship between the acidity of soils and their content of iron and aluminium compounds and concludes that these acid compounds are absorbed by soil colloids. In drawing conclusions from his work Rice (*7*) states:

much aluminum was present in extracts from soils of the highest acidity,

and Connor (*3*) also concludes that—

much of the harmful acidity of acid soils is due to the presence of toxic acid salts of aluminum and iron.

Ames and Schollenberger (*2*) state in their conclusions on the subject—

for the several soils investigated there was found to be an approximately quantitate · relationship between the bases soluble in fifth-normal nitric acid (excluding carbonates) and differences in lime requirement by the vacuum method.

A further list of references concerning effects and other causes of soil acidity may be found in the publications already cited and also in those of Frear (*5*) and Truog (*8*), and to avoid repetition the reader is referred to these papers for a more complete discussion of the subject.

THE PROBLEM

From studies on glacial-formed soils the author has for some time observed an apparent relationship between iron and aluminium salts in these soils and their reaction to litmus paper and phenolphthalein indicator. Yellow and brown colored soils of the Glacial Province are often acid to litmus paper. Small, well-marked, highly colored areas of soil

---

[1] Approved for publication in the Journal of Agricultural Research by the Director of the Michigan Agricultural Experiment Station.

[2] Reference is made by number (italic) to "Literature cited," p 672.

Journal of Agricultural Research,
Washington, D. C.

Vol. XI, No. 12
Dec. 17, 1917
Key No. Mich. 7

(659)

in these regions are invariably acid in reaction. Frequently areas of dark-colored acid soils are found having an exceptionally high content of iron or aluminium, or both. Recently an opportunity was given to study the problem from a chemical standpoint, and the results obtained from the investigation are presented in this paper as a contribution to our knowledge of the soil acidity question.

EXPERIMENTAL WORK

EXPERIMENT 1.—Twenty-nine samples of miscellaneous soils collected from the several glacial formations of Ingham County, Mich, were analyzed for calcium, iron, aluminium, and silicon by the solubility method, using $N/5$ hydrochloric acid. Averages of duplicate determinations are reported. The reaction of the soils was determined by means of litmus paper in contact with moist soil and also the Veitch qualitative test. The two methods gave similar results on these soils. A summary of the analytical results is given in Table I.

TABLE I.—*Soil classes and quantities of calcium, iron, and aluminium soluble in N/5 hydrochloric acid*

ACID SOILS

[Estimated as oxids. Results expressed as pounds per 2,000,000 pounds of soil]

Laboratory No	Soil class.	Calcium oxid.	Iron oxid.	Aluminium oxid.
2	Sandy loam	4,270	3,010	5,330
6	Loam	2,390	3,698	3,894
7	Sandy loam	3,100	3,182	2,776
9	Medium sand	1,390	1,462	2,288
10	....do	660	1,892	3,132
13	Sandy loam	2,270	3,956	2,144
14	....do	2,650	3,612	3,432
16	Medium sand	960	1,720	2,450
17	Clay loam	9,410	4,000	9,356
20	Medium sand	1,300	2,666	3,076
21	....do	1,780	2,332	2,790
22	Silt loam	4,170	3,010	3,100
23	Medium sand	1,760	2,708	2,944
24	Sandy loam	4,140	2,880	2,952
27	....do	2,040	2,706	1,916
28	....do	3,710	2,408	3,294

ALKALINE SOILS

Laboratory No	Soil class.	Calcium oxid.	Iron oxid.	Aluminium oxid.
1	Sandy loam	10,780	2,966	5,886
3	....do	8,840	2,666	3,220
4	Loam	4,060	3,182	452
5	....do	4,750	3,096	530
8	Silt loam	5,860	3,440	3,614
11	....do	8,140	3,526	4,908
12	Sandy loam	7,140	3,268	4,062
15	....do	6,210	2,966	3,118
18	....do	6,380	2,838	3,036
19	....do	10,000	3,182	1,988
25	Silt loam	36,000	4,042	9,406
26	....do	11,520	2,924	5,041
29	Loam	7,410	2,838	4,310

The results given in the foregoing table do not show striking differences except that the calcium-oxid content of the acid solutions from the alkaline soils is somewhat greater than in the case of the acid soils; but, when ratios between the soluble calcium oxid and iron oxid and aluminium oxid are determined, an interesting relationship is exposed. Table II gives these ratios.

TABLE II.—*Ratios between soluble calcium oxid and iron oxid and aluminium oxid*

ACID SOILS

Laboratory No.	Iron oxid.	Aluminium oxid.	Iron oxid and aluminium oxid.
2	0.70	1.25	1.95
6	1.55	1.63	3.18
7	1.03	.90	1.93
9	1.05	1.65	2.70
10	2.85	4.75	7.60
13	1.74	.94	2.68
14	1.36	1.30	2.66
16	1.77	2.53	4.30
17	.43	.99	1.42
20	2.05	2.37	4.42
21	1.35	1.55	2.90
22	.72	.74	1.46
23	1.54	1.67	3.21
24	.69	.95	1.64
27	1.33	.94	2.27
28	.65	.89	1.54

ALKALINE SOILS

	Iron oxid.	Aluminium oxid.	Iron oxid and aluminium oxid.
1	0.26	0.55	0.81
3	.30	.38	.68
4	.78	.11	.89
5	.65	.11	.76
8	.59	.62	1.20
11	.43	.53	96
12	.46	.57	1.03
15	.48	.50	.98
18	.43	.46	.89
19	.32	.30	.62
25	.11	.26	.37
26	.25	.44	.69
29	.38	.58	.96

These results show that, although the quantities of calcium, iron, and aluminium soluble in *N/5* hydrochloric acid under the conditions of the method vary considerably in the several soils, there is a marked relationship between the ratios of calcium oxid and iron oxid and aluminium oxid as expressed in Table II and the soil reaction. Considering the ratios CaO : (Fe$_2$O$_3$ + Al$_2$O$_3$), we find the lowest ratio for an acid soil is 1.42 and the highest ratio for an alkaline soil, 1.20. A neutral ratio may then be approximated by averaging these two numbers, and it is

1.31, or 1.3 nearly. From this standpoint we have a condition in which the ratios $CaO : (Fe_2O_3 + Al_2O_3)$ for the acid soils are all greater than 1 : 1.3, without exception. A similar relationship exists in case of the ratios $CaO : Al_2O_3$, the lowest ratio for an acid being 0.74 and the highest ratio for an alkaline soil 0.62, giving as an average 0.68 for a neutral ratio also without exception. The ratios $CaO : Fe_2O_3$, however, can not be placed in like order without having several exceptions, but if the other ratios can be considered correct because of the absence of exceptions, the correct ratio $CaO : Fe_2O_3$ can be computed and is 1.31 − 0.68, or 0.63. Three exceptions to this ratio are still apparent—namely, alkaline soils 4 and 5 have ratios greater than 0.63, and the acid soil 17 has a considerable lower ratio. It can be seen, however, that the two alkaline soils mentioned gave exceptionally low quantities of soluble aluminium oxid and the acid soil gave a very high quantity of soluble aluminium oxid. This point is also generally true in case of other soils—that is, alkaline soils having high $CaO : Fe_2O_3$ ratios gave low quantities and acid soils with low $CaO : Fe_2O_3$ ratios gave high quantities of soluble aluminium oxid with few exceptions. Therefore these exceptions can not be considered as invalidating the other results but rather tend to show the close relationship between soluble iron and aluminium in the soils and the soil reaction. There are no exceptions in the ratios between calcium oxid and iron oxid plus aluminium oxid.

The relationship between total calcium, iron, and aluminium was studied by making total analyses in duplicate of the soils for these elements. Ten gm. of soil ground to pass a 100-mesh sieve and 25 gm. of sodium peroxid were fused together in a nickel crucible. After cooling the mass was transferred to a 500-c.c. volumetric flask by means of hot, distilled water, acidified with concentrated hydrochloric acid, and allowed to stand on a steam bath for several hours. The contents of the flask were then cooled to room temperature, brought to a volume of 500 c.c., well mixed, and filtered. The solution was then analyzed according to official methods. The results are given in Table III and the ratios in Table IV.

TABLE III —*Total quantities of calcium, iron, and aluminium in soils*

[Estimated as oxids. Results expressed as pounds per 2,000,000 of soil]

ACID SOILS

Laboratory No.	Total calcium oxid.	Total iron oxid.	Total aluminium oxid.
2	17, 261	82, 421	120, 984
6	16, 154	45, 601	139, 239
7	14, 042	89, 531	99, 107
9	11, 808	39, 647	94, 184
10	9, 553	87, 801	99, 977
13	19, 352	74, 087	110, 029
14	13, 079	74, 071	136, 349

TABLE III.—*Total quantities of calcium, iron, and aluminum in soils*—Continued

ACID SOILS—continued

Laboratory No.	Total calcium oxid.	Total iron oxid.	Total aluminium oxid.
16	13,384	82,131	66,008
17	12,874	109,221	217,523
20	13,678	96,851	98,882
21	15,908	79,531	117,478
22	12,505	82,563	131,648
23	14,268	75,656	100,603
24	16,687	100,571	132,287
27	12,546	65,721	110,245
28	15,006	55,866	121,890

ALKALINE SOILS

Laboratory No.	Total calcium oxid.	Total iron oxid.	Total aluminium oxid.
1	16,236	78,241	155,076
3	19,926	83,391	120,501
4	16,236	41,631	132,972
5	31,652	68,601	139,204
8	15,129	67,455	139,502
11	20,582	73,756	163,434
12	19,106	50,887	129,846
15	27,101	61,221	134,820
18	15,990	75,403	129,359
19	18,163	61,495	127,930
25	44,217	63,483	166,471
26	21,443	53,553	147,943
29	17,056	35,673	113,958

TABLE IV.—*Ratios between total calcium oxid and iron oxid and aluminium oxid in soils*

ACID SOILS

Laboratory No.	Iron oxid.	Aluminium oxid.	Iron oxid + Aluminium oxid.
2	4.78	7.01	11.79
6	2.82	8.62	11.44
7	6.38	7.06	13.44
9	3.36	7.98	11.34
10	9.20	10.46	19.66
13	3.83	5.68	9.51
14	5.66	10.42	16.08
16	6.14	4.93	11.97
17	8.48	16.90	25.38
20	7.08	7.23	14.31
21	5.00	7.38	12.38
22	6.60	10.53	17.13
23	5.30	7.05	12.35
24	6.03	7.93	13.96
27	5.24	8.79	14.03
28	3.72	8.12	8.77

TABLE IV.—*Ratios between total calcium oxid and iron oxid and aluminium oxid soils*—Continued

ALKALINE SOILS

Laboratory No	Iron oxid.	Aluminium oxid.	Iron oxid + Aluminium oxid.
1.	4.82	9.55	14.37
3.	4.19	6.05	10.24
4.	2.56	8.19	10.75
5.	2.17	4.40	6.57
8.	4.46	9.22	13.68
11.	3.58	7.94	11.52
12.	2.66	6.80	9.46
15.	2.26	4.97	7.23
18.	4.72	8.09	12.81
19.	3.39	7.04	10.43
25.	1.44	3.76	5.20
26.	2.50	6.90	9.40
29.	2.09	6.68	8.77

While these results show that, in general, the total quantities of calcium and aluminium are lower and the total quantities of iron higher in the acid than in the alkaline soils, there is no distinct relationship between these total quantities and the soil reaction as there is in the case of the acid-soluble portions.

Total sulphur and magnesium were determined also, but no relationships were found between these elements and the soil reaction. The largest quantity of magnesium found in an acid soil was 3,226 pounds per 2,000,000 pounds of soil, and for an alkaline soil 4,288 pounds: consequently these soils may be characterized as low in magnesium.

As stated before, the acid-soluble silica was determined and these results, together with the "lime requirement" of the acid soils determined by the Veitch method, are given in Table V. A certain but not well-defined relationship exists between the quantities of acid-soluble silica and the "lime requirement." No other relationships concerning acid-soluble silica were discovered.

TABLE V.—*Quantity of acid-soluble silica (SiO_2) in 2,000,000 pounds of the soils and also the "lime requirement" of the acid soils by Veitch method*

Acid soils.			Alkaline soils.	
Laboratory No.	SiO₂	Lime requirement.	Laboratory No.	SiO₂
	Pounds.	*Pounds.*		*Pounds.*
2	1,020	1,560	1	1,570
6	960	1,320	3	1,330
7	670	1,440	4	1,230
9	620	1,440	5	1,430
10	520	1,440	8	1,380
13	730	1,440	11	2,280
14	960	1,560	12	1,490
16	540	1,440	15	1,010
17	2,520	3,600	18	1,370
20	780	1,200	19	1,460
21	640	1,080	25	3,480
22	1,110	1,560	26	2,060
23	640	1,440	29	850
24	1,020	1,800		
27	590	1,320		
28	730	1,560		

From the data given in Tables I and II it is evident that the relationship between the acid-soluble calcium and iron and aluminium may control the soil reaction. The question is, then, under what conditions can this phenomenon be produced. It is well known that many calcium salts hydrolyze in solution giving an excess of base, and reacting alkaline to litmus indicator, and that iron and aluminium salts generally produce acid solutions under similar conditions. The reaction of a solution containing a mixture of a salt that hydrolyzes and gives an excess of hydrion with a salt that produces an excess of hydroxidion on hydrolyzing depends upon the concentration and degree of hydrolysis of each salt in the presence of the other when equilibrium is established. The term "hydrolytic ratio" is used here to express this relationship. In other words, if the acid exceeds the alkali by chemical equivalents, the solution will react acid, and an alkaline solution results on reversing the conditions, while a certain hydrolytic ratio gives a neutral solution.

Results obtained from the analytical determinations with the solubility method and the theory of hydrolysis give a basis for the following hypothesis: that the reaction of these soils is probably determined chiefly by the hydrolytic ratio between salts of the alkaline earth elements on the one hand and salts of iron and aluminium on the other when these compounds are in equilibrium. From this standpoint any hydrolyzable salt in a soil or added to a soil, or any salt added to a soil that causes the formation of a hydrolyzable salt, may influence the soil reaction. In the following pages other data are given to support the above hypothesis.

EXPERIMENT II.—The quantity of lime required to neutralize these soils was determined by indirect titration with limewater in the following manner: 6–10, 300-c. c. Erlenmeyer flasks were used for each soil, 1 gm. of air-dried soil was placed into the first flask and the quantity increased in the remaining flasks by increments of 1 gm. each. Fifty c. c. of distilled water were then added to each flask and ten c. c. of a standardized lime-water solution. The flasks were sealed immediately with rubber stoppers, shaken several times, and allowed to stand about 16 hours. End points were determined in two ways: (1) by means of electrical resistance, and (2) with phenolphthalein indicator. Resistance measurements were taken by removing 25 c. c. of the supernatent solution with a pipette to a resistance cup and the readings taken with a Kohlrausch bridge. The highest resistance reading in the series was taken as the end point. A drop of indicator was then placed into each flask and the end point taken when the red color showed faintly. In all cases with the soils under discussion there was exact agreement in the end points as determined by the two methods.

On considering the ratio $CaO : Fe_2O_3 + Al_2O_3 :: 1 : 1.3$ to be correct for a neutral soil, the "lime requirement" of the acid soils may be computed by determining the quantities of calcium oxid necessary to add to the calcium oxid found soluble in the soils to bring all the ratios to 1:1.3. Table VI contains these figures, together with the "lime requirement" as determined by indirect titration. The quantities of acid-soluble calcium oxid plus the quantities of calcium oxid required to neutralize the soils by indirect titration are given in the fourth column, and the last column contains the ratios between these quantities of calcium oxid and the quantities of acid-soluble iron and aluminium.

TABLE VI.—*Lime requirement and ratios produced by neutralizing acid soils*

[Expressed as pounds per 1,000,000 pounds of soil.]

Laboratory No.	Calcium oxid required from neutral ratio.	Calcium oxid required by titration.	Acid-soluble calcium oxid + calcium oxid required by titration (=X).	$\dfrac{Fe_2O_3 + Al_2O_3}{X}$
	Pounds.	*Pounds.*	*Pounds.*	
2	2,140	2,040	6,310	1.32
6	3,450	3,240	5,630	1.35
7	1,480	1,600	4,700	1.27
9	1,490	1,600	2,990	1.25
10	3,210	3,140	3,800	1.32
13	2,420	2,400	4,670	1.31
14	2,770	2,800	5,450	1.29
16	2,250	2,200	3,160	1.30
17	860	940	10,350	1.29
20	3,120	2,800	4,100	1.40
21	2,120	1,900	3,680	1.38
22	530	660	4,830	1.27
23	2,590	2,160	3,920	1.44
24	1,220	1,160	5,300	1.29
27	1,520	1,400	3,440	1.34
28	580	660	4,370	1.30

In Table VII is shown the excess of acid-soluble iron oxid and aluminium oxid in the acid soils over the amount required to satisfy the ratio $CaO : Fe_2O_3 + Al_2O_3 :: 1 : 1.3$, and the ratios between the calcium oxid required by indirect titration and this excess of iron oxid plus aluminium oxid.

TABLE VII.—*Excess (in pounds) of iron oxid plus aluminium oxid over ratio 1 to 1.3 and ratios of calcium oxid required by titration to excess of iron oxid plus aluminium oxid*

Laboratory No.	Excess of iron oxid plus aluminium oxid.		Laboratory No	Excess of iron oxid plus aluminium oxid.	
	Over 1:1.3.	Ratio of calcium oxid by titration.		Over 1:1.3.	Ratio of calcium oxid by titration.
	Pounds.			*Pounds.*	
2	2,790	1.37	17	1,122	1.19
6	4,485	1.38	20	4,052	1.45
7	1,928	1.21	21	2,758	1.45
9	1,943	1.22	22	690	1.05
10	4,166	1.33	23	3,364	1.56
13	3,149	1.31	24	1,449	1.25
14	3,599	1.29	27	1,970	1.41
16	2,922	1.33	28	879	1.33

It may be seen from Table VI that the "lime requirement" as determined from the neutral-soil ratio and by indirect titration closely agree, causing the neutral ratio for all the acid soils to approximate 1:1.3 also, as shown by Table VII, when the lime required to neutralize the acid soils is compared to the excess of iron oxid plus aluminium oxid over the ratio 1:1.3, these ratios closely approach the same figure. The relationship between acid-soluble calcium and iron and aluminium compounds in these soils and the soil reaction is clearly shown, and it is also evident that we are dealing with equilibrium conditions.

EXPERIMENT III.—This experiment was performed to determine, if possible, the nature of the equilibrium—that is, whether physical or chemical. Soil 23 was chosen, the procedure being similar to that used in determining "lime requirements" by the indirect-titration method, except that after standing, the several portions were filtered, the soil washed with cold distilled water, and the filtrates titrated with an acid solution equivalent to the limewater solution used, with phenolphthalein as indicator. The experiment was performed in two ways: (1) By keeping the quantities of soil constant and varying the quantities of calcium oxid in contact with the soil, and (2) by varying the quantities of soil but keeping those of calcium oxid constant. The concentrations change, however, when the soil begins to take up lime. The results are given in Table VIII.

TABLE VIII.—*Effects of an excess of a solution of calcium hydroxid on the amount of calcium oxid used by acid soil 23*

VARYING QUANTITIES OF CALCIUM OXID, SOIL CONSTANT

Erlenmeyer flask No.	Weight of soil taken.	Quantity of calcium hydroxid taken.	Quantity of water added.	Titer of filtrate (cubic centimeters of hydrochloric acid).	Quantity of calcium hydroxid used by soil.	Percentage of applied calcium hydroxid.	Quantity of calcium oxid used per 2,000,000 pounds of soil.
	Gm.	*C. c.*	*C. c.*		*C. c.*		*Pounds.*
1	5	2	10	....	2.0	100	....
2	5	4	20	....	4.0	100	....
3	5	6	30	0.2	5.8	97	2,320
4	5	8	40	.6	7.4	92	2,960
5	5	10	50	2.4	7.6	76	3,040
6	5	12	60	3.6	8.4	70	3,360
7	5	14	70	5.5	8.5	61	3,400
8	5	16	80	7.7	8.3	52	3,320

VARYING QUANTITIES OF SOIL, CALCIUM OXID CONSTANT

1	1	3	20	0.75	2.25	75.0	4,500
2	2	3	20	.25	2.75	91.6	2,748
3	3	3	20	.1	2.9	96.6	1,930
4	4	3	20	....	3.0	100.0	....
5	5	3	20	....	3.0	100.0	....

It is evident that the soil continues to take up calcium oxid after the solution becomes alkaline to phenolphthalein, but the quantity used (Table VIII) soon reaches a maximum and remains quite constant. Soil acidity is undoubtedly neutralized when an excess of calcium oxid is found in the solution. A second reaction must then take place which is complete when the maximum amount of calcium oxid is taken. It will be observed from the table that this maximum amount increases with the concentrations of the solutions in contact with the soil. The quantity of calcium oxid required to neutralize 2,000,000 pounds of soil 23, as determined by the indirect-titration method with a different concentration of limewater than was used in Experiment III, is 2,160 pounds, an amount comparable to that used by the soil in this experiment when the solution becomes slightly alkaline.

Two generalizations can be drawn from the data given:

(1) A definite quantity of calcium oxid is required to neutralize an acid soil, with phenolphthalein as the indicator; and this quantity is independent of the concentration of the limewater used.

(2) Beyond the neutral point the amount of calcium oxid used varies directly with the concentration of the limewater solution bathing the soil; hence, it appears that the first generalization deals with a chemical reaction, and the second with a physical reaction. However, mass action may cause a chemical reaction when an excess of calcium oxid

At this point the possible combinations of the alkali earth elements and iron and aluminium with other elements in soils may well be considered. The principal mineral salts of the alkali earths are carbonates, silicates, phosphates, sulphates, chlorids, and nitrates. Of this list only the carbonates and silicates of calcium and magnesium give a red color on boiling their solutions with phenolphthalein. The principal iron and aluminium salts found in soils are silicates and phosphates, while sulphates, chlorids, and nitrates may be present. Theoretically these salts hydrolyze and give acid solutions, and will react with carbonates of silicates of the alkali earths; but these reactions proceed to an equilibrium and not to the limits of chemical equivalents.

EXPERIMENT IV.—Dilute solutions of the carbonate and silicate of calcium were boiled with phenolphthalein until the red color appeared. Small quantities of iron and aluminium salts were then added until the red color disappeared. Several salts of iron and aluminium were used in this manner, and the red color could be obtained or dispelled at will by increasing in order the calcium or iron and aluminium salts. Similar results were obtained by using solutions from alkaline and acid soils.

If a soil solution gives a red color with phenolphthalein when boiled and concentrated, a free alkali hydroxid is undoubtedly present, probably coming from the decomposition of a carbonate or silicate of an alkali earth element. On the other hand, if a soil solution gives no red color under the above-named conditions, an acid salt may be present in sufficient quantity to prevent the formation of free alkali.

From the ratios given in the preceding pages two general conclusions can be drawn: (1) The quantities of acid-soluble calcium are proportional to the quantities of acid-soluble iron and aluminium in neutral soils, and also to the unreacting acid-soluble iron and aluminium in the acid soils; and (2) the quantities of calcium required to neutralize the acid soils are proportional to the quantities of acid-soluble iron and aluminium in excess of the ratio 1:1.3 or to the reacting iron and aluminium. These results can not be explained on the basis of absorption, unless iron and aluminium compounds are the only absorbing substances in the soils, or that these compounds are proportional in acid-soluble quantities to the total amounts of absorbing compounds; also, if soils contain other absorbing substances, their absorbing power is satisfied to an equal degree; or that $N/5$ hydrochloric acid dissolves an amount of absorbed lime proportional to the unreacting iron and aluminium in the acid soils; and also soils exhibit two kinds of absorption, one kind independent of the concentration of limewater and the other dependent upon it. According to present knowledge of soil composition and absorption, these considerations are untenable.

The proportions found are in direct accord with the theory of hydrolysis and chemical equilibrium. When a neutral equilibrium is established between hydrolyzing compounds, the reacting substances bear certain fixed relationships to each other, which may be expressed in the form of a proportion. An equilibrium not neutral in reaction requires the addition of a quantity of acid or aklali proportional to the uncombined substances before the neutral point is obtained. If a solution of an acid salt of iron or aluminium is titrated with a standard solution of limewater until the reaction proceeds to a certain point but not to neutrality, the quantity of calcium oxid used is proportional to the quantity of iron or aluminium in the portion of the salt decomposed, and the amount of calcium oxid yet required to neutralize the solution is in the same proportion to the quantity of iron or aluminium in the portion of the salt yet to be acted upon, as is also the total quantity of iron or aluminium in the salt neutralized, a condition similar to that found in the acid-soluble portions of the soils studied.

When determinations for "acid consumed" were made on the soils preliminary to the solubility method, it was observed in case of the medium sands that more alkali was required to neutralize the acid solutions than was necessary to combine with the quantity of $N/5$ hydrochloric acid used; therefore these soils may have contained acid salts soluble in the acid in order to require this excess of alkali to neutralize the solutions.

The reddening of blue litmus paper by soils has been explained from the standpoint of selective absorption, but the phenomenon can be as logically explained on the basis that these soils contain acid salts absorbed by soil gels or relatively insoluble. Phosphates and hydrated silicates of iron or aluminium can be washed with water until the filtrates do not react acid to litmus paper, but when the paper is brought in contact with the solid material it is strongly reddened. It is also well known that soils which do not redden blue litmus paper, as well as neutralized acid soils, absorb large quantities of calcium oxid from limewater solutions. Silica gels also hold soluble-acid salts of iron or aluminium so tenaciously that these salts can not be washed out. The fact that a soil solution is not acid to litmus is no proof that an acid salt does not exist in the soil.

From the results obtained it is believed that the indirect-titration method as previously explained is adequate to determine the quantity of lime required to neutralize an acid soil. In determining the lime requirement of an acid soil due regard should be given to the factors influencing salt hydrolysis, solubilities of soil minerals, and mass action. Time, temperature, and concentrations of the reacting compounds are the principal factors involved. It is well known that in reactions concerning mass action, hydrolysis, and solubilities a certain length of time is required for equilibrium. Because of the effects of temperature on hydrolysis and solubility, a temperature considerably above or below that normal to field soils under plant-growing conditions should probably not be used. A point of importance in the procedure of the indirect-titration method is that the quantities of reacting substances, soil and limewater, are not changed during the reaction. Generally speaking, quantities of the reacting compounds are chosen that give a neutral equilibrium at a normal temperature. In this way the true point of neutrality is obtained, a fact undoubtedly known but not fully appreciated by many investigators, because in investigations on the subject and in developing lime-requirement methods an excess of soil or reagents is commonly used without apparent regard to the effects of mass action. Consideration should also be given to the nature of the products formed during the reaction. It will not be sufficient to determine the degree of acidity in a soil and calculate the amount of lime required to neutralize this acidity, because soil acidity undoubtedly results from the hydrolysis of soil compounds and enough reagent should be used to force the hydrolytic action to the neutral point. If calcium carbonate is used as a reagent, it should be in a very finely divided condition; otherwise the particles become coated with gels of iron, aluminium, or silica at the point of reaction with the soil, and their rate of solution is retarded. Considering all the aspects of the case, the logical reagent to use in a lime requirement method is calcium hydroxid.

SUMMARY

(1) A definite relationship was found between the ratios of calcium to iron and aluminium soluble in $N/5$ hydrochloric acid and the soil reaction. All the acid soils have ratios $CaO : (Fe_2O_3 + Al_2O_3)$ above $1 : 1.3$, and all the alkaline soils have ratios below this figure.

(2) It is believed that the reactions of the soils studied depend chiefly upon the hydrolytic ratios existing between hydrolyzing compounds of the alkali earths and iron and aluminium.

(3) A method for determining the calcium oxid required to neutralize a soil by indirect titration is described in which certain fixed quantities of a standardized calcium-hydroxid solution are allowed to react separately with varying amounts of soil. The concentrations of soil and solution giving a neutral reaction are chosen from the series by comparing the electrical resistance of the several solutions, also by means of phenolphthalein indicator. The greatest electrical resistance and faint color of the indicator is coincident with the concentration giving a neutral reaction.

(4) The quantities of lime required to neutralize the acid soils may be determined by computing the quantities of calcium oxid necessary to add to the acid soluble calcium oxid found in the soils to bring the ratios $CaO : Fe_2O_3 + Al_2O_3$ to $1 : 1.3$. The quantity of calcium oxid required by this factor method corresponds closely to the quantities required when determined by means of the indirect titration method, and it appears that the titration method is accurate and convenient.

LITERATURE CITED

(1) ABBOTT, J. B , CONNER, S. D., and SMALLEY, H. R.
    1913. THE RECLAMATION OF AN UNPRODUCTIVE SOIL OF THE KANKAKEE MARSH
    REGION. SOIL ACIDITY, NITRIFICATION, AND THE TOXICITY OF SOLUBLE
    SALTS OF ALUMINUM. Ind. Agr. Exp. Sta. Bul. 170, p. 327–374, 22 fig.
(2) AMES, J. W., and SCHOLLENBERGER, C. J.
    1916. LIMING AND LIME REQUIREMENT OF SOIL. Ohio Agr. Exp. Sta. Bul. 306,
    P. 281–396, 5 fig.
(3) CONNOR, S. D.
    1916. ACID SOILS AND THE EFFECT OF ACID PHOSPHATE AND OTHER FERTILIZERS
    UPON THEM. *In* Jour. Indus. and Engin. Chem., v. 8, no. 1, p. 35–40,
    2 fig.
(4) DAIKUHARA, G.
    1914 UEBER SAURE MINERALBÖDEN. *In* Japan Imp. Cent. Agr. Exp. Sta.
    Bul., v 2, no. 1, p. 1–40, 1 pl.
(5) FREAR, William.
    1915. SOUR SOILS AND LIMING. Pa. Dept. Agr. Bul. 261, 221 p.
(6) HARRIS, J. E.
    1914. SOIL ACIDITY. Mich. Agr. Exp. Sta. Tech. Bul. 19, p. 524–536.
(7) RICE, F. E.
    1916. STUDIES ON SOILS. I. *In* Jour. Phys. Chem., v. 20, no. 3, p. 214–227,
    1 fig.
(8) TRUOG, E.
    1916. THE CAUSE AND NATURE OF SOIL ACIDITY WITH SPECIAL REGARD TO COL-
    LOIDS AND ADSORPTION. *In* Jour. Phys. Chem., v. 20, no. 6, p. 457–484.
(9) VEITCH, F. P.
    1904. COMPARISON OF METHODS FOR THE ESTIMATION OF SOIL ACIDITY. *In* Jour.
    Amer. Chem. Soc., v. 26, no. 6, p. 637–662.

STRUCTURE OF THE POD AND THE SEED OF THE GEORGIA VELVET BEAN, STIZOLOBIUM DEERINGIANUM

By Charles V. Piper, *Agrostologist in Charge, Forage-Crop Investigations*, and J. Marion Shull, *Botanical Artist, Fruit Disease Investigations, Bureau of Plant Industry, United States Department of Agriculture*

INTRODUCTION

Since 1915 the culture of the velvet bean in the South has been very greatly increased, mainly owing to the introduction and development of short-season varieties. The early sorts are the Yokohama bean (*Stizolobium hassjoo* Piper and Tracy); the Chinese velvet bean (*S. niveum* (Roxb.) Kuntze var.); the Georgia velvet bean, an early variety of the old Florida velvet bean (*S. deeringianum* Bort); and also several varieties of hybrid origin. The Georgia velvet bean is at present the most largely grown. In 1916 the crop of beans reached such large proportions that numerous mills found it profitable to manufacture velvet-bean meal, prepared by grinding the dry ripe pods and seeds. This product has met with high favor as a rich protein feed. The acreage of velvet beans planted in 1917 is several times larger than in 1916.

No previous studies seem to have been made of the microscopic structure of the velvet bean, or, indeed, of any species of Stizolobium. The structure of other leguminous seeds has been described and illustrated by various botanists, particularly Kondo.[1] Inasmuch as velvet-bean meal promises to become an important commercial feed, the microscopic structures of the pod and seed are important as a basis of identifying the meal, either pure, adulterated, or in mixtures.[2]

STRUCTURE OF THE POD

HAIRS (Pl. 5⁰, G).—The hairs on the surface of the pod consist of a long terminal cell, empty or apparently so, and one to six very short basal cells filled with protoplasm. The hairs vary greatly in length,

---

[1] Kondo, M. Der Anatomische Bau einiger ausländischer Hülsenfrüchte, die jetzt in den Handel kommen. *In* Ztschr. Untersuch. Nahr. u. Genussmtl., Bd. 25, Heft 1, p. 1–56, 40 fig. 1913.
[2] The microscopic preparations of the material were made mostly by Dr Albert Mann, of the Bureau of Plant Industry, and the drawings are all by the junior writer.

Journal of Agricultural Research,
Washington, D. C.

Vol. XI, No. 13
Dec. 24, 1917
lg   Key No. G.—130

(673)

from 300 to 1,500 μ. The surface of the hairs is roughened with minute tubercles, increasingly so toward the apex. The smaller hairs are color-less, and on account of a relatively thicker wall remain cylindric, while the larger ones collapse and become blackish and twisted.

CARPELLARY WALL (Pl. 50, C, G).—The wall of the pod consists of an outer layer of epidermal cells, polyhedral in form, 15 to 30 μ thick, approximately cubical, and 15 to 30 μ in diameter. They are filled with dense protoplasm. The numerous basal cells of the hairs are conspicuous on account of the black base of the hairs. Stomata of typical form are about one-half as numerous as the hair bases.

Beneath the epidermis lie about five layers of peculiar sclerenchyma cells (Pl. 50, G, H, I, J) from 50 to 234 μ in length. The cells are spindle-shaped, and two to six times as long as broad, the lumen broadening at each end.

The inner half of the pod consists of small-celled parenchyma and numerous longitudinal bundles of thick-walled fibers (Pl. 50, D, E, F, G), 300 to 500 μ long. The lumen of these fibers is small, and in section numerous lateral extensions are apparent.

The innermost layer of the pod is satinlike and consists of thin-walled parenchyma (Pl. 50, A).

The fibrovascular bundles which traverse the pod require no particular mention. Their structure in cross section is illustrated in Plate 50, B.

STRUCTURE OF THE SEED

The most characteristic external feature of the seed of the velvet bean is the thick, white caruncle which surrounds the hilum (Pl. 51, A). In cross section (Pl. 51, B) this is seen to be composed of thick-walled parenchyma.

TESTA.—In cross section (Pl. 51, B, E) the testa is seen to be composed of an outer layer of long palisade cells, a second layer of the so-called "hour-glass" cells, the central portion of nearly empty parenchyma, and, between this and the small-celled inner epidermis, several layers of paren-chyma filled with protein.

The palisade cells (Pl. 51, C, D) are thick-walled cylinders, 100 to 180 μ long and 10 to 22 μ in diameter. In the neighborhood of the hilum the palisade cells are longest and broken near the middle by the so-called "light line," owing to the fact that these cells have the lumen much enlarged near the middle

The hour-glass cells (Pl. 51, F, G) are 50 to 100 μ long. They closely resemble those of related genera (Phaseolus, Vigna, Soja).

The cotyledons (Pl. 51, I, J) consist mainly of an outer epidermis of polyhedral cells about 10 μ in diameter; an inner epidermis of cells more than twice as large; and between, the much larger cells of the body of the cotyledons. All the cells are filled with grains of starch and proteids. The starch grains show no special peculiarities.

CHARACTERISTIC ELEMENTS

In velvet-bean meal the most abundant recognizable elements are: (1) The palisade cells of the testa; (2) the sclerenchyma cells of the pod; (3) the hairs of the pod, and (4) an occasional hour-glass cell. Most of the fibers are broken, but with careful examination nearly all of the structures may be found.

The most important elements to determine that the material is composed of velvet beans are the sclerenchyma cells and the hairs of the

PLATE 50

A.—Thin-walled parenchyma forming the satin tissue that lines the pod. ×80.

B.—Cross section of a vascular bundle of the pod   ×420.

C —Section of epidermis of pod, showing a stoma and several of the basal cells from which the hairs arise   ×80.

D —Fiber cells   ×80.

E —Fiber cell in longitudinal section. ×420.

F —Fiber cells in cross section. ×420

G —Cross section of velvet-bean pod, showing both conditions of surface hairs arising from the epidermal layer   Both types of hairs are of similar construction, being unicellular except at base, which consists variously of from one to five or six cells. The hairs are granulate or roughened, increasingly so toward the apex.   The longer, weaker-walled hairs have collapsed and become twisted or otherwise deformed and darkened in color, while the shorter, heavier-walled hairs remain stiff, tubular bristles, straight or sickle-shaped, and transparent.   Underlying the epidermis is the layer of sclerenchyma tissue several cells in depth, and below this appear fiber bundles in cross section.   ×80.

H —Sclerenchyma cells   ×80. These cells vary greatly in size, ranging from 50 to 234 μ in length.

I —Sclerenchyma cells   ×420

J.—Sclerenchyma cells in cross section   ×420.

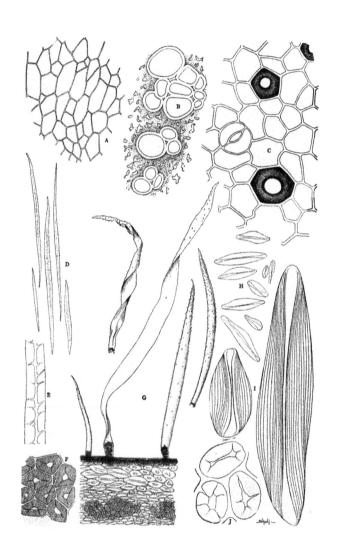

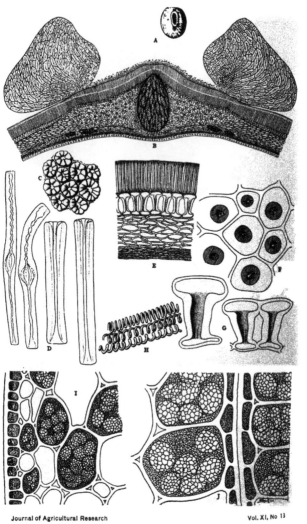

PLATE 51

A.—Georgia velvet bean. Natural size

B.—Cross section through hilum showing thick-walled parenchyma of the caruncle and the very short palisade cells of the hilar tissues. Beneath these are the palisade cells of the seed coat with the "light line" sinking to near the middle of the cells under the hilum. To right and left of the large vascular bundle consisting of scalariform ducts are the lesser spiral vascular bundles of the inner seed coat. ×27

C.—Palisade cells of seed coat, top view. ×420.

D.—Palisade cells of seed coat. The two cells on the left are from that part of the seed coat underlying the hilum and show the modification of the lumen that results in the so-called "light line." Palisade cells of the seed coat are 100 to 180 μ long and 10 to 22 μ wide. ×420.

E —Cross section through seed coat elsewhere than at the hilum showing relationship of palisade cells, hour-glass cells, and the underlying parenchyma. ×80.

F.—Hour-glass cells seen from above. ×420.

G.—Hour-glass cells, side view. ×420.

H.—Spirals from vascular bundles of the seed coat. ×420.

I —Cross section of cotyledon, showing outer epidermis, and smaller cells filled with starch grains and proteid granules. ×420.

J —Cross section of cotyledons, showing the medial line of the two cotyledons with inner epidermis and adjoining much larger cells with starch grains and proteid granules correspondingly larger. ×420.

DECOMPOSITION OF GREEN AND STABLE MANURES IN SOIL

By R. S. Potter, *Assistant Chief in Soil Chemistry,* and R. S. Snyder, *Assistant in Soil Chemistry, Iowa State College Experiment Station*

HISTORICAL REVIEW

This paper is the third of a series of reports whose principal object is the determination of the rate of decomposition of original or added organic material in soil. This is done by the measurements at frequent intervals of the carbon dioxid evolved from manured and unmanured soils. In the first of these papers (7)[1] the effects of lime, ammonium sulphate, and sodium nitrate on carbon-dioxid production were determined. In the second paper (8) several phases of the subject were taken up, the more important of which are: (1) The relation of the amount of air drawn over soils to the amount of carbon dioxid evolved, (2) a determination of the percentage of carbon dioxid in plots treated with varying amounts and kinds of organic matter, (3) a comparison of the results obtained by the laboratory method for the determination of carbon dioxid evolved from soils to the results of the determination of the percentage of carbon dioxid in the atmosphere of soils in the field, and (4) determinations of the amount of carbon dioxid evolved from soils treated with lime and varying amounts of stable manure. The conclusions were briefly as follows: Within limits there was not much variation in the amount of carbon dioxid evolved from soil with different amounts of air passed over the soil. The results obtained by the laboratory method agreed well with those obtained in the field plots. Calcium carbonate accelerated the rate of decomposition of both the original organic matter of soil and that added in amounts of stable manure varying from 10 to 50 tons per acre. There was less calcium carbonate decomposed in those soils receiving applications of manure than in those unmanured.

A complete review of the work done in American Experiment Stations on green manuring published quite recently (6) calls attention to the fact that, while there are very little definite data concerning the effects of green manuring upon succeeding crop yields, yet the conclusion to be drawn from all the work done upon the subject is that in general green manuring increases crop yields. A common recommendation to farmers is to plow in green manure to build up the organic matter in the soil, but there is no definite information available as to how rapidly one may expect such a practice to build up the soil or as to how long, for instance, such

---

[1] Reference is made by number (italic) to "Literature cited", pp 697-698

Journal of Agricultural Research, Washington, D C. li

Vol. XI, No. 13
Dec 24, 1917
Key No. Iowa—5

an application will last. It was to throw some light on this phase of the green manuring problem that the work reported herein was undertaken.

In one of the papers referred to above (8) a quite complete review of the literature on carbon-dioxid production in soils is given, so that here mention will be made of only those papers which have appeared since the preparation of that bibliography.

J. G. Lipman and associates (4) have at various times reported the results of a field experiment on the effect of a light application of stable manure on the decomposition of green manures. The essential points of the plan of their experiment are as follows: Cow manure in amounts varying from 1,000 to 4,000 pounds per acre is added to plots upon which in one series a nonlegume is growing and on the other series, a legume. The cow manure is plowed under with the accompanying green manure. This plan has been followed for several years. Crop yields and the amount of nitrogen in the crop are noted. From the fact that crop and nitrogen yields have been somewhat larger on those plots receiving the cow manure plus the green manure than on the ones receiving the green manure, they have concluded that the cow manure enhanced the rate of decomposition of the green manure. Also, those plots receiving the larger amounts of cow manure did not always give greater crop yields than those receiving the smaller. This, too, is interpreted as being indicative of an increase of the rate of decomposition of the green manure by the cow manure.

In 1916 Boltz (1) published the results of an interesting experiment with clover used as a green manure. Green clover at the rate of 7,744 pounds per acre was applied to two small areas, in one case merely put on the surface and in another spaded under. The clover and soil were analyzed for organic matter at the beginning of the experiment and the surface application of clover was collected at the end of the experiment and the amount of organic matter remaining was determined. The soil and soil and clover mixture were analyzed at the end of the experiment also. Throughout the course of the experiment, which ran from October 12 to May 5, 206 days, 66 per cent of the clover applied to the surface was lost, while 28.45 per cent of it was lost when it was incorporated with the soil. In a similar experiment, but carried out in a lysimeter and involving treatments of clover of 17,520 pounds per acre, 68.5 per cent of the organic matter of the clover was lost when applied to the surface and 58.5 per cent when spaded in with the soil. Only about 1 per cent of the organic matter of the clover was lost by drainage. The results obtained by Boltz are interesting when compared with those reported in this paper.

EXPERIMENT I

Part of the large sample of Miami silt loam soil that was used in the former experiments of this series (7, 8) was used. Practically the same apparatus was used as formerly, the only difference being that, instead of putting the soil in pots under bell jars, it was placed in the ordinary

2.5-liter acid bottles. This diminished chances for leaks. The treatment of the soils, in tons per acre, was as follows:

Pot No.	Treatment.
1, 2......	Control.
3, 4....	3 tons of calcium carbonate.
5, 6...	10 tons of manure.
7, 8.	10 tons of manure and 3 tons of calcium carbonate
9, 10.....	1 ton of oats.
11, 12.	1 ton of oats and 3 tons of calcium carbonate.
13, 14..	1 ton of oats and 10 tons of manure.
15, 16. ..	1 ton of oats, 10 tons of manure, and 3 tons of calcium carbonate.
17, 18	1 ton of clover.
19, 20	1 ton of clover and 3 tons of calcium carbonate.
21, 22	. 1 ton of clover and 10 tons of manure
23, 24.	. 1 ton of clover, 10 tons of manure, and 3 tons of calcium carbonate.

The manure used was well-rotted horse manure, air-dried and coarsely ground. The oats and clover were about two-thirds mature tops, air-dried and ground rather fine. The weights as given above refer to air-dried weights in all cases. Two and one-half pounds of soil were mixed with their respective treatments and then placed in the acid bottles. These were then connected with the apparatus which was freed from carbon dioxid. The soils were all made up to 22 per cent of moisture. This was done by adding 250 c. c. of "water" to each bottle. In every case 100 c. c. of this "water" were made up of an emulsion made from fresh soil; and, in the case of those soils receiving stable manure, 100 c. c. of fresh-manure emulsion were included in the 250 c. c. of "water." The soils, manures, and calcium carbonate were analyzed for total carbon at the beginning of the experiment. The experiment ran from July 10 to September 2, 1915, a total of 53 days.

At the end of the experiment all the soils were analyzed for carbonate carbon. Throughout the course of the experiment the carbon dioxid evolved was determined from time to time. In Table I will be found the total amount of carbon evolved from the soil as carbon dioxid, calculated in pounds per acre of 2,000,000 pounds. It will be observed that in general the duplicates agree very well. It may be stated that for each period the duplicates checked on the average about as closely as the summed-up amounts indicate. To put all the determinations which were made into a table would take up more space than we care to use here.

TABLE I.—*Amount of carbon evolved as carbon dioxid from soils in Experiment I*

[Results expressed as pounds of carbon per acre]

Soil No.	Original.	Duplicate.	Soil No.	Original.	Duplicate.
1, 2... ..	771. 9	767. 7	13, 14	1, 490. 1	1, 477. 8
3, 4....	1, 392. 5	1, 393. 3	15, 16..............	2, 153. 0	2, 123. 3
5, 6..... .	1, 070. 1	1, 071. 8	17, 18	1, 309. 6	1, 304. 2
7, 8	1, 625. 5	1, 625. 5	19, 20	1, 883. 1	1, 909. 8
9, 10.	1, 153. 8	1, 223. 0	21, 22..............	1, 567. 7	1, 551. 8
11, 12	1, 794. 1	1, 802. 3	23, 24............ ...	2, 188. 0	2, 204. 9

In Table II are found the number of pounds of carbon per acre given per day for the various periods. These were taken from the total carbon dioxid given off in the various periods. The data appearing in this table were used in plotting the curves of figures 1, 2, and 3.

TABLE II.—*Amount of carbon evolved as carbon dioxid per day from soils in Experiment I*

[Results expressed as pounds of carbon per acre per day]

Soil No.	Day.												
	1	2	3-4	5-8	9-11	12-15	16-18	19-22	23-25	26-32	33-39	40-46	47-53
1,2.....	57.0	51.0	33.3	18.3	12.3	18.4	18.7	12.5	7.0	9.8	a13.6	13.1	10.6
3,4.....	240.0	128.0	54.0	35.2	22.3	20.3	20.7	18.5	12.1	14.4	a21.0	13.1	12.5
5,6.....	81.2	78.0	47.6	25.0	17.2	17.7	18.8	16.4	10.4	13.1	a19.4	14.0	16.4
7,8.....	263.0	134.0	67.5	30.5	25.4	24.7	25.5	25.0	13.7	17.0	a24.2	a17.0	19.0
9,10....	81.8	146.0	73.5	34.0	22.5	20.5	19.7	16.2	9.3	11.9	b13.0	14.1	14.6
11,12...	283.0	196.0	97.5	49.5	32.9	29.5	27.8	23.9	14.0	16.7	b21.7	16.8	15.1
13,14...	136.0	156.0	86.0	42.5	28.6	26.5	25.5	20.7	11.7	14.8	a21.2	15.4	14.4
15,16...	317.8	196.0	121.0	57.2	38.7	35.2	34.0	29.7	17.7	20.8	a26.7	a14.0	19.7
17,18...	6710	215.0	92.5	34.7	21.0	19.7	19.3	16.0	9.1	11.9	a20.4	13.5	12.4
19,20 .	800.0	307.0	119.0	40.5	31.4	28.5	28.1	23.6	14.2	16.8	a23.2	27.8	16.9
21,22..	118 a	233.0	100.0	40.7	25.8	24.7	24.8	19.8	11.8	15.1	a22.4	15.0	15.8
23,24..	273.0	284.0	126.0	55.5	36.7	34.7	33.3	30.1	17.9	21.6	a29.0	20.7	20.3

aRepresents values for one soil, one determination having been lost.
bRepresents values taken from average of preceding and succeeding determinations, both duplicates having been lost.

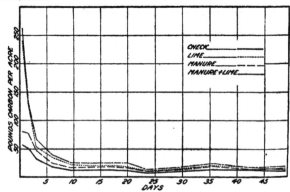

FIG. 1 —Graphs of the quantity of carbon evolved as carbon dioxid in the control, limed, manured, and manured and limed soils.

DISCUSSION OF THE CURVES

In figure 1 are given the curves for the carbon lost from the control, the limed, the manured, and the manured and limed soils. Throughout the course of the experiment it is seen that the limed soil is losing more carbon than the corresponding unlimed. It can not be said with certainty whether this is really due to an enhanced rate of decomposition of the organic matter or whether it is simply due to a gradual liberation of the carbon dioxid of the calcium carbonate by relatively insoluble acids or by acids being gradually formed from organic matter.

Figure 2 shows the curves for the soils treated with oats alone and in the combinations with manure and lime. The noteworthy thing here is the slow initial rate of decomposition of the green manure and then its rapid decomposition followed by a slowing up of the process. Anticipating the results of the experiment which follows, we may say that probably the slow action at first was due to the dry condition of the oats. The slowing up of the process is due, no doubt, simply to the relative exhaustion of the more easily decomposed material in the added green manure.

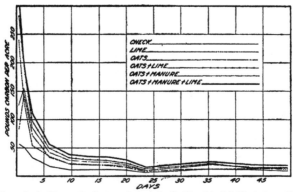

FIG. 2.—Graphs of the quantity of carbon evolved as carbon dioxid in soils treated with oats alone and in combination with manure and lime.

The curves for the green clover in its various combinations are shown in figure 3. In general the same may be said of these curves as of those for the oats. However, the clover decomposed much more rapidly than the oats in the early periods.

In Table III the summarized data for the whole experiment are given. An explanation of some of the columns of the table besides that appearing at the column head will probably make the table clearer. Unless otherwise indicated, all results are expressed in terms of pounds of carbon per acre of 2,000,000 pounds of soil. The figures in the second column were obtained from the analyses of the respective manure for total carbon. The values in the third column were found from the pounds of carbon in the soil originally and the added carbon. A sample of the air-dried calcium carbonate used was analyzed for carbon dioxid. From the amount obtained it was computed that the calcium carbonate was 94.42

per cent pure. Since practically no free alkali was found to be present, no doubt hygroscopic moisture made up the 5.58 per cent. Therefore the amount of carbon in the 3 tons of lime used per acre was 680 pounds, which number appears in the fourth column of the table. The figures in the next column were obtained directly from Table I. In the column headed "Inorganic carbon at end" are given the values obtained from the analysis of the soil for inorganic carbon by the MacIntire (5) method. These soils were analyzed while still moist, immediately after removal from the apparatus. No doubt the reason for finding inorganic carbon in the unlimed soil was that some carbon dioxid was retained in the soil moisture. No carbonates were found in the original air-dried soil. As is observed, only the averages of the duplicates are given. The greatest

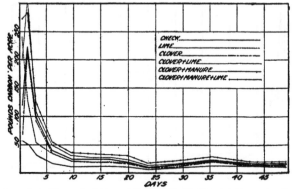

Fig. 3—Graphs of the quantity of carbon evolved as carbon dioxid in soils treated with green clover alone and in combination with manure and lime.

variation of the averages for the inorganic carbon in the unlimed soils was 2.5 pounds, while the greatest variation of the corresponding values for the limed soils was 5 pounds. This is thought to be a satisfactorily low experimental error. The weight of "Organic carbon lost" in the case of the unlimed soils were obtained simply by adding the total carbon lost to the inorganic found at the end of the experiment. For the limed soils, the following method was followed: The amount of inorganic carbon at the end of the experiment was subtracted from the amount added at the beginning. This, which is, of course, the number of pounds of inorganic carbon lost, was subtracted from the number of pounds of total carbon lost. This gives the desired figures listed in column 7. The next column is self-explanatory. In order to obtain the values found in column 9, it was assumed that in every case where organic matter was added to the soil the amount of carbon lost from the original organic matter of the soil was the same as where no organic matter was added. Hence, the value found for soils number 5 and 6 in column 9 was obtained by subtracting the value for 1 and 2 from 5 and 6, in column 7. To obtain the number for 7 and 8 in column 9, the value for soils 3 and 4 was subtracted from 7 and 8 in column 7. Columns 10 and 11 are self-explanatory. To obtain the values found in column 12, it was assumed that where stable and green manure were used in the soil in combination the stable manure and soil organic matter lost as much carbon as where the stable manure alone was added to the soil.

TABLE III.—*Summarised data of Experiment I*

[Unless otherwise given, results are expressed as pounds of carbon per acre]

Soil No.	Organic carbon added.	Total organic carbon at start.	Inorganic carbon added.	Total carbon lost.	Inorganic carbon at end.	Organic carbon lost.	Excess organic carbon lost from limed over unlimed soil.	Organic carbon from added manure.	Carbon lost from added manure.	Total organic carbon lost.	Organic carbon from green manure alone.	Quantity of calcium carbonate at end in limed soils.
1	2	3	4	5	6	7	8	9	10	11	12	13
									Per ct.	Per ct.		Lbs.
1, 2	None.	29,513	......	770	27	797	......	......		2.70	......	......
3, 4	None.	29,513	680	1,393	228	931	134	......		3.15	......	1,818
5, 6	7,500	37,013	......	1,075	44	1,115	......	318	4.23	3.02	......	......
7, 8	7,500	37,013	680	1,625	271	1,216	101	285	3.80	3.28	......	2,160
9, 10	861	30,374	......	1,188	31	1,219	......	422	49.00	4.01	......	......
11, 12	861	30,374	680	1,798	270	1,388	169	457	53.10	4.57	......	2,250
13, 14	8,361	37,874	......	1,484	71	1,555	......	758	9.07	4.10	440	......
15, 16	8,361	37,874	680	2,138	283	1,741	186	810	9.70	5.10	525	2,360
17, 18	898	30,411	......	1,306	36	1,342	......	545	60.80	4.41	......	......
19, 20	898	30,411	680	1,896	274	1,490	148	559	62.30	4.89	......	2,280
21, 22	8,398	37,911	......	1,560	69	1,629	......	832	9.91	4.29	515	......
23, 24	8,398	37,911	680	2,196	310	1,826	197	895	10.70	4.82	610	2,580

DISCUSSION OF RESULTS

The number of pounds of organic carbon at the end of the experiment (shown in column 6 of Table III) are worthy of study. These results will be considered with those for the number of pounds of residual calcium carbonate. It is observed that in every case the organic manures acted as conservers of lime. As to why that is, it is difficult to understand. One would think that the acids formed by the decomposition of the manure would decompose the lime. Perhaps this takes place in the early stages, and then these calcium salts of the acids are in turn acted on by the microorganisms with the re-formation of calcium carbonate. In this connection the work of Gimingham (2) may be pertinent. He stated that the soil organisms changed the calcium salts of such organic acids as oxalic found in plant tissues to calcium carbonate. This can hardly account for all of the excess carbonate in the green-manured soils, as the following considerations will show. According to data appearing in Hopkins's Soil Fertility and Permanent Agriculture (3, p. 603) 1 ton of young clover hay contains 34.2 pounds of calcium. If all of this were converted to calcium carbonate, it would give 85.5 pounds. The excess lime in the clover- and lime-treated soils over the soil receiving no treatment except lime is 460 pounds. Oat straw, according to the same authority, contains 6 pounds of calcium per ton; or in terms carbonate, 15 pounds. This accounts for only a small part of the excess carbonate in the oat-treated soils.

In a consideration of this question all the potential bases and acids in the manures should be considered, for after fairly complete decomposition the potassium, for instance, however originally combined with the organic material, would tend to combine either with an acid in the soil or with an acid from the organic material itself. This would relieve an equivalent amount of calcium carbonate from the necessity of having to neutralize acid.

Bearing directly upon this are the data found in Table IV, which is adapted directly from Hopkins (*3, p. 603*).

TABLE IV.—*Quantity (in pounds) of the elements in 1 ton of the clover hay and oat straw (dry material)*

Material.	Phos-phorus.	Potas-sium.	Magne-sium.	Cal-cium.	Sul-phur.	Sodi-um.	Chlo-ris.
Oats................................	2.4	27.0	2.8	6.0	1.2	3.0	5.4
Calcium carbonate equivalent.......	7.7	34.6	11.7	15.0	3.7	6.5	7.6
Clover..............................	9.0	50.6	9.4	34.2	1.4	2.8	6.6
Calcium carbonate equivalent......	29.0	64.8	39.1	85.5	4.4	6.1	9.3

In Table IV are given the pounds of the elements in clover hay and oat straw found in 1 ton. It should be noted that the values are for oat straw and not the immature oat plant which was used in this experiment. No data could be found for such material. In the table the calcium carbonate equivalent of the various elements appear. These were calculated on the supposition that the elements would be completely hydrolyzed or oxidized to their respective bases or acids and would then neutralize an equivalent of acid, in the case of the base, or an equivalent of base, probably lime, in the case of the acids. For the oats the excess lime equivalent of the bases over the acids is 48.1 pounds, and for the clover 152.8 pounds. It should be noted that these numbers include in them the calcium carbonate equivalent of the calcium as mentioned above. This, then, does not account for all of the excess lime remaining in the green-manured soils, but it does, at least with the clover, account for an appreciable percentage of it. Therefore this excess lime must come from the calcium silicates, but this does not explain why there is a greater amount left in the manured than in the unmanured soils.

Next to be considered are the values found in columns 7 and 8 of Table III. It is seen that in every case the lime causes an increased decomposition of organic matter. This of course has both a beneficial effect and a harmful effect from a practical standpoint. The greater rate of decomposition of the manure causes an increase in the rate of the making available of the plant food. It no doubt causes an increase in the rate of the destruction of possible toxic substances and probably in itself increases the air circulation in the soil. The harmful effect is the depletion of the organic matter in the soil, but since the main benefit derived from the organic matter of soil is its enhancement of microbial activity, anything within limits which aids in this activity is desirable.

Columns 9 and 10 (Table III) give the number of pounds of carbon lost from the added manure and the percentage loss, respectively. The most noteworthy thing in this connection is the large percentage of the green manure lost as compared with the stable manure. Thus, it is seen that 49 per cent of the carbon of the oats are lost, 60.8 per cent of that of the clover, and only 4.23 per cent of that of the manure. It seems surprising that about half of such an application of green manure would be lost in less than two months. It is believed that probably a somewhat comparable loss would take place in the field, because except for a lack of drainage these soils were held under conditions not greatly unlike those in the field. As to why, in the case of the stable manure in soils, a somewhat smaller amount of the manurial carbon was lost from the limed than the unlimed soils, we can not state. No doubt our assumption that the amount of carbon lost from the original soil organic matter in the manured soils is the same as that from the soil organic matter in the unmanured soils is not strictly true. As the manurial organic matter is more easily available, it is, no doubt, attacked largely at the expense of the soil organic matter. On the other hand, an enormous number, and doubtless a very great variety, of bacteria are introduced with the manure which of themselves serve to enhance the rate of decomposition of the organic matter and, hence, make a comparison of the manured, limed, and untreated soils of questionable significance. However, to arrive at some conclusion as to the amount of manure consumed in an experiment like this, some such an assumption must be taken. From column 11 (Table III) it is seen that in every case where total organic matter is considered, a larger percentage is decomposed from the limed than from the respective unlimed soil.

In column 12 (Table III) it is observed that the addition of stable manure has increased the rate of decomposition of the green manures to a small extent. According to some agriculturists, a dressing of stable manure on green manure before plowing under serves to enhance the rate of decomposition of the latter because of added bacteria. This experiment seems to confirm that belief. In anticipation of the consideration of the next experiment, it may be stated here that this result was not entirely confirmed when conditions were kept more typical of field

CONCLUSIONS FROM EXPERIMENT I

(1) Lime in the form of a carbonate under the conditions of this experiment appreciably enhances the rate of decomposition of both original soil organic matter and the organic matter of stable manure and the green manures oats and clover when added to soil. Two of the more important results of this are the increased availability of plant food and the more rapid depletion of the soil organic matter. This latter effect would be partially and perhaps entirely offset by the fact that with lime larger crops could be grown which would give more organic matter to return to the soil.

(2) The green manures oats and clover under the conditions of this experiment are decomposed much more completely than stable manure. Clover is decomposed somewhat more rapidly than oats.

(3) Stable manure increases the rate of decomposition of green manure when used in connection with the latter.

(4) Both stable and green manures act as conservers of lime.

EXPERIMENT II

In the experiment just described it will be remembered that the stable and green manures were applied in dry and ground condition. It was thought that perhaps the reason that such a large percentage of oats and clover was decomposed and their carbon given off as carbon dioxid was because of their finely divided condition. Of course the reason for adding manures in such a state was to make sure that a composite sample was added in each case. To test this and other points, a similar experiment was planned, but with some of these objectionable features eliminated The treatment of the soils, in tons per acre, is given in Table V.

TABLE V.—*Treatment of soils in Experiment II*

Pot No.	Air-dry basis.	Fresh basis.
1, 2......	Control.........................	Control.
3, 4 .	3 tons of lime	3 tons of lime.
5, 6 .	4 77 tons of stable manure	10 tons of stable manure.
7, 8	14 3 tons of stable manure........	30 tons of stable manure.
9, 10..	23 8 tons of stable manure.........	50 tons of stable manure.
11, 12 . .	4 77 tons of stable manure, 3 tons of lime.	10 tons of stable manure, 3 tons of lime.
13, 14...	14 3 tons of stable manure, 3 tons of lime	30 tons of stable manure, 3 tons of lime.
15, 16...	23.8 tons of stable manure, 3 tons of lime.	50 tons of stable manure, 3 tons of lime.
17, 18...	4.77 tons of stable manure, 1.29 tons of oats.	10 tons of stable manure, 4 tons of oats.
19, 20.	4.77 tons of stable manure, 1.29 tons of oats, 3 tons of lime.	10 tons of stable manure, 4 tons of oats, 3 tons of lime.
21, 22...	4.77 tons of stable manure, 1.18 tons of clover.	10 tons of stable manure, 4 tons of clover.
23, 24...	4 77 tons of stable manure, 1.18 tons of clover, 3 tons of lime.	10 tons of stable manure, 4 tons of clover, 3 tons of lime.
25, 26...	1.29 tons of oats...................	4 tons of oats.
27, 28...	1 29 tons of oats, 3 tons of lime.....	4 tons of oats, 3 tons of lime
29, 30 .	1 18 tons of clover......	4 tons of clover.
31, 32.. .	1 18 tons of clover, 3 tons of lime.....	4 tons of clover, 3 tons of lime.

The stable manure used was fresh horse manure, not dried. The oats and clover were taken from samples grown in the greenhouse to about two-thirds maturity. Material was cut into half-inch lengths and weighed out while still in an unwilted condition and thus added to the soil immediately. At the same time that the samples of manure were added to their respective soils, several samples of the green and stable manures were weighed into suitable containers and dried in the oven at 70° C. The weights were again taken, and from this the percentage of moisture in the samples as added was computed. The dried samples were finely ground and carbon determinations made on some of them. The results are given in Table VI.

TABLE VI —*Composition of the manures used in Experiment II*

Material.	No.	Weight taken.	Dry weight.	Average.	Quantity of carbon in 0.5 gm.	Average.	Percentage of carbon.	Average.
		Gm.	Gm.		Gm.			
Oats...............	1	4·5	1.52	....	0.2152	....	43.04	......
Do.............	2	4·5	1.39	.	.2110	....	42.20	......
Do.............	3	4·5	1.36	.	.2329	....	46.58	..
Do :	4	4·5	1.45	......	.2098	0.2172	41.96	43.45
Do.............	5	4·5	1.49	....	....	....	....	......
Do.............	6	4·5	1.51	1.454	....	....	....	......
Clover.............	1	4·5	1.32	....	.2281	....	45.62	......
Do.............	2	4·5	1.32	....	.2346	....	46.92	......
Do.............	3	4·5	1.30	.	.2184	....	43.68	......
Do.............	4	4·5	1.31	......	2243	2263	44.86	45.26
Do.............	5	4·5	1.31	....	....	....	....	......
Do.	6	4·5	1.37	1.322	....	....	....	......
Stable manure.......	1	11.3	5.3	....	.1930	....	38.60	......
Do.............	2	11.3	5.6	....	.1952	....	39.04	.....
Do.............	3	11.3	5 6	....	.1936	1939	38.72	38.7
Do.............	4	11.3	5.35	5.39	....	....	....	.....

An examination of the results in Table VI shows with what degree of uniformity in composition the samples of manures were as they were added to the soil. The experiment was started and carried through in all respects as the previous experiment.

In Table VII will be found the results for the total amount of carbon given off as carbon dioxid. Owing to breakage, the alkali from the alkali towers of pots 17, 19, and 30, ran over into their respective soils and spoiled them early in the experiment. Large molds appeared in pots 13 and 15 after several weeks, which accounts for the missing determination of those soils. A mold also appeared in No. 16, which was removed as soon as discovered. Since its duplicate had already been discarded, the values for the carbon dioxid of No. 16 were taken from the values found for the soils treated with the same amount of manure but without lime. It is of course realized that this does not have much justification, and, hence, the values for No. 16 should not be depended upon for any definite conclusion.

TABLE VII.—*Total carbon evolved as carbon dioxid from soils in Experiment II*

[Results expressed as pounds of carbon per acre]

Soil No.	Original.	Duplicate.	Soil No.	Original.	Duplicate.
1, 2	1,686	1,722	18	4,239	
3, 4	2,710	2,744	20	5,082	
5, 6	3,658	3,636	21, 22	4,195	4,242
7, 8	8,211	8,206	23, 24	5,077	4,969
9, 10	12,483	12,050	25, 26	2,657	2,704
11, 12	4,289	4,357	27, 28	3,215	3,407
14	8,769		29	2,688	
16	12,703		31, 32	3,245	3,274

In Table VIII appear the average amounts of carbon evolved as carbon dioxid per day in the successive periods. From this table the curves appearing in figures 4 to 9 were drawn.

TABLE VIII.—*Quantity of carbon evolved as carbon dioxid per acre per day in soils of Experiment II*

[Results expressed as pounds per acre]

Soil No.	1st day.	2d day.	3d day.	4th day.	5th day.	6th day.	7th day.	8th day.	9th day.	10th day.	11–12 day.	13–14 day.	15–16 days.	17–19 days.	20–22 days.
1, 2	78.5	87.7	46.4	58.1	43.3	34.1	29.9	30.4	26.7	25.0	25.0	15.3	15.9	20.0	14.6
3, 4	277.0	146.0	73.4	90.6	81.8	57.3	50.7	50.7	43.1	38.4	37.0	26.4	21.4	29.0	22.7
5, 6	120.0	137.0	91.7	160.0	139.0	120.0	126.0	113.0	104.0	86.7	88.5	56.5	42.2	49.7	40.0
7, 8	221.0	210.0	202.0	307.0	294.0	247.0	232.0	208.0	200.0	187.0	178.0	120.0	116.0	117.0	95.0
9, 10	328.0	312.0	303.0	468.0	412.0	338.0	305.0	289.0	260.0	262.0	239.0	169.0	137.0	177.0	121.0
11, 12	336.0	256.0	135.0	178.0	175.0	149.0	147.0	122.0	119.0	107.0	97.0	65.9	54.0	53.0	44.0
13, 14	445.0	256.0	256.0	408.0	383.0	266.0	232.0	265.0	208.0	186.0	193.0	126.0	103.0	119.0	90.3
15, 16	476.0	345.0	335.0	562.0	433.0	375.0	374.0	338.0	292.0	272.0	237.0	164.0	173.0	152.0	136.0
18	216.0	197.0	181.0	232.0	203.0	149.0	143.0	137.0	134.0	114.0	101.0	64.0	57.5	58.0	42.3
20	374.0	216.0	197.0	264.0	206.0	204.0	180.0	145.0	148.0	141.0	122.0	100.0	65.5	63.0	47.7
21, 22	192.0	174.0	191.0	266.0	186.0	150.0	149.0	158.0	160.0	120.0	103.0	63.5	52.5	54.0	38.7
23, 24	373.0	202.0	202.0	288.0	275.0	184.0	180.0	195.0	141.0	120.0	113.0	71.0	58.0	59.3	43.1
25, 26	156.0	157.0	125.0	136.0	108.0	72.2	69.2	65.1	51.7	44.6	45.3	30.3	28.3	34.7	32.7
27, 28	325.0	289.0	152.0	152.0	135.0	88.4	91.8	76.1	67.4	68.7	60.0	30.8	34.2	38.7	17.8
29	149.0	134.0	127.0	184.0	132.0	85.0	78.5	59.4	53.9	42.6	42.2	23.7	22.7	27.2	18.0
31, 32	322.0	159.0	277.0	187.0	126.0	105.0	87.6	80.9	77.9	58.9	55.0	37.2	33.3	33.3	24.0

Soil No.	23–26 days.	27–29 days.	30–33 days.	34–36 days.	37–44 days.	45–51 days.	52–58 days.	59–65 days.	66–72 days.	73–79 days.	80–86 days.	87–93 days.	94–100 days.	101–107 days.	108–114 days.
1, 2	10.8	8.1	12.0	8.4	6.4	7.9	5.0	5.1	4.6	4.3	4.0	4.5	4.7	4.6	3.6
3, 4	14.6	11.1	17.6	14.2	9.3	13.0	7.3	6.4	6.6	5.1	5.3	6.4	6.3	6.7	5.2
5, 6	23.1	20.4	29.8	22.2	16.9	19.0	12.3	11.5	9.3	10.6	9.7	7.7	8.5	9.4	7.4
7, 8	62.0	54.8	72.2	54.3	37.6	48.6	37.0	25.7	31.4	25.7	23.0	33.3	23.0	21.0	16.6
9, 10	86.0	76.3	105.0	84.3	73.5	97.8	76.4	81.4	61.3	39.9	35.0	33.6	32.3	31.4	24.3
11, 12	26.0	18.8	30.0	20.7	14.7	20.6	13.1	11.1	12.3	9.1	8.7	10.6	9.4	9.6	6.4
13, 14	68.0	59.7	73.2	82.0	54.6	56.1	33.6[a]	23.0[a]	19.7[a]	18.6[a]	17.0[a]	20.8[a]	15.8[a]	17.6[a]	10.9
15, 16	96.0	65.7	215.0	61.7	75.2										
18	26.0	20.2	40.3	20.5	14.4	17.0	16.7	9.1	9.2	8.6	8.6	11.2	10.1	8.7	7.1
20	27.7	23.7	42.7	23.9	18.1	18.9	12.7	9.6	10.9	8.8	8.7	11.2	9.2	8.9	8.1
21, 22	25.5	20.4	28.7	19.9	14.0	18.0	10.4	9.4	9.4	8.4	7.9	9.2	8.9	8.7	7.4
23, 24	27.0	20.6	42.5	32.1	17.6	22.0	12.5	9.7	11.6	9.2	7.7	10.3	10.1	9.9	7.8
25, 26	14.8	10.3	17.0	11.8	8.8	9.8	6.4	4.9	6.8	5.6	5.6	5.1	5.6	5.5	4.1
27, 28	11.8	14.1	24.1	17.8[a]	11.4	13.1	7.9	7.3	7.5	6.6	5.9	7.7	9.3	7.3	4.5
29	13.1	9.5	29.5	18.2	18.2	9.1	12.9	8.6	8.4	5.9	7.6	4.5	5.0	5.4	4.5
31, 32	19.1	14.8	19.2	16.2	9.2	18.6	6.5	9.3	6.9	6.7	7.2	8.6	4.7	7.9	5.3

TABLE VIII.—*Quantity of carbon evolved as carbon dioxid per acre per day in soils of Experiment II*—Continued

Soil No.	115–121 days.	122–128 days.	129–135 days.	136–142 days.	143–149 days.	150–156 days.	157–163 days.	164–170 days.	171–178 days.	179–185 days.	186–192 days.	195–199 days.	200–209 days.	210–216 days.
1, 2............	2.3	3.0	3.3	3.0	2.7	1.9	3.4	2.8	5.3	a 6.4	7.8	14.1	6.3	4.8
3, 4	5.0	4.0	4.2	3.5	3.0	2.8	4.5	4.0	6.1	13.9	9.3	16.4	14.7	6.6
5, 6	4.9	6.0	6.0	5.0	4.6	3.9	7.0	5.7	8.9	a 23.8	3.3	a 3.9	5.9	12.0
7, 8............	10.2	12.9	13.6	10.5	10.3	7.1	12.1	9.3	10.2	a 16.6	14.7	a 19.4	14.8	12.7
9, 10..........	14.1	18.0	17.7	13.7	13.4	9.7	15.6	12.6	21.2	a 26.9	4.7	13.2	17.5	13.5
11, 12..........	6.1	6.3	6.1	4.9	5.4	4.2	5.8	5.4	8.8	a 7.1	9.3	3.9	11.0	10.5
13, 14..........	a 7.6	a 9.3	a 8.1	a 12.2	a 8.9	a 5.1	a 9.9	a 8.8	a 12.2	a 13.6	a 14.0	b 14.4	a 15.0	a 12.0
15, 16..........	14.1	10.9	16.3	11.9	11.5	8.2	12.7	11.7	17.0	18.3	23.3	13.5	15.5	12.7
18............	4.4	5.5	5.8	4.6	4.7	3.5	5.7	4.5	8.4	8.5	11.0	b 9.5	7.8	5.9
20............	5.0	6.6	6.2	5.3	4.9	3.9	7.2	6.3	8.4	10.7	13.4	b 12.0	10.6	8.7
21, 22..........	4.5	5.7	5.8	4.7	4.7	3.8	6.2	4.5	8.7	7.9	8.8	a 8.5	8.8	6.6
23, 24..........	4.5	6.5	6.7	6.3	4.9	4.1	7.4	6.8	9.2	10.7	10.0	a 10.7	10.1	7.3
25, 26..........	2.9	3.2	3.7	2.7	3.2	2.0	3.8	2.6	6.0	9.2	10.0	14.8	10.9	11.4
27, 28 ..	3.2	5.2	5.5	3.1	4.5	3.2	4.5	4.5	6.9	a 5.9	7.0	13.7	a 8.4	7.0
29............	2.7	3.7	3.7	3.1	2.9	2.4	3.8	3.3	6.1	6.0	6.8	5.9	7.0	4.5
31, 32..........	3.6	4.7	4.9	3.8	3.5	2.7	5.9	4.5	6.8	8.4	6.3	a 10.0	8.9	10.2

a Represents value for one soil, one determination having been lost.
b Represents values taken from the average of preceding and succeeding determinations, both duplicates having been lost.

DISCUSSION OF THE CURVES

In figures 4 and 5 appear the curves for the soils receiving the various amounts of stable manure used alone and in combination with lime. The control and limed soils are also shown. Special attention is called to the manner of construction of the curves in the two different figures. In these two and the succeeding figures different standards have been used which are thought to bring out the important points to advantage. These changes can be followed easily by watching the notation on each figure. All of the curves are seen to be well above the curves for the control soil. With the exception of a few isolated points, the curves for the limed soils are well above those for their respective unlimed soils. The fact that the limed soils almost invariably gave off more carbon dioxid than their respective unlimed duplicates, even toward the end of the experiment, which ran more than half a year, shows that there must be greater bacterial activity induced by the lime. Of course it might be said that the excess carbon dioxid came from the lime itself, but it will be shown in Table IX that less lime was used in the manured than in the unmanured soils. This precludes that explanation of the fact. Also it would seem that after such an extended length of time all of the soil acid would be neutralized, so that the most reasonable conclusion is as stated above—namely, that lime increases bacterial activity to a marked extent.

When compared to the curves for the manure in the former experiment, it is seen that the manure curve is well above that where dry, well-rotted manure was used, in spite of the fact that in the case of the dry manure much more organic matter was present than in the case of the wet manure. As shown in Table VI, the 11.3 gm. of wet manure representing an application of 10 tons per acre, was only equivalent to 5.39 gm., or slightly less than 5 tons per acre. The reason for this is no doubt due to several causes. One of these is that many more bacteria were added with the fresh manure, and the second is that drying resulted in precipitation of considerable soluble organic matter in such a form that it redissolved rather slowly in the soil moisture, and drying also rendered the solid organic matter more impervious to moisture and,

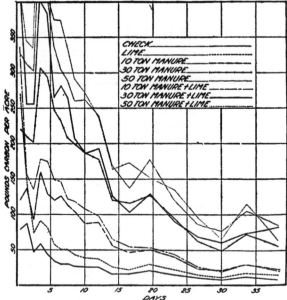

CHECK
LIME
10 TON MANURE
30 TON MANURE
50 TON MANURE
10 TON MANURE + LIME
30 TON MANURE + LIME
50 TON MANURE + LIME

POUNDS CARBON PER ACRE

DAYS

FIG. 4.—Graphs of the quantity of carbon evolved as carbon dioxid in soils receiving various amounts of manure used alone and in combination with lime. Period, 1–39 days.

hence, to bacterial action. Perhaps the main reason is that the easily available organic matter of the well-rotted manure had all been decomposed.

On comparing for the moment the curves in figures 6 and 7 with the curves for the similarly treated soils of the former experiment, shown in figure 2, it is seen that in the latter case where the fresh undried green manure was used, the rate of decomposition at first was enormously greater. Therefore the grinding had no effect on the green

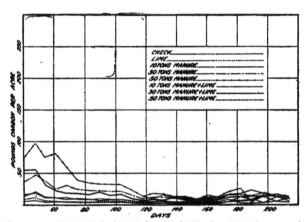

Fig. 5.—Graphs of the quantity of carbon evolved as carbon dioxid in soils receiving various amounts of manure used alone and in combination with lime. Period, 40–216 days.

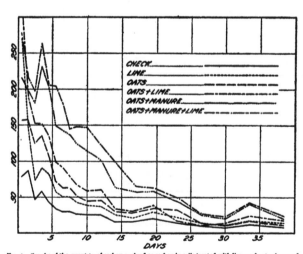

Fig. 6.—Graphs of the quantity of carbon evolved as carbon in soils treated with lime and oats alone and in combination with manure and lime. Period, 1–39 days.

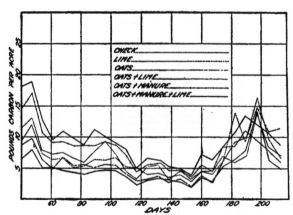

Fig. 7.—Graphs of the quantity of carbon evolved as carbon in soils treated with lime and oats alone and in combination with manure and lime. Period, 40-216 days.

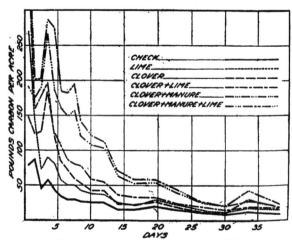

Fig. 8.—Graphs of the quantity of carbon evolved as carbon in soils treated with lime and clover alone and in combination with manure and lime. Period, 1-39 days.

manure, as was thought at first. Or, if it did, it was more than over-come by the effect of the drying. Another important thing to be noted in connection with the curves in figures 6 and 7 is that, after about the one-hundred-and-twentieth day of the experiment, the amount of car-bon being given off from the soils treated with the oats was about the same as the control soil. This means that most of the carbon of the green manure has been decomposed and evolved. The same may be said in regard to soil treated with lime and oats when compared to the soil treated with lime only.

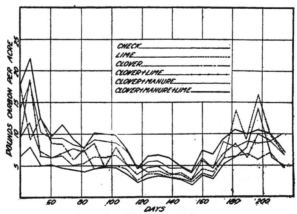

Fig 9.—Graphs of the quantity of carbon evolved as carbon in soils treated with lime and clover alone and in combination with manure and lime. Period, 40–216 days.

The curves for the clover-treated soil, with its various combinations, appear in figures 8 and 9. In general the same may be said in regard to the clover-treated as of the oat-treated soils. Here the fresh mate-rial decomposed much more rapidly than the dried and ground material of the former experiments. Also after the one-hundred-and-twentieth day, the rate of evolution of carbon as carbon dioxid is the same from the green-manured soil as the corresponding ungreen-manured. It is seen that here, as in the former experiment, the clover decomposed somewhat more rapidly than the oats.

In Table IX the summarized data for the whole experiment is given. Since this table is constructed exactly as Table III, no further explana-tion is needed in regard to it.

Table IX.—*Summarized results of Experiment II*

Soil No.	Organic carbon added.	Total organic carbon at start.	Inorganic carbon added.	Total carbon lost.	Inorganic carbon at end.	Organic carbon lost.	Excess organic carbon lost from limed over unlimed soil.	Organic carbon lost from added manure.	Carbon lost from added manure.	Total organic carbon lost.	Organic carbon from green manure alone.	Quantity of calcium carbonate at end in limed soils.
1	2	3	4	5	6	7	8	9	10	11	12	13
									P. ct.	P. ct.		Lbs.
1, 2........	None.	29,513	....	1,704	46	1,730	....	....	....	5.96	....	....
3, 4........	None.	29,513	680	2,727	182	2,229	479	....	....	7.55	....	1,516
5, 6........	3,686	33,199	....	3,647	36	3,683	....	1,933	52.4	11.10	....	....
7, 8........	11,058	40,571	....	8,308	34	8,242	....	6,492	58.8	20.70	....	....
9, 10........	18,430	47,943	....	12,967	55	12,322	166	10,632	57.7	25.80	....	....
11, 12........	3,686	33,199	680	4,343	208	3,849	79	1,500	44.0	11.60	1,715	....
14........	11,058	40,571	680	8,709	242	8,311	73	6,092	55.2	20.80	2,680	....
16........	18,430	47,943	680	12,703	226	12,249	73	10,020	54.4	25.60	2,885	....
18........	4,801	34,314	....	4,239	53	4,292	343	2,542	52.5	12.50	609	....
20........	4,801	34,314	680	5,088	230	4,638	306	2,409	50.2	13.50	779	1,915
21, 22........	4,740	34,253	....	4,218	47	4,263	....	2,515	57.8	12.50	582	....
23, 24........	4,740	34,253	680	5,033	218	4,571	306	2,342	49.4	13.40	712	1,825
25, 26........	1,115	30,628	....	2,680	35	2,735	....	980	76.5	8.93	....	....
27, 28........	1,115	30,628	680	3,310	230	2,860	125	631	56.6	9.34	....	1,985
29........	1,054	30,567	....	2,688	62	2,750	....	1,000	95.0	9.00	....	....
31, 32........	1,054	30,567	680	3,259	204	2,783	33	554	52.6	9.11	....	1,700

DISCUSSION OF THE RESULTS

The quantity of carbonate carbon left in the soil at the end of the experiment and shown in column 6 are particularly interesting when compared with the values of the preceding experiment shown in Table III. The amount of inorganic carbon in the unlimed soil of the two experiments does not show any striking differences. This is to be expected, since, as stated earlier, this carbon probably is simply the carbon dioxid dissolved in the moisture of the moist soils used for the analysis. But, when an examination of the results for the limed soil is made, consistency is observed. In the experiment which ran for 216 days, 182 pounds of carbon remained, while 218 pounds were left in the experiment which ran only 53 days. Thus, in the time from the fifty-third day to the two-hundred-and-sixteenth day 36 pounds of carbon were lost. In terms of calcium carbonate,this represents 300 pounds. With the 10-pound manure treatment there were 455 pounds of calcium carbonate decomposed from the fifty-third to the two-hundred-and-sixteenth day. However, not too much significance should be placed on this figure because of the different character and amount of the manure used. In the case of the oats the difference is 335 pounds of carbonate. For the clover the difference is 480 pounds. Somewhat more significance can be placed on these figures although no absolute dependence, for the green manures were in quite different states when added to the soils.

Here, as in the former experiment, the manures act as conservers of lime. A discussion of this question is given in connection with the first experiment. The oats seem to use less of the lime than the clover.

It is seen that in every case except where the soils were treated with 50 tons of manure the lime enhanced the rate of decomposition of the organic matter. As already mentioned, the soil treated with 50 tons of manure and lime developed a mold, which was later removed. Therefore, not much dependence should be placed on the results for that soil. The figures as a whole given in column 8 are not very consistent. The excess carbon from the soils treated with both stable manure and green manure is greater than where the similar amounts of manure were used separately. Yet, where no manure was used, the excess carbon was greater than in any other case. It seems that the original organic matter in a soil needs lime for its decomposition, while the organic matter of manures used here does not need lime.

The values given in columns 9 and 10 are extremely interesting. On considering the percentage of carbon lost from the various manures, it is seen that more than half of the carbon of the stable manures has been evolved in the course of the experiment. Thus, more than 10 times as much of the manure in this experiment has been decomposed as in the earlier experiment reported in this paper. This is a far greater difference than can be accounted for by a consideration of the time the two experiments ran—namely, 53 and 216 days, respectively. It should be recalled that in the first experiment dry, well-rotted manure was used, whereas in the latter fresh, moist manure was used. As stated above, in the discussion of the curves the reasons for this difference are that the physical condition of the fresh manure was in favor of its more rapid decomposition, more bacteria were added with it, and it contained more easily available organic matter.

Another noteworthy circumstance is that with the limed soils less organic carbon has been evolved than from the corresponding unlimed soil. Here, again the only explanation seems to be that the organic matter of the manures does not need lime for its decomposition. Possibly the lime, together with the organic matter, causes such a high percentage of carbon dioxid in the soil atmosphere and soil solution that it becomes slightly toxic to the bacteria. Our apparatus was of such a nature that cultivation of the soil during the course of the experiment was not feasible. Possibly with cultivation, which would have tended to diminish the concentration of carbon dioxid, the results would have been different. Particularly to be noticed is the very large percentage of the green manures which were decomposed. There were evolved as carbon dioxid 78.9 per cent of the carbon of the oats and 95 per cent of that of the clover. It should be recalled that the respective figures for the experiment which ran 53 days were 52.4 per cent and 62.3 per cent. Therefore it seems that the physical condition of the green manures did not greatly influence the rate of decomposition. While the manner of application of the oats and clover in this last experiment—

that is, in pieces not over ¼ inch in length—can not be said to represent conditions exactly comparable to field practice, yet it is believed that the rate of decomposition would have been approximately the same if they had been added in larger pieces. The results found here confirm those of Boltz already referred to. What shall we say, therefore, concerning the use of green manures in building up the organic matter of the soil? In the first place, it is to be remembered that neither in our experiments nor in those of others have the roots of the plants been used. Parts of the roots would probably decompose somewhat more slowly because of their toughness. However, it is doubtful if this would radically change the results. As Pieters (6) concludes from his review of the subject, the use of green manures generally gives greater crop yields to crops immediately following. This increase of crop would usually, if properly handled, give greater crop residues to plow under, which in turn would tend to enrich the soil. Then it must not be forgotten that the green manure would of itself add available plant food to the soil, taken partly or largely from the unavailable supply. And, of course, in the case of leguminous green manures, nitrogen would be added also. Then its decomposition would render an additional amount of plant food available.

In practice it is often observed that, where insufficient time is allowed between the plowing under of the green manure and the planting of a crop, in a dry season the crop suffers from lack of moisture. This probably is due to the turning under of the green manure in large bunches. From the results of this experiment it is seen that if some practical plan could be found to mix the material thoroughly with the soil, most of it would be decomposed in a short time. This must be left, of course, to field experiments.

When results for total organic matter are considered, it is observed from column 11 (Table IX) that the limed soils showed a greater percentage of decomposition than the unlimed. In some instances, however, the differences are so small as not to be significant.

In the twelfth column (Table IX) the carbon coming from the green manure alone is shown. In the case of the unlimed soils, it is less by nearly one-half than the corresponding amount from the green manure applied to the soil alone. From the limed soils, however, it is seen that the stable manure apparently quite appreciably increases the rate of decomposition of green manures. We are not prepared to give an explanation of these facts. However, we believe that the results of this experiment and the one preceding warrant the conclusion that green manures under most conditions will be decomposed sufficiently rapidly in the soil without the addition of stable manure.

CONCLUSIONS FROM EXPERIMENT II

The conclusions to be drawn from Experiment II are as follows: It is, of course, understood that it is not proposed that they hold for all soils under all conditions. However, it is believed that some generalizations are possible.

(1) Soil organic matter decomposes more rapidly under limed than under unlimed conditions.

(2) When organic matter in the form of stable manure and the green manures oats and clover is added to the soil, the total organic matter— that is, the organic matter of the soil plus the added organic matter— decomposes more rapidly under the influence of lime than without it. When the added organic matter alone is considered the rate of decomposition is lessened by the lime.

(3) The carbon of stable manures is evolved as carbon dioxid from soil under unlimed conditions to the extent of approximately 55 per cent. The carbon of oats under like conditions is evolved to the extent of 79 per cent and that of clover 95 per cent. Under unlimed conditions the amount of stable-manure carbon evolved is only slightly less than under limed conditions, while only about 57 per cent of the carbon of oats and 53 per cent of the carbon of clover is given off under limed conditions.

(4) All the manures tended to conserve the lime.

(5) Under unlimed conditions stable manure did not increase the rate of decomposition of the green manure as measured by the evolution of the carbon dioxid. With lime there was a slight increase in the amount of carbon given from the green manure when used with the stable manure over that given by the green manure when the latter was used alone. It should be recalled that in the former experiment stable manure enhanced the rate of decomposition of the green manure only to a slight extent.

(6) There is not a very great difference in the rate of decomposition of the green manure when added in a finely ground, dry state and when used fresh and in a relatively coarse state of subdivision.

LITERATURE CITED

(1) BOLTZ, G. E.
    1916. LOSS OF ORGANIC MATTER IN GREEN MANURING. In Ohio Agr. Exp. Sta.
        Mo. Bul., v. 1, no. 11, p. 347-350.
(2) GIMINGHAM, C. T.
    1911. THE FORMATION OF CALCIUM CARBONATE IN THE SOIL BY BACTERIA. In
        Jour. Agr. Sci , v. 4, pt. 2, p. 145-149.
(3) HOPKINS, C. G.
    1910. SOIL FERTILITY AND PERMANENT AGRICULTURE. 653 p., illus , pl., maps.
        New York.
(4) LIPMAN, J. G., et al.
    1913-16. THE INFLUENCE OF BACTERIA IN MANURE ON THE DECOMPOSITION OF
        GREEN MANURE (LEGUME AND NON-LEGUME). In N. J. Agr. Exp.
        Sta., 33d-36th Ann. Rpt., [1911]/12-[1914]/15.

(5) MacIntire, W. H., and Willis, L. G.

1913. SOIL CARBONATES, A NEW METHOD OF DETERMINATION. Tenn Agr. Exp. Sta. Bul. 100, p. 83–97, 1 fig. References, p. 97.

(6) Pieters, A. J.

1917. GREEN MANURING: A REVIEW OF THE AMERICAN EXPERIMENT STATION LITERATURE. *In* Jour. Amer. Soc. Agron., v. 9, no. 2, p. 62–82; no. 3, p. 109–126, no. 4, p. 162–190. Bibliography, p. 173–190.

(7) Potter, R. S., and Snyder, R. S.

1916. CARBON AND NITROGEN CHANGES IN SOIL VARIOUSLY TREATED: SOIL TREATED WITH LIME, AMMONIUM SULPHATE, AND SODIUM NITRATE. *In* Soil Science, v. 1, no. 1, p. 76–94, 2 fig. Literature cited, p. 93–94.

(8) ——— ———

1917. CARBON DIOXIDE PRODUCTION IN SOILS AND CARBON AND NITROGEN CHANGES IN SOILS VARIOUSLY TREATED. Iowa Agr. Exp. Sta. Research Bul. 39, p. 255–309, 21 fig. Bibliography, p. 306–309.

EUPATORIUM URTICAEFOLIUM AS A POISONOUS PLANT

By C. Dwight Marsh, *Physiologist in Charge*, and A. B. Clawson, *Physiologist, Poisonous Plant Investigations, Pathological Division, Bureau of Animal Industry, United States Department of Agriculture.*

INTRODUCTION

Among the suggested causes for the disease popularly known as milk sickness, trembles, slows, tires, etc., has been the plant commonly called white snakeroot (*Eupatorium urticaefolium*, the *Eupatorium ageratoides* of the older literature). Arguments have been advanced *in extenso* both for and against this theory. In later years the possibility of the connection of this plant with the disease has lessened since Crawford (4) [1] in 1908 published his negative pharmacological work and Jordan and Harris (7) in 1909 published their paper on the *Bacterium lactimorbi*. While there have still been authors who clung to the white snakeroot origin of the disease, it has been rather generally believed that the real cause was not a plant poison but more likely a disease-producing germ. Although the investigation of the subject has been carried on by the Department of Agriculture for some years, little attention has been paid to the question of the connection of this plant with the disease, as it was thought probable that it should be disregarded.

In the fall of 1914 there were some cases of milk sickness near Beecher City, Ill., and the attending physician, Dr. E. R. Brooks, made some experiments with the *E. urticaefolium* which led him to think it to be connected with the disease. At his solicitation a representative of the department visited the neighborhood, and arrangements were made to conduct a series of feeding experiments with a view to determine, if possible, whether or not the plant produced the disease. The experiments were carried on during the fall and winter of 1914 and 1915 at Washington and during the fall of 1915 near Beecher City. The material used in the first season was collected by Dr. Brooks and shipped to Washington, where it was fed both in fresh and dried condition. All the material used in the fall of 1915 was fed fresh from the immediate neighborhood of the experiment.

The general results of the experiments in their relation to the disease of milk sickness are not in form for publication, but it has been clearly demonstrated that *E. urticaefolium* must be counted as one of the rather important stock-poisoning plants, which produces serious losses of domestic animals. On this account it seems wise to publish the results so far

---

[1] Reference is made by number (italic) to "Literature cited," p. 714.

Journal of Agricultural Research,
Washington, D. C.

Vol XI, No. 13
Dec 24, 1917
Key No. A—35

lh

as they relate to *E. urticaefolium* as a poisonous plant, with the hope that some of the losses may be avoided; for it is evident that while in many localities this plant is suspected to be the cause of milk sickness, its importance as a stock-poisoning plant has received little recognition.

DESCRIPTION OF THE PLANT

Eupatorium urticaefolium Reichard (Pl. 52) quoted as *E. ageratoides* L. in the older botanies, is a slender, erect, perennial herb, belonging to the family Compositae. Its leaves, which are 3 to 5 inches long, are opposite, broadly ovate, pointed, sharply toothed and thin, and have rather long petioles. The stems are viscid-villous. The small white flowers are in compound corymbs of 8 to 30 flowers, appearing in the late summer and fall. The plants are from 1 to 4 feet in height.

It is found widely distributed in the eastern United States and as far west as Minnesota, Oklahoma, Nebraska, and Louisiana. Its favorite habitat is in rich, damp woods. It is abundant in the groves along the watercourses of the Middle States, and grows with special profusion in the so-called coves, or damp, shaded ravines, of the north slopes of the southern Appalachians. It is by no means, however, confined to shaded situations, for it sometimes grows in masses on cleared hillsides in the open.

It is commonly known as white snakeroot. Other names which have been applied to it are white sanicle, Indian sanicle, deerwort, bone set, poolwort, poolroot, richweed, squawweed, whitetop, and steria.

The root has been used in medicine as a diuretic and antispasmodic.

The common thoroughwort, *Eupatorium perfoliatum* L., which grows in similar localities, but usually on soils with more moisture, is readily distinguished from *E. urticaefolium*, as it is a coarser plant and the leaves are long, narrow, rugose, and, as the name indicates, are without petioles and are united around the stem.

EXPERIMENTAL WORK

In the experimental work of 1914 three head of cattle and two head of sheep were used. Following is the detail of the experiments.

CALF 668.—This animal was a bull calf, born in March, 1914, and kept during the summer in a pasture on the department farm at Arlington, Va. It was brought into the barn for experimental feeding October 7. It was in good condition and weighed 340 pounds.

The plants fed to this animal were shipped from Beecher City, Ill., and consisted of stems, leaves, and flowers, which were fed in as fresh a condition as possible, but had lost some moisture, so that the weights of material fed probably represented a somewhat larger quantity of the strictly fresh plant.

The animal received and ate the plants as follows· October 8, 4.5 pounds; October 10, 9.5 pounds; October 12, 7 pounds; and October 14,

7 pounds. He received, all told, on the basis of 1,000 pounds of animal weight, 82.4 pounds of *E. urticaefolium*, a daily ration of 11.8 pounds, during the time he was eating. During this time he was also fed hay in quantities not weighed on October 9, 13, 14, and 15. No symptoms of poisoning were shown until October 15. On this day the animal was depressed and constipated. He lay down much of the time, and sometimes refused to rise. He ate very little of the hay. On the 16th his condition was unchanged. On the 17th he could get up, but would do so only with considerable urging. He showed no desire to eat either hay or green grass. He was taken out of the pen, and after being led around a little, trembled violently and lay down. It was difficult to make him get up and return to the pen. On October 18 and 19 he grew gradually weaker. He could get upon his feet, but trembled violently when standing. The respiration was rapid. He ate nothing except a few handfuls of grass and a very little grain.

He died on the morning of October 20 without any struggle. Death occurred on the sixth day after the last feeding of *E. urticaefolium*. Observations upon the temperature during the illness showed nothing abnormal.

In the autopsy, petechiæ and hemorrhagic spots were found on the heart, both on the auricles and ventricles and also on the aorta and *vena cava*. The walls of the duodenum, jejunum, and ileum were congested, and there were homorrhagic spots on the cecum. The liver was light in color, possibly somewhat degenerated. There was nothing else abnormal.

BULL 663.—This animal, a bull, was brought in for feeding October 7, 1914, He had been in pasture on the Arlington farm since the preceding March and was in good condition, weighing 692 pounds.

The plants used for this animal had been shipped in from Beecher City, and autoclaved at a pressure of 7 pounds in order to destroy any possible germs of disease. This sterilized material was fed as follows: October 8, 5.2 pounds; October 10, 10.5 pounds; October 12, 26 pounds; October 14, 29 pounds; October 15, 15 pounds; making a total in 8 days of 85.7 pounds. On the basis of 1,000 pounds of animal weight, this would have been 123.8 pounds of *E. urticaefolium*, an average daily ration of 15.45 pounds per 1,000 pounds of animal weight. Hay was given on October 13 and 15 and on the days following. No symptoms were noted until October 15; then the animal exhibited constipation and a loss of appetite. On the 17th when he was led about, the exercise caused more or less labored breathing and there was some trembling of the muscles of the flanks, hips, and muzzle. On the 19th he was somewhat worse; he had little appetite, showed a disinclination to stand, and exhibited the same muscular trembling as on the 17th. On October 20, while still disinclined to stand, he, on the whole, showed improvement, and after this gained steadily. On the 23d he was returned to the pasture and seemed to be recovered from his indisposition. His temperature was normal during the experiment.

This bull received a second feeding of *E. urticaefolium* from January 7 to 21, 1915. This was of dried material which was ground up and mixed with grain. In the 15 days of the experiment he ate 103 pounds of this material. It was found by tests that the plant loses in drying 65 per cent, so that he ate an equivalent of 294.3 pounds of fresh plant. As the animal at this time weighed 760 pounds, this amounted to 387.2 pounds of *E. urticaefolium* per 1,000 pounds of animal weight, and the average daily ration on this basis was 25.8 pounds.

When this experiment was begun, the animal was very active. On October 15 he was rather dull, and on January 20 and 21 he was decidedly depressed and exhibited slight trembling. There were no further symptoms. The temperature remained normal during the experiment.

Cow 122.—This was a Jersey cow, 4 years old, that had been used as a milch cow during the preceding summer. She had been tested in the fall for tuberculosis and gave a positive reaction. At the time of the experiment she weighed 785 pounds.

From December 15, 1914, to January 8, 1915, she was fed on *E. urticae-folium* that had been collected and dried at Beecher City. This was fed with hay and grain. She received a total of 96 pounds, which would be equivalent to 274.3 pounds of fresh plant. On the basis of 1,000 pounds of animal weight, this cow received a total of 394.4 pounds, or an average daily ration of 14.1 pounds.

Except for some constipation in the early days of the experiment, there were no ill effects from the feeding. Her temperature remained normal during the experiment and she weighed 790 pounds on January 7.

Sheep 308.—A female spring lamb, weighing 80 pounds, was given *E. urticaefolium* from Beecher City on October 24, 26, and 28, 1914. None of that given on the 28th was eaten, and the total quantity eaten on the other days was 5 pounds. The animal was fed hay on October 25 and was later given what grain she would eat. The total quantity eaten was equivalent to 6.25 pounds per hundredweight of animal. In the forenoon of October 26 she exhibited a lack of appetite, eating nothing, and some mucus was running from the nose. At 3.30 p. m. she showed marked depression and was standing humped up, with the head held low, and brownish green material running from the mouth. When led about, she began to stagger and tremble, the trembling appearing first in the shoulders. When left alone, she lay down. In the forenoon of October 29 she was found lying on the left side with legs extended. She was raised, but could not stand alone. She remained in this condition, growing gradually weaker through the 30th and died between 2 and 4 a. m. on October 31.

Her temperature was normal during the experiment, except that it was rather high on October 30. In the autopsy the only positive abnormal condition was a congested duodenum.

SHEEP 309.—This was a female spring lamb weighing 84 pounds. She was fed upon autoclaved *E. urticaefolium.* This was given October 24 and 26, 1914, and the uneaten material was removed on the 29th. She ate in this time 4 pounds, or 4 4 pounds per hundredweight of animal. There was no effect from this feeding.

On November 2 she was given autoclaved material, the feeding being given by the balling gun. When she had received 1.38 pounds she commenced to salivate, and the feeding was terminated. She frothed at the mouth, was nauseated, and her breath had a bad odor. On November 3 she showed a lack of appetite and preferred to lie down. When led about, she walked unwillingly, showed marked trembling, and soon lay down. Plate 53, A, shows the weakness of the animal, and Plate 53, B, while apparently a poor picture, shows by the indistinctness of the sheep as compared with other objects the continuous trembling to which the animal was subject when on its feet. On November 4 the symptoms were more marked. When led about, she trembled violently and soon fell. This was repeated several times. The respiration and pulse were very rapid when attempts at movement were made. The temperature was 104 2° F. This condition continued through November 5, the temperature varying from 104.8 to 105.4. In the forenoon of the 6th the temperature was 105.8, and the respiration was labored and noisy. The animal was unable to stand, and there were occasional convulsive movements of the body muscles. She died at 4.30 p. m., and an autopsy was made on the morning of November 7. Nothing abnormal was found except possible congestion of the duodenum and small intestine and a similar condition of the liver and kidneys.

SHEEP 310.—This was a ewe about 6 months old, weighing 66 pounds. January 29, 1915, dried plants of *E. urticaefolium* ground to a powder and mixed with water were fed with a balling gun, 1.75 pounds being used. This was equivalent to 7.6 pounds of green material per 100 pounds of animal. It produced no bad effect.

On February 9 another feeding of 3 pounds of similar material was made in the same way. This was an equivalent of 13.82 pounds of green material per 100 pounds of animal. It, too, produced no effect besides evidence of discomfort from the large feeding. As will appear in the general discussion, the failure to produce intoxication was probably due to the fact that the plant in drying loses much of its toxicity.

Cow 699.—This animal, which was fed at Fancher, Ill., in the fall of 1915, was an old milch cow, estimated to weigh 1,000 pounds. She was in good condition, giving a good flow of milk. She was fed from September 21 to 30 on fresh material of *E. urticaefolium,* including stems, leaves, and flowers; this was mixed with grass and bran. She received during this time 106 pounds of *E. urticaefolium*, 138 pounds of grass, and 11.5 pounds of bran, an average daily ration of 10.6 pounds of Eupatorium, 12 5 pounds of grass,' and 1 pound of bran. The material was kept before her constantly in the attempt to make her eat as much as possible. She did not take kindly to the *E. urticaefolium*, and it was necessary to keep mixing fresh grass with it. A calf (No. 700) was allowed to take all her milk.

On September 30 it was evident that she was losing in activity and that her flow of milk was much reduced. She moved slowly and after a little exercise trembled noticeably in the shoulders, hips, and thighs. On October 1 she was much constipated and strained when defecating. She ate no *E. urticaefolium* and even refused to pick grass out of the mixture. On October 2 she was lying down and unable to get on her feet. Grass was offered to her, and she tried to eat it. The nose and lower jaw trembled violently. After attempting to rise, she went over on her side, with marked trembling about her head. Her flow of milk had nearly stopped. On October 8, in the morning, she was found dead.

Her temperature during the experiment was rather low, but hardly subnormal. At the autopsy the liver was somewhat yellowish, the duodenum was slightly congested, and the ileum showed many patches of congestion. Nothing else abnormal was noticed.

CALF 700.—This was a male calf which on September 21 was only 2 or 3 days old. He was put with cow 699 and took all her milk supply. He showed no ill effects from the milk, but suffered from lack of nourishment as the cow began to dry up.

From October 4 to 9 he was fed with *E. urticaefolium* by the balling gun, receiving a total of 5.6 pounds. This would be equivalent to 56 pounds for a 1,000-pound animal, or an average daily ration of 9.3 pounds for six days. During this time he was fed on skimmed milk. He showed nausea on October 8, and on the 9th this was accompanied by vomiting, with loss of appetite. On October 7 and 8 he shivered at times, but this may have been due to insufficient food rather than to any toxic effect of the *E. urticaefolium*. He was killed and autopsied on October 10. The result of the autopsy was entirely negative.

SHEEP 367.—This sheep, a yearling ewe weighing 60 pounds, was fed at Fancher, Ill., from September 22 to 28, inclusive, on fresh plants of *E. urticaefolium*. She ate 1.1 pounds on September 22, and on the remaining days, except September 24, she was given the material by a balling gun, receiving all told 4.3 pounds, or on the basis of 7.1 pounds per 100 pounds of animal, an average daily ration of 1 pound.

In the afternoon of September 25, after having received the plant at the rate of 3.8 pounds per 100 pounds animal weight, the sheep appeared inactive, preferring to lie down. On September 26 it was thought that she trembled some after being exercised, and there was some trembling of the jaw when she was being fed with the balling gun. On September

27 she was constipated; no further evidence of trembling. On September 28 she trembled after exercise in the morning, and the lower jaw trembled when she was being fed with the balling gun. She grew weaker during the day, lay down most of the time, ate very little, and trembled when on her feet. From September 29 to October 1 her condition continued much the same as on September 28, with gradually increasing weakness until on October 1 she was unable to rise or even to remain in an upright position, and lay on her side during the day. Plate 54 shows the condition of the animal. The urine and feces were bloody. She was found dead on the morning of October 2. There had been almost a complete loss of appetite, so that she had eaten very little during the experiment and weighed after death only 43 pounds.

The records of temperature, pulse, and respiration showed nothing unusual except a rather high pulse rate. The autopsy showed nothing distinctly abnormal, except that the liver had a spotted appearance and was possibly somewhat degenerated.

SHEEP 368 —This was a ewe about 4 years old, weighing 100 pounds. She was fed at Fancher, Ill., from September 22 to 24, 1915, with the balling gun, receiving a total of 5 pounds, an average daily ration of 1.7 pounds. She became nauseated during the feeding on September 23, but showed no further symptoms at that time. While being fed in the morning of the 24th some trembling of the nose was noted and the animal preferred to lie down. A little later the trembling and weakness became more marked, increasing until she was unable to stand and went over upon her side, salivating and vomiting. Plate 55 shows her condition at this time. She died at 1.30 p. m., having lost 12 pounds during the experiment. Her temperature remained normal during the experiment. In the autopsy the only marked change was in the liver, which presented a mottled appearance.

SHEEP 369.—This animal was a yearling ewe weighing 63 pounds. She was given *E. urticaefolium* with the balling gun, at Fancher, Ill., from October 3 to 7, 1915. She received 0.22 pound on October 3, and on the following days received four feedings of 0.22 pound each, thus getting all told 3.7 pounds. On the basis of an animal of 100 pounds this would be 5.9 pounds, or an average daily ration of 1.2 pounds.

While she was being fed in the afternoon of October 5, after about 1.5 pounds of the plant had been eaten, there was some trembling of the head. At the end of the first feeding on October 6 there was trembling of the jaw; at this time about 2.2 pounds had been eaten. There were no other symptoms until the morning of the 8th. When first seen on that day, she was lying down, and when made to rise walked a little way and lay down. The respiratory movements were deep and jerky, and she lay with the eyes half closed, grunting with each expiration. A little later when made to rise she stood humped up and moved about uneasily. The forelegs began to tremble, the trembling increasing in violence for about five minutes, when she lay down. She gradually grew weaker, exhibiting a repetition of the same symptoms. On the morning of October 10 she could still stand for a few minutes. She was killed at 9 a. m. on that day, and an autopsy was made. The autopsy showed nothing that could be considered abnormal. The records of temperature, pulse, and respiration showed nothing that could be considered specifically abnormal.

LETHAL DOSE

The experiments were too few in number to make it possible to speak positively in regard to the lethal dose. The results, however, are significant.

SHEEP.—Excluding No. 310, which was fed on the dry plant, considering No. 369 as a case of death, for this animal would have died, without doubt, and adding the amounts of the two feedings of No. 309 for reasons which will appear later, the average lethal dose was 6.05 pounds per hundredweight of animal, with a minimum of 5 pounds and a maximum of 7.1 pounds. The variation from the average is so little that it is probably safe to say that the lethal dose for sheep is about 6 per cent of the animal's weight.

CATTLE.—For the cattle the average lethal dose of the two fatal cases, Nos. 668 and 699, is 94.2 pounds per 1,000 pounds of animal. If it can be assumed that these are average cases, it would appear that the lethal dose for cattle is about 10 per cent of the animal's weight. Cattle apparently are somewhat less susceptible to the poison than sheep.

TOXIC DOSE

The experiments with the sheep failed to show any difference between the lethal and toxic doses

In the case of the cattle, No. 699 showed symptoms on 82.6 pounds per 1,000 pounds of animal, and succumbed after 106 pounds, and No. 663 showed symptoms after 123.8 pounds of autoclaved material. No. 700 exhibited symptoms after 56 pounds, but in this case there was a complication because of the animal's weakness from lack of food. It appears then that the margin between the toxic and lethal limits is quite narrow.

EUPATORIUM POISONING CUMULATIVE

While there was some evidence that the toxic dose was smaller when the material was given in a short time, this difference was comparatively slight. It appears from the experimental cases that the elimination of the toxic substance takes place very slowly, so that there is a distinct cumulative effect. This was evident both in the cattle and in the sheep. This may explain the fact that sheep 309 was killed in a forced feeding of only 1.6 pounds per hundredweight on November 2, 1914. From October 24 to 26 it had received 4.4 pounds with no symptoms. The combined feeding of the two periods was 6 pounds, the quantity which appears to be the average toxic or lethal dose; and if it is assumed that the toxic substance of the first period had not been eliminated the fatal effect of the small forced feeding may be explained.

SYMPTOMS

The most noticeable symptom and perhaps the most typical is trembling, which is seen especially in the muscles about the nose and in those of the legs. This becomes marked after exercise, and may be violent, ending in the animal falling to the ground.

In some cases trembling is the first symptom noted, but in others there is a preceding period of marked depression and inactivity.

In most cases the animals are constipated, and some have bloody feces.

Generally when fed intensively the animals are nauseated, and this is sometimes accompanied with vomiting

The respiration is normal, except that during exertion it is quickened and somewhat labored

In the average of cases there were no significant changes in temperature. In one or two cases it was slightly lower than normal, and in others somewhat higher than normal, during the latter part of the illness.

Weakness is very pronounced. The animals have difficulty in standing and sometimes remain down for a prolonged period before death. Sometimes they live for several days after the appearance of the first symptoms.

AUTOPSY FINDINGS

Generally speaking, there was more or less congestion in the duodenum and ileum. The liver in most cases showed evidence of a pathological condition, presenting the appearance of "nutmeg liver." Except in the liver and intestine, there was no evident abnormal condition.

PATHOLOGY

Microscopic examinations have been made of various tissue from animals poisoned by *E. urticaefolium*. Two organs, the liver and kidney, show pathological changes of a significant nature.

LIVER.—The livers from those animals which were autopsied shortly after death in all cases showed a condition of fatty infiltration, and most of them showed congestion, either acute or subacute. The fat globules were formed within the cells, in some cases occupying nearly the entire cell body. The globules were variable in size. They were least numerous in the case of calf 700 and most marked in the case of cow 669. It is to be noted that cattle 669 died from the effect of the plant, while calf 700 did not show positive symptoms.

Of the sheep, No. 368 showed very marked congestion, the cells of the lobules, more particularly near the *vena centralis*, being pushed far apart, giving the appearance of nutmeg liver. In cases Nos. 367, 368, and 369 the fat globules were more noticeable on the peripheral cells of each lobule   Sheep 369 showed pronounced necrosis of the cells in certain areas   In sheep 367 the liver had apparently been congested, as the cells about the *vena centralis* were pushed apart.

KIDNEY.—The kidney sections from all animals examined were very similar in pathological appearance   The lumina of the tubules, more particularly those near the periphery, and including the convoluted tubules and the loop of Henle, contained a serous exudate, showing that congestion had existed.   There were few normal cells remaining, most of them having begun to break down, the portion of the cells next the lumen having sloughed off, leaving the edge irregular and ragged.   The remains of such cells could often be seen in the lumina.   Most of the blood vessels excepting the capillaries were well filled with blood, which, taken with the serous exudate in the lumina of the tubules, indicates a subacute stage of congestion.   In the case of cow 699 the capillaries as well as other blood vessels were distended with blood.   The condition of degenerated epithelial cells would suggest the presence of some irritating substance.

While certain pathological conditions were found in the lung tissue of some of the cases, they were not constant enough to be considered significant.

ANIMALS SUSCEPTIBLE TO EUPATORIUM POISONING

The writers' experiments were with cattle and sheep.   Definite feeding experiments have been made by others, as will be shown later, which prove the toxicity of the plant for horses, and extracts have poisoned cats, dogs, and guinea pigs.   So far as reported experimentation goes it appears that no animals are immune to the toxic principle of the plant.

In regard to man, the definite evidence is hardly conclusive, in spite of a number of instances in which the plant or an extract is said to have been poisonous.   Some of these cases are clearly apochryphal.   But on the whole there seems to be little doubt that human beings may be poisoned by the plant if a sufficient dosage is used.

COMPARATIVE TOXICITY OF GREEN AND DRY PLANT

The dry plant was fed by the balling gun to sheep 310 on January 29 and February 9, the animal receiving the equivalent of 7 6 pounds of green plant per hundredweight on the first date and 13 82 pounds on the second.   The first feeding was more than the average toxic dose of the fresh plant and the second feeding more than twice that amount, but there was no evidence of poisoning in either case.

Two head of cattle were fed on dry material, No. 122 receiving the equivalent of 274.3 pounds of green plant per 1,000 pounds with no effect, while No. 663 received the equivalent of 294.3 pounds with only slight symptoms.   This latter amount was nearly three times the estimated toxic dose of the fresh plant. It seems that without doubt the plant loses a large part of its toxicity in drying.

ARE THE MILK AND FLESH OF POISONED ANIMALS POISONOUS?

Milk from cow 122, during the period in which she was eating *E urticaefolium*, was fed to cat 3 and dog 15. It should be noted in this connection that cow 122 received dry material and showed no symptoms of poisoning. While it is generally stated that "milk-sick" cows show no symptoms while giving milk, it is evident not only from the experiments of the Department of Agriculture but from the testimony of others that cows even when giving milk can show symptoms of Eupatorium poisoning.

Cat 3 was an old animal He drank the milk for about a week and then refused it, and some time later died. There was no evidence that the milk injured it, and its death was due to old age.

Dog 15 was fed milk from the same cow from December 20 to December 31 with no deleterious effect.

Meat from cattle 668 and sheep 309, both animals having died from Eupatorium poisoning, was fed to dog 11 from October 23 to November 16. During this time the dog ate over 60 pounds of meat, or an average of about 2.4 pounds a day. As he weighed only 15 pounds, he ate 4 times his weight of meat. He was in poor condition at the conclusion of the experiment, and died some days later There were no definite symptoms, and probably the death was not due to *E. urticaefolium* It must be recognized, however, that it is possible that the plant had some injurious effect.

Dog 12, weighing about 10 pounds, was fed like dog 11 from October 23 to November 16 on the meat from cattle 668 and sheep 309, eating 18.9 pounds, without any effect.

Dog 21, an old shepherd dog, was put in a pen with the meat from cattle 699, which had died of Eupatorium poisoning. He ate very little, and later the meat was cooked and fed to him from October 5 to 9, 1915, during which time he ate 6½ pounds. He refused to eat more, and was killed and autopsied on October 10. His lack of appetite was doubtless due to his age. The result of the feeding and of the autopsy was entirely negative.

During the time of the experiment with dog 21 two or three hounds from a neighboring farm broke into the pen where the meat was kept and practically cleaned up the skeleton. None of the dogs were known to be poisoned.

While the experiments with the flesh of poisoned animals can not be considered conclusive, everything points to the probability of the harmlessness of the meat.

In this connection it may be noted that Mr. George Walker, of Andrews, North Carolina, a region where trembles in cattle is common, has called attention to the fact that during the Civil War both armies, when in that neighborhood, took all the animals they could get, and, under the circumstances, the owners did not take pains to point out the animals affected with trembles, although doubtless many such were used, and yet there were, so far as known, no bad results.

BACTERIA NOT THE CAUSE OF ILLNESS IN ANIMALS FED WITH EUPATORIUM URTICAEFOLIUM

Cultures were made from the blood of bull 663 and calf 668 and from the milk of cow 122. In connection with the feeding experiments with cow 699 and calf 700 and sheep 368 and 369, cultures were made from the skin, liver, spleen, feces, and milk. From these cultures were isolated the individual organisms, where any were found. Out of all these cultures two of the bacteria bore some resemblance to the *Bacterium lactimorbi* of Jordan and Harris, but it seemed reasonably certain both from the morphological and cultural characters that they were not identical with it. The one that showed the most resemblance to *Bact. lactimorbi* was inoculated into dogs with no results. It should be noted, too, that bull 663 showed toxic symptoms of illness from autoclaved material, and sheep 309 was killed on plants treated in that manner.

These experiments, while not sufficient in number for positive proof, make it probable that it is a toxic substance in the plant which causes the illness, and not the presence of pathogenic bacteria.

RELATIONSHIP BETWEEN POISONING BY EUPATORIUM URTICAEFOLIUM AND MILK SICKNESS

There is no question that *E. urticaefolium* is poisonous and produces a line of symptoms closely resembling those said to be typical of milk sickness in cattle and other animals. Most of the so-called milk-sick cases in cattle occur in localities where *E. urticaefolium* grows. Field cases seen by the writers have the same symptoms as those which have been observed in experimental animals and have been diagnosed as intoxication by *E. urticaefolium*. There seems little doubt that many if not most cases of milk sickness or trembles in cattle are caused by this plant. It does not follow, however, that all cases of milk sickness are produced by *E. urticaefolium*.

Somewhat extended bacteriological investigations by the writers of this paper, the results of which will be reserved for another paper, appear to substantiate the claim of Jordan and Harris that there is a bacterium widely disseminated in the soil and on plants, which, under certain conditions—conditions for the most part unknown—produces a disease in man and animals, in which symptoms are exhibited which appear to be those of milk sickness.

Very much more work is necessary in order to clear up the subject, and definite plans have been made for extended work to this end; but it seems highly probable at this stage of the investigation that under the terms "milk sickness" or "trembles" are included at least two distinct things, one the poisoning of animals by *E. urticaefolium*, and the other a bacterial disease to which both animals and men are susceptible.

Although the question of the differential diagnosis of the two diseases will be discussed in detail in the report of the bacteriological investigations, it may be noted here that trembling appears to be more distinctly characteristic of Eupatorium poisoning, and that a subnormal temperature is one of the diagnostic symptoms of the bacterial diseases, while poisoning by *E. urticaefolium* produces no distinct effect on the temperature.

CONDITIONS UNDER WHICH ANIMALS MAY BE POISONED

It is known that stock may be pastured where *E. urticaefolium* is abundant, and that the use of these pastures may be continued for years with no harm. Most of the cases of poisoning occur in the late summer and fall and generally in years when there is a deficiency of moisture and a consequent shortage of forage grasses. Cases also occur when grazing animals are confined to a limited area on which the plant is abundant. The experimental work shows that *E. urticaefolium* is not palatable to our domestic animals and that they will avoid it in the presence of other foods. As in the case of most stock-poisoning plants, there is a direct relation between shortage of food and cases of poisoning, although, of course, sporadic cases may occur under other conditions.

REMEDIES

Sick animals should be treated with remedies to relieve the constipation and increase elimination. To this end purgatives may be used, of which perhaps Epsom salts is the best. This should be used in doses of 1 pound for a 1,000-pound animal. The feed should be laxative, like bran, oil meal, etc. In parts of North Carolina it is customary to feed milk-sick animals with green corn and pumpkins, and there is good reason to consider this a desirable diet.

Inasmuch as the toxic substance of the plant is eliminated very slowly, quick recovery must not be expected, and the animals should be given somewhat prolonged attention.

PREVENTION

In many places milk-sick areas have been fenced off with consequent prevention of losses. In some localities where *E. urticaefolium* is particularly abundant this evidently is advisable. The clearing of land and seeding to corn, grain, or grasses will of course stop the trouble. It has long been known that trembles affects animals pastured on unbroken land and that the disease disappears after cultivation.

So far as getting rid of the plant is concerned, however, it must be remembered that partial clearing is not sufficient. It is true that the plant favors damp, shaded places, but it sometimes grows most luxuriantly on partially cleared land.   It has been noticed in the mountains of North Carolina that the plant increases enormously on cleared land, and only disappears after the land has been seeded down.

If the poisonous character of the plant is recognized, much can be accomplished by so handling the animals that they do not graze largely in areas where *E. urticaefolium* is abundant.

PREVIOUS EXPERIMENTAL WORK ON EUPATORIUM URTICAEFOLIUM

The results of the experimental work undertaken by the writers were so definite and convincing that it seemed best to reexamine the literature of milk sickness with reference to the results obtained by other authors.   Although the literature is very extensive and *E. urticaefolium* has long been suspected as a possible cause of the disease, the published statements in regard to the plant are not very numerous.

Apparently the first published statement in regard to the effect of *E. urticaefolium* was by Rowe (*14*) in the Ohio State Journal of September 6, 1839.   He made a public experiment of feeding ''the weed'' on the farm of Francis Asbury, and published a statement by Mr. Asbury with affidavits by four other men.   ''The weed'' was fed to a cow for seven days, together with common feed.   A 3-months-old calf took the milk of the cow during that time, exhibited symptoms of trembles on the sixth day, and died on the ninth.   A yearling steer at the same time was fed with ''the weed'' and other necessary feed, had the trembles, and died at the end of seven days.   It is stated in the report of the Ohio State board of agriculture for 1858 (*9*) that the weed used was *Eupatorium ageratoides*.

A statement that *E. urticaefolium* (*ageratoides*) may be the cause of milk sickness was made by Barbee (*1*) in 1840.   He says that Dr. Owen told him that he had produced trembles in a calf by an extract of the plant.   Barbee himself gave a decoction to a dog and produced ''shaking palsy,'' vomiting, and death in three hours.

Drake in 1841 (*7, p. 215–216*) tells of two specific cases in which cattle were turned into pastures covered with *E. ageratoides* without any bad results.

Dewey in 1854 (*6*), thinks the cause of the disease is a

succulent plant, which grows in damp and thickly shaded bottoms, and bears a white blossom until late in the autumn.

Probably the plant thus described is *E. urticaefolium*.   He states that he knows all the symptoms of the disease were produced in calves by experimental feeding of this plant.

In the Thirteenth Annual Report of the Ohio State Board of Agriculture (*9*), printed in 1859, is an unsigned article, probably written by Townshend, in which is a letter from W. J. Vermilya.   Vermilya in 1855 put 200 sheep in a field containing *E. ageratoides*, saw them eat the plant, and lost 8 lambs that had developed the trembles In 1856 he fed the plant to a mare, beginning the feeding on October 12. Symptoms of trembles were very pronounced on October 21, and on the 23d the animal died During the illness the animal was examined by three doctors, two farriers, and other citizens, and all agreed that the mare had the trembles.

William Jerry, of Madison County, Ill , in 1867 published in the Missouri Republican a statement which was reprinted in the Medical and Surgical Reporter of Philadelphia (*13*), that he had been made violently ill with symptoms of milk sickness by eating E. *ageratoides* prepared as greens. He also produced sickness in a dog by a decoction of the plant. Sawyer in the same number of the magazine says that he has experimental evidence that the plant will produce in animals a disease very similar to milk sickness.

Moseley in 1906, in a detailed article (*10*), in which he states positively that E. *ageratoides* is the cause of trembles in animals and of milk sickness in man, relates many cases of poisoning of animals by the plant in Ohio pastures, and gives details of his own successful experiments with the plant and with extracts, using as experimental animals cats, dogs, rabbits, and sheep.

Crawford in 1908 (*4*) discusses Moseley's investigations and records a series of experiments on rabbits, cats, dogs, sheep, and man, for the most part made with extracts, from which he draws the conclusion that E. *ageratoides* is not the cause of milk sickness.[1]

Brooks in 1914 (*2*) states that his experiments show that both cattle and sheep are poisoned by the plant, the sheep being the more susceptible. He states that an animal must eat about 10 per cent of its weight in order to be poisoned.

Clay (*3*) fed two head of cattle and one sheep on fresh-cut material of the plant, and all died within three days with symptoms of trembles.

The results obtained by these authors were so concordant that it now seems strange that more importance was not attached to them. The experiments of Vermilya were especially conclusive. The fact that in the coves mentioned by Drake no bad results followed the pasturing of cattle is, of course, only negative evidence.

Even in the experiments by Crawford, from which he concluded that the plant was not harmful, an examination of the cases shows that of the eight rabbits tested with extracts four died and one other exhibited trembling of the muscles. The cat used was sick. The dog and sheep used showed no symptoms, but the doses given were very small. The same thing was true of the dose given to himself, for, granted that he weighed 150 pounds, the dosage was only o 44 of 1 per cent of his weight.

[1] In 1909 (*11*) and 1910 (*12*), in papers intended to prove that aluminium phosphate in plants is the cause of milk sickness, Moseley details other successful experiments with feeding the plant to rabbits and cows and claims to have produced the disease in rabbits and cats from the milk of diseased animals, in cats by butter, and in cats from the meat of diseased rabbits.

Inasmuch as the experiments detailed in this paper show that the toxic dose for animals is from 6 per cent to 10 per cent of their weight, symptoms would hardly be expected from the small quantity which Crawford administered to himself, even if man were vastly more susceptible than the lower animals.

Apparently experimenters were so interested in discovering the cause of milk sickness that the fact was overlooked that, whether milk sickness were produced by *E. urticaefolium* or not, actual proof had been made repeatedly of the poisonous character of the plant. The experiments of this paper have confirmed the former work and given definite information in regard to symptoms and dosage.[1]

SUMMARY

(1) *Eupatorium urticaefolium* has for many years been considered by many people as the cause of milk sickness, or trembles, in cattle.

(2) Experimental work shows conclusively that the plant is toxic, and produces a definite line of symptoms bearing a close resemblance to those considered characteristic of trembles.

(3) Probably many, possibly most, cases of trembles in cattle and sheep are due to poisoning by *E. urticaefolium*.

(4) Under the term "milk sickness," or "trembles," are probably grouped at least two distinct things: one poisoning by *E. urticaefolium* and the other a bacterial disease.

LITERATURE CITED

(1) BARBEE, W J
    1840 FACTS RELATIVE TO THE ENDEMIC DISEASE CALLED, BY THE PEOPLE OF THE WEST, MILK-SICKNESS *In* West Jour. Med. and Surg , v. 1, p. 178-190

(2) BROOKS, E. W
    1914. MILK-SICKNESS *In* Altamont News, Altamont, Ill., v 33, no 44, p 1, 5.

(3) CLAY, A J
    1914 PERSONAL AND CLINICAL EXPERIENCES WITH MILK SICKNESS *In* Ill. Med Jour , v 26, no 2, p. 103-108.

(4) CRAWFORD, A. C
    1908 THE SUPPOSED RELATIONSHIP OF WHITE SNAKEROOT TO MILK SICKNESS, OR "TREMBLES " *In* U S Dept. Agr. Bur. Plant Indus , Bul. 121, p 5-20, 1 pl

(5) CURTIS, R. S., and WOLF, F A.
    1917 EUPATORIUM AGERATOIDES, THE CAUSE OF TREMBLES *In* Jour. Agr. Research, v. 9, no. 11, p. 397-404, pl. 22-24. Literature cited, p. 404.

(6) DEWEY, J S.
    1854. [MILK SICKNESS.] *In* Northwest. Med and Surg Jour , v 11 (n s , v. 3), no 12, p 541-547

[1] Since this paper was offered for publication, Curtis and Wolf (5) have published the details of careful feeding experiments at the North Carolina Experiment Station with *Eupatorium urticaefolium* upon sheep, by which they produced 13 fatal cases of "trembles." Had their work been known when this paper was prepared, it would have necessitated some slight changes in the introductory part. As, however, their results do not conflict in any way with those obtained by the writers, it has seemed best to leave the paper in its original form.

(7) DRAKE, Daniel.

1841. A MEMOIR ON THE DISEASE CALLED BY THE PEOPLE "TREMBLES" AND THE "SICK-STOMACH" OR "MILK SICKNESS", AS THEY HAVE OCCURRED IN THE COUNTIES OF FAYETTE, MADISON, CLARK, AND GREEN IN THE STATE OF OHIO *In* West. Jour. Med and Surg , v. 3, no. 3, p. 161–226.

(8) JORDAN, E. O., and HARRIS, N. M.

1909 MILK SICKNESS. *In* Jour Infect. Diseases, v 6, no. 4, p 401–491, 6 fig. Bibliography, p 485–491.

(9) 1859. MILK SICKNESS. *In* 13th Ann. Rpt. Ohio State Bd Agr., 1858, p. 670–675 Reprinted in 28th Ann. Rpt Ohio State Bd. Agr., 1873, p. 489–494 1874.

(10) MOSELEY, E. L.

1906. THE CAUSE OF TREMBLES IN CATTLE, SHEEP AND HORSES, AND OF MILK-SICKNESS IN PEOPLE. *In* Ohio Nat , v. 6, no 4, p 463–470, no. 5, p 477–483

(11) 1909. THE CAUSE OF TREMBLES AND MILKSICKNESS *In* Med. Rec., N. Y., v. 75, no 20, p. 839–844

(12) 1910 ANTIDOTE FOR ALUMINUM PHOSPHATE, THE POISON THAT CAUSES MILK-SICKNESS. *In* Med. Rec., N. Y., v 77, no. 15, p. 620–622.

(13) 1867. THE PLANT THAT CAUSES MILK SICKNESS. *In* Med. and Surg. Reporter, v. 16, no. 3 (no. 526), p. 270–271. Reprinted from Missouri Republican.

(14) ROWE, John.

1839. [MILK SICKNESS.] *In* Ohio State Jour , Sept. 6. Cited in 13th Ann. Rpt. Ohio State Bd. Agr., 1858, p. 675. 1859.

PLATE 52

Eupatorium urticaefolium, white snakeroot.

(716)

PLATE 54

Poisoning by *Eupatorium urticaefolium:*

A.—Sheep 367 at 11.17 a.m. on September 30.
B.—Sheep 367 at 11.20 a.m. on September 30.

PLATE 55

Poisoning by *Eupatorium urticaefolium:*

A.—Sheep 368 on September 24.
B.—Sheep 368 a little later the same day.

23719°—17——4

INDEX

ADDITIONAL COPIES
OF THIS PUBLICATION MAY BE PROCURED FROM
THE SUPERINTENDENT OF DOCUMENTS
GOVERNMENT PRINTING OFFICE
WASHINGTON, D. C.
AT
5 CENTS PER COPY
▽

Lightning Source UK Ltd.
Milton Keynes UK
UKHW020843110119
335238UK00009B/1004/P